JACARANDA MATHS QUEST

FOUNDATION MATHEMATICS 12

VCE UNITS 3 AND 4 | FIRST EDITION

JACARANDA MATHS QUEST

FOUNDATION MATHEMATICS 12

VCE UNITS 3 AND 4 | FIRST EDITION

MARK BARNES

CONTRIBUTING AUTHORS

Pauline Holland

Christine Utber

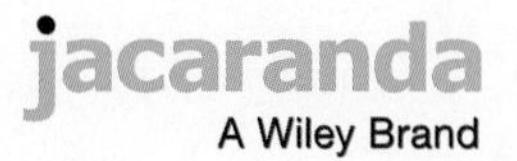

First edition published 2023 by
John Wiley & Sons Australia, Ltd
155 Cremorne Street, Cremorne, Vic 3121

Typeset in 10.5/13 pt TimesLTStd

ISBN: 978-1-119-87614-4

The covers of the *Jacaranda Maths Quest VCE Mathematics* series are the work of Victorian artist Lydia Bachimova.

Lydia is an experienced, innovative and creative artist with over 10 years of professional experience, including five years of animation work with Walt Disney Studio in Sydney. She has a passion for hand drawing, painting and graphic design.

Illustrated by diacriTech and Wiley Composition Services

Typeset in India by diacriTech

A catalogue record for this book is available from the National Library of Australia

Printed in Singapore
M121035_290822

Contents

About this resource

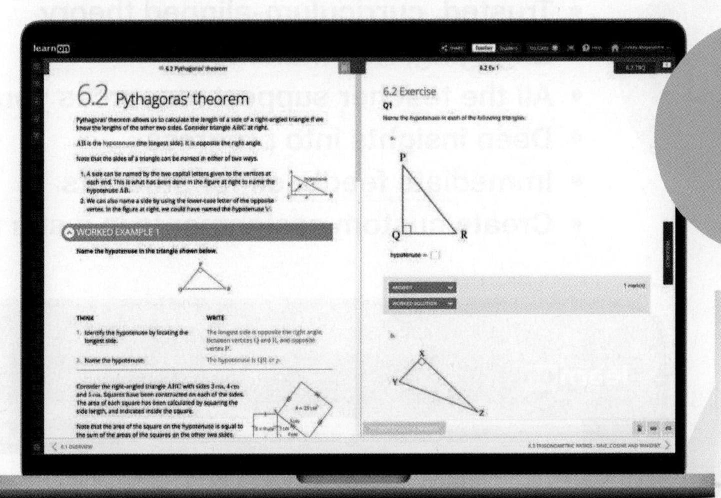

Everything you need for your students to succeed

JACARANDA MATHS QUEST
FOUNDATION MATHEMATICS 12 VCE UNITS 3 AND 4 | FIRST EDITION

Developed by expert Victorian teachers for VCE students

Tried, tested and trusted. The NEW Jacaranda VCE Mathematics series continues to deliver curriculum-aligned material that caters to students of all abilities.

Completely aligned to the VCE Mathematics Study Design

Our expert author team of practising teachers ensures 100 per cent coverage of the new VCE Mathematics Study Design (2023–2027).

Everything you need for your students to succeed, including:

- NEW! Access carefully scaffolded question sets, to ensure that all students can experience success and take the next steps. Ensure assessment preparedness with practice SACs.
- NEW! Be confident your students can get unstuck and progress, in class or at home. For every question online they receive immediate feedback and fully worked solutions.
- NEW! Teacher-led videos to unpack concepts and worked examples to fill learning gaps after COVID-19 disruptions.

Learn online with Australia's most

Everything you need for each of your lessons in one simple view

- **Trusted, curriculum-aligned theory**
- **Engaging, rich multimedia**
- **All the teacher support resources you need**
- **Deep insights into progress**
- **Immediate feedback for students**
- **Create custom assignments in just a few clicks.**

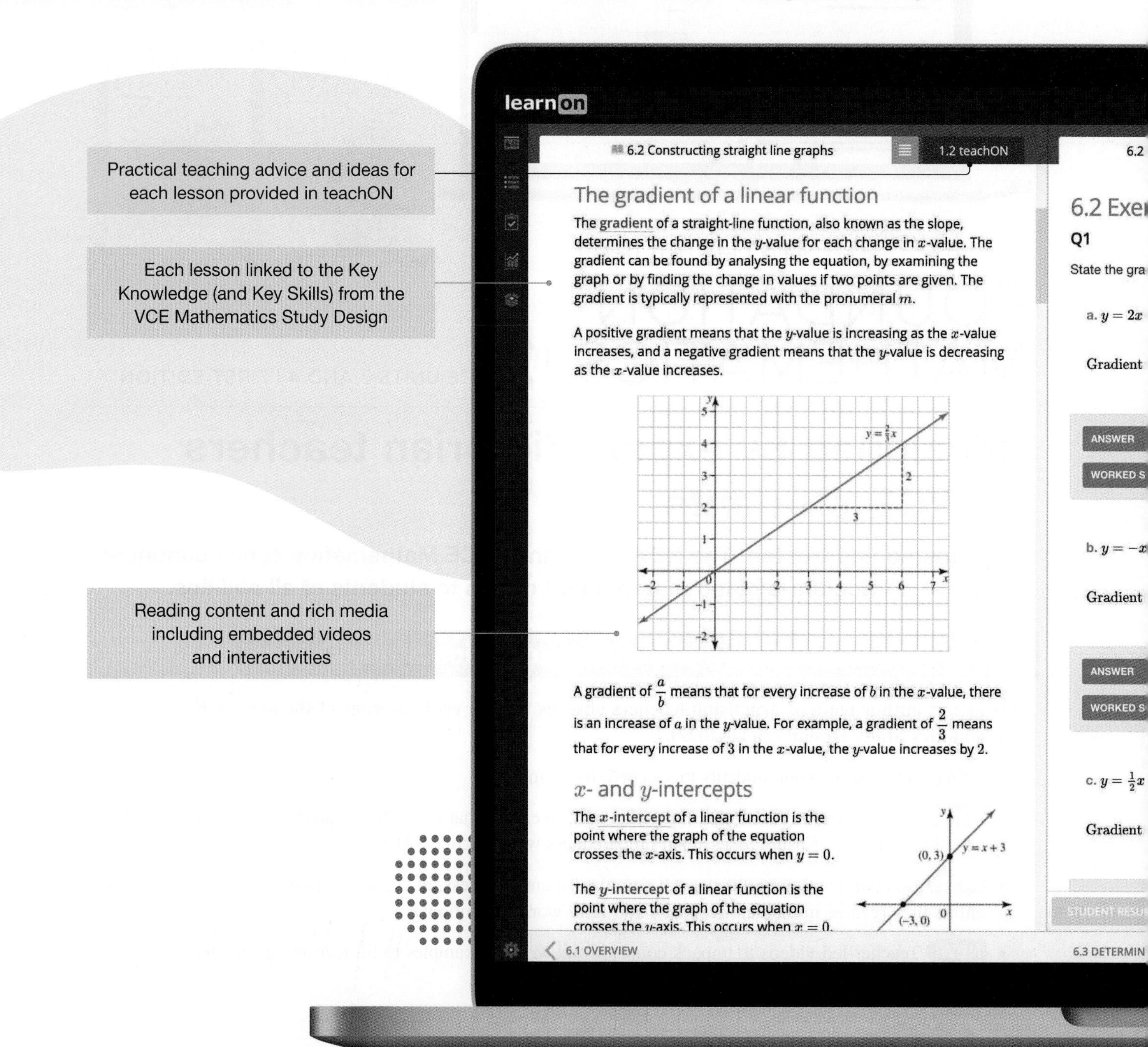

powerful learning tool, learnON

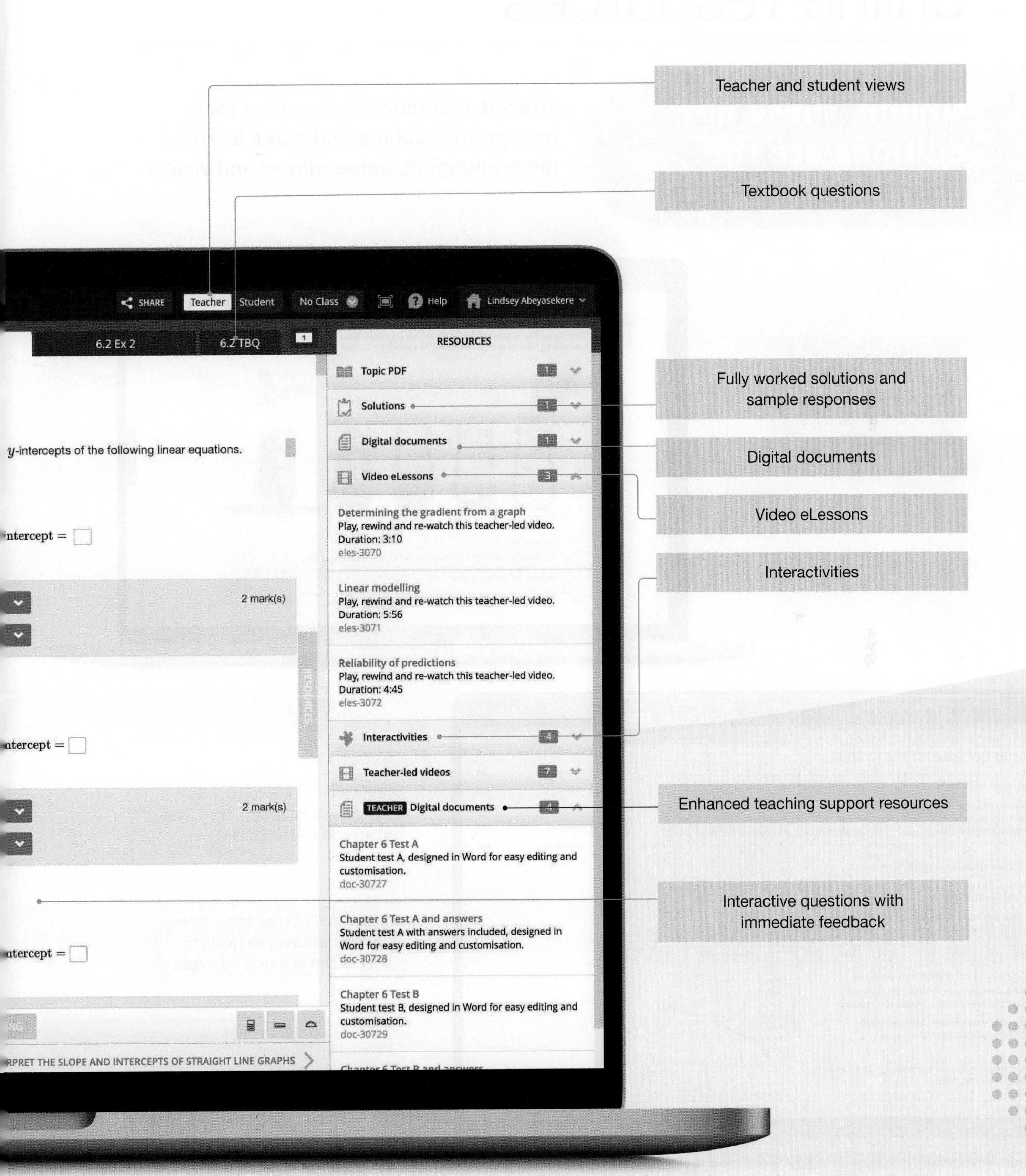

Get the most from your online resources

Online, these new editions are the **complete package**

Trusted Jacaranda theory, plus tools to support teaching and make learning more engaging, personalised and visible.

Each topic is linked to Key Knowledge (and Key Skills) from the VCE Mathematics Study Design.

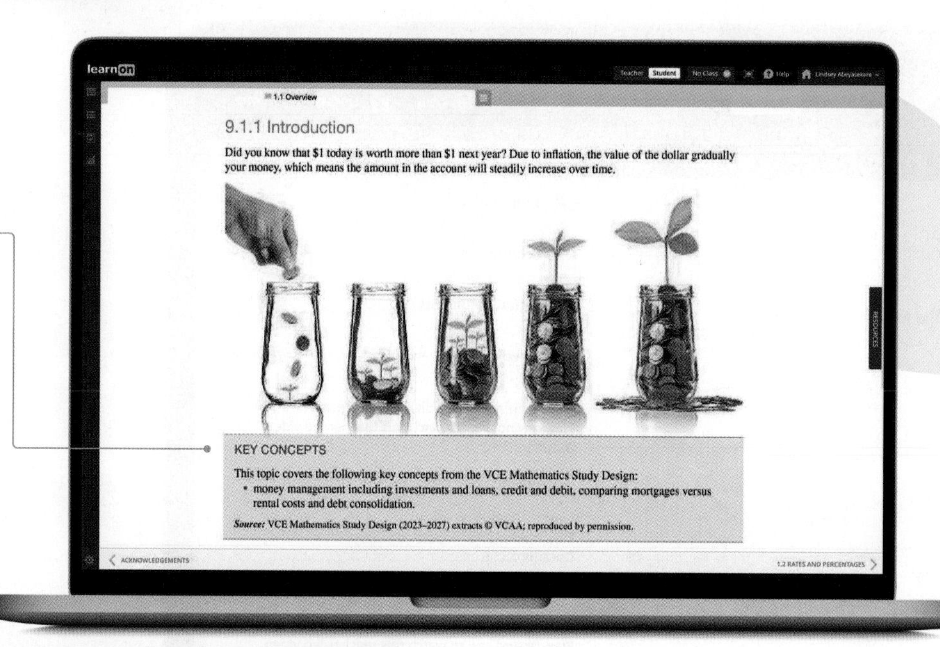

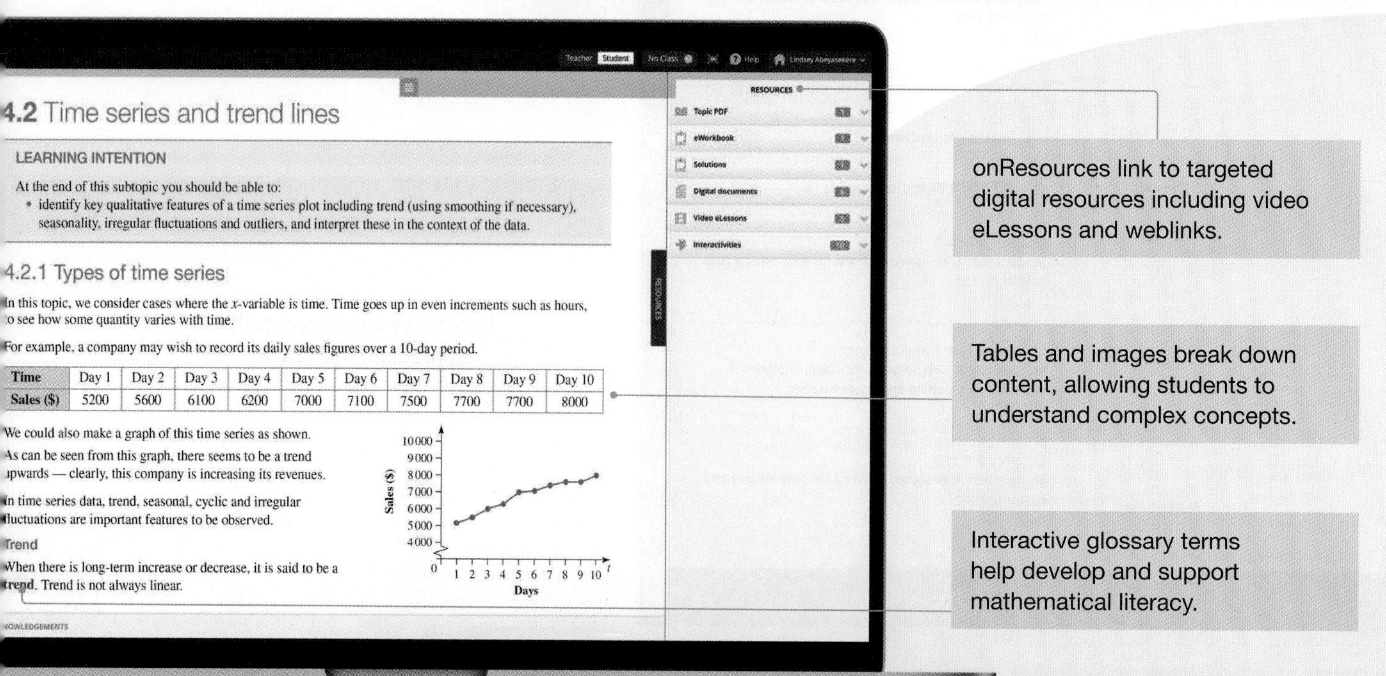

onResources link to targeted digital resources including video eLessons and weblinks.

Tables and images break down content, allowing students to understand complex concepts.

Interactive glossary terms help develop and support mathematical literacy.

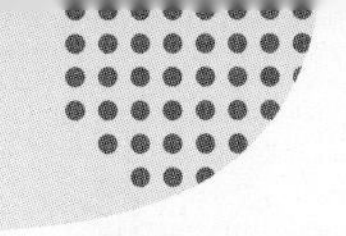

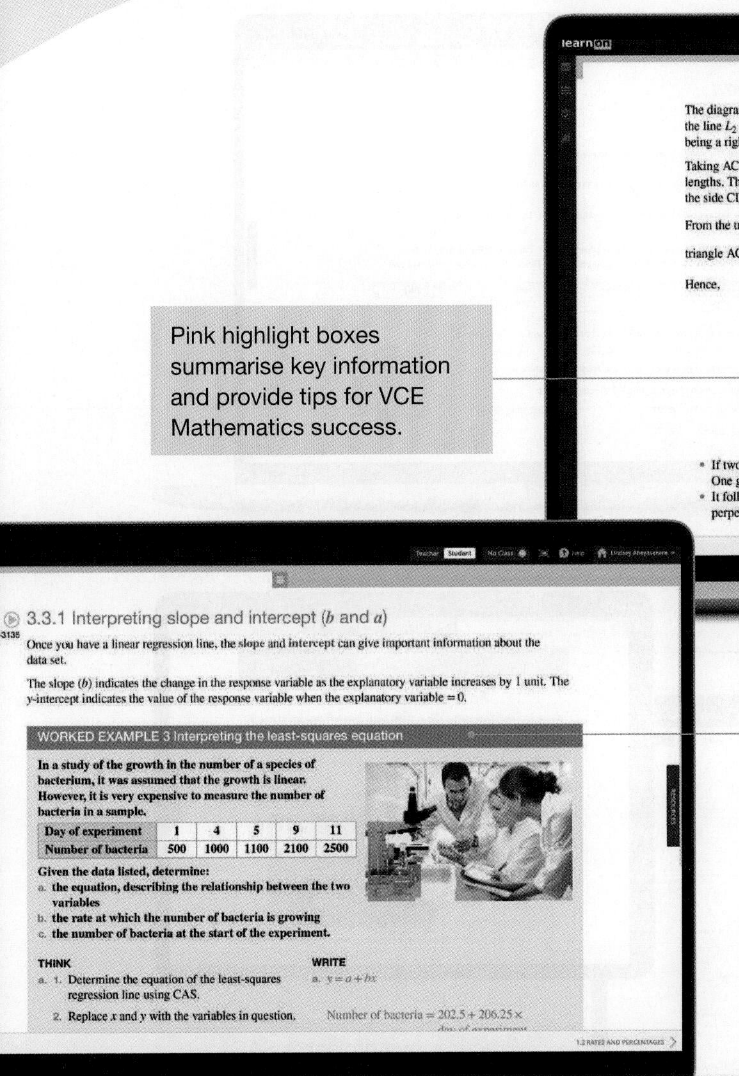

The diagram shows the line L_1 passing through the points A and B and the line L_2 passing through the points A and D, with the angle BAD being a right angle.

Taking AC as 1 unit, the sides in the diagram are labelled with their lengths. The side CB has length m_1. Because lengths must be positive, the side CD is labelled as $-m_2$, since $m_2 < 0$.

From the triangle ABC in the diagram, $\tan\theta = \frac{m_1}{1} = m_1$, and from the triangle ACD in the diagram, $\tan\theta = \frac{1}{-m_2}$.

Hence,

$$m_1 = \frac{1}{-m_2}$$
$$\therefore m_1 m_2 = -1$$

Gradients of perpendicular lines

$$m_1 m_2 = -1 \text{ or } m_2 = -\frac{1}{m_1}$$

- If two lines with gradients m_1 and m_2 are perpendicular, then the product of their gradients is -1. One gradient is the negative reciprocal of the other.
- It follows that if $m_1 m_2 = -1$, then the two lines are perpendicular. This can be used to test for perpendicularity.

Pink highlight boxes summarise key information and provide tips for VCE Mathematics success.

Worked examples, supported by teacher-led videos, break down the process of answering questions using a think/write format.

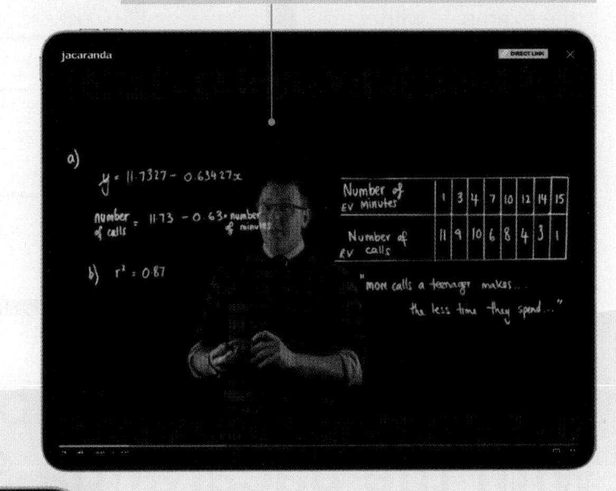

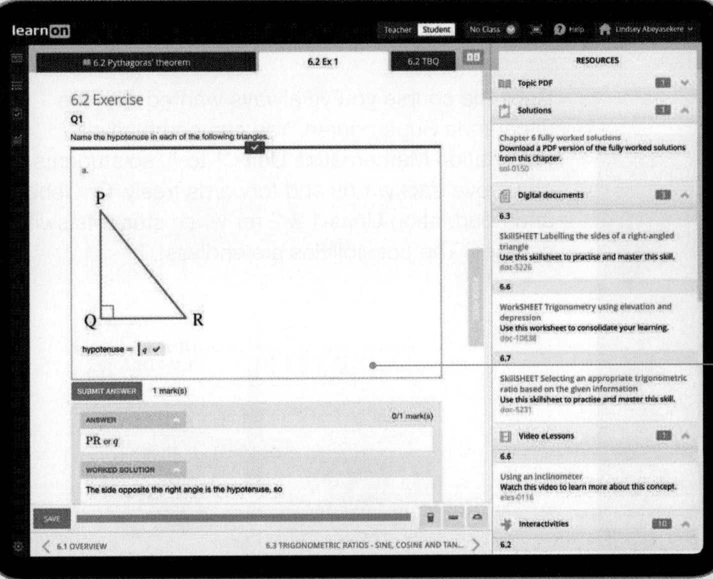

- Online and offline question sets contain practice questions, with exemplary responses.
- Every question has immediate, corrective feedback to help students to overcome misconceptions as they occur and to study independently — in class and at home.

Topic reviews

Topic reviews include online summaries and topic level review exercises that cover multiple concepts.

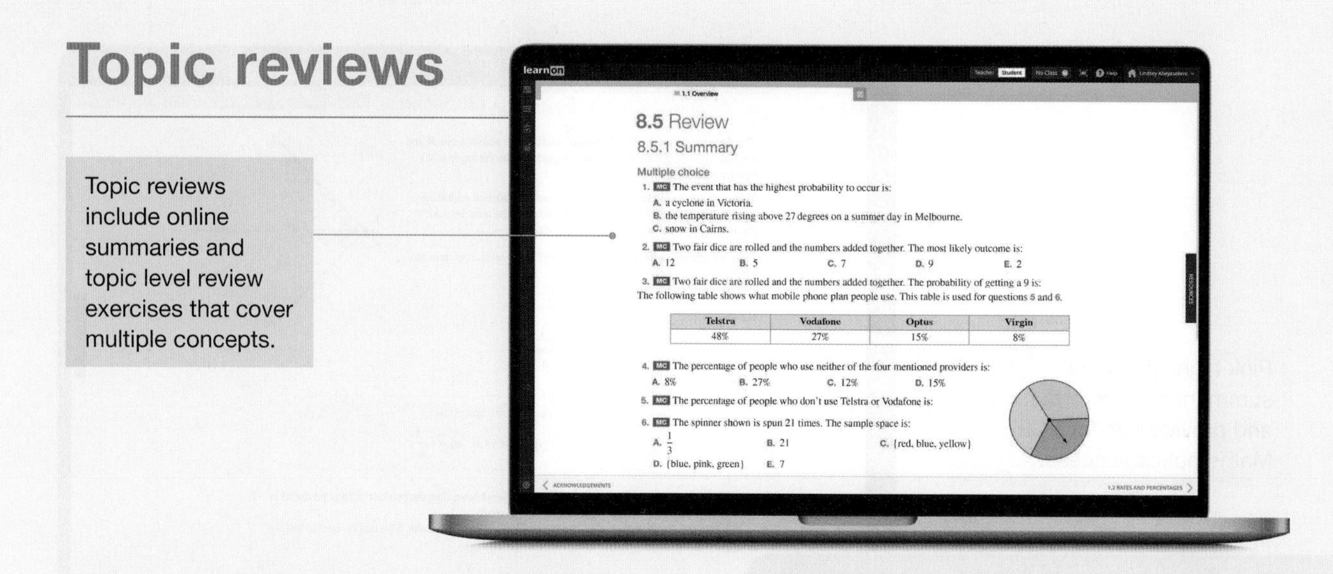
8.5 Review

8.5.1 Summary

Multiple choice

1. MC The event that has the highest probability to occur is:
A. a cyclone in Victoria.
B. the temperature rising above 27 degrees on a summer day in Melbourne.
C. snow in Cairns.

2. MC Two fair dice are rolled and the numbers added together. The most likely outcome is:
A. 12 B. 5 C. 7 D. 9 E. 2

3. MC Two fair dice are rolled and the numbers added together. The probability of getting a 9 is:

The following table shows what mobile phone plan people use. This table is used for questions 5 and 6.

Telstra	Vodafone	Optus	Virgin
48%	27%	15%	8%

4. MC The percentage of people who use neither of the four mentioned providers is:
A. 8% B. 27% C. 12% D. 15%

5. MC The percentage of people who don't use Telstra or Vodafone is:

6. MC The spinner shown is spun 21 times. The sample space is:
A. $\frac{1}{3}$ B. 21 C. {red, blue, yellow}
D. {blue, pink, green} E. 7

Get exam-ready!

Topic level review questions are structured just like the exams — with multiple choice, short answer and extended response questions.

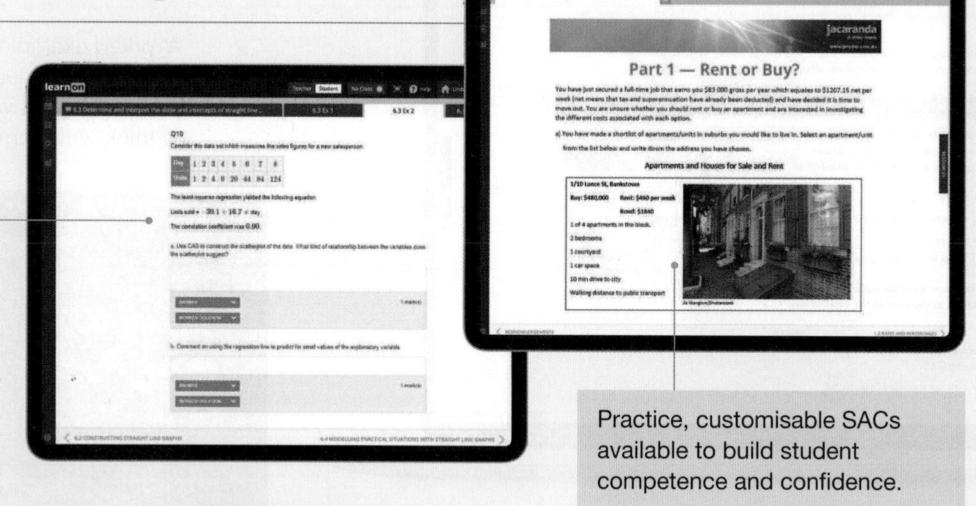

Practice, customisable SACs available to build student competence and confidence.

Combine units flexibly with the Jacaranda Supercourse

Build the course you've always wanted with the Jacaranda Supercourse. You can combine all Foundation Mathematics Units 1 to 4, so students can move backwards and forwards freely. Or General and Foundation Units 1 & 2 for when students switch courses. The possibilities are endless!

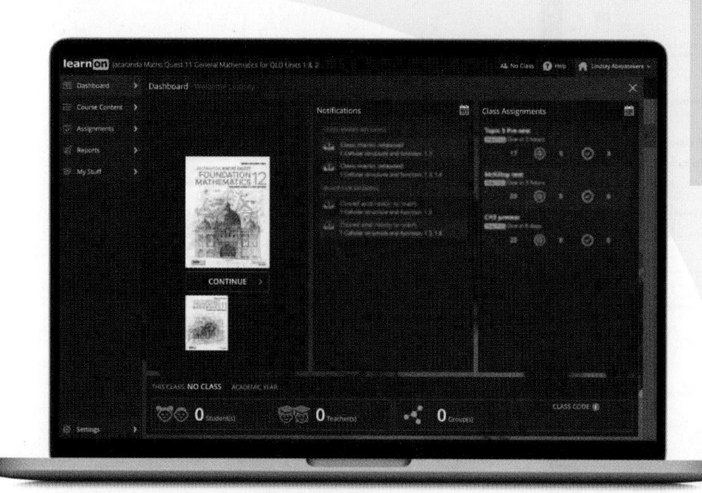

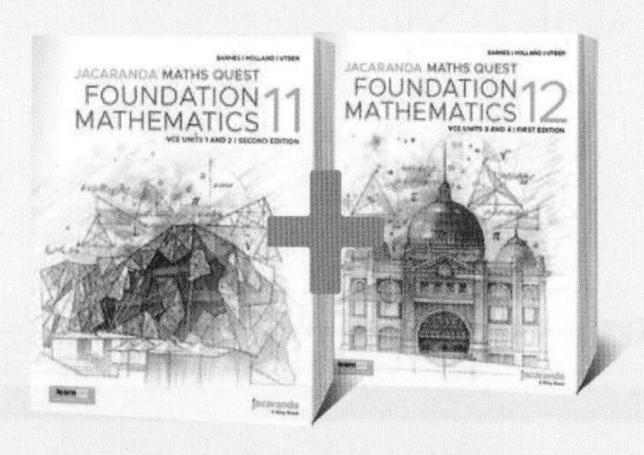

A wealth of teacher resources

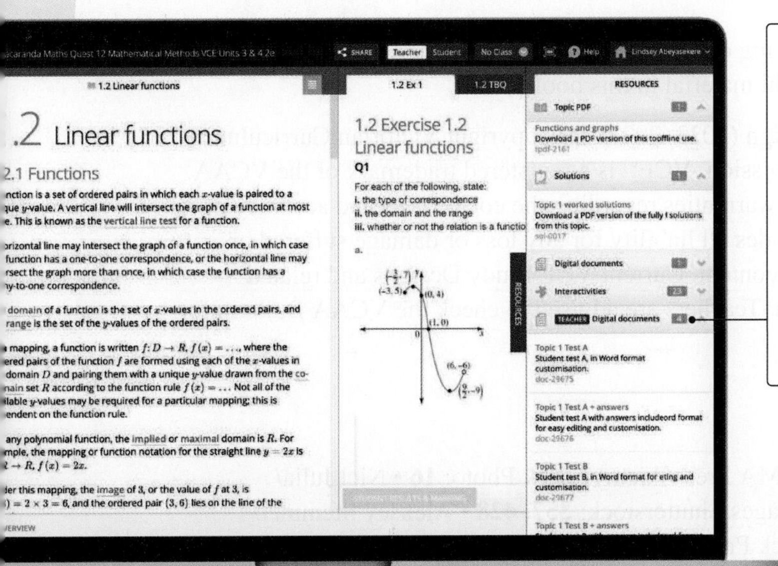

Enhanced teacher support resources, including:

- work programs and curriculum grids
- teaching advice and additional activities
- quarantined topic tests (with solutions)
- quarantined SACs (with worked solutions and marking rubrics)

Customise and assign

A testmaker enables you to create custom tests from the complete bank of thousands of questions.

Reports and results

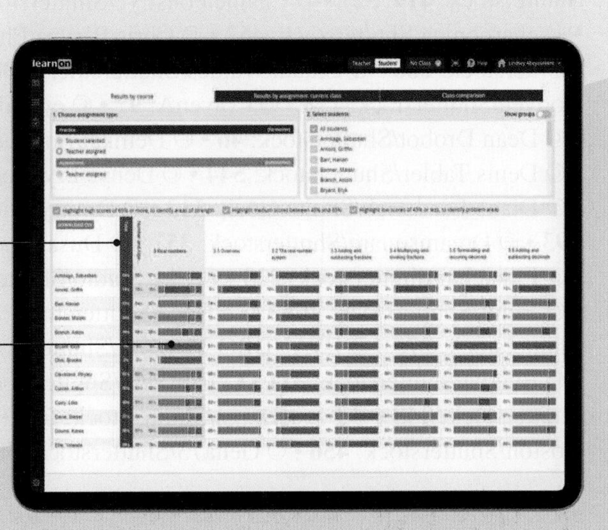

Data analytics and instant reports provide data-driven insights into performance across the entire course.

Show students (and their parents or carers) their own assessment data in fine detail. You can filter their results to identify areas of strength and weakness.

Acknowledgements

The authors and publisher would like to thank the following copyright holders, organisations and individuals for their assistance and for permission to reproduce copyright material in this book

Selected extracts from the VCE Mathematics Study Design (2023–2027) are copyright Victorian Curriculum and Assessment Authority (VCAA), reproduced by permission. VCE® is a registered trademark of the VCAA. The VCAA does not endorse this product and makes no warranties regarding the correctness and accuracy of its content. To the extent permitted by law, the VCAA excludes all liability for any loss or damage suffered or incurred as a result of accessing, using or relying on the content. Current VCE Study Designs and related content can be accessed directly at www.vcaa.vic.edu.au. Teachers are advised to check the VCAA Bulletin for updates.

Images

• © Kristian Dowling/Stringer/Getty Images: **17** • © ZUMA Press/Alamy Stock Photo: **16** • NickJulia/Shutterstock: **328** • Shutterstock: **436** • © Akhenaton Images/Shutterstock: **357, 428** • Aleksey Stemmer/Shutterstock: **508** • ArTDi101/Shutterstock: **247** • Atstock Productio/Shutterstock: **427** • Blue Planet Studi/Shutterstock: **413** • Canoneer/Shutterstock: **470** • Cat Box/Shutterstock: **431** • © Eivaisla/Shutterstock: **334, 341** • © elfishes/Shutterstock: **464, 470** • freesoulproductio/Shutterstock: **469** • Gorodenkoff/Shutterstock: **574** • GoTaR/Shutterstock: **508** • hddigital/Shutterstock: **341** • © Irina Rogova/Shutterstock: **469, 483** • Ljupco Smokovski/Shutterstock: **483** • NatalyaBond/Shutterstock: **335** • Oleksandr_Delyk/Shutterstock: **470** • Orla/Shutterstock: **508** • © Rawpixel.com/Shutterstock: **17 38, 410, 422** • S-F/Shutterstock: **449** • Seregam/Shutterstock: **473** • Sheila Say/Shutterstock: **440** • © spass/Shutterstock: **274, 341** • TZIDO SUN/Shutterstock: **366** • Zyn Chakrapong/Shutterstock: **441** • © 2009fotofriends/Shutterstock: **490** • © 3DConcepts/Shutterstock: **411** • © 63ru78/Shutterstock: **523** • © A.J. Pictures/Shutterstock: **310** • © Aarcady/Shutterstock: **470** • © Africa Studio/Shutterstock: **44, 148, 155, 224, 243, 351, 416, 557** • © Ailisa/Shutterstock: M. Unal Ozmen/Shutterstock, **340** • © Aleksandar Todorovic/Adobe Stock: **536** • © Aleksandr Simonov/Shutterstock: **103** • © Alex Farias/Shutterstock: **479** • © Alex Stemmer/Shutterstock: **294** • © alison1414/Shutterstock: **237** • © Alliya2/Shutterstock: **336** • © alphaspirit.it/Shutterstock: **408** • © Ana Vasileva/Shutterstock: **470, 484** • © Andrey_Popov/Shutterstock: **412** • © Andy Dean Photography/Shutterstock: **243, 457** • © Anne Mathiasz/Shutterstock: **32** • © Artazum/Shutterstock: **445** • © arturasker/Shutterstock: **305** • © Atstock Productions/Shutterstock: **381** • © B Cas/Shutterstock: **36** • © badahos.com/Shutterstock: **304** • © bbernard/Shutterstock: **350** • © bekulnis/Shutterstock: **334** • © bezikus/Shutterstock: **5** • © bikeriderlondon/Shutterstock: **273** • © Bikeworldtravel/Shutterstock: **236** • © bmphotographer/Shutterstock: **442** • © BobrinSKY: **94** • © BONNINSTUDIO/Shutterstock: **445** • © Brett Rabideau/Shutterstock: **496** • © Brian A Jackson/Shutterstock: **419, 527** • © ChameleonsEye/Shutterstock: **137** • © ChameleonsEye.com/Shutterstock: **247** • © charobnica/Shutterstock: **462** • © Chris Parypa Photography/Shutterstock: **252** • © Christine Pedretti/Shutterstock: **266** • © ClaudioValdes/Shutterstock: **138** • © cobalt88/Shutterstock: **268** • © Cornelis JC Van Rooyen/Shutterstock: **178** • © CravenA: **97** • © cybrain/Shutterstock: **482** • © David Gyung/Shutterstock: **145** • © Dean Drobot/Shutterstock: **46** • © Demkat/Shutterstock: **220** • © Denis Belitsky/Shutterstock: **303** • © Denis Tabler/Shutterstock: **544** • © Denys Prykhodov/Shutterstock: **250** • © Devenorr/Shutterstock: **529** • © Diane Diederich/Shutterstock: **513** • © Dmitry Sheremeta/Shutterstock: **305** • © dragon_fang/Shutterstock: **593** • © Dreamsquare/Shutterstock: **357** • © Dusan Pavlic/Shutterstock: **26** • © eakasarn/Shutterstock: **455** • © Ekkachai/Shutterstock: **228** • © Eky Studio/Shutterstock: **277** • © Elena Elisseeva/Shutterstock: **18** • © Elnur/Shutterstock: **268** • © elRoce/Shutterstock: **327** • © ElRoi/Shutterstock: **522** • © ESB Professional/Shutterstock: **275** • © EVANDRO MAZER/Shutterstock: **43** • © Evgeny Karandaev/Shutterstock: **472** • © Evikka/Shutterstock: **311** • © FamVeld/Shutterstock: **269** • © Farknot Architect/Shutterstock: **86** • © fizkes/Shutterstock: **448** • © FocusDzign/Shutterstock: **601** • © Francesco Abrignani/Shutterstock: **331** • © Franck Boston/Shutterstock: **456** • © Gena73/Shutterstock: **582** • © Gitanna/Shutterstock: **558** • © glebchik/

Shutterstock: **257** • © Hannamariah/Shutterstock: **273** • © Herbert Kratky/123RF: **280** • © Holli/Shutterstock: **276** • © HomeStudio/Shutterstock: **58** • © HSNphotography/Shutterstock: **478** • © HST6/Shutterstock: **480** • © Hurst Photo/Shutterstock: **246** • © Iakov Filimonov/Shutterstock: **219** • © Ilin Sergey/Shutterstock: **25** • © Inna Ogando/Shutterstock: **465, 472** • © Inspiring/Shutterstock: **213** • © Iryna Inshyna/Shutterstock: **284** • © Monkey Business Images/Shutterstock: **1, 2, 55, 222, 267, 340, 416** • © Iulian Valentin/Shutterstock: **293** • © James Watts/Shutterstock: **238** • © Jim David/Shutterstock: **109** • © Jimmie48 Photography/Shutterstock: **15** • © Joe Gough/Adobe Stock: **351** • © Judy Kennamer/Shutterstock: **223** • © Kardasov Films/Shutterstock: **449** • © karegg/Shutterstock: **59** • © karen roach: **95** • © Katarina Christenson/Shutterstock: **536** • © Kateryna Mashkevych/Shutterstock: **36** • © Katherine Welles/Shutterstock: **248** • © klikkipetra/Shutterstock: **582** • © Kolonko: **82** • © Kovaleva_Ka/Shutterstock: **10** • © Ksenia Palimski/Shutterstock: **461** • © Kucher Serhii/Shutterstock: **237** • © Kzenon/Shutterstock: **514** • © LanKS/Shutterstock: **114** • © Lemau Studio/Shutterstock: **245** • © Leonard Zhukovsky/Shutterstock: **154** • © Lev Kropotov/Shutterstock: **481** • © Lopolo/Shutterstock: **417, 449** • © loveaum/Shutterstock: **246** • © M-SUR/Shutterstock: **212** • © M. Unal Ozmen/Shutterstock: **233, 250** • © Mai Groves/Shutterstock: **187** • © Malyugin/Shutterstock: **342** • © mangostock/Shutterstock: **248** • © MaraZe/Shutterstock: **265** • © mariocigic/Shutterstock: **365** • © Marlon Lopez MMG1 Design/Shutterstock: **274** • © Max kegfire/Shutterstock: **99** • © Max Topchii/Shutterstock: **299, 374** • © Maxx-Studio/Shutterstock: **151, 461** • © Mehmet Cetin/Shutterstock: **237** • © Michael Smolkin/Shutterstock: **482** • © michaeljung/Shutterstock: **414, 425** • © Microgen/Shutterstock: **554, 243** • © Mikbiz/Shutterstock: **242** • © Milleflore Images/Shutterstock: **48** • © Minerva Studio/Shutterstock: **302** • © mtlapcevic/Shutterstock: **439** • © MvanCaspel/Shutterstock: **479** • © NadyaEugene/Shutterstock: **269** • © NaMaKuKi/Shutterstock: **280** • © Nata Bene/Shutterstock: **200** • © nata-lunata/Shutterstock: **265** • © Nattawit Khomsanit/Shutterstock: **95** • © NavinTar/Shutterstock: **35** • © Nejron Photo/Shutterstock: **296** • © nikamo/Shutterstock: **264** • © Nikita Shchavelev/Shutterstock: **445** • © Nils Versemann/Shutterstock: **272** • © oksana2010/Shutterstock: **484** • © Olaf Speier/Shutterstock: **581** • © Oleggg/Shutterstock: **439** • © Olesia Bilkei/Shutterstock: **332** • © Olga Dmitrieva/Shutterstock: **215** • © Olga Kashubin/Shutterstock: **49** • © Olga Listopad/Shutterstock: **351** • © OtmarW/Shutterstock: **481** • © Panumas Yanuthai/Shutterstock: **436** • © Patryk Kosmider/Shutterstock: **344** • © paulista/Shutterstock: **250** • © Pavel L Photo and Video/Shutterstock: **568** • © Pavel Nesvadba/Shutterstock: **350** • © Pavel1964/Shutterstock: **235** • © Peshkova/Shutterstock: **461** • © petcharaPJ/Shutterstock: **495** • © Peterfz30/Shutterstock: **69** • © Phattana Stock/Shutterstock: **444** • © PHILIPPE MONTIGNY/Shutterstock: **263** • © Phongphan/Shutterstock: **440** • © Phonlamai Photo/Shutterstock: **592** • © photoiconix/Shutterstock: **586** • © Photology1971/Shutterstock: **303** • © Photos BrianScantlebury/Shutterstock: **164** • © Photostriker/Shutterstock: **38** • © pikselstock/Shutterstock: **104** • © pisaphotography/Shutterstock: **301** • © Pixel Embargo/Shutterstock: **358** • © Pressmaster/Shutterstock: **407** • © Prostock-studio: **81** • © Quang Ho/Shutterstock: **278** • © Rido/Shutterstock: **270** • © Rinelle/Shutterstock: **410** • © Robyn Mackenzie/Shutterstock: **301, 413** • © Roman Samborskyi/Shutterstock: **373, 418, 434** • © Romolo Tavani/Shutterstock: **528** • © RossHelen/Shutterstock: **481** • © ryota.www/Shutterstock: **160** • © S_Photo/Shutterstock: **374** • © Sander van Sinttruye/Shutterstock: **561** • © Sararoom Design/Shutterstock: **581** • © Sararwut Jaimassiri/Shutterstock: **252** • © Sergey Novikov/Adobe Stock: **110** • © Sergii Korshun/Shutterstock: **519** • © shooarts/Shutterstock: **459** • © Shuang Li/Shutterstock: **165, 409** • © Shutter Baby photo/Shutterstock: **460** • © Gen Savina/Shutterstock: **135** • © Juice Flair/Shutterstock: **135** • © TY Lim/Shutterstock: **623** • © SIAATH/Shutterstock: **357** • © simez78/Shutterstock: **223** • © Soloviova Liudmyla/Shutterstock: **146** • © Sorbis/Shutterstock: **47** • © SpeedKingz/Shutterstock: **352** • © Stefan Schurr/Shutterstock: **294** • © Syda Productions/Shutterstock: **283, 479** • © Tanee Sawasdee/Shutterstock: **598** • © TK Kurikawa/Shutterstock: **288, 391** • © Toa55/Shutterstock: **309** • © Travnikov Studio/Shutterstock: **249** • © UfaBizPhoto/Shutterstock: **391** • © Vadim Sadovski/Shutterstock: **565** • © Vereshchagin Dmitry/Shutterstock: **345** • © VGstockstudio/Shutterstock: **38** • © Victoria Chudinova/Shutterstock: **100** • © virtu studio/Shutterstock: **72** • © visivastudio/Shutterstock: **463** • © VladKol/Shutterstock: **222** • © VLADYSLAV DANILIN/Shutterstock: **280** • © wavebreakmedia/Shutterstock: **211, 241** • © WAYHOME studio/Shutterstock: **369** • © Wirestock Creators/Shutterstock: **8** • © xpixel/Shutterstock: **232** • © yousang/Shutterstock: **426** • © Zdenka Darula: **98** • © Zety Akhzar/Shutterstock: **10** • © Zhukova Valentyna/Shutterstock: **519** • © Zolnierek/Shutterstock: **489** • © southstarcommunities/Flickr: **71** • © Getty Images: **258** • © Ingram Publishing/Alamy Stock Photo: **138** • © 9-Kilo/Shutterstock: **586** • © C Squared Studios/Photodisc/Getty Images: **530** • © ChameleonsEye/Shutterstock: Daniel Vine Photography/Shutterstock,

Brian Jackson/Adobe Stock Photos: **4** • © ET-ARTWORKS/Getty Images: **608** • © Fotyma/Shutterstock: **151** • © fritz16/Shutterstock: **5** • © Gino Santa Maria/Shutterstock: **573** • © Krakenimages.com/Adobe Stock: **110** • © Kym Smith/Newspix: **246** • © mama_mia/Shutterstock: **151** • © manaemedia/Shutterstock: **5** • © No change to MS: **17, 25** • © Oleksiy Mark/Shutterstock: **165** • © Photodisc: **5, 247, 523, 529** • © Photodisc/Getty Images: Artazum/Shutterstock, Designs Stock/Shutterstock: **587** • © SharonPhoto/Shutterstock: **597** • © Smart Vectors/Shutterstock: **54** • © Tatiana Koroleva/Alamy Stock Photo: **90** • © weim/Adobe Stock: **564** • © WooGraphics/Shutterstock: **89** • © xy/Adobe Stock: **98** • © Zeljko Radojko/Shutterstock: **569** • © Andrei Shumskiy/Shutterstock: **535** • © Antonio S/Shutterstock: **535** • © Kosorukov Dmitry/Shutterstock: **568** • © Madcat_Madlove/Shutterstock: **568** • © Mak3t/Shutterstock: **618** • © ppart/Shutterstock: **618** • © Ryan Fletcher/Shutterstock: **568** • © Science Photo Library/Shutterstock: **535** • © ThomasLENNE/Shutterstock: **535** • © Tim UR/Shutterstock: **535** • © tr3gin/Shutterstock: **618** • © Source: Redrawn by Spatial Vision: **458** • © Maths xpress 10 student resource book 9780731406272: **10B**, page **340, 50** • © Maths xpress 8 student resource book 9780731406159: **3D** Ratios, page **90, 60** • © Maths xpress 8 student resource book 9780731406159: **3F**, page **101, 70** • © Maths xpress 8 student resource book 9780731406159: Exercise 3D, page **94, 64** • © Source: Data from Cases by vicdhhs, Tableau public. https://public.tableau.com/app/profile/vicdhhs/viz/Cases_15982342702770/DashCasesGSG: **300** • © Source: Example 2: Tax invoice for a sale of more than $1,000, Tax invoices. © Australian Taxation Office for the Commonwealth of Australia: **420** • © Source: Example: Business activity statement - front, Example activity statement. Retrieved from https://www.ato.gov.au/business/business-activity-statements-(bas)/in-detail/instructions/payg-withholding---how-to-complete-your-activity-statement-labels/?page=2. © Australian Taxation Office for the Commonwealth of Australia: **421** • © Source: Example: Business activity statement - rear, Example activity statement. Retrieved from https://www.ato.gov.au/business/business-activity-statements-(bas)/in-detail/instructions/payg-withholding---how-to-complete-your-activity-statement-labels/?page=2. © Australian Taxation Office for the Commonwealth of Australia: **422**

Text

© Source: Table extract from Department of Education and Training, *Completion Rates of Higher Education Students — cohort analysis*, 2005–2014: **353**

Every effort has been made to trace the ownership of copyright material. Information that will enable the publisher to rectify any error or omission in subsequent reprints will be welcome. In such cases, please contact the Permissions Section of John Wiley & Sons Australia, Ltd.

1 Calculations

LEARNING SEQUENCE

Fully worked solutions for this topic are available online.

1.1 Overview

Hey students! Bring these pages to life online

Watch videos

Engage with interactivities

Answer questions and check results

Find all this and MORE in jacPLUS

1.1.1 Introduction

Using addition, subtraction, multiplication and division is essential in understanding mathematics, since these operations are the building blocks of calculations. These calculations involve the use of integers, decimals and fractions, and help us answer questions like 'how much?', 'how far?' and 'how many?'. These essential building blocks allow us to answer these everyday questions.

Examples of where these calculations are used in the real world are everywhere. A carpet layer uses calculations to determine how much carpet is required to carpet a house, and a fashion designer calculates the amount of fabric needed to make clothes. These calculations are the building blocks in mathematics and will be covered in this chapter.

KEY CONCEPTS

This topic covers the following key concepts from the VCE Mathematics Study Design:

- mathematical conventions and notations for number and number operations
- rational numbers and irrational numbers related to measurement, ratios and proportions in a practical context
- estimation and approximation including interval estimates, rounding, significant figures, leading-digit approximations, floor and ceiling values and percentage error.

Note: *Concepts shown in grey are covered in other topics.*

1.2 Number operations

LEARNING INTENTION

At the end of this subtopic you should be able to:

- solve practical numerical problems using various number operations.

1.2.1 Practical numerical problems

We constantly solve practical numerical problems in our head without realising that we are doing it. It could be calculating what time you need to leave home to get to school on time or how much money you need to purchase food from the school canteen. Life constantly requires us to solve these numerical problems, so it is important that we can do so quickly and accurately.

WORKED EXAMPLE 1 Solving practical numerical problems

A USB stick can hold 512 MB of data. If 386 MB of the USB stick is already filled, determine how much space is left.

THINK	WRITE
1. State the total storage space.	Space available = 512 MB
2. State how much space has been used.	Space used = 386 MB
3. The space left is the difference between the total space and the space used.	Space left = 512 − 386 = 126 MB
4. State the answer.	The space left on the USB stick is 126 MB.

WORKED EXAMPLE 2 Solving real-world problems using calculations

Nathan has a part-time job that pays $15.50 per hour. Nathan gets paid time and a half for hours worked on Saturdays and double time for hours worked on Sundays. Determine how much Nathan gets paid in a week in which they work 5 hours on Friday, 4 hours on Saturday and 5.5 hours on Sunday.

THINK	WRITE
1. Calculate the amount Nathan earned on Friday.	Money earned Friday = 5 × $15.50 = $77.50
2. Calculate the amount Nathan earned on Saturday.	Money earned Saturday = 4 × $15.50 × 1.5 = $93.00
3. Calculate the amount Nathan earned on Sunday.	Money earned Sunday = 5.5 × $15.50 × 2 = $170.50
4. Calculate Nathan's weekly pay by adding the amounts earned on Friday, Saturday and Sunday.	Total pay = $77.50 + $93.00 + $170.50 = $341.00
5. State Nathan's weekly pay.	Nathan earned $341.00 that week.

1.2 Exercise

Students, these questions are even better in jacPLUS

Receive immediate feedback and access sample responses

Access additional questions

Track your results and progress

Find all this and MORE in jacPLUS

1. WE1 The monthly data allowance included in your mobile phone plan is 15 GB. If you have already used 12 GB of data this month, determine how much data you have left to use for the rest of the month.
2. A UHD television was priced at $8999. It has a sale sign on it that reads: 'Take a further $1950 off the marked price.' Determine the sale price of the television.
3. A Year 12 class held a car wash that raised $345. The detergent, buckets and sponges cost $28 in total. Determine the profit from the car wash.
4. You went to the shops to buy a birthday present for your friend. You took $25 to the shops and spent a total of $19 on a present and a card.
   a. If the card cost $2, determine the cost of the present.
   b. Determine the amount of money left over from the $25 you had initially.
5. The photographs show three tall structures.

   a. Determine how much taller the Eiffel Tower is than the Chrysler Building.
   b. Calculate the height difference between the Infinity Tower and the Eiffel Tower.
   c. Determine the difference in height between the Chrysler Building and the Infinity Tower.
   d. Explain how you obtained your answers to parts a, b and c.

6. This sign shows the distances to a number of locations in the Northern Territory and Queensland.
   a. Determine how much further Tobermorey is from Harts Range.
   b. Calculate the distance between Boulia and Tobermorey.

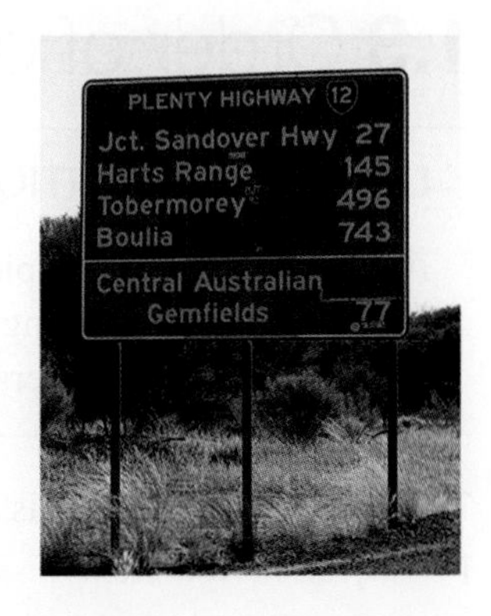

7. There are 400 students in total at a high school. If there are 70 students in Year 7, 73 students in Year 8, 68 students in Year 9, 65 students in Year 10 and 72 students in Year 11, determine the number of students in Year 12.

8. Consider the digits 2, 3, 4 and 5.
   a. Construct the largest possible number using these digits.
   b. Construct the smallest possible number using these digits.
   c. Calculate the difference between the numbers from parts **a** and **b**.

9. Assume your seventeenth birthday is today.
   a. Determine your age in weeks. (Assume 52 weeks in a year.)
   b. Determine your age in days. (Assume 365 days in a year.)

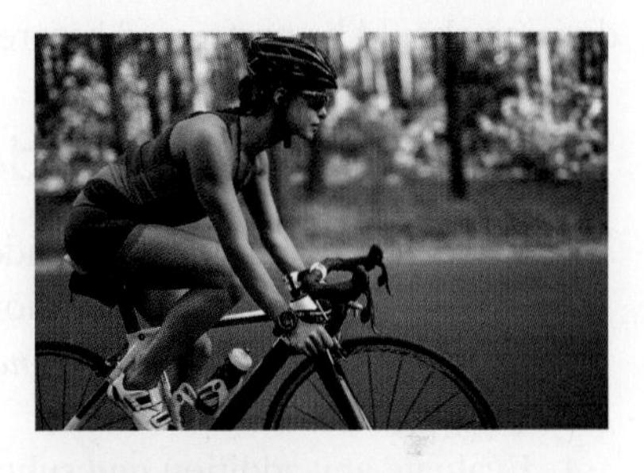

10. While training for a triathlon, your friend follows a specific training program which includes running 15 km, cycling 25 km and swimming 10 km each week. If your friend trains for 8 weeks, calculate the total distance she will travel in her training.

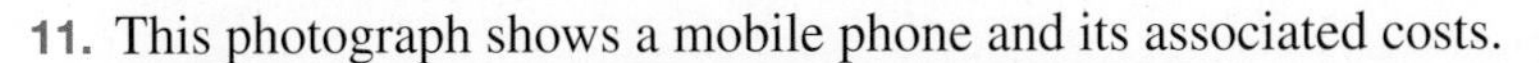

11. This photograph shows a mobile phone and its associated costs.
   a. Determine the cost of a 1-minute call on this phone.
   b. Determine the cost of a 35-minute call on this phone.
   c. If you use the phone to make five 3-minute calls every day, calculate how much the calls will cost you for a year. (Assume 365 days in a standard year.)
   d. Determine the cost of sending 20 text messages from this phone.
   e. Calculate the difference in cost between sending 100 text messages and sending 50 picture messages using this phone.

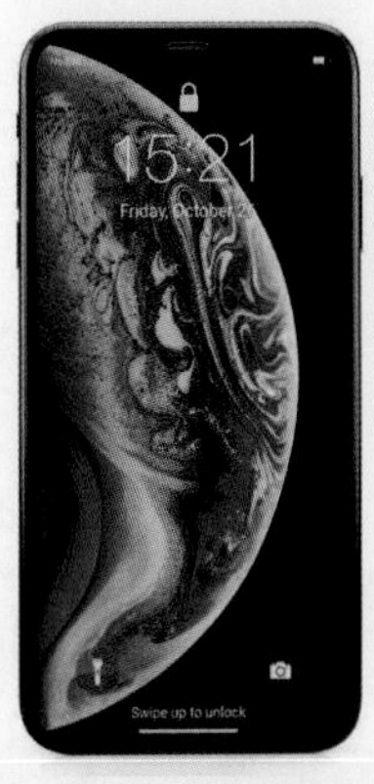

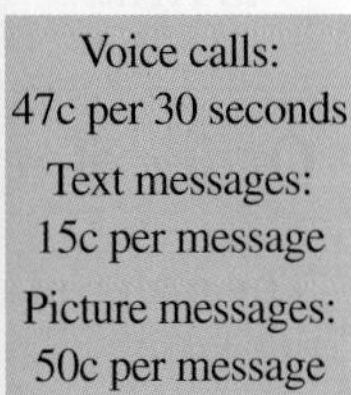

12. **WE2** Sarah has a part-time job that pays $18.50 per hour. Sarah gets paid time and a half for hours worked on Saturdays and double time for hours worked on Sundays. Use your calculator to calculate how much Sarah gets paid in a week in which she works 4 hours on Friday, 5 hours and 30 minutes on Saturday, and 8 hours on Sunday.

13. Assuming that each egg in the photograph is the same size, determine the mass of each egg.

14. In a class of 24 students, each student plays cricket, soccer or both. If 16 students play cricket and 10 students play soccer, determine the number of students who play both sports.

1.3 Order of operations

LEARNING INTENTION

At the end of this subtopic you should be able to:
- solve problems using the order of operations
- use the order of operations to solve real-world problems.

The **order of operations** is a set of rules that determines the order in which mathematical operations are to be performed.

Consider $8 + 12 \div 4$. If you perform the addition first, the answer is $20 \div 4 = 5$. If you perform the division first, the answer is $8 + 3 = 11$. The correct answer to $8 + 12 \div 4$ is 11, because the order of operations specifies that division should be performed before addition.

1.3.1 Rules for the order of operations

1. Evaluate any calculations inside brackets.
2. Evaluate any indices calculations (powers or roots).
3. Evaluate any multiplication *and* division calculations, working in the order in which they appear *from left to right.*
4. Evaluate any addition *and* subtraction calculations, working in the order in which they appear *from left to right.*

Order of operations

The order of operations rules can be remembered as BIDMAS:

B	I	D	M	A	S
Brackets	Indices	Divide	Multiply	Add	Subtract
( )	$\sqrt{x}$ x^2	$\div$ or	$\times$	$+$ or	$-$

WORKED EXAMPLE 3 Calculating using the order of operations rules

Use the order of operations rules to calculate each of the following.

a. $6 \div 3 \times 9 + 3$

b. $12 \div 2 + 4 \times (4 + 6)$

c. $\{4 + [3 \times 2 + (15 - 8)]\} + 5$

d. $\left[(4 + 2)^2 + 6\right] \div 7$

e. $\dfrac{3 \times \left(\sqrt{40 - 4} - 3\right)^2 + (10 \div 2)}{12 \div 3}$

THINK	WRITE
a. 1. Write the question.	a. $6 \div 3 \times 9 + 3$
2. No brackets or powers are included in this question. Evaluate the division and multiplication from left to right.	$= 2 \times 9 + 3$ $= 18 + 3$

3. Evaluate the addition to calculate the answer. $= 21$

b. 1. Write the question. **b.** $12 \div 2 + 4 \times (4 + 6)$

2. Evaluate the addition inside the brackets first. $= 12 \div 2 + 4 \times 10$

3. Evaluate the division and multiplication, working from left to right. $= 6 + 4 \times 10$
$= 6 + 40$

4. Evaluate the addition to calculate the answer. $= 46$

c. 1. Write the question. **c.** $\{4 + [3 \times 2 + (15 - 8)]\} + 5$

2. Evaluate the innermost brackets by calculating the answer to 15 − 8. $= \{4 + [3 \times 2 + 7]\} + 5$

3. Evaluate the next pair of brackets by first calculating the multiplication and then the addition. $= \{4 + [6 + 7]\} + 5$
$= \{4 + 13\} + 5$

4. Evaluate the last pair of brackets by calculating the answer to 4 + 13. $= 17 + 5$

5. Evaluate the addition to calculate the answer. $= 22$

d. 1. Write the question. **d.** $\left[(4 + 2)^2 + 6\right] \div 7$

2. Evaluate the inner brackets first by adding 4 and 2. $= [6^2 + 6] \div 7$

3. Evaluate the indices by calculating 6^2. $= [36 + 6] \div 7$

4. Evaluate the addition in the remaining brackets. $= 42 \div 7$

5. Evaluate the division to calculate the answer. $= 6$

e. 1. Write the question. **e.** $\dfrac{3 \times \left(\sqrt{40 - 4} - 3\right)^2 + (10 \div 2)}{12 \div 3}$

2. Evaluate any expressions in brackets. In the first bracket, evaluate the expression under the square root first, then complete the subtraction.

$= \dfrac{3 \times \left(\sqrt{36} - 3\right)^2 + (10 \div 2)}{12 \div 3}$

$= \dfrac{3 \times (6 - 3)^2 + (10 \div 2)}{12 \div 3}$

$= \dfrac{3 \times 3^2 + 5}{12 \div 3}$

3. Evaluate the power term by squaring 3. $= \dfrac{3 \times 9 + 5}{12 \div 3}$

4. Evaluate the multiplication and then the addition in the numerator. Evaluate the division in the denominator. $= \dfrac{32}{4}$

5. Calculate the answer. $= 8$

Using a calculator to evaluate expressions involving the order of operations

Many scientific calculators can input and evaluate expressions in one line.

In Worked example 3d, we are asked to evaluate $\left[(4+2)^2+6\right] \div 7$.

To determine its value on a calculator, follow the steps below.

- Input the expression $\left((4+2)^2+6\right) \div 7$ in one continuous line into your calculator using only round brackets.
- Press = or ENTER to evaluate.

$$\therefore \left[(4+2)^2+6\right] \div 7 = 6$$

In Worked example 3e, we are asked to evaluate $\dfrac{3 \times \left(\sqrt{40-4}-3\right)^2 + (10 \div 2)}{12 \div 3}$.

To determine its value on a calculator, follow the steps below.

- If the calculator has a fraction template, access this function. This will set up two entry boxes — one for the numerator and one for the denominator.
- Input $3 \times \left(\sqrt{40-4}-3\right)^2 + (10 \div 2)$ into the entry box for the numerator.
- Input $12 \div 3$ into the entry box for the denominator.
- Press = or ENTER to evaluate.

$$\therefore \frac{3 \times \left(\sqrt{40-4}-3\right)^2 + (10 \div 2)}{12 \div 3} = 8$$

WORKED EXAMPLE 4 Applying the order of operations to real-world problems

Sean buys 4 oranges at \$0.70 each and 10 pears at \$0.55 each.

a Calculate the price Sean paid for the oranges and pears.

b Calculate the average price Sean paid for a piece of fruit.

THINK	WRITE
a. 1. Calculate the total amount of money Sean spent on fruit.	a. $(4 \times \$0.70) + (10 \times \$0.55) = \$2.80 + \$5.50 = \$8.30$
2. Write the answer.	The total amount Sean spent on fruit is \$8.30.
b. 1. Count the total number of pieces of fruit that Sean purchased.	b. 4 oranges + 10 pears = 14 pieces of fruit
2. The average price Sean paid for a piece of fruit is equal to the total cost divided by the total number of pieces.	$\text{Average price} = \dfrac{\text{total cost}}{\text{total number of pieces}} = \dfrac{\$8.30}{14} = \$0.59$
3. Write the answer.	The average price Sean paid for a piece of fruit is \$0.59.

 Resources

Video eLesson BIDMAS (eles-1883)

Interactivity Order of operations (int-3707)

1.3 Exercise

Students, these questions are even better in jacPLUS

 Receive immediate feedback and access sample responses

 Access additional questions

 Track your results and progress

Find all this and MORE in jacPLUS

1. WE3a–c Use the order of operations rules to calculate each of the following.
   a. $3 \times 4 \div 2$
   b. $6 \div 3 \times 3 \div 2$
   c. $24 \div (12 - 4)$
   d. $12 \div 4 + 36 \div 6 + 2$
   e. $12 \times (20 - 12)$
2. Use the order of operations rules to calculate each of the following.
   a. $(18 - 15) \div 3 \times 27$
   b. $52 \div 13 + 75 \div 25$
   c. $(12 - 3) \times 8 \div 6$
   d. $\{[(16 + 4) \div 4] - 2\} \times 6$
   e. $5^2 - 15 + 10 \div 5$
3. WE3d–e Use the order of operations rules to calculate each of the following.
   a. $13 + 2^3 \div \left(\sqrt{64} - 1\right) + 2$
   b. $50 - \left(\sqrt{100 \div 4} + 2\right)^2$
   c. $[(5^3 - 25) \div 5^2]^2$
   d. $\dfrac{36 + \sqrt{36} \div 6}{7}$
   e. $\{[55 - (16 \div 4)^2] \div 13\} + 10$
4. Your friend is having trouble with the order of operations. A sample of her work is shown.
   a. Explain what your friend is doing incorrectly.
   b. Show the correct solutions to the questions.

$120 \div 6 \times 2$	$18 - 6 + 5$
$= 120 \div 12$	$= 18 - 11$
$= 10$	$= 7$

5. Use BIDMAS to evaluate each of the following.
   a. $(6 + 3) - (2 + 5)$
   b. $72 + \left(\dfrac{1}{2} \times 40\right) - 3 \times 4$
   c. $56 - 6 \times 4 \div (13 - 7)$
   d. $23 \times 3 - 12 \times 3$
   e. $\dfrac{32}{8} \times 3 + 6^2$
6. Use BIDMAS to evaluate each of the following.
   a. $(4 + 5)^2 - (3 \times 2)$
   b. $\dfrac{[7 + 2 + (3 \times 4)]}{3}$
   c. $\sqrt{(12 - 7) \times 5}$
   d. $\left[(27 \div 3)^2 + 4\right] \div 7$
7. Evaluate each of the following, leaving your answer as a fraction where appropriate.
   a. $\dfrac{13 - 4}{7 + 1}$
   b. $\dfrac{(2 + 3) \times 5}{(12 - 3) \div 3}$
   c. $\dfrac{(6 - 3)^3 - 20}{6^2}$
   d. $\dfrac{4 \times \left[(5 - 3)^3 - (4 \div 2)\right]}{15 \div 3 \times 2}$

8. Insert one set of brackets in the appropriate place to make each of these statements true.

a. $12 - 8 \div 4 = 1$
b. $4 + 8 \times 5 - 4 \times 5 = 40$
c. $3 + 4 \times 9 - 3 = 27$
d. $3 \times 10 - 2 \div 4 + 4 = 10$
e. $10 \div 5 + 5 \times 9 \times 9 = 81$
f. $18 - 3 \times 3 \div 5 = 9$

9. For a birthday party, you buy two packets of paper plates at \$2 each, three bags of chips at \$4 each and three boxes of party pies at \$5 each.
   a. State how the order of operations helps you calculate the correct total cost of these items.
   b. Write an equation to show the operations required to calculate the total cost.
   c. Calculate the total cost.

10. Use the digits 1, 2, 3 and 4 and the operators +, −, × and ÷ to construct equations that result in the numbers 1 to 5 (the numbers 2 and 4 are already done for you). You must use each digit in each expression, and you may not use any digit more than once. You may combine the digits to form larger numbers (like 21). You may also use brackets to make sure the operations are done in the correct order.

$$1 =$$
$$2 = 4 - 3 + 2 - 1$$
$$3 =$$
$$4 = 4 - 2 + 3 - 1$$
$$5 =$$

11. **WE4** Pat buys 5 bananas at \$0.30 each and 8 apples at \$0.25 each.
   a. Calculate the price Pat paid for the bananas and apples.
   b. Calculate the average price Pat paid for a piece of fruit (rounded to the nearest cent).

12. Carrol purchases chocolate for her daughter's party. She buys 6 packets of Mars bars at \$4.85 each and 8 packets of Freddo Frogs at \$3.55 each. Calculate the average price that Carrol paid for each packet of chocolate.

1.4 Rounding

LEARNING INTENTION

At the end of this subtopic you should be able to:
- round integers to the nearest 10, 100 and 1000
- round numbers using leading digit approximation
- use floor and ceiling values to round decimals
- round decimals to any decimal place.

1.4.1 Rounding integers

Numbers can be rounded to different degrees of accuracy.

Rounding to the nearest 10

To round to the nearest 10, think about which multiple of 10 the number is closest to. For example, if 34 is rounded to the nearest 10, the result is 30 because 34 is closer to 30 than it is to 40.

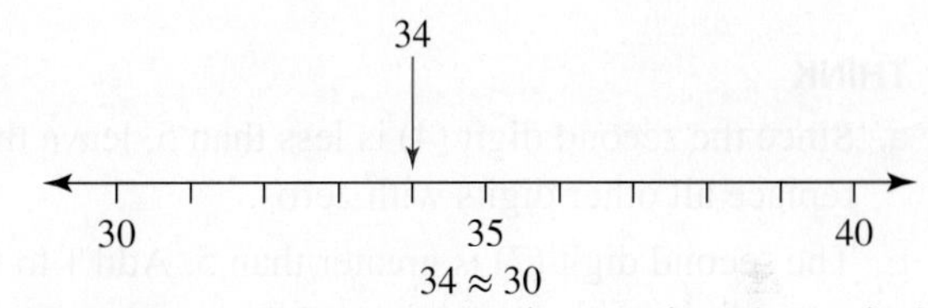

$34 \approx 30$

(*Note:* The symbol ≈ represents 'is approximately equal to'.)

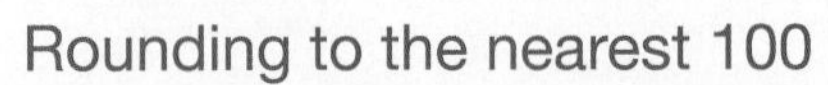

Rounding to the nearest 100

To round to the nearest 100, think about which multiple of 100 the number is closest to.

Leading-digit approximation (rounding to the first digit)

To round to the first (or leading) digit, use the following guidelines:
- Consider the next digit after the leading digit (i.e. the second digit).
- If the second digit is 0, 1, 2, 3 or 4, the first digit stays the same and all the following digits are replaced with zeros.
- If the second digit is 5, 6, 7, 8 or 9, the first digit is raised by 1 (rounded up) and all the following digits are replaced with zeros.

For example, if 2345 is rounded to the first digit, the result is 2000 because the second digit is a 3. On a number line, you can see that 2345 is closer to 2000 than it is to 3000.

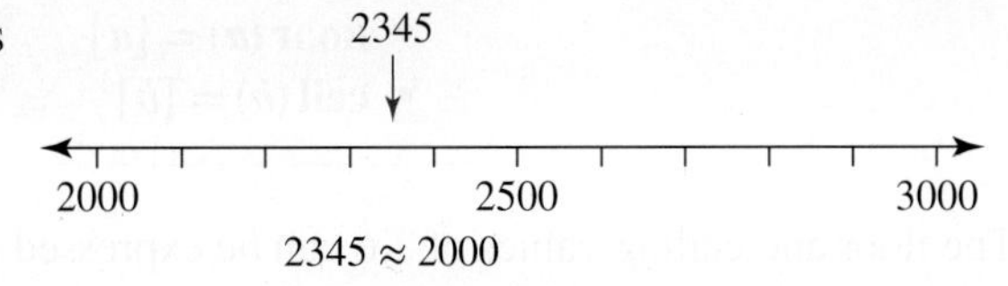

$2345 \approx 2000$

WORKED EXAMPLE 5 Rounding to the nearest 10, 100 or 1000

Round the number 23 743 to the:

a. nearest 10 **b. nearest 100** **c. nearest 1000.**

THINK	WRITE
a. 1. Consider the number starting in the tens position. The number is 43.	a. 23 743
2. Decide what multiple of 10 the number 43 is closest to.	43 is closest to 40.
3. Write the rounded number by replacing 43 with 40.	$23\,743 \approx 23\,740$

b. 1. Consider the number starting in the hundreds position. The number is 743. — b. 23 743

2. Decide what multiple of 100 the number 743 is closest to. — 743 is closest to 700.

3. Write the rounded number by replacing 743 with 700. — $23\,743 \approx 23\,700$

c. 1. Consider the number starting in the thousands position. The number is 3743. — c. 23 743

2. Decide what multiple of 1000 the number 3743 is closest to. — 3743 is closest to 4000.

3. Write the rounded number by replacing 3743 with 4000. — $23\,743 \approx 24\,000$

WORKED EXAMPLE 6 Rounding to the leading digit

Round each of the following numbers to the first (or leading) digit.

a. **5498**

b. **872**

THINK	WRITE
a. Since the second digit (4) is less than 5, leave the leading digit unchanged and replace all other digits with zeros.	a. $5498 \approx 5000$
b. The second digit (7) is greater than 5. Add 1 to the leading digit and replace all other digits with zeros.	b. $872 \approx 900$

1.4.2 Floor and ceiling values

Floor and ceiling values give us the nearest integer.

- The floor value gives us the nearest integer below our value.
- The ceiling value gives us the nearest integer above our value.

Floor and ceiling values

We represent floor and ceiling values as:

- **floor (a) = $\lfloor a \rfloor$**
- **ceil (b) = $\lceil b \rceil$**

The floor and ceiling values of 7.6 can be expressed as:

$$\text{floor}(7.6) \text{ or } \lfloor 7.6 \rfloor \text{ and } \text{ceil}(7.6) \text{ or } \lceil 7.6 \rceil$$

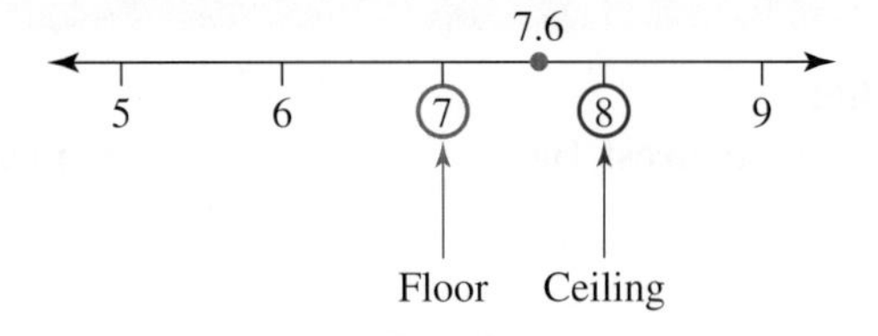

From the diagram:

$$\text{floor}(7.6) \text{ or } \lfloor 7.6 \rfloor = 7$$

$$\text{ceiling}(7.6) \text{ or } \text{ceil}(7.6) \text{ or } \lceil 7.6 \rceil = 8$$

WORKED EXAMPLE 7 Calculating floor and ceiling values

Calculate each of the following.

a. **floor(3.8)** **b.** $\lceil 9.1 \rceil$ **c.** $\lceil 5.3 \rceil + \lfloor 6.8 \rfloor$

THINK	WRITE
a. The floor is the nearest integer below 3.8.	**a.** floor(3.8) = 3
b. 1. The brackets with a 'ceiling' indicate we need to determine the ceiling of 9.1.	**b.** ceil(9.1)
2. The ceiling is the nearest integer above 9.1.	$\lceil 9.1 \rceil = 10$
c. 1. Understand the expression.	**c.** $\lceil 5.3 \rceil + \lfloor 6.8 \rfloor = \text{ceil}(5.3) + \text{floor}(6.8)$
2. Calculate each of the values.	$\lceil 5.3 \rceil + \lfloor 6.8 \rfloor = 6 + 6$
3. Evaluate the expression.	$\lceil 5.3 \rceil + \lfloor 6.8 \rfloor = 6 + 6$ $= 12$

1.4.3 Rounding decimals

When rounding a decimal, a similar approach is used. If a number is to be rounded to 2 decimal places, use the following guidelines:

- Look at the *third* decimal place.
- If the digit in the *third* decimal place is less than 5, leave the second decimal value unchanged and drop off all digits after the second decimal place.
- If the digit in the third decimal place is greater than or equal to 5, add 1 to the second decimal value, then drop off all digits after the second decimal place.

WORKED EXAMPLE 8 Rounding to decimal places

Round 23.1846 to:

a. **1 decimal place** **b.** **2 decimal places.**

THINK	WRITE
a. 1. Consider the digit in the second decimal place, 8.	**a.** 23.1846
2. Since it is greater than or equal to 5, add 1 to the first decimal value and drop off any digits after the first decimal place.	23.2
3. State the answer.	$23.1846 \approx 23.2$
b. 1. Consider the digit in the third decimal place, 4.	**b.** 23.1846
2. Since it is less than 5, leave the second decimal value unchanged and drop off any digits after the second decimal place.	23.18
3. State the answer.	$23.1846 \approx 23.18$

on Resources

Video eLesson Estimating and rounding (eles-0822)

Interactivities Rounding (int-3980)

Rounding (int-3932)

1.4 Exercise

Students, these questions are even better in jacPLUS

Receive immediate feedback and access sample responses

Access additional questions

Track your results and progress

Find all this and MORE in jacPLUS

1. WE5a Round each of the following to the nearest 10.

a. 6 b. 67 c. 173
d. 1354 e. 56 897 f. 765 489

2. WE5b Round each of the following to the nearest 100.

a. 41 b. 91 c. 151
d. 3016 e. 42 578 f. 345 291

3. WE5c Round each of the following to the nearest 1000.

a. 503 b. 1385 c. 6500
d. 12 287 e. 452 999 f. 2 679 687

4. WE6 Round each of the following to the first (or leading) digit.

a. 6 b. 45 c. 1368
d. 12 145 e. 168 879 f. 4 985 452

5. WE7a Calculate each of the following.

a. floor(2.2) b. floor(7.8) c. ceil(2.2) d. ceil(7.8)

6. WE7b Calculate each of the following.

a. $\lfloor 5.3 \rfloor$ b. $\lceil 9.1 \rceil$ c. $\lceil 6.6 \rceil$ d. $\lfloor 1.8 \rfloor$

7. WE7c Calculate each of the following.

a. $\lfloor 3.5 \rfloor + \lfloor 1.8 \rfloor$ b. $\lfloor 7.9 \rfloor + \lceil 4.3 \rceil$ c. $\lceil 4.8 \rceil - \lceil 3.5 \rceil$ d. $\lceil 12.3 \rceil - \lceil 7.3 \rceil$

8. Evaluate each of the following.

a. $\lceil 18.4 \rceil - \lceil 10.6 \rceil$ b. $\lceil 27.3 \rceil + \lceil 1.3 \rceil$ c. $\lfloor 12.7 \rfloor \times \lceil 4.3 \rceil$ d. $\lfloor 36.9 \rfloor \div \lfloor 12.6 \rfloor$

9. WE8a Round the following to 1 decimal place.

a. 0.410 b. 0.87 c. 9.27
d. 25.25 e. 300.06 f. 12.82

10. Round the following to 1 decimal place.

a. 99.91 b. 8.88 c. 17.610 27
d. 0.8989 e. 93.994 f. 0.959 027

11. WE8b Round the following to 2 decimal places.

a. 0.3241 b. 0.863 c. 1.246 10
d. 13.049 92 e. 7.128 63 f. 100.813 82

12. Round the following to 2 decimal places.

a. 71.260 39 b. 0.0092 c. 0.185 00
d. 19.6979 e. 0.3957 f. 0.999

13. Round the following to the number of decimal places shown in the brackets.

a. 2.386 214 (2) b. 14.034 59 (1) c. 0.027 135 (2)
d. 0.876 490 3 (4) e. 64.295 18 (4) f. 0.382 04 (3)

14. Round the following to the number of decimal places shown in the brackets.

a. 96.280 49 (1) b. 3.040 9 (2) c. 8.902 (2)
d. 47.879 69 (3) e. 0.099 498 632 (2) f. 0.486 259 0 (2)

15. **MC** 13.179 rounded to 2 decimal places is equal to:

A. 13.17 B. 13.20 C. 13.18 D. 13.19 E. 13.2

16. **MC** 1.7688 rounded to 3 decimal places is equal to:

A. 1.768 B. 1.770 C. 1.778 D. 1.769 E. 1.800

17. **MC** 2.998 rounded to 1 decimal place is equal to:

A. 3.0 B. 2.9 C. 2.8 D. 3.1 E. 2.99

18. Round the following to the nearest unit.

a. 10.7 b. 8.2 c. 3.6
d. 92.7 e. 112.1 f. 21.76

19. Round the following to the nearest unit.

a. 42.0379 b. 2137.50 c. 0.12
d. 0.513 e. 0.99 f. 40.987

20. Luke is shopping at a local clothes shop. He is interested in buying a pair of jeans marked with a price tag of $129.97. He has cash but also has a debit card. Use this example to explain how rounding works for both cash and debit card in this instance.

1.5 Estimation and approximation strategies

LEARNING INTENTION

At the end of this subtopic you should be able to:

- make estimates and calculations using mental and by-hand methods
- use estimation and other approaches to check for accuracy and reasonableness of results.

Estimating is useful when an accurate answer is not necessary. When you do not need to know an exact amount, an estimate or **approximation** is enough.

An estimate is based on information, so it is not the same as a guess. You can estimate the number of people in attendance at a sportsground based on an estimate of the fraction of seats filled.

Estimation using rounding can help you check if your calculations are correct.

WORKED EXAMPLE 9 Estimating in the real world

Estimate the number of people in attendance at Rod Laver Arena using the information in the photograph.

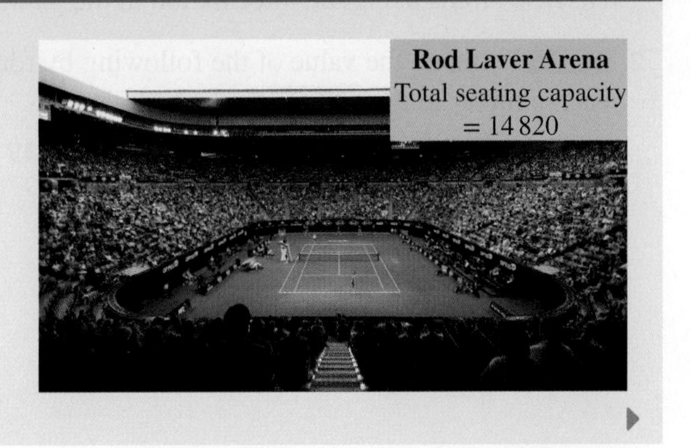

THINK	WRITE
1. Estimate the fraction of seats that are filled in the photograph.	Approximately 4 tenths of the seats are filled.
2. Calculate 4 tenths of the total seating capacity of Rod Laver Arena.	$4 \text{ tenths of } 14\,820 = \frac{4}{10} \times 14\,820$ $= 0.4 \times 14\,820$ $= 5928$
3. State the answer.	The estimated number of people in attendance is 5928.

WORKED EXAMPLE 10 Estimation by rounding

Estimate the value of 85 697 × 248 by rounding each number to its first (or leading) digit.

THINK	WRITE
1. Write the question.	$85\,697 \times 248$
2. Round the first number to its first digit. The second digit in 85 697 is 5, so round 85 697 up to 90 000.	$\approx 90\,000 \times 248$
3. Round the second number to its first digit. The second digit in 248 is 4, so round 248 down to 200.	$\approx 90\,000 \times 200$
4. Complete the multiplication.	$\approx 18\,000\,000$

Resources

Interactivity Estimation (int-4318)

1.5 Exercise

Students, these questions are even better in jacPLUS

Receive immediate feedback and access sample responses

Access additional questions

Track your results and progress

Find all this and MORE in jacPLUS

1. WE9 Estimate the number of people in attendance at Hisense Arena using the information in the photograph.

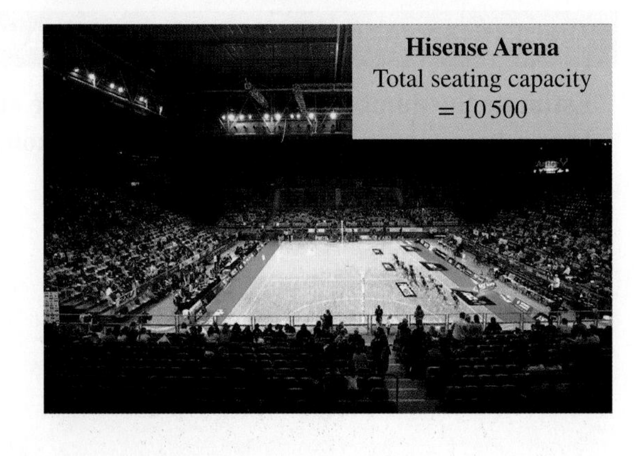

Hisense Arena
Total seating capacity
= 10 500

2. WE10 Estimate the value of the following by rounding each number to the first (or leading) digit.
   a. $487 + 962$
   b. $184\,029 + 723\,419$
   c. $942\,637 - 389\,517$
   d. 87×432
   e. $623 \times 12\,671$
   f. $69\,241 \div 1297$

3. Give three examples of situations in which it is suitable to use an estimate or a rounded value instead of an exact value.

4. An online map tells you that it will take 30 minutes to drive from your house to your cousin's house.
   a. State whether it will take *exactly* 30 minutes.
   b. Determine how the map designers might have estimated this amount of time.
5. In your own words, describe the difference between an estimate and a guess.
6. Describe one way to estimate how many students there are at your school.
7. Sports commentators often estimate the crowd size at sports events. They do this by estimating the percentage of the seats that are occupied, and then use the venue's seating capacity to estimate the crowd size.
   a. The photograph shows the crowd at a soccer match at the Brisbane Cricket Ground (the Gabba). Use the information given with the photograph to calculate the estimated number of people in attendance.
   b. Choose two well-known sports venues and research the maximum seating capacity for each.
   c. If you estimate that three out of every four seats are occupied at each of the venues you chose in part b, determine how many people are in attendance at each venue.

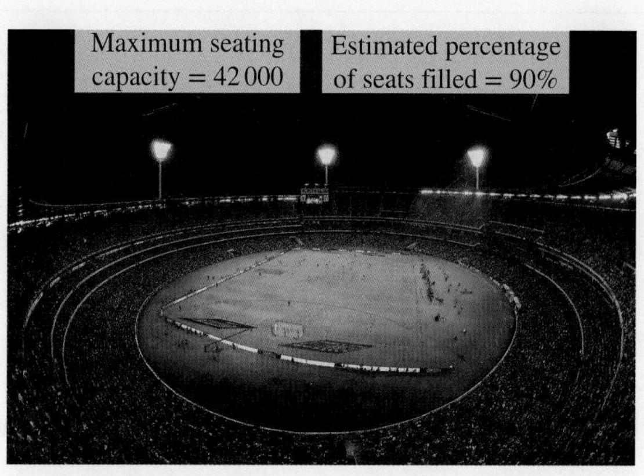

8. i. Estimate the answer to each of the following expressions by rounding to the first (or leading) digit.
   ii. Calculate the exact answer using a calculator.
   a. $46 + 85$
   b. $478 + 58 + 2185$
   c. $37 - 25$
   d. $54 - 28$
   e. $2458 - 1895$
9. i. Estimate the answer to each of the following expressions by rounding to the first (or leading) digit.
   ii. Calculate the exact answer using a calculator.
   a. 25×58
   b. 197×158
   c. $10\,001 - 572$
   d. $23\,547 \times 149$
10. i. Estimate the answer to each of the following by rounding to the nearest 10.
    ii. Calculate the exact answer using a calculator, writing answers as mixed numbers where appropriate.
    a. $68 \div 8$
    b. $158 \div 8$
    c. $425 \div 11$
    d. $3694 \div 6$
    e. $\frac{245}{5}$
    f. $\frac{168}{7}$
11. In the photo shown, estimate:
    a. the number of apples
    b. the number of grapes
    c. the number of peaches and nectarines.

12. Three 97-cm lengths of wood are needed to make a bookcase. When buying the wood at a hardware store, determine if the lengths should be rounded in any way. If so, state how they should be rounded.

13. Estimate the following by first rounding each number to its leading digit.

a. 957×13 b. $3634 \div 161$

14. Estimate the following by first rounding each number to its leading digit.

a. $\frac{(389+429)}{374}$ b. $\frac{(238 \times 275)}{(12+48)}$

1.6 Decimals and fractions

LEARNING INTENTION

At the end of this subtopic you should be able to:
- identify and compare place value
- add and subtract decimals
- multiply by powers of 10.

1.6.1 Decimal place value

In a decimal number, the whole number part and the fractional part are separated by a **decimal point**.

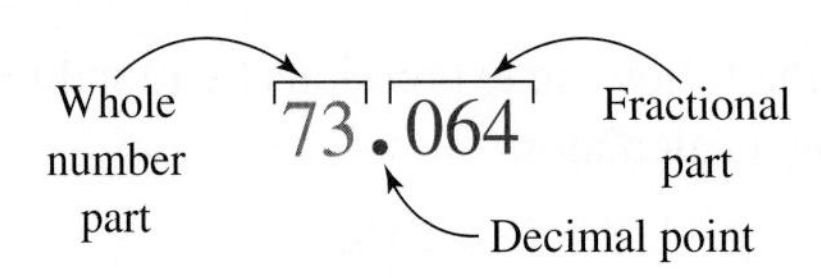

A place-value table can be extended to include decimal place values. It can be continued to an infinite number of **decimal places**.

Thousands	Hundreds	Tens	Ones	•	Tenths	Hundredths	Thousandths	Ten-thousandths
1000	100	10	1	•	$\frac{1}{10}$	$\frac{1}{100}$	$\frac{1}{1000}$	$\frac{1}{10\,000}$

The decimal number 73.064 represents 7 tens, 3 ones, 0 tenths, 6 hundredths and 4 thousandths.

In **expanded form**, 73.064 is written as:

$$(7 \times 10) + (3 \times 1) + \left(0 \times \frac{1}{10}\right) + \left(6 \times \frac{1}{100}\right) + \left(4 \times \frac{1}{1000}\right)$$

When reading decimals, the whole number part is read normally and each digit to the right of the decimal point is read separately. For example, 73.064 is read as 'seventy-three point zero six four'.

The number of decimal places in a decimal is the number of digits after the decimal point. The number 73.064 has 3 decimal places.

The zero (0) in 73.064 means that there are no tenths.

Any whole number can be written as a decimal number by showing the empty decimal places after the decimal point. For example, the number 2 can be written as 2.0, 2.00 or with any number of zeros after the decimal point. These are not place-holding zeros; they are trailing zeros.

WORKED EXAMPLE 11 Place value

Write the value of the digit 5 in each of the following decimal numbers.

a. 2.005 **b. 498.57** **c. 0.05**

THINK	WRITE
a. The digit 5 is the third decimal place to the right of the decimal point, so the value is thousandths.	a. $\frac{5}{1000}$ or 5 thousandths
b. The digit 5 is the first decimal place to the right of the decimal point, so the value is tenths.	b. $\frac{5}{10}$ or 5 tenths
c. The digit 5 is the second decimal place to the right of the decimal point, so the value is hundredths.	c. $\frac{5}{100}$ or 5 hundredths

1.6.2 Comparing decimals

Decimals are compared using digits with the same place value.

The decimal with the largest number in the highest place-value column is the largest decimal, *regardless of the number of decimal places.*

For example, 15.71 is larger than 15.702 because the first place value with different digits (moving from left to right) is hundredths, and 1 is greater than 0; that is, 15.71 > 15.702.

Tens	Ones •	Tenths	Hundredths	Thousandths
1	5 •	7	1	
1	5 •	7	0	2

1.6.3 Adding and subtracting decimals

A decimal number has an infinite number of decimal place values in its fractional part.

Trailing zeros are zeros that appear to the right of the decimal number and have no digits following them other than zero.

When decimals with different numbers of decimal places are added or subtracted, trailing zeros can be written so that both decimals have the same number of decimal places.

12.300 000

↖

Trailing zeros

Decimal numbers can be added and subtracted in a similar manner to whole numbers. Set out the numbers to be added or subtracted in vertical columns so that the decimal points are lined up.

Answers to decimal addition and subtraction may be checked mentally by rounding each decimal number to the nearest whole number and then adding or subtracting the rounded numbers.

WORKED EXAMPLE 12 Adding decimals

Calculate the value of 197.76 + 52.9. Check your answer using a calculator.

THINK	WRITE
1. Set out the problem in vertical columns so that the decimal points are lined up. Fill in the empty place values with trailing zeros, as shown in pink.	$\begin{array}{r} 197.76 \\ +52.90 \\ \hline \end{array}$
2. Add the digits as you would for whole numbers. Write the decimal point in the answer in line with the decimal points in the question.	$\begin{array}{r} {}^{1}1{}^{1}9{}^{1}7.76 \\ +\ \ 52.90 \\ \hline 250.66 \\ \hline \end{array}$
3. Write the answer and check using a calculator.	$197.76 + 52.9 = 250.66$

WORKED EXAMPLE 13 Subtracting decimals

Calculate the value of 125.271 – 85.08. Check your answer using a calculator.

THINK	WRITE
1. Set out the subtraction problem in vertical columns. Line up the decimal points so that the digits of the same place value are underneath each other. Fill in the empty place values with trailing zeros, as shown in pink.	$\begin{array}{r} 125.271 \\ -\ \ 85.080 \\ \hline \end{array}$
2. Subtract the digits as you would for whole numbers. Write the decimal point in the answer in line with the decimal point in the question.	$\begin{array}{r} {}^{0}\not{1}{}^{1}25.{}^{1}\not{2}{}^{1}71 \\ -\ \ 85.080 \\ \hline 40.191 \\ \end{array}$
3. Write the answer and check using a calculator.	$125.271 - 85.08 = 40.191$

1.6.4 Multiplying by powers of 10

Powers of 10 are 10, 100, 1000, 10 000 and so on. When you multiply a decimal by a power of 10, you move the position of the decimal point to the right by the number of zeros in the power of 10.

Multiples of 10 are 10, 20, 30, 40, ... , 120, ... , 400, ... , 16 000 and so on. Multiplying by 400 is the same as multiplying by 4 and then by 100.

Powers

$$10^1 = 10$$
$$10^2 = 100$$
$$10^3 = 1000$$
$$10^4 = 10\,000$$

Multiples

$$1 \times 100 = 100$$
$$2 \times 100 = 200$$
$$3 \times 100 = 300$$
$$4 \times 100 = 400$$

WORKED EXAMPLE 14 Multiplying by powers of 10

Calculate the following.

a. $\mathbf{1.34 \times 100}$ **b.** $\mathbf{64.7 \times 1000}$

THINK	WRITE
a. 1. Write the number 1.34 and identify where the decimal point is.	**a.** 1.34
2. There are 2 zeros in 100, so move the decimal point 2 places to the right.	1.34 $\rightarrow$ 134
3. Write the answer.	$1.34 \times 100 = 134$
b. 1. Write the number 64.7 and identify where the decimal point is.	**b.** 64.7
2. There are 3 zeros in 1000, so move the decimal point 3 places to the right. Add zeros where necessary.	64.700 $\rightarrow$ 64 700
3. State the answer.	$64.7 \times 1000 = 64\,700$

WORKED EXAMPLE 15 Multiplying by multiples of 10

Calculate the value of 2.56 $\times$ 7000. Check your answer using a calculator.

THINK	WRITE
1. Multiplying by 7000 is the same as multiplying by 7 and then by 1000.	$2.56 \times 7000 = 2.56 \times 7 \times 1000$
2. Complete the first multiplication, 2.56×7.	2.56 $\times$ 7 17.92
3. Rewrite the question showing the result of the first multiplication.	$2.56 \times 7 \times 1000 = 17.92 \times 1000$
4. Complete the second multiplication, 17.92×1000, by moving the decimal point in 17.92 three places to the right. Add zeros where necessary.	$= 17\,920$
5. Write the answer and check using a calculator.	$2.56 \times 7000 = 17\,920$

1.6.5 Dividing by powers of 10

When you divide a decimal by a power of 10, you move the position of the decimal point to the left by the number of zeros in the power of 10.

WORKED EXAMPLE 16 Dividing by powers of 10

Calculate the value of the following.

a. $\mathbf{234.25 \div 100}$ **b.** $\mathbf{4.28 \div 1000}$

THINK	WRITE
a. 1. Write the number 234.25 and identify where the decimal point is.	**a.** 234.25

2. There are 2 zeros in 100, so move the decimal point 2 places to the left.

234.25
→ 2.3425

3. Write the answer.

$234.25 \div 100 = 2.3425$

b. 1. Write the number 4.28 and identify where the decimal point is.

b. 4.28

2. There are 3 zeros in 1000, so move the decimal point 3 places to the left. Add zeros before 4 where necessary.

0004.28
→ 0.004 28

3. Write the answer.

$4.28 \div 1000 = 0.004\,28$

1.6.6 Converting decimals to fractions

Decimals can be written as fractions by using place values.

WORKED EXAMPLE 17 Writing decimals as fractions

Convert each of the following decimals to fractions in simplest form.

a. **0.7** b. **3.25**

THINK

WRITE

a. The digit 7 is in the tenths column, so it represents 7 tenths, which is written as $\frac{7}{10}$.

Ones	• Tenths	Hundredths
0	• 7	0

a. $0.7 = \frac{7}{10}$

b. 1. The number 3.25 consists of 3 ones, 2 tenths $\left(\frac{2}{10}\right)$ and 5 hundredths $\left(\frac{5}{100}\right)$.

Ones	• Tenths	Hundredths
3	• 2	5

b. $3.25 = 3 + \frac{2}{10} + \frac{5}{100}$

2. Add 2 tenths and 5 hundredths by writing $\frac{2}{10}$ as a fraction with a denominator of 100.

$$3.25 = 3 + \frac{20}{100} + \frac{5}{100}$$
$$= 3 + \frac{25}{100}$$

3. Simplify $\frac{25}{100}$ by cancelling by a factor of 25.

$$= 3 + \frac{{}^{1}\cancel{25}}{\cancel{100}^{4}}$$
$$= 3 + \frac{1}{4}$$
$$= 3\frac{1}{4}$$

1.6.7 Converting fractions to decimals

A fraction can be expressed as a decimal by dividing the numerator by the denominator.

$$\frac{5}{8} = 5 \div 8$$

When a division results in a remainder, add a trailing zero and continue the division until there is no remainder or until a pattern can be seen.

If a division reaches a stage where there is no remainder, the decimal is called a **finite decimal**.

If a division continues endlessly with a **repeating pattern**, it is called an **infinite recurring decimal**.

Infinite recurring decimals can be written in an abbreviated form by placing a dot above the repeating pattern. When more than one digit is repeated, a dot is placed above the first digit and last digit in the repeating pattern, or a bar is placed above the entire repeating pattern.

Common recurring decimals

$$\frac{1}{3} = 0.333\,333\ldots = 0.\dot{3}$$

$$\frac{1}{6} = 0.166\,666\ldots = 0.1\dot{6}$$

$$\frac{1}{7} = 0.142\,871\,4\ldots = 0.\dot{1}4285\dot{7}$$

tlvd-5952

WORKED EXAMPLE 18 Writing fractions as decimals

Write $\frac{5}{8}$ as a decimal. Check your answer using a calculator.

THINK	WRITE
1. The fraction $\frac{5}{8}$ can be expressed as $5 \div 8$. Write the division.	$8\overline{)5}$
2. Divide 8 into 5; the result is 0 remainder 5. Write 0 above 5 and write the remainder, as shown in pink. Add a decimal point and trailing zero next to 5.	$\begin{array}{r} 0 \\ 8\overline{)5.{}^{5}0} \end{array}$
3. Divide 8 into 50; the result is 6 remainder 2. Write the decimal point in the answer above the decimal point in the question. Write 6 above 0 and write the remainder, as shown in red. Add another trailing zero.	$\begin{array}{r} 0.\,6 \\ 8\overline{)5.{}^{5}0{}^{2}0} \end{array}$
4. Divide 8 into 20; the result is 2 remainder 4. Write 2 above 0 and write the remainder, as shown in purple. Add another trailing zero.	$\begin{array}{r} 0.\,6\,2 \\ 8\overline{)5.{}^{5}0{}^{2}0{}^{4}0} \end{array}$
5. Divide 8 into 40; the result is 5. Write 5 above 0, as shown in pink. There is no remainder, so the division is finished.	$\begin{array}{r} 0.\,6\,2\,5 \\ 8\overline{)5.{}^{5}0{}^{2}0{}^{4}0} \end{array}$
6. Write the answer and check using a calculator.	$\frac{5}{8} = 0.625$

tlvd-3505

WORKED EXAMPLE 19 Writing mixed fractions as decimals

Write $6\frac{2}{3}$ as a decimal. Check your answer using a calculator.

THINK	WRITE
1. The mixed number $6\frac{2}{3}$ can be converted to a decimal by converting the fractional part.	$6\frac{2}{3} = 6 + \frac{2}{3}$
2. The fraction $\frac{2}{3}$ can be expressed as $2 \div 3$. Write the division.	$3\overline{)2}$
3. Divide 3 into 2; the result is 0 remainder 2. Write 0 above 2 and write the remainder beside the next smallest place value, as shown in pink. Write the decimal point in the answer above the decimal point in the question and add a trailing zero.	$\begin{array}{r} 0 \\ 3\overline{)2.{}^{2}0} \end{array}$
4. Divide 3 into 20; the result is 6 remainder 2. Write 6 above 0 and write the remainder, as shown in red. Add a trailing zero.	$\begin{array}{r} 0.\ 6 \\ 3\overline{)2.{}^{2}0{}^{2}0} \end{array}$
5. Divide 3 into 20; the result is 6 remainder 2. This remainder is the same as the previous remainder. The remainder will continue to be the same. The decimal is a recurring decimal. Write the fraction and its equivalent decimal.	$\begin{array}{r} 0.\ 6\ 6 \\ 3\overline{)2.{}^{2}0{}^{2}0{}^{2}0} \end{array}$ $\frac{2}{3} = 0.\dot{6}$
6. Write the fractional part of the mixed number as a decimal.	$6\frac{2}{3} = 6 + \frac{2}{3}$ $= 6 + 0.\dot{6}$
7. Add the whole number and the decimal part. Write the answer and check using a calculator.	$= 6.\dot{6}$

on Resources

Video eLesson Place value (eles-0004)

Interactivities Addition of decimals (int-3982)
Subtraction of decimals (int-3984)
Comparing decimals (int-3976)
Conversion of decimals to fractions (int-3978)
Decimal parts (int-3975)
Division of decimals by a multiple of 10 (int-3990)
Place value and comparing decimals (int-4337)

1.6 Exercise

Students, these questions are even better in jacPLUS

Receive immediate feedback and access sample responses

Access additional questions

Track your results and progress

Find all this and MORE in jacPLUS

1. State the number of decimal places in each of the following decimals.

a. 548.5845 b. 0.007 c. 1.1223
d. 15.001 e. 4.1 f. 13.42

2. a. State how many decimal places we use in our currency.
b. Write the amount of money shown in the photograph:
i. in words
ii. in numbers.

3. Identify the place-holding zeros in each of the following.

a. 10.23 b. 105.021
c. 11.010 d. 0.001
e. 282.0001 f. 15.00

4. Write each of the following in expanded form.

a. 16.02 b. 11.046 c. 222.03
d. 15.11 e. 4.701 f. 68.68

5. Write the decimal number represented by each of the following.

a. $(4 \times 10) + (2 \times 1) + \left(5 \times \frac{1}{10}\right) + \left(0 \times \frac{1}{100}\right) + \left(4 \times \frac{1}{1000}\right)$

b. $\left(2 \times \frac{1}{10}\right) + \left(7 \times \frac{1}{100}\right) + \left(2 \times \frac{1}{1000}\right)$

c. $1 + \left(9 + \frac{1}{1000}\right)$

d. $(3 \times 10) + \frac{9}{10} + \frac{3}{100} + \frac{4}{1000}$

6. **WE11** Write the value of the digit 7 in each of the following decimal numbers.

a. 2.075 b. 15.701 c. 12.087
d. 93.1487 e. 17.16 f. 73.064

7. Use an appropriate method to determine which decimal number is larger in each of the following pairs.

a. 16.273 and 16.2 b. 137.02 and 137.202
c. 95.89 and 95.98 d. 0.001 and 0.0001
e. 0.123 and 0.2 f. 0.0101 and 0.012

8. **WE12** Calculate each of the following. Check your answers using a calculator.

a. $14.23 + 254.52$ b. $79.58 + 18.584$
c. $99.999 + 0.01$ d. $58.369 + 86.12 + 78$
e. $485.5846 + 5 + 584.58 + 0.57$ f. $34.2 + 7076 + 20.5604 + 1.53$

9. WE13 Calculate each of the following. Check your answers using a calculator.

a. $25.3458 - 25.2784$
b. $848.25 - 68.29$
c. $58.8 - 24.584$
d. $470 - 28.57$
e. $15.001 - 0.007$
f. $35.1 - 9.007\,51$

10. Calculate each of the following to 2 decimal places by:

i. rounding all numbers first, then adding or subtracting
ii. adding or subtracting the numbers first, then rounding.

a. $4.457 + 5.386$
b. $47.589 - 35.410$
c. $126.917 - 35.492$
d. $168.268 + 21.253$

11. WE14 Calculate each of the following.

a. 6.284×100
b. 5.3×1000

12. WE15 Calculate each of the following. Check your answers using a calculator.

a. 1.2345×50
b. 1.2345×600
c. 1.2345×7000
d. $1.2345 \times 20\,000$

13. WE16 Calculate each of the following.

a. $3.45 \div 10$
b. $123.98 \div 100$
c. $1245.37 \div 1000$
d. $3.569 \div 10$
e. $0.246 \div 1000$
f. $2.48 \div 100$

14. WE17 Convert each of the following decimals to fractions in simplest form.

a. 0.1
b. 0.5
c. 0.8
d. 0.12
e. 0.21
f. 0.84

15. Convert each of the following decimals to fractions in simplest form.

a. 0.05
b. 0.625
c. 3.8
d. 2.13
e. 12.42
f. 10.0035

16. WE18 & 19 Write each of the following fractions as a decimal. Check your answers using a calculator.

a. $\frac{3}{8}$
b. $\frac{4}{5}$
c. $\frac{5}{4}$
d. $\frac{4}{9}$

17. Write each of the following fractions as a decimal. Check your answers using a calculator.

a. $2\frac{1}{4}$
b. $3\frac{1}{20}$
c. $\frac{13}{2}$
d. $\frac{1}{5}$

1.7 Rational and irrational numbers

LEARNING INTENTION

At the end of this subtopic you should be able to:

- evaluate square and cube roots
- multiply, divide and simplify surd expressions
- round rational and irrational numbers to decimal places and significant figures.

1.7.1 Rational and irrational numbers

Ancient mathematicians found that rational numbers could not be used to label every point on the number line. In other words, they discovered lengths that could not be expressed as fractions. These numbers are called irrational numbers.

A number is irrational if it is not rational — that is, if it cannot be written as a fraction, nor as a terminating or recurring decimal.

Irrational numbers are denoted by the letter I.

Rational and irrational numbers

A rational number is any number that can be written as a fraction.

An irrational number is any number that cannot be written as a fraction.

1.7.2 Roots

The nth root of any positive number can be found. That is, $\sqrt[n]{b}=a$.

For example, $\sqrt{25}=5$ because $25=5\times 5$.

Note: 2 is usually not written in square roots; for example, $\sqrt{25}=\sqrt[2]{25}$.

$\sqrt[3]{64}=4$ because $64=4\times 4\times 4$.

Any scientific calculator can calculate a square root using the square root button ($\sqrt{\ }$) or any root using the nth root button ($\sqrt[n]{\ }$).

tlvd-3506

WORKED EXAMPLE 20 Evaluating square, cube and other roots

Evaluate each of the following.

a. $\sqrt{16}$ b. $\sqrt[3]{27}$ c. $\sqrt[5]{32}$

THINK	WRITE
a. Determine the number that when multiplied by itself gives 16: $16=4\times 4$.	a. $\sqrt{16}=4$
b. Determine the number that when multiplied by itself 3 times gives 27: $27=3\times 3\times 3$.	b. $\sqrt[3]{27}=3$
c. Determine the number that when multiplied by itself 5 times gives 32: $32=2\times 2\times 2\times 2\times 2$.	c. $\sqrt[5]{32}=2$

1.7.3 Surds

When the square root of a number is an irrational number, it is called a *surd*. For example, $\sqrt{10}$ cannot be written as a fraction, nor as a recurring or terminating decimal. It is therefore irrational and is called a surd.

$$\sqrt{10}\approx 3.162\,277\,660\,17\ldots$$

The value of a surd can be approximated using a number line.

For example, $\sqrt{21}$ will lie between 4 and 5, because $\sqrt{16}=4$ and $\sqrt{25}=5$.

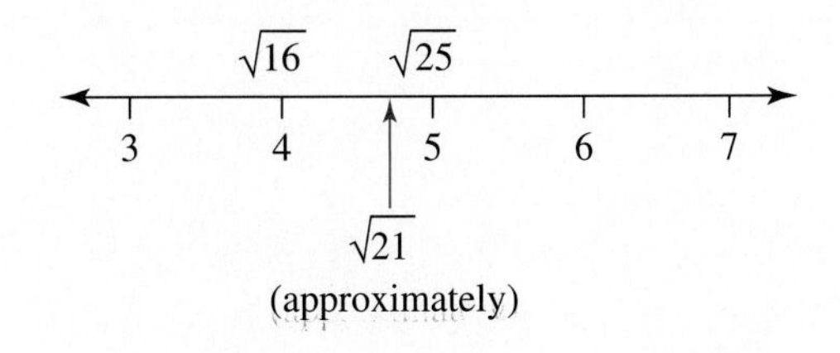

WORKED EXAMPLE 21 Determining surds

Determine which of the following are surds.

a. $\sqrt{0}$ b. $\sqrt{20}$ c. $-\sqrt{9}$ d. $\sqrt[3]{6}$

THINK	WRITE
a. $\sqrt{0}=0$. This is a rational number and therefore not a surd.	a. $\sqrt{0}=0$, which is not a surd.
b. $\sqrt{20}\approx 4.472\,135\,955\ldots$ This is an irrational number and therefore a surd.	b. $\sqrt{20}$ is a surd.
c. $-\sqrt{9}=-3$. This is a rational number and therefore not a surd.	c. $-\sqrt{9}=-3$, which is not a surd.
d. $\sqrt[3]{6}\approx 1.817\,120\,592\,83\ldots$ This is an irrational number and therefore a surd.	d. $\sqrt[3]{6}$ is a surd.

1.7.4 Multiplying and dividing surds

Consider that $3=\sqrt{9}$, $2=\sqrt{4}$, and $6=\sqrt{36}$.

The multiplication $3\times 2=6$ could be written as:

$$\sqrt{9}\times\sqrt{4}=\sqrt{36}$$

In general, $\sqrt{a}\times\sqrt{b}=\sqrt{ab}$.

For example, $\sqrt{7}\times\sqrt{3}=\sqrt{21}$.

Similarly, $\sqrt{a}\div\sqrt{b}=\sqrt{\frac{a}{b}}$.

For example, $\sqrt{18}\div\sqrt{3}=\sqrt{6}$.

Multiplying and dividing surds

Multiplying surds:

$$\sqrt{a}\times\sqrt{b}=\sqrt{ab}$$

Dividing surds:

$$\sqrt{a}\div\sqrt{b}=\sqrt{\frac{a}{b}}$$

tlvd-3507

WORKED EXAMPLE 22 Multiplying surds

Evaluate the following, leaving your answer in surd form.

a. $\sqrt{7}\times\sqrt{2}$ b. $5\times\sqrt{3}$ c. $\sqrt{5}\times\sqrt{5}$ d. $-2\sqrt{3}\times4\sqrt{5}$

THINK	WRITE
a. Apply the rule $\sqrt{a}\times\sqrt{b}=\sqrt{ab}$.	a. $\sqrt{7}\times\sqrt{2}=\sqrt{14}$
b. Only $\sqrt{3}$ is a surd. It is multiplied by 5, which is not a surd.	b. $5\times\sqrt{3}=5\sqrt{3}$
c. Apply the rule $\sqrt{a}\times\sqrt{b}=\sqrt{ab}$.	c. $\sqrt{5}\times\sqrt{5}=\sqrt{25}$ $=5$
d. Multiply the whole numbers by each other. Multiply the surds by each other.	d. $-2\sqrt{3}\times4\sqrt{5}=-2\times4\times\sqrt{3}\times\sqrt{5}$ $=-8\times\sqrt{15}$ $=-8\sqrt{15}$

tlvd-3508

WORKED EXAMPLE 23 Dividing surds

Evaluate the following, leaving your answer in surd form.

a. $\dfrac{\sqrt{10}}{\sqrt{5}}$ b. $\sqrt{\dfrac{10}{5}}$ c. $\dfrac{-6\sqrt{8}}{4\sqrt{4}}$ d. $\dfrac{\sqrt{20}}{\sqrt{5}}$

THINK	WRITE
a. Apply the rule $\sqrt{a}\div\sqrt{b}=\sqrt{\frac{a}{b}}$.	a. $\frac{\sqrt{10}}{\sqrt{5}}=\sqrt{\frac{10}{5}}$ $=\sqrt{2}$
b. Simplify the fraction first.	b. $\sqrt{\frac{10}{5}}=\sqrt{2}$
c. Simplify the whole numbers. Then apply the rule $\sqrt{a}\div\sqrt{b}=\sqrt{\frac{a}{b}}$.	c. $\frac{-6\sqrt{8}}{4\sqrt{4}}=\frac{-3\sqrt{2}}{2}$
d. Apply the rule $\sqrt{a}\div\sqrt{b}=\sqrt{\frac{a}{b}}$.	d. $\frac{\sqrt{20}}{\sqrt{5}}=\sqrt{4}$ $=2$

1.7.5 Simplifying surds

Just as a rational number can be written in many different ways $\left(\text{e.g. } \frac{1}{2}=\frac{5}{10}=\frac{7}{14}\right)$, so can a surd, and it is expected that surds should normally be written in simplest form.

A surd is in simplest form when the number inside the radical sign has the smallest possible value with no roots inside.

Note that $\sqrt{24}$ can be factorised in several ways. For example:

$$\sqrt{24} = \sqrt{2} \times \sqrt{12}$$
$$\sqrt{24} = \sqrt{3} \times \sqrt{8}$$
$$\sqrt{24} = \sqrt{4} \times \sqrt{6}$$

In the last case notice a square number, $\sqrt{4} = 2$.

$$\begin{aligned}\sqrt{24} &= 2 \times \sqrt{6} \\ &= 2\sqrt{6}\end{aligned}$$

$2\sqrt{6}$ is equal to $\sqrt{24}$, and $2\sqrt{6}$ is written in simplest form with no remaining square factors under the root symbol.

To simplify a surd you must determine a factor that is also a perfect square: 4, 9, 16, 25, 36, 49 and so on.

WORKED EXAMPLE 24 Reducing surds to their simplest form

Simplify the following surds.

a. $\sqrt{18}$ **b.** $6\sqrt{20}$

THINK	WRITE
a. 1. Rewrite 18 as the product of two numbers, one of which must be square.	**a.** $\sqrt{18} = \sqrt{9} \times \sqrt{2}$
2. Simplify.	$= 3 \times \sqrt{2}$ $= 3\sqrt{2}$
b. 1. Rewrite 20 as the product of two numbers, one of which must be square.	**b.** $6\sqrt{20} = 6 \times \sqrt{4} \times \sqrt{5}$
2. Simplify.	$= 6 \times 2 \times \sqrt{5}$ $= 12\sqrt{5}$

The surd $\sqrt{22}$ cannot be simplified because 22 has no perfect square factors. Surds can be simplified using more than one factor:

$$\begin{aligned}\sqrt{72} &= \sqrt{4} \times \sqrt{18} \\ &= 2\sqrt{18} \\ &= 2 \times \sqrt{9} \times \sqrt{2} \\ &= 2 \times 3\sqrt{2} \\ &= 6\sqrt{2}\end{aligned}$$

1.7.6 Entire surds

The surd $\sqrt{45}$, when simplified, is written as $3\sqrt{5}$.

The surd $\sqrt{45}$ is called an entire surd, because it is written entirely inside the radical sign, whereas $3\sqrt{5}$ is not.

WORKED EXAMPLE 25 Converting to an entire surd

Write $3\sqrt{7}$ as an entire surd.

THINK	WRITE
1. In order to place 3 inside the radical sign, it has to be written as $\sqrt{9}$.	$3\sqrt{7} = \sqrt{9} \times \sqrt{7}$
2. Apply the rule $\sqrt{a} \times \sqrt{b} = \sqrt{ab}$.	$= \sqrt{63}$

Surds should be simplified before adding or subtracting like terms.

tlvd-3509

WORKED EXAMPLE 26 Simplifying surds

Simplify each of the following.

a. $5\sqrt{75} - 6\sqrt{12} + \sqrt{8} - 4\sqrt{3}$

b. $\dfrac{5}{\sqrt{5}}$

THINK	WRITE
a. 1. Simplify $5\sqrt{75}$.	a. $5\sqrt{75} = 5 \times \sqrt{25} \times \sqrt{3}$ $= 5 \times 5 \times \sqrt{3}$ $= 25\sqrt{3}$
2. Simplify $6\sqrt{12}$.	$6\sqrt{12} = 6 \times \sqrt{4} \times \sqrt{3}$ $= 6 \times 2 \times \sqrt{3}$ $= 12\sqrt{3}$
3. Simplify $\sqrt{8}$.	$\sqrt{8} = \sqrt{4} \times \sqrt{2}$ $= 2\sqrt{2}$
4. Rewrite the original expression and simplify by adding like terms.	$5\sqrt{75} - 6\sqrt{12} + \sqrt{8} - 4\sqrt{3}$ $= 25\sqrt{3} - 12\sqrt{3} + 2\sqrt{2} - 4\sqrt{3}$ $= 9\sqrt{3} + 2\sqrt{2}$
b. Rewrite the numerator as the product of two surds and then simplify.	b. $\dfrac{5}{\sqrt{5}} = \dfrac{\sqrt{5} \times \sqrt{5}}{\sqrt{5}}$ $= \sqrt{5}$

1.7.7 The irrational number π

Surds are not the only irrational numbers. Most irrational numbers cannot be written as surds. A few of these numbers are so important to mathematics and science that they are given special names. The best known of these numbers is π (pi). π is irrational, so it cannot be written as a fraction. If you tried to write π as a decimal, you would be writing forever, as the digits never recur and the decimal does not terminate. In the 20th century, computers were used to determine the value of π to 1 trillion decimal places. The value of π is very close (but not equal) to 3.14 or $\frac{22}{7}$. Most calculators store an approximate value for π.

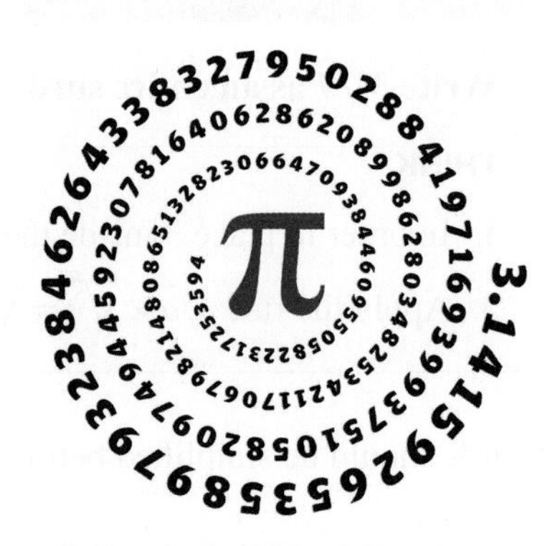

1.7.8 Rounding rational and irrational numbers to decimal places

Because numbers such as $\sqrt{3}$ or π cannot be written exactly as decimals, approximate values are often used. These values can be found using a calculator and then rounded off to the desired level of accuracy. When we round numbers, we write them to a certain number of decimal places or significant figures.

1.7.9 Rounding to significant figures

Another method of rounding decimals is to write them correct to a certain number of significant figures. In a decimal number, the significant figures start with the first non-zero digit.

Consider the approximate value of $\sqrt{2}$.

$\sqrt{2} \approx 1.414$. This approximation is written correct to 3 decimal places, but it has 4 significant figures. When counting significant figures, the non-zero digits before the decimal point are included.

The decimal 0.0302 is written to 4 decimal places, but it has only 3 significant figures. This is because significant figures start with the first non-zero digit. In this case, the first significant figure is 3, and the significant figures are 302.

WORKED EXAMPLE 27 Determining significant figures

Determine how many significant figures there are in each of the following numbers.

a. **25** b. **0.04** c. **3.02** d. **0.100**

THINK	WRITE
a. The first significant figure is 2.	a. 25 There are 2 significant figures.
b. The first significant figure is 4.	b. 0.04 There is 1 significant figure.
c. The first significant figure is 3.	c. 3.02 There are 3 significant figures.
d. The first significant figure is 1.	d. 0.100 There are 3 significant figures.

WORKED EXAMPLE 28 Rounding to significant figures

Round these numbers correct to 5 significant figures.

a. π b. $\sqrt{200}$ c. $0.\dot{0}\dot{3}$ d. 2530.166

THINK	WRITE
a. $\pi \approx 3.141\,592\ldots$ The first significant figure is 3. Write 4 more digits and round using the next digit.	a. 3.1416
b. $\sqrt{200} \approx 14.142\,13\ldots$ The first significant figure is 1. Write 4 more digits.	b. 14.142
c. $0.\dot{0}\dot{3} = 0.030\,303\,0\ldots$ The first significant figure is 3. Write 4 more digits.	c. 0.030 303
d. 2530.16 ... The first significant figure is 2. Write 4 more digits and round using the next digit.	d. 2530.2

Resources

Interactivity Balance surds (int-2762)

1.7 Exercise

Students, these questions are even better in jacPLUS

Receive immediate feedback and access sample responses

Access additional questions

Track your results and progress

Find all this and MORE in jacPLUS

1. WE20 Evaluate each of the following.

a. $\sqrt{36}$
b. $\sqrt[3]{8}$
c. $\sqrt{121}$
d. $\sqrt[3]{64}$
e. $\sqrt{64}$
f. $\sqrt[4]{81}$

2. WE21 Determine which of the following are surds.

a. $\sqrt{10}$
b. $\sqrt[3]{9}$
c. $-\sqrt{2}$
d. $\sqrt{50}$
e. $-\sqrt{16}$
f. $\sqrt{9} + \sqrt{49}$

3. WE22 Evaluate the following, leaving your answer in surd form.

a. $\sqrt{3}\times\sqrt{7}$
b. $2\times3\sqrt{7}$
c. $2\sqrt{7}\times5\sqrt{2}$
d. $\sqrt{7}\times9$
e. $\sqrt{3}\times\sqrt{3}$
f. $2\sqrt{3}\times4\sqrt{3}$

4. WE23 Evaluate the following, leaving your answer in surd form.

a. $\dfrac{\sqrt{18}}{\sqrt{3}}$
b. $\dfrac{15\sqrt{6}}{5\sqrt{2}}$
c. $\dfrac{15\sqrt{6}}{10}$
d. $\dfrac{5}{3}\sqrt{\dfrac{15}{3}}$
e. $\dfrac{\sqrt{9}}{\sqrt{3}}$
f. $\dfrac{\sqrt{28}}{\sqrt{7}}$

5. WE24a Simplify each of the following.

a. $\sqrt{18}$
b. $\sqrt{50}$
c. $\sqrt{28}$
d. $\sqrt{108}$
e. $\sqrt{48}$
f. $\sqrt{500}$

6. WE24b Simplify each of the following.

a. $5\sqrt{27}$
b. $6\sqrt{64}$
c. $7\sqrt{50}$
d. $10\sqrt{24}$
e. $4\sqrt{42}$
f. $9\sqrt{45}$

7. WE25 Write each of the following in the form $\sqrt{a}$ (as an entire surd).

a. $5\sqrt{7}$
b. $6\sqrt{3}$
c. $8\sqrt{6}$
d. $3\sqrt{10}$
e. $4\sqrt{2}$
f. $10\sqrt{6}$

8. MC $\sqrt{80}$ in simplest form is equal to:

A. $4\sqrt{5}$
B. $2\sqrt{20}$
C. $8\sqrt{10}$
D. $5\sqrt{16}$
E. $10\sqrt{8}$

9. MC Determine which of the following surds is in simplest form.

A. $\sqrt{60}$
B. $\sqrt{147}$
C. $\sqrt{105}$
D. $\sqrt{117}$
E. $\sqrt{72}$

10. MC $6\sqrt{5}$ is equal to:

A. $\sqrt{900}$
B. $\sqrt{30}$
C. $\sqrt{150}$
D. $\sqrt{180}$
E. $\sqrt{65}$

11. Simplify each of the following.

a. $4\sqrt{5}-6\sqrt{5}-2\sqrt{5}$

b. $10\sqrt{11}-6\sqrt{11}+\sqrt{11}$

c. $4\sqrt{2}+6\sqrt{2}+5\sqrt{3}+2\sqrt{3}$

d. $10\sqrt{5}-2\sqrt{5}+8\sqrt{6}-7\sqrt{6}$

12. **WE26** Simplify each of the following.

a. $\sqrt{45}-\sqrt{80}+\sqrt{5}$

b. $\sqrt{7}+\sqrt{28}-\sqrt{343}$

c. $\sqrt{24}+\sqrt{180}+\sqrt{54}$

d. $\sqrt{12}+\sqrt{20}-\sqrt{125}$

e. $3\sqrt{45}+2\sqrt{12}+5\sqrt{80}+3\sqrt{108}$

f. $6\sqrt{44}+4\sqrt{120}-\sqrt{99}-3\sqrt{270}$

13. **MC** $\sqrt{2}+6\sqrt{3}-5\sqrt{2}-4\sqrt{3}$ is equal to:

A. $-5\sqrt{2}+2\sqrt{3}$ B. $-3\sqrt{2}+23$ C. $6\sqrt{2}+2\sqrt{3}$ D. $-4\sqrt{2}+2\sqrt{3}$ E. $6\sqrt{5}-1$

14. **MC** $4\sqrt{8}-6\sqrt{12}-7\sqrt{18}+2\sqrt{27}$ is equal to:

A. $-7\sqrt{5}$ B. $29\sqrt{2}-18\sqrt{3}$ C. $-13\sqrt{2}-6\sqrt{3}$ D. $-13\sqrt{2}+6\sqrt{3}$ E. $7\sqrt{5}$

15. **MC** $2\sqrt{20}+5\sqrt{24}-\sqrt{54}+5\sqrt{45}$ is equal to:

A. $19\sqrt{5}+7\sqrt{6}$ B. $9\sqrt{5}-7\sqrt{6}$ C. $-11\sqrt{5}+7\sqrt{6}$ D. $-11\sqrt{5}-7\sqrt{6}$ E. $12\sqrt{35}$

16. **WE27** State how many significant figures there are in each of the following numbers.

a. 1631

b. 5.04

c. 95.00

d. 0.000 316

e. 0.1007

f. 0.010

17. **WE28** Write each number correct to 5 significant figures.

a. $\sqrt{15}$

b. $5.\dot{1}$

c. 5.15

d. 11.72^2

e. $\frac{3}{7}$

f. $2\frac{3}{7}$

18. Explain whether each of the following statements is True or False.

a. Every surd is a rational number.

b. Every surd is an irrational number.

c. Every irrational number is a surd.

d. Every surd is a real number.

19. Explain whether each of the following statements is True or False.

a. $1.\dot{3}\dot{1}$ is a rational number.

b. $1.\dot{3}\dot{1}$ is an irrational number.

c. $1.\dot{3}\dot{1}$ is a surd.

d. $1.\dot{3}\dot{1}$ is a real number.

20. Ryan is laying floor tiles in his bathroom. His bathroom floor is rectangular with a width of $3\sqrt{5}$ and a length of $4\sqrt{3}$. Calculate the area of the tiles Ryan lays in his bathroom.
$(A_{\text{rectangle}} = \text{length} \times \text{width})$

21. A landscaper is designing a triangular garden bed. He wants to put border edging all around the garden bed. The side lengths of the garden bed are $\left(3+\sqrt{3}\right)$ m, $\left(2\sqrt{3}+\sqrt{5}\right)$ m and $\left(5\sqrt{5}-\sqrt{3}\right)$ m. Calculate the length of the border edging required:

a. as a surd
b. rounded to 3 significant figures.

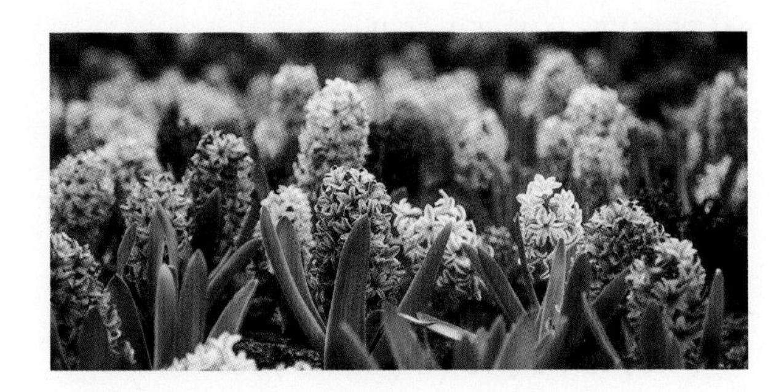

22. Explain why $\sqrt{36}$ and $-\sqrt{36}$ are defined, but $\sqrt{-36}$ is not.

23. The area of a circle is calculated using the formula $A = \pi \times r^2$, where r is the radius of the circle. π is sometimes rounded to 2 decimal places (3.14). A particular circle has a radius of 7 cm.

a. Use $\pi = 3.14$ to calculate the area of the circle to 4 significant figures.
b. Use the π key on your calculator to calculate the area of the circle to 6 significant figures.
c. Round your answer for part **b** to 2 decimal places.

24. The volume of a sphere (a ball shape) is calculated using the formula $V = \frac{4}{3} \times \pi \times r^3$, where r is the radius of the sphere. A beach ball with a radius of 25 cm is bouncing around the crowd at the MCG during the Boxing Day Test.

a. Calculate the volume of the beach ball to 4 decimal places.
b. When the volume is calculated to 4 decimal places, state how many significant figures it has.
c. State whether the calculated volume is a rational number. Give reasons for your answer.

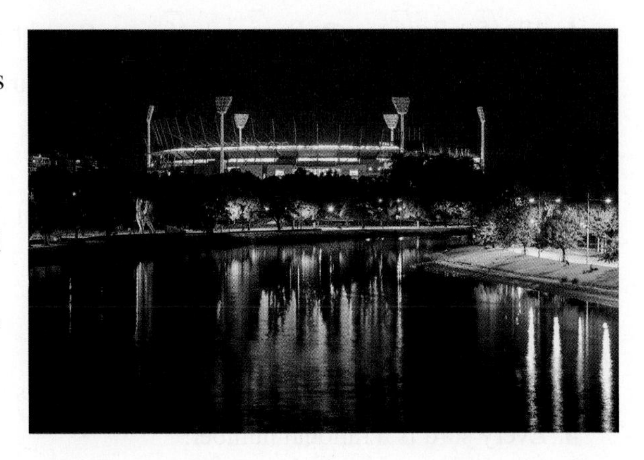

1.8 Review

1.8.1 Summary

doc-38009

Hey students! Now that it's time to revise this topic, go online to:

Review your results

Watch teacher-led videos

Practise questions with immediate feedback

Find all this and MORE in jacPLUS

1.8 Exercise

Multiple choice

1. MC Evaluating the expression $6 + 4 \times 5$ gives:
   A. 50 B. 15 C. 26 D. 24 E. 34
2. MC Evaluating the expression $\frac{3 \times (5+7)}{2^2}$ gives:
   A. $\frac{11}{2}$ B. 9 C. $\frac{13}{2}$ D. 3 E. $\frac{33}{4}$
3. MC Rounding 432 679 to the first (or leading) digit gives:
   A. 430 000 B. 400 000 C. 500 000 D. 440 000 E. 420 000
4. MC Rounding 6.957 21 to 3 decimal places gives:
   A. 6.957 B. 6.96 C. 6.958 D. 6.95 E. 7.000
5. MC The estimate of $251 + 978$ by rounding each number to its first (or leading) digit is:
   A. 1229 B. 900 C. 1200 D. 1300 E. 1100
6. MC The estimate of $3333 - 967 - 1545$ by first rounding each number to its nearest 100 is:
   A. 800 B. 900 C. 700 D. 10 000 E. 1000
7. MC The value of 2.56×2000 is:
   A. 5120 B. 2560 C. 4120 D. 5600 E. 5210
8. MC The place value of 7 in 26.07 is:
   A. Tens B. Units C. Tenths D. Hundredths E. Hundreds
9. MC The expression $2\sqrt{2} + 3\sqrt{18} + 5\sqrt{2} - 4\sqrt{18}$ simplified is:
   A. $7\sqrt{2} - \sqrt{18}$ B. $7\sqrt{2} + \sqrt{18}$ C. $4\sqrt{2}$ D. $28\sqrt{2}$ E. $8\sqrt{2}$
10. MC Rounding the irrational number $\frac{\pi}{4}$ to 5 significant figures gives:
    A. 0.7853 B. 0.785 40 C. 0.785 39 D. 0.7854 E. 0.785 40

Short answer

11. For each of the following expressions:
    i. estimate the answer by rounding to the first (or leading) digit
    ii. calculate the exact answer using a calculator.
    a. $58 + 89$ b. $211 - 58$ c. $169 + 239$ d. $1234 - 456$
12. Round each of the following to the nearest unit.
    a. 121.60 b. 0.512 c. 79.4 d. 9.6

13. a. Determine the value of the following, writing your answer as a fraction.

$$\frac{2\left[\left(4^3 \div 2^3\right) + 3\,(24 - 18)\right]}{(3 + 6 \div 2)}$$

b. Using a calculator, write your answer from part a as a decimal.

14. Convert each of the following decimals to fractions in simplest form.

a. 0.6   b. 5.75

15. Write $3\frac{7}{8}$ as a decimal.

Extended response

16. Diego fills his car with petrol once a week. Over the last 4 weeks, he paid the following prices per litre for his petrol: \$1.24, \$1.39, \$1.46 and \$1.40.
Determine the average price, to 2 decimal places, that Diego paid for petrol over the 4 weeks.

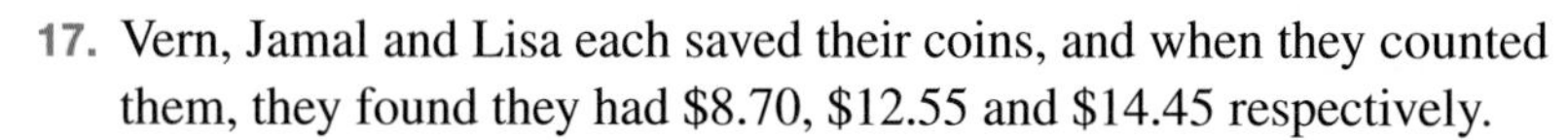

17. Vern, Jamal and Lisa each saved their coins, and when they counted them, they found they had \$8.70, \$12.55 and \$14.45 respectively.

a. Estimate their total savings by first rounding each amount to the nearest dollar.

b. Calculate their actual total savings.

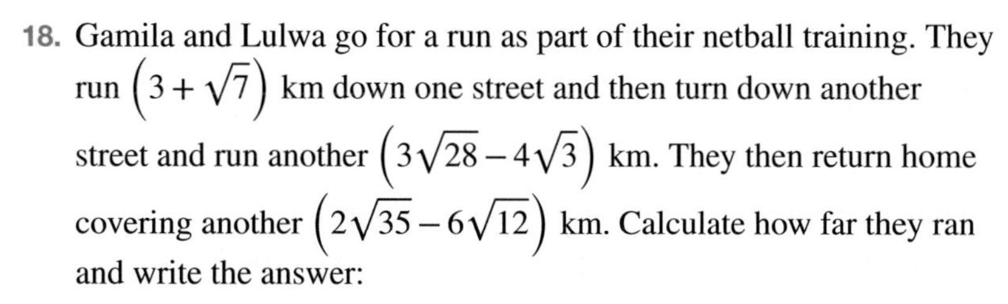

18. Gamila and Lulwa go for a run as part of their netball training. They run $\left(3 + \sqrt{7}\right)$ km down one street and then turn down another street and run another $\left(3\sqrt{28} - 4\sqrt{3}\right)$ km. They then return home covering another $\left(2\sqrt{35} - 6\sqrt{12}\right)$ km. Calculate how far they ran and write the answer:

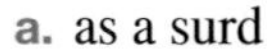

a. as a surd

b. as a decimal rounded to 2 decimal places.

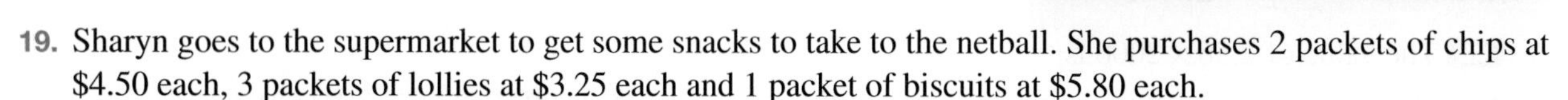

19. Sharyn goes to the supermarket to get some snacks to take to the netball. She purchases 2 packets of chips at \$4.50 each, 3 packets of lollies at \$3.25 each and 1 packet of biscuits at \$5.80 each.

a. Calculate how much Sharyn spent.

b. If Sharyn paid with a \$30 note, calculate how much change she received.

20. Kingston and Mariana are going to paint sections of their house. They will use three different types of paint. The prices and quantities of each type required are:

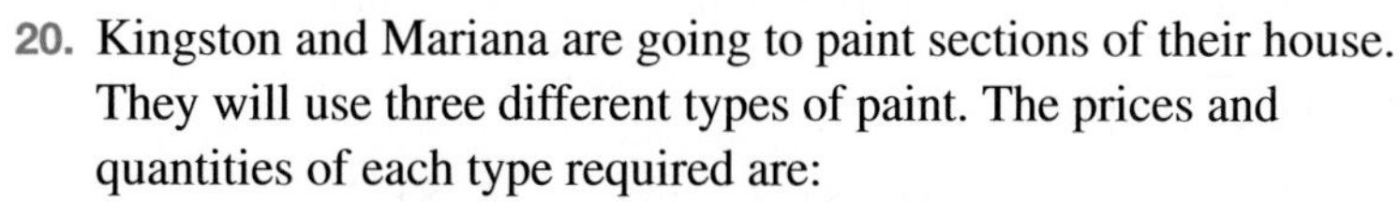

- 3 litres of paint A at \$21.75 per litre
- 0.75 litres of paint B at \$72.50 per 2 litres
- 7.5 litres of paint C at \$12.75 per 0.5 litre.

Determine the total cost of the paint required to complete the project.

Answers

Topic 1 Calculations

1.2 Number operations

1.2 Exercise

1. 3 GB
2. $7049
3. $317
4. a. $17 b. $6
5. a. 5 m
   b. 75 m
   c. 70 m
   d. To calculate the difference, subtract the height of the smaller building from the height of the taller building.
6. a. 351 km b. 247 km
7. 52
8. a. 5432 b. 2345 c. 3087
9. a. 884 b. 6205
10. 400 km
11. a. $0.94 b. $32.90 c. $5146.50
    d. $3 e. $10
12. $522.63
13. $54\frac{1}{6}$ grams
14. 2 students

1.3 Order of operations

1.3 Exercise

1. a. 6 b. 3 c. 3
   d. 11 e. 96
2. a. 27 b. 7 c. 12
   d. 18 e. 12
3. a. $16\frac{1}{7}$ b. 1 c. 16
   d. $\frac{37}{7}$ e. 13
4. a. When the only operations are multiplication and division, they should be done from left to right. When the only operations are addition and subtraction, they should be done from left to right.
   b. 40 and 17
5. a. 2 b. 80 c. 52
   d. 33 e. 48
6. a. 75 b. 7 c. 5 d. $12\frac{1}{7}$
7. a. $1\frac{1}{8}$ b. $8\frac{1}{3}$ c. $\frac{7}{36}$ d. $2\frac{2}{5}$
8. a. $(12 - 8) \div 4 = 1$
   b. $(4 + 8) \times 5 - 4 \times 5 = 40$
   c. $3 + 4 \times (9 - 3) = 27$
   d. $3 \times (10 - 2) \div 4 + 4 = 10$
   e. $10 \div (5 + 5) \times 9 \times 9 = 81$
   f. $(18 - 3) \times 3 \div 5 = 9$
9. a. Use multiplication first to calculate the cost of each type of food, and then add these costs together to calculate the total cost of all the food.
   b. Cost = $(2 \times \$2) + (3 \times \$4) + (3 \times \$5)$
   c. $31
10. $1 = (4 - 3) \div (2 - 1)$
    $2 = 4 - 3 + 2 - 1$
    $3 = 21 \div (3 + 4)$
    $4 = 4 - 2 + 3 - 1$
    $5 = 3 + 4 - (2 \times 1)$
11. a. $3.50 b. $0.27
12. $4.11

1.4 Rounding

1.4 Exercise

1. a. 10 b. 70 c. 170
   d. 1350 e. 56 900 f. 765 490
2. a. 0 b. 100 c. 200
   d. 3000 e. 42 600 f. 345 300
3. a. 1000 b. 1000 c. 7000
   d. 12 000 e. 453 000 f. 2 680 000
4. a. 6 b. 50 c. 1000
   d. 10 000 e. 200 000 f. 5 000 000
5. a. 2 b. 7 c. 3 d. 8
6. a. 5 b. 10 c. 7 d. 1
7. a. 4 b. 12 c. 1 d. 5
8. a. 8 b. 30 c. 60 d. 3
9. a. 0.4 b. 0.9 c. 9.3
   d. 25.3 e. 300.1 f. 12.8
10. a. 99.9 b. 8.9 c. 17.6
    d. 0.9 e. 94.0 f. 1.0
11. a. 0.32 b. 0.86 c. 1.25
    d. 13.05 e. 7.13 f. 100.81
12. a. 71.26 b. 0.01 c. 0.19
    d. 19.70 e. 0.40 f. 1.00
13. a. 2.39 b. 14.0 c. 0.03
    d. 0.8765 e. 64.2952 f. 0.382
14. a. 96.3 b. 3.04 c. 8.90
    d. 47.880 e. 0.10 f. 0.49
15. C
16. D
17. A
18. a. 11 b. 8 c. 4
    d. 93 e. 112 f. 22
19. a. 42 b. 2138 c. 0
    d. 1 e. 1 f. 41
20. If Luke pays for the jeans with a debit card, the exact amount will be debited from his account. Australian currency has the smallest denomination of $0.05, which means that the price would be rounded to the nearest 5 cents. If Luke pays for the jeans with cash, the price would be rounded to $129.95.

1.5 Estimation and approximation strategies

1.5 Exercise

1. 6300
2. a. 1500 b. 900 000 c. 500 000
   d. 36 000 e. 6 000 000 f. 70
3. Adding the cost of groceries when shopping; calculating the cost of petrol for a trip; determining the size of a crowd
4. a. Possibly, but most likely not
   b. They may have divided the distance by the average speed.
5. An estimate is based on information.
6. Multiply the number of students in an average class by the number of classes in the school.
7. a. 37 800
   b. Sample responses can be found in the worked solutions in the online resources.
   c. Sample responses can be found in the worked solutions in the online resources.
8. a. i. 140 ii. 131
   b. i. 2560 ii. 2721
   c. i. 10 ii. 12
   d. i. 20 ii. 26
   e. i. 0 ii. 563
9. a. i. 1800 ii. 1450
   b. i. 40 000 ii. 31 126
   c. i. 9400 ii. 9429
   d. i. 2 000 000 ii. 3 508 503
10. a. i. 7 ii. $8\frac{1}{2}$
    b. i. 16 ii. $19\frac{3}{4}$
    c. i. 43 ii. $38\frac{7}{11}$
    d. i. 369 ii. $615\frac{2}{3}$
    e. i. 25 ii. 49
    f. i. 17 ii. 24
11. a. Approximately 90
    b. Approximately 2000
    c. Approximately 80
12. If material lengths are to be rounded, it should be to a longer length than required or the purchased wood will be too short.
13. a. 10 000 b. 20
14. a. 2 b. 1000

1.6 Decimals and fractions

1.6 Exercise

1. a. 4 b. 3 c. 4 d. 3 e. 1 f. 2
2. a. 2
   b. i. Seventy-eight dollars and fifty-five cents
   ii. $78.55
3. a. Ones
   b. Tens, tenths
   c. Tenths
   d. Ones, tenths, hundredths
   e. Tenths, hundredths, thousandths
   f. There are no place-holding zeros.
4. a. $(1\times 10)+(6\times 1)+\left(0\times \frac{1}{10}\right)+\left(2\times \frac{1}{100}\right)$
   b. $(1\times 10)+(1\times 1)+\left(0\times \frac{1}{10}\right)+\left(4\times \frac{1}{100}\right)+\left(6\times \frac{1}{1000}\right)$
   c. $(2\times 100)+(2\times 10)+(2\times 1)+\left(0\times \frac{1}{10}\right)+\left(3\times \frac{1}{100}\right)$
   d. $(1\times 10)+(5\times 1)+\left(1\times \frac{1}{10}\right)+\left(1\times \frac{1}{100}\right)$
   e. $(4\times 1)+\left(7\times \frac{1}{10}\right)+\left(0\times \frac{1}{100}\right)+\left(1\times \frac{1}{1000}\right)$
   f. $(6\times 10)+(8\times 1)+\left(6\times \frac{1}{10}\right)+\left(8\times \frac{1}{100}\right)$
5. a. 42.504 b. 0.272 c. 1.009 d. 30.934
6. a. $\frac{7}{100}$ b. $\frac{7}{10}$ c. $\frac{7}{1000}$
   d. $\frac{7}{10\,000}$ e. 7 f. 70
7. a. 16.273 b. 137.202 c. 95.98
   d. 0.001 e. 0.2 f. 0.012
8. a. 268.75 b. 98.164 c. 100.009
   d. 222.489 e. 1075.7346 f. 7132.2904
9. a. 0.0674 b. 779.96 c. 34.216
   d. 441.43 e. 14.994 f. 26.092 49
10. a. i. 9.85 ii. 9.84
    b. i. 12.18 ii. 12.18
    c. i. 91.43 ii. 91.43
    d. i. 189.52 ii. 189.52
11. a. 628.4 b. 5300
12. a. 61.725 b. 740.7 c. 8641.5 d. 24 690
13. a. 0.345 b. 1.2398 c. 1.245 37
    d. 0.3569 e. 0.000 246 f. 0.0248
14. a. $\frac{1}{10}$ b. $\frac{1}{2}$ c. $\frac{4}{5}$
    d. $\frac{3}{25}$ e. $\frac{21}{100}$ f. $\frac{21}{25}$
15. a. $\frac{1}{20}$ b. $\frac{5}{8}$ c. $3\frac{4}{5}$
    d. $2\frac{13}{100}$ e. $12\frac{21}{50}$ f. $10\frac{7}{2000}$
16. a. 0.375 b. 0.8 c. 1.25 d. 0.75
17. a. 2.25 b. 3.05 c. 6.5 d. 0.2

1.7 Rational and irrational numbers

1.7 Exercise

1. a. 6 b. 2 c. 11
   d. 4 e. 8 f. 3
2. a. Yes b. Yes c. Yes
   d. Yes e. No f. No
3. a. $\sqrt{21}$ b. $6\sqrt{7}$ c. $10\sqrt{14}$
   d. $9\sqrt{7}$ e. 3 f. 24
4. a. $\sqrt{6}$ b. $3\sqrt{3}$ c. $\frac{3\sqrt{6}}{2}$
   d. $\frac{5}{3}\sqrt{5}$ e. $\sqrt{3}$ f. 2
5. a. $3\sqrt{2}$ b. $5\sqrt{2}$ c. $2\sqrt{7}$
   d. $6\sqrt{3}$ e. $4\sqrt{3}$ f. $10\sqrt{5}$
6. a. $15\sqrt{3}$ b. 48 c. $35\sqrt{2}$
   d. $20\sqrt{6}$ e. $4\sqrt{42}$ f. $27\sqrt{5}$
7. a. $\sqrt{175}$ b. $\sqrt{108}$ c. $\sqrt{384}$
   d. $\sqrt{90}$ e. $\sqrt{32}$ f. $\sqrt{600}$
8. A
9. C
10. D
11. a. $-4\sqrt{5}$ b. $5\sqrt{11}$
    c. $10\sqrt{2}+7\sqrt{3}$ d. $8\sqrt{5}+\sqrt{6}$
12. a. 0 b. $-4\sqrt{7}$
    c. $5\sqrt{6}+6\sqrt{5}$ d. $2\sqrt{3}-3\sqrt{5}$
    e. $29\sqrt{5}+22\sqrt{3}$ f. $9\sqrt{11}-\sqrt{30}$
13. D
14. C
15. A
16. a. 4 b. 3 c. 4
    d. 3 e. 4 f. 2
17. a. 3.8730 b. 5.1111 c. 5.1515
    d. 137.36 e. 0.428 57 f. 2.4286
18. a. False b. True c. False d. True
19. a. True b. False c. False d. True
20. $12\sqrt{15}$ m
21. a. $3+2\sqrt{3}+6\sqrt{5}$ m b. 19.9
22. $\sqrt{36}$ means the positive square root of 36, which equals 6 since $6^2 = 36$. Similarly, $-\sqrt{36}$ means the negative square root of 36, which equals -6. But $\sqrt{-36}$ is undefined since you cannot square a number to get a negative number.
23. a. 153.9 cm^2 b. 153.938 cm^2 c. 153.94 cm^2
24. a. 65 449.8470 cm^3
    b. 9
    c. Since the calculated value has a finite number of decimal places, it is rational.

1.8 Review

1.8 Exercise

Multiple choice

1. C
2. B
3. B
4. A
5. D
6. A
7. A
8. D
9. C
10. B

Short answer

11. i. a. 150 b. 140
    c. 400 d. 500
    ii. a. 147 b. 153
    c. 408 d. 778
12. a. 122 b. 1 c. 79 d. 10
13. a. $\frac{26}{3}$ b. 8.6
14. a. $\frac{3}{5}$ b. $5\frac{3}{4}$
15. 3.875

Extended response

16. $1.37
17. a. $36 b. $35.70
18. a. $3+13\sqrt{7}-16\sqrt{3}$ km
    b. 9.68 km
19. a. $24.55 b. $5.45
20. $283.69

2 Ratios, proportion and variation

LEARNING SEQUENCE

Fully worked solutions for this topic are available online.

2.1 Overview

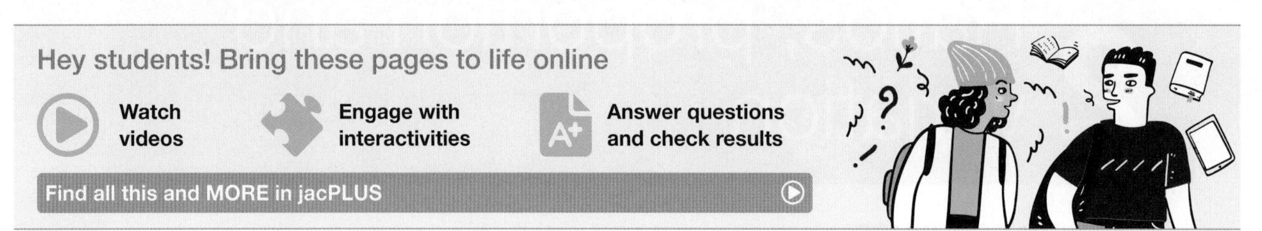

2.1.1 Introduction

Ratios are used to compare different quantities. For example, they can be used to describe money exchanges, ingredients used in cooking or perspective in scaled drawings. Some ratios have been found to occur in nature, and in ancient times some philosophers believed that numbers could explain the nature of the universe. Proportion is a mathematical comparison between two numbers. A proportion is an equation in which two ratios are set equal to each other. And variation is defined by any change in some quantity due to change in another. All of these concepts will be explored in this topic.

KEY CONCEPTS

This topic covers the following key concepts from the VCE Mathematics Study Design:
- rational numbers and irrational numbers related to measurement, ratios and proportions in a practical context
- direct and indirect variation.

2.2 Direct variation (proportion)

LEARNING INTENTION

At the end of this subtopic you should be able to:
- identify direct variation
- calculate the constant of proportionality
- create equations in the form $y = kx$ to solve problems.

A direct variation, also called *direct proportion*, is a relationship between two variables x and y that can be written as $y = kx, k \neq 0$.

If two quantities vary directly, then doubling one will double the other.

Direct variation produces a linear graph that passes through the origin.

The statement 'quantity a varies directly with quantity b' can also be written as 'quantity a is directly proportional to quantity b'.

Direct variation is written as an equation of the form:

$$y = kx$$

which is often shortened to:

$$y \propto x$$

The **proportion sign** $\propto$ is equivalent to '$= k$'.

k is called the **constant of proportionality** or the constant of variation (proportion) and is the gradient of the graph.

For example, $P = 4l$ represents the perimeter of a square with side length l. The constant of proportionality in this case is 4.

tlvd-3510

WORKED EXAMPLE 1 Direct variation and proportionality

The graph shows the relationship between the mass and cost of apples purchased at the supermarket.

a. Determine if the cost of the apples varies directly with the mass.

b. Calculate the constant of proportionality.

c. Write a rule connecting cost, C, and mass, m.

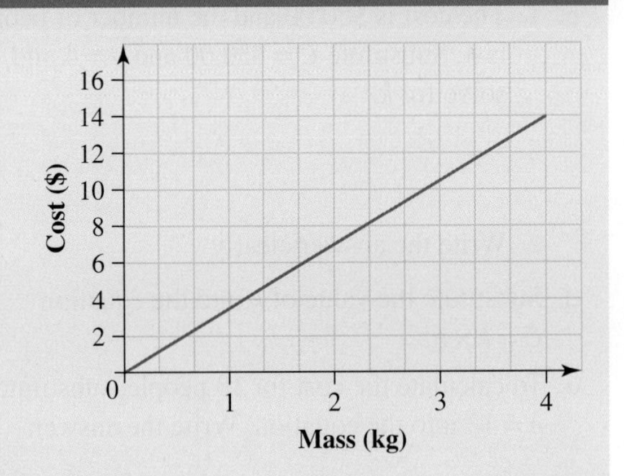

THINK	WRITE
a. The graph is a straight line that passes through the origin, so the cost of apples varies directly with the mass.	a. The cost of the apples varies directly with the mass: $C \propto m$.

b. The constant of proportionality, k, is the same as the gradient. Use the formula for calculating the gradient to determine k.

b. $\text{Gradient} = \dfrac{\text{rise}}{\text{run}}$

$$k = \frac{14}{4}$$

$$k = 3.5$$

c. Substitute the value of k into the equation $C = km$.

c. $C = km$

$$C = 3.5m$$

tlvd-3511

WORKED EXAMPLE 2 Direct variation and proportionality 2

The cost for a group of people to go to the movies varies directly with the number of people.

a. Let n equal the number of people and let C equal the cost. Use the proportionality sign ($\propto$) to write the relationship between n and C as a mathematical statement of direct variation.

b. Write the answer to part a using an equals sign and the constant k.

c. If it costs \$50.00 for 4 people to go to the movies, calculate k.

d. Write an equation connecting C and n.

e. Use the equation to calculate the cost for 13 people to go to the movies.

f. Use the equation to calculate how many people could go to the movies for \$112.50.

THINK	WRITE
a. The cost varies directly with the number of people.	**a.** $C \propto n$
b. Replace $\propto$ with '$= k$'.	**b.** $C = k \times n$
c. 1. The cost is \$50.00 and the number of people is 4. Substitute $C = \$50.00$ and $n = 4$ and solve for k.	**c.** $C = k \times n$ $50 = k \times 4$ $\frac{50}{4} = \frac{k \times 4}{4}$ $k = 12.5$
2. Write the answer clearly.	The constant of proportionality, $k = 12.5$.
d. Substitute the value of k into the equation $C = k \times n$.	**d.** $C = 12.5n$
e. To calculate the cost for 13 people, substitute $n = 13$ into the equation. Write the answer.	**e.** $C = 12.5 \times 13$ $= 162.50$ It costs \$162.50 for 13 people to go to the movies.

f. 1. To calculate the number of people who could go to the movies for \$112.50, substitute $C = 112.5$ into the equation.	**f.** $C = 12.5n$ $112.5 = 12.5n$
2. Solve by dividing both sides by 12.5.	$\dfrac{112.5}{12.5} = \dfrac{12.5n}{12.5}$ $n = 9$
3. Write the answer.	Nine people could go to the movies for \$112.50.

2.2 Exercise

Students, these questions are even better in jacPLUS

Receive immediate feedback and access sample responses

Access additional questions

Track your results and progress

Find all this and MORE in jacPLUS

1. Explain what is meant by direct variation.

2. **WE1** The graph shows how the cost of a wedding reception varies with the number of people attending.
   a. Determine if the cost is directly proportional to the number of people attending.
   b. Calculate the constant of variation.
   c. Write a rule connecting the cost, C, and the number of people attending, n.

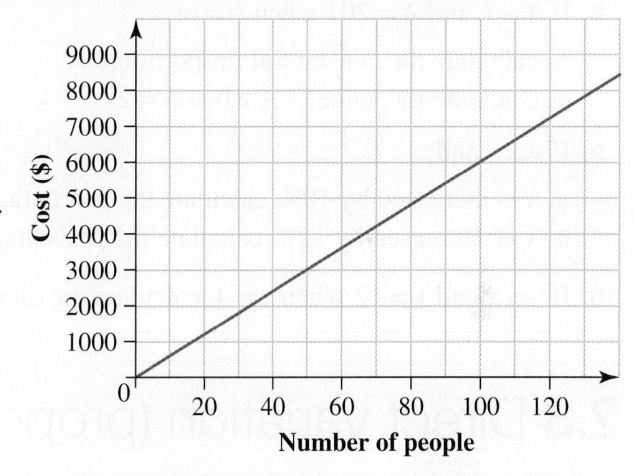

3. The wages earned by a worker for different numbers of hours are shown in the table.

Time (hours)	1	2	3	4
Wages (\$)	25	50	75	100

   a. Draw a graph of wages (vertical axis) versus time (horizontal axis).
   b. Determine if the amount earned varies directly with the amount of time worked.

4. a. Draw a graph of the data in the table.

x	0.7	1.2	3	4.1
y	2.8	4.8	12	16.4

   b. State whether y is directly proportional to x.

5. a. Draw a graph of the data in the table.

x	0.4	1.5	2.2	3.4
y	0.84	3.15	4.62	7.14

b. State whether y varies directly with x.

6. **WE2** The amount of interest earned by an investment is proportional to the amount of money invested.

a. Use the symbols I for the amount of interest and A for the amount of money invested, together with the proportionality sign ($\propto$), to write this in mathematical shorthand.

b. Write the answer to part **a** using an equals sign and the constant k.

c. If an investment of \$20 000 earns \$10 000 interest, calculate k.

d. Write an equation connecting I and A.

e. Use the equation to calculate the interest earned by an investment of \$45 000.

f. Use the equation to calculate the investment needed to earn \$37 000 in interest.

7. If $H \propto n$ and $H = 27.9$ when $n = 6.2$:

a. calculate the constant of proportionality

b. determine the rule connecting H and n

c. calculate the value of H when $n = 72$

d. calculate the value of n when $H = 450$.

8. If $A \propto L$ and $A = 201$ when $L = 6$:

a. calculate the constant of proportionality

b. determine the rule connecting A and L

c. calculate the value of A when $L = 25$

d. calculate the value of L when $A = 385.25$.

9. If $y \propto x$ and:

a. x is increased by 10%, calculate the percentage increase in y

b. x is decreased by 15%, calculate the percentage decrease in y.

10. If $y \propto x$ and $y = 12$ when $x = 4$, calculate the increase in x if y increases from 120 to 960.

2.3 Direct variation (proportion) and ratios

LEARNING INTENTION

At the end of this subtopic you should be able to:

- calculate the constant of proportionality using $\frac{y}{x} = k$
- create equations in the form $y = kx$ to solve problems.

If two quantities x and y are directly proportional, then:

$$y \propto x$$

$$\text{or } y = kx$$

This can be expressed as $\frac{y}{x} = k$. The constant of variation or constant of proportionality, k, is equal to the **ratio of y to x** for any data pair.

The constant of variation, k, is the **rate at which y varies with x**.

Some examples of direct linear variation are:

- The length that a flexible spring extends varies directly with the mass of an object attached to one end.
- Distance varies directly with time for an object travelling at constant speed.

WORKED EXAMPLE 3 Using the constant of variation to calculate distance and time

The distance travelled by a vehicle over 4 hours is shown in the table. The distance is directly proportional to the time: $d \propto t$.

Time (hours)	1	2	3	4
Distance (km)	90	180	270	360

a. **Write the relationship between d and t as an equation with a constant of variation, k.**

b. **Calculate the constant of variation.**

c. **Add a row to the bottom of the given table and calculate the rate $\frac{\text{distance}}{\text{time}}$ for each pair of values given.**

d. **Comment on the connection between your answers to parts b and c.**

THINK

a. Given $d \propto t$, write the relationship between d and t as an equation with a constant of variation, k.

WRITE

a. $d \propto t$

$d = kt$

THINK

b. Choose a data pair, such as $t = 2$ hours and $d = 180$ km. Substitute these values into the equation and solve for k by dividing both sides by 2.

WRITE

b. $180 = k \times 2$

$\frac{180}{2} = \frac{k \times 2}{2}$

$90 = k$

THINK

c. Draw the table and add a row for the rate $\frac{\text{distance}}{\text{time}}$.

In each case, divide the distance by the time.

WRITE

c.

Time (hours)	1	2	3	4
Distance (km)	90	180	270	360
$\frac{\text{distance}}{\text{time}}$ (km/h)	90	90	90	90

THINK

d. The rate $\frac{\text{distance}}{\text{time}}$ is the same as the constant of variation. Write a statement to answer the question.

WRITE

d. The constant of variation is the rate at which the distance changes over time.

WORKED EXAMPLE 4 Using the constant of variation to calculate price and weight

Consider the image shown.

a. Letting C equal the cost and m equal the mass of potatoes, write an equation connecting the quantities.

b. Use the equation to calculate the cost of 2.3 kg of potatoes.

c. Use the equation to calculate the mass of potatoes that can be purchased for $7.00.

THINK	WRITE
a. 1. The cost will vary directly with the mass because the rate is constant at $2.50 per kg. Also, if no potatoes are purchased, the cost will be $0. Write this relationship as an equation.	a. $C \propto m$ $C = k \times m$
2. The rate is the constant of variation, k. Substitute that rate into the equation.	$C = 2.5 \times m$
b. Substitute the value $m = 2.3$ and simplify. Write the answer.	b. $C = 2.5 \times 2.3$ $= 5.75$ 2.3 kg of potatoes costs $5.75.
c. 1. Substitute $C = 7$ and solve for m by dividing both sides by 2.5.	c. $C = 2.5 \times m$ $7 = 2.5 \times m$ $\frac{7}{2.5} = \frac{2.5 \times m}{2.5}$ $2.8 = m$
2. Write the answer.	2.8 kg of potatoes can be purchased for $7.00.

2.3 Exercise

1. Explain, in your own words, the connection between the ratio of y to x $\left(\frac{y}{x}\right)$ and the constant of variation, k.

2. WE3 For the data shown in the table, y is directly proportional to x ($y \propto x$).

y	1.6	2.6	3.1	4.9
x	11.2	18.2	21.7	34.3

a. Write the relationship between y and x as an equation with a constant of variation, k.
b. Calculate the constant of variation.
c. Add another row to the table and calculate the rate, $\frac{y}{x}$.
d. Comment on the connection between your answers to parts **b** and **c**.

3. An employee is paid by the hour. The table shows the number of hours worked in a five-day week and the wages earned each day. The wages earned vary directly with the number of hours worked.

	Monday	Tuesday	Wednesday	Thursday	Friday
Hours worked	6.5	7.0	8.5	9.0	7.0
Wages	$110.50	$119.00	$144.50	$153.00	$119.00

a. Using W for wages, H for hours worked, and the constant of variation, k, write an equation for the relationship between the variables.
b. Calculate the constant of variation, k.
c. For each day of the week, calculate the value of $\frac{\text{wages}}{\text{hours}}$.
d. Interpret the results in part **c**.

4. **WE4** Consider the image shown.

a. Using the symbol D for distance and L for the number of litres of fuel, write an equation connecting the two quantities.
b. Use the equation to calculate the distance that can be travelled on 4.6 L of fuel.
c. Use the equation to calculate the number of litres of fuel needed to travel 130 km.

Fuel consumption rate 10.6 km/L

5. The gravitational force F (Newtons, N) varies directly with the mass m (kg). A mass of 10 kg experiences a force of 98 N.
a. Calculate the proportionality constant k in the expression $F = k \times m$.
b. Interpret the result in part **a**.

6. For the table of results shown, calculate the ratio $\frac{y}{x}$ for each data pair and state whether y is directly proportional to x.

x	3.5	6.1	9.7	11.2
y	7.35	11.59	23.28	24.64

7. Chemists have found that there is a direct variation relationship between the temperature of a gas and its pressure in atmospheres — $P \propto T$. However, the temperature has to be measured on the Kelvin scale (K). To convert from °C to K, add 273°.

The lid is put on an empty soft-drink bottle when the temperature is 10 °C. The initial pressure inside the container is 1 atmosphere. It is left out in the sun on a day when the temperature reaches 40 °C.

a. Because $P \propto T$, the ratio $\frac{P}{T}$ is constant. Use this to calculate the final pressure in the bottle.
b. Calculate the percentage by which the pressure in the bottle has increased.

2.4 Partial variation (proportion)

LEARNING INTENTION

At the end of this subtopic you should be able to:
- create partial variation equations in the form $y = kx + c$ to solve problems
- calculate the constant of variation using $k = \dfrac{y_2 - y_1}{x_2 - x_1}$.

Partial variation involves quantities that produce a linear graph that *does not* pass through the origin.

Constant of variation

The equation for the relationship is in the form $y = kx + c$ where k is the constant of variation (the gradient of the graph) and c is the y-intercept. This is described as 'y varies partly as x and is partly constant'.

The rule for calculating the gradient of a line, $k = \dfrac{y_2 - y_1}{x_2 - x_1}$, is used to determine the constant of variation.

tlvd-3512

WORKED EXAMPLE 5 Partial variation in the real world

The cost of electricity is partly constant and varies partly with the number of kilowatt hours used.

a. Using C for the cost and n for the number of kilowatt hours (kWh), together with the constant c and the constant of variation k, write a rule for the relationship between the quantities.

b. If the cost is \$90 for 200 kWh of usage and \$105 when the usage is 350 kW, sketch a graph showing this information.

c. Calculate the constant of variation k.

d. Interpret the result in part c.

e. Calculate the value of c, the vertical intercept.

f. Write the relationship between C and n.

THINK	WRITE
a. Since the cost of electricity is partly constant and varies partly with the number of kilowatt hours used, the relationship will be in the form $y = kx + c$. Write the rule.	a. $C = kn + c$
b. Cost is the dependent variable, as the cost depends on the number of kilowatt hours used, so it will be on the vertical axis. The graph is a sketch and does not need to be drawn to scale. Plot the points (200, 90) and (350, 105).	b. 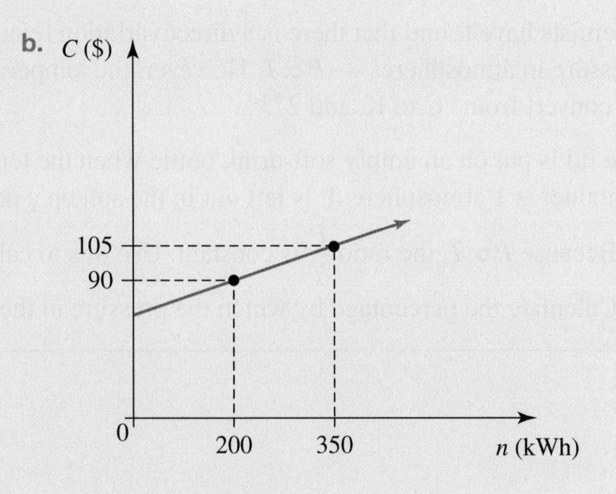

c. The constant of variation is the gradient of the graph.
Use the formula to calculate its value.

c. $k = \frac{y_2 - y_1}{x_2 - x_1}$

$k = \frac{105 - 90}{350 - 200}$

$= \frac{15}{150}$

$= 0.1$

d. The gradient k is the rate or cost per kWh.

d. The cost of electricity is \$0.10 per kilowatt hour.

e. Substitute $k = 0.1$ and the coordinates of one point on the graph into the equation from part a. In this case use (200, 90).

e. $C = kn + c$

$C = 0.1n + c$

$90 = 0.1 \times 200 + c$

$90 = 20 + c$

$70 = c$

f. Substitute the values of k and c into the equation $C = kn + c$.

f. $C = 0.1n + 70$

WORKED EXAMPLE 6 Using partial variation equations to solve problems

Use the equation $C = 0.1n + 70$ to calculate:

a. the cost when 475 kWh of electricity is used

b. the amount of electricity used in the bill shown.

Page 1 of 2

SYNERGY ENERGY

Account number	**123456**
Amount due	**\$147.20**
Due date	**22 Jun 2022**

ELECTRICITY ACCOUNT

Mr Scott Fitzroy
155 Cremorne St
Melbourne Vic. 3121

ELECTRICITY SUMMARY

Usage and service charges (see over for details)	\$147.20
Total other charges	\$0.00
Current charges (including GST)	\$147.20
Total amount due	**\$147.20**

THINK

a. 1. Write the equation.
Substitute the value $n = 475$ into the equation to calculate the cost when 475 kWh of electricity is used.

2. Write the answer.

WRITE

a. $C = 0.1n + 70$

$C = 0.1 \times 475 + 70$

$= 117.50$

475 kWh of electricity costs \$117.50.

b. 1. • Read the total amount owing from the bill.
• Substitute the value $C = 147.20$.
• Solve for n.

b.
$$C = 0.1n + 70$$
$$147.20 = 0.1n + 70$$
$$77.20 = 0.1n$$
$$\frac{77.20}{0.1} = \frac{0.1n}{0.1}$$
$$772 = n$$

2. Write the answer.

A bill of \$147.20 means that the amount of electricity used was 772 kWh.

2.4 Exercise

1. In your own words, describe the graph resulting from partial variation.
2. y varies partly with x and is partly constant. If $y = 56$ when $x = 4$ and $y = 243$ when $x = 15$:
   a. calculate the value of the constant of variation, k
   b. calculate the y-intercept, c
   c. state the rule for y in terms of x
   d. use the rule to calculate y when $x = 11.5$
   e. use the rule to calculate x when $y = 404.5$.
3. The relationship between y and x is one of partial variation. It is found that $y = 41$ when $x = 2$ and $y = 97$ when $x = 9$.
   a. Calculate the value of the constant of variation, k.
   b. Calculate the y-intercept, c.
   c. Determine the rule for y in terms of x.
   d. Use the rule to calculate y when $x = 6.75$.
   e. Use the rule to calculate x when $y = 114.6$.
4. WE5 For any polygon, the sum of the interior angles, S, varies partly as the number of sides, n, and is partly constant.
   a. Using the symbols S and n, together with a constant of variation, k, and a constant c, write a rule for the relationship between the quantities.
   b. In any triangle, the sum of the angles is 180°, while in a quadrilateral, the sum of the angles is 360°. Sketch a graph of the number of sides versus the sum of the angles.
   c. Calculate the constant of variation, k.
   d. Interpret the result in part c.
   e. Calculate the value of c, the vertical intercept.
   f. Write the relationship between S and n.

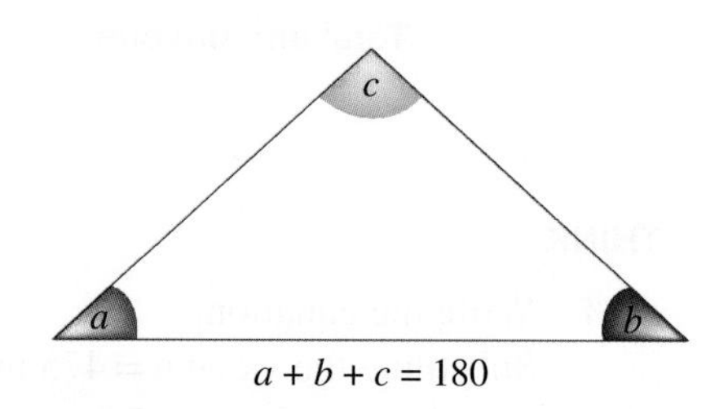

5. **WE6** The rule for the relationship between the sum of the angles, S, and the number of sides, n, is $S = 180n - 360$.

a. Use the rule to calculate the sum of the angles in a decagon (10 sides).
b. Use the rule to calculate the number of sides if the sum of the angles was 2520°.

6. When hiring a car for a day, the cost, C, is partly constant and varies partly with the distance, d, travelled.

a. Using the symbols C and d, together with the constant of variation, k, and the constant c, write a rule for the relationship between the quantities.
b. One day's hire to travel 200 km costs \$90, while a day when 550 km is travelled costs \$142.50. Sketch a graph of cost versus distance travelled.
c. Calculate the constant of variation, k.
d. Interpret the result in part **c**.
e. Calculate the value of c, the vertical intercept, and interpret the result.
f. Write the rule for the relationship between C and d.
g. Use the rule from part **f** to calculate the cost to hire the car for a day and travel 360 km.

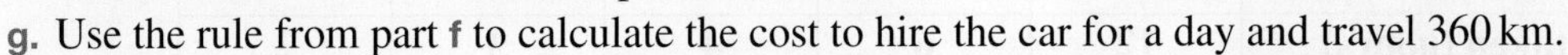

h. Use the rule to calculate the distance travelled if the cost for the day was \$123.

7. **MC** If Q varies partly with x and is partly constant, and the rule connecting Q and x is $Q = 0.4x - 7$, when $x = 1.4$, Q equals:

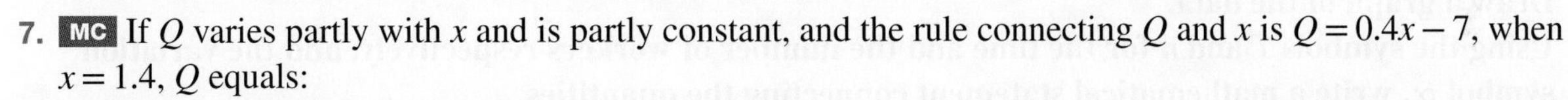

A. 0.56 B. −6.6 C. −6.44 D. 1.4 E. 11.2

8. **MC** A graph of y versus x is shown. Select the correct statement.

A. y varies directly with x and the constant of variation, k, is 6.
B. y varies directly with x and the constant of variation, k, is 5.
C. y varies partly with x and is partly constant. The constant of variation, k, is 6.
D. y varies partly with x and is partly constant. The constant of variation, k, is 5.
E. y varies partly with x and is partly constant. The constant of variation, k, is 10.

2.5 Inverse variation (proportion)

LEARNING INTENTION

At the end of this subtopic you should be able to:
- identify when two quantities show an inverse variation (proportion)
- create inverse variation equations in the form $y = \frac{k}{x}$ to solve problems.

If two quantities **vary inversely**, then increasing one variable decreases the other.

Inverse variation produces the graph of a **hyperbola**.

The statement 'quantity a varies inversely with quantity b' can also be written as 'quantity a is **inversely proportional** to quantity b'.

If y varies inversely to x, it is written as: $y \propto \frac{1}{x}$ or $y = \frac{k}{x}$.

As in direct variation, k is called the *proportionality constant* or the *constant of variation*.

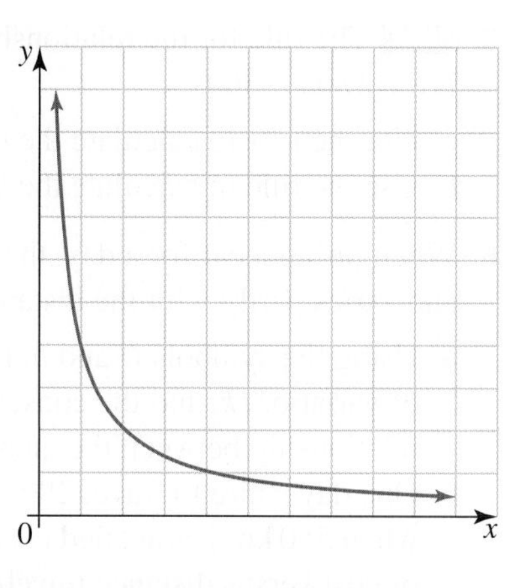

tlvd-3513

WORKED EXAMPLE 7 Inverse variation 1

The table shows how the time that it takes to complete a task varies with the number of workers.

Number of workers	1	2	3	4	6
Time taken (hours)	12	6	4	3	2

a. **Draw a graph of the data.**
b. **Using the symbols T and n for the time and the number of workers respectively, and the variation symbol $\propto$, write a mathematical statement connecting the quantities.**
c. **Rewrite the statement from part b using an equals sign and the constant of variation k.**
d. **Choose one of the data points and calculate the value of k.**
e. **Restate the equation using the value of k.**

THINK	WRITE
a. The amount of time taken to complete the task depends on the number of workers; time is the dependent variable, so place it on the vertical axis. The smallest number of hours is 2 and the highest is 12, so use a scale of 2 on the vertical axis. The smallest number of workers is 1 and the highest is 6. Use a scale of 1 on the horizontal axis.	a. 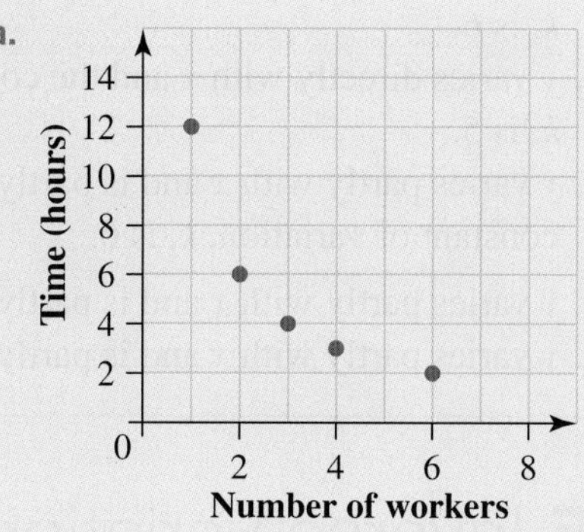
b. The graph looks like a hyperbola, so assume that the variables are inversely proportional. Write this mathematically.	b. $T \propto \frac{1}{n}$
c. The proportion sign can be replaced with '$= k$'. Write the equation.	c. $T = \frac{k}{n}$
d. Choose one pair of coordinates to help calculate k — say (4, 3). Substitute the values $T = 3$ and $n = 4$. Solve for k by multiplying both sides by 4.	d. $3 = \frac{k}{4}$ $3 \times 4 = \frac{k}{4} \times \frac{4}{1}$ $12 = k$
e. Replace k with its value of 12 in the equation. Write the equation.	e. $T = \frac{12}{n}$

WORKED EXAMPLE 8 Inverse variation 2

Using the result from Worked example 7, calculate:

a. the time that it would take 7 people to complete the task

b. the number of people required to complete the task in half an hour.

THINK	WRITE
a. 1. To calculate the time taken for 7 workers to complete the task, substitute the value $n = 7$ into the equation $T = \frac{12}{n}$ and solve for T.	a. $T = \frac{12}{n}$ $= \frac{12}{7}$ $= 1.7$
2. State the result.	It would take 7 people 1.7 hours to complete the task.
b. 1. To determine the number of workers required to complete the task in half an hour, substitute $T = \frac{1}{2}$ into the formula.	b. $T = \frac{12}{n}$ $\frac{1}{2} = \frac{12}{n}$
2. Solve for n by: • multiplying both sides by n • multiplying both sides by 2.	$\frac{1}{2} \times \frac{n}{1} = \frac{12}{n} \times \frac{n}{1}$ $\frac{n}{2} = 12$ $\frac{n}{2} \times \frac{2}{1} = 12 \times 2$ $n = 24$
3. Write the answer.	24 people are required to complete the task in half an hour.

2.5 Exercise

1. Explain, in your own words, some of the differences between direct and inverse variation.

2. WE7 The table shows the time taken to travel 100 km at different speeds.

Speed (km/h)	100	50	25	10	5	1
Time (h)	1	2	4	10	20	100

a. Plot the information on a graph on paper or on a calculator.

b. Using the symbols s for speed and t for time, together with the variation sign, write a mathematical shorthand statement for the connection between the quantities.

c. Write the statement in part **b** as a rule with an equals sign.

d. Choose one of the data points and calculate the value of k.

e. Restate the equation using the value of k.

3. A packet of sweets contains 20 pieces. They are to be divided equally among several friends.

a. Copy and complete the table to show the number of sweets that each friend receives for various numbers of friends.

Number of friends	20	10		4	
Number of sweets each	1		4		20

b. Plot the information on a graph on paper or on a calculator.

c. If n is the number of friends and S is the number of sweets each friend receives, write a rule connecting the quantities, using an equals sign and a constant of variation, k.

4. **WE8** In question 3, if there were 20 friends, each received 1 sweet.

a. Use this data to calculate the constant of variation, k.

b. Choose any of the other four data pairs and recalculate the value of k.

c. Write the rule connecting S and n using an equals sign.

5. In question 4, it was found that the relationship between the number of people, n, and the number of sweets, S, was $S = \frac{20}{n}$. Use this result to calculate:

a. the number of sweets each person receives if there were 12 people

b. the number of people present if each receives $1\frac{3}{7}$ sweets each.

6. A company that makes wotzits determines that the number sold depends on the price. If the price is higher, fewer are sold. The market research gives the following expected sales results:

Price of the wotzit	\$1	\$5	\$20	\$50	\$100	\$200	\$400
Number sold (thousands)	400	80	20	8	4	2	1

a. Produce a graph of the number sold versus the price.

b. Using P for price and n for the number sold, state what equation connects the quantities.

c. Use the equation to predict the number sold if the price is \$25.

d. Use the equation to calculate the price if the required sales are 250 000 wotzits.

7. A process to sterilise surgical instruments is tested and its success is found to depend on the temperature used. The results are shown in the table.

Temperature of steriliser (°C)	10	20	40	50	80	90	100
Microbes remaining alive (%)	95	47.5	23.75	19	11.875	10.55	9.5

a. Produce a graph on paper or on a calculator of the percentage of microbes remaining versus the temperature of the steriliser.
b. Using M for the percentage of microbes remaining and T for the temperature, state what equation connects the quantities.
c. Use the equation to predict the percentage of microbes remaining if a temperature of 75 °C is used.
d. Use the equation to calculate the temperature required to ensure that no more than 2% of microbes remain.

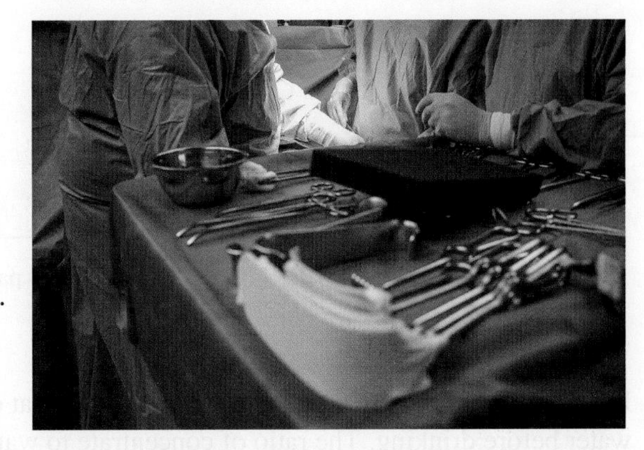

8. MC If $y \propto \frac{1}{x}$ and $y = 12$ when $x = 50$, when $x = 25$:

A. $y = 6$
B. $y = 18$
C. $y = 12\frac{1}{2}$
D. $y = 24$
E. $y = 11\frac{1}{2}$

9. MC When an amount of gas is enclosed in a container, the pressure is inversely proportional to the volume. This is noticeable when using a bicycle pump. As the volume inside the pump is reduced by pushing the plunger, the pressure increases. Using P for pressure and V for volume, determine which of the following is *not* True.

A. P varies inversely with V.
B. As V increases, P decreases.
C. If V is halved, P is doubled.
D. $P \propto \frac{1}{V}$
E. $\frac{P}{V} = k$

10. If $y \propto \frac{1}{x}$ then $y = \frac{k}{x}$ and $yx = k$.

a. Copy and complete the following table.

x	48	24	16	12	8
y	2	4	6	8	12
yx					

b. State if it is true in this case that $y \propto \frac{1}{x}$.

2.6 Ratios

LEARNING INTENTION

At the end of this subtopic you should be able to:
- write ratios in simplest form
- use ratios to calculate fractions of quantities
- determine equivalent ratios.

Ratios are used to compare two or more quantities that are measured in the same unit. Ratios are written without units.

Ratios can be written in words or in numerical form using a colon (:).

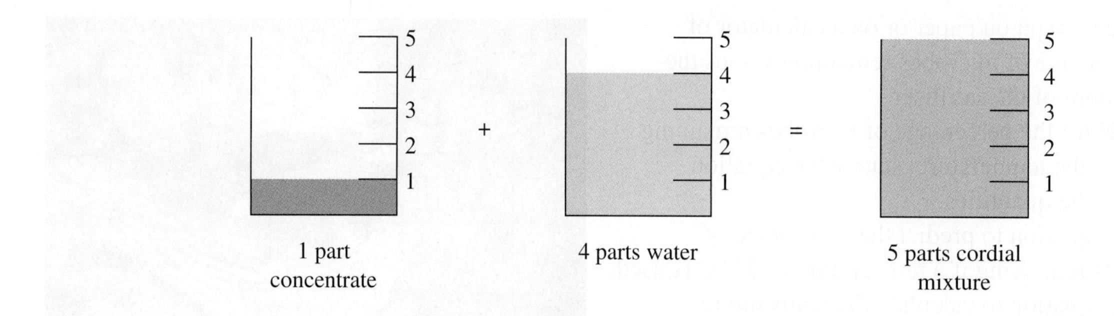

For example, cordial manufacturers recommend that one part of cordial concentrate be mixed with four parts of water before drinking. The ratio of concentrate to water is 1 to 4 and is written as 1 : 4. It means there is one part of concentrate for every four parts of water.

The order of the numbers in a ratio follows the order of the description of the ratio. The ratio of water : concentrate = 4 : 1.

WORKED EXAMPLE 9 Determining ratios

From the picture shown, write a ratio for:

a. blue M&M's to red M&M's to yellow M&M's

b. green M&M's to the total number of M&M's.

THINK	WRITE
a. 1. Count the blue, red and yellow M&M's.	a. There are seven blue, six red and four yellow M&M's.
2. Write this as a ratio listing blue, then red and then yellow M&M's.	blue : red : yellow = 7 : 6 : 4
b. 1. Count the number of green M&M's and total M&M's.	b. There are five green M&M's and 24 in total.
2. Write this as a ratio of green and total M&M's.	green : total = 5 : 24

Ratios in simplest form

One part of a ratio can be written as a fraction of the other part of the ratio. If the fraction can be simplified by cancelling a common factor, then the ratio can be simplified by dividing both numbers by the same common factor.

For example, consider a cordial mixture where concentrate : water is 2 : 4.

To simplify, divide by the common factor, 2:

$$\frac{2}{2} : \frac{4}{2} = 1 : 2$$

Ratios in simplest form should contain whole numbers. If a ratio contains fractions or decimals, these should be converted to whole numbers in simplest form.

WORKED EXAMPLE 10 Simplifying ratios

Rewrite the following as ratios in simplest form.

a. $7\text{ kg} = 140\text{ g}$ **b.** $\frac{1}{3} : \frac{1}{4}$

THINK	WRITE
a. 1. Write the ratio so that the amounts are in the same unit. There are 1000 g in 1 kg, so $7\text{ kg} = (7 \times 1000)\text{ g}$.	**a.** 7 kg to 140 g $= 7000$ g to 140 g
2. Write the two numbers as a ratio. (Ratios have no units.)	$= 7000 : 140$
3. Divide the ratio by the highest common factor. The highest common factor of 7000 and 140 is 140.	$= \frac{7000}{140} : \frac{140}{140}$ $= 50 : 1$
b. 1. Write the ratio.	**b.** $\frac{1}{3} : \frac{1}{4}$
2. Make both fractions whole numbers by multiplying by the lowest common multiple. The lowest common multiple of 3 and 4 is 12.	$\frac{1}{\cancel{3}_1} \times \frac{\cancel{12}^4}{1} : \frac{1}{\cancel{4}_1} \times \frac{\cancel{12}^3}{1}$
3. Write the answer.	$4 : 3$

Using ratios to calculate fractions of amounts

A ratio can be used to form a fraction that compares one part of a ratio with another part or with the whole amount.

Ratios that compare parts of a whole

Using the ratio of concentrate : water $= 1 : 4$, it is possible to describe the amount of concentrate as a fraction of the amount of water, and vice versa. Using the values from the cordial and water example:

$$\frac{\text{amount of concentrate}}{\text{amount of water}} = \frac{1}{4}$$

This means that the amount of concentrate is $\frac{1}{4}$ the amount of water. Similarly, the amount of water is 4 times the amount of concentrate.

Ratios that compare a part with a whole amount

The ratio of concentrate : cordial mixture $= 1 : 5$. Using the values from the ratio:

$$\frac{\text{amount of concentrate}}{\text{amount of cordial mixture}} = \frac{1}{5}$$

This means that the amount of concentrate is $\frac{1}{5}$ of the amount of cordial mixture. Similarly, the amount of cordial mixture is 5 times the amount of concentrate.

Equivalent ratios

To double the amount of cordial, we keep the ratio of concentrate to water the same by doubling the amounts of both concentrate and water. The relationship between the amount of concentrate to water is now 2 : 8.

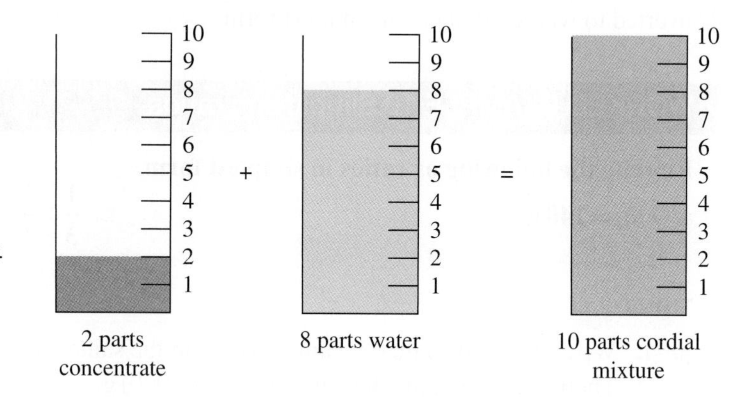

Two ratios are **equivalent** when they describe the same relationship. The ratios 2 : 8 and 1 : 4 are equivalent. In both ratios, there is 1 part of concentrate for every 4 parts of water.

If two ratios are equivalent, their parts are combined in the same **proportion**.

The ratios 1 : 4 and 2 : 8 are equivalent, so $1:4=2:8$. This statement is a proportion.

If we match up **corresponding** numbers in the two ratios shown, we can see that both pairs of corresponding numbers have been multiplied or divided by the same amount (the same **multiplicative factor**). In the proportions shown here, the multiplicative factor is 4.

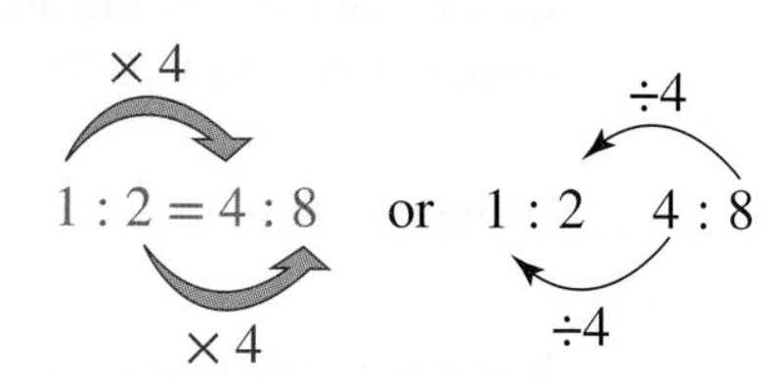

To go from the ratio with the smaller values to the ratio with the larger values, multiply by the multiplicative factor.

To go from the ratio with the larger values to the ratio with the smaller values, divide by the multiplicative factor.

tlvd-3514

WORKED EXAMPLE 11 Equivalent ratios

Determine whether the following pairs of ratios are equivalent.

a. 2 : 3 and 24 : 36 **b. 5 : 6 and 81 : 89**

THINK	WRITE
a. 1. • Check for a common multiplicative factor by comparing corresponding numbers in the two ratios. • Divide the numbers in the ratio with the larger values by the corresponding values in the other ratio. • Divide 24 by 2 and 36 by 3.	**a.** 2 : 3 and 24 : 36 $\frac{24}{2}=12 \quad \frac{36}{3}=12$
2. The two answers are the same and show the common multiplicative factor of 12. The two ratios are equivalent.	The ratios are equivalent. $2:3=24:36$
b. 1. Divide 81 by 5 and 89 by 6 to check for a common multiplicative factor.	**b.** 5 : 6 and 81 : 89 $\frac{81}{5}=16.2 \quad \frac{89}{6}=14.83$
2. The two answers are different so the multiplicative factors are different. The ratios are not equivalent.	The ratios are not equivalent. $5:6\neq 81:89$

Proportion statements can also be used to determine the missing value of a ratio.

WORKED EXAMPLE 12 Calculating proportions

Calculate the value of x that makes the following ratios proportional.

a. **$2:5$ and $26:x$**

b. **$x:3$ and $14:21$**

THINK	WRITE
a. 1. We know the ratios are proportional, which means that they have a common multiplicative factor.	**a.** $2:5 = 26:x$
2. To determine the common multiplicative factor, divide 26 by 2.	$\frac{26}{2} = 13$
3. The multiplicative factor is 13. To calculate the value of x, multiply 5 by the multiplicative factor (13).	$2:5 = 26:x$ $x = 5 \times 13$ $= 65$
b. 1. We know the ratios are proportional, which means that they have a common multiplicative factor.	**b.** $x:3 = 14:21$
2. To find the multiplicative factor, divide 21 by 3.	$\frac{21}{3} = 7$
3. The multiplicative factor is 7. To calculate the value of x, divide 14 by 7.	$14 \div 7 = 2$ $x = 2$

2.6 Exercise

1. Explain what a ratio is and what it is used for.
2. WE9 A bag contains five red marbles, eight green marbles and three blue marbles. Write the ratio of:
   a. blue marbles to green marbles
   b. red marbles to blue marbles
   c. green marbles to red marbles to blue marbles
   d. blue marbles to the total number of marbles.
3. The ratio of boys to girls in a class is $7:5$.
   a. State the ratio of girls to boys.
   b. State the number of girls as a fraction of the total.
   c. State the number of boys as a fraction of the total.

4. **MC** In the photograph, the ratio of ballpoint pens to pencils is:

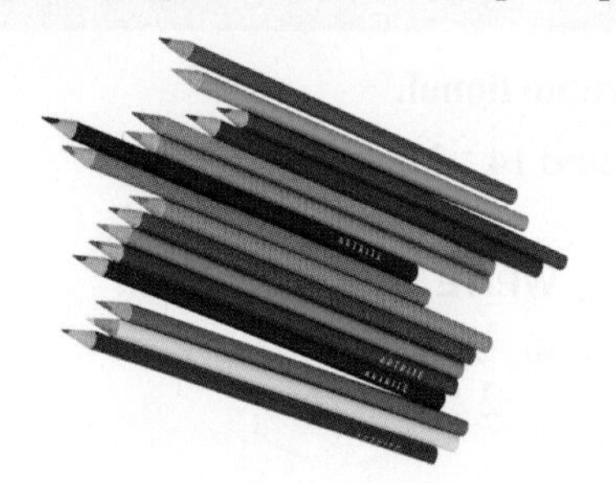

A. 16 : 7 B. $\frac{8}{16}$ C. 16 : 31 D. 7 : 16 E. 8 : 17

5. **WE10** Write each of the following ratios in simplest form.
   a. 6 : 16 b. 27 : 9 c. 56 : 4 d. 1.5 : 2 e. 5 : 15 : 35 f. 1.6 : 2.8
6. Rewrite each of the following statements as a ratio in simplest form.
   a. 12 cm to 9 mm b. 70 minutes to 2 hours c. 1 week to 4 days
   d. 3 hours to 2 days e. 120 seconds to 1 hour f. 1 kg to 150 g
7. The photograph shows a model used to represent a new building to be built. If the new building is to be 110 metres high, write the ratio of the model height to the actual height of the building in simplest form.

8. **WE11** Determine whether the following pairs of ratios are equivalent.
   a. 2 : 3 and 10 : 15 b. 8 : 3 and 168 : 63
   c. 4 : 7 and 32 : 60 d. 5 : 9 and 110 : 198
   e. 12 : 16 and 4 : 3 f. $\frac{10}{3}$ and $\frac{25}{7}$
9. **WE12** Calculate the value of x that makes each of the following pairs of ratios equivalent.
   a. 4 : 7 and 8 : x b. x : 9 and 2 : 3
   c. 10 : x and 3 : 15 d. 12 : 10 and x : 5
   e. x : 6 and 25 : 30 f. 7 : 49 and x : 1
10. The wing of a model aeroplane is 4 cm long, and the scale of the model is 1 : 300.
    a. Determine the ratio of the length of the model's wing to the length of the wing of the actual aeroplane.
    b. Calculate the length of the wing of the actual aeroplane.
11. The nutritional panel for a certain cereal is shown. Sugar is one form of carbohydrate. Calculate the ratio of sugar to total carbohydrates in the cereal.

	Quantity per 30-g serving	Percentage daily intake per 30-g serving	Quantity per 30-g serving with $\frac{1}{2}$ cup skim milk
Energy	480 kJ	5.5%	670 kJ
Protein	6.6 g	13.1%	11.2 g
Fat			
– total	0.2 g	0.3%	0.3 g
– saturated	0.1 g	0.1%	0.2 g
Carbohydrate			
– total	20.8 g	6.7%	27.3 g
– sugars	9.6 g	10.7%	16.1 g

12. There are apples, oranges and bananas in a fruit bowl. The ratio of apples to oranges is 3 : 4, and the number of apples is $\frac{3}{8}$ of the total number of pieces of fruit. State how many bananas there are.

2.7 Ratio and proportion

LEARNING INTENTION

At the end of this subtopic you should be able to:
- calculate proportions using ratios
- use proportions to solve real-world problems.

In mathematics, a **proportion** is an equation that shows equality of two ratios.

When objects are in proportion to each other, they have the same fractional relationships between the size of their parts and the size of the whole. For example, the rectangles below are in proportion to each other.

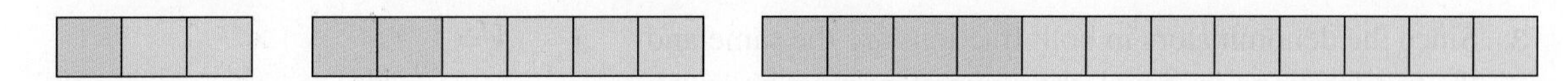

In each rectangle:
- when you compare the parts with the whole rectangle, the green squares form $\frac{1}{3}$ of the whole rectangle, while the yellow squares form the other $\frac{2}{3}$
- when you compare the two coloured parts, there is 1 green square for every 2 yellow squares; the number of green squares is $\frac{1}{2}$ of the number of the yellow squares.

Equivalent ratios have the same relationships between the two parts in each ratio. In the rectangles above, the green : yellow ratios are 1 : 2, 2 : 4 and 4 : 8. All of these ratios are equivalent to each other and, in each of these ratios, the fraction $\frac{\text{green}}{\text{yellow}} = \frac{1}{2}$.

Equivalent ratios

The common fractional relationship between the parts of two equivalent ratios can be represented as shown.

$$\text{If } a : b = c : d,$$

$$\text{then } \frac{a}{b} = \frac{c}{d}.$$

In mathematics, an equation showing two equal ratios is called a *proportion*.

WORKED EXAMPLE 13 Using fractions to calculate proportions

Write the following proportions as equal fractions and calculate the value of x.

a. $3 : x = 18 : 42$

b. $x : 2 = 29 : 58$

THINK	WRITE
a. 1. Write the numbers from each ratio as a fraction, making sure that corresponding numbers from each ratio are in the same position in both fractions.	**a.** $3 : x = 18 : 42$ $\frac{3}{x} = \frac{18}{42}$
2. To calculate the value of x, simplify the fraction $\frac{18}{42}$ by a factor of 6 so that it has the same denominator as the fraction containing x.	$\frac{3}{x} = \frac{\cancel{18}^{3}}{\cancel{42}^{7}}$

3. Since the numerators in both fractions are the same, and the two fractions are equal, the denominators must be equal.	$x=7$
b. 1. Write the proportion as two equal fractions.	**b.** $x:2=29:58$ $\frac{x}{2}=\frac{29}{58}$
2. To calculate the value of x, simplify the fraction $\frac{29}{58}$ so that it has the same denominator as the fraction containing x. Simplify by a factor of 29 ($58 \div 2 = 29$).	$\frac{x}{2}=\frac{29^{1}}{58^{2}}$
3. Since the denominators in both fractions are the same and the two fractions are equal, the numerators must be equal.	$x=1$

Proportion can be used to compare two objects.

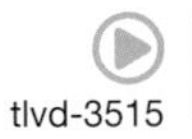
tlvd-3515

WORKED EXAMPLE 14 Comparing quantities using proportions

Two jugs contain mixtures of orange juice and water.

- **Mixture 1 contains 3 cups of orange juice and 5 cups of water.**
- **Mixture 2 contains 2 cups of orange juice and 4 cups of water.**

Determine whether the mixtures have the same orange flavour. If the flavour is not the same, determine which mixture has the stronger orange flavour.

THINK	WRITE
1. Write the quantities for each mixture as a ratio of orange juice to water.	Mixture 1: orange juice : water = 3 : 5 Mixture 2: orange juice : water = 2 : 4
2. Find the fractional relationship between the two parts of each ratio by writing the numbers from each ratio as a fraction. Make sure that corresponding numbers from each ratio are in the same position in both fractions.	Mixture 1: $\frac{\text{orange juice}}{\text{water}}=\frac{3}{5}$ Mixture 2: $\frac{\text{orange juice}}{\text{water}}=\frac{2}{4}$ $=\frac{1}{2}$
3. If the mixtures are the same, the fractional relationship between the two parts of each ratio will be the same. Compare the two fractions.	Mixture 1: $\frac{3}{5}\times\frac{2}{2}=\frac{6}{10}$ Mixture 2: $\frac{1}{2}\times\frac{5}{5}=\frac{5}{10}$ Therefore, $\frac{3}{5}\neq\frac{1}{2}$.
4. Since the fractions are not the same, the mixtures are not in the same proportion and will not taste the same.	The mixtures do not have the same orange flavour.

5. The mixture with the stronger orange flavour will have more orange juice in it. Calculate the amount of orange juice in each mixture as a fraction of the amount of water.	Mixture 1: $(\text{orange juice}) = \frac{6}{10} \times (\text{water})$ Mixture 2: $(\text{orange juice}) = \frac{5}{10} \times (\text{water})$
6. In mixture 1, orange juice is a larger fraction of water than in mixture 2.	Mixture 1 has the stronger orange flavour.

2.7 Exercise

Students, these questions are even better in jacPLUS

Receive immediate feedback and access sample responses

Access additional questions

Track your results and progress

Find all this and MORE in jacPLUS

1. Explain what it means when two objects are proportional to each other.

2. **WE13** Write the following proportions as equal fractions and calculate the value of a.

a. $a : 2 = 8 : 16$
b. $a : 6 = 8 : 12$
c. $3 : a = 45 : 105$
d. $7 : a = 14 : 48$
e. $24 : 16 = a : 4$
f. $17 : a = 51 : 81$

3. **MC** A classmate was having a party and invited 30 people. They were going to spend about $120 on food for the party, but then 10 people cancelled. The proportion equation to work out how much money should now be spent on food is:

A. $\frac{30}{120} = \frac{10}{x}$
B. $\frac{30}{120} = \frac{20}{x}$
C. $\frac{10}{120} = \frac{x}{30}$
D. $\frac{20}{30} = \frac{120}{x}$
E. $\frac{30}{120} = \frac{x}{20}$

4. Solve the following for x.

a. $x : 4 = \frac{30}{12}$
b. $10 : x = \frac{80}{104}$

5. A classmate's work is shown at right. Fix the error in your classmate's work.

$$3 : 5 = 51 : x$$
$$\frac{3}{5} = \frac{51}{x}$$
$$3 \times 5 = 51 \times x$$
$$153 = 5x$$
$$x = 30.6$$

6. A powdered sports drink is prepared by dissolving two scoops of powder in 1 litre of water.

a. State the ratio of the number of powder scoops to the volume of drink.
b. If an athlete had a 750-mL bottle, determine how much powder would be needed to fill the bottle with the prepared sports drink in the correct proportion.

7. A recipe to make enough spaghetti bolognaise to feed four people needs 500 g of mincemeat. According to this recipe, calculate how much mincemeat would be needed to make enough spaghetti bolognaise to feed a group of nine people.

8. In class 8A, there are 4 boys and 5 girls. In class 8B, there are 12 boys and 15 girls. In class 8C, there are 6 girls and 8 boys.
   a. For each class, write the ratio of boys to girls.
   b. State in which two classes the proportions of boys and girls are the same.
   c. Class 8A joins class 8B to watch a program for English. State the resulting boy-to-girl ratio.
   d. Class 8C joins class 8B for PE lessons. State the boy-to-girl ratio in the combined PE class.
   e. Determine if the ratio of boys to girls watching the English program is equivalent to the ratio of boys to girls in the combined PE class.

9. **WE14** An internet search for a homemade lemonade recipe yielded the following results. All of the recipes had water added, but the sweetness of the lemonade is determined by the ratio of lemon juice to sugar. For each of the recipe ratios shown below, determine the mixtures that have the same taste. If they do not have the same taste, determine which website has the sweeter lemonade recipe.
   a. Website 1: 3 tablespoons of sugar for every 15 tablespoons of lemon juice
      Website 2: 4 tablespoons of sugar for every 20 tablespoons of lemon juice
   b. Website 3: 3 tablespoons of sugar for every 9 tablespoons of lemon juice
      Website 4: 5 tablespoons of sugar for every 16 tablespoons of lemon juice
   c. Website 5: 2 cups of sugar for every 3 cups of lemon juice
      Website 6: 5 tablespoons of sugar for every 8 tablespoons of lemon juice
   d. Website 7: 3 tablespoons of sugar for every 8 tablespoons of lemon juice
      Website 8: 7 tablespoons of sugar for every 12 tablespoons of lemon juice

10. The ingredients for a chocolate pudding are shown. The cook follows the recipe for the pudding but makes $1\frac{1}{2}$ times the quantity of the sauce, thinking that the pudding may be too dry otherwise.

Pudding		**Sauce**
1 cup self-raising flour	$\frac{1}{2}$ teaspoon salt	$\frac{3}{4}$ cup brown sugar
$\frac{3}{4}$ cup sugar	2 tablespoons cocoa	$\frac{1}{4}$ cup cocoa
$\frac{1}{2}$ cup milk	1 teaspoon vanilla	$1\frac{1}{4}$ cups boiling water
2 tablespoons butter		

   a. For each sauce ingredient, write the ratio of the amount used to the amount in the original recipe, in simplest form.
   b. The modified recipe makes enough pudding for four people. If the cook wants to make enough to feed 10 people, calculate how much of each ingredient is required.

2.8 Dividing in a given ratio

LEARNING INTENTION

At the end of this subtopic you should be able to:
- calculate amounts split into a given ratio.

When something is shared, we often use ratios to ensure that the sharing is fair.

Consider this situation. Two people buy a lottery ticket for $3. They win a prize of $60. How is the prize divided fairly?

Each person contributes $1.50.	One person contributes $1 and the other $2.
• The contribution for the ticket is in the ratio 1 : 1. • The prize is divided in the ratio 1 : 1. • Each person receives $30.	• The contribution for the ticket is in the ratio 1 : 2. • The prize is divided in the ratio 1 : 2. • The person who contributed $1 receives $20 and the other person receives $40.
\$30 | \$30	\$20 | \$20 | \$20

In the situation:
- the 1 : 1 ratio has $1 + 1 = 2$ total parts. Each person receives $\frac{1}{2}$ of the prize money ($30).
- the 1 : 2 ratio has $1 + 2 = 3$ total parts. Person 1 paid for 1 part of the ticket and therefore receives $\frac{1}{3}$ of the prize money ($20); person 2 paid for 2 parts of the ticket and therefore receives $\frac{2}{3}$ of the prize money ($40).

WORKED EXAMPLE 15 Sharing amounts using ratios

Share the amount of $250 in the ratio 2 : 3.

THINK	WRITE
1. Determine the total number of equal parts of the ratio.	Total number of parts $= 2 + 3$ $= 5$
2. The first part of the ratio represents $\frac{2}{5}$ of the total amount.	$\frac{2}{5}$ of \$250 $= \frac{2}{5} \times \frac{250}{1}$ $= 100$

3. The second part represents $\frac{3}{5}$ of the total amount.

$\frac{3}{5}$ of \$250

$= \frac{3}{5} \times \frac{250}{1}$

$= 150$

4. Answer the question.

The first part of the ratio equates to \$100 and the second part of the ratio equates to \$150.

The amount that the second part should receive can also be calculated by subtracting the amount the first part received from the original amount.

In Worked example 15, this would be $\$250 - \$100 = \$150$.

tlvd-3516

WORKED EXAMPLE 16 Sharing amounts using ratios 2

Three people buy a raffle ticket for a \$2400 shopping spree. If they put in \$5, \$3 and \$2 each, determine how the prize should be divided if they win.

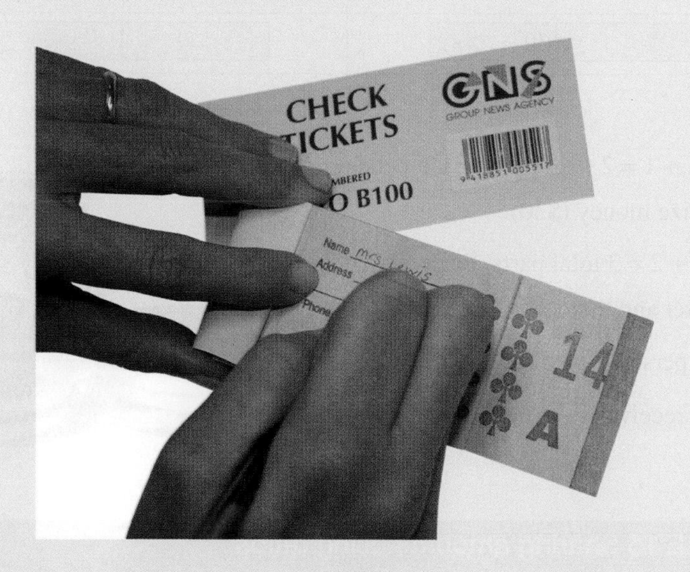

THINK

1. Determine the total number of equal parts of the ratio. (This is the total amount of money put in.)

WRITE

Total number of parts $= 5 + 3 + 2$
$= 10$

2. The first part of the ratio represents $\frac{5}{10}$ of the amount, the second part represents $\frac{3}{10}$ of the amount, and the third part represents $\frac{2}{10}$ of the amount.

$\frac{5}{10}$ of \$2400
$= \frac{5}{10} \times \frac{2400}{1}$
$= \$1200$

$\frac{3}{10}$ of \$2400
$= \frac{3}{10} \times \frac{2400}{1}$
$= \$720$

$\frac{2}{10}$ of \$2400
$= \frac{2}{10} \times \frac{2400}{1}$
$= \$480$

3. Answer the question.

The person who put in \$5 should receive \$1200.
The person who put in \$3 should receive \$720.
The person who put in \$2 should receive \$480.

2.8 Exercise

Students, these questions are even better in jacPLUS

Receive immediate feedback and access sample responses

Access additional questions

Track your results and progress

Find all this and MORE in jacPLUS

1. Explain how ratios can be used to share an amount between different parts.
2. Write the total number of parts for each of the following ratios.

   a. 3 : 4 b. 1 : 8 c. 6 : 7 d. 9 : 2 e. 3 : 11 f. 1 : 3 : 4
3. WE15 Share $2000 in the following ratios.

   a. 1 : 1 b. 1 : 3 c. 1 : 6 d. 3 : 4 e. 8 : 3 f. 3 : 7
4. The ratio of left-handers to right-handers in a survey of 390 people was 2 : 11.

   a. State what fraction of the people were left-handed.

   b. Calculate how many of those surveyed were right-handed.
5. WE16 Three people contribute to an $8 raffle ticket for a $3200 cash prize. If they put in $5, $2 and $1, determine how the prize should be divided if they win.
6. A triangle has sides in the ratio 2 : 3 : 4. Calculate the length of each side if the perimeter of the triangle is 45 cm.
7. The sides of a quadrilateral are in the ratio 1 : 3 : 3 : 4. If the perimeter of this shape is 121 cm, calculate the length of each of the sides.
8. The ratio of the three angles in a particular triangle is 2 : 4 : 5. Correct to 2 decimal places, calculate the sizes of the three angles.
9. Three friends bought a lottery ticket costing $10. Determine how they should share the prize of $350 000 if they each contribute:

   a. $1, $1 and $8 b. $3, $3 and $4

   c. $5.50, $1.50 and $3 d. $4.65, $1.15 and $4.20.
10. Three friends in the photo have been working in a garden. The garden owner paid $500 for the work that was done. Determine how this money should be divided fairly among the three friends.

11. One week, your friend worked for 38 hours at his normal rate of pay and then an extra three hours for one and a half times his normal rate of pay.

    a. Determine the ratio of the pay he received at normal rates to the pay he received for the extra hours.

    b. If he earned $1870 during the week, calculate his normal rate of pay per hour.
12. Three friends bought a raffle ticket and agreed to share the $1500 prize fairly if they won. If the first contributed half of the cost and the second contributed one-third of the cost, determine how much prize money each of the three friends should receive.

13. Three people who shared in a lottery ticket won the jackpot. The first person received \$250 000, the second person received \$175 000 and the third person received \$775 000. If the ticket cost \$12, calculate how much each contributed towards the ticket. (*Hint:* Simplify the ratio.)

14. Your elderly neighbour likes to tell stories. In particular, he likes to tell the story of how he won the lottery when he was younger. The thing is, he can never quite remember the prize pool. He knows that he received \$21 000, and that he contributed \$2.00 to a \$7.50 ticket. Help your neighbour figure out the total prize pool.

15. A pizza is purchased by Kamil and his two friends. Kamil put in \$3, one friend put in \$4 and the other put in \$5.
    a. State how many pieces the pizza should be cut into.
    b. Determine how many pieces of pizza each friend should get.
    c. Calculate the fraction of the pizza each person gets.

16. A triangle has one angle that is double the size of another angle. If the third angle is 60°:
    a. determine the sizes of the other two angles
    b. write the three angles as a ratio.

17. Write at least *three* different ways that you could share out \$240 among four friends as:
    a. ratios
    b. money shared.

18. You and four friends were pooling money to bid for 3 hours at a function where your favourite football star was going to appear. If your friends contributed \$12, \$15, \$10 and \$8, calculate how much you should contribute to guarantee you at least $\frac{1}{2}$ hour there (assuming that you won the bid).

2.9 Review

2.9.1 Summary

doc-38010

Hey students! Now that it's time to revise this topic, go online to:

Review your results

Watch teacher-led videos

Practise questions with immediate feedback

Find all this and MORE in jacPLUS

2.9 Exercise

Multiple choice

1. MC Determine which of the following represents an example of direct proportion.
   A. Hours of sleep and the type of breakfast you eat
   B. Thickness of a book and the number of pages in the book
   C. Distance travelled to work and the type of car you own
   D. The football team you support and the phone you have
   E. The school you go to and the gaming console you use

2. MC If $y \propto x$ and $y = 12$ when $x = 3$, the proportionality constant is:
   A. 12 B. 3 C. 36 D. 4 E. 15

3. MC If $y \propto \frac{1}{x}$ and $y = 2$ when $x = 5$, the value of y when $x = 20$ is:
   A. 2 B. 5 C. $\frac{1}{2}$ D. $\frac{1}{5}$ E. 10

Use the following graph to answer questions 4 to 6.

The graph shows the connection between the number of revolutions of a bicycle wheel and the distance travelled.

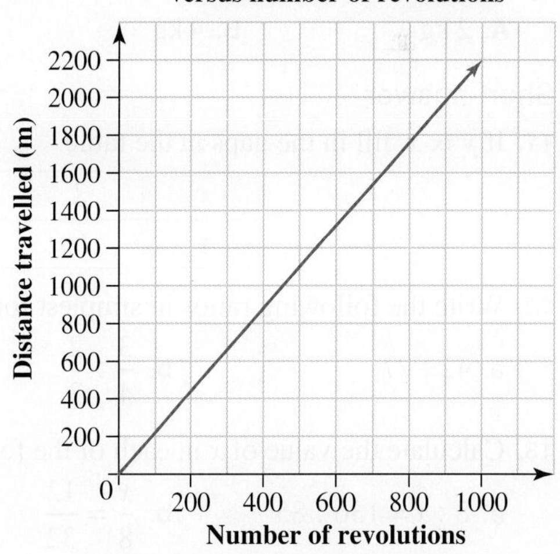

4. MC If d is the distance travelled and r is the number of revolutions, determine which of the following is True.
   A. $d \times r = k$ B. $\frac{d}{r} = k$ C. $d \propto k$ D. $r \propto k$ E. $r = kk$

5. **MC** Determine which of the following is True.

A. $k = 2.2$

B. For any point on the graph, $\frac{r}{d} = 2.2$.

C. The units for k are revolutions per metre.

D. $r = 2.2 \times d$

E. $k = 22$

6. **MC** Determine which of the following is True.

A. If the number of revolutions is 250, then the distance travelled is 114 m.

B. If the distance was 2420 m, then the wheels completed 1100 revolutions.

C. If the distance was 3000 m, then the wheels completed 6600 revolutions.

D. The wheels completed 2.2 revolutions per metre travelled.

E. The wheels completed 22 revolutions per minute.

7. **MC** The ratio 48 : 72 in its simplest form is:

A. 24 : 36

B. 6 : 9

C. 2 : 3

D. 48 : 72

E. 1 : 2

8. **MC** In the equation $3 : 7 = x : 42$, the value of x is:

A. 18 B. 3 C. 24 D. 38 E. 15

9. **MC** Three families order fish and chips and the total bill comes to $168. They split the bill according to how many people there are in each family. The three families have 3, 5 and 6 members respectively. The family with 6 members paid:

A. $60 B. $70 C. $36 D. $50 E. $72

10. **MC** Concrete is mixed with 1 part cement, 2 parts sand and 3 parts stones. If 24 kg of concrete was made, the amount of sand used would be:

A. 2 kg B. 4 kg C. 6 kg D. 8 kg E. 12 kg

Short answer

11. If $y \propto x$, fill in the gaps in the table.

x	2	7		15
y	6	21	30	

12. Write the following ratios in simplest form.

a. 42 : 77 b. $\frac{3}{4} : \frac{2}{3}$ c. 0.1 : 0.05 d. 5 cm : 1 m

13. Calculate the value of x in each of the following proportion equations.

a. $8 : x = 136 : 85$ b. $\frac{x}{8} = \frac{12}{32}$ c. $\frac{16}{21} = \frac{x}{168}$ d. $\frac{11}{7} = \frac{209}{x}$

14. Three people buy a lottery ticket for a prize of $875 000. Determine how much they would each win if they contributed:

a. $10, $7 and $8 b. $2, $2.50 and $0.50 c. $10, $6 and $4.

Extended response

15. The strength of an electric field (Vm^{-1}) is inversely proportional to the distance the plates are apart (m). Determine the effect on the electric field if the distance between the plates is:
    a. three times the distance apart   b. half the distance apart.
16. The variable y is inversely proportional to the variable x. When $x = 4, y = 6$.
    a. Write a variation statement using the $\propto$ sign for the relationship between y and x.
    b. Calculate the constant of variation and hence write a rule for the relationship.
    c. Use the rule to calculate the value of y when $x = 3$.
    d. Calculate x when $y = 12$.
17. The time t (hours) it takes to complete a 100-km journey varies inversely with the speed v (m/s).
    a. Write a proportionality statement connecting t and v.
    b. Write a rule connecting t and v using a constant of variation, k.
    c. Calculate the time needed to complete the 100 km at 50 km/h.
    d. Calculate the constant of variation.
    e. State the rule connecting t and v.
18. A mathematician is attempting to model the probability of an AFL player successfully kicking a goal under various conditions. They suggest that the probability p of kicking the goal varies inversely with the distance d (m) from the goal and the angle, θ (°), from directly in front.
    a. Write a proportionality statement for this situation.
    b. They find that a player 50 m out from the goal and at a 30° angle has a probability of 0.5 of scoring the goal. Calculate the constant of variation.
    c. Write the rule that models the probability of kicking the goal in terms of the distance and angle.
    d. Use the rule to calculate the probability for a shot at goal from 45 m out at a 45° angle.
19. Two friends are sharing a house. When the gas bill of $136 arrives, they discuss how much each should pay. In the end, they decide that since gas is used only for hot water in the house, they should pay according to the length of their showers. One usually spends 3 minutes in the shower. The other usually spends 5 minutes. Calculate how much of the bill each should pay.
20. A Year 12 student holding a birthday party has invited 25 friends and 15 family members.
    a. State the ratio of friends to family.
    b. Express the number of friends as a percentage of the total number of people.
    c. For catering purposes, 20% of people are vegetarians and 75% of people like chips. Calculate how many people are vegetarians and how many people like chips.
    d. The finger food comprised samosas, rice-paper rolls and mini dim sims in the ratio 1 : 2 : 5 (in that order). If there were 160 individual items of finger food, calculate how many of each type there were.

Answers

Topic 2 Ratios, proportion and variation

2.2 Direct variation (proportion)

2.2 Exercise

1. When two variables are related so that their ratio is constant, the relationship between them is called direct variation or proportion.
2. a. Yes, because it is a straight line that goes through the origin.
   b. $k = 60$
   c. $C = 60n$
3. a. 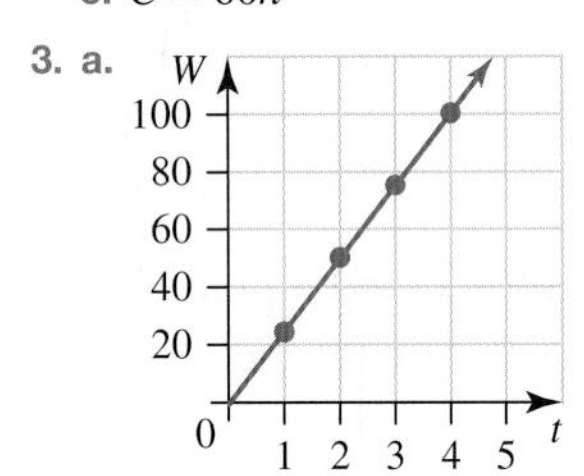

   b. Yes
4. a. b. Yes
5. a. b. Yes
6. a. $I \propto A$ b. $I = kA$ c. $k = 0.5$
   d. $I = 0.5A$ e. $I = \$22\,500$ f. $A = \$74\,000$
7. a. $k = 4.5$ b. $H = 4.5n$ c. $H = 324$
   d. $n = 100$
8. a. $k = 33.5$ b. $A = 33.5L$ c. $A = 837.5$
   d. $L = 11.5$
9. a. $y \propto x$
   Since they are directly proportional to each other, if x increases by 10% then y will also increase by 10%.
   b. $y \propto x$
   Since they are directly proportional to each other, if x decreases by 15% then y will also decrease by 15%.
10. x increases from 40 to 320 when y increases from 120 to 960.

2.3 Direct variation (proportion) and ratios

2.3 Exercise

1. The constant of variation k is equal to the ratio $\frac{y}{x}$. It is the rate at which y varies with x.
2. a. $y = kx$
   b. $k = \frac{1}{7}$
   c.

y	1.6	2.6	3.1	4.9
x	11.2	18.2	21.7	34.3
$\frac{y}{x}$	$\frac{1}{7}$	$\frac{1}{7}$	$\frac{1}{7}$	$\frac{1}{7}$

   d. $k = \frac{y}{x} = \frac{1}{7}$
3. a. $W = kH$
   b. $k = 17$
   c. $\frac{\text{wages}}{\text{hours}} = 17$ for each day of the week
   d. An employee is paid \$17 per hour.
4. a. $D = 10.6L$ b. $D = 48.76\text{ km}$
   c. $L = 12.264$ litres
5. a. $k = 9.8\text{ N/kg}$
   b. This is the gravitational constant — the force per kg of mass.
6.

y	3.5	6.1	9.7	11.2
x	7.35	11.59	23.28	24.64
$\frac{y}{x}$	2.1	1.9	2.4	2.2

   y is not directly proportional to x.
7. a. $P = 1.106$ atmospheres
   b. 10.6% increase

2.4 Partial variation (proportion)

2.4 Exercise

1. The graph is a straight line but does not go through the origin.
2. a. $k = 17$ b. $c = -12$ c. $y = 17x - 12$
   d. $y = 183.5$ e. $x = 24.5$
3. a. $k = 8$ b. $c = 25$ c. $y = 8x + 25$
   d. $y = 79$ e. $x = 11.2$
4. a. $S = kn + c$
   b.
   c. $k = 180$
   d. For every increase in the number of sides, there is an increase of 180 in the sum of the angles.
   e. $c = -360$
   f. $S = 180n - 360$
5. a. $S = 1440°$ b. $n = 16$ sides
6. a. $C = kd + c$

b. 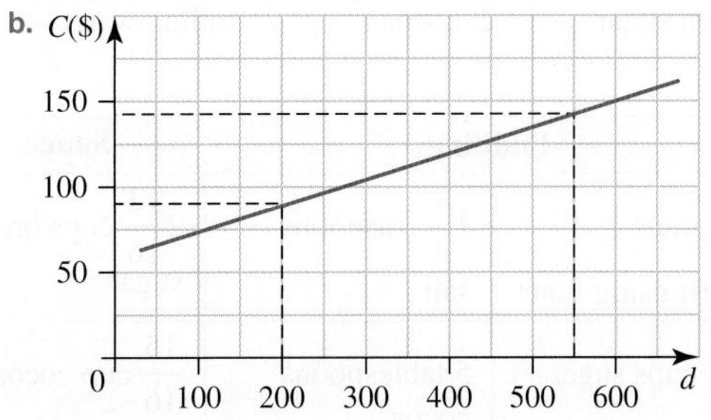

c. $k = 0.15$

d. The cost for the distance component is $0.15 per km.

e. $c = 60$; it costs $60 to hire the car before any distance is travelled.

f. $C = 0.15d + 60$

g. $114

h. 420 km

7. C

8. D

2.5 Inverse variation (proportion)

2.5 Exercise

1.

Direct variation	Inverse variation
Increase in x means an increase in y.	Increase in x means a decrease in y.
Graph is a straight line through (0, 0).	Graph is a hyperbola.
$y = kx$ $\Rightarrow k = \frac{y}{x}$	$y = \frac{k}{x}$ $\Rightarrow k = xy$

2. a. 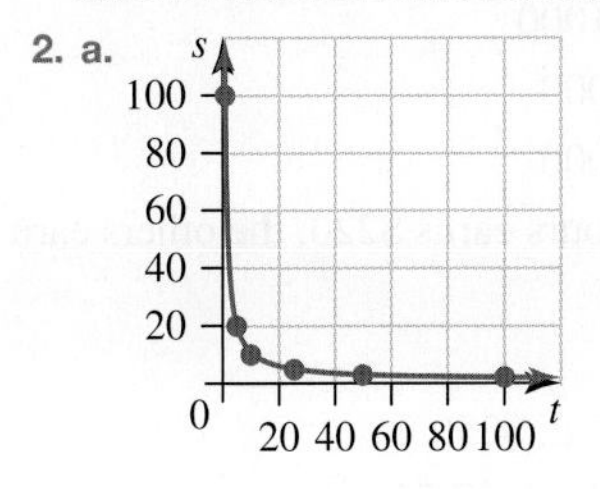

b. $t \propto \frac{1}{s}$

c. $t = \frac{k}{s}$

d. $k = 100$

e. $t = \frac{100}{s}$

3. a.

Number of friends	20	10	5	4	1
Number of sweets	1	2	4	5	20

b. 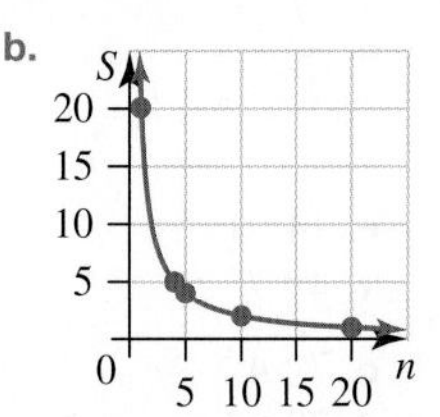

c. $S = \frac{k}{n}$

4. a. $k = 20$ b. $k = 20$ c. $S = \frac{20}{n}$

5. a. Each receives $1\frac{2}{3}$ sweets.

b. There are 14 people present.

6. a. 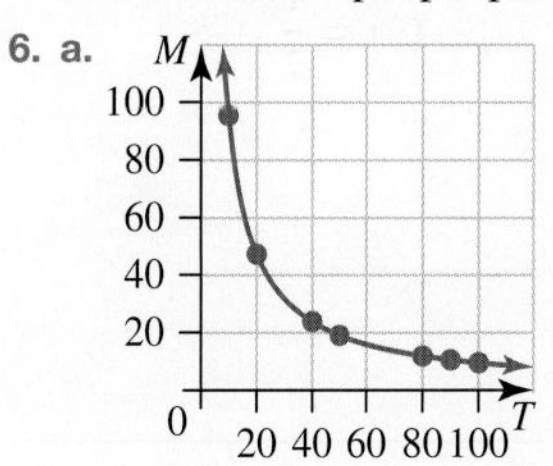

b. $n = \frac{400}{P}$

c. $n = 16\,000$ sold

d. $P = \$1.60$

7. a. 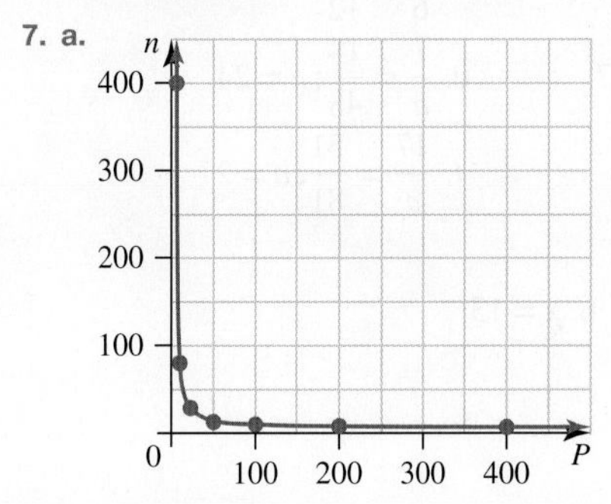

b. $m = \frac{950}{T}$

c. 12.67%

d. More than 475 °C

8. D

9. E

10. a.

x	48	24	16	12	8
y	2	4	6	8	12
yx	96	96	96	96	96

b. Yes

2.6 Ratios

2.6 Exercise

1. A ratio is used to compare two or more quantities that are measured in the same unit.

2. a. 3 : 8 b. 5 : 3 c. 8 : 5 : 3 d. 3 : 16

3. a. 5 : 7 b. $\frac{5}{12}$ c. $\frac{7}{12}$

4. D

5. a. 3 : 8 b. 3 : 1 c. 14 : 1
d. 3 : 4 e. 1 : 3 : 7 f. 4 : 7

6. a. 40 : 3 b. 7 : 12 c. 7 : 4
d. 1 : 16 e. 1 : 30 f. 20 : 3

7. Model height : actual height = 55 cm : 110 m
= 55 : 11 000
= 1 : 200

8. a. Equivalent b. Equivalent c. Not equivalent
d. Equivalent e. Not equivalent f. Not equivalent

9. a. $x = 14$ b. $x = 6$ c. $x = 50$
d. $x = 6$ e. $x = 5$ f. $x = \frac{1}{7}$

10. a. 1 : 300 b. 1200 cm

11. 9.6 : 20.8 = 6 : 13

12. 1 banana

2.7 Ratio and proportion

2.7 Exercise

1. They have the same fractional relationship between the size of their parts and the size of the whole.

2. a. $\frac{a}{2} = \frac{8}{16}; a = 1$ b. $\frac{a}{6} = \frac{8}{12}; a = 4$
c. $\frac{3}{a} = \frac{45}{105}; a = 7$ d. $\frac{7}{a} = \frac{14}{48}; a = 24$
e. $\frac{24}{16} = \frac{a}{4}; a = 6$ f. $\frac{17}{a} = \frac{51}{81}; a = 27$

3. B

4. a. $x = 10$ b. $x = 13$

5. $\frac{3}{5} = \frac{51}{x}$
$3 \times x = 5 \times 51$
$3x = 225$
$x = 85$

6. a. 2 : 1
b. 1 litre = 1000 millilitres so $\frac{x}{2} = \frac{750}{1000}$; $x = 1.5$.

7. 1125 g

8. a. Class 8A — 4 : 5; class 8B — 4 : 5; class 8C — 4 : 3
b. Classes 8A and 8B
c. 4 : 5
d. 20 : 21
e. No

9. a. $\frac{3}{15} = \frac{4}{20}$; same taste
b. $\frac{3}{9} \neq \frac{5}{16}$; website 3 has the sweeter recipe.
c. $\frac{2}{3} \neq \frac{5}{8}$; website 5 has the sweeter recipe.
d. $\frac{3}{8} \neq \frac{7}{12}$; website 8 has the sweeter recipe.

10. a. Brown sugar — 3 : 2; cocoa — 3 : 2; boiling water — 3 : 2
b.

Pudding		Sauce
$2\frac{1}{2}$ cups self-raising flour	$1\frac{1}{4}$ teaspoons salt	$2\frac{13}{16}$ cups brown sugar
$1\frac{7}{8}$ cups sugar	5 tablespoons cocoa	$\frac{15}{16}$ cup cocoa
$1\frac{1}{4}$ cups milk	$2\frac{1}{2}$ teaspoons vanilla	$4\frac{11}{16}$ cups boiling water
5 tablespoons butter		

2.8 Dividing in a given ratio

2.8 Exercise

1. Ratios can be used to share an amount fairly between different parts.

2. a. 7 b. 9 c. 13
d. 11 e. 14 f. 8

3. a. \$1000 : \$1000 b. \$500 : \$1500
c. \$285.71 : \$1714.29 d. \$857.14 : \$1142.86
e. \$1454.55 : \$545.45 f. \$600 : \$1400

4. a. 60 people b. 330 people

5. \$2000 : \$800 : \$400

6. 10 cm, 15 cm, 20 cm

7. 11 cm, 33 cm, 33 cm, 44 cm

8. 32.73° : 65.45° : 81.82°

9. a. \$35 000, \$35 000, \$280 000
b. \$105 000, \$105 000, \$140 000
c. \$192 500, \$52 500, \$105 000
d. \$162 750, \$40 250, \$147 000

10. The friend who worked 11 hours earns \$220; the others earn \$140 each.

11. a. 2 : 3 b. \$44

12. First: \$750; second: \$500; third: \$250

13. First: \$2.50; second: \$1.75; third: \$7.75

14. \$78 750

15. a. 12
b. Kamil gets 3 pieces. The friend who put in \$4 gets 4 pieces; the friend who put in \$5 gets 5 pieces.
c. Kamil gets $\frac{1}{4}$ of the pizza. His friends each get $\frac{1}{3}$ and $\frac{5}{12}$ of the pizza.

16. a. 80° and 40° b. 3 : 4 : 2

17. Answers will vary. Sample responses can be found in the worked solutions in the online resources.

18. \$9

2.9 Review

2.9 Exercise

Multiple choice

1. B
2. D
3. C
4. B
5. A
6. B
7. C
8. A
9. E
10. D

Short answer

11.

x	2	7	10	15
y	6	21	30	45

12. a. 6 : 11 b. 9 : 8 c. 2 : 1 d. 1 : 20

13. a. 5 b. 3 c. 128 d. 133

14. a. \$10: \$350 000
\$7: \$245 000
\$8: \$280 000

b. \$2: \$350 000
\$2.50: \$437 500
\$0.50: \$87 500

c. \$10: \$437 500
\$6: \$262 500
\$4: \$175 000

Extended response

15. a. E is $\frac{1}{3}$ of its original strength.

b. E is 2 times its original strength.

16. a. $y \propto \frac{1}{x}$ b. $y = \frac{24}{x}$ c. 8

d. 2

17. a. $t \propto \frac{1}{v}$ b. $t = \frac{k}{v}$ c. 2 h

d. 100 e. $t = \frac{100}{v}$ f. 54.5 min

18. a. $p \propto \frac{1}{d\theta}$ b. 750 c. $p = \frac{750}{d\theta}$

d. 0.37

19. 3 minutes: \$51
5 minutes: \$85

20. a. 25 : 15

b. 62.5%

c. 20%: 8 vegetarians
75%: 30 people like chips

d. Samosas: 20
Rice-paper rolls: 40
Mini dim sims: 100

3 Percentages

LEARNING SEQUENCE

Fully worked solutions for this topic are available online.

3.1 Overview

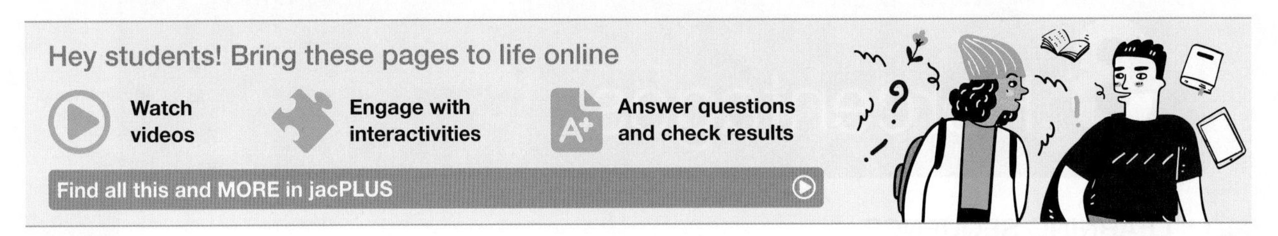

3.1.1 Introduction

Percentages are used extensively in everyday life. Percentage increases or decreases, percentage discounts, percentage profit or loss, the goods and services tax, and percentage errors are all examples of how we use percentages every day. This is why an understanding of percentages is important — so you can understand if you are paying the correct price after a percentage discount, or what percentage profit you may have made after selling a particular item. Percentages are used extensively in many different professions, such as hospitality, finance, statistics and journalism. It is an essential skill to have in any profession.

KEY CONCEPTS

This topic covers the following key concepts from the VCE Mathematics Study Design:
- rational numbers and irrational numbers related to measurement, ratios and proportions in a practical context
- estimation and approximation including interval estimates, rounding, significant figures, leading-digit approximations, floor and ceiling values and percentage error.

Note: Concepts shown in grey are covered in other topics.

3.2 Converting percentages

LEARNING INTENTION

At the end of this subtopic you should be able to:
- convert percentages to fractions and decimals
- convert fractions and decimals to percentages.

3.2.1 Percentages

Per cent means 'for every hundred'. The total amount of anything is 100%.

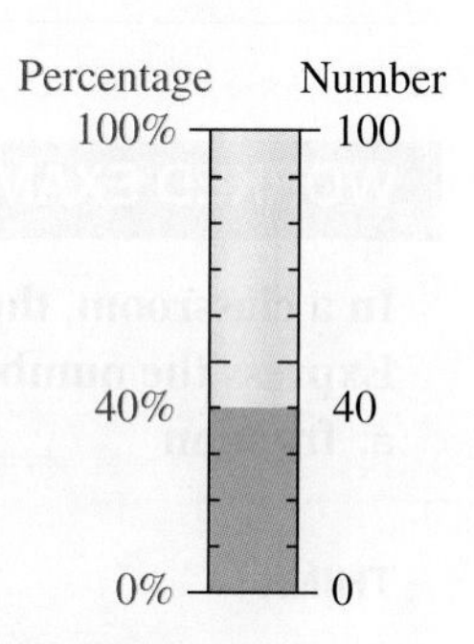

A percentage can be written as a fraction with a denominator of 100. For example, 40% is the same as the fraction $\frac{40}{100}$. It is shown as 40 out of 100 (or 40%) on the percentage comparison scale.

If the original amount is not exactly 100, the percentage can be worked out by relating the original amount to 100 using a percentage comparison chart.

3.2.2 Converting between fractions, decimals and percentages

Percentages, decimals and fractions can all be used to describe the same relationship between the part and the whole. The shaded portion of the square shown is 35%, 0.35, $\frac{35}{100}$ or $\frac{7}{20}$.

The following chart can be used to help you convert between fractions, decimals and percentages.

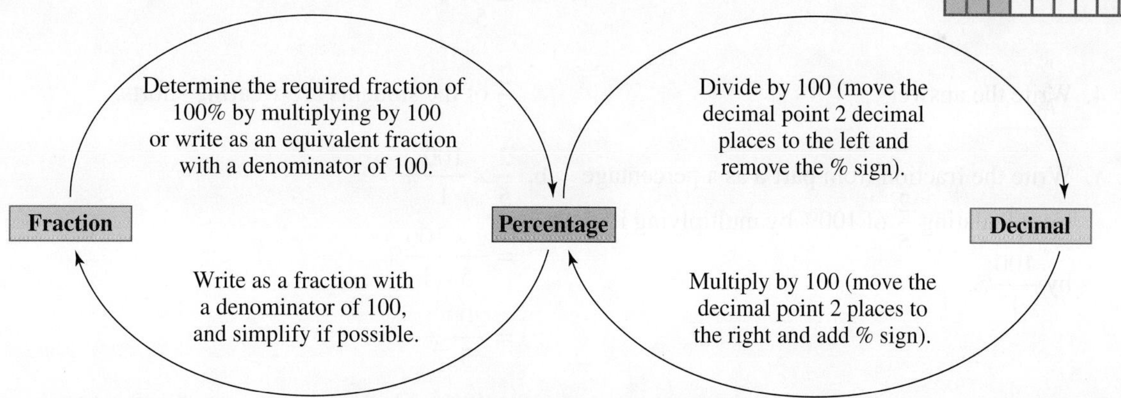

WORKED EXAMPLE 1 Converting between fractions, decimals and percentages

In a class, 48% of students arrived late to school today. Express the percentage of students who arrived on time as a:

a. fraction **b. decimal.**

THINK	WRITE
a. 1. If 48% of students arrived late then (100 − 48)% or 52% arrived on time. 52% is 52 in every 100.	a. $52\% = \frac{52}{100}$

2. Cancel by the highest common factor. (The highest common factor of 52 and 100 is 4.)	$\frac{52}{100} = \frac{\cancel{52}^{13}}{\cancel{100}_{25}}$
3. Write the fraction in simplest form.	$52\% = \frac{13}{25}$
b. 1. Write the percentage as a fraction with a denominator of 100 (52% is 52 hundredths).	b. $52\% = \frac{52}{100}$
2. Remove the % sign and move the decimal point 2 places to the left.	$52\% = 0.52$

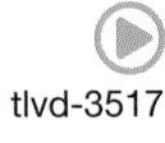

tlvd-3517

WORKED EXAMPLE 2 Calculating fractions, decimals and percentages

In a classroom, there are nine students wearing skirts, ten wearing shorts and six wearing long pants. Express the number of students wearing shorts out of the total number of students as a:

a. **fraction** b. **percentage** c. **decimal.**

THINK	WRITE
a. 1. Calculate the total number of students in the class.	a. The total number of students $= 9 + 10 + 6$ $= 25$
2. Express the number of students who are wearing shorts as a fraction of the total number of students in the class.	There are 10 students wearing shorts. The fraction of students wearing shorts is 10 out of 25.
3. Simplify the fraction by cancelling by the highest common factor of 5.	$\frac{10}{25}$ $= \frac{2}{5}$
4. Write the answer.	$\frac{2}{5}$ of the students are wearing shorts.
b. 1. Write the fraction from part **a** as a percentage by calculating $\frac{2}{5}$ of 100% by multiplying it by $\frac{100}{1}\%$.	b. $\frac{2}{5} \times \frac{100}{1}\%$ $= \frac{2 \times 100}{5 \times 1}\%$ $= \frac{200}{5}\%$ $= 40\%$
2. Write the answer.	Forty per cent of the students are wearing shorts.
c. 1. Write the percentage from part **b** as a fraction with a denominator of 100.	c. $40\% = \frac{40}{100}$
2. Write the fraction as a decimal.	$= 0.40$
3. Write the answer.	0.4 of the total number of students are wearing shorts.

on Resources

 **Interactivity** Percentages, fractions and decimals (int-3741)

3.2 Exercise

Students, these questions are even better in jacPLUS

 Receive immediate feedback and access sample responses

 Access additional questions

Track your results and progress

Find all this and MORE in jacPLUS

1. In your own words, describe how to convert:
   a. a fraction to a percentage
   b. a percentage to a fraction
   c. a decimal to a percentage
   d. a percentage to a decimal.
2. Copy and complete the following table, writing all fractions in simplest form.

Fraction	Decimal	Percentage
$\frac{1}{3}$	$0.\dot{3}$	$33\frac{1}{3}\%$
		64%
	0.75	
$\frac{3}{8}$		
	0.92	
$\frac{2}{3}$		
		52.5%

3. Convert each of the following fractions to percentages. Where appropriate, round your answer to 2 decimal places.
   a. $\frac{32}{100}$
   b. $\frac{43}{50}$
   c. $\frac{7}{250}$
   d. $\frac{9}{10}$
   e. $\frac{5}{67}$
   f. $\frac{56}{72}$
4. Express the following percentages as fractions in simplest form.
   a. 11%
   b. 82%
   c. 12.5%
   d. 202%
5. Express the following decimals as percentages.
   a. 0.15
   b. 0.85
   c. 3.10
   d. 0.024
6. Express the following fractions as percentages. Round your answer to 2 decimal places where appropriate.
   a. $\frac{7}{8}$
   b. $\frac{3}{5}$
   c. $\frac{5}{6}$
   d. $2\frac{1}{3}$
7. **WE1** In a tennis club, 63% of players are right-handed. Express the percentage of left-handed players as a:
   a. fraction
   b. decimal.

8. WE2 A student surveyed the teacher car park and counted 15 white cars, 10 silver cars, 2 green cars, 7 red cars and 6 blue cars. Express the number of cars of each colour out of the total number of cars as a:

a. fraction b. percentage c. decimal.

9. Year 12 students were surveyed about the type of computer they had at home. 84% said that they had PCs and the rest had Macs.

a. Calculate the percentage of students who had Macs.

b. Calculate the fraction of students who had PCs.

10. Air contains 21% oxygen, 0.9% argon and 0.1% trace gases, and the remainder is nitrogen.

a. Calculate the percentage of nitrogen in the air.

b. Calculate how many litres of oxygen are present in 100 litres of air.

c. Calculate how many litres of oxygen are present in 250 litres of air.

11. A group of students was practising their basketball free throws. Each student had four shots and the results are displayed in the table.

Free-throw results	Number of students	Percentage of students
No shots in	3	
One shot in, three misses	11	
Two shots in, two misses	10	
Three shots in, one miss	4	
All shots in	2	

a. Identify how many students participated in the game.

b. Complete the table to show the percentage of students for each result.

c. Calculate how many students made exactly 25% of their shots.

d. Calculate the percentage of students who made less than 50% of their shots.

12. The table shows the percentage of households with 0 to 5 children. Calculate:

a. the percentage of households that have 6 or more children

b. the percentage of households that have fewer than 2 children

c. the fraction of households that have no children

d. the fraction of households that have 1, 2 or 3 children.

Number of children	Percentage (%)
0	56
1	16
2	19
3	6
4	2
5	1

13. Use the bunch of flowers shown to answer these questions.

a. Calculate the percentage of flowers that are yellow.

b. Calculate the fraction of flowers that are pink.

14. Survey your classmates on the brand of mobile phone that they have. Present your results in a table showing the percentage, fraction and decimal amount of each brand.

3.3 Percentage quantities and percentage error

LEARNING INTENTION

At the end of this subtopic you should be able to:
- calculate percentages of quantities
- calculate percentage error.

3.3.1 Calculating the percentage of an amount

Quantities are often expressed as the percentage of an amount. For example, 82% of 1000 Year 12 students believe that they get too much homework. This statement gives an indication of the proportion of the Year 12 population that feels this way.

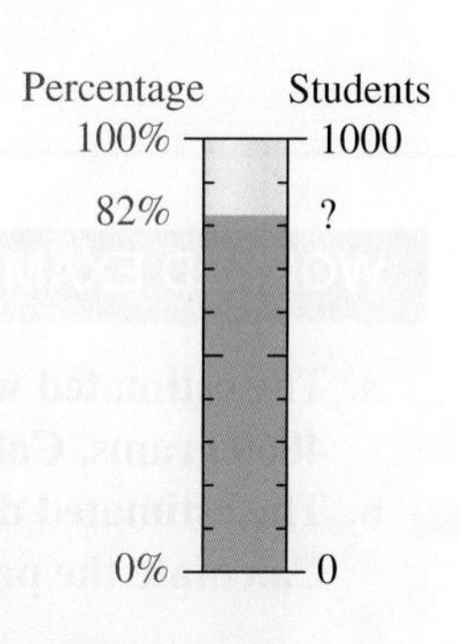

Calculations involving a percentage of an amount can be performed by changing the percentage to a fraction and finding this fraction of the amount.

WORKED EXAMPLE 3 Calculating percentages of quantities

Calculate 15% of $140.

THINK	WRITE
1. Convert 15% to a fraction and change 'of' to ×.	$15\%\text{ of }\$140 = \frac{15}{100} \times \frac{140}{1}$
2. Simplify by cancelling the highest common factors and then multiply the fractions.	$= \frac{{}^{3}\cancel{15}}{{}_{20}\cancel{100}} \times \frac{140}{1}$ $= \frac{3}{20} \times \frac{140}{1}$
3. Simplify the answer if possible.	$= \frac{420}{20}$ $= 21$
4. Answer the question.	15% of \$140 is \$21.

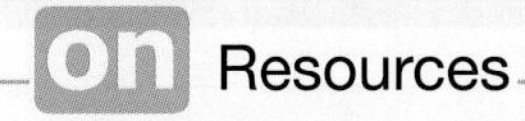

Video eLesson Percentages of an amount (eles-3621)

Interactivity Percentage of an amount (int-3743)

3.3.2 Percentage error

Percentage error is used to compare the difference between an estimate of a quantity and its actual value. For example, manufacturers and scientists use percentage error to determine the reliability of their equipment and processes, and the validity of their experiments. The closer the percentage error is to zero, the better the estimate.

Calculating percentage error

If the approximate value is greater than the exact value:

$$\text{percentage error} = \frac{\text{approximate value} - \text{exact value}}{\text{exact value}} \times 100\%$$

If the approximate value is less than the exact value:

$$\text{percentage error} = \frac{\text{exact value} - \text{approximate value}}{\text{exact value}} \times 100\%$$

WORKED EXAMPLE 4 Calculating percentage error

a. The estimated weight of a newborn baby was 3500 grams, but the baby's actual weight was 4860 grams. Calculate the percentage error.

b. The estimated distance between two towns was 70 km, but the actual distance was 65.4 km. Calculate the percentage error.

THINK	WRITE
a. 1. The estimated weight was less than the actual weight.	a. $\text{Percentage error} = \frac{\text{exact value} - \text{approximate value}}{\text{exact value}} \times 100$
2. Calculate the percentage error.	$\text{Percentage error} = \frac{4860 - 3500}{4860} \times 100$ $= 27.98\%$
3. Write the answer.	The percentage error is 27.98%.
b. 1. The estimated distance was greater than the actual distance.	b. $\text{Percentage error} = \frac{\text{approximate value} - \text{exact value}}{\text{exact value}} \times 100$
2. Calculate the percentage error.	$\text{Percentage error} = \frac{70 - 65.4}{65.4} \times 100$ $= 7.03\%$
3. Write the answer.	The percentage error is 7.03%.

3.3 Exercise

Students, these questions are even better in jacPLUS

Receive immediate feedback and access sample responses

Access additional questions

Track your results and progress

Find all this and MORE in jacPLUS

1. In your own words, describe how to calculate the percentage of an amount.
2. WE3 Calculate the value of each of the following.
   a. 20% of 50 b. 25% of 44 c. 80% of 70 d. 4% of 250 e. 5% of 40 f. 93% of 150

3. Calculate the following.
   a. 50% of 20 b. 20% of 80 c. 5% of 60 d. 10% of 30
4. Calculate the following.
   a. 70% of 110 b. 52% of 75 c. 90% of 70 d. 80% of 5000
5. Calculate the following.
   a. 95% of 200 b. 110% of 50 c. 150% of 8 d. 125% of 20
6. An online retailer has a 30% off end-of-year sale. Explain how you would determine the total amount of money to deduct from the advertised price.

7. Broadcasting regulations specify that 55% of television programs shown between 6 pm and midnight must be Australian and that, between 6 pm and midnight, there should be no more than 13 minutes per hour of advertising.
   a. Calculate how many minutes of advertising are allowed between 6 pm and midnight.
   b. Calculate for how many minutes programs are screened between 6 pm and midnight.
   c. Calculate the percentage of time spent screening advertising.
   d. Calculate how many minutes of Australian content must be screened between 6 pm and midnight.
8. In a Maths competition, the top 8% of students across the state achieve a score of 40 or more out of a possible 50.
   a. In a school where 175 students entered the Maths competition, state how many scores of more than 40 you would expect.
   b. In one school, there were 17 scores of 40 or more, and 204 scores that were less than 40. Determine if the students performed better than the state average.
9. **MC** 60% of 30 is:
   A. $19\frac{4}{5}$ B. $\frac{31}{5}$ C. 186 D. 18 E. 30
10. Thirty per cent of residents in the suburb Hunters Hill are over the age of 65. If there are 180 000 residents, calculate how many are over the age of 65.
11. The grocery bill for Mika's shopping was \$250. The following week, Mika spent 7% more on his groceries. Calculate how much Mika spent in the following week.
12. A classmate attempted to solve a problem, as shown. Help them calculate the correct answer.

Question: **Calculate 21% of \$22.05.**
$21 \div 22.05 \times 100$ = \$0.95??

13. When I am 10% older than I am now, I will be 22. Calculate how old I am now.
14. **WE4** The estimated grocery bill budgeted for the week was \$250, but the actual bill was \$262.20. Calculate the percentage error.
15. A long-distance runner estimated her run took 120 minutes, but the official time recorded was 118.3 minutes. Calculate the percentage error.
16. Answer the following questions.
   a. The estimated grocery bill budgeted for the week was \$250, but the actual bill was \$262.20. Calculate the percentage error.
   b. A long-distance runner estimated that her run took 120 minutes, but the official time recorded was 118.3 minutes. Calculate the percentage error.

17. In supermarkets, potatoes are frequently sold in 2-kg bags. As potatoes are discrete objects, the bags rarely weigh exactly 2 kg. For reasons relating to both customer satisfaction and profit, the warehouse supervisor knows that a percentage error of more than 10% is unacceptable.
Two bags of potatoes are chosen at random and weighed. Bag A weighs 2.21 kg and bag B weighs 1.88 kg. Calculate the percentage error for each of these bags and determine if either or both will pass the inspection.

18. Ninety per cent of students at a school were present for school photographs. If the school has 1100 students, calculate how many were absent on the day the photographs were taken.

3.4 Application of percentages

LEARNING INTENTION

At the end of this subtopic you should be able to:

- write an amount as a percentage of another
- use percentages to solve real-world problems
- determine percentage increases and decreases.

Percentages are used in a variety of real-world contexts. One that we see in our day-to-day lives is shopping.

3.4.1 Discount, sale prices and profit

The amount that a price is reduced by is called a **discount**. For the bicycle shown, the discount can be calculated as:

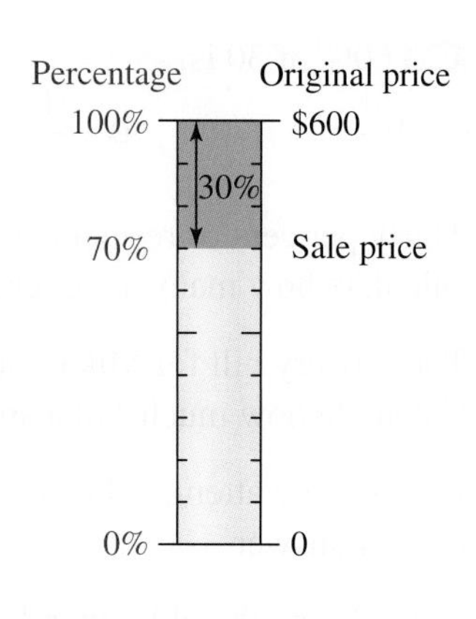

$$\begin{aligned}\text{Discount amount} &= 30\% \text{ of } \$600\\ &= \frac{30}{100} \times \frac{600}{1}\\ &= \$180\end{aligned}$$

The **sale price** is the price of the item after the discount has been removed. The sale price can be calculated using two different methods.

Method 1: Subtracting the discount amount from the original price	Method 2: Determining the remaining percentage of the original price
Sale price $= \text{original price} - \text{discount}$ $= 600 - (30\% \text{ of } 600)$ $= 600 - \left(\frac{30}{100} \times \frac{600}{1}\right)$ $= 600 - \left(\frac{30}{{}_{1}\cancel{100}} \times \frac{\cancel{600}^{6}}{1}\right)$ $= 600 - 180$ $= 420$ The sale price is \$420.	Sale price $= (100\% - 30\%) \text{ of original price}$ $= 70\% \text{ of } 600$ $= \frac{70}{100} \times \frac{600}{1}$ $= \frac{70}{{}_{1}\cancel{100}} \times \frac{\cancel{600}^{6}}{1}$ $= 420$ The sale price is \$420.

When you buy something, you pay more for it than the shop did. So that they can make a **profit**, shops purchase items for one price and then add a certain percentage on top of what they paid.

tlvd-3519

WORKED EXAMPLE 5 Calculating percentage increase or decrease

A shop attendant has to label items with price tags. If the shop normally increases prices by 15%, calculate the price that the shop attendant should put on an item that the shop bought for \$55.

THINK	WRITE
1. Calculate 15% of \$55 by first expressing both as fractions.	$15\% \text{ of } 55 = \frac{15}{100} \times \frac{55}{1}$
2. Simplify and then multiply the two fractions. Simplify the answer.	$= \frac{{}^{3}\cancel{15}}{{}_{20}\cancel{100}} \times \frac{55}{1}$ $= \frac{3}{20} \times \frac{55}{1}$ $= \frac{165}{20}$ $= \$8.25$
3. Add this answer to the original price.	$\$55 + \$8.25 = \$63.25$
4. Answer the question.	The item should be priced at \$63.25.

on Resources

Interactivity Percentage increase and decrease (int-3742)

3.4.2 Expressing one amount as a percentage of another

Percentages can be used to express the proportion of an amount.

Expressing one amount as a percentage of another amount allows comparisons to be made with other similar situations. For example, the number of questions correct in a test is often expressed as a percentage. The score from one test can then be compared with the score from another test, even if they contain a different total amount of marks.

WORKED EXAMPLE 6 Determining percentages

Write the test score '30 out of 50' as a percentage.

THINK	WRITE
1. 30 out of 50 is the same as $\frac{30}{50}$.	$\frac{30}{50}$
2. Write $\frac{30}{50}$ as a percentage by calculating $\frac{30}{50}$ of $\frac{100}{1}\%$. This could also be done by writing $\frac{30}{50}$ as $\frac{60}{100}$ using equivalent fractions $\left(\frac{30}{50} \times \frac{2}{2} = \frac{60}{100}\right)$.	$= \frac{30}{50} \times \frac{100}{1}\%$
3. Cancel any common factor (50), then multiply the two fractions.	$= \frac{30}{{}_1\cancel{50}} \times \frac{\cancel{100}^2}{1}\%$ $= \frac{60}{1}\%$
4. Answer the question.	$= 60\%$

WORKED EXAMPLE 7 Calculating percentage discounts or profits

An electronics store is having a sale. A gaming console originally priced at $600 is on sale for $510. Calculate the percentage discount.

THINK	WRITE
1. Calculate the discount.	$\$600 - \$510 = \$90$
2. Calculate the percentage the discount is out of the original amount.	$= \frac{90}{600} \times \frac{100\%}{1}$
3. Reduce the fraction by cross-dividing.	$= \frac{90}{{}_6\cancel{600}} \times \frac{{}^1\cancel{100}\%}{1}$ $= \frac{90}{6}\%$ $= 15\%$
4. State your answer.	The percentage discount on the gaming console is 15%.

3.4 Exercise

Students, these questions are even better in jacPLUS

Receive immediate feedback and access sample responses

Access additional questions

Track your results and progress

Find all this and MORE in jacPLUS

1. WE5 A shop attendant at a sports store has to label items with price tags. If the store normally increases prices by 35%, calculate the price that should be put on a tennis racquet that the store bought for $250.
2. Calculate the amount of money saved (the discount) on each of the following items.
   a. A skirt originally priced at $64 is discounted by 25%.
   b. A CD valued at $29 is discounted by 10%.
   c. Rollerblades originally priced at $149 are discounted by 15%.
   d. A digital camera has a marked price of $99 before it is discounted by 40%.
3. Without using a calculator, calculate the percentage discount of the following.

	Marked price	Discount
a.	$100	$10
b.	$250	$125
c.	$90	$30
d.	$80	$20

Calculate the sale price of each item with the following marked prices and percentage discounts.

	Marked price	Discount
e.	$1000	15%
f.	$250	20%

4. a. Calculate the discount you would receive on the guitar shown.
   b. Determine the sale price of the guitar.
   c. Subtract the percentage amount from 100%. Calculate this percentage of the guitar's marked price. Explain what you notice about the answer.

5. In your own words, explain two different methods that can be used to calculate the sale price of an item, when you are given the original price and the discount amount as a percentage.
6. WE6 Write the following test scores as percentages.
   a. 15 out of 20
   b. 120 out of 200
   c. 10 out of 200
   d. 9 out of 45
7. Express, to the nearest whole number:
   a. 70 as a percentage of 80
   b. 12 as a percentage of 42
   c. 11 as a percentage of 30
   d. 14 as a percentage of 18
   e. 80 as a percentage of 450
   f. 30 as a percentage of 36.
8. WE7 Headphones originally priced at $200 are on sale for $175. Calculate the percentage discount.

9. For each of the following, express the first amount as a percentage of the second amount, correct to the nearest whole number. In each case, make sure that the two amounts are in the same unit.
   a. 50c and \$3
   b. 300 m and 5 km
   c. 5 mm and 4 cm
   d. 3 days and 3 weeks
   e. 20 minutes and 1 hour
   f. 250 mL and 4 L
10. Decrease the following amounts by the percentages given.
   a. \$50 by 10%
   b. \$90 by 50%
   c. \$45 by 20%
11. A sale discount of 20% was offered by the music store Solid Sound. Calculate:
   a. the cash discount allowed on a \$350 sound system
   b. the sale price of the system.
12. Conduct a survey of the students in your class to determine the method of transport they use to get to school.
   a. Determine how many different methods of transport to school are used by the students in your class.
   b. Express the numbers of students using each method of transport to school as a percentage of the total number of students in your class.
13. A classmate was completing a discount problem where she needed to calculate a 25% discount on \$79. She misread the question and calculated a 20% discount to get \$63.20. She then realised her mistake and took a further 5% off from \$63.20. Explain if this is the same as taking 25% off \$79, using calculations to support your answer.
14. A store had to increase its prices by 10% to cover increasing expenses. A particular soundbar was originally priced at \$220. Use the following questions to help you calculate the new price of the soundbar using two different methods.
   a. i. Calculate the cost increase and add it to the original price.
      ii. Add the percentage amount to 100% and multiply your answer by the original price.
   b. Explain what you notice about the answers to part a.

15. State what you would multiply the original prices of items by to get their new prices with:
   a. a 20% discount
   b. a 15% discount
   c. a 25% increase
   d. a 5% increase
   e. a 35% discount
   f. an 11% increase.

3.5 Profit and loss

LEARNING INTENTION

At the end of this subtopic you should be able to:

- use percentages to calculate cost price, selling price, and profit and loss.

3.5.1 Cost price, selling price and percentages

When a manufacturer makes a product, it is usually sold to a wholesaler who subsequently sells the product on to retail outlets. At each stage, the product is marked up by a certain percentage.

When a retailer calculates the price to be marked on an article (the *selling price*, SP), many overhead costs must be taken into account (staff wages, rent, store improvements, electricity, advertising and so on).

The total price the retail shop owner pays for the product, including overhead costs, is the *cost price*, CP.

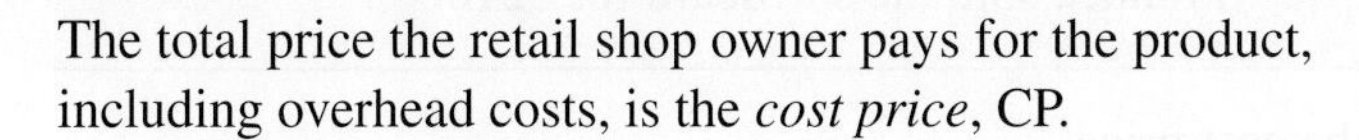

The *profit* is the difference between the total of the retailer's costs (cost price) and the price for which the goods actually sell (selling price).

Profit and loss

If selling price (SP) > cost price (CP), a profit is made.

$$\textbf{profit} = \textbf{selling price} - \textbf{cost price}$$

If selling price (SP) < cost price (CP), a loss is made.

$$\textbf{loss} = \textbf{cost price} - \textbf{selling price}$$

3.5.2 Selling price

- To calculate the selling price of an item given the cost price and the percentage profit, increase the cost price by the given percentage.

$$\text{selling price} = (100\% + \text{ percentage profit}) \text{ of cost price}$$

- To calculate the selling price of an item given the cost price and the percentage loss, decrease the cost price by the given percentage.

$$\text{selling price} = (100\% - \text{percentage loss}) \text{ of cost price}$$

WORKED EXAMPLE 8 Calculating selling price

Raimi operates a sports store at a fixed profit margin of 55%. Calculate the price he would mark on a pair of running shoes that cost him $120.

THINK	WRITE
1. Determine the selling price by first adding the percentage profit to 100%, then calculating this percentage of the cost price.	$\text{Selling price} = 155\% \text{ of } \$120$ $= 1.55 \times \$120$ $= \$186$
2. Write the answer as a sentence.	The running shoes would sell for \$186.

WORKED EXAMPLE 9 Calculating selling price 2

Frankee bought a surfboard for \$300 and sold it at a 30% loss a year later. Calculate the selling price.

THINK	WRITE
1. Determine the selling price by first subtracting the percentage loss from 100%, then calculating this percentage of the cost price.	$\text{Selling price} = 70\% \text{ of } \$300$ $= 0.70 \times \$300$ $= \$210$
2. Write the answer as a sentence.	Frankee sold the surfboard for \$210.

Profit or loss is usually calculated as a percentage of the cost price.

Percentage profit and loss

$$\textbf{percentage profit} = \frac{\textbf{profit}}{\textbf{cost}} \times 100\%$$

$$\textbf{percentage loss} = \frac{\textbf{loss}}{\textbf{cost}} \times 100\%$$

tlvd-3520

WORKED EXAMPLE 10 Percentage profit

A clothing store buys T-shirts at \$15 each and sells them for \$28.95 each. Calculate the percentage profit made on the sale of one T-shirt.

THINK	WRITE
1. Calculate the profit on each T-shirt: selling price − cost.	$\text{Profit} = \$28.95 - \$15$ $= \$13.95$
2. Calculate the percentage profit: $\frac{\text{profit}}{\text{cost}} \times 100\%$.	$\text{Percentage profit} = \frac{13.95}{15} \times 100\%$ $= 93\%$
3. Write the answer as a sentence, rounding to the nearest per cent if applicable.	The profit is 93% of the cost price.

Modern accounting practice favours calculating profit or loss as a percentage of the selling price. This is because commissions, discounts, taxes and other items of expense are commonly based on the selling price.

$$\text{percentage profit} = \frac{\text{profit}}{\text{selling price}} \times 100\%$$

$$\text{percentage loss} = \frac{\text{loss}}{\text{selling price}} \times 100\%$$

3.5 Exercise

Students, these questions are even better in jacPLUS

Receive immediate feedback and access sample responses

Access additional questions

Track your results and progress

Find all this and MORE in jacPLUS

1. WE8, 9 Calculate the selling price for each of the following items.

	Cost price	%	Profit/loss
a.	\$18	40%	profit
b.	\$116	25%	loss
c.	\$1300	30%	profit
d.	\$213	75%	loss
e.	\$699	$33\frac{1}{3}\%$	profit

2. WE10 For each of the following items, calculate the percentage profit or loss.

	Cost price	Selling price
a.	\$15	\$20
b.	\$40	\$50
c.	\$40	\$30
d.	\$75	\$85
e.	\$38.50	\$29.95

3. A supermarket buys frozen chickens for \$3.50 each and sells them for \$5.60 each. Calculate the percentage profit made on the sale of each chicken.

4. A restored motorbike was bought for \$1250 and later sold for \$4900.
   a. Calculate the profit made.
   b. Calculate the percentage of the profit. Give your answer correct to the nearest whole number.

5. A retailer bought a laptop for \$1200 and advertised it for \$1525.
   a. Calculate the profit made.
   b. Calculate the percentage profit (to the nearest whole number) on the cost price.
   c. Calculate the percentage profit (to the nearest whole number) on the selling price.
   d. Compare the differences between the answers to parts **b** and **c**.

6. Rollerblades bought for \$139.95 were sold after six months for \$60.
   a. Calculate the loss made.
   b. Calculate the percentage loss. Give your answer to the nearest whole number.

7. A sports card collection costing \$80 was sold for \$65. Calculate the percentage loss.

8. Camila runs a jewellery business that uses a fixed profit margin of 98%. Calculate the price she would charge for a necklace that cost her $830.

9. Calculate the selling price for each item.
   a. Jeans costing $20 are sold with a profit margin of 95%.
   b. A soccer ball costing $15 is sold with a profit margin of 80%.
   c. A sound system costing $499 is sold at a loss of 45%.
   d. A skateboard costing $30 is sold with a profit margin of 120%.

10. A fruit-and-veg shop bought 500 kg of tomatoes for $900 and sold them for $2.80 per kg.
    a. Calculate the profit per kilogram.
    b. Calculate the profit as a percentage of the cost price (round to 1 decimal place).
    c. Calculate the profit as a percentage of the selling price (round to 1 decimal place).
    d. Compare the answers to parts **b** and **c**.

11. To produce a set of crockery consisting of a dinner plate, soup bowl, bread plate and coffee mug, the costs per item are $0.98, $0.89, $0.72 and $0.69 respectively. These items are packaged in boxes of sets and sell for $39. If the company sells 4000 boxes in a month, calculate its total profit.

12. Copy and complete the table below.

Cost per item	Items sold	Sale price	Total profit
$4.55	504	$7.99	
$20.00		$40.00	$8040.00
$6.06	64 321		$225 123.50
	672	$89.95	$28 425.60

13. Determine the cost price for the following items.
    a. A diamond ring sold for $2400 with a percentage profit of 60% of the selling price
    b. A cricket bat sold for $69 with a percentage profit of 25% of the selling price
    c. A 3-seater sofa sold for $1055 with a percentage profit of 35% of the selling price

14. Sonja bought an old bike for $20. She spent $47 on parts and paint and renovated it. She then sold it for $115 through her local newspaper. The advertisement cost $10.
    a. Calculate Sonja's total costs.
    b. Calculate the percentage profit (to the nearest whole number) that Sonja made on costs.
    c. Calculate the percentage profit (to the nearest whole number) that Sonja made on the selling price.

15. Max bought a car for $6000. He sold it to Janine for 80% of the price he paid for it. Janine sold it to Jennifer at a 10% loss. Jennifer then sold the car to James for 75% of the price she paid.
    a. Calculate the price James paid for the car.
    b. Calculate the total percentage loss on the car from Max to James.

3.6 Review

3.6.1 Summary

doc-38011

Hey students! Now that it's time to revise this topic, go online to:

Review your results

Watch teacher-led videos

Practise questions with immediate feedback

Find all this and MORE in jacPLUS

3.6 Exercise

Multiple choice

1. MC The fraction $\frac{1}{4}$ as a percentage is:

   A. 20% B. 25% C. 30% D. 35% E. 45%

2. MC 15% of 450 is:

   A. 72.5 B. 85 C. 65 D. 67.5 E. 75

3. MC 36 as a percentage of 90 is:

   A. 10% B. 20% C. 25% D. 30% E. 40%

4. MC 33% as a decimal is:

   A. 3.3 B. 0.033 C. 0.33 D. 33.33 E. 3.33

5. MC If 68% of the 64 500 crowd at the MCG supported Collingwood, the number of Collingwood supporters was:

   A. 43 860 B. 20 640 C. 44 640 D. 32 250 E. 48 680

6. MC If an NBA collectors' card was purchased for $8.50 and sold later for $25.50, the profit made was:

   A. $12 B. $16.50 C. $15 D. $17 E. $15.50

7. MC For the same NBA collectors' card in question 6, the percentage profit was:

   A. 20% B. 200% C. 2% D. 17% E. 150%

8. MC Callum bought a car for $7500 and sold it for $6200. The percentage loss, to the nearest percent, was:

   A. 12% B. 14% C. 17%
   D. 19% E. 21%

9. MC If the original price of a record was $18.95 and it increased in value over time and sold for $46.50, the profit was:

   A. $25.55 B. $27.55 C. $27.50
   D. $25.50 E. $46.50

10. MC Due to a lack of supply, the price of timber is going up by 15%. If timber originally cost $6.65 per metre, the new increased price is closest to:

    A. $7.55 B. $6.80 C. $7.60 D. $7.65 E. $8.15

Short answer

11. Fill in the following table.

Fraction	Decimal	Percentage
$\frac{1}{5}$		
		87.5%
	0.4	

12. Calculate:
    a. 43% of $230
    b. 110% of 80.

13. Calculate the percentage discount on an item originally priced at $49.99 and now priced at $35.

14. Express:
    a. 54 as a percentage of 68
    b. 42 as a percentage of 162
    c. 19 as a percentage of 17.

Extended response

15. Jill purchased a handbag for $250 and later sold it on eBay for $330.
    a. Calculate the percentage profit on the cost price.
    b. Calculate the percentage profit on the selling price.
    c. Compare the answers to parts **a** and **b**.

16. Jacques' furniture shop had a sale with 30% off the original price. If the original price of a lounge suite was $5745, calculate the new sale price.

17. William owns a hairdressing salon and raises the price of haircuts from $26.50 to $29.95. Calculate the percentage increase of the price of haircuts.

18. The price of milk increased by 8%. If the original price of milk was $2.65, calculate the new price of milk.

19. A pair of jeans bought for $129 was later sold for $85. Calculate the percentage loss made on the sale.

20. Tia works in a sports shop. She purchased wholesale golf shirts for $55 each. Calculate how much she sold the shirts for if she made a 163% profit.

Answers

Topic 3 Percentages

3.2 Converting percentages

3.2 Exercise

1. a. Write the fraction as an equivalent fraction with a denominator of 100.
   b. Write the percentage as a fraction with a denominator of 100, and simplify if possible.
   c. Multiply the decimal by 100%.
   d. Write the percentage as a fraction with a denominator of 100 and convert to a decimal.

2.

Fraction	Decimal	Percentage
$\frac{1}{3}$	$0.\dot{3}$	$33\frac{1}{3}\%$
$\frac{16}{25}$	0.64	64%
$\frac{3}{4}$	0.75	75%
$\frac{3}{8}$	0.375	37.5%
$\frac{23}{25}$	0.92	92%
$\frac{2}{3}$	$0.\dot{6}$	$66\frac{2}{3}\%$
$\frac{21}{40}$	0.525	52.5%

3. a. 32% b. 86% c. 2.8%
   d. 90% e. 7.46% f. 77.78%
4. a. $\frac{11}{100}$ b. $\frac{41}{50}$ c. $\frac{1}{8}$ d. $\frac{101}{50}$
5. a. 15% b. 85% c. 310% d. 2.4%
6. a. 87.5% b. 60% c. 83.33% d. 233.33%
7. a. $\frac{37}{100}$ b. 0.37
8.

	White	Silver	Green	Red	Blue
a.	$\frac{3}{8}$	$\frac{1}{4}$	$\frac{1}{20}$	$\frac{7}{40}$	$\frac{3}{20}$
b.	37.5%	25%	5%	17.5%	15%
c.	0.375	0.25	0.05	0.175	0.15

9. a. 16% b. $\frac{21}{25}$
10. a. 78% b. 21 litres c. 52.5 litres
11. a. 30
    b.

Free throw results	Number of students	Percentage of students
No shots in	3	10%
One shot in, three misses	11	$36.\dot{6}\%$
Two shots in, two misses	10	$33.\dot{3}\%$
Three shots in, one miss	4	$13.\dot{3}\%$
All shots in	2	$6.\dot{6}\%$

    c. 11
    d. $46.\dot{6}\%$
12. a. 0% b. 71% c. $\frac{14}{25}$ d. $\frac{41}{100}$
13. a. 38.10% b. $\frac{8}{21}$
14. Answers will vary. They should reflect the students' knowledge of the different forms (percentages, fractions and decimals) of numbers.

3.3 Percentage quantities and percentage error

3.3 Exercise

1. Change the percentage to a fraction and find this fraction of the amount.
2. a. 10 b. 11 c. 56
   d. 10 e. 2 f. 139.5
3. a. 10 b. 16 c. 3 d. 3
4. a. 77 b. 39 c. 63 d. 4000
5. a. 190 b. 55 c. 12 d. 25
6. Multiply the advertised price by 30% or $\frac{3}{10}$.
7. a. 78 minutes
   b. 282 minutes or 4 hours 42 minutes
   c. $21\frac{2}{3}\%$
   d. 198 minutes
8. a. 14
   b. 7.69% of students achieved a score of 40 or more, which is just below the state average.
9. D
10. 54 000
11. $267.50
12. 21% of 22.05 $= 0.21 \times 22.05$
    $= \$4.65$
13. 110% of my age = 22; I am now 20 years old.
14. 4.65%
15. 1.44%
16. a. 4.65% b. 1.44%
17. Bag B will pass. Bag A will not pass.
18. 110

3.4 Application of percentages

3.4 Exercise

1. $337.50
2. a. $16 b. $2.90 c. $22.35 d. $39.60

3. a. 10% b. 50% c. $33\frac{1}{3}\%$ d. 25%

e. $850 f. $200

4. a. $78

b. $442

c. 100% − 15% = 85%
85% of $520 = $442
The answer is the cost price of the guitar after the discount.

5. • Find the discount amount and subtract this from the original price.
• Subtract the discount percentage from 100 and find this percentage of the original price.

6. a. 75% b. 60% c. 5% d. 20%

7. a. 88% b. 29% c. 37% d. 78% e. 18% f. 83%

8. 12.5%

9. a. 17% b. 6% c. 13% d. 14% e. 33% f. 6%

10. a. $45 b. $45 c. $36

11. a. $70 b. $280

12. a. Answers will vary depending on the survey responses.
b. Answers will vary. They should demonstrate the students' knowledge of conversion to percentages.

13. 95% of $63.20 = $60.05
75% of $79 = $59.25
The two methods calculate percentages of different amounts and therefore result in different answers.

14. a. i. 10% of $220 = $22
$220 + $22 = $242
ii. 110% of $220 = $242
b. The answers are the same.

15. a. 80% b. 85% c. 125%
d. 105% e. 65% f. 111%

3.5 Profit and loss

3.5 Exercise

1. a. $25.20 b. $87 c. $1690
d. $53.25 e. $931.77

2. a. $33\frac{1}{3}\%$ profit
b. 25% profit
c. 25% loss
d. $13\frac{1}{3}\%$ profit
e. 22.21% loss

3. 60%

4. a. $3650 b. 292%

5. a. $325
b. 27%
c. 21%
d. The percentage profit is greater on the cost price.

6. a. $79.95 b. 57%

7. 18.75%

8. $1643.40

9. a. $39 b. $27 c. $274.45 d. $66

10. a. $1.00 profit per kg
b. 55.6%
c. 35.7%
d. The percentage profit is greater on the cost price.

11. $142 880

12.

Cost per item	Items sold	Sale price	Total profit
$4.55	504	$7.99	$1733.76
$20.00	402	$40.00	$8040.00
$6.06	64 321	$9.56	$225 123.50
$47.65	672	$89.95	$28 425.60

13. a. $960 b. $51.75 c. $685.75

14. a. $77 b. 49% c. 33%

15. a. James paid $3240.
b. The total percentage loss was 46%.

3.6 Review

3.6 Exercise

Multiple choice

1. B
2. D
3. E
4. C
5. A
6. D
7. B
8. C
9. B
10. D

Short answer

11.

Fraction	Decimal	Percentage
$\frac{1}{5}$	**0.2**	**20%**
$\frac{7}{8}$	**0.875**	87.5%
$\frac{2}{5}$	0.4	**40%**

12. a. $98.90 b. 88

13. 29.99%

14. a. 79.41% b. 25.93% c. 111.76%

Extended response

15. a. 32%
b. 24.24%
c. The percentage profit is greater on the cost price since this is the lower value.

16. $4021.50

17. 13.02%

18. $2.86 ≈ $2.85

19. 34.11%

20. $144.65

4 Linear and simultaneous equations

LEARNING SEQUENCE

Fully worked solutions for this topic are available online.

4.1 Overview

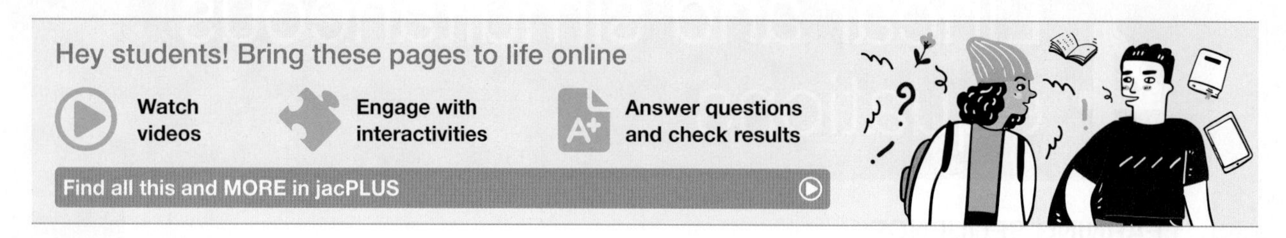

4.1.1 Introduction

Formulas and equations have been developed for a wide variety of theoretical and practical situations, from calculating part-time pay for your week's work to calculating the volume of water required to fill your pool for summer. Once we have a formula, we can use it to calculate required unknown values by substituting known values into the correct formula.

Often in life, we are presented with multiple options and must decide on a combination of outcomes that provides you with the best result.

Consumers face these decisions every day. For example, if a person was looking to hire a car there are multiple options available. Most car companies rent cars with either a flat daily fee, cost per kilometre driven or a combination of both. Each option can be the cheapest, depending on how far the consumer is going to drive the car. Therefore, the consumer can use simultaneous equations to calculate which option would best suit them.

In this topic, we will use our previous knowledge of linear equations to solve equations simultaneously to determine a break-even point, and to learn how to choose the best option in everyday situations.

KEY CONCEPTS

This topic covers the following key concepts from the VCE Mathematics Study Design:

- symbolic expressions, equations and formulas
- graphical and algebraic analysis of relations including transposition of formulas and finding a break-even point using simultaneous equations.

Source: VCE Mathematics Study Design (2023–2027) extracts © VCAA; reproduced by permission.

4.2 Algebraic substitution and transposition

LEARNING INTENTION

At the end of this subtopic you should be able to:
- evaluate algebraic expressions using substitution
- rearrange formulas and use them to solve problems.

4.2.1 Substitution

Unknown numbers are represented with letters and are called **variables**.

To **evaluate** an expression, **substitute** a number in place of the variable and then use the order of operations to work out the value of the expression.

Recall the order of operations (BIDMAS) from Topic 1 to evaluate expressions.

WORKED EXAMPLE 1 Algebraic substitution

Evaluate the following expressions by substituting $a = 2$ and $b = 5$.

a. $3ab$ b. $7(a + b)$ c. $\sqrt{a^2 + b}$

THINK	WRITE
a. 1. Write the expression and substitute 2 for a and 5 for b.	a. $3ab = 3 \times a \times b$ $= 3 \times 2 \times 5$
2. Evaluate the expression.	$= 30$
b. 1. Write the expression and substitute 2 for a and 5 for b.	b. $7(a + b) = 7 \times (a + b)$ $= 7 \times (2 + 5)$
2. Evaluate the expression. Remember to work out the part of the expression in brackets first.	$= 7 \times 7$ $= 49$
c. 1. Write the expression and substitute 2 for a and 5 for b.	c. $\sqrt{a^2 + b} = \sqrt{2^2 + 5}$
2. Evaluate the expression. Remember to follow the order of operations.	$= \sqrt{4 + 5}$ $= \sqrt{9}$ $= 3$

4.2.2 Formulas

Equations are mathematical statements that show two equal **expressions** — that is, the left-hand side and the right-hand side of an equation are equal. In the equation $a + 2 = 6$, the expression on the left-hand side $(a + 2)$ is equal to that on the right-hand side (6).

A **formula** is a special equation or rule that describes the relationship between different quantities.

The plural of *formula* is *formulas* or *formulae*.

To be able to use a formula, you need to know:

- what the formula is used for
- what the variables represent
- any specific unit requirements of the variables.

Formula	What it is used for	What the pronumerals represent	Unit requirements
$A = lw$	Calculating the area of a rectangle	A represents the area of the rectangle. l represents the length of the rectangle. w represents the width of the rectangle.	l and w must be in the same unit.
$E = mc^2$	Calculating the energy contained in a given mass	E represents the amount of energy. m represents the mass. c represents the speed of light.	m must be in kg. c must be in m/s.

The variable written by itself in the formula is the subject of the formula. For example, A is the subject of the formula $A = lw$.

If you are given the values of the other variables in a formula, you can evaluate the subject of the formula by substituting these values.

WORKED EXAMPLE 2 Calculating values using substitution

In the formula $S = ut + \frac{1}{2}at^2$, determine the value of S if $a = 10$, $u = 25$ and $t = 6$.

THINK	WRITE
1. Write the formula.	$S = ut + \frac{1}{2}at^2$
2. Substitute the values of a, u and t. Calculate the value of S using the order of operations (BIDMAS).	$S = 25 \times 6 + \frac{1}{2} \times 10 \times 6^2$ $= 150 + 180$ $= 330$
3. Answer the question.	$S = 330$

4.2.3 Changing the subject of a formula

Sometimes you may need to change the subject of a formula. You can do this by rearranging or **transposing** the formula to get the variable you want as the subject of the equation. This can be done by using backtracking techniques, or by performing inverse (opposite) operations on both sides of the formula.

Inverse operations

+ and − are inverse operations.

× and ÷ are inverse operations.

WORKED EXAMPLE 3 Rearranging a formula

Rearrange $v = at + u$ to make the variable a the subject of the formula.

THINK	WRITE
1. Think about how the formula was built up from a.	a was multiplied by t, then u was added to the result, giving v.
2. Backtrack by performing inverse operations one at a time and simplifying.	$v = a \times t + u$
3. Subtract u from both sides.	$v - u = a \times t + u - u$ $v - u = a \times t$
4. Divide both sides by t.	$\frac{v-u}{t} = \frac{a \times \not{t}}{\not{t}}$ $\frac{v-u}{t} = a$
5. Write the new formula with a as the subject.	$a = \frac{v-u}{t}$

4.2.4 Solving problems involving formulas

We can solve problems involving formulas using two methods.

Method 1: Substitute the known values into the formula and solve for the unknown.

Method 2: Change the subject of the formula, then substitute the known values into the rearranged formula and evaluate for the unknown.

Both methods usually require solving an equation. However, rearranging (or transposing) the formula to make the unknown the subject is often more difficult.

tlvd-3516

WORKED EXAMPLE 4 Rearranging formulas to solve problems

The perimeter, P, of a rectangle is given by the formula $P = 2(l + w)$, where l is the length and w is the width of the rectangle.

a. Calculate the length of a rectangle with a perimeter of 3 metres and a width of 45 centimetres.

b. i. Transpose the formula to make l the subject.

ii. Use the transposed formula to calculate the length of the rectangle given in part a.

THINK	WRITE
a. 1. Write the formula and substitute the values of P and w.	a. $P = 2(l + w)$
2. Remember to always use the same units ($P = 300$ cm).	$300 = 2(l + 45)$
3. Solve using inverse operations.	
4. Divide both sides by 2.	$\frac{300}{2} = \frac{2(l+45)}{2}$ $150 = l + 45$
5. Subtract 45 from both sides.	$150 - 45 = l + 45 - 45$ $105 = l$
6. Answer the question.	The length of the rectangle is 105 cm.

b. i. 1. Write the formula.	**b. i.** $P = 2(l + w)$
2. Change the subject to l by using inverse operations one at a time.	
3. Divide both sides by 2.	$\frac{P}{2} = \frac{2(l+w)}{2}$ $\frac{P}{2} = l + w$
4. Subtract w from both sides.	$\frac{P}{2} - w = l + w - w$ $\frac{P}{2} - w = l$
5. Answer the question.	$l = \frac{P}{2} - w$
ii. 1. Substitute for P and w (note the units need to be the same) and evaluate.	**ii.** $l = \frac{300}{2} - 45$ $= 150 - 45$ $= 105$
2. Answer the question.	The length of the rectangle is 105 cm.

Resources

Interactivity Transposing linear equations (int-6449)

4.2 Exercise

1. **WE1** Evaluate the following expressions by substituting $x = 3$, $y = 7$ and $z = 15$.
   a. $x + 3y$ **b.** $3x + 2z$ **c.** $4z - 15$ **d.** $3(2z - y)$ **e.** $5x + 2y - z$ **f.** $4z - 2y + x$
2. Evaluate the following expressions by substituting $x = 3$, $y = 7$ and $z = 15$.
   a. $3x + 7z - 2x + 7$ **b.** $x^2 + y^2$ **c.** $\frac{z}{x}$
   d. $\frac{x + y}{z}$ **e.** $x(x^2 + y)$ **f.** $\sqrt{2(z - y) + x^2}$
3. **a.** Substitute 5 for m in the expression $m + 6$.
   b. If the value of k is 15, calculate the value of $2k - 7$.
   c. Given that $l = 3$, calculate the value of $10l - 27$.
   d. If $j = 15$, calculate the value of $11j - 62$.
   e. If $s = 21$, evaluate $3s - 7$.
4. **WE2** In the formula $S = ut + \frac{1}{2}at^2$, calculate the value of S if $a = 8$, $u = 12$ and $t = 9$.

5. The cost of hiring a kayak ($C) can be calculated using the formula $C = 50 + 10h$, where h is the number of hours that the kayak is hired for.

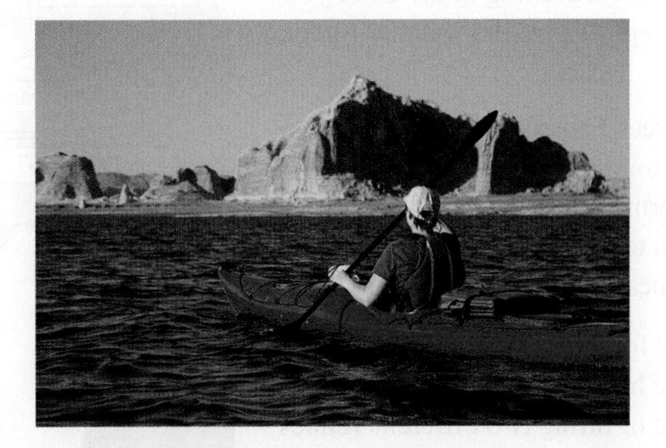

Calculate the cost of hiring the kayak for:

a. 1 hour
b. 3 hours
c. 6.5 hours
d. 90 minutes.

6. The perimeter of a rectangle (P) is calculated using the formula $P = 2(l + w)$, where l is the length and w is the width of the rectangle.
   a. Calculate the perimeter of a rectangle that has:
      i. a length of Time = 2 hours and a width of 2 cm
      ii. a length of 15 cm and a width of 3 cm.
   b. Write the formula for the area of a rectangle and use it to calculate the areas of the two rectangles in part a.

7. **WE3** Rearrange the following formula to make the variable n the subject of the formula.

$$T = a + nd$$

8. Transpose each of the following formulas to make the variable shown in brackets the subject of the formula.

a. $C = AB + D \ (A)$
b. $v = u + at \ (t)$
c. $V = \dfrac{Ah}{3} \ (h)$

9. **WE4** a. Write the formula for the perimeter of the rectangle shown.

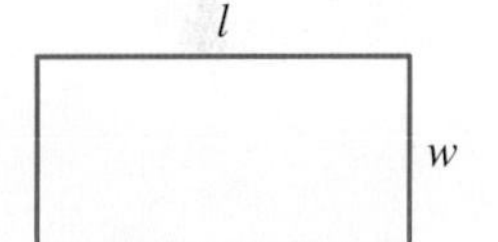

   b. Calculate the length of the rectangle with a perimeter of 2 metres and a width of 28 centimetres.
   c. i. Transpose the formula to make l the subject.
      ii. Use the transposed formula to calculate the length of the rectangle given in part b.

10. The total surface area of a cylinder can be found using the formula $A = 2\pi r(r + h)$ where r is the radius and h is the height of the cylinder.
    a. Rewrite the formula to make h the subject of the formula.
    b. Use your formula to calculate the height of a cylinder with a radius of 3 cm and a total surface area of 160 cm^2. Give your answer to 1 decimal place.

11. The distance, D km, travelled at a speed of s km/h for t hours is given by the formula $D = st$.
    a. Calculate the distance travelled at 80 km/h for:
       i. 4.5 hours
       ii. 150 minutes.
    b. Calculate the average speed if a car travels 120 kilometres in:
       i. 1.6 hours
       ii. 75 minutes.
    c. Calculate the time needed to travel a distance of 640 kilometres at an average speed of:
       i. 80 km/h
       ii. 60 km/h.

12. The formula $F = \frac{9C}{5} + 32$ converts degrees Celsius, °C, to degrees Fahrenheit, °F. While travelling in the USA you discover that the weather reports give the weather forecast in degrees Fahrenheit.

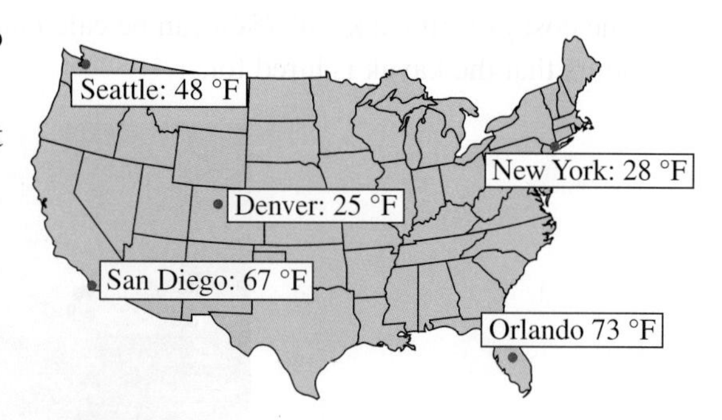

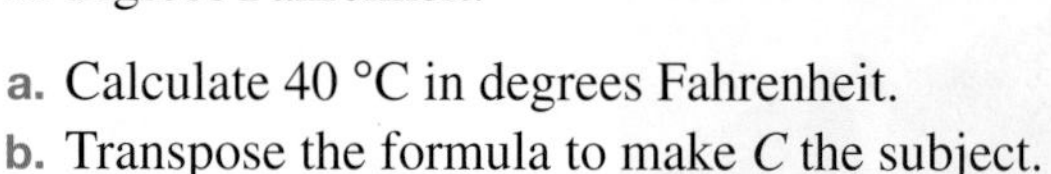

a. Calculate 40 °C in degrees Fahrenheit.
b. Transpose the formula to make C the subject.
c. Use your transposed formula from part **b** to convert the temperatures shown to degrees Celsius. Give your answers correct to the nearest whole number.

13. The cost of hiring a venue for a party includes a set amount of money, called a hiring fee, and a cost for each person attending. The cost of hiring two different venues for p people is given below.
Cost at venue A = \$100 + \$5.00p
Cost at venue B = \$50 + \$5.75p

a. Calculate the hiring fee for each venue.
b. Calculate the cost per person for each venue.
c. If 50 people are attending a party, determine which venue it would be cheaper to hire.

14. The volume of a sphere, V, is given by the formula $V = \frac{4}{3}\pi r^3$, where r is the radius of the sphere.
A basketball is shaped like a sphere.

a. State the radius of the basketball shown in the diagram.
b. Calculate the volume of the basketball shown, giving your answer correct to 2 decimal places.

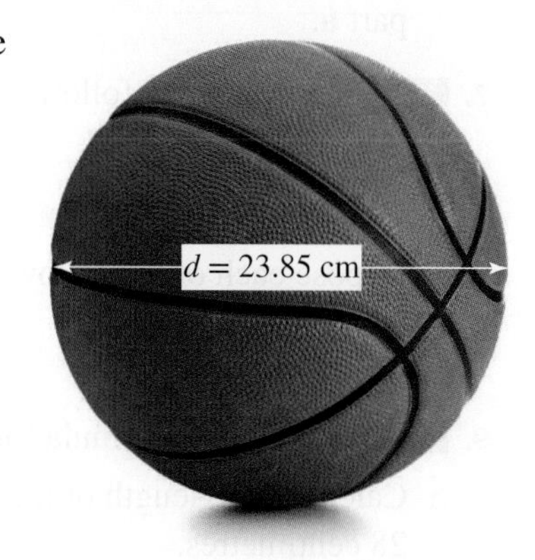

15. Stopping distance is the distance required to bring a moving vehicle to a stop from the moment the brakes are applied. The formula to calculate the stopping distance, d metres, when travelling at a speed of v m/s is:

$$d = \frac{v^2}{2\mu g}$$

where μ is the coefficient of friction between the wheels and the road surface and g is the gravitational force ($g = 9.8\ \text{m/s}^2$).
Calculate the stopping distances under the following conditions. Give your answers to 2 decimal places.

a. A car is travelling at a speed of 50 km/h on a dry road with $\mu = 0.8$.
b. Road and weather conditions, such as rain, change the value of μ.
   i. If $\mu = 0.45$, calculate the stopping distance when travelling at a speed of 100 km/h.
   ii. If $\mu = 0.15$, calculate the stopping distance when travelling at a speed of 100 km/h.
c. Comment on your findings in part **b**.

4.3 Linear equations

LEARNING INTENTION

At the end of this subtopic you should be able to:
- solve linear equations including those with variables on both sides
- construct linear equations from word problems.

4.3.1 Solving linear equations

A **linear equation** has variables whose highest power is 1 — for example, $3a + 5 = 10$ or $6 - \frac{x}{2} = 8$.

Linear equations can be solved using **inverse operations** and creating equivalent equations.

When solving an equation, the last operation performed on the variable when building the equation is the first operation undone by applying inverse operations to both sides of the equation.

WORKED EXAMPLE 5 Solving linear equations

Solve $6(a - 3) = 72$.

THINK	WRITE
1. To create the equation, a had 3 subtracted from it and the result was multiplied by 6 to get 72. The last operation performed was multiplying by 6. To undo the $\times 6$ operation, divide both sides by 6, as shown in red, and simplify.	$\frac{{}^{1}\cancel{6}(a-3)}{{}^{1}\cancel{6}} = \frac{72}{6}$ $a - 3 = 12$
2. Looking at the equivalent equation ($a - 3 = 12$), a has had 3 subtracted from it. To undo the -3 operation, add 3 to both sides, as shown in red, and simplify.	$a - 3 + 3 = 12 + 3$ $a = 15$
3. Alternatively, the expression can be solved by applying the distributive law and then rearranging the equation to solve for a.	$6(a-3) = 72$ $6a - 18 = 72$ $6a - 18 + 18 = 72 + 18$ $\frac{6a}{6} = \frac{90}{6}$ $a = 15$

Solutions to equations can be checked by substituting the solution back into the equation and comparing the left- and right-hand sides of the equation. For example, $a = 15$ is the solution for $6(a - 3) = 72$.

$$\begin{aligned} \text{LHS} &= 6(a-3) \qquad \text{RHS} = 72 \\ &= 6(15-3) \\ &= 6(12) \\ &= 72 \end{aligned}$$

Therefore $a = 15$ is the solution for $6(a - 3) = 72$.

4.3.2 Equations with variables on both sides

When a variable appears on both sides of an equation, inverse operations are used to collect the variable into a single term.

WORKED EXAMPLE 6 Equations with variables on both sides

Solve $4+2x=5x-8$.

THINK	WRITE
1. The variable appears on both sides of the equation. Subtracting from both sides of the equation will leave only one term containing the variable: the term $3x$.	$4+2x=5x-8$ $4+2x-2x=5x-2x-8$ $4=3x-8$
2. Looking at the new equation formed, $4=3x-8$, the last operation performed on x was to subtract 8. Add 8 to both sides to undo this operation, as shown in red.	$4+8=3x-8+8$ $12=3x$
3. Looking at the new equation formed, $12=3x$, the last operation performed on x was to multiply by 3. • Divide both sides by 3 to undo this operation, as shown in red. • Simplify the fraction, as shown in pink. • Rearrange the equation so x is on the LHS.	$\frac{{}^{4}\cancel{12}}{{}^{1}\cancel{3}}=\frac{{}^{1}\cancel{3}x}{{}^{1}\cancel{3}}$ $4=x$ $x=4$

4.3.3 Developing equations from a word description

To solve problems presented in words, turn the problem into an equation, solve the equation and use the solution to answer the problem.

- **Step 1:** Read the question carefully and identify the unknown variables.
- **Step 2:** Construct an appropriate equation using your variable.
- **Step 3:** Solve the equation.
- **Step 4:** Check your solution to see if it makes sense in the context of the question.
- **Step 5:** Answer the question.

WORKED EXAMPLE 7 Creating equations from word problems

The ages of three sisters add to 35 years. If the eldest is 4 years older than the middle sister and the youngest is 2 years younger than the middle sister, calculate the ages of the three sisters.

THINK	WRITE
1. Read the question carefully and identify the unknown variables.	Let m be the middle sister's age. Then the eldest sister is $m+4$ years old and the youngest sister is $m-2$ years old.
2. The sum of the sisters' ages is 35 years. Construct an appropriate equation using your variable.	Middle sister's age + eldest sister's age + youngest sister's age = 35 years $m+m+4+m-2=35$
3. Solve the equation by: • subtracting 2 from both sides • dividing both sides by 3.	$3m+2=35$ $3m+2-2=35-2$ $3m=33$ $\frac{3m}{3}=\frac{33}{3}$ $m=11$

4. Check your solution to see if it makes sense in the context of the question. It is reasonable for the middle sister to be 11 years old, which is the solution to the equation. This makes the eldest sister $11 + 4 = 15$ years old and the youngest sister $11 - 2 = 9$ years old. The ages of the three sisters add to 35 years.	If $m = 11$, $\text{LHS} = 3m + 2 \quad \text{RHS} = 35$ $= 3 \times 11 + 2$ $= 33 + 2$ $= 35$ LHS = RHS Therefore, $m = 11$ is a solution.
5. Answer the question.	The ages of the three sisters are 15, 11 and 9.

WORKED EXAMPLE 8 Creating equations from word problems 2

A gardener constructs a fence with three strands of wire around a rectangular garden bed. She bought a roll of wire that was 100 m long and has 4 m left over. If the length of the garden bed is 2 m longer than its width, calculate the dimensions of the garden bed.

THINK	WRITE
1. Read the question carefully and identify the unknown variables.	Let $w =$ the width of the garden bed. The length will be $w + 2$.
2. There are three strands of wire around the garden bed and 4 m left over. Construct an equation to represent this using the variables.	One strand (perimeter) $= 2w + 2(w + 2)$ $= 2w + 2w + 4$ $= 4w + 4$ For three strands: $3(4w + 4)$ $= 12w + 12$ Add leftover wire: $12w + 12 + 4$ $= 12w + 16$ Total length of wire = 100 Equation: $12w + 16 = 100$
3. Solve the equation by: • subtracting 16 from both sides • dividing both sides by 12.	$12w + 16 - 16 = 100 - 16$ $12w = 84$ $\frac{12w}{12} = \frac{84}{12}$ $w = 7$
4. Check your answer to see if it makes sense in the context of the question. If $w = 7$, the length is $9(w + 2)$. One strand of wire would be $14 + 18 = 32$ m long. Three strands would be 96 m long. That leaves 4 m left of the 100 m of wire. So the answer is reasonable.	If $w = 7$, $\text{LHS} = 12w + 16 \quad \text{RHS} = 100$ $= 12(7) + 16$ $= 100$ LHS = RHS Therefore, $w = 7$ is a solution.
5. Answer the question.	The garden bed is 7 m wide and 9 m long.

on Resources

Interactivity Solving linear equations (int-6450)

4.3 Exercise

Students, these questions are even better in jacPLUS

Receive immediate feedback and access sample responses

Access additional questions

Track your results and progress

Find all this and MORE in jacPLUS

1. WE5 Solve the following equations.
 a. $3(a+6)=24$ b. $2(b-7)=22$ c. $4(c+7)=51$ d. $7(d-15)=61$
2. Solve the following equations by using inverse operations and performing the same operations on both sides.
 a. $2a=18$ b. $b-4=48$ c. $\frac{c}{8}=2$
 d. $4f=16$ e. $g+7=11$ f. $\frac{u}{3}=4$
3. State what operation is the first to be undone when solving the following equations.
 a. $3a+9=87$ b. $(b-7)+3=5$ c. $\frac{(c-2)}{5}+11=8$ d. $\frac{3(2d+7)}{11}=-1$
4. Solve the following equations by using inverse operations.
 a. $2x+4=10$ b. $3p-2=7$ c. $7a+15=57$
 d. $11a-13=119$ e. $\frac{c}{2}+6=28$ f. $\frac{d+8}{7}=1$
5. WE6 Solve each of the following equations and check your solutions.
 a. $10a-4=3a+3$ b. $5b+1=7b+5$ c. $6+2c=5c-3$ d. $12+5d=3+4d$
6. Solve the following equations and check your solutions.
 a. $2k+8=-4-k$ b. $g-4=3g+8$ c. $10r+2=8r-4$
 d. $4w+2=3w+7$ e. $-2q+7=4q+19$ f. $2-3v=-2v-11$
7. WE7 The ages of three sisters add to 45 years. The eldest sister is 5 years older than the middle sister. The middle sister is 5 years older than the youngest sister. Calculate how old the sisters are.
8. A mother is five times as old as her son. Their ages add to 36 years. Calculate how old they both are.
9. WE8 A farmer constructs a fence with four strands of wire around a rectangular paddock. She bought a roll of wire that was 500 m long and has 180 m left over. If the length of the paddock is 6 m longer than its width, calculate the dimensions of the paddock.

10. A pair of shoes and matching bag cost $175. If the shoes cost $50 more than the bag, calculate the prices of the bag and the shoes.
11. The length of the rectangle below is 7 centimetres less than three times its width. If the perimeter of the rectangle is the same as the perimeter of the triangle, calculate the side lengths of the rectangle and the triangle.

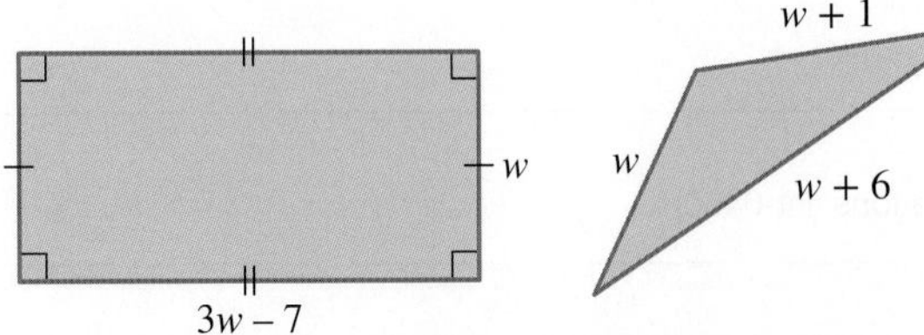

12. You have 12 more than 3 times the number of marker pens in your pencil case that your friend has in his pencil case. The teacher has 5 more than 4 times the number of marker pens in your friend's pencil case. Let your friend have x marker pens in his pencil case.
   a. Write an expression for the number of marker pens in:
      i. your pencil case
      ii. your teacher's pencil case.
   b. You have the same number of marker pens as the teacher. Write an equation to show this.
   c. Calculate the number of marker pens in your friend's pencil case by solving the equation from part **b**.

13. If a number is halved and then increased by 12, it will be six more than the original number. Calculate the value of this original number.

14. During the walkathon around the school oval, you walk five laps more than your friend, while your sister walks 10 laps more than you. The three of you raise $96 at a rate of $3 per lap. Calculate the number of laps each person walked.

Sponsor name	Number of laps	Total ($)	Paid

15. The two rectangles pictured have the same area but different dimensions.
   a. Calculate the value of x.
   b. Calculate the dimensions of the rectangles.
   c. State the area of each rectangle.

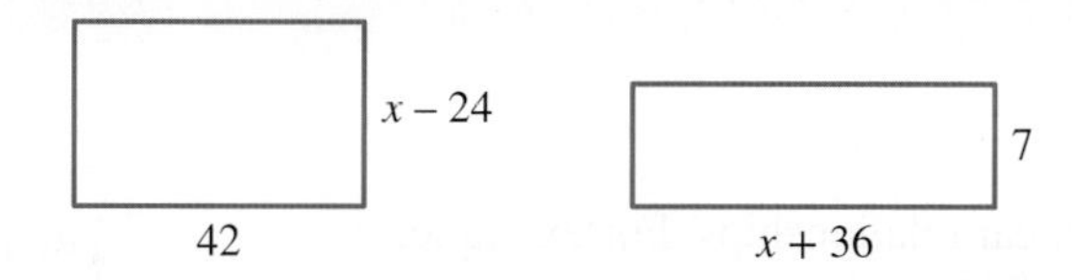

16. For each of the following, write a linear equation and then solve it. In each case, let x equal the original number.
   a. When I multiply this number by 3, add 2 and divide by 4, the result is 5. Calculate the value of the original number.
   b. When I divide by 3, subtract 4 and multiply by 9, my answer is 9. Calculate the value of the original number.
   c. Three consecutive whole numbers add to 81. Calculate the values of the three numbers.

17. Consider the image shown.
    a. Write a rule that could be used to calculate the cost of hiring a windsurfer for n hours.
    b. Use your rule from part **a** to calculate the cost of hiring a windsurfer for 6 hours.
    c. You have \$100 to spend. Write an equation to work out the number of hours you can hire a windsurfer for.
    d. Solve the equation from part **c** to calculate how many hours you can hire a windsurfer for. (*Note:* You can hire a windsurfer for a whole number of hours only.)
    e. After windsurfing for as long as you can with \$100, you return the windsurfer. Calculate how much money you will have left over.

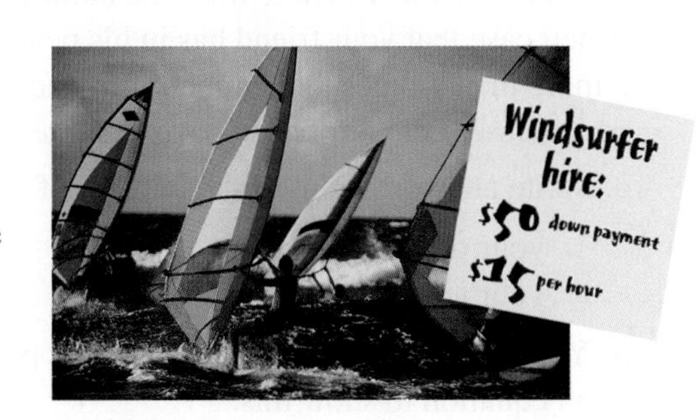

18. Consider the image shown.
    a. Calculate how many rides you need so that option 1 and option 2 shown cost the same.
    b. If you were planning on having only two rides, explain which option you would choose and why.

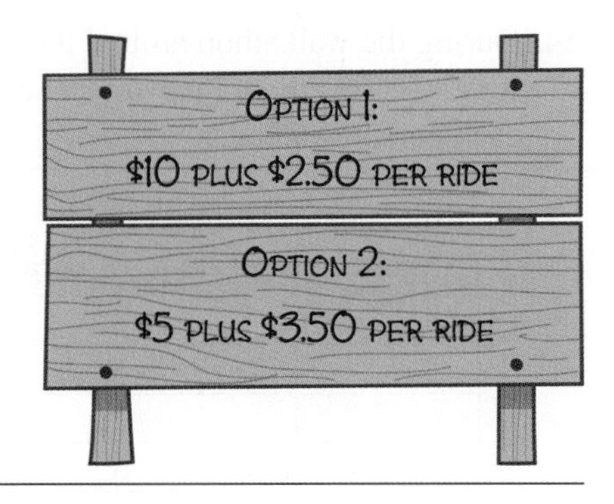

4.4 Linear modelling

LEARNING INTENTION

At the end of this subtopic you should be able to:
- create and solve problems using linear modelling
- graph linear relationships.

4.4.1 Introduction

Many real-life situations have linear relationships. For example, a mobile phone bill might be modelled by the linear relationship $C = 12 + 0.25t$, where C is the cost in dollars and t is the number of text messages sent. This equation is in the form $y = mx + c$.

On a graph, the explanatory variable is on the horizontal axis and the response variable is on the vertical axis. The value on the vertical axis 'depends' on the value on the horizontal axis.

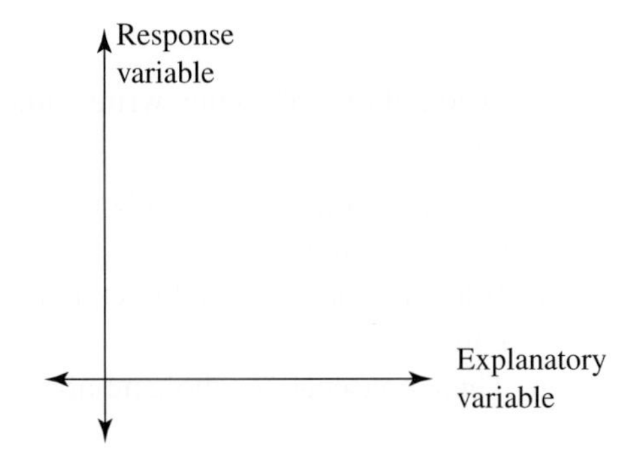

In practical applications, it is important to determine which variable is the explanatory variable and which is the response variable so that they can be graphed appropriately.

In practical experiments, the explanatory variable is the variable that is changed or manipulated to determine how this affects another variable (the response variable). For example, you could change the temperature of an oven and determine how this affects the time taken for a clay model to dry.

Time (hours, minutes or seconds) is almost always the explanatory variable in a practical problem.

tlvd-3521

WORKED EXAMPLE 9 Solving problems using linear modelling

A car hire company rents cars for $60 up front and a further cost of 50 cents per km.

a. Determine which variable is explanatory and which is response.
b. Write an equation relating the cost to the distance travelled.
c. Calculate the cost of travelling 200 km.
d. Sketch the graph of the relationship.
e. Calculate how far you have travelled if the cost is $335.

THINK	WRITE
a. The cost depends on the distance travelled, so the cost is the response variable.	a. Cost is response. Distance is explanatory.
b. 1. Read the question carefully and identify the unknown variables.	b. Let $C =$ cost (in dollars). Let $d =$ distance in km.
2. • Initially the cost is $60 before any travelling is done. This is a constant. • The cost per km is 50 cents (or 0.5 dollars as the cost is in dollars). This represents the gradient. • Write the equation.	$C = 0.50d + 60$
c. • The distance travelled is d. • Substitute 200 for d in the equation. • Write the answer.	c. Let $d = 200$. $C = 0.50(200) + 60$ $= 100 + 60$ $= 160$ The cost is $160.
d. Sketch a graph of the equation $C = 0.50d + 60$. • Use the y-intercept from the equation (0, 60) and the point (200, 160) that was found in the previous step to draw the graph. • Check for restrictions on the domain: ∘ The minimum number of kilometres that can be travelled is zero, so the point (0, 60) is an endpoint. ∘ There is no maximum distance that can be travelled, so the right end of the line is an arrowhead.	d. 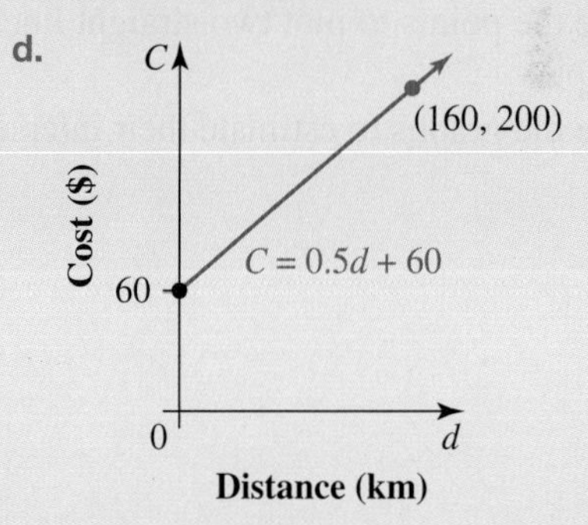
e. 1. Determine the distance travelled for a cost of $335 by substituting 335 for C. • Solve the equation for d.	e. Let $C = 335$. $C = 0.50d + 60$ $335 = 0.50d + 60$ $335 - 60 = 0.50d + 60 - 60$ $275 = 0.50d$ $\frac{{}^{1}\cancel{0.5}\,d}{{}^{1}\cancel{0.5}} = \frac{275}{0.5}$ $d = 550$
2. Write the solution.	The distance travelled is 550 km.

tlvd-3522

WORKED EXAMPLE 10 Solving problems using linear modelling 2

Two delivery companies, Express Delivery and HotWire Delivery, charge the following rates depending on the number of kilometres travelled to deliver goods.

Express Delivery	HotWire Delivery
Call-out fee of **\$10** plus **10** cents per km	Call-out fee of **\$5** plus **20** cents per km

a. **Use the information from the table to form two equations.**
b. **Compare the cost for travelling 60 km. Explain.**
c. **Sketch the graphs to determine where the graphs intersect.**
d. **Solve algebraically to confirm your answer in part c.**
e. **Explain when it would be cheaper to use Express Delivery.**

THINK	WRITE
a. 1. Identify the unknown variables.	a. Let C = cost in dollars. Let d = distance in km.
2. Write an equation for the cost of each company.	Express: $C = 10 + 0.10d$ HotWire: $C = 5 + 0.20d$
b. 1. To determine the cost, substitute 60 for d in each equation.	b. $C = 10 + 0.10(60)$ $= 16$ Express charges \$21. HotWire: $C = 5 + 0.20(60)$ $= 17$ HotWire charges \$17.
2. State the answer.	It is cheaper to use Express Delivery for 60 km.
c. Create points using information from part b. d = explanatory variable C = response variable – Express Delivery: (60, 16) and (0, 10) – HotWire Delivery: (60, 17) and (0, 5) Use the points to plot two straight lines on a graph. Use the graphs to estimate their intersection.	c. 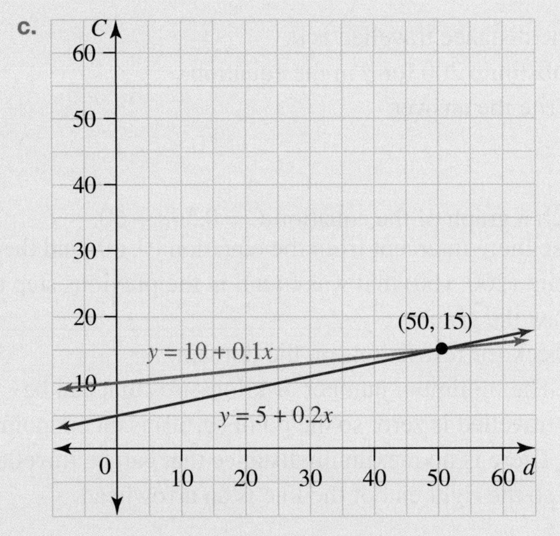
d. 1. To determine when the cost is the same, equate the two equations.	d. $10 + 0.10d = 5 + 0.20d$ $10 - 5 = 0.20d - 0.10d$ $5 = 0.10d$ $d = \frac{5}{0.10}$ $d = 50$
2. Substitute $d = 50$ into one of the equations to find C.	$C = 10 + 0.10(50)$ $= 10 + 5$ $= 15$ Therefore, the point of intersection is (50, 15).

e. Use the graph to explain when it would be cheaper to use Express Delivery.	**e.** Express Delivery becomes cheaper when the distance is above 50 km, as the line is below the HotWire line.

4.4 Exercise

Students, these questions are even better in jacPLUS

Receive immediate feedback and access sample responses

Access additional questions

Track your results and progress

Find all this and MORE in jacPLUS

1. Explain the difference between the explanatory and the response variable.

2. **MC** Choose the situation that is modelled by the graph shown.
   - **A.** You have $100 and plan to spend $5 each week.
   - **B.** You have $100 and plan to save $5 each week.
   - **C.** You need $100 for a new mobile and plan to save $5 each week.
   - **D.** You need $100 for a new mobile and plan to spend $5 each week.
   - **E.** You have 5 weeks to save for a new mobile.

3. **WE9** An electronics shop has 30 OLED TVs in stock. The manufacturer ships the TVs in containers of 12.
   - **a.** Write a linear equation that relates the number of containers to the total number of TVs.
   - **b.** Graph the function.
   - **c.** Explain how the gradient and y-intercept relate to the total number of TVs.

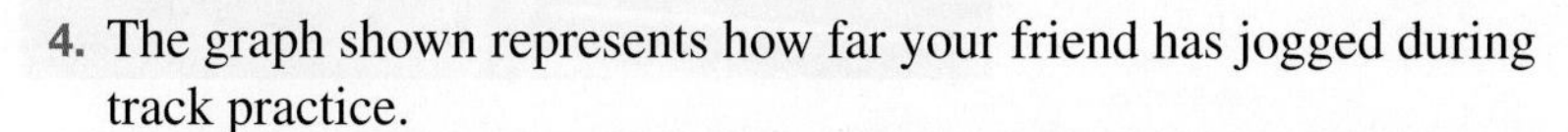

4. The graph shown represents how far your friend has jogged during track practice.
   - **a.** Explain what the slope and the y-intercept of the graph mean in terms of jogging.
   - **b.** If the graph had a slope of 5, explain if your friend would be travelling faster or slower.

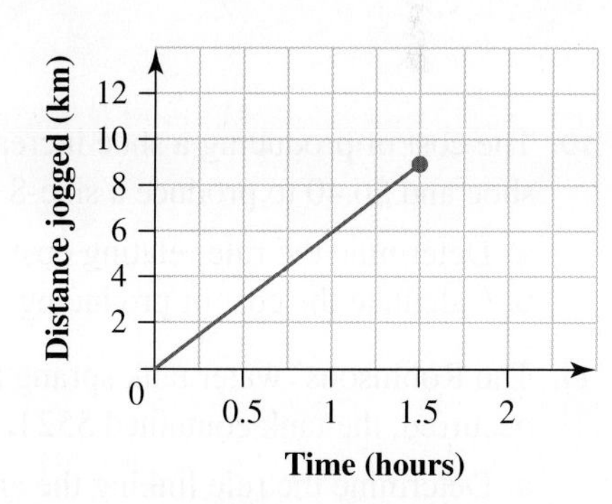

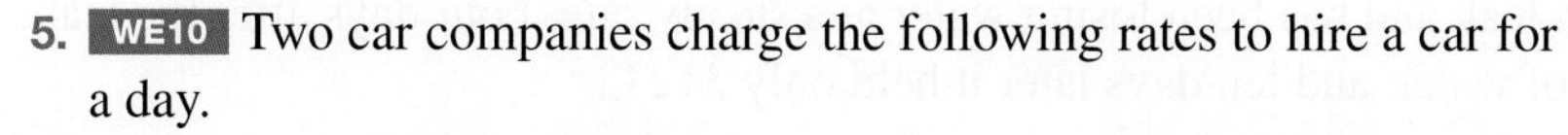

5. **WE10** Two car companies charge the following rates to hire a car for a day.

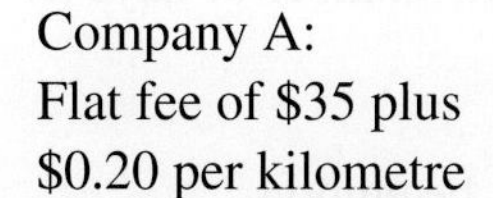

Company A:	Company B:
Flat fee of $35 plus	Flat fee of $25 plus
$0.20 per kilometre	$0.30 per kilometre

   - **a.** Use the information given to form two equations.
   - **b.** Compare the cost for travelling 50 km. Explain.
   - **c.** Sketch the graphs and locate where they intersect.
   - **d.** Solve algebraically to confirm your answer in part **c.**
   - **e.** Explain when it would be cheaper to use company A.

6. Optican, a telephone company, is trying to attract new customers. Its new plan charges a monthly fee of \$25 for unlimited texts, calls and 10 GB of data. However, when a customer uses all of their data for the month, Optican charges 30 cents per megabyte (MB). Let C be the total cost and s one MB of excess data.
   a. Write an equation to represent this information.
   b. Calculate the total cost if you used 50 MB more than your allowed data this month.
   c. Sketch a graph to show the relationship.
   d. Determine how much excess data you used if you were sent a monthly bill of \$106.
   e. A start-up telephone company, Primeus, offers new customers \$5 a month for unlimited texts, calls and 10 GB of data. However, they charge 70 cents per megabyte for excess data. Determine when each company is the better deal.
7. Your friend says the following to her sister: 'If you multiply my age in years by five and triple your age in years, the result is 83.' How old are your friend and her sister?
8. Write linear equations for each of the following statements, using x to represent the unknown. (Do not attempt to solve the equations.)
   a. When 3 is added to a certain number, the answer is 5.
   b. Subtracting 9 from a certain number gives a result of 7.
   c. Seven times a certain number is 24.
   d. A certain number divided by 5 gives a result of 11.
   e. Dividing a certain number by 2 equals −9.
   f. Three subtracted from five times a certain number gives a result of −7.
9. Nathan's bank balance has increased in a linear manner since he started his part-time job. If after 20 weeks of work his bank balance was at \$560 and after 21 weeks of work it was at \$585, determine:
   a. the rule that relates the size of his bank balance, A, and the time (in weeks) worked, t
   b. the amount in his account after 200 weeks
   c. the initial amount.

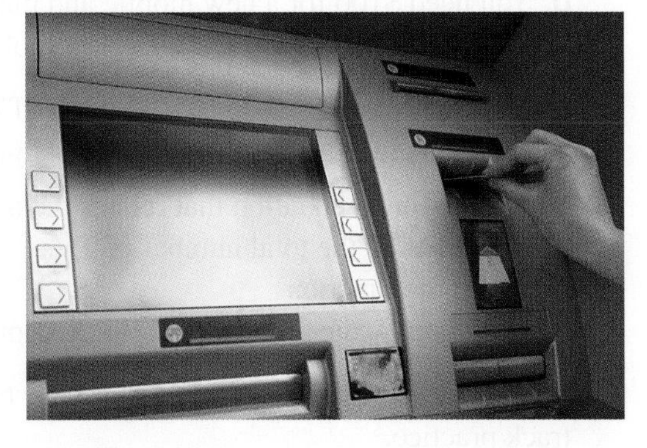

10. The cost of producing a shoe increases as the size of the shoe increases. It costs \$5.30 to produce a size-6 shoe and \$6.40 to produce a size-8 shoe. Assume that a linear relationship exists.
    a. Determine the rule relating cost (C) to shoe size (s).
    b. Calculate the cost of producing a size-12 shoe.
11. The Robinsons' water tank sprang a leak and has been losing water at a steady rate. Four days after the leak occurred, the tank contained 552 L of water, and ten days later it held only 312 L.
    a. Determine the rule linking the amount of water in the tank (w) and the number of days (t) since the leak occurred.
    b. Calculate the initial amount of water in the tank.
    c. If water loss continues at the same rate, calculate when the tank will be empty.
12. Maria is paid \$11.50 per hour, plus \$7 for each jacket that she sews. If she earned \$176 for one 8-hour shift, calculate how many jackets she sewed.
13. Mai hired a car for a fee of \$120 plus \$30 per day. Casey's rate for his car hire was \$180 plus \$26 per day. If their final cost and rental period were the same, determine the length of the rental period.

14. A Year 12 class are raising money for a local charity. So far, they have raised \$1200 in donations from various events. The class want to reach their goal of \$2100. They decide to sell cupcakes for 95c each at lunch time for the next two weeks. Determine how many cupcakes they need to sell to reach their goal.

15. Joseph wishes to have some flyers delivered for his grocery business. Post Quick quotes a price of \$200 plus 50 cents per flyer, while Fast Box quotes \$100 plus 80 cents per flyer.

    a. If Joseph needs to order 1000 flyers, determine which distributor would be cheaper to use.
    b. Calculate the number of flyers for which the cost will be the same if he uses either distributor.

4.5 Solving simultaneous equations graphically

LEARNING INTENTION

At the end of this subtopic you should be able to:

- solve simultaneous equations graphically by hand
- rearrange formulas and use them to solve problems
- identify special cases of simultaneous equations that are parallel, perpendicular or have multiple solutions.

4.5.1 Introduction

Simultaneous means occurring at the same time.

When a point belongs to more than one line, the coordinates of the point satisfy both equations. The equations of the lines are called **simultaneous equations**.

A **system of equations** is a set of equations that can all be satisfied by the same values for variables.

When the system of equations is sketched, the coordinates of the point of intersection of all of the lines are the values of the variables that satisfy all of the equations.

To solve simultaneous equations is to calculate the values of the variables that satisfy all equations in the system and, hence, determine the coordinates of the point (or points) of intersection for all of the lines.

tlvd-3523

WORKED EXAMPLE 11 Solving simultaneous equations graphically

Solve the following simultaneous equations graphically:

a. $y = 3x + 1$
$y - 2 = 2x$

b. $2x - 3y = 6$
$x + y = 3$

THINK	WRITE
a. 1. Rearrange both equations into the gradient-intercept form, $y = mx + c$.	a. $y = 3x + 1$ $\quad$ $y - 2 = 2x$ $y - 2 + 2 = 2x + 2$ $y = 2x + 2$

2. Graph both equations using the gradient m and y-intercept c.

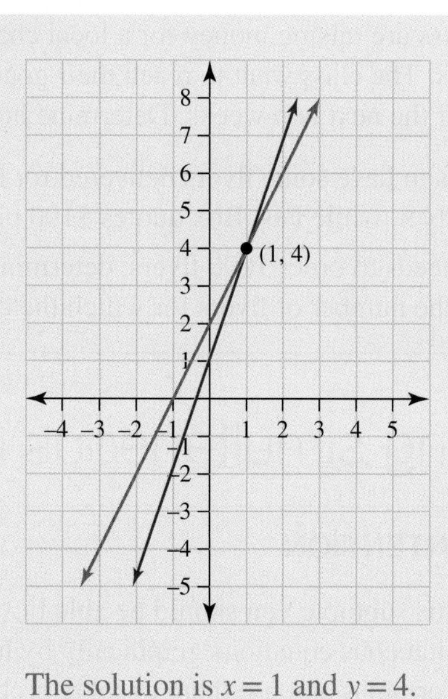

3. Write the solution.

The solution is $x = 1$ and $y = 4$. This is the point $(1, 4)$.

4. Check that the solution satisfies both equations by substituting the values $x = 1$ and $y = 4$ into both equations.

$$y = 3x + 1$$
$$(4) = 3(1) + 1$$
$$4 = 4$$
$$y = 2x + 2$$
$$(4) = 2(1) + 2$$
$$4 = 4$$

The solution is correct.

b. 1. Determine the intercepts for each equation.

- The y-intercept occurs when $x = 0$.
- The x-intercept occurs when $y = 0$.
- State the coordinates of the x- and y-intercepts.

b. $2x - 3y = 6$ $\quad$ $x + y = 3$

y-intercept: let $x = 0$

$2(0) - 3y = 6$ $\quad$ $(0) + y = 3$

$-3y = 6$ $\quad$ $y = 3$

$y = -2$

x-intercept: let $y = 0$

$2x - 3(0) = 6$ $\quad$ $x + (0) = 3$

$2x = 6$ $\quad$ $x = 3$

$x = 3$

$(0, -2), (3, 0)$ $\quad$ $(0, 3), (3, 0)$

2. On grid paper, plot the x- and y-intercepts for each line and join them to graph each line.

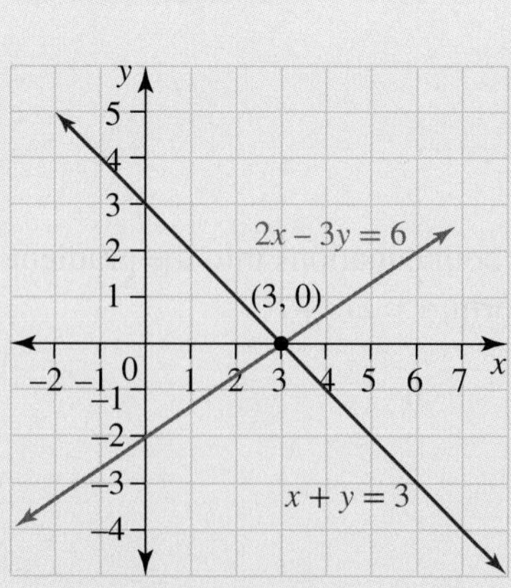

3. Carefully read the coordinates of the point of intersection.

The point of intersection appears to be $(3, 0)$, as shown in the graph.

4. Write the solution.	The solution is $x = 3$ and $y = 0$. This is the point $(3, 0)$.
5. Check that the solution satisfies both equations by substituting the values $x = 3$ and $y = 0$ into each equation.	$2x - 3y = 6$ $x + y = 3$ $2(3) - 3(0) = 6$ $3 + 0 = 3$ $6 - 0 = 6$ $3 = 3$ $6 = 6$ The solution is correct.

4.5.2 Special cases of simultaneous equations

Equations with multiple solutions

Two lines are **coincident** if they lie one on top of the other. For example, the blue line and the red line in the graph shown are coincident.

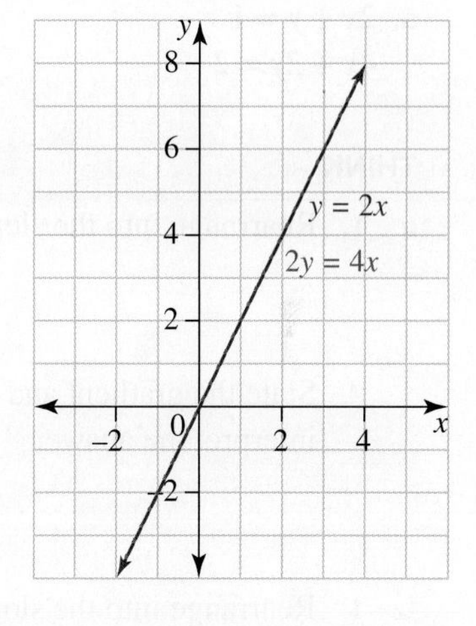

There are an infinite number of solutions to coincident equations. Every point where the lines coincide satisfies both equations and hence is a solution to the simultaneous equations.

Coincident equations have the same equation, although the equations may have been transposed so they look different. For example, $y = 2x$ and $2y = 4x$ are coincident equations.

Equations with no solutions

If two lines do not intersect, there is no simultaneous solution to the equations. For example, the lines shown in the graph do not intersect, so there is no point that belongs to both lines.

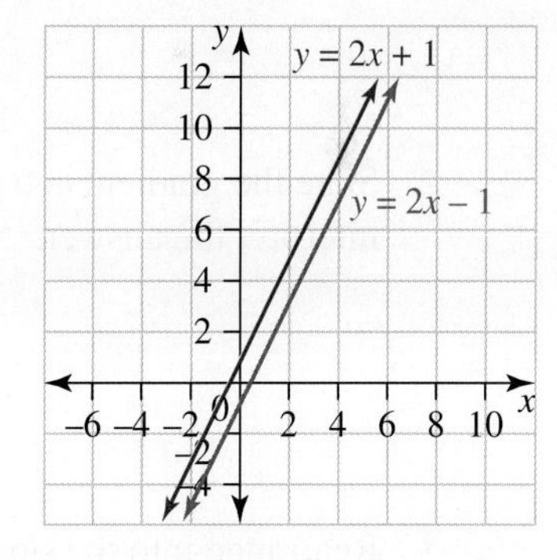

Parallel lines have the same gradient but a different y-intercept.

For straight lines, the only situation in which the lines do not cross is if they are parallel *and* not coincident.

Perpendicular lines

Perpendicular lines meet at right angles (90°).

Perpendicular lines have negative reciprocal gradients, $m_1 = \dfrac{-1}{m_2}$ or $m_1 m_2 = -1$, where m_1 is the gradient of the first line and m_2 is the gradient of the second line. For example, for the two lines shown in the graph, $m_1 = 2$ and $m_2 = \dfrac{-1}{2}$.

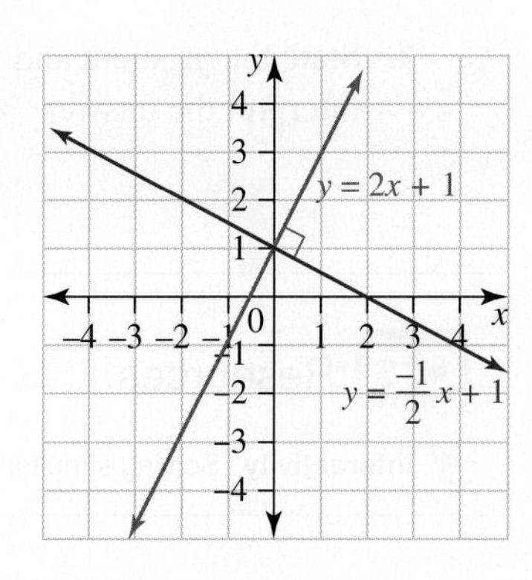

Parallel and perpendicular lines

Two parallel lines have the same gradient but different y-intercepts.

The gradients of two perpendicular lines multiplied together equal -1.

$$m_1 m_2 = -1$$

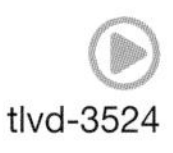

tlvd-3524

WORKED EXAMPLE 12 Identifying equations with multiple solutions, no solutions or one solution

For each pair of equations, determine whether the lines are parallel, perpendicular or neither.

a. $2x + y = 1$
$4x + 2y = 2$

b. $y = 3x - 2$
$3y + x = 6$

c. $y - x = 3$
$2y = 2x - 4$

THINK	WRITE
a. 1. Rearrange into the slope-intercept form.	**a.** $2x + y = 1$, $y = -2x + 1$ $4x + 2y = 2$, $2y = -4x + 2$, $y = -2x + 1$
2. State the gradient and y-intercept and interpret the answer.	$m_1 = -2,\ c = 1$ $m_2 = -2,\ c = 1$ Because they are the same line, they have multiple solutions. Therefore, they are neither.
b. 1. Rearrange into the slope-intercept form.	**b.** $y = 3x - 2$ $3y + x = 6$, $3y = -x + 6$, $y = \frac{-1}{3}x + 2$
2. State the gradient and y-intercept and interpret the answer.	$m_1 = 3,\ c = -2$ $m_2 = \frac{-1}{3},\ c = 2$ The gradients multiply together to give -1. $m_1 m_2 = -1$ $3 \times \frac{-1}{3} = -1$ Therefore, the equations are perpendicular.
c. 1. Rearrange into the slope-intercept form.	**c.** $y - x = 3$, $y = x + 3$ $2y = 2x - 4$, $y = x - 2$
2. State the gradient and y-intercept and interpret the answer.	$m_1 = 1,\ c = 3$ $m_2 = 1,\ c = -2$ The gradients are the same and the y-intercepts are different. Therefore, they are parallel with no solution.

on Resources

Interactivity Solving simultaneous equations graphically (int-6452)

4.5 Exercise

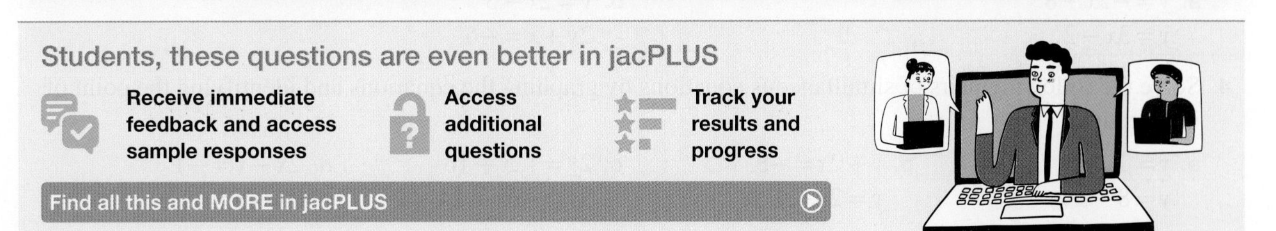

1. Explain in your words what simultaneous equations are.
2. For each of the following graphs, determine the coordinates of the point of intersection. Justify your answers.

a.

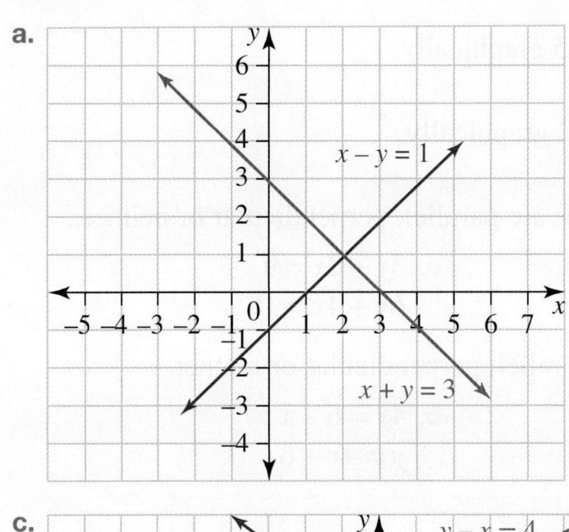

b.

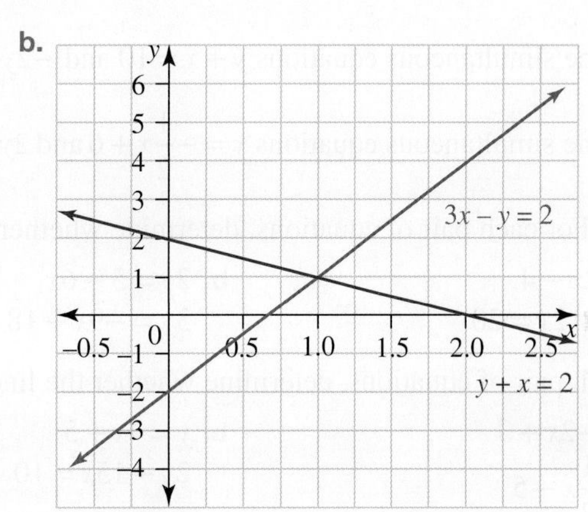

c.

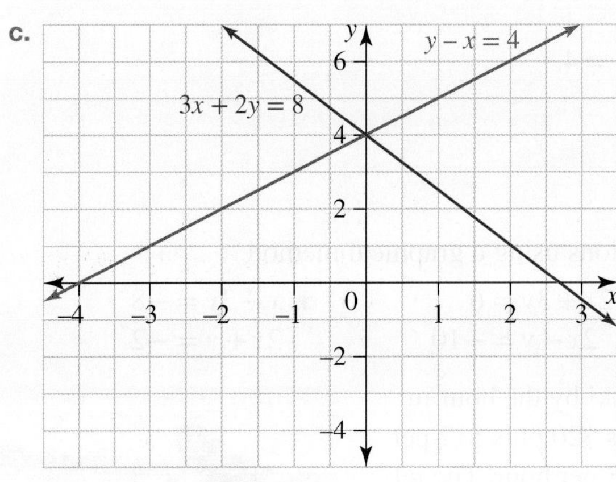

d.

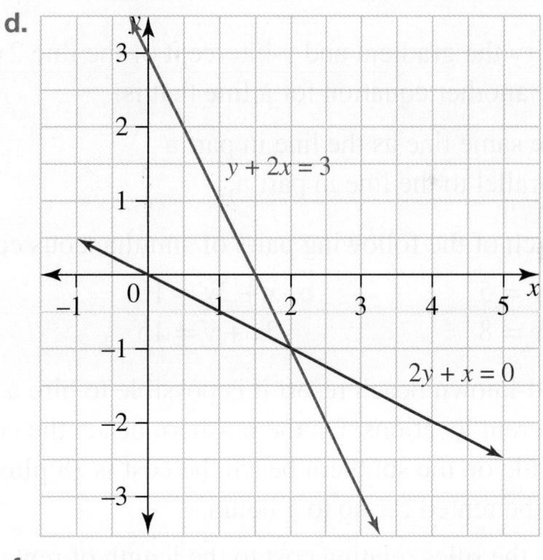

e.

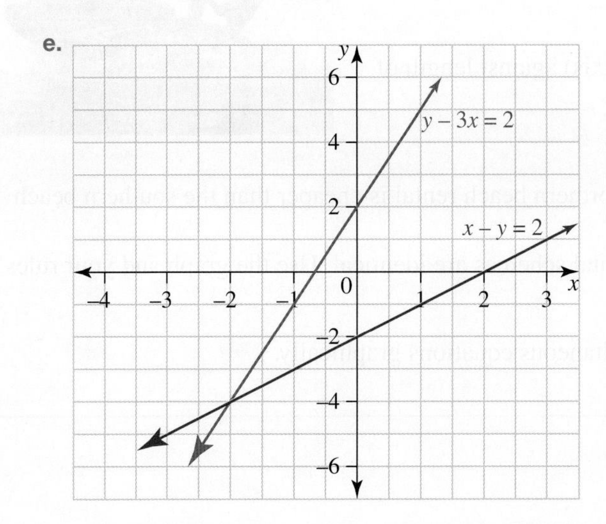

f.

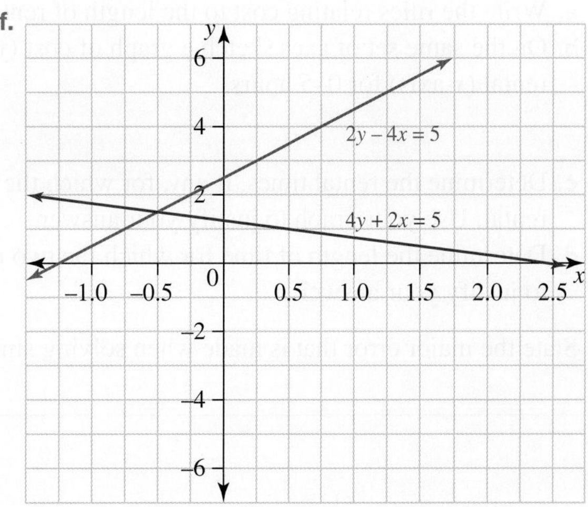

3. **WE11** Solve the following simultaneous equations graphically:

a. $y = -2x + 8$
$y = 3x - 2$

b. $y = 2x - 3$
$2y + x = -6$

4. Solve the following pairs of simultaneous equations by graphing the equations and identifying the point of intersection.

a. $y = 2x + 3$
$y = 8 - 3x$

b. $y + 2x = -8$
$y = 2x + 4$

c. $2y = 12x + 16$
$3y = -6x - 24$

d. $2x - y = -1$
$3x + y = 11$

5. Solve the following pairs of simultaneous equations by graphing the equations and identifying the point of intersection.

a. $31 = y - 2x$
$2x + 2y = 14$

b. $x + 2y = 4$
$y = 2x - 3$

c. $x + y = 6$
$2x + 4y = 20$

d. $y = \frac{2x}{3} + 2$
$y = 2x - 2$

6. Solve the simultaneous equations $y + x = 10$ and $-2y + x = -5$ graphically.

7. Solve the simultaneous equations $y = -\frac{1}{2}x + 6$ and $2y - x = 12$ graphically.

8. **WE12** For each pair of equations, determine whether the lines are parallel, perpendicular or neither.

a. $y = 2x - 4$
$6y - 12x = 20$

b. $2y = 5 - 6x$
$3y = -9x + 18$

c. $3y + 2x = 9$
$6y + 4x = 18$

9. For each pair of equations, determine whether the lines are parallel, perpendicular or neither.

a. $y = -2x + 3$
$y = \frac{1}{2}x - 5$

b. $y = 4x + 5$
$2y - 15x = 10$

c. $4y = 6 - x$
$y = 4x + 6$

10. a. Identify the gradient and y-intercept of the line $2x - y = 4$.
b. Write another equation for a line that is:
   i. the same line as the line in part a
   ii. parallel to the line in part a.

11. Solve each of the following pairs of simultaneous equations using a graphical method.

a. $x + y = 5$
$2x + y = 8$

b. $x + 2y = 10$
$3x + y = 15$

c. $2x + 3y = 6$
$2x - y = -10$

d. $x - 3y = -8$
$2x + y = -2$

12. At a well-known beach resort it is possible to hire a jet ski by the hour in two different locations. On the northern beach the cost is \$20 plus \$12 per hour, while on the southern beach the cost is \$8 plus \$18 per hour. The jet skis can be rented for up to 5 hours.

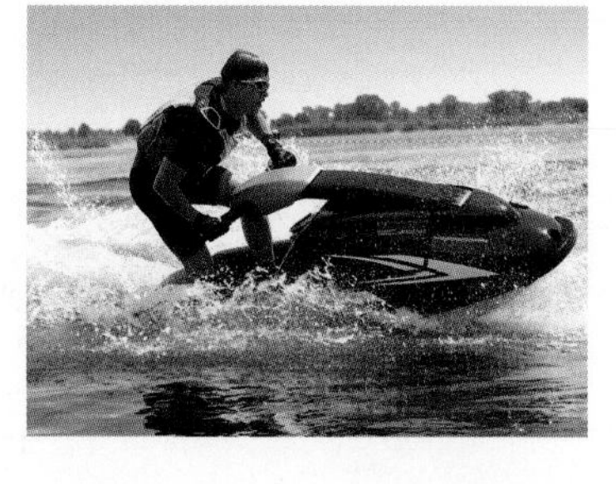

a. Write the rules relating cost to the length of rental.
b. On the same set of axes sketch a graph of cost (y-axis) against length of rental (x-axis) for 0–5 hours.

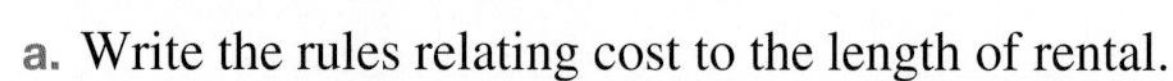

c. Determine the rental times, if any, for which the northern beach rental is cheaper than the southern beach rental. Use your graph to justify your answer.
d. Determine the length of time for which the two rental schemes are identical. Use the graph and your rules to justify your answer.

13. State the major error that is made when solving simultaneous equations graphically.

4.6 Solving simultaneous equations by substitution

LEARNING INTENTION

At the end of this subtopic you should be able to:
- solve simultaneous equations using the substitution method.

4.6.1 Introduction

Substitution involves replacing a variable in an equation or formula with an equivalent value or expression.

At the point of intersection of two lines, the x-coordinate gives the same y-value when substituted into each equation. The values of x and y are the same in both equations: $(x_1, y_1) = (x_2, y_2)$ at the point of intersection.

Substituting the expression for one y-value into the other equation for its y-value creates a single equation with one variable.

The same substitutions can be made using x-values to give a single equation with y as the variable.

This method of solving simultaneous equations is known as the substitution method.

The substitution method is normally used when one or both equations are written in the form $y = mx + c$.

tlvd-3525

WORKED EXAMPLE 13 Solving simultaneous equations by substitution

Solve the system of equations $y = x + 3$ and $y = 2x + 5$.

THINK	WRITE
1. Write the two equations, labelling one as equation 1 and the other as equation 2.	$y=x+3$ [1] $y=2x+5$ [2]
2. At the point of intersection, the two y-values are equal, so [1] = [2].	At point of intersection: [1] = [2] $x+3 = 2x+5$
3. Solve the equation for x. • x appears on each side of the equation, so subtract x from both sides. • Subtract 5 from both sides.	$x+3-x = 2x+5-x$ $3 = x+5$ $3-5 = x+5-5$ $-2 = x$
4. The x-coordinate of the point of intersection is −2. To determine the value of the y-coordinate, substitute the value of x into either of the equations.	$y = x+3$ $= (-2)+3$ $= 1$
5. Write the solution as the pair of coordinates representing the point of intersection.	The solution is $x = -2$ and $y = 1$, or $(-2, 1)$.

Interactivity Solving simultaneous equations using substitution (int-6453)

tlvd-3526

WORKED EXAMPLE 14 Rearranging equations to solve by substitution

Solve the system of equations $x + 3y = 1$ and $y = 2x + 5$ by substitution.

THINK	WRITE
1. Label the two equations as equation 1 and equation 2, as shown in purple and red.	$x + 3y = 1$ [1] $y = 2x + 5$ [2]
2. At the point of intersection, the two y-values are equal, so replace y with $2x + 5$ in equation [1].	$y_1 = y_2$ $x + 3(2x+5) = 1$
3. Solve the equation for x. • Expand the brackets. • Collect like terms of x. • Subtract 15 from both sides. • Divide both sides by 7.	$x + 6x + 15 = 1$ $7x + 15 = 1$ $7x = -14$ $x = -2$
4. The x-coordinate of the point of intersection is -2. To determine the value of the y-coordinate, substitute $x = -2$ into one of the equations (it does not matter which).	Substitute $x = -2$ into [2]: $y = 2x + 5$ $= 2(-2) + 5$ $= 1$
5. Write the solution as the pair of coordinates representing the point of intersection.	The solution is $x = -2$ and $y = 1$, or $(-2, 1)$.
6. Check by substituting $(-2, 1)$ into the other equation.	Check: substitute $(-2, 1)$ into [1]: $x + 3y = 1$ LHS $= -2 + 3(1)$ RHS $= 1$ $= 1$ LHS = RHS The solution is correct.

4.6 Exercise

1. Explain the advantage of solving simultaneous equations by substitution rather than by looking at the graphs.

2. WE13 Solve the system of equations $y = 5x$ and $y = -3x + 8$.

3. WE14 Solve the system of equations $x + 3y = 5$ and $y = 2x + 4$ by substitution.

4. Solve the following pairs of equations simultaneously using the substitution method.

a. $2x + y = 17$
$x = 2y + 1$

b. $y + 4x = 6$
$y = 2x - 3$

c. $3x - 2y = 5$
$y = 3 - 4x$

d. $y = 1 - 2x$
$y = 3 - 4x$

5. Solve the following pairs of equations simultaneously using the substitution method.

a. $y = 2x - 5$
$5x - y = 1$

b. $y = -x + 4$
$5x = 3y$

c. $4x - 5y = 17$
$y = 2x - 1$

d. $x = -5y + 1$
$3x + 2y = 16$

6. Use substitution to solve the following pair of linear equations.

$$2x + 2y = 6$$
$$y = -x + 3$$

a. Explain what you notice about this solution.
b. Comment on the values of the gradients and intercepts for these equations.
c. Sketch the two equations.
d. Explain what you observe about these lines.

7. For each of the pairs of simultaneous equations below, determine whether they are the same line, parallel lines or intersecting lines.

a. $2x - y = -9$
$-4x - 18 = -2y$

b. $x - y = 7$
$x + y = 7$

c. $x + 6 = y$
$x + y = 6$

d. $x + y = -2$
$x + y = 7$

8. Determine which of the following problems has one solution, an infinite number of solutions or no solution. Explain your answers.

a. $x - y = 1$
$2x - 3y = 2$

b. $2x - y = 5$
$4x - 2y = -6$

c. $x - 2y = -8$
$4x - 8y = -16$

9. Use substitution to solve each of the following pairs of simultaneous equations.

a. $5x + 2y = 17$
$y = \dfrac{3x - 7}{2}$

b. $2x + 7y = 17$
$x = \dfrac{1 - 3y}{4}$

c. $2x + 3y = 13$
$y = \dfrac{4x - 15}{5}$

d. $-2x - 3y = -14$
$x = \dfrac{2 + 5y}{3}$

e. $3x + 2y = 6$
$y = 3 - \dfrac{5x}{3}$

f. $-3x - 2y = -12$
$y = \dfrac{5x - 20}{3}$

10. A small farm has sheep and chickens. There are twice as many chicken as sheep, and there are 104 legs between the sheep and the chickens. Determine how many chickens there are.

11. Solve the following pairs of simultaneous equations using the substitution method.

a. $y = 2x - 11$ and $y = 4x + 1$
b. $y = 3x + 8$ and $y = 7x - 12$
c. $y = 2x - 10$ and $y = -3x$

12. a. For the pair of simultaneous equations below, state which equation is the logical choice to make x the subject of the equation.
$8x - 7y = 9$
$x + 2y = 4$
b. Use the substitution method to solve the system of equations. Show all your working.

13. A particular Chemistry book costs $6 less than a particular Physics book. Two such Chemistry books and three such Physics books cost a total of $123. Construct two simultaneous equations and solve them using the substitution method. Show your working.

14. Use substitution to solve each of the following pairs of simultaneous equations for x and y in terms of m and n.

a. $mx + y = n$
$y = mx$

b. $x + ny = m$
$y = nx$

c. $mx - y = n$
$y = nx$

d. $mx - ny = n$
$y = x$

4.7 Solving simultaneous equations by elimination

LEARNING INTENTION

At the end of this subtopic you should be able to:

- rearrange and solve equations simultaneously using the elimination method.

4.7.1 The elimination method

The **elimination method** is an algebraic method to solve simultaneous equations without graphing.

If two balanced equations contain the same variables, the equations can be added or subtracted to eliminate one of the variables. For example, the equations $2x + y = 5$ and $x + y = 3$ can be solved by elimination.

Looking at the balance scale diagram, if the left-hand side of the second equation is subtracted from the left-hand side of the first equation, and the right-hand side of the second equation is subtracted from the right-hand side of the first equation, the variable y is eliminated, leaving $x = 2$.

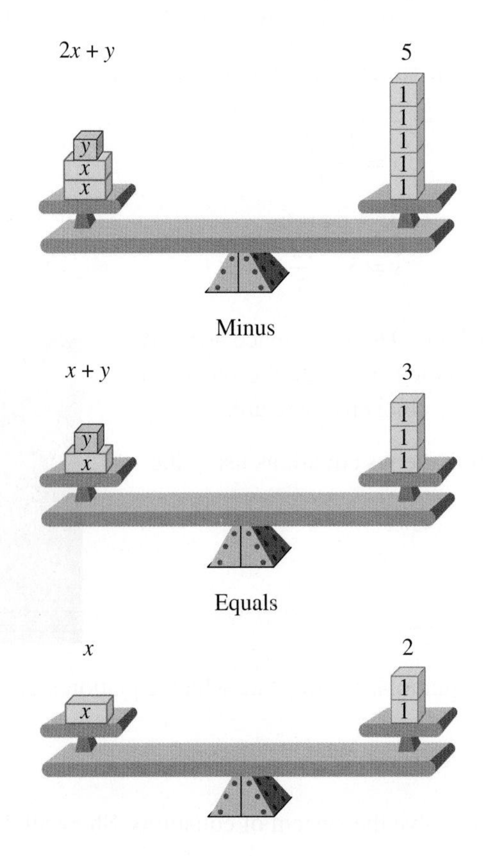

Another way to represent this situation is:

$$\begin{array}{r} 2x + y = 5 \\ -(x + y = 3) \\ \hline x = 2 \end{array}$$

In this example, the variable is eliminated by subtraction to reveal the value of x. The value of y can then be calculated by substituting $x = 2$ into either equation.

$$\begin{aligned} 2(2) + y &= 5 \\ y &= 1 \end{aligned}$$

WORKED EXAMPLE 15 Identifying when to add or subtract with elimination

For each of the following pairs of simultaneous equations, state whether you should add or subtract to eliminate a variable. State which variable will be eliminated.

a. $3x + y = 5$
$2x - y = 3$

b. $2x + 3y = 9$
$2x + y = 7$

THINK	WRITE
a. 1. Label the equations as equation 1 and equation 2, as shown in purple and red.	a. $3x + y = 5$ [1] $2x - y = 3$ [2]
2. Both the coefficients of y have the same value; one is positive and the other is negative.	
When these equations are added, the y-terms will cancel each other out.	Add the equations to eliminate y.
b. 1. Label the equations as equation 1 and equation 2, as shown in purple and red.	b. $2x + 3y = 9$ [1] $2x + y = 7$ [2]
2. Both the coefficients of x have the same value, but both are positive. If the equations are subtracted, the x-terms will cancel each other out.	Subtract equation 2 from equation 1 to eliminate x.

tlvd-3527

WORKED EXAMPLE 16 Solving simultaneous equations with elimination

Solve the simultaneous equations $3x + 2y = -12$ and $5x - 2y = -4$ by elimination.

THINK	WRITE
1. Label the equations as equation 1 and equation 2, as shown in purple and red. When these equations are added, the y-terms will cancel each other out.	$3x + 2y = -12$ [1] $5x - 2y = -4$ [2]
2. Add the two equations together by: • adding the two left-hand sides • adding the two right-hand sides.	$3x + 2y = -12$ $+(5x - 2y = -4)$ $8x = -16$
3. Solve for x by dividing both sides by 8.	$x = -2$
4. Substitute the x-coordinate into either of the equations and solve for y.	Substitute $x = -2$ into [1]: $3(-2) + 2y = -12$ $-6 + 2y = -12$ $2y = -6$ $y = -3$
5. Write the solution as the pair of coordinates for the point of intersection.	The solution is $x = -2$ and $y = -3$, or $(-2, -3)$.

6. Check by substituting $x = -2$ and $y = -3$ into the other equation.

Check: substitute $(-2, -3)$ into [2]:
$5x - 2y = -4$
$\text{LHS} = 5(-2) - 2(-3)$ $\text{RHS} = -4$
$= -10 + 6$
$= -4$
LHS = RHS
The solution is correct.

4.7.2 Multiplying equations to solve using the elimination method

If two equal quantities are multiplied by the same number, the results remain equal. Double both sides and it remains balanced. If a variable is not eliminated when the equations are simply added or subtracted, it may be necessary to multiply one or both equations so that when the equations are added, one of the variables is then eliminated.

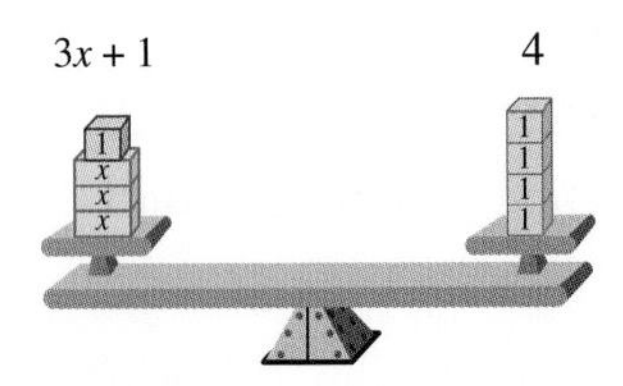

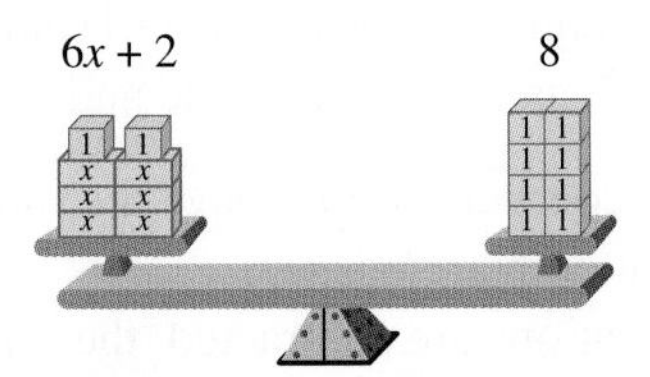

tlvd-3528

WORKED EXAMPLE 17 Solving harder simultaneous equations with elimination

Use the elimination method to solve the simultaneous equations $5x + 3y = 12$ and $7x + 2y = 19$.

THINK	WRITE
1. Choose a variable to eliminate. In this case, choose to eliminate the y-variable.	$5x + 3y = 12$ [1] $7x + 2y = 19$ [2]
2. The lowest common multiple of the coefficients of y is 6. • Multiply equation [1] by 2 to create equation [3]. • Multiply equation [2] by 3 to create equation [4].	$[1] \times 2: 10x + 6y = 24$ [3] $[2] \times 3: 21x + 6y = 57$ [4]
3. To eliminate y, subtract equation [3] from equation [4] by subtracting the left-hand sides and then subtracting the right-hand sides.	$21x + 6y = 57$ $-(10x + 6y = 24)$ $11x = 33$
4. Solve for x.	$x = 3$
5. Substitute $x = 3$ into one of the equations to find y.	Substitute $x = 3$ into [1]: $5x + 3y = 12$ $5(3) + 3y = 12$ $3y = -3$ $y = -1$
6. Write the solution as the pair of coordinates representing the intersection point.	Solution is $x = 3$ and $y = -1$, or $(3, -1)$.

7. Check the solution by substituting the point (3, −1) into the other equation.	To check, substitute (3, −1) into [2]: $7x + 2y = 19$ $\text{LHS} = 7(3) + 2(-1) \quad \text{RHS} = 19$ $= 21 - 2$ $= 19$ LHS = RHS The solution is correct.

Resources

Interactivity Solving simultaneous equations using elimination (int-6127)

4.7.3 Problem-solving with simultaneous equations

When solving practical problems, the following steps can be useful:

1. Determine the variables and allocate them a variable name.
2. Set up two equations using the variables.
3. Solve these equations using an appropriate method.
4. Write your solution in words.
5. Check your solution.

tlvd-3529

WORKED EXAMPLE 18 Using simultaneous equations to solve problems

Charley is passionate about Mathematics and Science. In two recent tests, the difference between Charley's two marks was 13 and the sum of the two marks was 183. Assuming Charley's Mathematics mark was their better result, determine their test results for each subject.

THINK	WRITE
1. Determine the unknowns and allocate a variable to each of them.	Let M = Mathematics result. Let S = Science result.
2. Set up equations for the sum (+) of the test results and the difference (–) of the results.	$M + S = 183$ [1] $M - S = 13$ [2]
3. Use the elimination technique to solve the simultaneous equations by adding the two equations to eliminate S.	$M + S = 183$ [1] $+ \quad + \quad +$ $M - S = 13$ [2] $2M = 196$
4. Solve for M.	$M = \frac{196}{2}$ $M = 98$
5. Substitute $M = 98$ into one of the equations to solve for S.	$M + S = 183$ $98 + S = 183$ $98 - 98 + S = 183 - 98$ $S = 85$
6. Write the answer.	The Mathematics result is 98% and the Science result is 85%.
7. Check your solution.	$98 + 85 = 183$ $98 - 85 = 13$

4.7 Exercise

Students, these questions are even better in jacPLUS

Receive immediate feedback and access sample responses

Access additional questions

Track your results and progress

Find all this and MORE in jacPLUS

1. Explain how you know whether to add or subtract two equations together to eliminate one of the variables.
2. **WE15** For each of the following pairs of simultaneous equations, state whether you should add or subtract to eliminate a variable. State which variable will be eliminated.
   a. $x+y=-5$
   $2x-y=-1$
   b. $5x+3y=19$
   $5x+y=13$
   c. $-3x+2y=19$
   $-3x+y=11$
   d. $3x-4y=7$
   $5x-4y=21$
3. **WE16** Solve the simultaneous equations $5x+3y=12$ and $6x-3y=21$.
4. Solve the following pairs of equations by subtracting equations to eliminate either x or y.
   a. $3x+2y=13$
   $5x+2y=23$
   b. $2x-5y=-11$
   $2x+y=7$
   c. $-3x-y=8$
   $-3x+4y=13$
5. Solve each of the following equations using the elimination method.
   a. $6x-5y=-43$
   $6x-y=-23$
   b. $x-4y=27$
   $3x-4y=17$
   c. $-4x+y=-10$
   $4x-3y=14$
6. **WE17** Use the elimination method to solve the following simultaneous equations.
   a. $3x+2y=13$
   $-x+3y=-8$
   b. $4x-2y=7$
   $3x+2y=14$
   c. $y-x=5$
   $3x-5y=-21$
   d. $x+y=20$
   $3x+11y=100$
7. Use the elimination method to solve the following simultaneous equations.
   a. $3x+2y=10$
   $12x-5y=14$
   b. $2x+5y=4$
   $7x+15y=9$
   c. $3x+4y=5$
   $5x+2y=-1$
   d. $2x-7y=-4$
   $3x-5y=5$
8. Use the elimination method to solve the following simultaneous equations.
   a. $3x-2y=5$
   $2x-3y=5$
   b. $5x+3y=11$
   $3x+2y=6$
   c. $3x+7y=8$
   $15x+2y=-26$
   d. $12x-3y=6$
   $3x+2y=7$
9. The points $(2, 3)$ and $(-2, 6)$ are on the line $ax+by=18$. Determine the values of a and b by:
   a. writing two equations in terms of a and b using the two points
   b. solving the simultaneous equations created in part **a**.

10. WE18 A student had a Physics test and a Chemistry test in the same week. The difference between the two tests' results is 18 and the sum is 144.
Assuming the Physics result is the better result, calculate each of the test results.

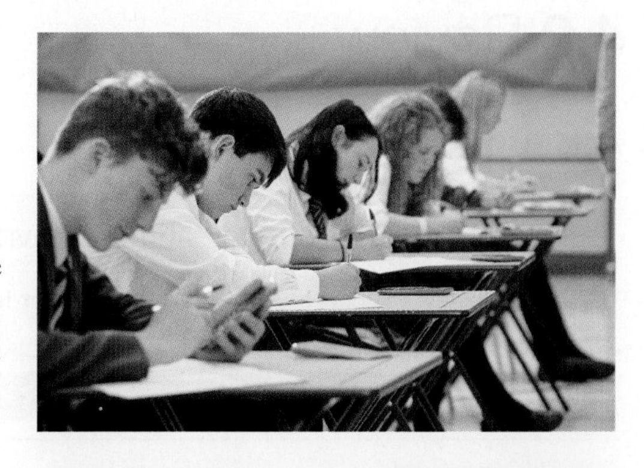

11. A Year 12 student buys some fruit for the week. They choose to buy 6 bananas and 4 apples. The total cost was \$6.24 and they spent the same amount of money on bananas as they did on apples.
Calculate the cost of each banana and apple.

12. A person goes out to buy some doughnuts and coffees to share at work. If they buy 5 coffees and 8 doughnuts, it would cost \$37.50, but if they had to buy 6 coffees and 10 doughnuts it would cost \$45.50.
Determine the cost of one coffee and the cost of one doughnut.

13. Colleen goes out to buy milk and bread for the week. If she buys 3 cartons of milk and 4 loaves of bread it would cost \$26.40, but if she bought 4 cartons of milk and 5 loaves of bread it would cost \$33.80.
Determine the price of a loaf of bread and the price of a carton of milk.

14. A triangle is formed by the graph of the equations $x + y = 13$, $x + 2y = 9$ and $2x + y = 9$.
Determine the vertices of the triangle by using simultaneous equations to identify the coordinates of the points at which each pair of lines cross.

4.8 Review

4.8.1 Summary

doc-38012

Hey students! Now that it's time to revise this topic, go online to:

Review your results

Watch teacher-led videos

Practise questions with immediate feedback

Find all this and MORE in jacPLUS

4.8 Exercise

Multiple choice

1. MC For the equation $2(3x-7)=10$, the value of x is:
   A. $\frac{-11}{6}$
   B. 5
   C. $\frac{17}{5}$
   D. 4
   E. $\frac{-2}{3}$

2. MC An equation of motion is given by $v=u+at$. Changing the subject of the formula to t gives:
   A. $t=\frac{u-v}{a}$
   B. $t=\frac{v-u}{a}$
   C. $t=\frac{v}{a}-u$
   D. $t=u-\frac{v}{a}$
   E. $t=a(v-u)$

3. MC The area of a trapezium is given by the formula $A=\frac{1}{2}(a+b)h$. If $a=3, b=7$ and $h=4$, A is:
   A. 40
   B. 20
   C. 24.5
   D. 37.5
   E. 30

4. MC Declan the electrician charges a call-out fee of \$80 and then \$65 per hour on top of that call-out fee. If Declan works n hours, the total cost equation is:
   A. $C=15n+80$
   B. $C=80n+65$
   C. $C=80n$
   D. $C=65n+80$
   E. $C=65n$

5. **MC** The graph shown plots the simultaneous equations $y = 2x - 1$ and $y = -2x + 7$. The solution to the simultaneous equations is:

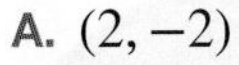

A. $(2, -2)$
B. $(3, 2)$
C. $(2, 3)$
D. $(2, 7)$
E. $(2, 2)$

6. **MC** The solution to the simultaneous equations $y = 5x$ and $y = 3x - 4$ using substitution is:

A. $(5, 3)$ B. $(-1, 5)$ C. $(1, -7)$
D. $(-1, 5)$ E. $(-2, -10)$

7. **MC** To eliminate x in the simultaneous equations $3y - 2x = 5$ and $2y - 2x = -3$, you need to:

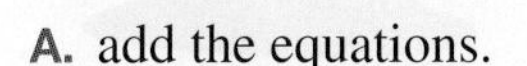

A. add the equations. B. multiply the equations.
C. divide the equations. D. subtract the equations.
E. square the equations.

8. **MC** The solution to the simultaneous equations $3y - 5x = 7$ and $-3y + x = 1$ using elimination is:

A. $(2, -1)$ B. $(-2, -1)$ C. $(-2, 1)$
D. $(3, -5)$ E. $(5, -3)$

9. **MC** The solution to the simultaneous equations $2y - 3x = 12$ and $y = \frac{x}{2}$ is:

A. $(-6, -3)$ B. $(6, -3)$ C. $(-6, 3)$ D. $(-3, -6)$ E. $(3, 6)$

10. **MC** Which of the following pairs of simultaneous equations have a solution of $(-4, 5)$:

A. $y - 3x = 8$ and $y = x$ B. $3y + 2x = 7$ and $-2y + x = -14$
C. $2y - 3x = 8$ and $-y + 2x = -6$ D. $2y - 2x = 14$ and $y - 2x = 13$
E. $2y - 3x = 22$ and $3x - 2y = 2$

Short answer

11. If $m = 3$ and $n = 5$, evaluate the following expressions.

a. $m + n$ b. $4m + 2n$ c. $6(m + 4)$

12. If $m = 3$ and $n = 5$, evaluate the following expressions.

a. $4m^2 - 2n$ b. $\frac{m + n}{2}$ c. $\sqrt{m^2 + n^2 + (n - m)}$

13. Use inverse operations to solve the following equations.

a. $4a = 24$ b. $2d - 6 = 14$ c. $\frac{e + 7}{2} = 15$ d. $3(f - 2) = 24$

14. The cost of taking a taxi (C) in dollars includes a flag fall of \$3.50 plus \$2 per km. The equation for this is $C = 2k + 3.50$, where k is the distance travelled in km.
Calculate how much it would cost to travel:

a. 22 km
b. 15.5 km.

Extended response

15. The surface area of a sphere, A, is given by the formula $A = 4\pi r^2$, where r is the radius of the sphere. A globe of Earth is approximately spherical, with a diameter of 45 cm.
    a. State the radius of the globe.
    b. Calculate the surface area of the globe, giving your answer to the nearest 100.
    c. The globe is to be protected with a special clear paint finish. The cost of the special finish is \$5.67 per 100 cm^2 or part thereof. Calculate the cost, to the nearest 5 cents, of protecting the globe.

16. You have a summer job delivering pamphlets. You are paid \$22 and a further 12 cents per pamphlet delivered.
    a. Write an equation to represent the information.
    b. Sketch the graph of the equation.
    c. Calculate what you would earn if you delivered 1150 pamphlets.
17. Use the substitution method to solve each of the following simultaneous equations.
    a. $y = x - 2$
       $x + y = -6$
    b. $4x + 4y = 2$
       $-5x + y = 2$
18. Use the elimination method to solve each of the following simultaneous equations.
    a. $x - y = 2$
       $x + y = -6$
    b. $6x + 8y = 2$
       $3x - 2y = 4$
19. Determine the point of intersection for $2x + y = 1$ and $y = 4 + x$.
20. A pharmacy is having a sale on sun hats and sunscreen to promote skin care this summer. Two bottles of sunscreen and a hat cost \$33, and four bottles of sunscreen and three hats cost \$72.
    a. Write a system of equations that describe the situation.
    b. Solve the system by elimination to determine the cost of one bottle of sunscreen and a hat.

Answers

Topic 4 Linear and simultaneous equations

4.2 Algebraic substitution and transposition

4.2 Exercise

1. a. 24 b. 39 c. 45
   d. 69 e. 14 f. 49
2. a. 115 b. 58 c. 5
   d. $\frac{2}{3}$ e. 48 f. 5
3. a. 11 b. 23 c. 3
   d. 103 e. 56
4. 432
5. a. \$60 b. \$80 c. \$115 d. \$65
6. a. i. 14 cm ii. 36 cm
   b. i. 10 cm^2 ii. 45 cm^2
7. $n = \frac{T-a}{d}$
8. a. $A = \frac{C-D}{B}$ b. $t = \frac{v-u}{a}$
   c. $h = \frac{3V}{A}$
9. a. $P = 2(l+w)$
   b. 72 cm
   c. i. $l = \frac{P-2w}{2}$ ii. 72 cm
10. a. $h = \frac{A-2\pi r^2}{2\pi r}$ b. 5.5 cm
11. a. i. 360 km ii. 200 km
    b. i. 75 km/h ii. 96 km/h
    c. i. 8 hours
    ii. 10 hours and 40 minutes
12. a. 104 °F
    b. $C = \frac{5(F-32)}{9}$
    c. Seattle: 9 °C
    Denver: −4 °C
    San Diego: 19 °C
    New York: −2 °C
    Orlando: 23 °C
13. a. Venue A: \$100 Venue B: \$50
    b. Venue A: \$5 Venue B: \$5.75
    c. Venue B
14. a. 11.925 cm b. 7103.36 cm^3
15. a. 12.30 metres
    b. i. 87.48 metres ii. 262.45 metres
    c. When μ (the coefficient of friction between the wheels and the road) is smaller, the stopping distance at the same speed is a lot longer.

4.3 Linear equations

4.3 Exercise

1. a. $a = 2$ b. $b = 18$
   c. $c = 5.75$ d. $d = \frac{166}{7}$
2. a. $a = 9$ b. $b = 52$ c. $c = 16$
   d. $f = 4$ e. $g = 4$ f. $u = 12$
3. a. Subtract 9.
   b. Remove the brackets and simplify.
   c. Subtract 11.
   d. Multiply by 11.
4. a. $x = 3$ b. $p = 3$ c. $a = 6$
   d. $a = 12$ e. $c = 44$ f. $d = -1$
5. a. $a = 1$ b. $b = -2$
   c. $c = 3$ d. $d = -9$
6. a. $k = -4$ b. $g = -6$ c. $r = -3$
   d. $w = 5$ e. $q = -2$ f. $v = 13$
7. 10, 15 and 20 years of age
8. The son is 6 years old and his mother is 30 years old.
9. 23 metres long and 17 metres wide
10. Bag: \$62.50; shoes: \$112.50
11. Rectangle: 4.2 cm, 5.6 cm
    Triangle: 4.2 cm, 5.2 cm, 10.2 cm
12. a. i. $3x + 12$, where your friend has x marker pens
    ii. $4x + 5$
    b. $3x + 12 = 4x + 5$
    c. 7 marker pens
13. 12
14. Your friend walked 4 laps, you walked 9 laps and your sister walked 19 laps.
15. a. $x = 36$
    b. Dimensions of first rectangle: 42 cm by 12 cm
    Dimensions of second rectangle: 72 cm by 7 cm
    c. 504 cm^2
16. a. 6 b. 15 c. 26, 27, 28
17. a. $C = 50 + 15n$, where C = cost of hiring, n = hours hired
    b. \$140
    c. $100 = 50 + 15n$
    d. 3 hours
    e. \$5
18. a. 5 rides
    b. Option 2: it costs \$12 for the two rides, which is \$3 cheaper than option 1.

4.4 Linear modelling

4.4 Exercise

1. The response variable depends on the values of the explanatory variable.
2. B
3. a. $T = 30 + 12b$, where T = total number of TVs and b = number of containers

b.

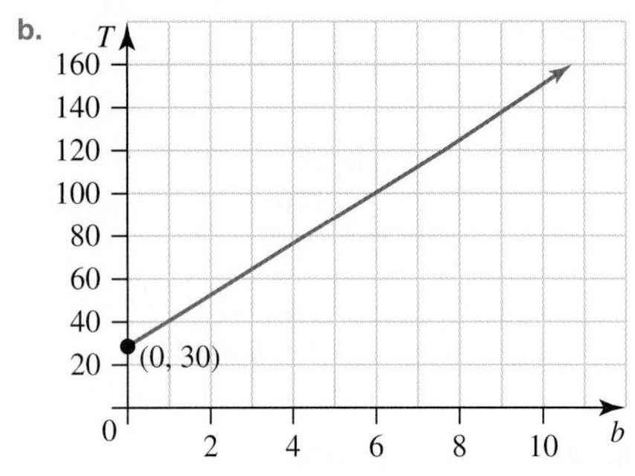

c. The y-intercept is the initial number of TV sets; the gradient is the number of TVs in each container.

4. a. The slope (gradient) means that the friend jogs at 6 km per hour; the y-intercept means he started at 0.

b. Slower

5. a. $y = 0.2x + 35$
$y = 0.3x + 25$

b. Company A = \$45
Company B = \$40
Company B is cheaper for travelling 50 km.

c.

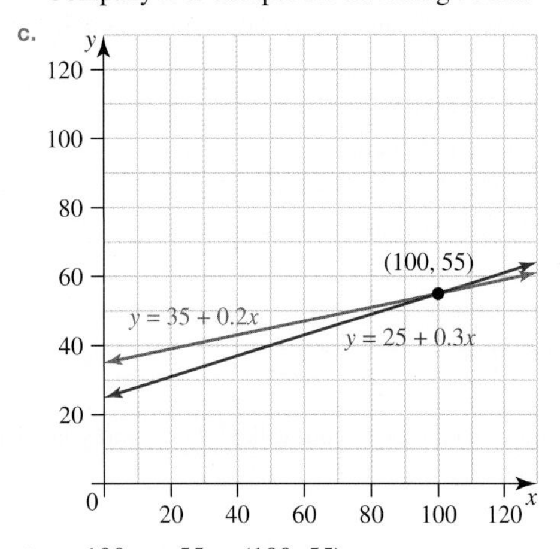

d. $x = 100$, $y = 55$ or (100, 55)

e. If you are driving over 100 km a day

6. a. $C = 0.3s + 25$

b. \$40

c.

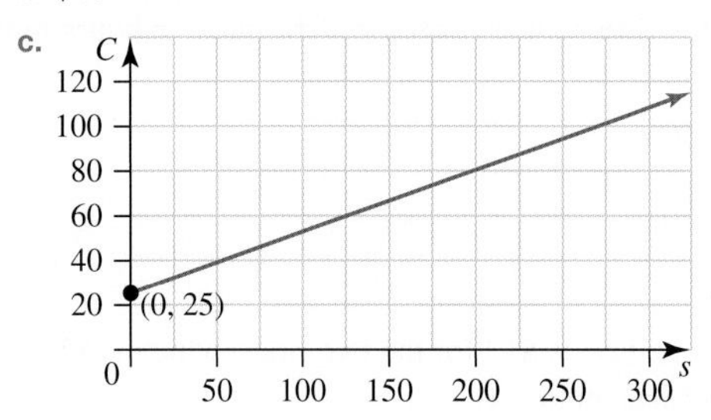

d. 270 MB

e. Primeus offers a better deal when customers don't go over their plan or use less than 50 MB more than their limit. Over 50 MB, Optican is the better deal.

7. Let the friend's age $= x$ and her sister's age $= y$.
Multiplying the friend's age by 5 gives $5x$; trebling the sister's age gives $3y$.
The total of the ages is $5x + 3y = 83$.
There are a number of possible integer answers.

Friend's age, x	1	4	7	10	13	16
Friend's sister's age, y	26	21	16	11	6	1

8. a. $x + 3 = 5$ b. $x - 9 = 7$ c. $7x = 24$

d. $\frac{x}{5} = 11$ e. $\frac{x}{2} = -9$ f. $5x - 3 = -7$

9. a. $A = 25t + 60$ b. \$5060 c. \$60

10. a. $C = 0.55s + 2.0$ b. \$8.60

11. a. $W = -40t + 712$

b. 712 L

c. 18 days

12. 12 jackets

13. 15 days

14. 947 cupcakes

15. a. Post Quick (cost = \$700)

b. The cost is nearly the same for 333 flyers (\$366.50 and \$366.40).

4.5 Solving simultaneous equations graphically

4.5 Exercise

1. Simultaneous equations are two or more equations that occur at the same time and generally intersect one another.

2. a. $x - y = 1$, $x + y = 3$; (2, 1)

b. $x + y = 2$, $3x - y = 2$; (1, 1)

c. $y - x = 4$, $3x + 2y = 8$; (0, 4)

d. $y + 2x = 3$, $2y + x = 0$; $(2, -1)$

e. $y - 3x = 2$, $x - y = 2$; $(-2, -4)$

f. $2y - 4x = 5$, $4y + 2x = 5$; $\left(-\frac{1}{2}, 1\frac{1}{2}\right)$

3. a.

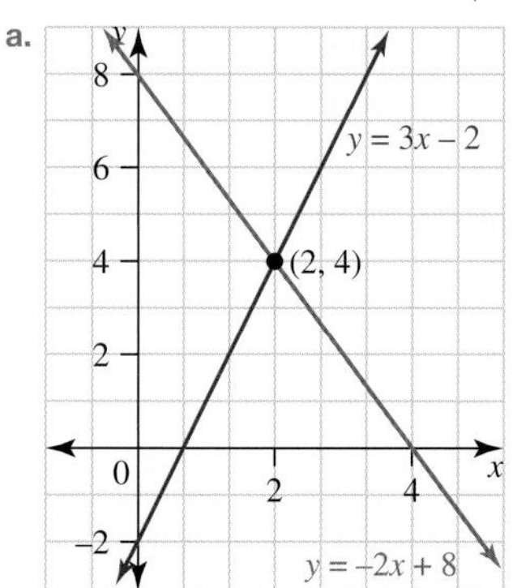

b.

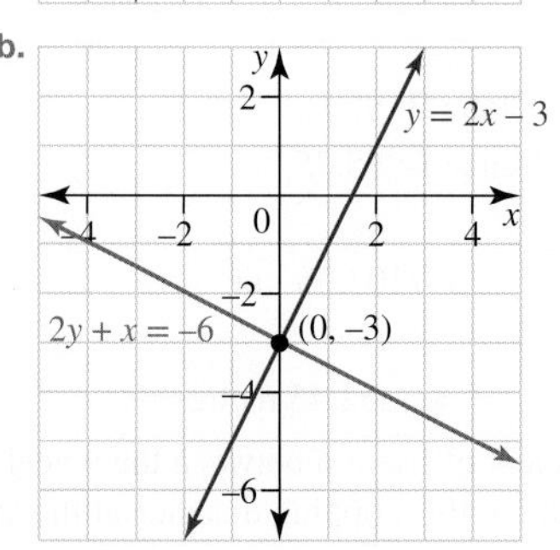

4. a.

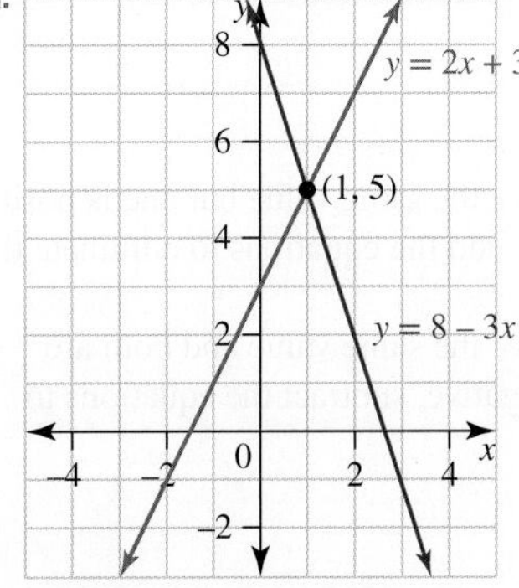

Intersection point $(1, 5)$

b.

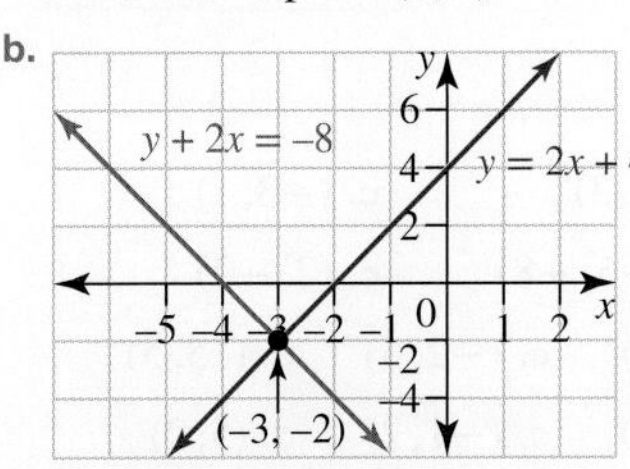

Intersection point $(-3, -2)$

c.

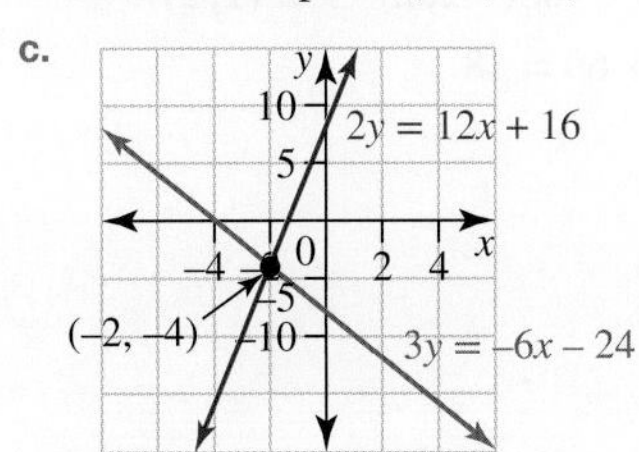

Intersection point $(-2, -4)$

d.

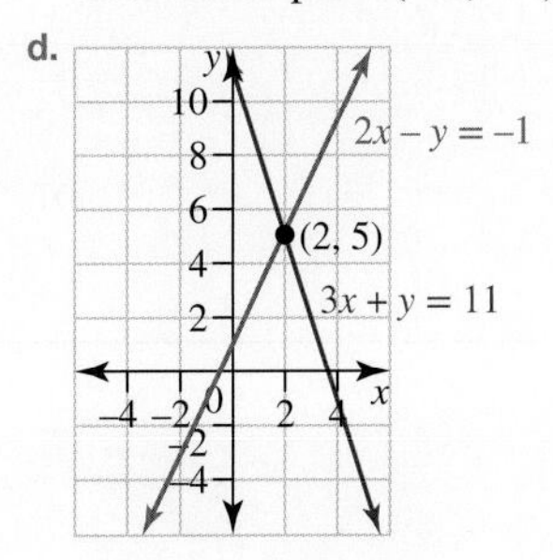

Intersection point $(2, 5)$

5. a.

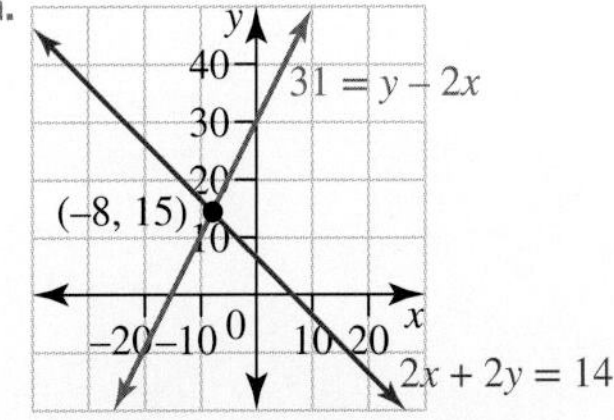

Intersection point $(-8, 15)$

b.

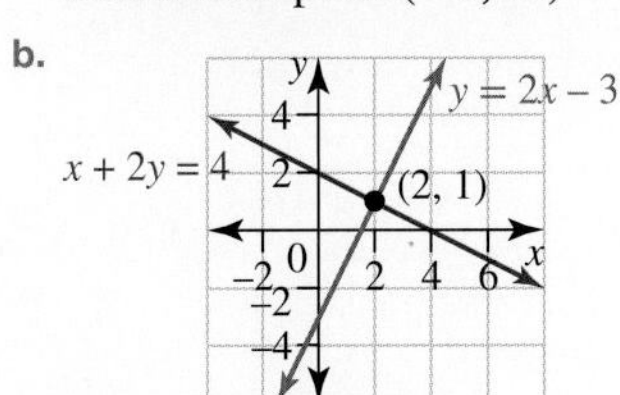

Intersection point $(2, 1)$

c.

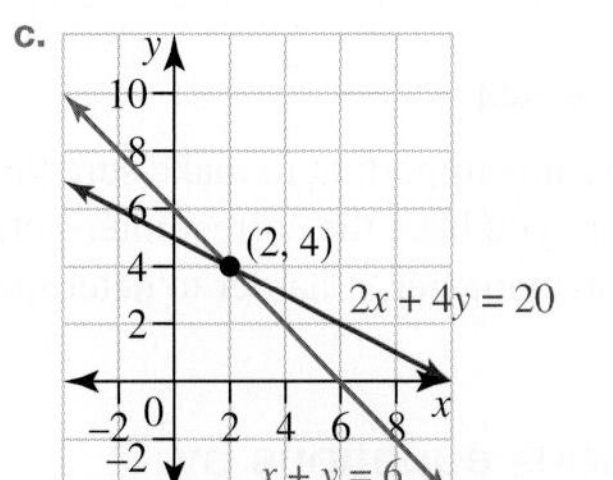

Intersection point $(2, 4)$

d.

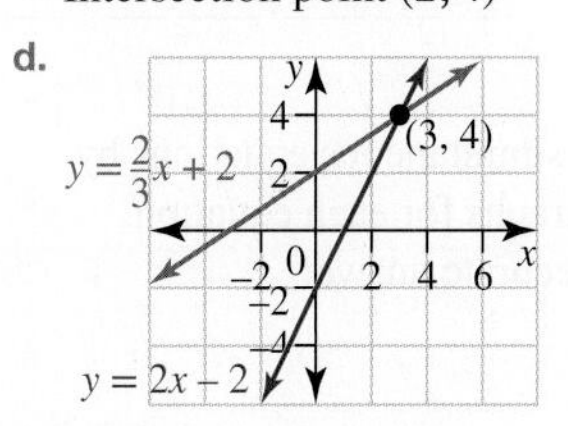

Intersection point $(3, 4)$

6. $(5, 5)$

7. $(0, 6)$

8. a. $m_1 = 2$, $m_2 = 2$; parallel lines
   b. $m_1 = -3$, $m_2 = -3$; parallel lines
   c. $m_1 = -\frac{2}{3}$, $m_2 = -\frac{2}{3}$; parallel lines

9. a. $m_1 = -2$, $m_2 = \frac{1}{2}$; perpendicular lines
   b. $m_1 = 4$, $m_2 = \frac{15}{2}$; neither parallel nor perpendicular lines
   c. $m_1 = -\frac{1}{4}$, $m_2 = 4$; perpendicular lines

10. a. $m = 2$, $c = -4$
    b. i. Answers will vary. An example is $2y = 4x - 8$.
       ii. Answers will vary. An example is $y = 2x - 2$.

11. a. The given point is not a solution.
    b. The given point is a solution.
    c. The given point is a solution.
    d. The given point is not a solution.

12. a. Northern beach:

$$C = 20 + 12t,\ 0 \le t \le 5$$

Southern beach:

$$D = 8 + 18t,\ 0 \le t \le 5$$

b. Northern beaches in red; southern beaches in blue

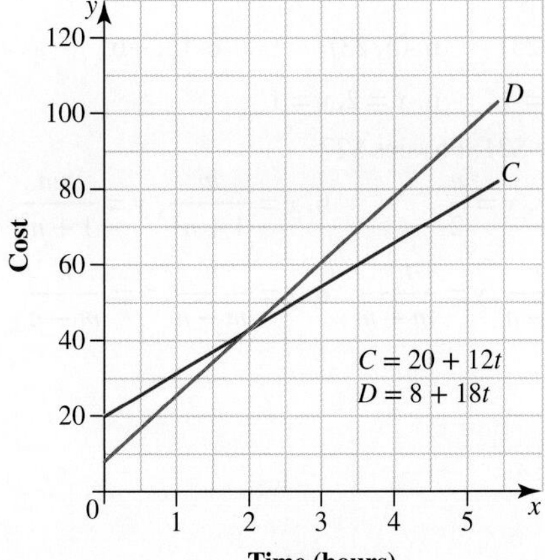

c. Time > 2 hours

d. Time = 2 hours; cost = \$44

13. When solving graphically it is important to make sure your graph is accurate to ensure you have the correct intersection point. If the answer is not an integer, it harder to determine the exact value.

4.6 Solving simultaneous equations by substitution

4.6 Exercise

1. It is often quicker to solve simultaneous equations by substitution than to draw graphs for each equation. Substitution will give an accurate answer.
2. $(1, 5)$
3. $(-1, 2)$
4. a. $(7, 3)$ b. $\left(\frac{3}{2}, 0\right)$
   c. $(1, -1)$ d. $(1, -1)$
5. a. $\left(-\frac{4}{3}, -\frac{23}{3}\right)$ b. $\left(\frac{3}{2}, \frac{5}{2}\right)$
   c. $(-2, -5)$ d. $(6, -1)$
6. a. There is no solution.
   b. The two equations have the same gradient ($m = -1$) and y-intercept ($c = 3$).
   c.

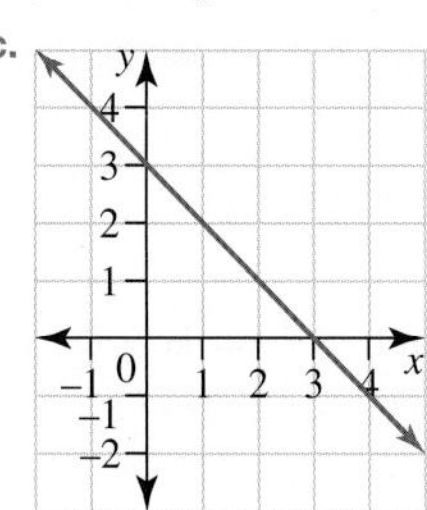

   d. The lines are coincident.
7. a. Same line b. Intersecting
   c. Intersecting d. Parallel
8. a. 1 solution
   b. No solution (parallel lines)
   c. No solution (parallel lines)
9. a. $(3, 1)$ b. $(-2, 3)$ c. $(5, 1)$
   d. $(4, 2)$ e. $(0, 3)$ f. $(4, 0)$
10. 26 chickens
11. a. $(-6, -23)$ b. $(5, 23)$ c. $(2, -6)$
12. a. $x + 2y = 4$ b. $x = 2, y = 1$
13. Chemistry \$21; Physics \$27
14. a. $x = \frac{n}{2m}, y = \frac{n}{2}$ b. $x = \frac{m}{1 + n^2}, y = \frac{mn}{1 + n^2}$
    c. $x = \frac{n}{m - n}, y = \frac{n^2}{m - n}$ d. $x = \frac{n}{m - n}, y = \frac{n}{m - n}$

4.7 Solving simultaneous equations by elimination

4.7 Exercise

1. If both coefficients have the same value but one is positive and the other negative, add the equations to eliminate the variable.
   If both coefficients have the same value and both are positive or both are negative, subtract the equations to eliminate the variable.
2. a. Add; eliminate y.
   b. Subtract; eliminate x.
   c. Subtract; eliminate x.
   d. Subtract; eliminate y.
3. $(3, -1)$
4. a. $(5, -1)$ b. $(2, 3)$ c. $(-3, 1)$
5. a. $(-3, 5)$ b. $(-5, -8)$ c. $(2, -2)$
6. a. $(5, -1)$ b. $(3, 2.5)$ c. $(-2, 3)$ d. $(15, 5)$
7. a. $(2, 2)$ b. $(-3, 2)$ c. $(-1, 2)$ d. $(5, 2)$
8. a. $(1, -1)$ b. $(4, -3)$ c. $(-2, 2)$ d. $(1, 2)$
9. a. $2a + 3b = 18, -2a + 6b = 18$
   b. $a = 3, b = 4$
10. Physics = 81%
    Chemistry 63%
11. Banana = \$0.52
    Apple = \$0.78
12. Donut = \$1.25
    Coffee = \$5.50
13. Bread = \$4.20
    Milk = \$3.20
14. $(3, 3), (-4, 17), (17, -4)$

4.8 Review

4.8 Exercise

Multiple choice

1. D
2. B
3. B
4. D
5. C
6. E
7. D
8. B
9. A
10. B

Short answer

11. a. 8 b. 22 c. 42

12. a. 26 b. 4 c. 6

13. a. 6 b. 10 c. 23
 d. 10 e. 5 f. 10

14. a. $47.50 b. $34.50

Extended response

15. a. 22.5 cm b. 6400 cm^2 c. $362.90

16. a. $P = 0.12d + 22$
 where P = pay and d = number of delivered pamphlets.

 b. 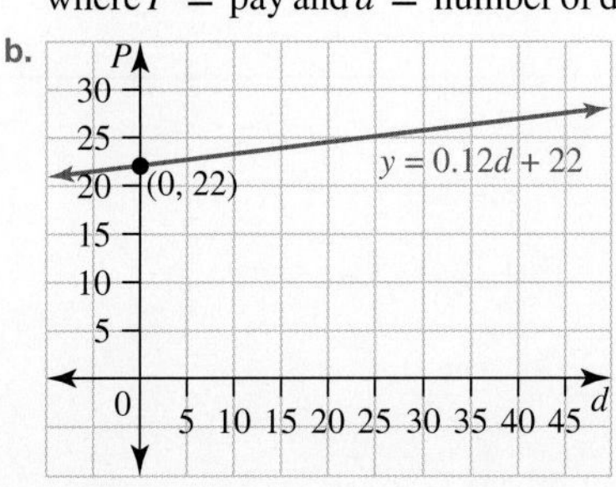

 c. $160

17. a. $(-2, -4)$ b. $\left(-\frac{1}{4}, \frac{3}{4}\right)$

18. a. $(-2, -4)$ b. $\left(1, -\frac{1}{2}\right)$

19. $(-1, 3)$

20. a. S = cost of a sunscreen bottle
 H = cost of a hat
 $2S + H = 33$
 $4S + 3H = 72$

 b. 1 sunscreen = $13.50
 1 hat = $6

5 Data collection and organisation

LEARNING SEQUENCE

Fully worked solutions for this topic are available online.

5.1 Overview

Hey students! Bring these pages to life online

Watch videos

Engage with interactivities

Answer questions and check results

Find all this and MORE in jacPLUS

5.1.1 Introduction

Data is an integral part of our daily lives. It's used to both inform us and inform others about us.

Our purchasing decisions, both online and instore using loyalty cards, can be used to influence our future consumer choices. On social media, the people we follow or the articles we read provide information about our personal lives — for example, our political allegiances or our entertainment and travel interests. This allows companies and other groups to target us with more of the same. Mobile phones can be used to track the movements of people and assist law enforcement to solve crimes. Education departments, both at state and federal level, collect information about school attendance, NAPLAN results, Year 12 results, school retention rates and much more. This information can be used to determine the needs of students and schools and to compare students nationally and internationally.

The media provides us with information determined from data collections about our health, employment, housing, public transport, road usage, the stock market, politics and many other issues that are important to us. The Bureau of Meteorology collects vast amounts of weather data to make accurate forecasts, and is used on a daily basis by millions of Australians to plan our daily activities. Governments and companies use data to meet current needs and plan for the future. The power of technology to collect vast amounts of data has had a significant impact on sport. Coaching and training methods for top athletes and sporting teams use data to improve individual performance and game plans. An understanding of the power and purpose of data enables us to make informed decisions about many aspects of our lives.

In this chapter we will investigate data collection and data representation. The representation of data through tables or graphs is a visual way to display the story the data is telling. This chapter will develop your skills in the collection and appropriate representation of data through tables, graphs and the use of technology.

KEY CONCEPTS

This topic covers the following key concepts from the VCE Mathematics Study Design:

- development and specification of data collection requirements and methods, including consideration of audience and purpose of data collection, errors and misrepresentations in statistics
- collection and modelling of data, including the construction of tables or spreadsheets and graphs to represent data and correct representations
- contemporary representations of data and graphs derived from technology including reviewing appropriateness of graphical representations, including pictograms, bubble, Mekko, radar, sunburst, heat map and stacked area charts.

5.2 Types of data

LEARNING INTENTION

At the end of this subtopic you should be able to:
- classify types of data.

5.2.1 Types of data

Before we collect any data, we need to understand the various types of data that can be gathered. The diagram shown distinguishes these types.

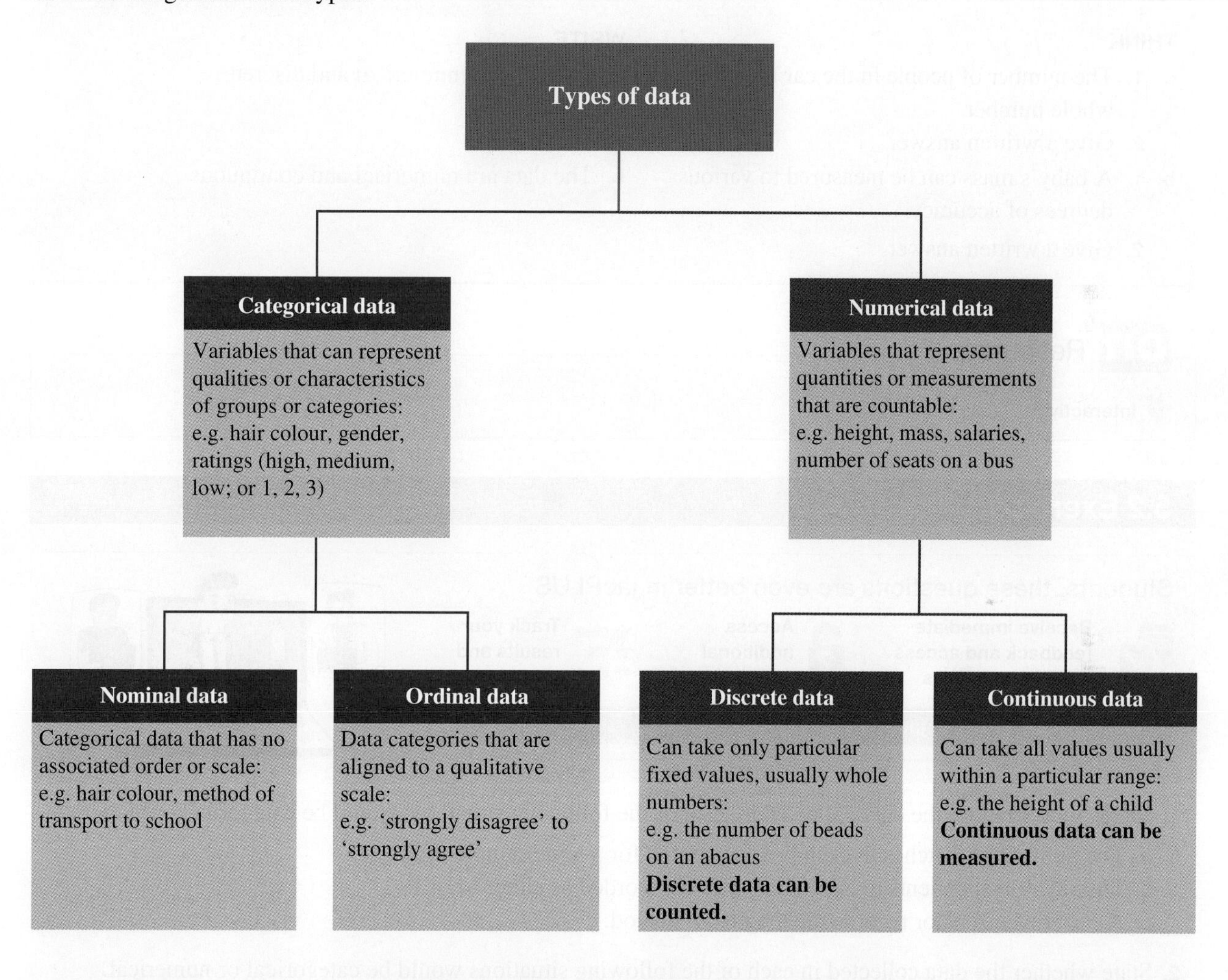

WORKED EXAMPLE 1 Determining whether data is categorical or numerical

State whether the following examples of data are categorical or numerical.
a. The value of sales recorded at each branch of a fast-food outlet
b. The breeds of dog at a show

THINK	WRITE
a. The value of sales at each branch can be measured.	a. The value of sales is numerical data.
b. The breeds of dog at a show cannot be measured.	b. The breeds of dog are categorical data.

WORKED EXAMPLE 2 Determining whether numerical data is discrete or continuous

State whether each of the following records of numerical data is discrete or continuous.

a. The number of people in each car that passes through a tollgate

b. The mass of a baby at birth

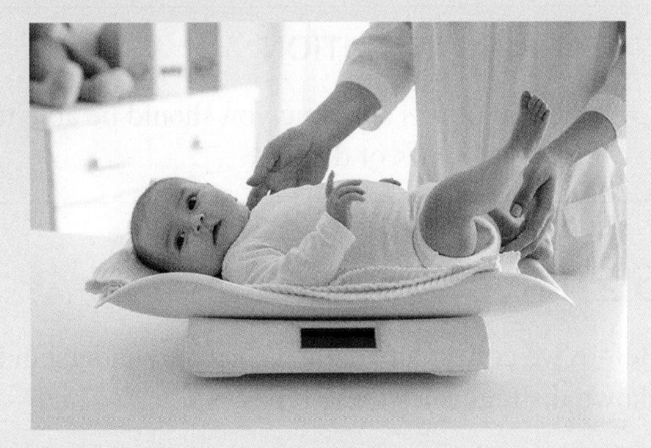

THINK	WRITE
a. 1. The number of people in the car must be a whole number.	
2. Give a written answer.	a. The data are numerical and discrete.
b. 1. A baby's mass can be measured to various degrees of accuracy.	
2. Give a written answer.	b. The data are numerical and continuous.

 Resources

 **Interactivity** Types of data (int-6086)

5.2 Exercise

1. WE1 State whether the data collected in each of the following situations would be categorical or numerical.
   a. The number of matches in each box is counted for a large sample of boxes.
   b. The sex of respondents to a questionnaire is recorded as either M or F.
   c. A fisheries inspector records the lengths of 40 cod.
2. State whether the data collected in each of the following situations would be categorical or numerical.
   a. The occurrence of hot, warm, mild and cool weather for each day in January is recorded.
   b. The actual temperature for each day in January is recorded.
   c. Cinema critics are asked to judge a film by awarding it a rating from one to five stars.
3. WE2 State whether the numerical data gathered in each of the following situations are discrete or continuous.
   a. The heights of 60 tomato plants at a plant nursery
   b. The number of jelly beans in each of 50 packets
   c. The time taken for each student in a class of six-year-olds to tie their shoelaces
   d. The petrol consumption rate of a large sample of cars
   e. The intelligence quotient (IQ) of each student in a class

4. For each of the following, state if the data are categorical or numerical. If numerical, state if they are discrete or continuous.
   a. The number of students in each class at your school
   b. The teams people support at a football match
   c. The brands of peanut butter sold at a supermarket
   d. The heights of people in your class
   e. The interest rate charged by each bank
   f. A person's heart rate
5. An opinion poll was conducted. A thousand people were given the statement 'Euthanasia should be legalised'. Each person was offered five responses: strongly agree, agree, unsure, disagree and strongly disagree. Describe the data type in this example.
6. A teacher marks her students' work with the grade A, B, C, D, or E. Describe the data type used.
7. A teacher marks his students' work using a mark out of 100. Describe the data type used.
8. **MC** The number of people who are using a particular bus service are counted over a two-week period. The data formed by this survey would best be described as:
   A. categorical data.
   B. numerical and discrete data.
   C. numerical and continuous data.
   D. quantitative data.
   E. qualitative data.
9. The graph shows the number of days of each weather type for Melbourne in January.

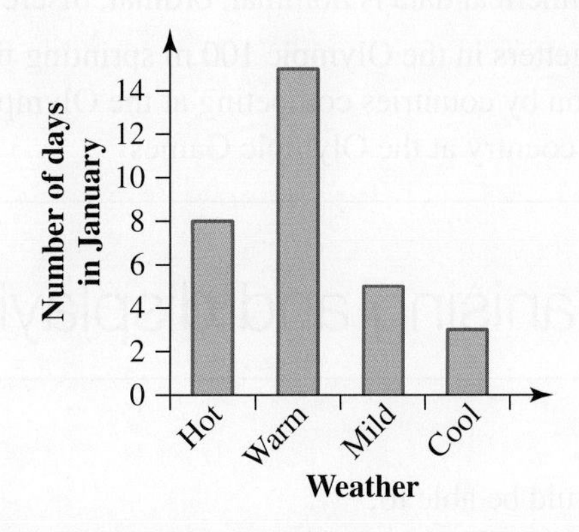

   Describe the data in this example.
10. The graph shows a girl's height each year for 10 years.

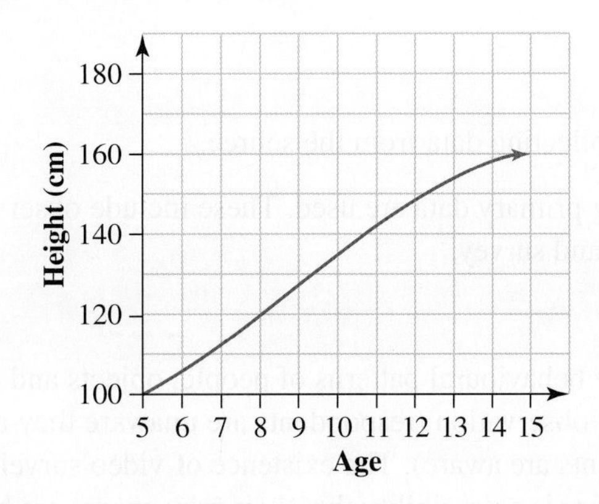

   Describe the data in this example.

11. Data on the different types of cereal on supermarket shelves is collected.
Verify that the collected data is categorical, and determine whether it is ordinal or nominal.

12. Data on the rating of hotels from 'one star' to 'five star' is collected.
Verify that the collected data is categorical, and determine whether it is ordinal or nominal.
13. Identify whether the following numerical data is discrete or continuous.
    a. The amount of daily rainfall in Geelong
    b. The heights of players in the National Basketball League
    c. The number of children in families
14. Identify whether the following numerical data is nominal, ordinal, discrete or continuous.
    a. The times taken for the place getters in the Olympic 100 m sprinting final
    b. The number of gold medals won by countries competing at the Olympic Games
    c. The types of medals won by a country at the Olympic Games

5.3 Collecting, organising and displaying data

LEARNING INTENTION

At the end of this subtopic you should be able to:
- consider the development of data collection requirements and represent data through the construction of tables, spreadsheets and graphs.

5.3.1 Data collection

Primary data collection involves collecting data from the source.

A variety of methods for collecting primary data are used. These include observation, digital footprints, measurement, experiment, census and survey.

Observation

Observation involves recording the behavioural patterns of people, objects and events in a systematic manner. Data can be collected by disguised observation (respondents are unaware they are being observed) or undisguised observation (respondents are aware). The existence of video surveillance cameras, for example, can mean that people know that there is a possibility that their movements are being recorded, but they are not always aware of when the recording takes place.

Observations can be in a natural environment (for example, small children may be observed at play to assist in understanding their cognitive development) or a contrived environment (for example, at a food-tasting session for a food company). Mechanical devices (for example video cameras, closed circuit television or counting devices across a road) can also be used.

Digital footprint

Vast amounts of data are collected through our **digital footprints**. Search engines, emails, social media websites, online shopping websites, apps on our smart phones and loyalty cards are just examples of a whole range of means by which data collection agencies are able to learn about individuals. Each click of a mouse or tap on a tablet adds to the trillions of pieces of data that are being generated around the world every second, all of which can potentially be mined and analysed.

Measurement

Measurement involves using a measuring device to collect data. This generally involves conducting an experiment of some type.

- The height of everyone in your class can be measured.
- The mass of all newborn babies can be collected.
- A pedometer measures the number of steps the wearer takes.

Experiment

Generally, when an experiment is conducted, the data collected are quantitative. Particular care should be taken to ensure that the experiment is conducted in a manner that would produce similar results if repeated. Care must be taken with the recording of results, and those results must be in a form that can readily be analysed.

All results need to be recorded, including any weird or unexpected outcomes.

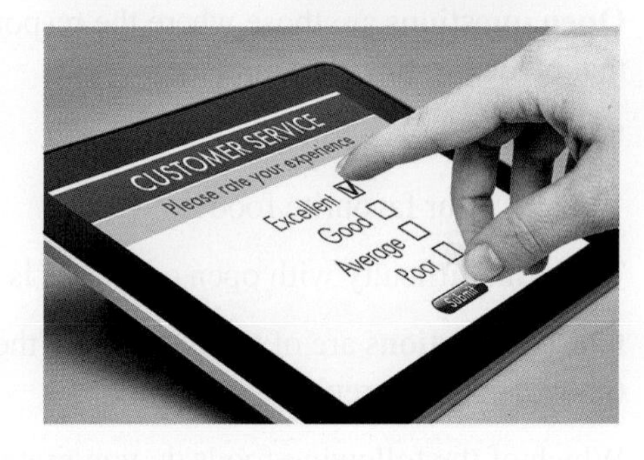

Census

A census is conducted by systematic collection of data from an entire population. Every five years, the Australian Bureau of Statistics conducts a census. In this census, data is collected from every household in Australia on a designated evening. It is called the *Census of Population and Housing* and not only includes households, but also motels, caravan parks and campgrounds. It provides a snapshot of Australia's population on that date, but can be used to compare changes to the population from the past and to predict possible changes in the future.

A census is different from a survey because a survey takes a sample of a population, whereas a census uses the entire population.

Survey

A survey is a questionnaire in which people are interviewed. Often a survey needs to be rewritten many times to ensure it is clear and unbiased.

The interview can be in person (face to face), by video call or by telephone. Advantages of these methods are that the interviewer is able to gauge the reactions of the respondents, and that they can explain particular questions if necessary.

Email can also be used to survey participants. There are advantages and disadvantages to using this type of survey.

Advantages:
- Email can cover a large number of people or organisations.
- Wide geographic coverage is possible.
- It avoids embarrassment on the part of the respondent.
- Interviewer bias is avoided.
- The respondent has time to consider responses.
- It is relatively cheap.

Disadvantages:
- The questions have to be relatively simple.
- The response rate is often quite low (and inducements are often given as an incentive to answer).
- The reliability of the answers is questionable.
- There is no control over who actually completes the questionnaire.
- The returned questionnaires might be incomplete.

Questionnaires

As mentioned previously, it is important to take care in designing the questionnaire in order to obtain relevant and reliable data. In formulating the questions, we must also keep in mind that we require the data collected to be in a form that is not difficult to analyse.

The questions posed can be either open or closed.

Open questions are those where the respondent has no guided boundaries within which to answer. Questions that belong to this class include:

'Who is your favourite singer?'

'What is your favourite food?'

The main difficulty with open questions is that they are often difficult to classify and analyse.

Closed questions are of the type where the respondent must answer within a category. The question about food (above) could be rephrased as:

Which of the following foods do you prefer?
☐ Meat
☐ Seafood
☐ Poultry
☐ Vegetables
☐ Fruit

Analysis of these answers would be easier than trying to fit the open-ended answers into a category.

Avoid options such as:
☐ None of the above
☐ Don't know

Such options provide the respondents with an excuse not to answer.

tlvd-4060

WORKED EXAMPLE 3 Classifying survey responses

Twenty people answered the open question:
'What do you particularly *dislike* about the way the news is presented on TV?'
They provided the following responses.

1. **The ads interrupt too often.**
2. **Have to wait too long for the headline items.**
3. **The violent scenes are too graphic.**
4. **It was better when it was only half an hour instead of a full hour.**
5. **There's too much violence.**
6. **The reporters are politically biased.**
7. **The accident scenes are not sensitively handled.**
8. **Some of the reports are too long.**
9. **It's all about politics.**
10. **It mixes up local, interstate and overseas news.**
11. **I'd like it to be shorter.**
12. **There are more advertisements than news.**
13. **It seems to concentrate on murder and death.**
14. **I find it far too long.**
15. **The newsreaders don't pronounce names correctly.**
16. **It's too informal.**
17. **The newsreader is far too old.**
18. **I don't like ads interrupting the news.**
19. **Some reports show too much blood and gore.**
20. **The interviews by the reporters are too long.**

Classify the responses into categories to identify the main reasons given.

THINK	WRITE
1. Look for four or five categories under which they could be classified.	Four main categories are apparent: 1. Length 2. Violence 3. Advertisements 4. Newsreader
2. Classify the responses under these headings.	Length: Responses No. 4, 8, 11, 14, 20 Violence: Responses No. 3, 5, 7, 13, 19 Advertisements: Responses No. 1, 12, 18 Newsreader: Responses No. 15, 16, 17
3. Identify those not classified.	Those not classified are: No. 2. Too long for headline items No. 6. Reporters politically biased No. 9. All about politics No. 10. Mixes up news
4. Identify the main reasons.	The main reasons for people disliking the TV news presentations seem to be centred around the length of the news and the violence portrayed.

Note: It is possible to classify these responses into different categories. No single way is correct. In practice, the responses to open-ended questions are generally classified in several ways until the researcher is satisfied with the classifications.

Preparing a questionnaire

In preparing good questions, it is advisable to keep the following points in mind:

1. The questions should flow smoothly from one to the next.
2. Include introductory remarks outlining the aim and purpose of the questionnaire, along with any necessary instructions.
3. Avoid jargon, slang and abbreviations.
4. Do not ask questions that are vague or ambiguous.
5. Avoid bias and emotional language.
6. Avoid double-barrelled questions.
7. Do not pose leading questions (that is, those that lead to an expected response).
8. Make sure your questions can be answered by your respondents.
9. Avoid questions with double negatives.
10. At the conclusion, thank the respondent for answering.

5.3.2 Organising data

Once data have been collected and checked for errors, they need to be put into an organised form. This involves tallying the responses to a questionnaire, accurately recording your observations or tabulating the results of your research.

This task is made easier if the questionnaire is designed with ease of tabulation in mind. Usually the results are first organised into a table and the number of responses in each category recorded. This is often done with tally marks and using the gatepost method.

WORKED EXAMPLE 4 Creating a table of survey responses

A survey is conducted among 24 students who were asked to name their favourite spectator sport. Their responses are recorded below.

AFL	**Cricket**	**Cricket**	**Soccer**	**Rugby League**
Cricket	**Tennis**	**Cricket**	**AFL**	**Rugby League**
AFL	**AFL**	**Rugby Union**	**Soccer**	**Netball**
Basketball	**Basketball**	**Netball**	**AFL**	**Cricket**
Cricket	**AFL**	**Rugby League**	**Cricket**	

Put these results into a table.

THINK

Draw a table and beside each sport put a tally mark for each response. Every fifth tally mark becomes a gatepost.

WRITE

Sport	Tally	Frequency
AFL	~~IIII~~ I	6
Basketball	II	2
Cricket	~~IIII~~ II	7
Netball	II	2
Rugby League	III	3
Rugby Union	I	1
Soccer	II	2
Tennis	I	1

tlvd-4061

WORKED EXAMPLE 5 Creating a table of survey responses 2

A Year 12 class was surveyed on their weekly income. The responses are shown below.

\$75	**\$115**	**\$60**	**\$54**	**\$88**	**\$0**	**\$98**	**\$102**
\$56	**\$45**	**\$83**	**\$71**	**\$40**	**\$37**	**\$87**	**\$117**
\$43	**\$79**	**\$58**	**\$89**	**\$70**	**\$105**	**\$99**	**\$55**

Complete the table below.

Income	Tally	Frequency
0–20		
21–40		
41–60		
61–80		
81–100		
101–120		

THINK

Count the number of responses within each category and put a tally mark in the column.

WRITE

Income	Tally	Frequency
0–20	I	1
21–40	II	2
41–60	~~IIII~~ II	7
61–80	IIII	4
81–100	~~IIII~~ I	6
101–120	IIII	4

5.3.3 Displaying data

The most common way of displaying data is by using a graph. Graphs provide a visual illustration of relationships and patterns within a set of data. Different graphs have different purposes. We will now look briefly at column graphs and sector graphs, then look at histograms, stem plots and boxplots.

Column graphs

A **column graph** (or **bar graph**) is used when we wish to show a quantity. Categories are written on the horizontal axis and frequencies on the vertical axis.

tlvd-4062

WORKED EXAMPLE 6 Creating a column graph for survey responses

The table below shows the results of a survey on favourite sports. Show this information in a column graph.

Sport	Frequency
AFL	6
Basketball	2
Cricket	7
Netball	2
Rugby League	3
Rugby Union	1
Soccer	2
Tennis	1

THINK

1. Draw the horizontal axis showing each sport.
2. Draw a vertical axis to show frequencies up to 7.
3. Draw the columns. Make sure that they all have the same width, and that there are gaps between them.
4. Use a ruler.
5. Label the axes.
6. Give the graph a title.

WRITE

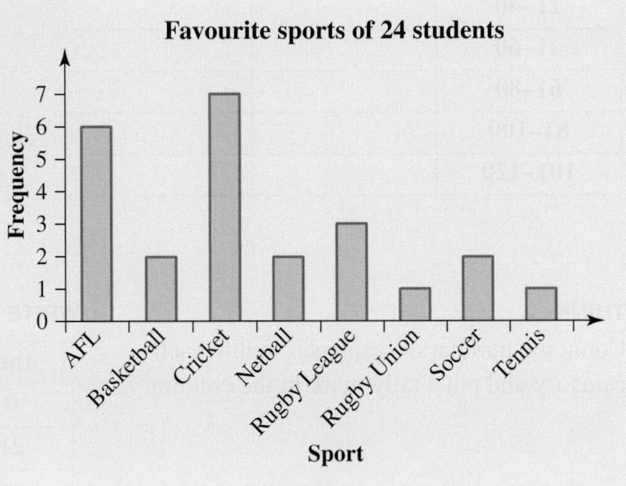

Sector graphs

A **sector graph** (also known as circle graph or pie graph) is used when we want the graph to display a comparison of quantities. An angle is drawn at the centre of the circle that is the same fraction of 360° as the fraction of people making each response.

tlvd-4063

WORKED EXAMPLE 7 Creating a sector graph for survey responses

Using the same table as in Worked example 6, draw a sector graph.

Sport	Frequency
AFL	6
Basketball	2
Cricket	7
Netball	2
Rugby League	3
Rugby Union	1
Soccer	2
Tennis	1

THINK

1. Calculate each result as a fraction of 360°.

WRITE

$\text{AFL} = \frac{6}{24} \times 360° = 90°$

$\text{Basketball} = \frac{2}{24} \times 360° = 30°$

$\text{Cricket} = \frac{7}{24} \times 360° = 105°$

$\text{Netball} = \frac{2}{24} \times 360° = 30°$

$\text{Soccer} = \frac{2}{24} \times 360° = 30°$

$\text{Tennis} = \frac{1}{24} \times 360° = 15°$

$\text{Rugby League} = \frac{3}{24} \times 360° = 45°$

$\text{Rugby Union} = \frac{1}{24} \times 360° = 15°$

2. Draw the graph.

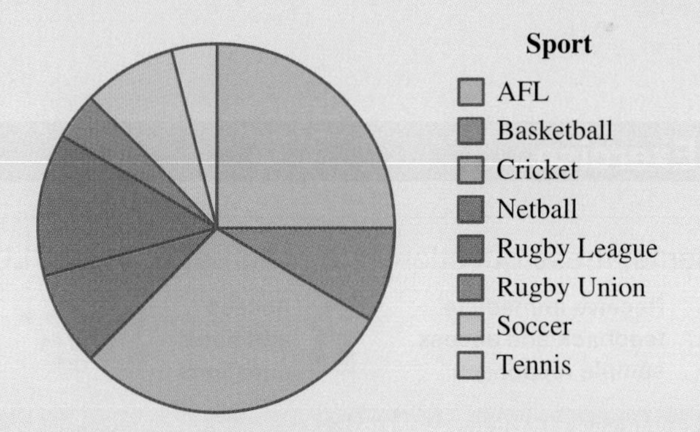

3. Label each sector or provide a legend.

These graphs can also be drawn using a spreadsheet.

5.3.4 Spreadsheets — Displaying numerical data

Worked examples 6 and 7 explained the calculations needed to manually display the same numerical data as a column graph and as a sector graph. In this activity, we look at entering the data into a spreadsheet, then exploring some of the graphical tools available. The instructions provided refer to the Excel spreadsheet.

1. Enter the heading 'Sports' and the categories of sport in Column A, as shown.
2. Enter the heading 'Frequency' and the relative frequencies in Column B, as shown.
   - Highlight the sports and the frequencies, then select the 'Insert' tab and the 'Column graph' option, and then select 'Clustered Column'. Follow the instructions, remembering to label the axes and title the graph, to produce the column graph shown.
   - Select the data again and follow the same process but select 'Pie graph' to produce the pie graph shown.

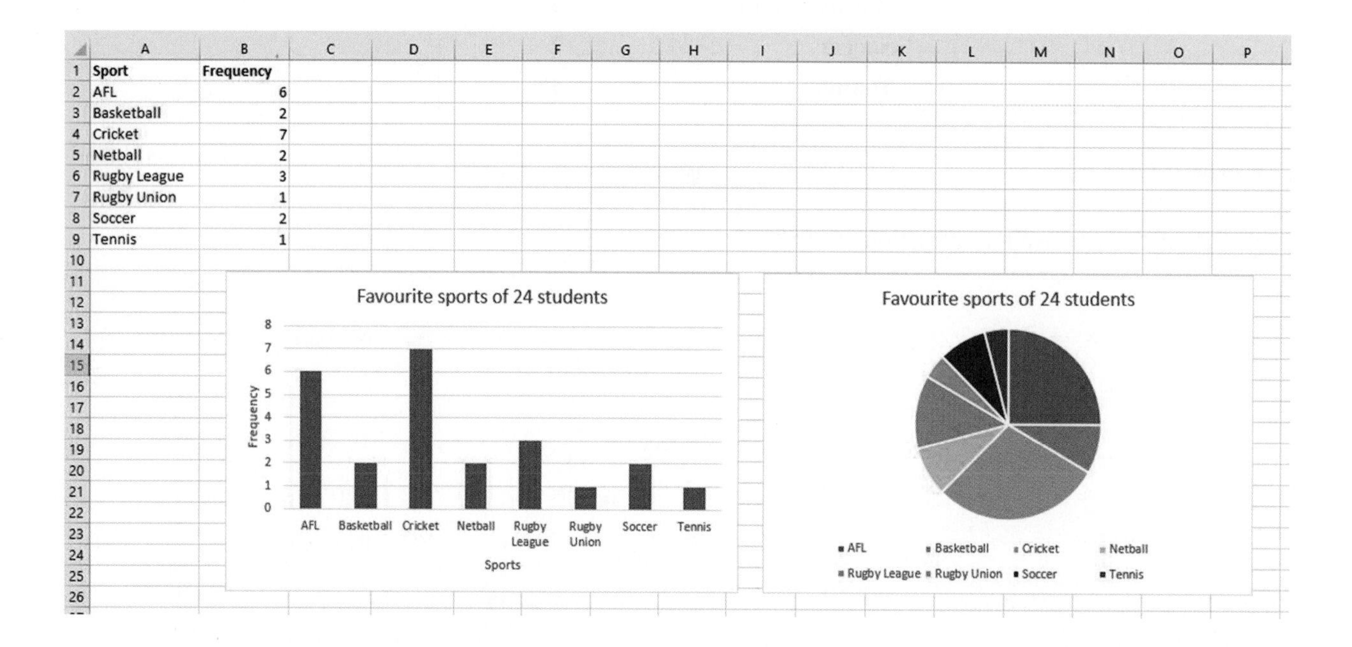

5.3 Exercise

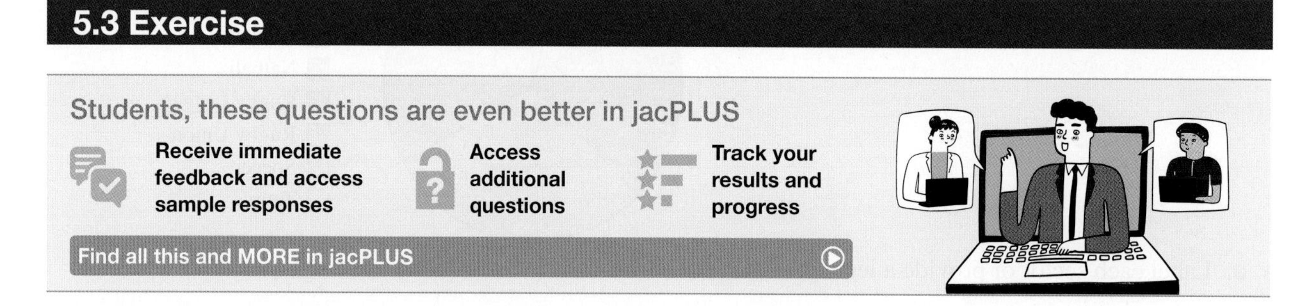

1. Explain what you understand by the terms *open* and *closed* questions. Give an example of an open question. Rewrite your question in a closed format.

2. **WE3** Twenty students were asked their opinions about the cause of congestion at the school's front gate. Analyse their responses below, suggest categories into which they could be classified and identify the most commonly stated reasons for the congestion.

   1. The cars shouldn't come up the front driveway.
   2. The front entrance is too small.
   3. There should be another entrance.

4. The buses are the problem.
5. The students get in the way of the cars.
6. Bike riders should have a separate entrance.
7. The senior school and the junior school should start and finish at different times.
8. The cars block the gate.
9. The bike riders don't know the road rules.
10. The buses should stop further down the road.
11. Too many students.
12. Parents don't care where they park.
13. The front gates should be wider.
14. Bike riders should go out the back gate.
15. Kids block the cars.
16. Kids just sit around talking there.
17. There should be a traffic control officer there to direct the traffic.
18. Students should not just sit around there.
19. The buses all arrive at the same time.
20. The road is too narrow.

3. Thirty students were asked to identify one thing in their Maths course that they particularly disliked. Classify the responses below into appropriate categories, then identify the main reasons.
    1. It's too hard.
    2. There's too much homework.
    3. I can't understand the teacher.
    4. It's boring.
    5. The boys are too distracting.
    6. The teacher doesn't like girls.
    7. It's too much work.
    8. We get homework every night.
    9. I can't understand it.
    10. The boys show off.
    11. The boys always get better marks.
    12. I can't concentrate for that length of time.
    13. We do something new every lesson.
    14. I don't like working in groups.
    15. The teacher expects too much.
    16. Our teacher is too strict.
    17. The teacher doesn't help us enough with our problems.
    18. There's too much work to cover.
    19. We're expected to do assignments.
    20. I don't like giving presentations to the class.
    21. Our class is too big.
    22. The work is not interesting.
    23. I just don't understand maths.
    24. I don't like the teacher.
    25. The teacher expects too much of us.
    26. There's too much work to cover.
    27. We're expected to remember too much.
    28. It won't help me later in life.
    29. The teacher picks on me because I don't understand the work.
    30. The course is not relevant.

4. Identify the areas of concern in the following questions, then rewrite each so that the meaning is clear and understandable.
   a. How much do you earn?
   b. Do you exercise regularly?
   c. Is the GST in Australia smaller than the VAT in England?
   d. Do you generally support the causes of murderous terrorists who threaten the lives of peace-loving people?
   e. Do you support the Prime Minister's policy on wildlife preservation?
   f. What is your height in inches?
   g. Did you buy your sneakers for comfort and quality?
   h. You don't agree with charging more for skim milk (where they've taken out the cream) than for full cream milk, do you?
   i. Do you agree that we should do more for our diggers, who risked their lives during the war so that we could be free?
5. Write the following open questions in closed format.
   a. What is your age?
   b. How much pocket money do you get each week?
   c. How do you travel to school?
   d. What type of destination do you prefer for a holiday?
6. Primary data from the students at a school is required to determine the internet access students have at home. The data collected is to provide support for opening the computer room for student use at night.
   a. State what data should be collected.
   b. Outline possible methods to collect the data.
   c. Decide which method you consider to be the best option and discuss its advantages and disadvantages.
7. **MC** Select the method most suitable to collect the primary data for the height of trees along a tree-lined street.
   A. Measurement
   B. Observation
   C. Survey
   D. Interview
   E. Experiment

8. **MC** Select the method most suitable to collect the primary data on student opinion regarding the length of lessons.
   A. Measurement
   B. Observation
   C. Survey
   D. Interview
   E. Experiment
9. **WE4** A class of students was asked to identify the make of car their family owned. Their responses are shown below.
   a. Put these results into a table.

Holden	Ford	Nissan	Mazda	Toyota	Holden
Ford	Holden	Ford	Mitsubishi	Toyota	Toyota
Nissan	Holden	Holden	Ford	Toyota	Mazda
Mazda	Toyota	Ford	Holden	Holden	Ford
Mitsubishi	Toyota	Holden	Ford	Ford	Toyota

   b. **WE6** Draw a column graph to display the data.
   c. **WE7** Draw a sector graph to display the data.

10. The results of a spelling test done by 30 students are shown below.

6	7	6	8	4	6	6	7	5	9
5	7	8	10	5	9	7	7	7	6
4	7	8	8	7	8	6	5	9	7

a. Put these results into a table.
b. Draw a column graph to display the data.

11. a. WE5 The marks scored on a Maths exam, out of 100, by 25 Year 12 students are shown below.

87	44	95	66	78	69	66	92	78
54	60	66	69	66	77	79	66	71
71	83	74	81	69	70	57		

Copy and complete the following table.

Mark	Tally	Frequency
40–49		
50–59		
60–69		
70–79		
80–89		
90–99		

b. Draw a column graph to display the data.
c. Draw a sector graph to compare the number of people in each category.

12. The data below show the number of customers who entered a shop each day in a certain month.
a. Choose suitable groupings to tabulate these data.

114	195	175	163	180	120	204	199
178	216	200	147	168	173	102	150
169	185	173	164	130	119	158	163
141	155	132	143	190	179	200	

b. Draw a column graph to display the data.

5.4 Histograms and frequency polygons

LEARNING INTENTION

At the end of this subtopic you should be able to:
- display data using frequency histograms and frequency polygons.

We shall now consider three other forms of graphical display of data: histograms, stem plots and boxplots.

5.4.1 Displaying data using frequency histograms

A **frequency histogram** is similar to a column graph with the following essential features.

1. Gaps are never left between the columns, except for a half-unit space before the first column.
2. If the chart is coloured or shaded, then it is done all in one colour. (The columns essentially all represent different levels of the same thing.)
3. Frequency is always plotted on the vertical axis.
4. For ungrouped data, the horizontal scale is marked so that the data labels appear under the centre of each column. For grouped data, the horizontal scale is marked so that the class centre of each class appears under the centre of the column.

tlvd-4064

WORKED EXAMPLE 8 Creating a frequency histogram from a table

The following table shows the number of people living in each house in a street. Show this information in a frequency histogram.

No. of people	Frequency
1	1
2	4
3	10
4	15
5	8

THINK

1. Draw a set of axes with the number of people living in a house on the horizontal axis and frequency on the vertical axis.
2. Draw the graph, leaving a half-column-width space before the first column.

WRITE

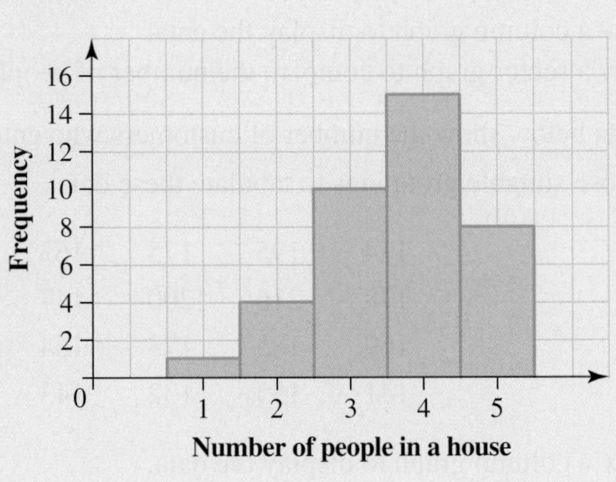

A **frequency polygon** is a line graph that can be drawn by joining the centres of the tops of each column of the histogram.

Frequency polygons are a graphical device for understanding the shapes of distributions. They serve the same purpose as histograms, but are especially helpful for comparing sets of data.

The **polygon** starts and finishes on the horizontal axis a half-column-width space from the group boundary of the first and last groups.

The figure shows the frequency polygon drawn on top of the histogram from Worked example 8.
It is common practice to draw the histogram and the polygon on the same set of axes.

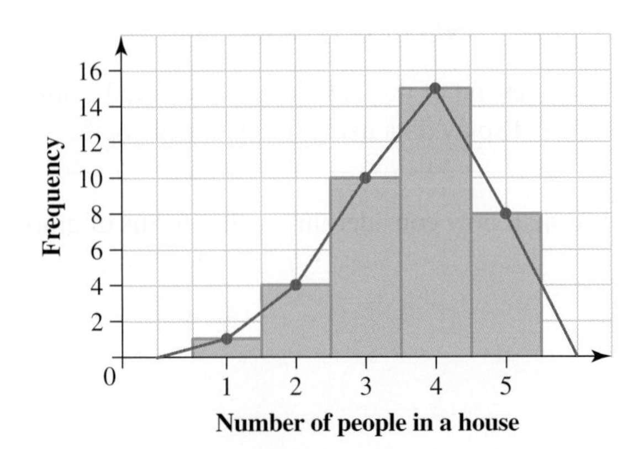

WORKED EXAMPLE 9 Creating a frequency histogram with a polygon from a table

The frequency table below shows a class set of marks on an exam. Draw a frequency histogram and polygon on the same set of axes.

Mark	Class centre	Frequency
51–60	55.5	3
61–70	65.5	5
71–80	75.5	12
81–90	85.5	7
91–100	95.5	3

THINK

1. Draw a set of axes with the exam mark on the horizontal axis and frequency on the vertical axis. Show the class centres for the exam marks.
2. Draw the columns, leaving a half-column-width space before the first column.
3. Draw a line graph to the centre of each column.
4. Make sure the line graph begins and ends on the horizontal axis.

WRITE

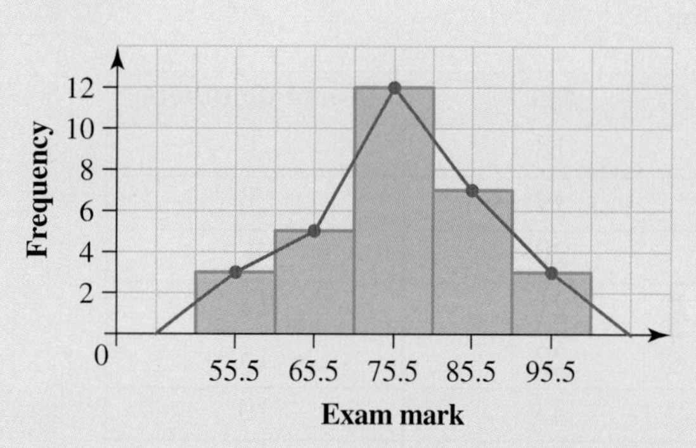

5.4 Exercise

Students, these questions are even better in jacPLUS

Receive immediate feedback and access sample responses

Access additional questions

Track your results and progress

Find all this and MORE in jacPLUS

1. WE8 A survey is done on young drivers taking the written test for their licence. The number of mistakes each makes is recorded and the results are shown in the following frequency distribution table. Show this information in a frequency histogram.

No. of mistakes (score)	No. of drivers (frequency)
0	5
1	8
2	11
3	4
4	3
5	1

2. Students in a class were asked how many children there were in their family. The results are shown in the following frequency distribution table. Show this information in a frequency histogram and polygon.

No. of children in a family	Frequency
0	3
2	5
3	8
4	4
5	2
6	1

3. The table below shows the age in years of the members of a surf club. Show this information in a frequency polygon.

Age	No. of members
18	3
19	5
20	8
21	13
22	15
23	10
24	8
25	5

4. The label on a box of matches states that on average each box contains 50 matches. Quality control surveyed 50 boxes for the number of matches. The results are shown below.

48	50	50	51	50	49	53	52	48	51
50	50	51	49	48	53	52	50	49	49
49	50	50	51	53	52	54	47	50	49
48	49	47	53	49	52	50	51	50	50
50	48	47	50	51	49	50	49	52	51

a. Put this information into a frequency table.
b. Show the results in a frequency histogram and polygon.

5. **WE9** The table below shows the length of 71 fish caught in a competition. Show this information in a frequency histogram and polygon.

Length of fish (mm)	Class centre	Frequency
300–309	304.5	9
310–319	314.5	15
320–329	324.5	20
330–339	334.5	12
340–349	344.5	8
350–359	354.5	7

6. Sixty people were involved in a psychology experiment. The following frequency table shows the times taken for the 60 people to complete a puzzle for the experiment.

Time taken (seconds)	Class centre	Frequency
6 to almost 8		1
8 to almost 10		4
10 to almost 12		15
12 to almost 14		18
14 to almost 16		12
16 to almost 18		8
18 to almost 20		2

a. Copy the frequency table and complete the class centre column.
b. Show the information in a frequency histogram and polygon.

7. WE10 In a suburb of roughly 1500 houses, a random sample of 40 households were surveyed to find the number of children living in each. The data were collected and recorded as follows.

0, 3, 2, 4, 1, 2, 3, 2, 2, 2, 2, 1, 3, 4, 5, 2, 3, 1, 1, 1,
0, 0, 2, 3, 4, 1, 3, 4, 2, 2, 0, 1, 2, 3, 2, 0, 2, 4, 5, 1

Construct a frequency table for this data.

8. A quality control officer selected 25 boxes of smart watches at random from a production line. She tested every smart watch in each box and recorded the number of defective smart watches in each box as follows:

1, 3, 2, 5, 2, 2, 1, 5, 2, 1, 2, 4, 3, 0, 5, 3, 2, 1, 3, 2, 1, 3, 4, 2, 1

Construct a frequency table for this data.

9. WE11 This table shows the number of hours of sport played per week by a group of Year 11 students.

Score (hours of sport played)	Frequency (f)
1	3
2	8
3	10
4	12
5	16
6	8
7	7
Total	64

Draw a combined histogram and frequency polygon for this data.

10. A block of houses in a suburb was surveyed to find the size of each house (in m^2). These are results.

Size of house (m^2)	Frequency
100–< 150	13
150–< 200	18
200–< 250	19
250–< 300	17
300–< 350	14
350–< 400	11
Total	92

Draw a combined histogram and frequency polygon for this data.

11. Forty people in a shopping centre were asked about the number of hours per week they spent watching TV.

The result of the survey is shown below.

10, 13, 7, 12, 16, 11, 6, 14, 6, 11, 5, 14, 12, 8,
27, 17, 13, 8, 14, 10, 13, 7, 15, 10, 16, 8, 18,
14, 21, 28, 9, 12, 11, 13, 9, 13, 29, 5, 24, 11

a. Completed the following frequency table for the given data.

Class interval	Midpoint (class centre)	Frequency
5–< 10		
10–< 15		
15–< 20		
20–< 25		
25–< 30		
Total		

b. Draw a combined histogram and polygon to suit the data.

12. The amount of money (in dollars) spent on snacks each week by a random sample of 17-year-olds was found to be as shown below.

10, 15, 5, 4, 8, 10, 4, 15, 5, 6, 10, 6, 5, 10, 8, 10, 5, 10, 10, 6

a. Organise the data into class intervals of 0–< 5, 5–< 10 and so on, and display it as a frequency table.
b. Display the graph as a histogram.
c. Discuss whether you feel these sample results reflect those of 17-year-olds generally.

5.5 Contemporary representation of data

LEARNING INTENTION

At the end of this subtopic you should be able to:
- display data using contemporary representations such as pictograms and bubble, Mekko, radar, sunburst, heat map and stacked area charts.

In subtopics 5.3 and 5.4 we have seen some methods of representing data. In this subtopic, we will explore more contemporary digital representations.

5.5.1 Pictograms

A pictogram is a simplistic but very popular way of representing data in a visual form. They are graphs that use icons and images to represent the data. They have the advantage of making data look more compelling and interesting to the reader.

tlvd-4059

WORKED EXAMPLE 10 Using pictograms

Sketch the following data as a pictogram using Excel.

Student	Number of apples eaten per week
John	7
Paula	10
Minh	3
Serena	5
Tony	2
Claire	6

THINK

1. a. Enter the data in Excel.
   b. Highlight the data.
   c. Click 'Insert', then 'Column or Bar Chart'.
   d. Select your Column or Bar Chart. This example has used a 2D Horizontal Graph.

WRITE

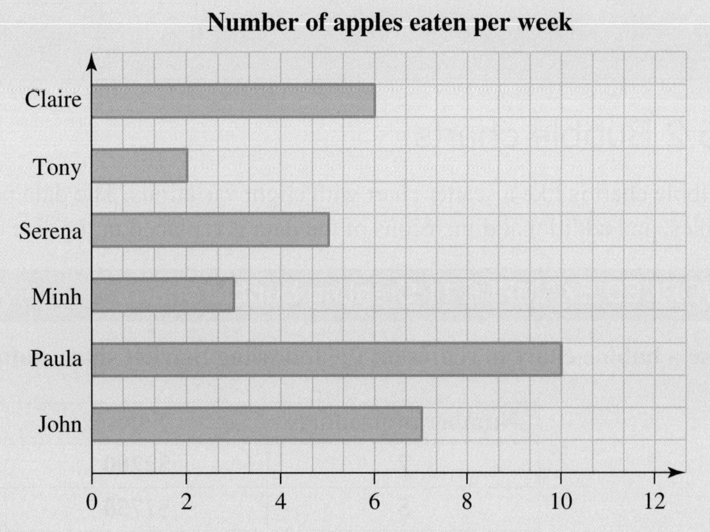

2. a. Click one of the horizontal bars.
   b. Right-click and select 'Format plot area'.
   c. On the right of the screen click 'Fill and line'.
   d. Click 'Fill'.
   e. Click 'Picture or texture fill'.

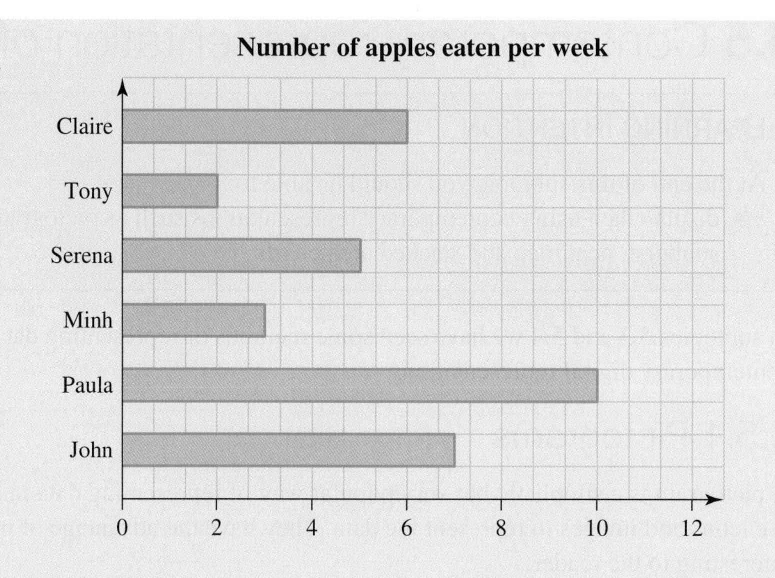

3. a. Click 'Insert'.
   b. Click 'Online Picture'.
   c. Click the apple option.
   d. Select the apple of choice.
   e. Click 'Insert'.
4. a. Make sure you clicked the bar graph.
   b. Click 'Stack and scale with'.
   c. Click 'Series options'.
   d. Use the gap width slider to reduce it to 0%.
   e. Tidy up the graph and complete labelling if required.

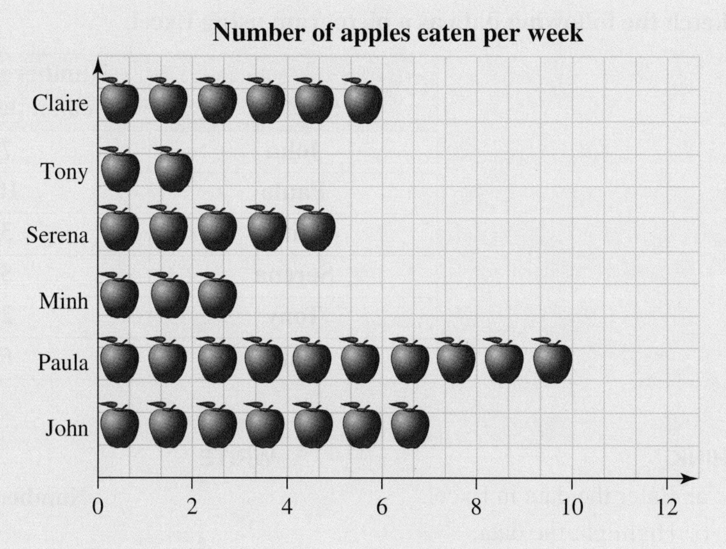

5.5.2 Bubble charts

A bubble chart is like a scatter chart with slight variations. The data points on a scatter chart are replaced with bubbles, and additional dimensions of the data is replaced in the size of the bubble.

WORKED EXAMPLE 11 Creating bubble charts

Use a bubble chart to represent the following market share data.

Number of products	Sales	% market share
7	$2200	14%
5	$1750	10%
12	$5100	25%
15	$7200	32%
9	$2700	19%

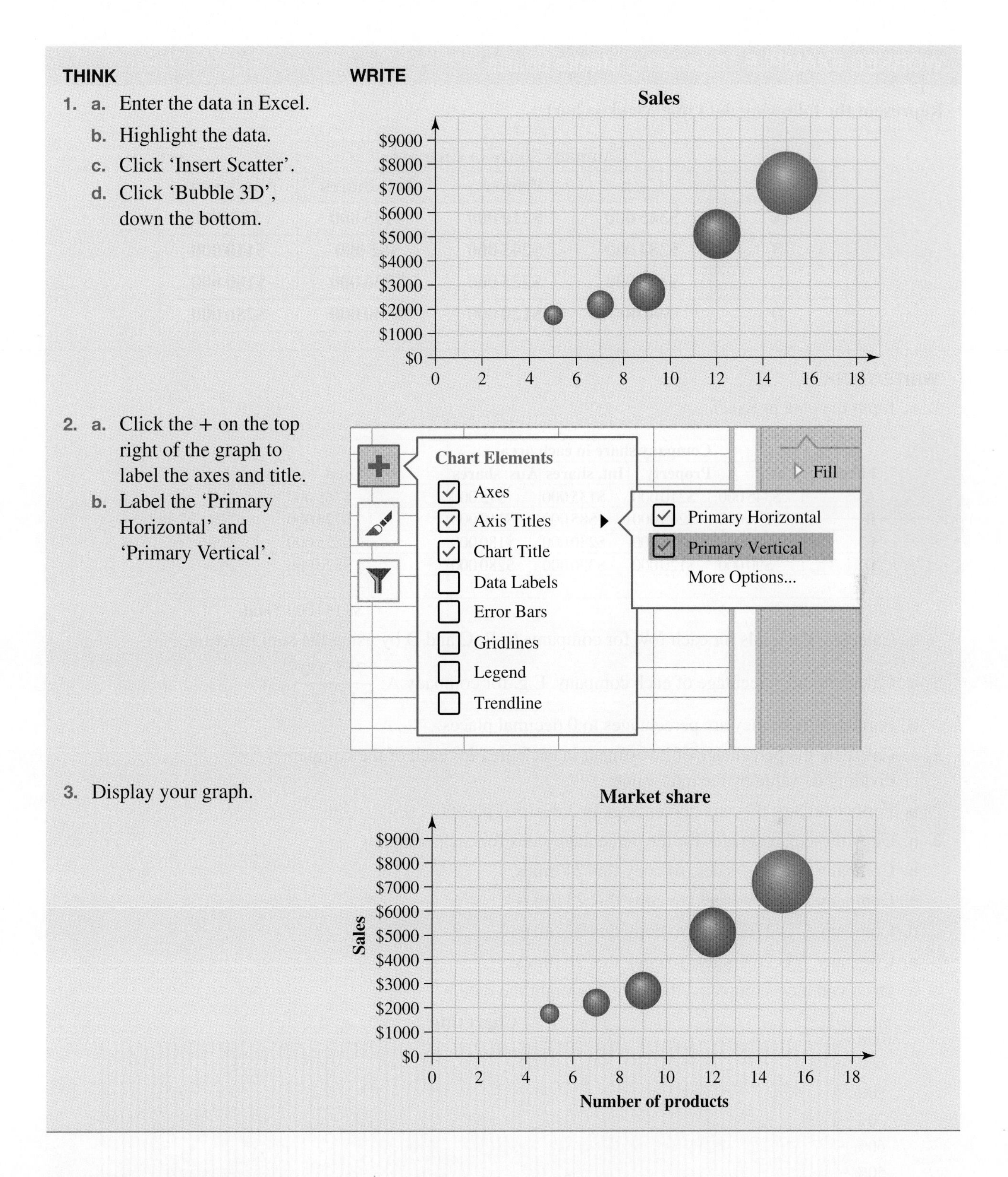

5.5.3 Mekko charts

A Marimekko (Mekko) chart is a two-dimensional stacked chart that is most commonly used by business analysts in finance and consulting companies.

WORKED EXAMPLE 12 Creating Mekko charts

Represent the following data in a Mekko chart.

Company share in each area				
Fund	**Cash**	**Property**	**Int. shares**	**Aus. shares**
A	**$345 000**	**$210 000**	**$135 000**	**$75 000**
B	**$284 000**	**$245 000**	**$85 000**	**$110 000**
C	**$120 000**	**$325 000**	**$230 000**	**$180 000**
D	**$90 000**	**$120 000**	**$330 000**	**$280 000**

WRITE/THINK

1. a. Input the data in Excel.

		Company share in each area					
Fund	Cash	Property	Int. shares	Aus. shares		Total	% sales
A	$345 000	$210 000	$135 000	$75 000		$765 000	24%
B	$284 000	$245 000	$85 000	$110 000		$724 000	23%
C	$120 000	$325 000	$230 000	$180 000		$855 000	27%
D	$90 000	$120 000	$330 000	$280 000		$820 000	26%
						$3 164 000	Total

   b. Calculate the totals for each row for company A, B, C and D by using the sum function.

   c. Calculate the percentage of each company. E.g. for company A: $\frac{765\,000}{3\,164\,000}$.

   d. Format cells so they are percentages to 0 decimal places.

2. a. Calculate the percentage of investment in each area for each of the companies by dividing its value by the total value.

   b. Format cells so they are percentages to 2 decimal places.

3. a. Copy these percentages for the percentage sales for each company.

   b. Company A is 24% sales, so copy this 24 times.

   c. Company B is 23% sales, so copy this 23 times.

   d. Company C is 27% sales, so copy this 27 times.

   e. Company A is 26% sales, so copy this 26 times.

4. a. Once you have completed the table, highlight the data.

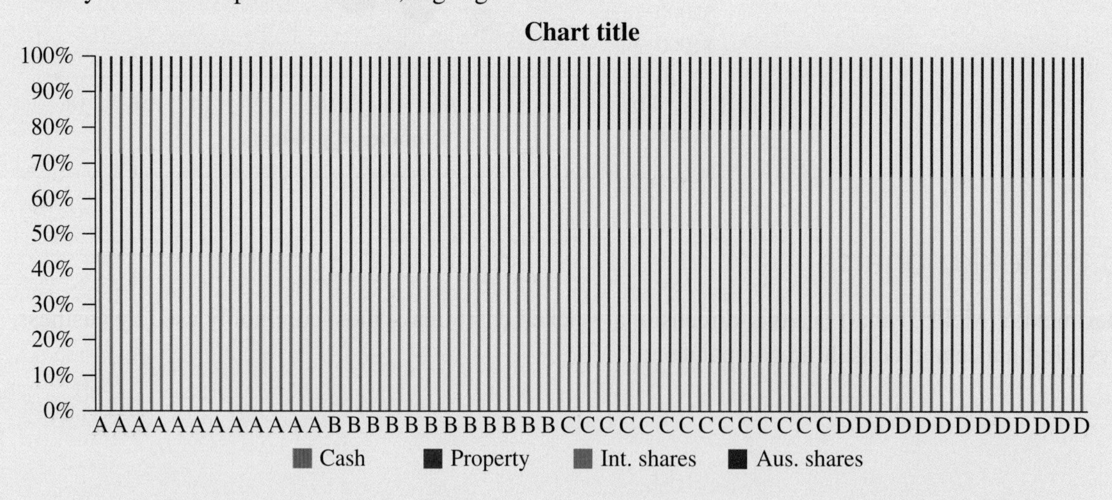

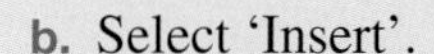

b. Select 'Insert'.

c. Select 'Column or bar chart'.

d. Select the 2D stacked column graph.

5. a. Click anywhere on the graph.

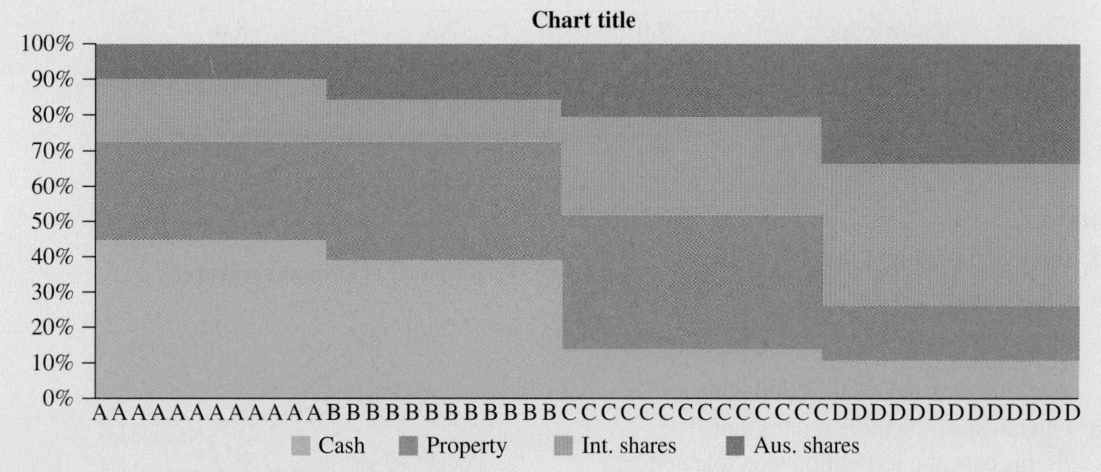

b. Right-click and select 'Format data series'.

c. Use the slider to reduce gap width from 150% to 0%.

6. a. To reduce all the A/B/C/Ds from your data, delete them all except the middle one for each company.

b. Label the axes and title on your graph.

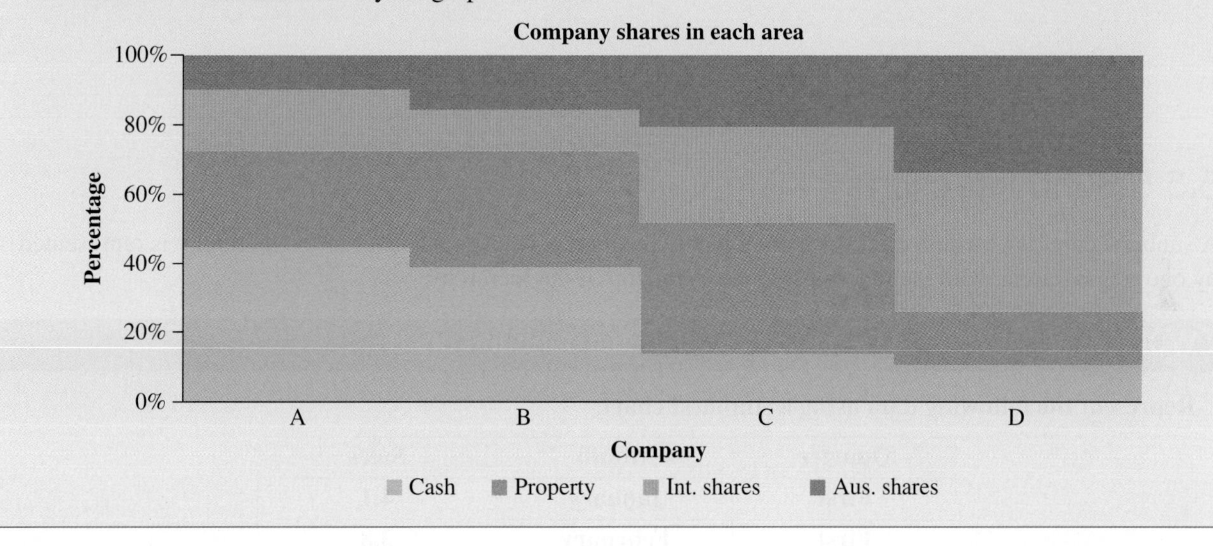

5.5.4 Radar charts

A radar (or spider) chart is used when you need to represent multivariable information in a 2D plane. This could assist in comparing, analysing and decision-making of data. It is a graph used to compare data, but it can get clustered if there are too many variables.

WORKED EXAMPLE 13 Representing data in a radar chart

Represent the following data on the performance of keynote speakers in a radar chart.

	Paul	Claire	Ross
Engagement	8.1	8.9	6.5
Knowledge	9.0	7.3	8.0
Usefulness	5.0	7.5	6.8
Delivery	6.7	7.8	7.2
Enjoyment	8.3	9.2	7.0

THINK

1. a. Input and highlight the data in Excel.
   b. Click 'Insert'.
   c. Click the graph icon: Different versions may have a different icon.
   d. Click one of the three radar graphs. In this case the left option was selected.
   e. Add a title to the graph.

WRITE

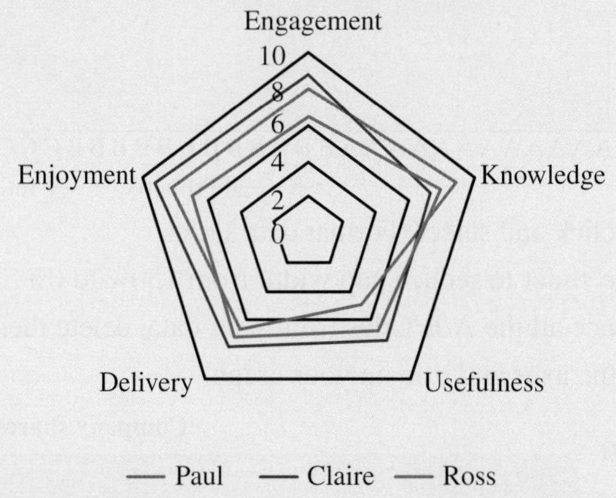

5.5.5 Sunburst charts

A sunburst chart is best suited for displaying hierarchical data, where each level of the hierarchy is represented by one ring or circle, with the innermost circle as the top of the hierarchy.

WORKED EXAMPLE 14 Representing data in a sunburst chart

Represent the following data using a sunburst chart.

Quarter	Month	Sales
First	January	4.1
First	February	3.8
First	March	1.8
Second	April	1.2
Second	May	1.5
Second	June	0.4
Third	July	0.7
Third	August	0.3
Third	September	1.1
Fourth	October	1.5
Fourth	November	2.1
Fourth	December	4

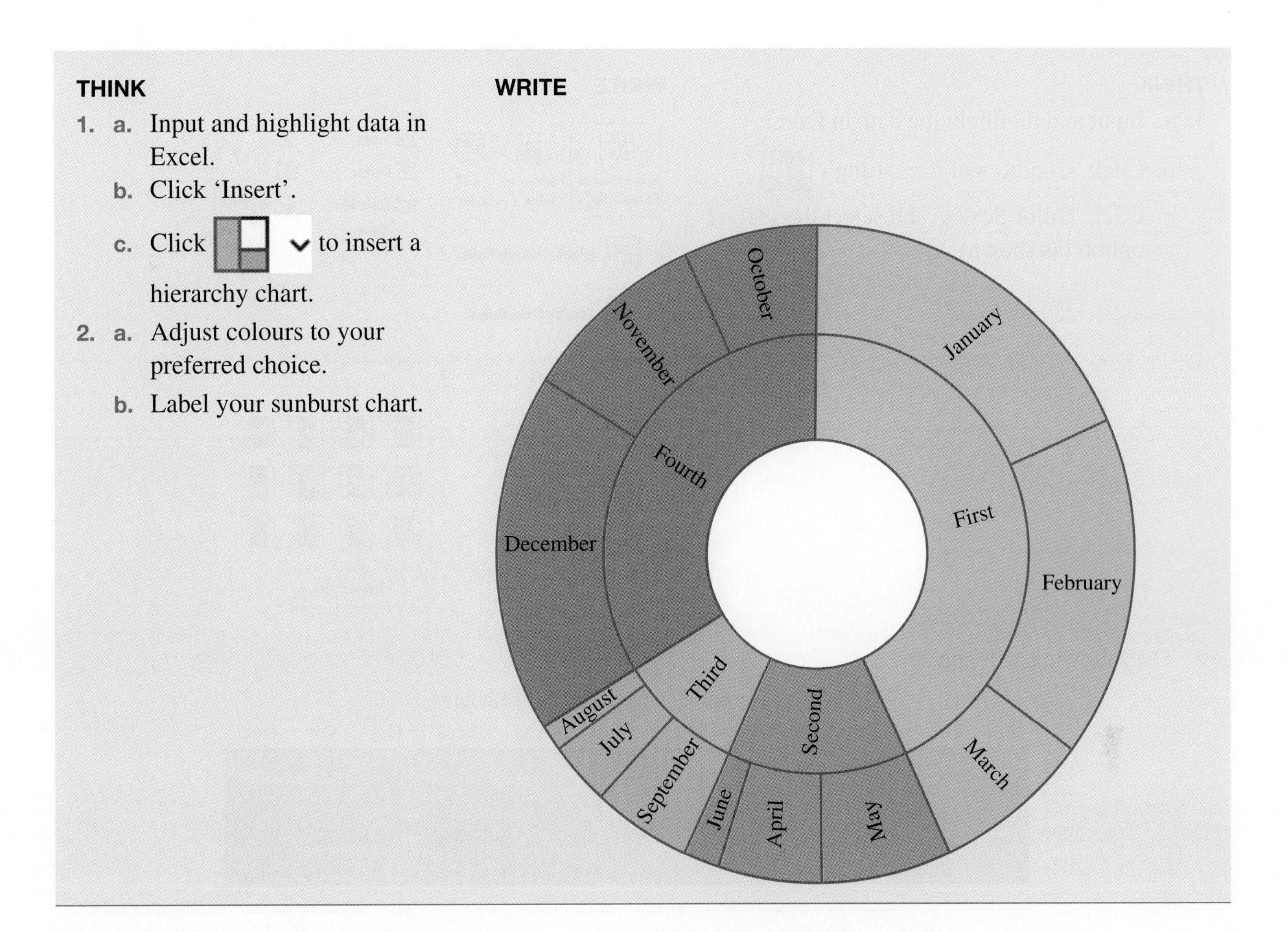

THINK	WRITE
1. a. Input and highlight data in Excel. b. Click 'Insert'. c. Click [icon] to insert a hierarchy chart. 2. a. Adjust colours to your preferred choice. b. Label your sunburst chart.	

5.5.6 Heat maps

A heat map is a method of visually representing numerical data where the value of each data point is represented using colour. Generally, warmer colours are used together with red.

WORKED EXAMPLE 15 Using heat maps

Represent the average monthly temperatures in Melbourne using a heat map.

	Average monthly temperature in Melbourne											
	Jan	**Feb**	**March**	**April**	**May**	**June**	**July**	**Aug**	**Sept**	**Oct**	**Nov**	**Dec**
2018	26	25.4	23.9	19.8	16.1	13.5	12.9	13.8	16.3	18.9	21.6	23.7
2019	26	25.7	23.9	19.9	16.2	13.4	12.8	13.8	16.3	18.9	21.5	23.8
2020	26	25.8	23.9	20	16.1	13.5	13	13.9	16.3	18.8	21.7	23.9
2021	26	25.8	24	19.8	16.1	13.6	21.9	13.8	16.3	19	21.6	23.7

THINK	WRITE
1. a. Input and highlight the data in Excel. b. Click 'Conditional formatting'. c. Click 'Color Scales' and select the second option (as shown).	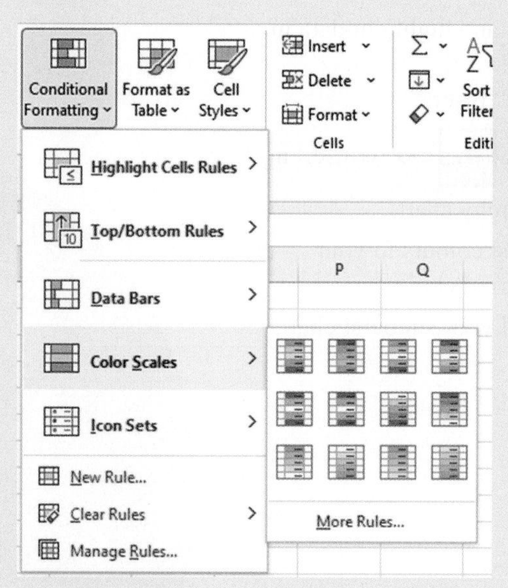

2. The following will appear.

Average monthly temperature in Melbourne

	Jan	Feb	March	April	May	June	July	Aug	Sept	Oct	Nov	Dec
2018	26	25.4	23.9	19.8	16.1	13.5	12.9	13.8	16.3	18.9	21.6	23.7
2019	26	25.7	23.9	19.9	16.2	13.4	12.8	13.8	16.3	18.9	21.5	23.8
2020	26	25.8	23.9	20	16.1	13.5	13	13.9	16.3	18.8	21.7	23.9
2021	26	25.8	24	19.8	16.1	13.6	12.9	13.8	16.3	19	21.6	23.7

5.5.7 Area charts

An area chart is similar to a line chart, but includes colour under the plotted line. Area charts are useful when there are multiple time-series data, and you want to determine the contribution of each.

WORKED EXAMPLE 16 Using an area chart to represent data

Represent the data on the monthly phone sales below using an area chart.

Phone sales per month			
	iPhone	**Samsung**	**Google**
November	**65**	**54**	**27**
December	**89**	**69**	**33**
January	**78**	**66**	**31**
February	**56**	**41**	**23**
March	**48**	**32**	**18**

THINK	WRITE
1. a. Enter and highlight the data in Excel. b. Click 'Insert'. c. Select 'Line and Area Chart'. 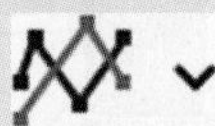 d. Select 'Stacked Area' 2D as shown.	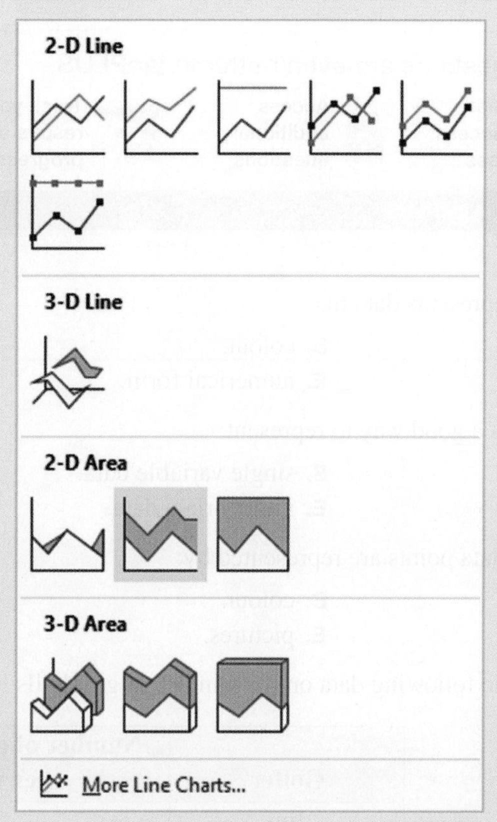
2. The following graph will appear. Label your graph.	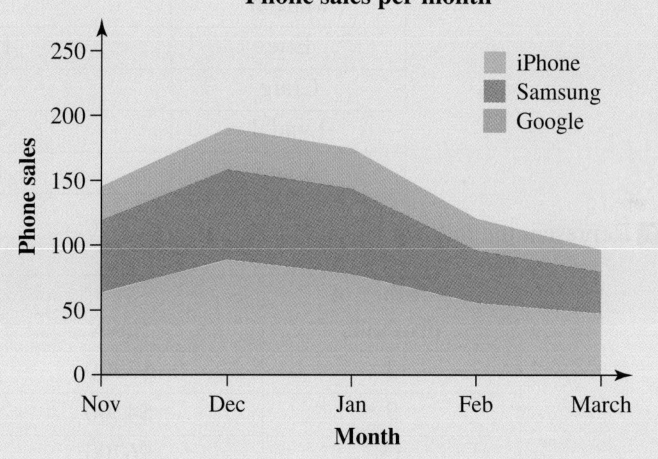

5.5 Exercise

Students, these questions are even better in jacPLUS

Receive immediate feedback and access sample responses

Access additional questions

Track your results and progress

Find all this and MORE in jacPLUS

1. **MC** A pictogram represents data in:
   A. a table.
   B. colour.
   C. visual form.
   D. black and white.
   E. numerical form.
2. **MC** A radar chart is a good way to represent:
   A. categorical data.
   B. single variable data.
   C. ordinal data.
   D. numerical data.
   E. multivariate data.
3. **MC** In heat maps, data points are represented by:
   A. bars.
   B. colour.
   C. dots.
   D. columns.
   E. pictures.
4. **WE10** Represent the following data on the number of golf balls lost per week using a pictogram.

Golfer	Number of golf balls lost per week
Jim	8
Marion	5
Eric	12
Craig	3
Lyndal	5
Adam	1

5. **WE11** Represent the following data in a bubble graph.

Number of products	Sales	% market share
3	$3300	5%
9	$4300	11%
15	$6700	28%
21	$7100	40%
7	$3900	16%

6. **WE12** Represent the following data in a Mekko graph.

Managed fund breakdown of investments				
Managed fund	% cash	% property	% gold	% shares
A	34	22	10	34
B	8	32	15	45
C	20	15	15	50
D	4	44	16	36

7. **WE13** Use a radar chart to represent the following data on how real estate agents were rated by their clients.

	Brent	Jane	Alice
Polite	6.2	7.8	9.1
Trustworthy	4.8	6.1	7.3
Helpful	6.4	7.1	6.4
Organised	8.4	5.8	7.9
Communicative	5.6	8.1	7.2

8. **WE14** Represent the following data using a sunburst chart.

Quarter	Month	Week	Sales
1	Jan	1	4.7
		2	8.2
		3	2.8
		4	5.2
1	Feb	1	0.6
		3	2.4
1	March	2	3.8
		3	9.4
		4	3.6
2	April	3	0.4
2	May	2	1.8
		3	7.2
		4	1.1
3	July	1	1.8
		2	8.3
		3	0.5
		4	6.6

9. **WE15** Represent the average sales of AFL tickets per game using a heat map.

	Round 1	Round 2	Round 3	Round 4	Round 5	Round 6	Round 7	Round 8	Round 9	Round 10
Fri	65 424	43 793	58 938	78 284	60 324	58 362	43 569	82 892	33 720	48 598
Sat	35 792	45 923	69 367	25 793	54 783	39 760	48 268	27 680	44 324	31 769
Sun	28 572	22 475	31 580	36 782	41 890	30 210	28 450	36 456	32 480	29 960

10. **WE16** Display the monthly TV sales from a particular store using a stacked area chart.

	Monthly TV sales		
	Samsung	LG	Hisense
October	43	38	21
November	69	54	60
December	116	78	68
January	78	51	44
February	32	22	18

5.6 Stem-and-leaf plots

LEARNING INTENTION

At the end of this subtopic you should be able to:
- display and interpret stem-and-leaf plots
- calculate medians and quartiles from stem-and-leaf plots.

5.6.1 Stem-and-leaf plots

- As an alternative to a frequency table, a *stem-and-leaf plot* may be used to group and summarise data.
- A stem is made using the first part of each piece of data. The second part of each piece of data forms the leaves.
- The following data show the mass (in kg) of 20 possums trapped, weighed, then released by a wildlife researcher.

1.8	0.9	0.7	1.4	1.6	2.1	2.7	2.2	1.8	2.3
2.3	1.5	1.1	2.2	3.0	2.5	2.7	3.2	1.9	1.7

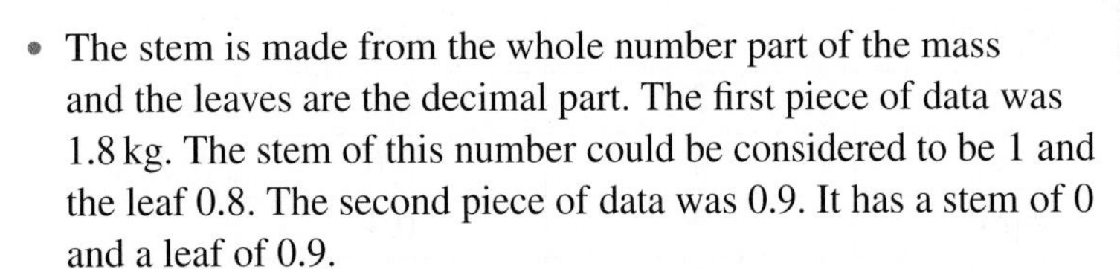

- The stem is made from the whole number part of the mass and the leaves are the decimal part. The first piece of data was 1.8 kg. The stem of this number could be considered to be 1 and the leaf 0.8. The second piece of data was 0.9. It has a stem of 0 and a leaf of 0.9.
- To compose the stem-and-leaf plot, rule a vertical column of stems, then enter the leaf of each piece of data in a neat row beside the appropriate stem. The first row of the stem-and-leaf plot records all data from 0.0 to 0.9.
- The second row records data from 1.0 to 1.9, etc.
- Attach a key to the plot to show the reader the meaning of each entry.
- The data should be arranged in order of size, so this stem-and-leaf plot should be written in such a way that the numbers in each row of 'leaves' are in ascending order.

  Key: 0|7 = 0.7 kg

Stem	*Leaf*
0	7 9
1	1 4 5 6 7 8 8 9
2	1 2 2 3 3 5 7 7
3	0 2

- When preparing a stem-and-leaf plot, it is important to try to keep the numbers in neat vertical columns, because a neat plot gives the reader an idea of the distribution of scores. The plot itself looks a bit like a histogram turned on its side.

WORKED EXAMPLE 17 Using a stem-and-leaf plot to represent data

The information below shows the mass, in kilograms, of twenty 17-year-old boys. Show this information in a stem-and-leaf plot.

65	45	56	57	58	54	61	72	70	69
61	58	49	52	64	71	66	65	66	60

THINK	WRITE
1. Make the 'tens' the stem and the 'units' the leaves.	Key: 5|6 = 56 kg
2. Write a key.	

Stem	*Leaf*
4	5 9
5	6 7 8 4 8 2
6	5 1 9 1 4 6 5 6 0
7	2 0 1

3. Complete the plot with the leaves in each row in ascending order.

Stem	*Leaf*
4	5 9
5	2 4 6 7 8 8
6	0 1 1 4 5 5 6 6 9
7	0 1 2

5.6.2 Class size of a stem-and-leaf plot

It is also useful to be able to represent data with a class size of 5. This could be done for the previous stem-and-leaf plot by choosing stems 0*, 1, 1*, 2, 2*, 3, where the class with stem 1 contains all the data from 1.0 to 1.4 and stem 1* contains the data from 1.5 to 1.9, etc. If stems are split in this way, it is a good idea to include two entries in the key. The stem-and-leaf plot for the 'possum' data would appear as follows:

Key: 1|1 = 1.1 kg
1*|5 = 1.5

Stem	*Leaf*
0*	7 9
1	1 4
1*	5 6 7 8 8 9
2	1 2 2 3 3
2*	5 7 7
3	0 2

Advantages of a stem-and-leaf plot

A stem-and-leaf plot has the following advantages over a frequency distribution table.

- The plot itself gives a graphical representation of the spread of data. (It is like a histogram turned on its side.)
- All the original data are retained, so there is no loss of accuracy when calculating statistics such as the mean and standard deviation. In a grouped frequency distribution table, some generalisations are made when these values are calculated.

tlvd-4065

WORKED EXAMPLE 18 Using a stem-and-leaf plot with class sizes

The following data give the length of gestation in days for 24 mothers. Prepare a stem-and-leaf plot of the data using a class size of 5.

280	**287**	**285**	**276**	**266**	**292**	**288**	**273**	**295**	**279**	**284**	**271**
292	**288**	**279**	**281**	**270**	**278**	**281**	**292**	**268**	**282**	**275**	**281**

THINK

1. A group size of 5 is required. The smallest piece of data is 266 and the largest is 295, so make the stems: 26*, 27, 27*, 28, 28*, 29, 29*.
   The key should give a clear indication of the meaning of each entry.
2. Enter the data piece by piece. Enter the leaves in pencil first, so that they can be rearranged in order of size. Check that 24 pieces of data have been entered.
3. Now arrange the leaves in order of size.

WRITE

Key: $26^*|6 = 266$ days
$27|0 = 270$ days

Stem	*Leaf*
26*	6 8
27	0 1 3
27*	5 6 8 9 9
28	0 1 1 1 2 4
28*	5 7 8 8
29	2 2 2
29*	5

5.6.3 Stem-and-leaf plot: medians and quartiles

Since all the original data are recorded on the stem-and-leaf plot and are conveniently arranged in order of size, the plot can be used to locate the upper and lower quartiles and the median.

- The *median* is the middle score or the average of the two middle scores.
- The *lower quartile* is the median of the lower half of the data.
- The *upper quartile* is the median of the upper half of the data. Using the 'possum' mass data from subtopic 5.6.1 as an example:
  Key: $0|7 = 0.7$ kg

Stem	*Leaf*
0	7 9
1	1 4 5 6 7 8 8 9
2	1 2 2 3 3 5 7 7
3	0 2

There were 20 records, so the median is the average of the 10th and 11th scores. Counting each score as it appeared in the stem-and-leaf plot, we can see that the 10th score is the number 1.9 and the 11th score is the number 2.1.

$$\text{Median} = \frac{1.9 + 2.1}{2}$$
$$= 2.0\,\text{kg}$$

The median divides the data into halves.

The lower quartile is the median of the lower half, which has ten scores in it. So the position of the lower quartile is given by the average of the 5th and 6th scores.

The 5th score is the number 1.5. The 6th score is the number 1.6.

$$\text{The lower quartile} = \frac{1.5 + 1.6}{2}$$
$$= 1.55\text{ kg}$$

The upper quartile is the median of the upper half, which also has ten scores in it.

The 5th score in this half is the number 2.3. The 6th score is the number 2.5.

$$\text{The upper quartile} = \frac{2.3 + 2.5}{2}$$
$$= 2.4\text{ kg}$$

The *interquartile* range is the difference between the upper and lower quartiles.

tlvd-4066

WORKED EXAMPLE 19 Calculating the interquartile range from a stem-and-leaf plot

Determine the interquartile range of the data presented in the following stem-and-leaf plot.

Key: 15|7 = 157 kg

Stem	***Leaf***
15	**488**
16	**13368**
17	**00147999**
18	**123357889**
19	**278**
20	**02**

THINK	WRITE
1. There are 30 scores and so the median will be the average of the 15th and 16th scores.	$\text{Median} = \frac{179 + 179}{2}$ $= 179\text{ kg}$
2. There are 15 scores in each half, so the lower and upper quartiles will be the 8th score in each half.	The lower quartile = 168 kg The upper quartile = 188 kg
3. The interquartile range is the difference between the upper and lower quartiles.	Interquartile range = upper quartile − lower quartile = 188 − 168 = 20 kg

Note: Remember to provide appropriate units for the answer.

5.6 Exercise

Students, these questions are even better in jacPLUS

Receive immediate feedback and access sample responses

Access additional questions

Track your results and progress

Find all this and MORE in jacPLUS

1. **WE17** The data below give the number of errors made each week by 20 machine operators. Prepare a stem-and-leaf diagram of the data using a stem of 0, 1, 2, and so on.

6	15	20	25	28	18	32	43	52	27
17	26	38	31	26	29	32	46	13	20

2. The data below give the time taken (in minutes) for each of the 40 runners on a 10-km fun run. Prepare a stem-and-leaf diagram for the data using a class size of 10 minutes.

36	42	52	38	47	59	72	68	57	82
66	75	45	42	55	38	42	46	48	39
42	58	40	41	47	53	68	43	39	48
71	42	50	46	40	52	37	54	48	52

3. **WE18** The typing speed (in words per minute) of 30 word processors is recorded below. Prepare a stem-and-leaf diagram of the data using a class size of 5.

96	102	92	96	95	102	95	115	110	108
88	86	107	111	107	108	103	121	107	96
124	95	98	102	108	112	120	99	121	130

4. Twenty transistors are tested by applying increasing voltage until they are destroyed. The maximum voltage that each could withstand is recorded below. Prepare a stem-and-leaf plot of the data using a class size of 0.5.

14.8	15.2	13.8	14.0	14.8	15.7	15.5	15.6	14.7	14.3
14.6	15.2	15.9	15.1	14.3	14.6	13.9	14.7	14.5	14.2

5. **WE19** The stem-and-leaf plot shown gives the exact mass of 24 packets of biscuits. Calculate the interquartile range of the data.

Key: 248|4 = 248.4 g

Stem	*Leaf*
248	4 7 8
249	2 3 6 6
250	0 0 1 1 6 9 9
251	1 5 5 5 6 7
252	1 5 8
253	0

6. The time taken (in seconds) for a test vehicle to accelerate from 0 to 100 km/h is recorded during a test of 24 trials. The results are represented by the stem-and-leaf plot below.

Key: 7|2 = 7.2 s
7*|6 = 7.6 s

Stem	*Leaf*
7	2 4 4
7*	5 5 7 9
8	0 0 1 2 4 4 4
8*	5 5 6 8 9
9	2 2 3
9*	5 7

a. Determine the median of the data.
b. Determine the upper and lower quartiles of the data.
c. Determine the interquartile range of the data.

Questions** 7 **to** 10 **refer to the following information.

Each student in a class has been assigned a newly planted tree to look after, and must provide a weekly report on its growth and condition. From the latest reports, the teacher recorded the height of each tree (in mm), and entered these in the stem-and-leaf plot shown.

Key: 12|1 = 1210 mm
12*|5 = 1250 mm

Stem	*Leaf*
12*	1 2 4
12*	5 7 7 9 9
13*	0 1 1 2 3 4 4
13*	5 6 6 7 9 9
14*	0 2 3 4
14*	6 7

7. **MC** The class size used in the stem-and-leaf plot is:
A. 1 B. 10 C. 33 D. 50 E. 100

8. **MC** The number of scores that have been recorded is:
A. 21 B. 27 C. 33 D. 1210 E. 1410

9. **MC** The median of the data is:
A. 13.4 B. 14 C. 134 D. 1335 E. 1340

10. **MC** The interquartile range of the data (in mm) is:
A. 10 B. 14 C. 100 D. 1290 E. 1390

11. The maximum hand spans (in cm) of 20 male concert pianists are recorded as follows.

23.6	20.2	22.8	21.4	25.1	24.8	23.2	21.6	20.7	23.6
22.8	24.6	21.8	22.8	23.1	24.6	21.7	24.7	22.2	23.0

a. Complete a stem-and-leaf plot to represent the data.
b. Determine the median of the data.
c. Determine the upper and lower quartiles of the data.
d. Determine the interquartile range of the data.

12. The heights (in cm) of a sample of 30 plants are recorded as follows.

93	88	94	99	91	85	126	107	110	111
98	96	117	101	97	92	101	132	103	82
114	84	96	103	108	115	90	110	126	85

a. Complete a stem-and-leaf plot to represent the data.
b. Determine the median of the data.
c. Determine the upper and lower quartiles of the data.
d. Determine the interquartile range of the data.

5.7 Five-number summary and boxplots

LEARNING INTENTION

At the end of this subtopic you should be able to:
- develop a five-number summary and construct a boxplot for a data set.

5.7.1 Boxplots (box-and-whisker plots)

In drawing a boxplot, five values are extracted from the data set. These values summarise the data and are therefore referred to as the **five-number summary**.

Five-number summary

The five-number summary consists of:
- **lower extreme — the lowest score in the data set**
- **lower quartile — the middle score of the lower half of the data set**
- **median — the middle score**
- **upper quartile — the middle score of the upper half of the data set**
- **upper extreme — the highest score in the data set.**

tlvd-4067

WORKED EXAMPLE 20 Developing a five-number summary

For the set of scores below, develop a five-number summary.

12 15 46 9 36 85 73 29 64 50

THINK	WRITE
1. Re-write the list in ascending order.	9 12 15 29 36 46 50 64 73 85
2. Write the lowest score.	Lower extreme = 9
3. Calculate the lower quartile.	Lower quartile = 15
4. Calculate the median.	$\text{Median} = \frac{36+46}{2}$ $= 41$
5. Calculate the upper quartile.	Upper quartile = 64
6. Write the upper extreme.	Upper extreme = 85 Five-number summary = 9, 15, 41, 64, 85

Once a five-number summary has been developed, it can be graphed using a box-and-whisker plot, a powerful way to display the spread of the data.

The box-and-whisker plot consists of a central divided box with attached whiskers. The box spans the interquartile range, the vertical line inside the box marks the median and the whiskers indicate the range. Each section of the box-and-whisker plot represents one quarter of the scores of the data set.

Box-and-whisker plot

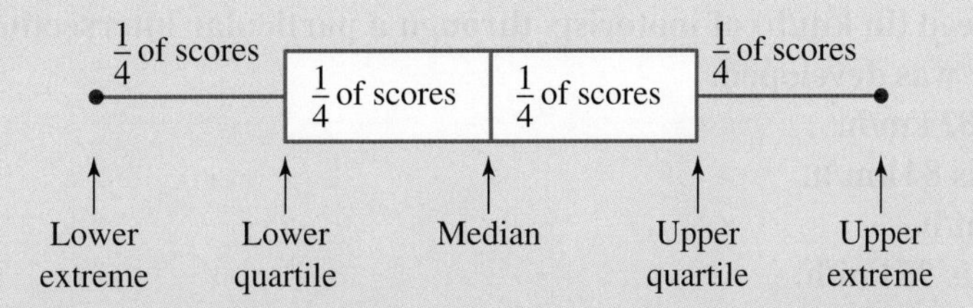

Box-and-whisker plots are always drawn to scale. This can be drawn with the five-number summary attached as labels:

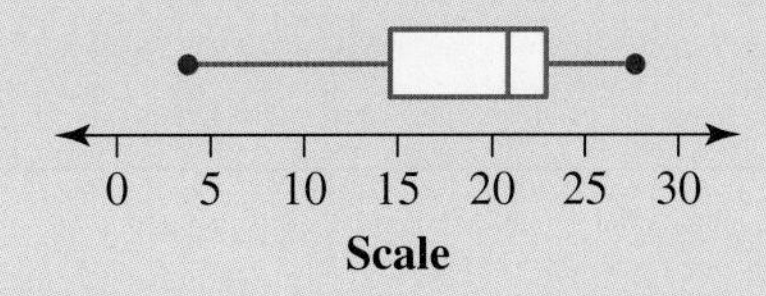

It can also be drawn with a scale presented alongside the box-and-whisker plot. (This representation is preferable.)

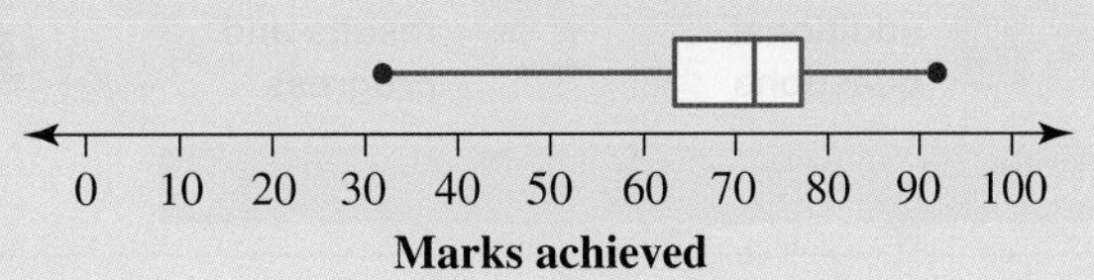

WORKED EXAMPLE 21 Interpreting data from a box-and-whisker plot

The following box-and-whisker plot shows the marks achieved by students on their end-of-year exam.

0 10 20 30 40 50 60 70 80 90 100

Marks achieved

a. **State the median.**
b. **Determine the interquartile range.**
c. **Determine the highest mark in the class.**

THINK	WRITE
a. The mark in the box shows the median (72).	a. Median = 72 marks
b. 1. The lower end of the box shows the lower quartile (63).	b. Lower quartile = 63 marks
2. The upper end of the box shows the upper quartile (77).	Upper quartile = 77 marks
3. Subtract the lower quartile from the upper quartile.	Interquartile range = 77 − 63 = 14 marks
c. The top end of the whisker gives the top mark (92).	c. Top mark = 92 marks

on Resources

Interactivity Box and whisker plots (int-6245)

tlvd-4068

WORKED EXAMPLE 22 Showing five-number-summary data in a box-and-whisker plot

After analysing the speed (in km/h) of motorists through a particular intersection, the following five-number summary was developed.

- **The lowest score is 82 km/h.**
- **The lower quartile is 84 km/h.**
- **The median is 89 km/h.**
- **The upper quartile is 95 km/h.**
- **The highest score is 114 km/h.**

Show this information in a box-and-whisker plot.

THINK	WRITE
1. Draw a scale from 70 to 120 using 1 cm = 10 km/h.	70 80 90 100 110 120 **Speed in km/h**
2. Draw the box from 84 to 95.	
3. Mark the median at 89.	
4. Draw the whiskers to 82 and 114.	

5.7 Exercise

1. **WE20** Write a five-number summary for the following data set.

5 17 16 8 25 18 20 15 17 14

2. For each of the following data sets, write a five-number summary.
   a. 23 45 92 80 84 83 43 83
   b. 2 6 4 2 5 7 1
   c. 60 75 29 38 69 63 45 20 29 93 8 29 93

3. **WE21** From the five-number summary 6, 11, 13, 16, 32, determine:
   a. the median b. the interquartile range c. the range.

4. From the five-number summary 101, 119, 122, 125, 128, determine:
   a. the median b. the interquartile range c. the range.

5. **WE22** A five-number summary is given below.
   - Lower extreme = 39.2
   - Lower quartile = 46.5
   - Median = 49.0
   - Upper quartile = 52.3
   - Upper extreme = 57.8

   Draw a box-and-whisker plot of the data.

6. The following box-and-whisker plot shows the distribution of final points scored by a Rugby team over a season's roster.
   a. Determine the team's greatest score.
   b. Determine the team's smallest score.
   c. Determine the team's median score.
   d. Determine the range of points scored.
   e. Determine the interquartile range of points scored.

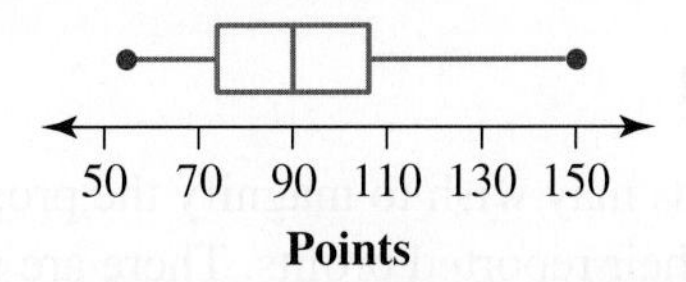

7. The following box-and-whisker plot shows the distribution of data formed by counting the number of honey bears in each pack.
   In any pack, determine:
   a. the largest number of honey bears
   b. the smallest number of honey bears
   c. the median number of honey bears
   d. the range of numbers of honey bears
   e. the interquartile range of honey bears.

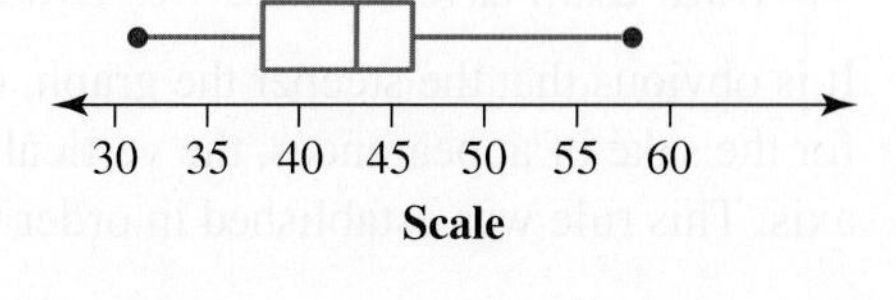

Questions* 8–10 *refer to the following box-and-whisker plot.

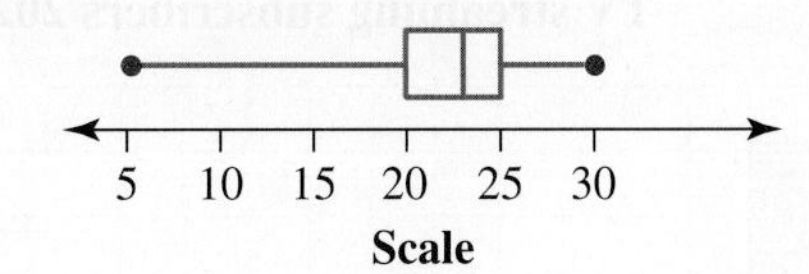

8. **MC** The median of the data is:
   A. 5 B. 20 C. 23 D. 25 E. 31
9. **MC** The interquartile range of the data is:
   A. 5 B. 20 C. 25 D. 20 to 25 E. 31
10. **MC** Determine which of the following is *not* true of the data represented by the box-and-whisker plot.
    A. One-quarter of the scores is between 5 and 20.
    B. One-half of the scores is between 20 and 25.
    C. The lowest quarter of the data is spread over a wide range.
    D. Most of the data are contained between the scores of 5 and 20.
    E. The top half of the scores is spread over a smaller range than the bottom half of the scores.
11. The number of sales made each day by a salesperson is recorded over a fortnight:

    25, 31, 28, 43, 37, 43, 22, 45, 48, 33

    a. Write a five-number summary of the data.
    b. Draw a box-and-whisker plot of the data.
12. The data below show monthly rainfall in millimetres.

Jan	Feb	Mar	Apr	May	June	July	Aug	Sept	Oct	Nov	Dec
10	12	21	23	39	22	15	11	22	37	45	30

    a. Provide a five-number summary of the data.
    b. Draw a box-and-whisker plot of the data.

5.8 Graphical methods of misrepresenting data

LEARNING INTENTION

At the end of this subtopic you should be able to:
- identify errors and misrepresentations in data analysis.

5.8.1 Methods of misrepresenting data

Many people have reasons for misrepresenting data: politicians may wish to magnify the progress achieved during their term, or businesspeople may wish to accentuate their reported profits. There are numerous ways of misrepresenting data. In this subtopic, only graphical methods of misrepresentation are considered.

Vertical axis and horizontal axis

It is obvious that the steeper the graph, the better the growth appears. A 'rule of thumb' for statisticians is that for the sake of appearances, the vertical axis should be two-thirds to three-quarters the length of the horizontal axis. This rule was established in order to have some comparability between graphs.

This graph shows the number of subscribers of TV streaming companies. The information is distorted as the y-axis starts at 20 million, so it looks like Apple and Discovery have little or no subscribers when they actually have around 20 and 22 million subscribers respectively.

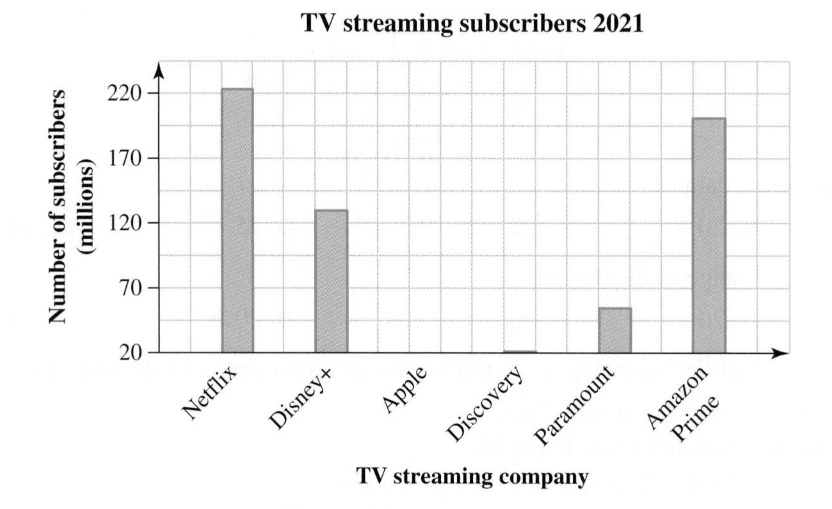

Changing the scale on the vertical axis

The following table gives the holdings of ROPE Corporation during 2001.

Quarter	Holdings (× $1 000 000)
Jan–Mar	200
Apr–June	200
July–Sept	201
Oct–Dec	202

Here is one way of representing this data:

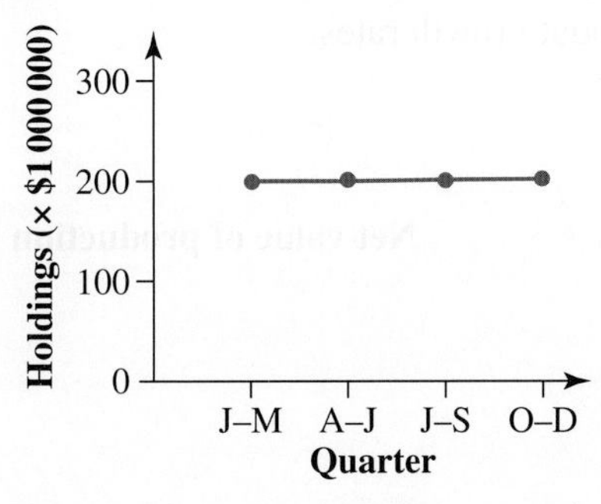

This graph doesn't accurately show the change in value of the quarters. Now look at the following graph.

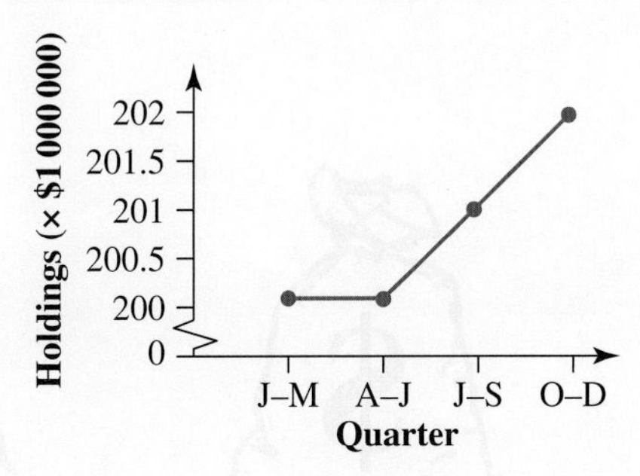

Omitting certain values

If we chose to ignore the second quarter's value, which shows no increase, then the graph would show even better growth.

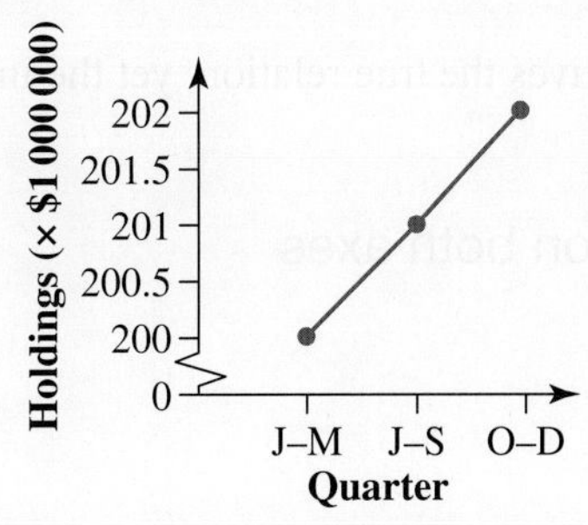

Foreshortening the vertical axis

Look at the following figures. Notice in graph (a) that the numbers from 0 to 4000 on the y-axis have been omitted.

In graph (b) these numbers have been inserted. The rate of growth of the company looks far less spectacular in graph (b) than in graph (a).

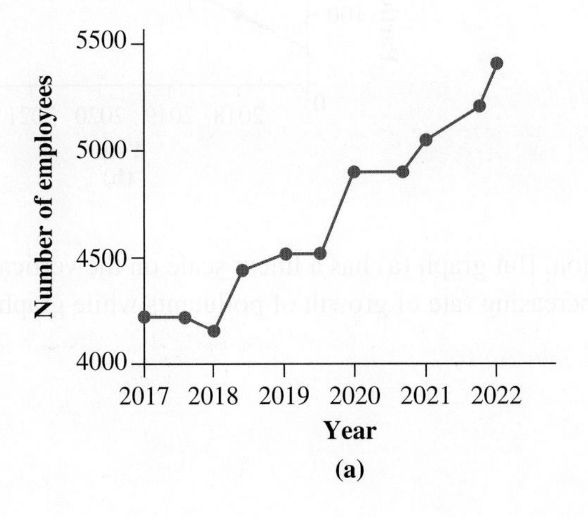

(a)

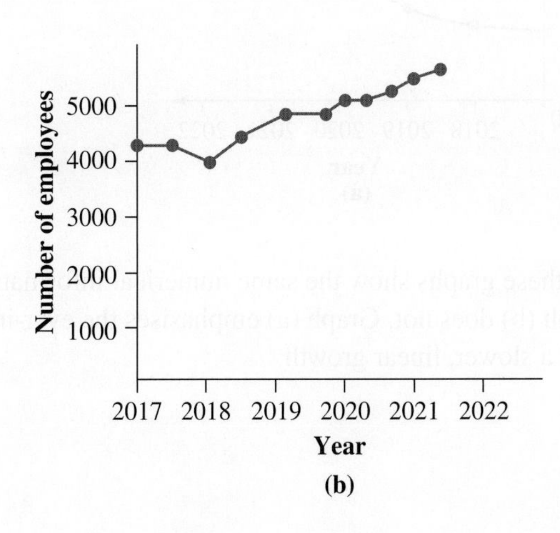

(b)

Foreshortening the vertical axis is a very common procedure. It does have the advantage of giving extra detail but it can give the wrong impression about growth rates.

Visual impression

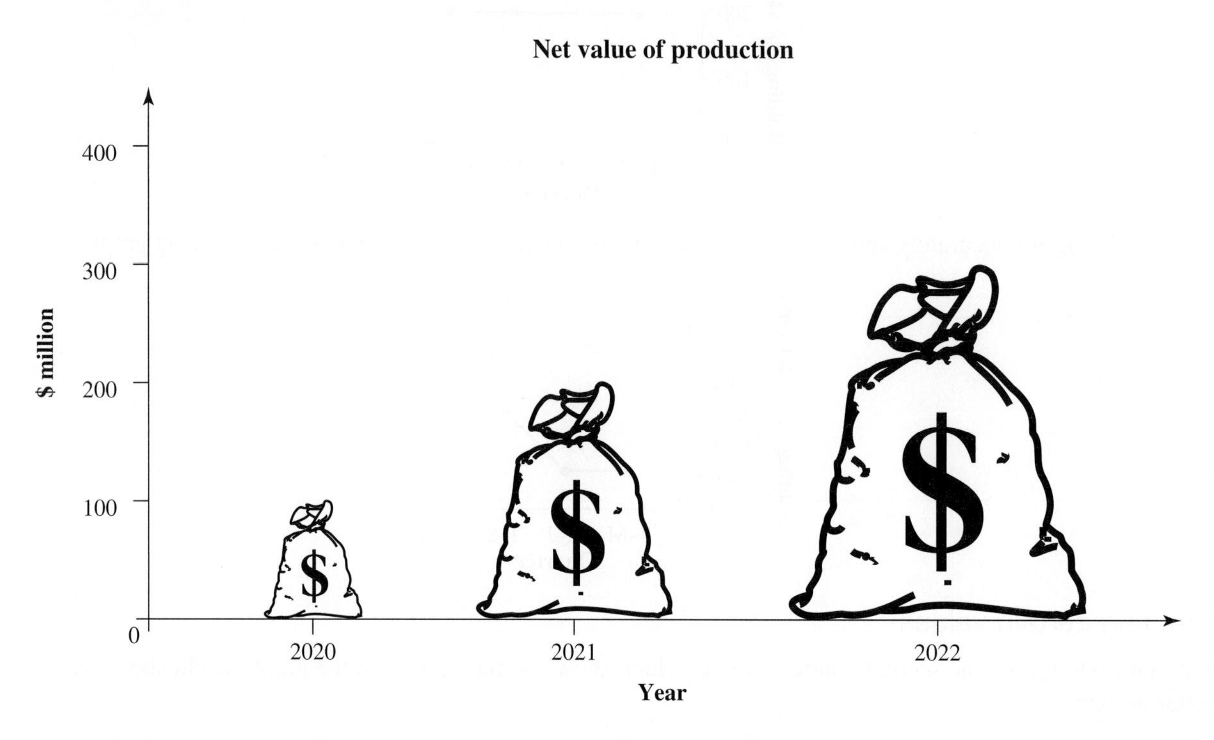

In this graph, height is the property that gives the true relation, yet the impression of a much greater increase is given by the volume of each money bag.

A non-linear scale on an axis or on both axes

Consider the following two graphs.

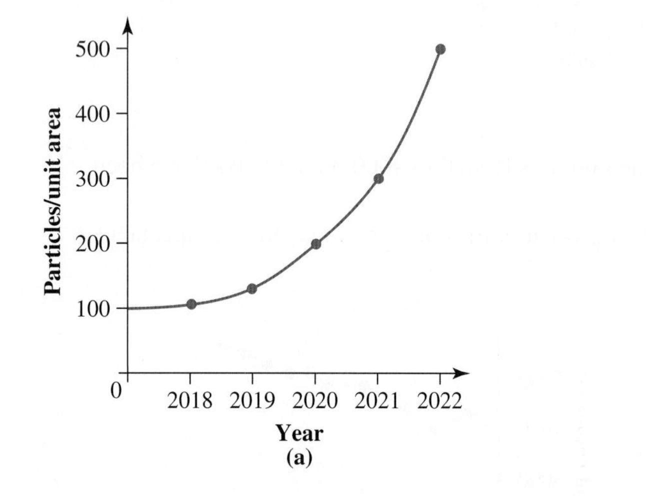

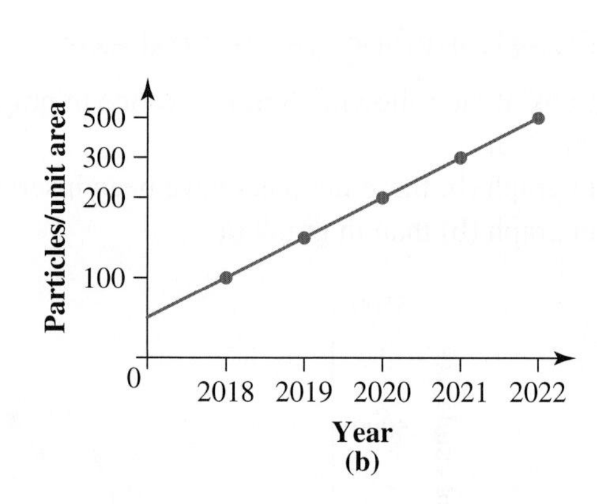

Both of these graphs show the same numerical information. But graph (a) has a linear scale on the vertical axis and graph (b) does not. Graph (a) emphasises the ever-increasing rate of growth of pollutants while graph (b) suggests a slower, linear growth.

tlvd-4069

WORKED EXAMPLE 23 Interpreting survey graphs

The following data show wages and profits of a certain company. All figures are in millions of dollars.

Year	**2005**	**2010**	**2015**	**2020**
Wages	**6**	**9**	**13**	**20**
% increase in wages	**25**	**50**	**44**	**54**
Profits	**1**	**1.5**	**2.5**	**5**
% increase in profits	**20**	**50**	**66**	**100**

Now consider the graphs:

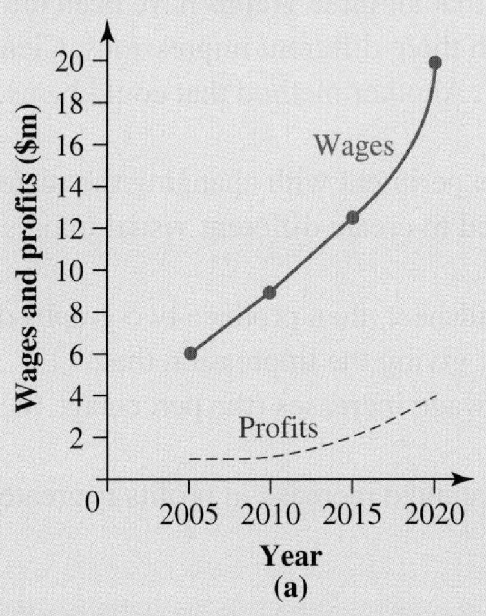

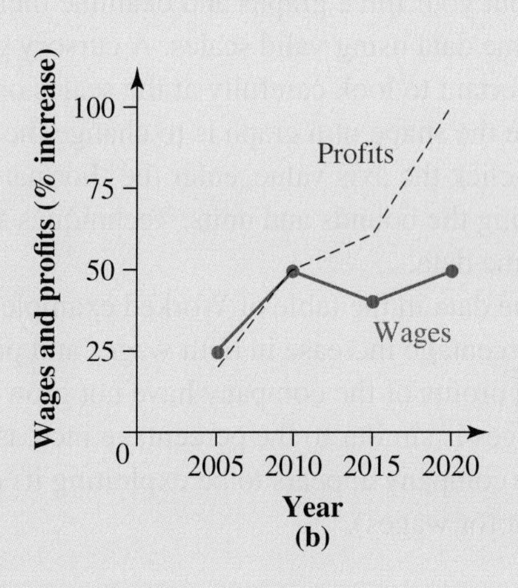

a. **Explain whether the graphs accurately reflect the data.**
b. **Explain which graph you would rather have published if you were:**
 i. **an employer dealing with employees requesting pay increases**
 ii. **an employee negotiating with an employer for a pay increase.**

THINK	WRITE
a. 1. Look at the scales on both axes. All scales are linear.	a. Graphs do represent data accurately. However, quite a different picture of wage and profit increases is painted by graphing with different units on the y-axis.
2. Look at the units on both axes. Graph (a) has the y-axis in $ while graph (b) has the y-axis in %.	
b. i. 1. Compare wage increase with profit increase.	b. i. The employer would prefer graph (a) because they could argue that employees' wages were increasing at a greater rate than profits.
2. The employer wants high profits.	
ii. 1. Consider again the increases in wages and profits.	ii. The employees would choose graph (b), arguing that profits were increasing at a great rate while wage increases clearly lagged behind.
2. The employees don't like to see profits increasing at a much greater rate than wages.	

5.8.2 Spreadsheets creating misleading graphs

Let us return to Worked example 22 and use a spreadsheet to draw graph (a).

1. Enter the data as indicated in the spreadsheet.
2. Select the data and select 'Insert', then 'Insert Scatter' and select 'Scatter with smooth lines'. This will create Graph 1.
3. Copy and paste the graph twice within the spreadsheet.
4. Graph 2 gives the impression that the wages are a great deal higher than the profits. This effect was obtained by reducing the length of the horizontal axis. Experiment with shortening the horizontal length and lengthening the vertical axis.
5. In Graph 3, we get the impression that the wages and profits are not very different. This effect was obtained by lengthening the horizontal axis and shortening the vertical axis. Experiment with various combinations.
6. Print out your three graphs and examine their differences. Note that all three graphs have been drawn from the same data using valid scales. A cursory glance leaves us with three different impressions. Clearly, it is important to look carefully at the scales on the axes of graphs. Another method that could be used to change the shape of a graph is to change the scale of the axes.
7. Right-click the axis value, enter the 'Format axis' option, then experiment with changing the scale by adjusting the bounds and units. Techniques such as these are used to create different visual impressions of the same data.
8. Use the data in the table of Worked example 23 to create a spreadsheet, then produce two graphs depicting the percentage increase in both wages and profits over the years, giving the impression that:
   a. the profits of the company have not grown at the expense of wage increases (the percentage increase in wages is similar to the percentage increase in profits)
   b. the company appears to be exploiting its employees (the percentage increase in profits is greater than that for wages).

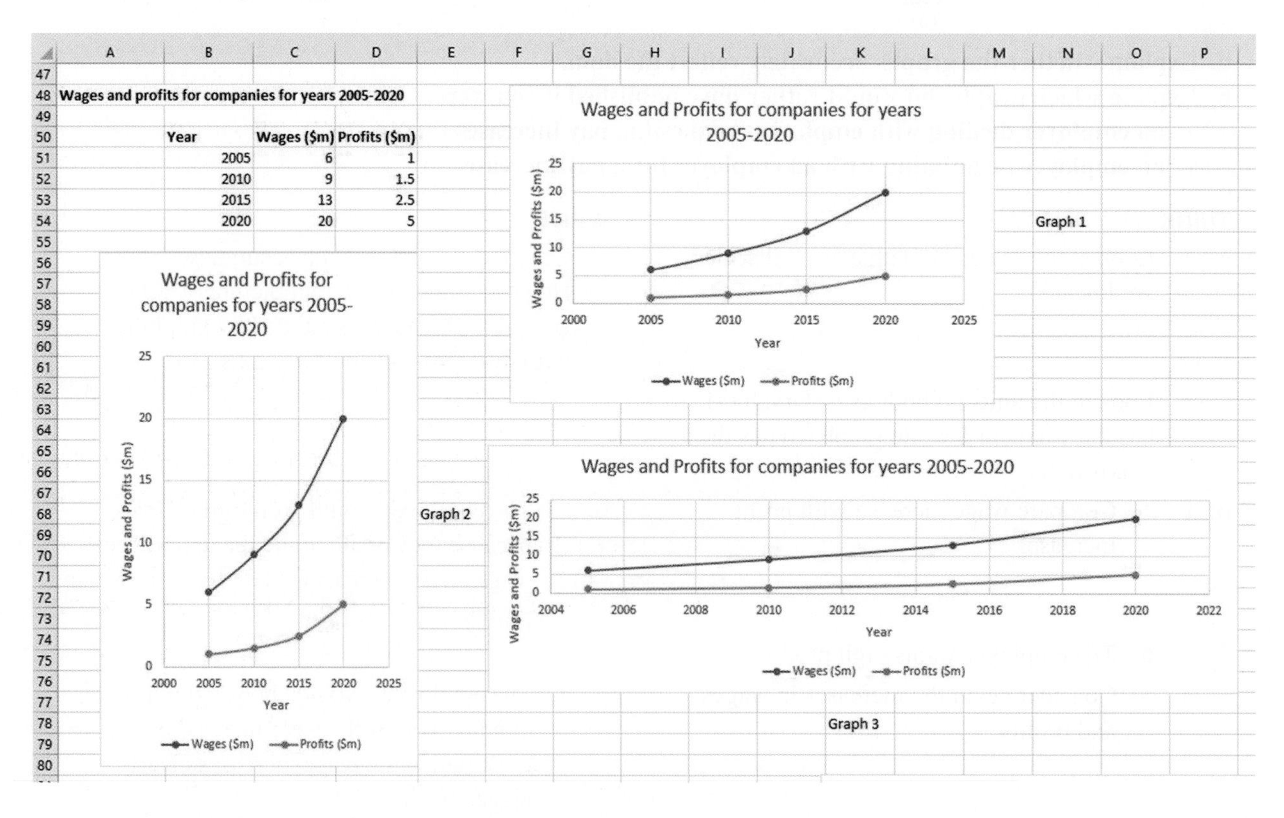

5.8 Exercise

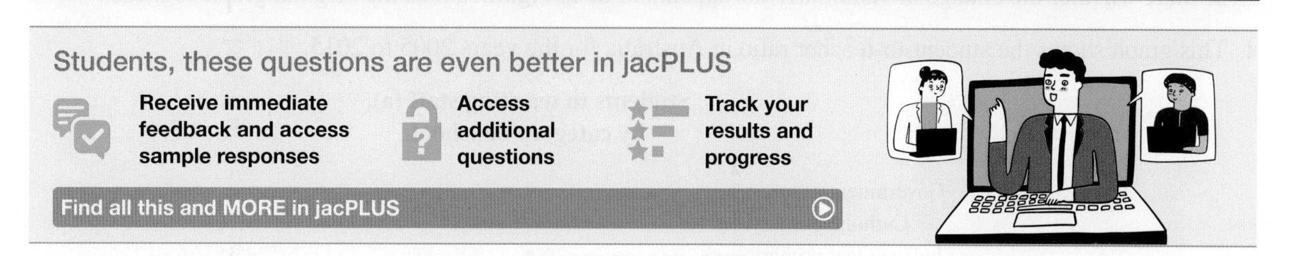

1. This graph shows the money spent on research by a company in the years 2014, 2018 and 2022. Draw another bar graph that minimises the appearance of the fall in research funds.

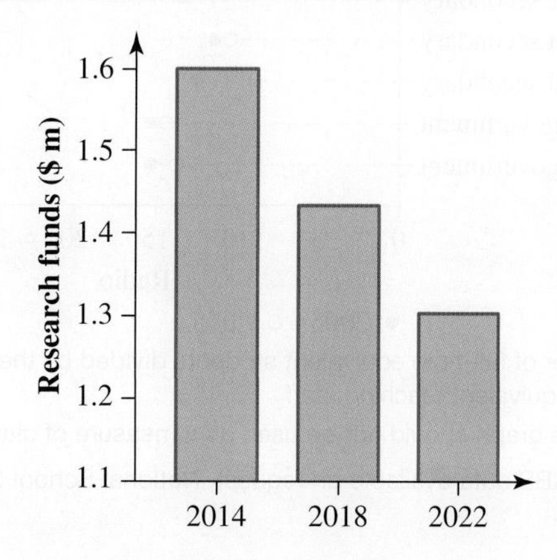

2. WE23 Examine this graph of employment growth in a company.

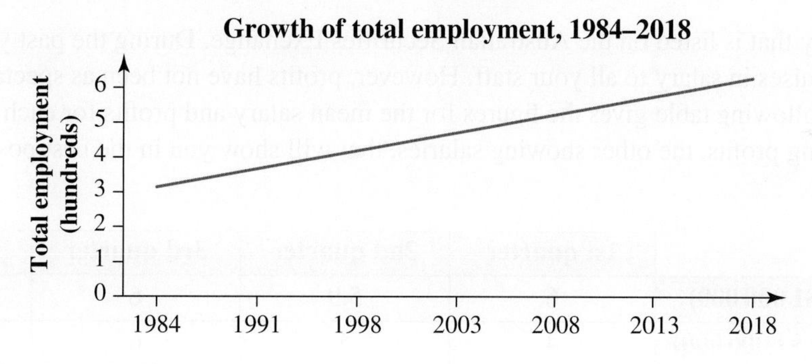

Explain why the graph is misleading.

3. Examine this graph.

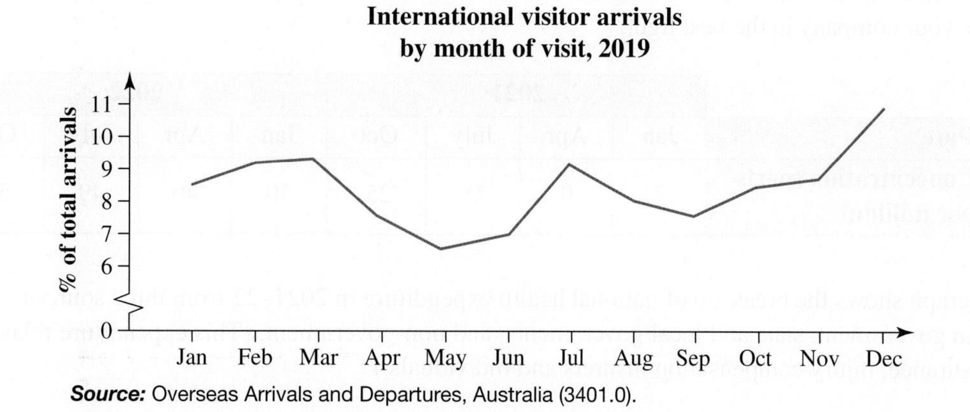

Source: Overseas Arrivals and Departures, Australia (3401.0).

a. Redraw this graph with the vertical axis showing percentage of total arrivals starting at 0.
b. State whether the change in visitor arrivals appears to be as significant as the original graph suggests.

4. This graph shows the student-to-teacher ratio in Australia for the years 2005 to 2015.

Students to teaching staff (a), by category of school

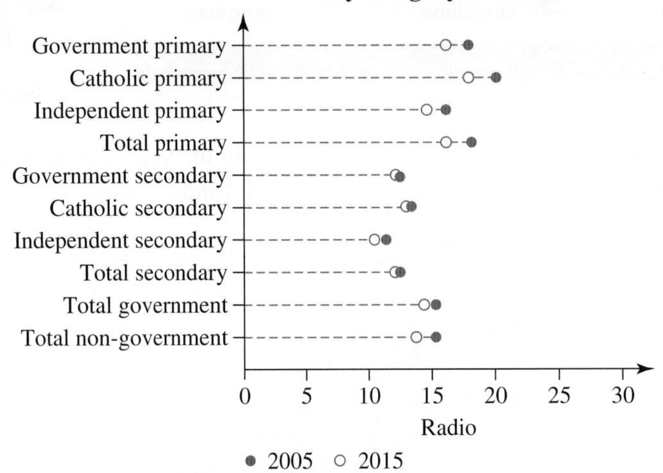

(a) Number of full-time equivalent students divided by the number of full-time equivalent teaching staff

Note: This graph should not be used as a measure of class size.

Source: ABS data available on request, National School Statistics collection.

a. Describe what has generally happened to the ratio of students to teaching staff over the 10-year period.
b. A note says that the graph should not be used as a measure of class size. Explain why.

5. You run a company that is listed on the Australian Securities Exchange. During the past year you have given substantial raises in salary to all your staff. However, profits have not been as spectacular as in the year before. The following table gives the figures for the mean salary and profits for each quarter. Draw two graphs, one showing profits, the other showing salaries, that will show you in the best possible light to your shareholders.

	1st quarter	2nd quarter	3rd quarter	4th quarter
Profits (× \$1 000 000)	6	5.9	6	6.5
Salaries (× \$1 000 000)	4	5	6	7

6. You are a manufacturer and your plant is discharging heavy metals into a waterway. Your own chemists do tests every three months and the following table gives the results for a period of two years. Draw a graph that will show your company in the best light.

	2021				2022			
Date	Jan	Apr	July	Oct	Jan	Apr	July	Oct
Concentration (parts per million)	7	9	18	25	30	40	49	57

7. This pie graph shows the break-up of national health expenditure in 2021–22 from three sources: Australian government, state and local governments, and non-government. (This expenditure relates to private health insurance, injury compensation insurers and individuals.)

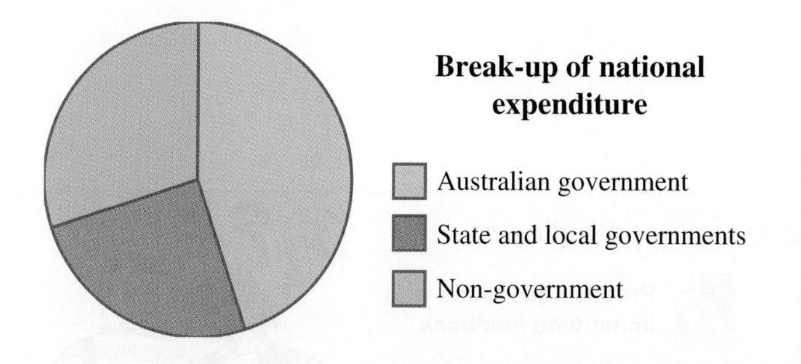

Expenditure source ($m)%	($m)	%
Australian government	37 229	45
State and local governments	21 646	25
Non-government	28 004	30

a. Comment on the claim that $87 000 million was spent on health from these three sources.
b. Determine which source contributes least to the national health expenditure. Comment on its quoted percentage.
c. Determine which source contributes the next greatest amount to the national health expenditure. Comment on its quoted percentage.
d. The Australian government contributes the greatest amount. Comment on its quoted percentage.
e. Consider the pie chart.
  i. Based on the percentages shown in the table, state what the angles should be.
  ii. Based on the actual expenditures, state what the angles should be.
  iii. Measure the angles in the pie chart and comment on their values.

8. This graph shows how the $27 that a buyer pays for a new album is distributed among the departments involved in its production and marketing.

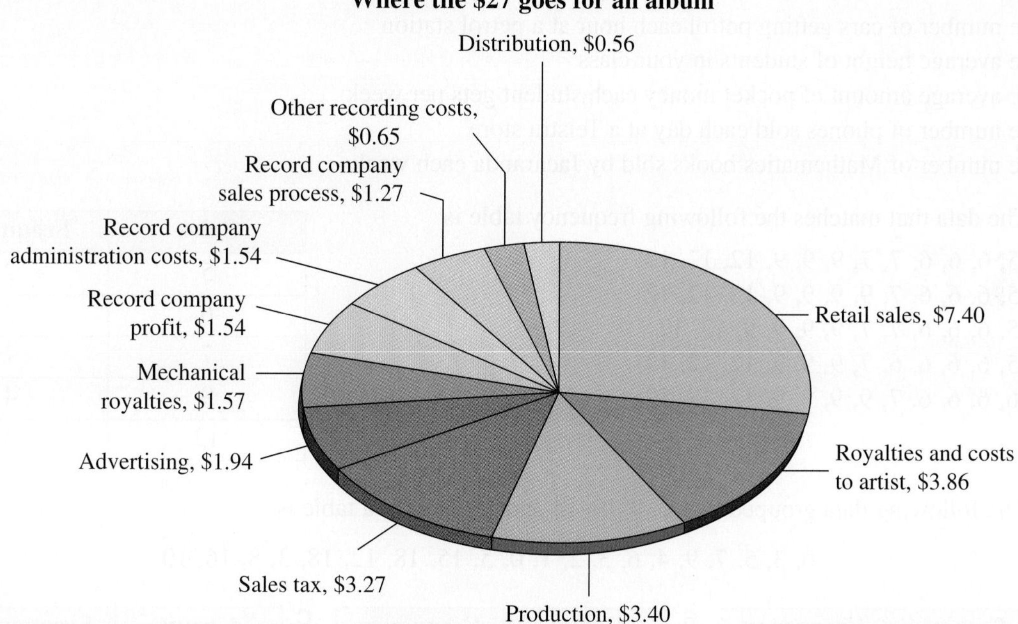

Determine whether or not the graph is misleading. Fully explain your reasoning and support any statements that you make. Also:

a. comment on the shape of the graph and how it could be obtained
b. explain whether your visual impression of the graph supports the figures.

5.9 Review

5.9.1 Summary

doc-38013

Hey students! Now that it's time to revise this topic, go online to:

Review your results

Watch teacher-led videos

Practise questions with immediate feedback

Find all this and MORE in jacPLUS

5.9 Exercise

Multiple choice

1. MC Determine which of the following describes categorical data.
   - A. The number of people at Bunnings each day
   - B. The goals kicked by Richmond Tigers each week
   - C. Your classmate's favourite movie this year
   - D. The time each student spends on homework each night
   - E. The amount of rubbish dumped at the tip each day.

2. MC Determine which of the following represents continuous data.
   - A. The number of cars getting petrol each hour at a petrol station
   - B. The average height of students in your class
   - C. The average amount of pocket money each student gets per week
   - D. The number of phones sold each day at a Telstra store
   - E. The number of Mathematics books sold by Jacaranda each week

3. MC The data that matches the following frequency table is:
   - A. 5, 5, 6, 6, 6, 7, 7, 9, 9, 9, 12, 12, 12
   - B. 5, 5, 6, 6, 6, 7, 9, 9, 9, 9, 12, 12, 12
   - C. 5, 5, 6, 6, 6, 7, 7, 9, 9, 9, 9, 12, 12
   - D. 5, 5, 6, 6, 6, 6, 7, 9, 9, 9, 12, 12, 12
   - E. 5, 6, 6, 6, 6, 7, 9, 9, 9, 9, 12, 12, 12

Sales	Frequency
5	2
6	3
7	1
9	4
12	3

4. MC The following data grouped 0–4, 5–9, 10–14 and 15–19 into a table is:

   0, 3, 5, 7, 9, 4, 6, 3, 2, 1, 0, 5, 15, 18, 12, 18, 3, 8, 16, 10

A.

Group	Frequency
0–4	7
5–9	6
10–14	2
15–19	4

B.

Group	Frequency
0–4	8
5–9	6
10–14	2
15–19	4

C.

Group	Frequency
0–4	7
5–9	6
10–14	2
15–19	4

D.

Group	Frequency
0–4	8
5–9	5
10–14	3
15–19	4

E.

Group	Frequency
0–4	8
5–9	5
10–14	2
15–19	5

5. MC The data that matches the histogram shown is:

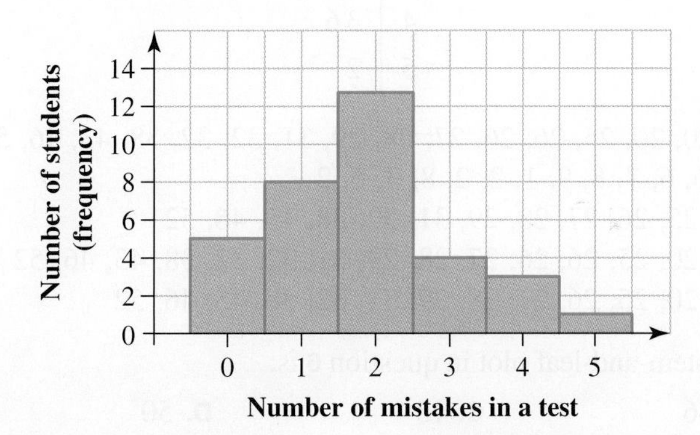

A.

Number of mistakes in a test	Frequency
0	4
1	8
2	13
3	4
4	3
5	1

B.

Number of mistakes in a test	Frequency
0	4
1	8
2	13
3	4
4	4
5	1

C.

Number of mistakes in a test	Frequency
0	5
1	8
2	13
3	4
4	3
5	1

D.

Number of mistakes in a test	Frequency
0	5
1	8
2	12
3	4
4	3
5	1

E.

Number of mistakes in a test	Frequency
0	5
1	8
2	12
3	4
4	4
5	1

6. **MC** The data that matches the following stem-and-leaf plot is:

Key: 0|6 = 6 errors

Stem	*Leaf*
0	6
1	3 5 7 8
2	0 0 5 6 6 7 8 9
3	1 2 2 8
4	3 6
5	2

A. 0, 6, 13, 15, 17, 18, 20, 20, 25, 26, 26, 27, 28, 29, 31, 32, 32, 38, 43, 46, 52
B. 6, 3, 5, 7, 8, 0, 0, 5, 6, 6, 7, 8, 9, 1, 2, 2, 8, 3, 6, 2
C. 6, 13, 15, 17, 18, 20, 25, 26, 27, 28, 29, 31, 32, 38, 43, 46, 52
D. 6, 13, 15, 17, 18, 20, 20, 25, 26, 26, 27, 28, 29, 31, 32, 32, 38, 43, 46, 52
E. 6, 13, 15, 17, 18, 20, 20, 25, 26, 27, 28, 29, 31, 32, 38, 43, 46, 52

7. **MC** The range from the stem-and-leaf plot in question 6 is:

A. 52 **B.** 46 **C.** 13 **D.** 50 **E.** 48

8. **MC** For the following data set, the five-number summary is:

3, 15, 16, 10, 8, 9, 21, 16, 12, 23

A. 3, 9, 13.5, 16, 23 **B.** 3, 9, 13, 16, 23 **C.** 3, 10, 13.5, 16, 23
D. 3, 9, 13.5, 21, 23 **E.** 3, 9, 14, 18, 23

The following box-and-whisker plot is to be used for questions 9 and 10.

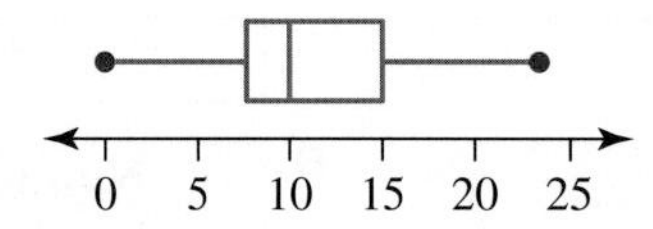

9. **MC** The median of the box-and-whisker plot is:

A. 0 **B.** 8 **C.** 10 **D.** 15 **E.** 24

10. **MC** The upper quartile (Q_3) value of the box-and-whisker plot is:

A. 0 **B.** 8 **C.** 10 **D.** 15 **E.** 24

Short answer

11. State whether each of the following data types is categorical or numerical.
   a. The television program that people watch at 7:00 pm
   b. The number of pets in each household
   c. The amount of water consumed by athletes in a marathon run
   d. The average distance that students live from school
   e. The mode of transport used between home and school

12. For each of the numerical data types below, determine if the data are discrete or continuous.
   a. The dress sizes of Year 11 girls
   b. The volume of backyard swimming pools
   c. The amount of water used in households
   d. The number of viewers of a particular television program
   e. The amount of time Year 11 students spend studying

13. Identify the problems with the following questions, then rewrite each so that the meaning is clear and understandable.
   a. Do you think that the school holidays are too long at Christmas time and that they should be spread evenly over the year?
   b. Have you never said that you don't like chocolate?
   c. Do you support the building of a fence around the duck pond to protect the poor defenceless ducklings from the savage dogs in the area?
   d. Did you know that the VCAA administer the VCE exams?
   e. You support the police in their endeavour to reduce the road toll, don't you?

14. a. A group of Year 11 students were asked to state the number of books that they had purchased in the last year. The results are shown below.

12	1	13	20	5	22	35	12	17	20
9	5	11	0	14	25	3	8	10	9
12	6	18	7	10	9	6	23	14	19

   Put the results into a table using the categories 0–4, 5–9, 10–14, etc.
   b. Draw a column graph and a sector graph to represent the results.

15. The table below shows the number of sales made each day over a month in a car yard.

Number of sales	Frequency
0	2
1	5
2	12
3	6
4	2
5	0
6	1

   Show this information in a frequency histogram and polygon.

Extended response

16. Represent the number of coffees drunk by a person during the week on a pictogram.

Day	Coffees drunk
Monday	3
Tuesday	2
Wednesday	1
Thursday	3
Friday	4
Saturday	6
Sunday	5

17. Sketch a radar chart of AFL players' key performance indicators shown.

KPI	Scott	Patrick	Christian
Marks	7	6	8
Tackles	6	8	6
Running	9	5	6
Clearances	8	10	9

18. For the box-and-whisker plot shown:

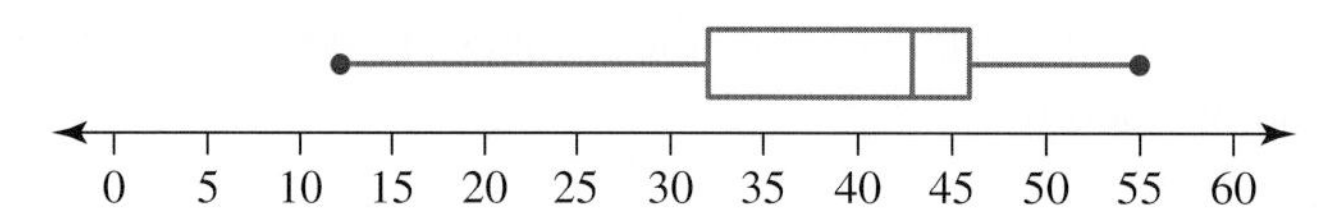

a. state the median
b. calculate the range
c. calculate the interquartile range.

19. Consider the following stem-and-leaf plot. It represents the number of typing errors recorded by a class of students in one page of typing.

Key: 1|2 = 12
1*|5 = 15

Stem	*Leaf*
0	0 1 4
0*	6 7 8 9
1	0 0 1 1 2 3 3 4
1*	5 6 8 9
2	3

a. Determine the number of students in the class.
b. Determine the median number of errors.

20. The following box-and-whisker plot shows the sales of two different brands of washing powder at a supermarket each day.

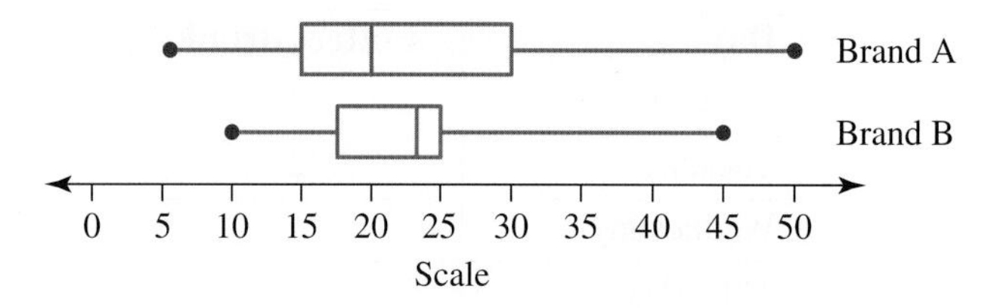

a. State the range for brand A.
b. State the interquartile range for brand A.

Answers

Topic 5 Data collection and organisation

5.2 Types of data

5.2 Exercise

1. a. Numerical b. Categorical c. Numerical
2. a. Categorical b. Numerical c. Numerical
3. a. Continuous b. Discrete c. Continuous
   d. Continuous e. Discrete
4. a. Numerical and discrete
   b. Categorical
   c. Categorical
   d. Continuous and numerical
   e. Continuous and numerical
   f. Numerical and discrete
5. Categorical and ordinal
6. Categorical and ordinal
7. Numerical and discrete
8. B
9. Categorical
10. Continuous and numerical
11. Nominal
12. Ordinal
13. a. Continuous b. Continuous c. Discrete
14. a. Continuous b. Discrete c. Ordinal

5.3 Collecting, organising and displaying data

5.3 Exercise

1. Open questions are those where the respondent has no guided boundaries within which to answer. Closed questions are those where the respondent must answer within a category.
   Open question:
   How long do you spend exercising at the gym each week?
   Closed question:
   Which of the following categories describes your weekly gym exercise program?
   ☐ no gym sessions
   ☐ less than 30 minutes
   ☐ about 1 hour
   ☐ about 2 hours
   ☐ more than 2 hours
2. One classification could be:
   1. The entrance: 1, 2, 3, 6, 8, 13, 14
   2. Student congestion: 5, 11, 15, 16, 18
   3. Vehicle parking: 4, 10, 12, 19
   Not classified: 7, 9, 17, 20
3. One classification could be:
   1. The work: 1, 2, 7, 8, 9, 13, 18, 19, 20, 23, 26, 27
   2. Teacher: 3, 15, 16, 17, 24, 25, 29
   3. Gender: 5, 6, 10, 11
   4. Interest: 4, 22, 28, 30
   Not classified: 12, 14, 21
4. a. This question is vague. What period of time does it refer to? Is it just for the main job?
   What is your yearly income?
   ☐ less than \$30 000
   ☐ \$30 000 – < \$50 000
   ☐ \$50 000 – < \$70 000
   ☐ \$70 000 – < \$100 000
   ☐ \$100 000 or more
   b. The terms should be defined.
   There are various forms of exercise — sport, walking, gym …
   Regularly could be twice every day, three times every week, once every month, …
   Do you walk at least 10 000 steps each day?
   ☐ yes
   ☐ no
   c. Explain the abbreviations.
   You are familiar with the Goods and Services Tax (GST) of 10% in Australia. The equivalent tax in England is called the Value Added Tax (VAT). Is the GST:
   ☐ less than the VAT
   ☐ equal to the VAT
   ☐ greater than the VAT?
   d. Emotional language can persuade respondents to answer in a biased manner.
   Do you generally support the causes of terrorists?
   ☐ yes
   ☐ no
   e. Explain the Prime Minister's policy on wildlife preservation.
   The Prime Minister does not support a policy of culling kangaroos. Do you agree with his policy?
   ☐ yes
   ☐ no
   f. Inches is an old imperial unit of length. Respondents may not know the conversion from cm to inches.
   One inch is equivalent to approximately 2.5 cm. What is your height in inches?
   g. This is a double-barrelled question. When you buy sneakers, what do you generally look for?
   ☐ comfort
   ☐ quality
   ☐ neither of these
   h. This is a leading question which tends to cause the respondent to answer 'no'.
   Do you agree with the policy of charging more for skim milk than for full cream milk?
   ☐ yes
   ☐ no
   i. Emotional language causes respondents to be sympathetic to a cause.
   Do you think that our diggers have been treated well since they fought during the war?
   ☐ yes
   ☐ no

5. a. How old are you?
- ☐ less than 10 years
- ☐ 10 years – < 15 years
- ☐ 15 years – < 20 years
- ☐ 20 years – < 30 years
- ☐ 30 years or older

b. How much pocket money do you get each week?
- ☐ less than \$5
- ☐ \$5 – < \$10
- ☐ \$10 – < \$15
- ☐ \$15 or more

c. How do you travel to school?
- ☐ car
- ☐ bus
- ☐ train
- ☐ bike
- ☐ walk
- ☐ other

d. What type of destination do you prefer for a holiday?
- ☐ seaside
- ☐ city
- ☐ outback
- ☐ snow
- ☐ other

6. a. As many students as possible should be asked about their internet access.

b. Possible answers are: online survey, interview, letters to parents/guardians.

c. An online survey would be the most efficient.

7. A

8. C

9. a.

Make	Tally	Frequency
Holden	𝍸 |||	8
Ford	𝍸 |||	8
Nissan	||	2
Mazda	|||	3
Toyota	𝍸 ||	7
Mitsubishi	||	2

b.

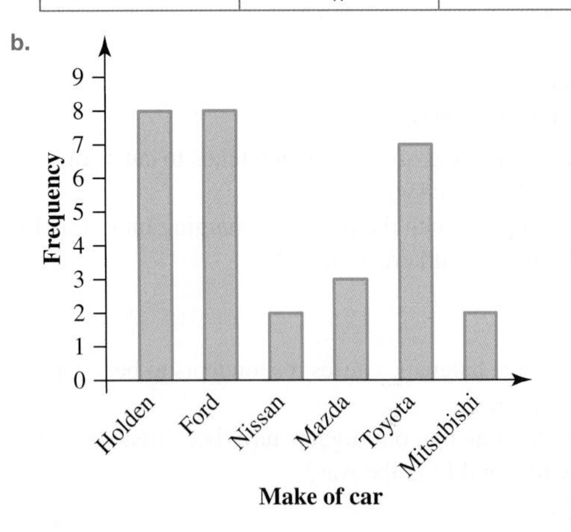

c.

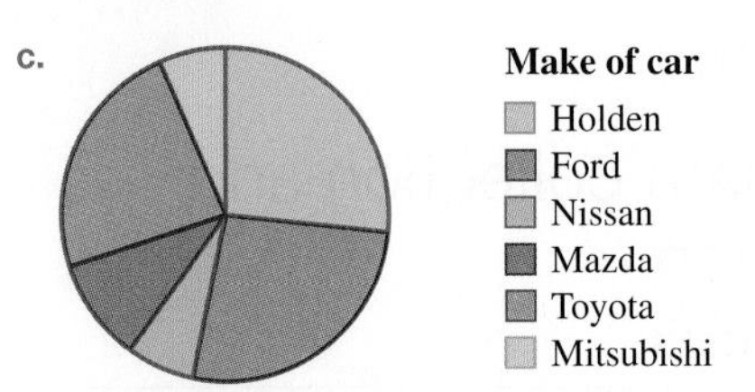

10. a.

Make	Tally	Frequency
4				2
5						4
6	𝍸		6
7	𝍸 ||||	9
8	𝍸	5
9					3
10			1

b.

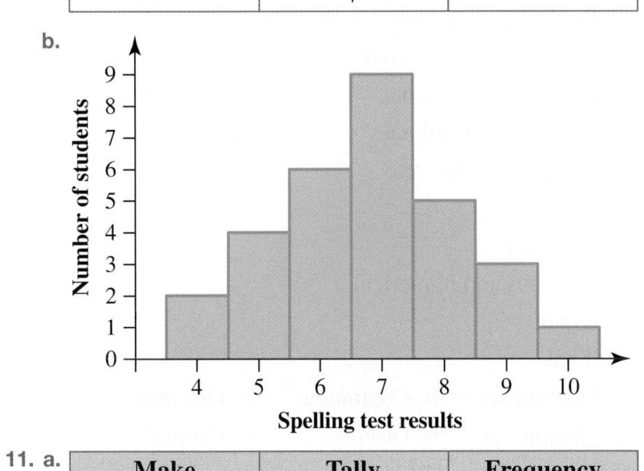

11. a.

Make	Tally	Frequency
40–49			1
50–59	||	2
60–69	𝍸 ||||	9
70–79	𝍸 |||	8
80–89	|||	3
90–99	||	2

b.

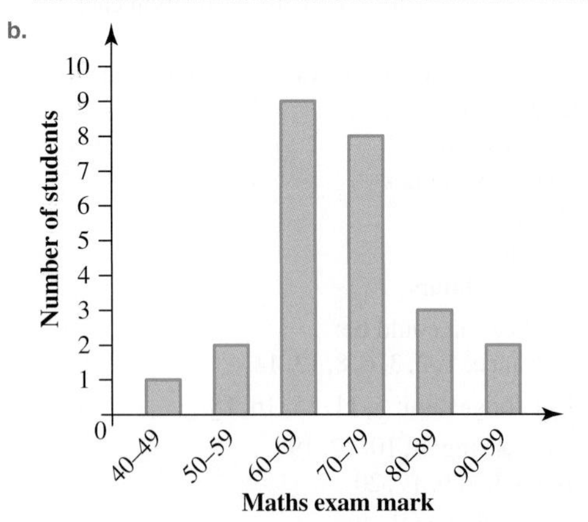

c.

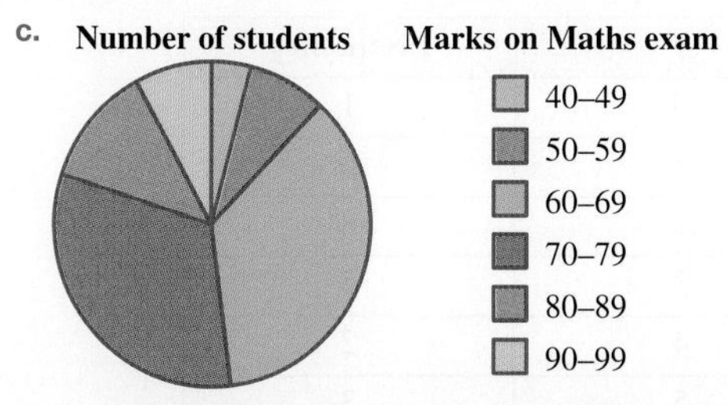

12. a.

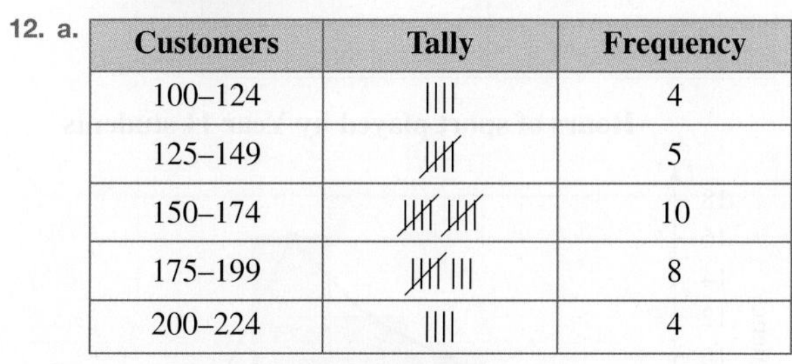

Customers	Tally	Frequency
100–124	||||	4
125–149	𝍸	5
150–174	𝍸 𝍸	10
175–199	𝍸 |||	8
200–224	||||	4

b. See the graph at the bottom of the page.*

5.4 Histograms and frequency polygons

5.4 Exercise

1.

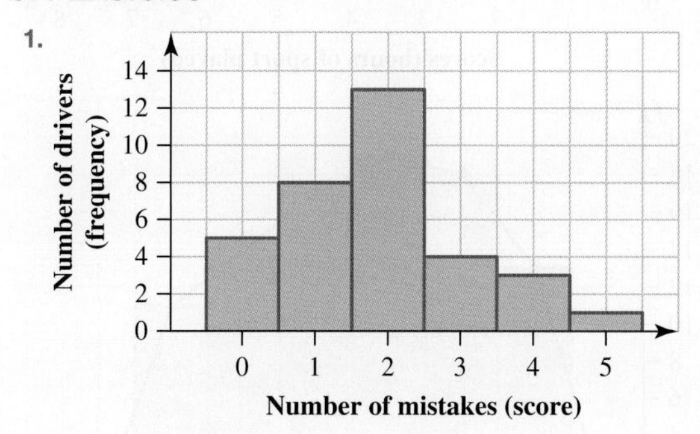

2.

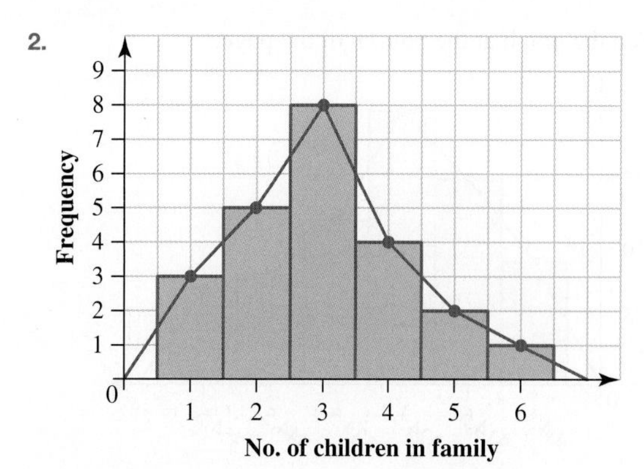

3.

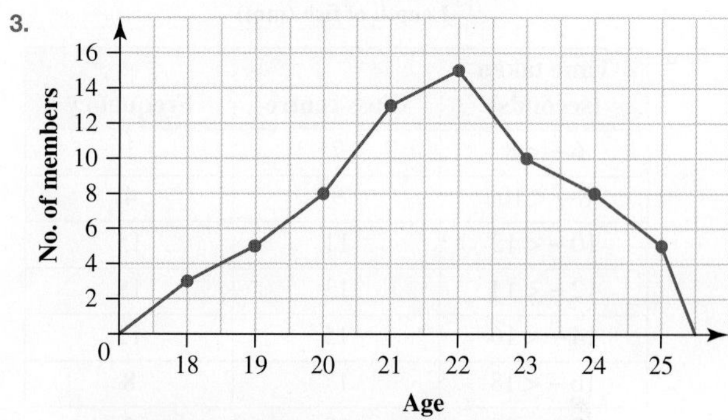

4. a.

Number of matches in a box	Frequency
47	3
48	5
49	10
50	15
51	7
52	5
53	4
54	1

*12. b.

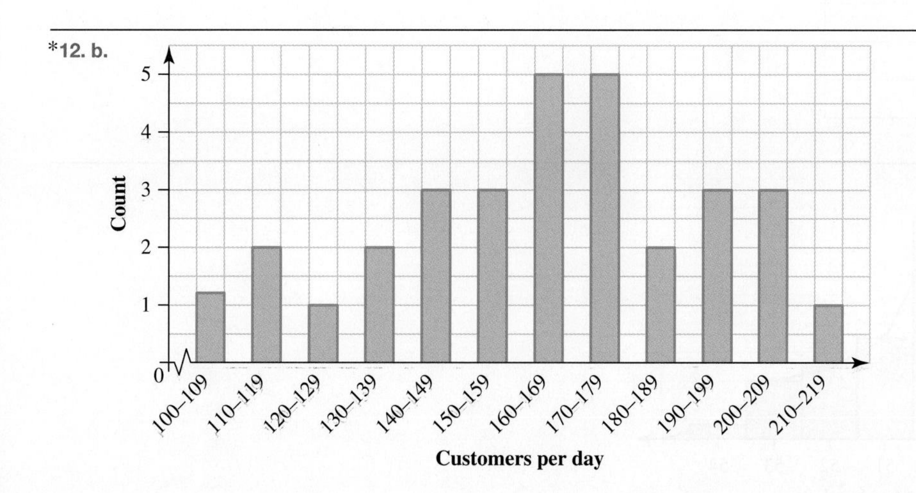

b. See the graph at the bottom of the page.*

5.

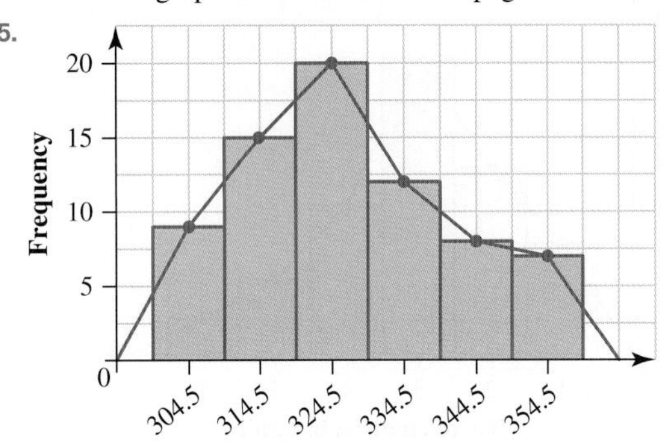

6. a.

Time taken (seconds)	Class centre	Frequency
6– < 8	7	1
8– < 10	9	4
10– < 12	11	15
12– < 14	13	18
14– < 16	15	12
16– < 18	17	8
18– < 20	19	2

b.

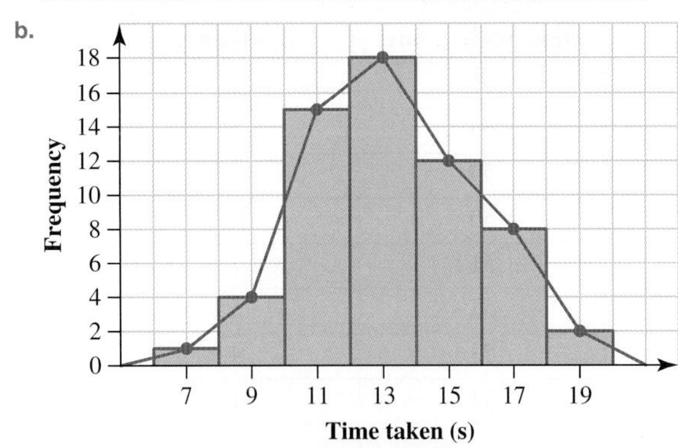

7.

x	Frequency
0	5
1	8
2	13
3	7
4	5
5	2
Total	40

8.

x	Frequency
0	1
1	6
2	8
3	5
4	2
5	3
Total	25

9.

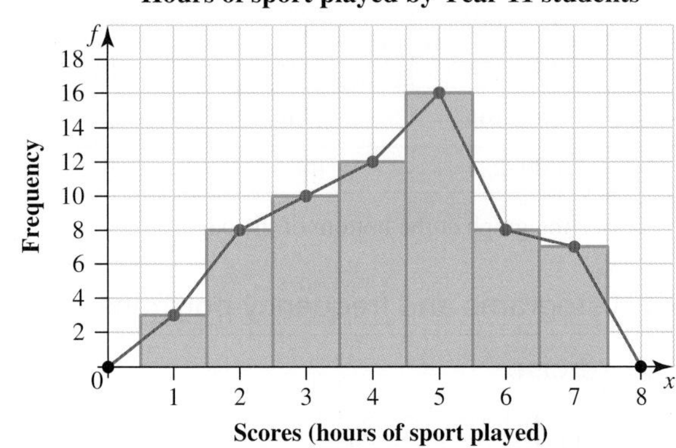

10.

f
Frequency
20
18
16
14
12
10
8
6
4
2
0
100 150 200 250 300 350 400
x
Size of house (m^2)

*4. b.

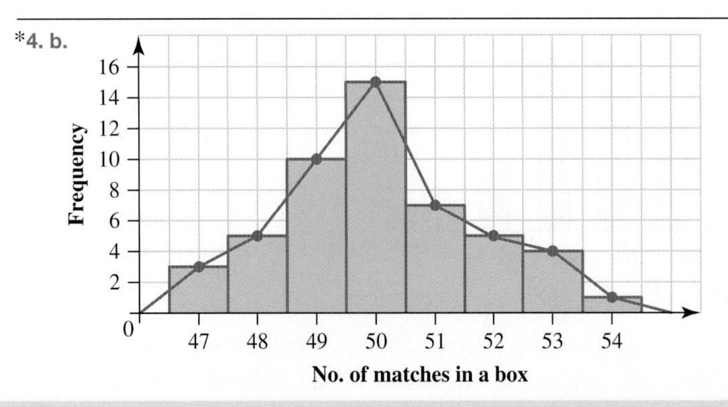

11. a.

Class interval	Midpoint (class centre)	Frequency
5– < 10	7.5	11
10– < 15	12.5	19
15– < 20	17.5	5
20– < 25	22.5	2
25– < 30	27.5	3
Total		40

b.

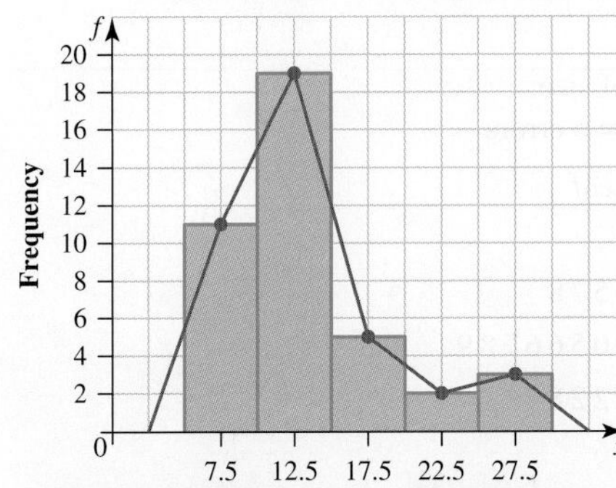

12. a.

Money on snacks	Frequency
0– < 5	2
5– < 10	9
10– < 15	7
15– < 20	2
Total	20

b.

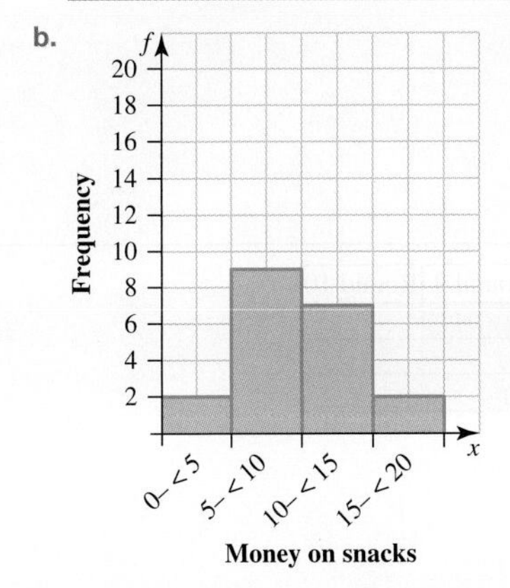

c. This is a very small sample of 17-year-olds, so it could not be said that this reflects their spending on snacks.

5.5 Contemporary representation of data

5.5 Exercise

1. C
2. E
3. B
4.

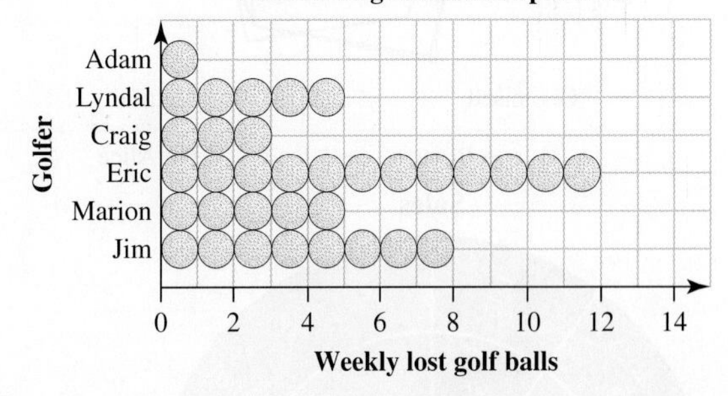

5.

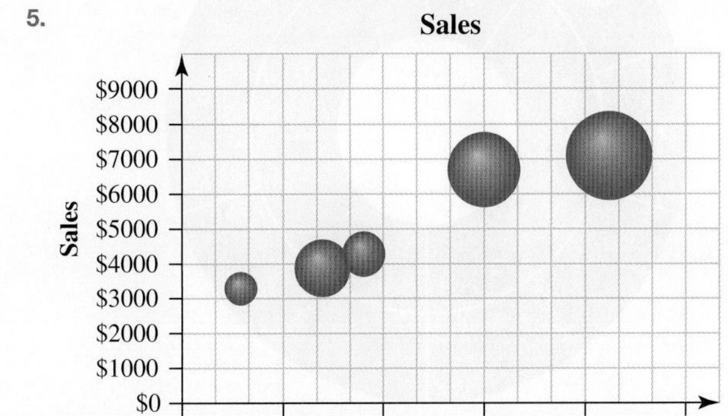

6. See the graph at the bottom of the page.*

*6.

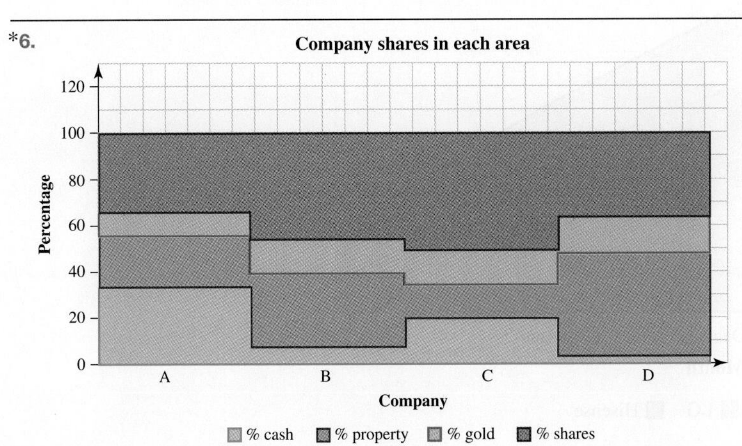

7.

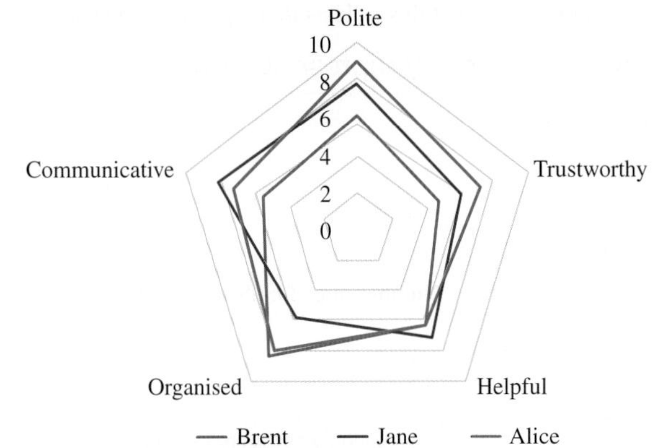

8.

Sales

9. See the table at the bottom of the page.*

10. See the graph at the bottom of the page.**

5.6 Stem-and-leaf plots

5.6 Exercise

1. Key: 0|6 = 6 errors

Stem	*Leaf*
0	6
1	5 8 7 3
2	0 5 8 7 6 6 9 0
3	2 8 1 2
4	3 6
5	2

In order of size:
Key: 0|6 = 6 errors

Stem	*Leaf*
0	6
1	3 5 7 8
2	0 0 5 6 6 7 8 9
3	1 2 2 8
4	3 6
5	2

*9.

Day	Round 1	Round 2	Round 3	Round 4	Round 5	Round 6	Round 7	Round 8	Round 9	Round 10
Fri										
Sat										
Sun										

**10.

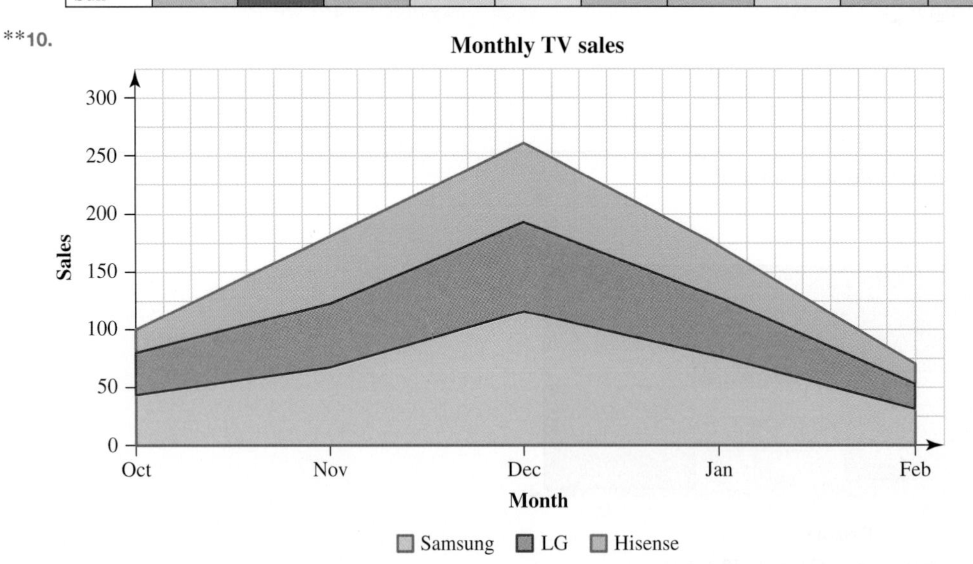

2. Key: 3|6 = 36 min

Stem	*Leaf*
3	6 8 8 9 9 7
4	2 7 5 2 2 6 8 2 0 1 7 3 8 2 6 0 8
5	2 9 7 5 8 3 0 2 4 2
6	8 6 8
7	2 5 1
8	2

In order of size:
Key: 3|6 = 36 min

Stem	*Leaf*
3	6 7 8 8 9 9
4	0 0 1 2 2 2 2 2 3 5 6 6 7 7 8 8 8
5	0 2 2 2 3 4 5 7 8 9
6	6 8 8
7	1 2 5
8	2

3. Key: 10|1 = 101 wpm
10*|6 = 106 wpm

Stem	*Leaf*
8*	8 6
9	2
9*	6 6 5 5 6 5 8 9
10	2 2 3 2
10*	8 7 7 8 7 8
11	0 1 2
11*	5
12	1 4 0 1
12*	
13	0

In order of size:
Key: 10|1 = 101 wpm
10*|6 = 106 wpm

Stem	*Leaf*
8*	6 8
9	2
9*	5 5 5 6 6 6 8 9
10	2 2 2 3
10*	7 7 7 8 8 8
11	0 1 2
11*	5
12	0 1 1 4
12*	
13	0

4. Key: 14|3 = 14.3 V
14*|8 = 14.8 V

Stem	*Leaf*
13*	8 9
14	0 3 3 2
14*	8 8 7 6 6 7 5
15	2 2 1
15*	7 5 6 9

In order of size:
Key: 14|3 = 14.3 V
14*|8 = 14.8 V

Stem	*Leaf*
13*	8 9
14	0 2 3 3
14*	5 6 6 7 7 8 8
15	1 2 2
15*	5 6 7 9

5. The median is halfway between the 12th and 13th scores.

Lower quartile = 249.6
Upper quartile = 251.55
IQR = 1.95 g

6. a. Median = 8.4s

b. Lower quartile = 7.8s
Upper quartile = 8.85s

c. IQR = 1.05s

7. D

8. B

9. E

10. C

11. a. Key: 20|6 = 20.6 cm

Stem	*Leaf*
20	2 7
21	4 6 7 8
22	2 8 8 8
23	0 1 2 6 6
24	6 6 7 8
25	1

b. Median = 22.9 cm

c. Lower quartile = 21.75 cm
Upper quartile = 24.1 cm

d. IQR = 2.35 cm

12. a. Key: 8|2 = 82 cm

Stem	*Leaf*
8	2 4 5 5 8
9	0 1 2 3 4 6 6 7 8 9
10	1 1 3 3 7 8
11	0 0 1 4 5 7
12	6 6
13	2

b. Median = 100 cm

c. Lower quartile = 92 cm
Upper quartile = 110 cm

d. IQR = 18 cm

5.7 Five-number summary and boxplots

5.7 Exercise

1. 8, 15, 16.5, 18, 25
2. a. 23, 44, 81.5, 83.5, 92 b. 1, 2, 4, 6, 7
   c. 8, 29, 45, 72, 93
3. a. 13 b. 5 c. 26
4. a. 122 b. 6 c. 27
5. 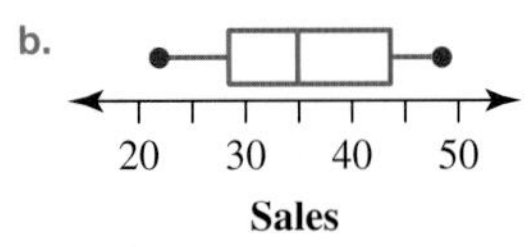

6. a. 147 b. 56 c. 90 d. 27 e. 8
7. a. 58 b. 31 c. 43 d. 27 e. 8
8. C
9. A
10. D
11. a. 22, 28, 35, 43, 48
    b. 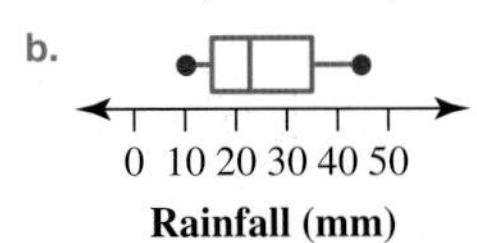

12. a. 10 mm, 13.5 mm, 22 mm, 33.5 mm, 45 mm
    b. 0 10 20 30 40 50
    Rainfall (mm)

5.8 Graphical methods of misrepresenting data

5.8 Exercise

1. 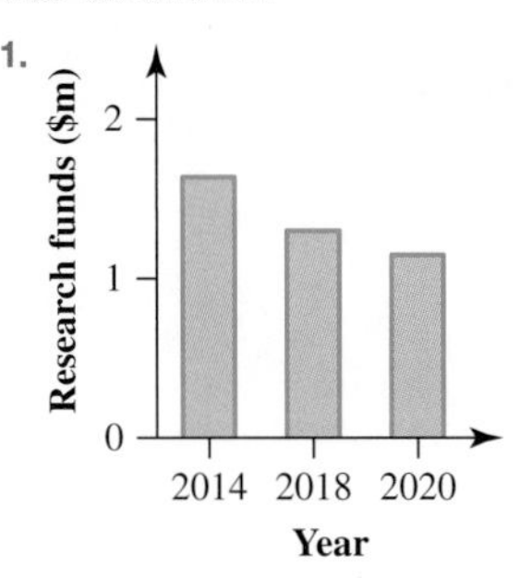

2. The horizontal axis uses the same division for both 5- and 7-year periods.
3. a. See the graph at the bottom of the page.*
   b. The change in visitor arrival rates throughout the year is still apparent; however, the new scale minimises these changes.
4. a. The student-to-teacher ratio has generally decreased over the 10-year period in all types of schools. This means that there are more teachers to cater for the number of students throughout Australia. The greatest change has occurred in Catholic primary schools, while the smallest change is noted in government secondary schools.
   b. These figures have been collected from large city schools to small country schools. An average of the numbers from this wide variety of schools is not an indication of average class sizes.

*3. a.

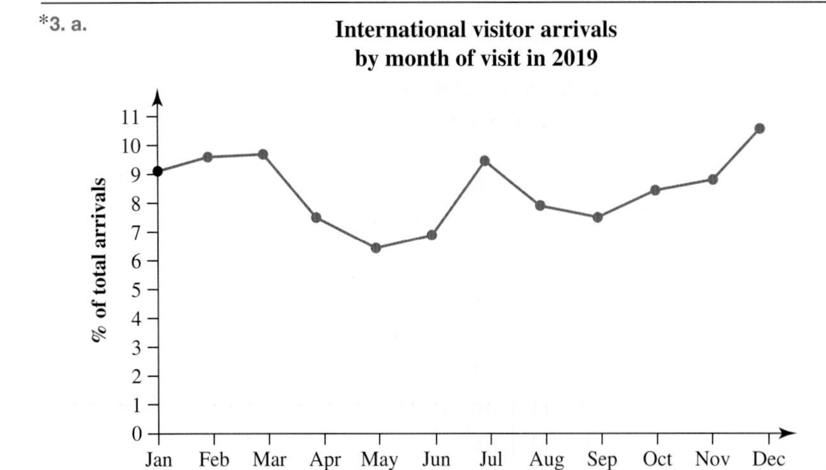

5. Company profits

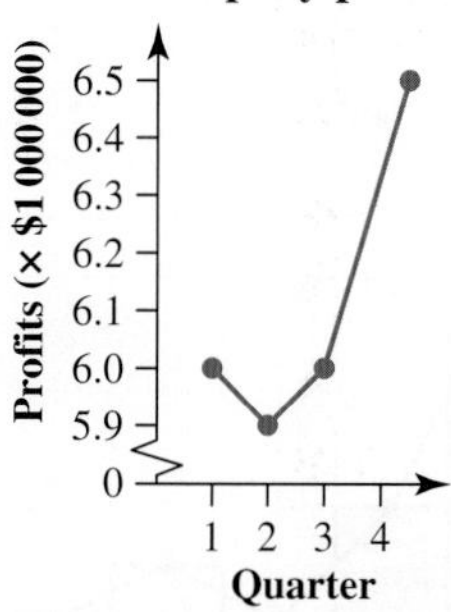

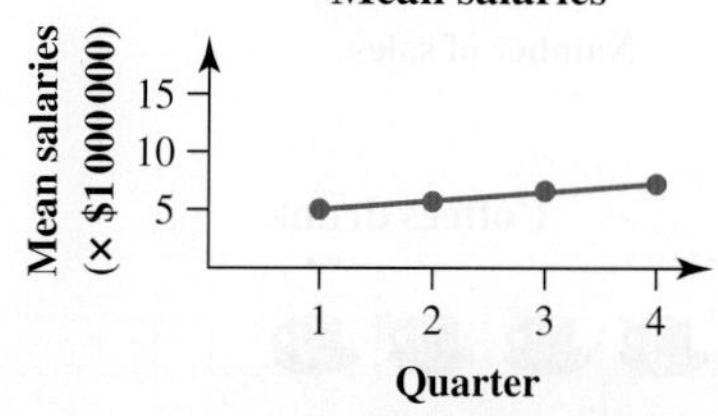

6.

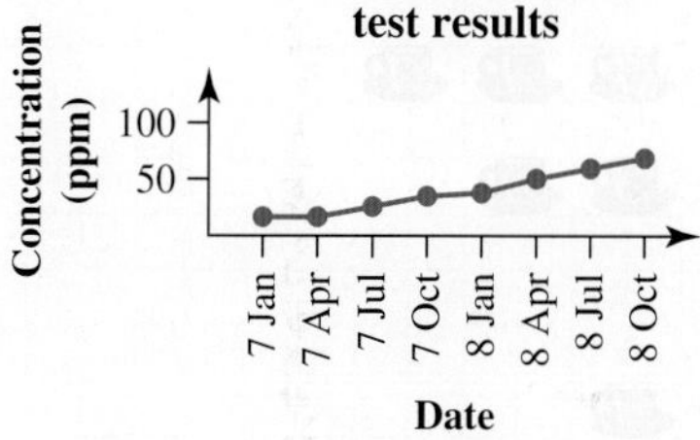

7. a. Total spent (in million dollars)

$= \$37\,229 + \$21\,646 + \$28\,004$
$= \$86\,879$ m

The claim of \$87 000 m is accurate enough in the context.

b. State and local governments contribute least to national health expenditure.

$$\text{Contribution \%} = \frac{21\,646}{86\,879} \times 100\%$$
$$= 24.9\%$$

This value has been rounded to 25%, showing that the quoted value is correct.

c. Non-government organisations are the next greatest contributor.

$$\text{Contribution \%} = \frac{28\,004}{86\,879} \times 100\%$$
$$= 32.2\%$$

This figure has been rounded down to 30%. The percentages being quoted seem to be rounded to the nearest 5%.

d. The Australian government contributes \$37 229 m.

$$\text{Contribution \%} = \frac{37\,229}{86\,879} \times 100\%$$
$$= 42.9\%$$

This figure has been rounded up to 45%, and this could be considered misleading in some contexts.

e. Angles for the pie graph:

i. Using quoted percentages:

$$\text{Australian government} = \frac{45}{100} \times 360° = 162°$$

$$\text{State and local governments} = \frac{25}{100} \times 360° = 90°$$

$$\text{Non-government} = \frac{30}{100} \times 360° = 108°$$

ii. Using actual expenditures:

$$\text{Australian government} = \frac{37\,229}{86\,879} \times 360° = 154°$$

$$\text{State and local governments} = \frac{21\,646}{86\,879} \times 360° = 90°$$

$$\text{Non-government} = \frac{28\,004}{86\,879} \times 360° = 116°$$

iii. Measuring pie graph angles:

Australian government = 154°
State and local governments = 78°
Non-government = 128°

Even though the pie graph gives a rough picture of the relative contributions of the three sectors, it has not been carefully drawn.

8. a. It is a circle viewed on an angle to produce an ellipse.

b. Because of the view, some of the sectors with roughly the same values appear to have greatly different central angles. This means that the visual impression does not support the figures in the graph.

5.9 Review

5.9 Exercise

Multiple choice

1. C
2. B
3. B
4. B
5. C
6. D
7. B
8. A
9. C
10. D

Short answer

11. a. Categorical b. Numerical c. Numerical
d. Numerical e. Categorical

12. a. Discrete b. Continuous c. Continuous
d. Discrete e. Continuous

13. a. This is a double-barrelled question.
Because of the length of the Christmas school holidays, do you think they should instead be spread evenly over the year?

b. There are two negatives in this question.
Has there ever been an occasion when you said that you don't like chocolate?

c. This question uses emotive language.
Would you support the building of a fence around the duck pond to protect the ducklings from the dogs in the area?

d. Explain the abbreviations used.
Did you know that the Victorian Curriculum and Assessment Authority is responsible for administering the VCE exams?

e. Don't lead the respondent to answer in a particular way.

f. Do you support the police in their endeavour to reduce the road toll?

14. a.

Number of books	Tally	Number of students
0–4	III	3
5–9	卌 IIII	9
10–14	卌 IIII	9
15–19	III	3
20–24	IIII	4
25–29	I	1
30–34		0
35–39	I	1

b.

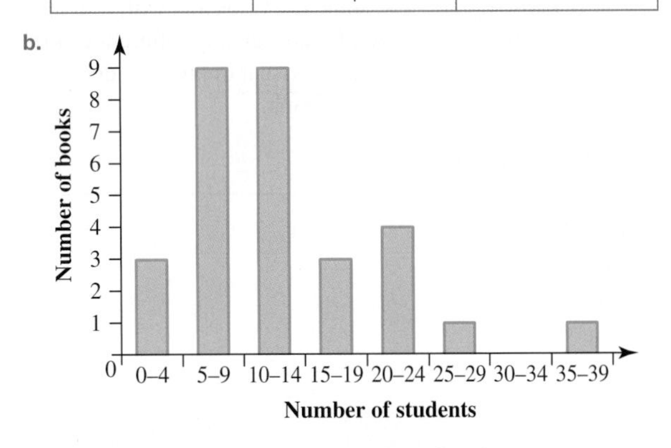

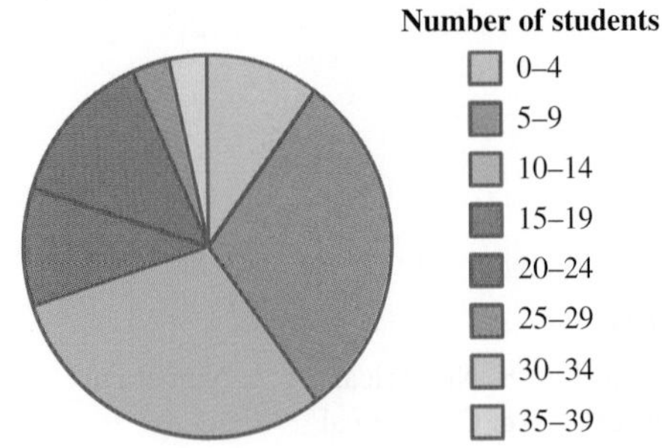

Number of students
- 0–4
- 5–9
- 10–14
- 15–19
- 20–24
- 25–29
- 30–34
- 35–39

15.

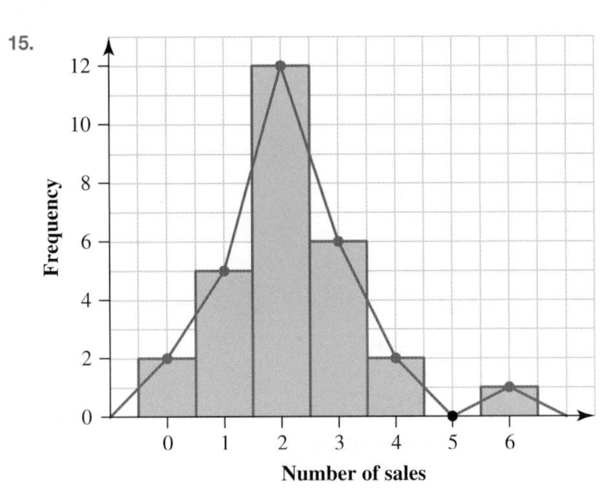

Extended response

16.

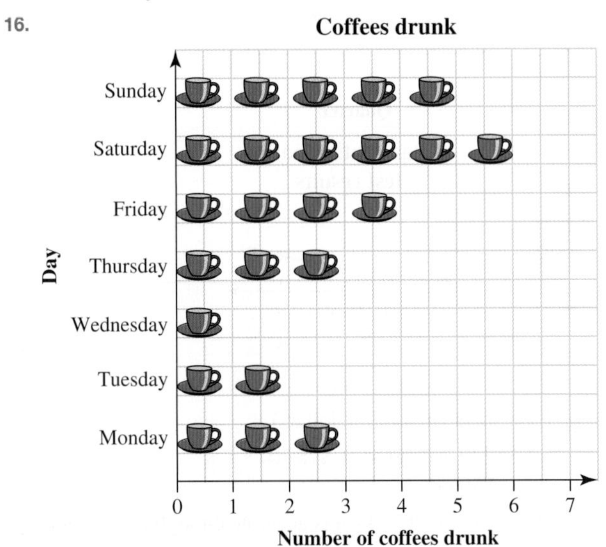

17.

Key performance indicators

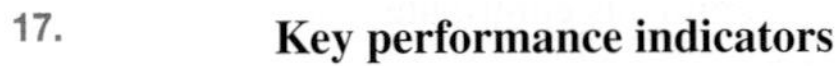

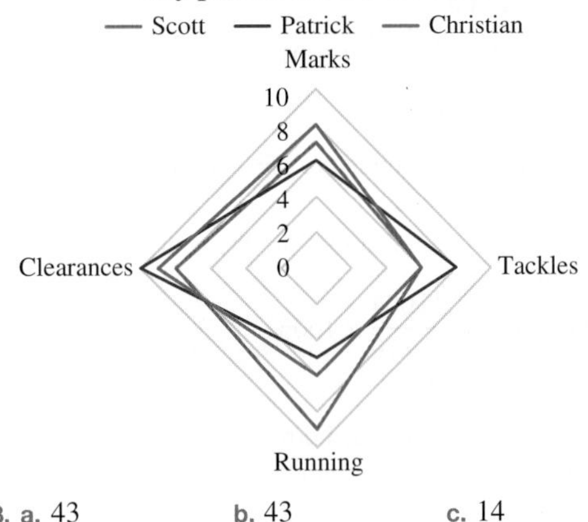

18. a. 43 b. 43 c. 14

19. a. 20 b. 11

20. a. 45 b. 15

6 Measures of central tendency and spread

LEARNING SEQUENCE

Fully worked solutions for this topic are available online.

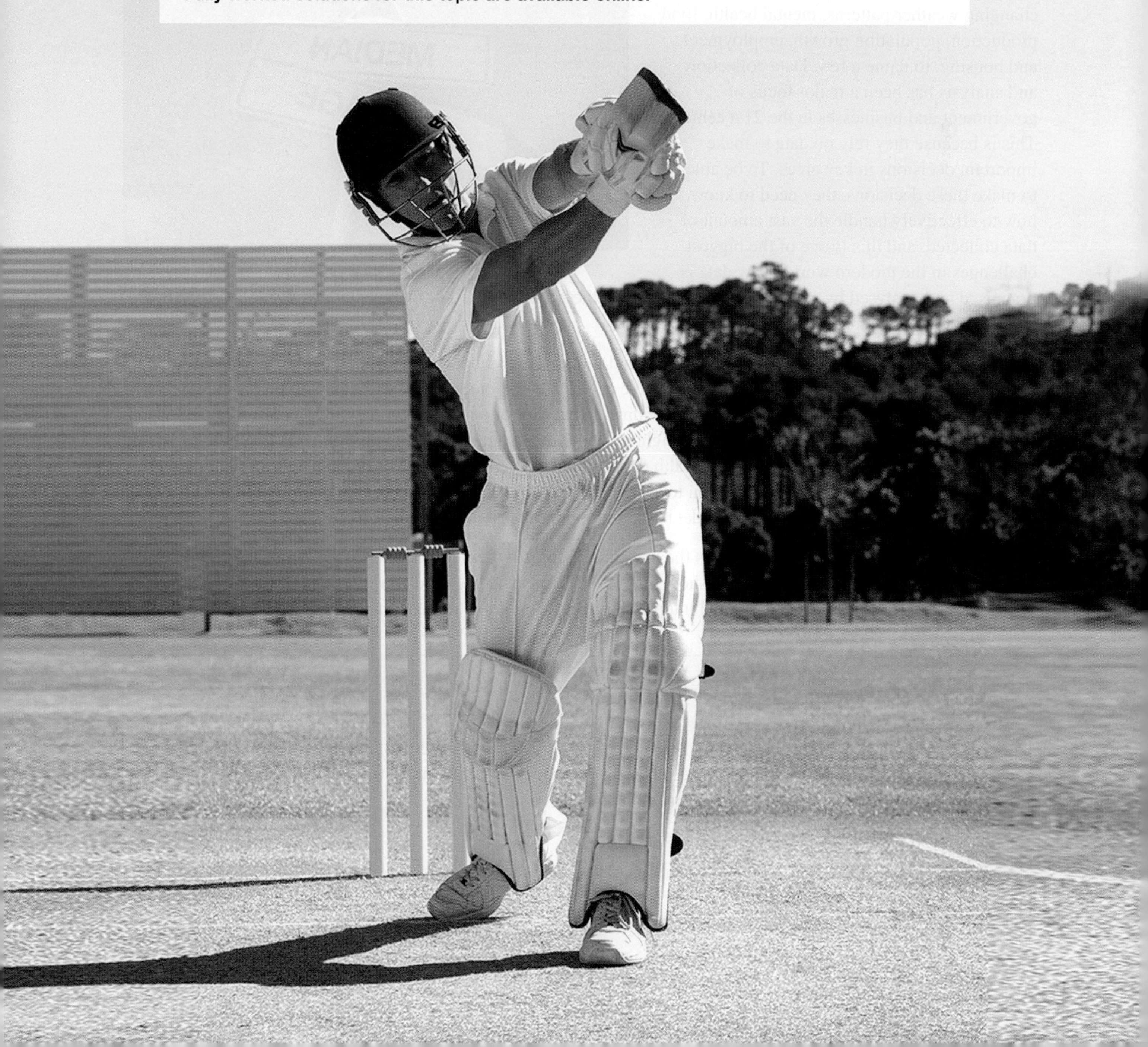

6.1 Overview

Hey students! Bring these pages to life online

Watch videos

Engage with interactivities

Answer questions and check results

Find all this and MORE in jacPLUS

6.1.1 Introduction

The analysis of data can improve our knowledge and understanding of the world and beyond. This includes issues such as environmental change, disease control, changing weather patterns, mental health, food production, population growth, employment and housing, to name a few. Data collection and analysis has been a major focus of government and businesses in the 21st century. This is because they rely on data to make important decisions in key areas. To be able to make these decisions, they need to know how to effectively handle the vast amount of data collected, and this is one of the biggest challenges in the modern world. Raw data is organised and analysed by statisticians to determine if there are expected or unexpected results and to discover short- and long-term trends.

The measures of central tendency (mean, median and mode) are an important tool that can be used to describe a whole set of data. The mean and median give us the centre and the mode gives us the most typical value of the data set. These values can be very useful in providing a snapshot of a large set of data. For example, a person looking to buy a home in an unfamiliar city might consider the mean or median house price in a suburb to decide if this suburb is one that they can afford. Health departments may use the mean or median age at which bowel cancer is first diagnosed to determine when to send early detection test kits to citizens. Manufacturers may use the mode to decide how many items to make based on popularity of size or function.

Another useful tool used to analyse data is the measure of spread: the range, interquartile range and standard deviation. These measures of spread can show how well the mean or the median represents the data. This topic will develop your skills in calculating and interpreting measures of central tendency and spread.

KEY CONCEPTS

This topic covers the following key concepts from the VCE Mathematics Study Design:

- interpolation and extrapolation of data, predictions, limitations, inferences and conclusions comparing and interpreting data sets and graphs, including using measures of central tendency and spread (percentiles and standard deviation) and cumulative frequency.

Note: Concepts shown in grey are covered in other topics.

6.2 Measures of central tendency

LEARNING INTENTION

At the end of this subtopic you should be able to:
- calculate the mean, median and mode of various data sets.

6.2.1 Mean and median

In statistics, a population is the entire set of subjects or objects being studied or investigated.

A sample is a smaller selection of subjects or objects taken from the population.

Statistical analysis on a sample can often be used to make generalisations about the population, as long as the subjects or objects in the sample are selected at random.

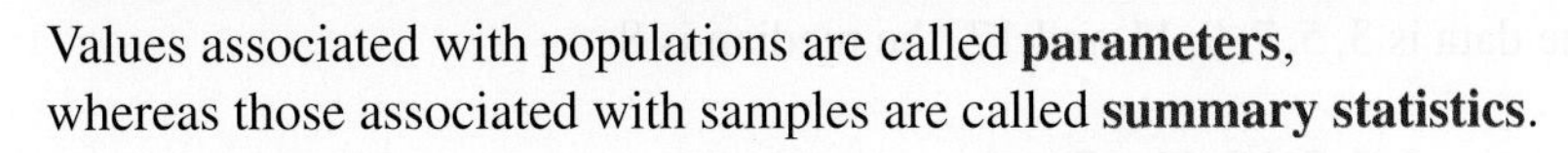

Values associated with populations are called **parameters**, whereas those associated with samples are called **summary statistics**.

The mean

The **mean** of a population is a theoretical measure of the centre of the entire population. The mean is referred to as the *average* in everyday language. It is not always a value in a data set; for example, the mean number of dogs in an Australian household is 1.5. This indicates that the mean average of dogs in an Australian household is between one and two dogs.

Obtaining the mean of a whole population is not always practicable, so the mean of a sample is often calculated instead. The mean of a population, μ, and the mean of a sample, $\bar{x}$, are calculated using the same formula.

Population and sample means

$$\mu = \frac{\textbf{sum of all the values in the population}}{\textbf{number of values in the population}} = \frac{\sum x}{N}$$

and

$$\bar{x} = \frac{\textbf{sum of all the values in the sample}}{\textbf{number of values in the sample}} = \frac{\sum x}{n},$$

where x is a data value, N is the total number of data values in the population and n is the total number of data values in the sample.

The summation symbol, $\sum$ (the Greek capital letter sigma), is the total value of the formula to the right of the summation symbol. Since the data sets in this chapter are samples, we will use the formula $\bar{x} = \frac{\sum x}{n}$ to calculate the mean.

The median

The **median** is the value in the middle position of the data set. It is the value that half the observations are less than and half are greater than. The median is also a *theoretical measure* of the centre of the set of data. It is not always a value in the set of data.

> **Median**
>
> **To calculate the median, arrange the values in the set of data in *ascending* order (smallest to largest). Then determine the middle number.**

For an *odd* number of values, the median is the *middle* value. For example, if the data is 6, 7, 9, 10 and 15, then the median is 9.

$$\begin{gathered}6, 7, 9, 10, 15\\ \uparrow\\ \text{Median} = 9\end{gathered}$$

For an *even* number of values, the median is the *mean* of the two middle values. In this case, the median may not be an actual data value. For example, if the data is 3, 5, 7, 9, 11 and 17, the median is 8.

$$\begin{gathered}3, 5, 7, \mid 9, 11, 17\\ \uparrow\\ \text{Median} = \frac{7+9}{2}\\ = 8\end{gathered}$$

If the set of data contains n values, the position of the median when the values are arranged in numerical (ascending) order can be calculated using the following formula.

$$\text{median position} = \frac{n+1}{2}$$

If $n = 6$, then $\frac{n+1}{2} = \frac{6+1}{2} = \frac{7}{2} = 3.5$

The median is halfway between the third and fourth values.

If $n = 5$, then $\frac{n+1}{2} = \frac{5+1}{2} = \frac{6}{2} = 3$

The median is the third value.

6.2.2 The mode

The **mode** is the value that occurs the most often. It is often described as the typical value of the observations. The mode can be found for both numerical and categorical data. (Recall that **categorical data** are data that can be grouped or classified, and **numerical data** are data that can be counted or measured.)

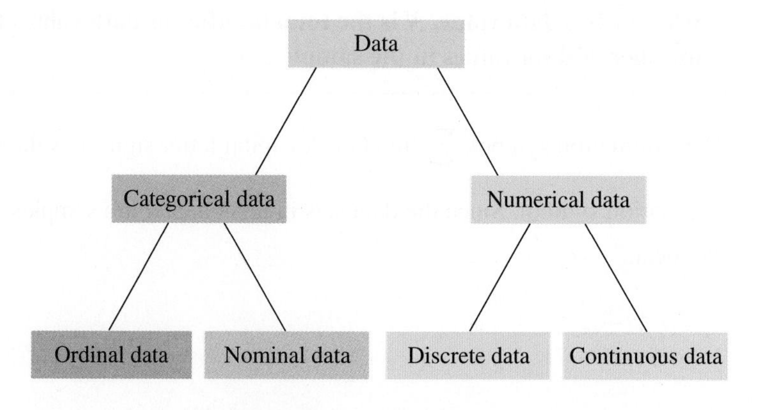

Mode

Mode is the data value that occurs the most often.

Some sets of data have no mode at all. If all of the values occur only once, there is no mode. It is possible for a data set to have more than one mode. A data set with two modes is referred to as 'bimodal'. A data set with three modes is referred to as 'trimodal'. A data set with more than one mode is referred to as 'multimodal'.

tlvd-3715

WORKED EXAMPLE 1 Calculating measures of central tendency

Ten Year 12 students were asked how many hours they spent competing in sport out of school hours. The results of the survey are:

2, 4, 5, 6, 7, 2, 2, 0, 5, 1

Calculate and intepret the values of:

a. **the mean**
b. **the median**
c. **the mode.**

THINK	WRITE
a. 1. Calculate the sum of all the values in the data set.	a. Sum of all the values $= 2+4+5+6+7+2+2+0+5+1$ $= 34$
2. Count the number of values in the data set (0 must be included).	Number of values = 10
3. The mean is the sum of all the values divided by the number of values.	$\text{Mean } (\bar{x}) = \dfrac{\text{sum of all the values}}{\text{number of values}} = \dfrac{\sum x}{n} = \dfrac{34}{10} = 3.4$
4. Explain what the mean tells us about the data.	If the total number of hours spent competing in sport was shared equally between the 10 students, each student would spend 3.4 hours on average competing in sport.
b. 1. Arrange the values in the data set in ascending order (smallest to largest).	b. 0, 1, 2, 2, 2, 4, 5, 5, 6, 7
2. The median is the middle value. Locate the position of the median.	There are 10 values in the set of data, so the median is the $\dfrac{10+1}{2} = 5\dfrac{1}{2}$th value (mean of the 5th and 6th values). 0, 1, 2, 2, 2, | 4, 5, 5, 6, 7

3. Determine the median value by calculating the mean of 2 and 4 or determine the value halfway between the two values.

$$\text{Median} = \frac{2+4}{2} = \frac{6}{2} = 3$$

4. Explain what the median tells us about the data.

Half of the students compete in more than 3 hours of sport and half of the students compete in less than 3 hours of sport.

c. 1. The mode is the most common value in the set of data.

c. 2, 4, 5, 6, 7, 2, 2, 0, 5, 1
Mode = 2

2. Explain what the mode tells us about the data.

The most common amount of time that students spend competing in sport is 2 hours.

6.2.3 Measures of central tendency in dot plots and stem plots

Recall that a list of data may be represented in different formats, such as a dot plot or stem (stem-and-leaf) plot. In a dot plot, each data value is represented as a dot on a number line. In a stem-and-leaf plot, each data value is split into two components, the stem and the leaf. Data is then grouped according to its stem.

WORKED EXAMPLE 2 Calculating the mean in dot plots and stem plots using technology

Calculate the mean of each of the following sets of data using technology. (Give your answer to 1 decimal place.)

a. 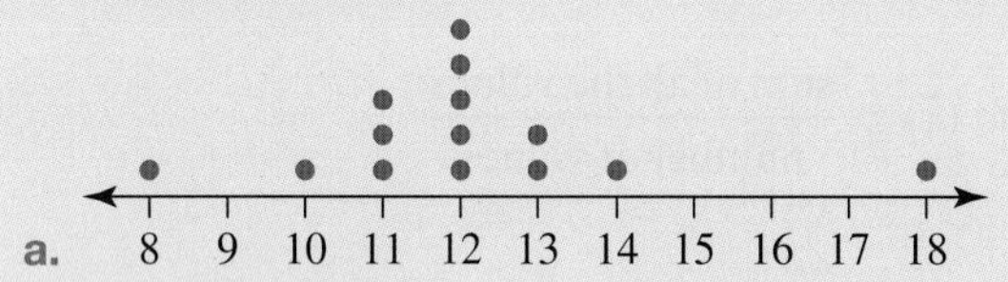

b. **Key: 12|1 = 121**

Stem	*Leaf*
12	1 5 9
13	4 7 9
14	8 8
15	2 8
16	3 8
17	2

THINK

a. 1. On the calculator keypad, press **data** and enter the list of data represented in the dot plot into L1.

DISPLAY/WRITE

a. 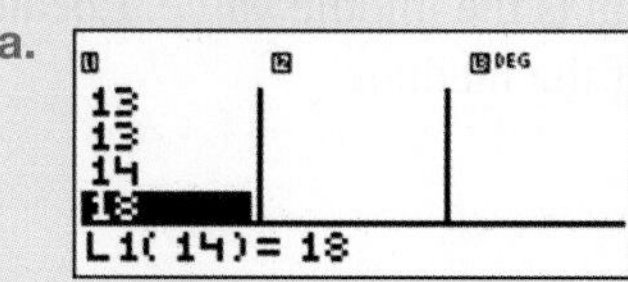

2. Press **2nd** and then **data**.
Select:
1:1-Var Stats, then press ENTER.
Select:
DATA: **L1**
FRQ: **ONE**
Select:
CALC
Then press ENTER.

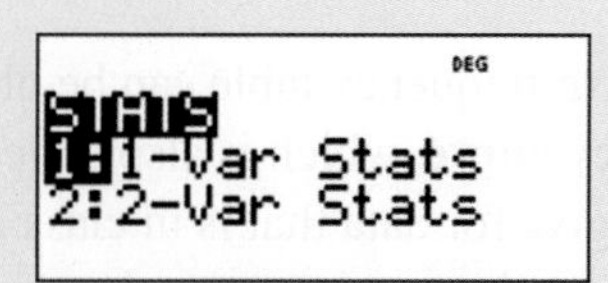

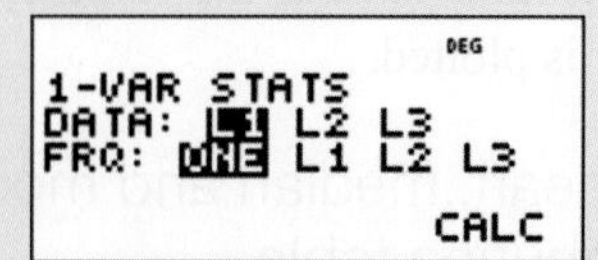

3. The answer appears on the screen.

4. Write the answer. $\bar{x} = 12.1$ correct to 1 decimal place.

b. 1. On the calculator keypad, press **data**, and enter the list of data represented in the stem and leaf plot into L1.

b.

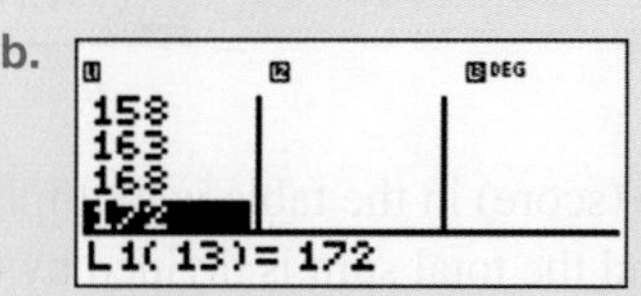

2. Press **2nd** and then **data**.
Select:
1:1-Var Stats, then press ENTER.
Select:
DATA: **L1**
FRQ: **ONE**
Select:
CALC
Then press ENTER.

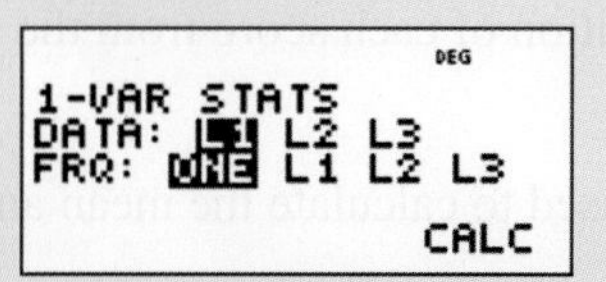

3. The answer appears on the screen.

4. Write the answer. $\bar{x} = 145.6$ correct to 1 decimal place.

6.2.4 Measures of central tendency in frequency distribution tables

Cumulative frequency

Cumulative frequency of a data value is the number of observations that are above or below the particular value. Cumulative frequency is recorded in a **cumulative frequency table**. The final value in a cumulative frequency table will always equal the total number of observations in the data set. This cumulative frequency table shows the number of movies watched in the last month by a group of 30 Year 12 students.

Data (x)	Frequency (f)	Cumulative frequency (cf)
3	3	3
4	5	$3 + 5 = 8$
5	7	$8 + 7 = 15$
6	10	$15 + 10 = 25$
7	0	$25 + 0 = 25$
8	5	$25 + 5 = 30$

Ogives

Data from a cumulative frequency table can be plotted to form a **cumulative frequency curve**, which is also called an **ogive** (pronounced *oh-jive*). To plot an ogive for data that is in class intervals, the maximum value for the class interval is used as the value against which the cumulative frequency is plotted.

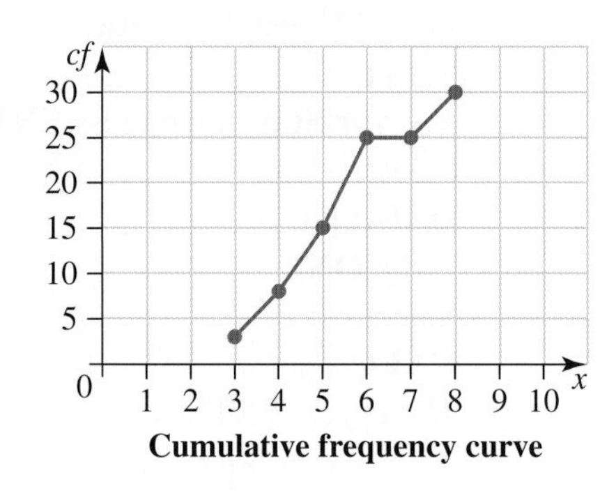

Cumulative frequency curve

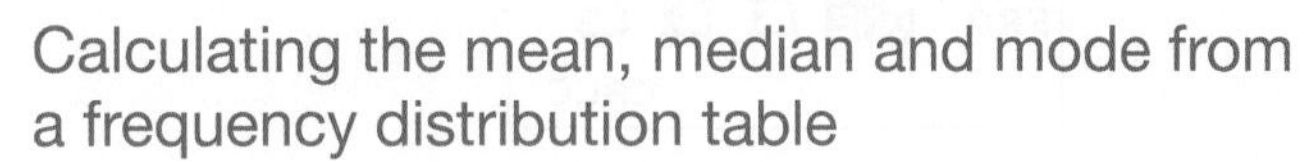

Calculating the mean, median and mode from a frequency distribution table

If data are presented in a frequency distribution table, the formula used to calculate the mean is:

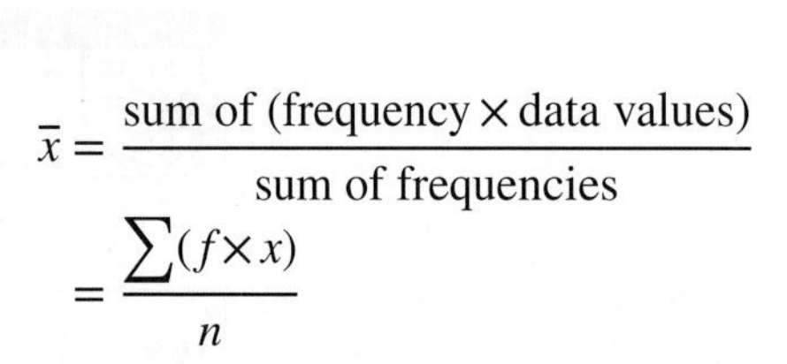

$$\bar{x} = \frac{\text{sum of (frequency} \times \text{data values)}}{\text{sum of frequencies}}$$
$$= \frac{\sum(f \times x)}{n}$$

Here, each data value (score) in the table is multiplied by its corresponding frequency; then all the $f \times x$ products are added together, and the total sum is divided by the number of observations in the set. To determine the median, find the position of each score from the cumulative frequency column. The mode is the score with the highest **frequency**.

Technology can be used to calculate the mean and median from a frequency distribution table.

WORKED EXAMPLE 3 Determining the mean, median and mode from a frequency distribution table

For the data shown in the frequency distribution table, calculate:

a. the mean using technology
b. the median using technology
c. the mode.

Score (x)	Frequency (f)
4	1
5	2
6	5
7	4
8	3
Total	15

THINK

a. 1. On the calculator keypad, press **data**. Enter the list of data from the column labelled Score (x) into L1. Enter the list of data in the column labelled Frequency (f) into L2.

2. Press **2nd** and then **data**.
Select:
1: 1-Var Stats, then press ENTER.
Select:
DATA: **L1**
FRQ: **L2**, then press ENTER.
Select:
CALC
Then press ENTER.

DISPLAY/WRITE

a.

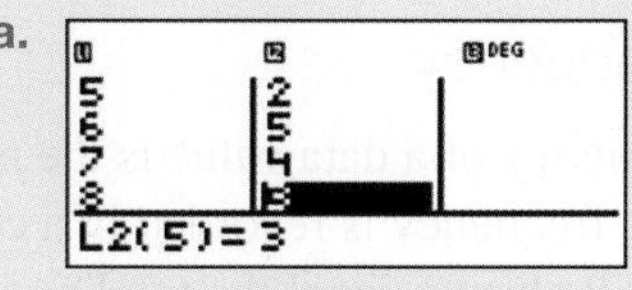

3. The answer appears on the screen.	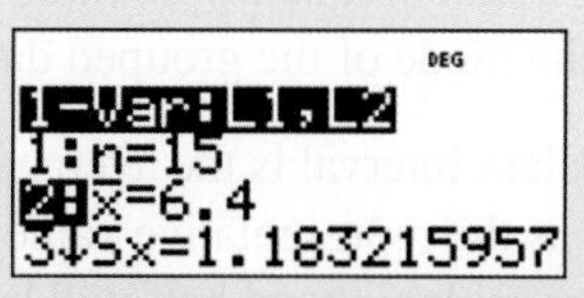
4. Write the answer.	$\bar{x} = 6.4$ correct to 1 decimal place.
b. 1. Locate the position of the median using the rule $\frac{n+1}{2}$, where $n = 15$ numbers.	**b.** $\frac{15+1}{2} = 8$ The median is the 8th score.
2. Use the frequency distribution table to count to the 8th number.	The median of the data set is 6.
c. 1. The mode is the score with the highest frequency.	**c.** The score with the highest frequency is 6.
2. Answer the question.	The mode of the data set is 6.

6.2.5 Measures of central tendency in real-world contexts

The mean is often called the average in real-world contexts. It is used in scientific research to describe many different phenomena, from weather patterns to sports statistics. The mean is also used to analyse the results of students in schools and universities. Examples of the mean being used in real-world contexts include the average number of children per household in Australia, average daily rainfall, average number of goals scored in a game and average mark on a Mathematics exam.

The median is the middle value of a data set by position and is often used when the information being investigated has some extremely high or extremely low values in the data set. The median is not greatly affected by these extreme values, as it looks at the number of data points, not their values. For this reason, the median is often used in economics. Examples include median household income and median house price.

The mode is less commonly used in real-world contexts, but it can still be a practical and useful way of describing information. For example, clothing shop owners will look at the highest-selling clothing size to make sure they have the most floor stock in that particular size.

Always remember to relate summary statistics back to real-world contexts, ensuring they have relevance and meaning. For example, a shoe-shop owner might determine that the shop sells a mean of 26.7 pairs of shoes a day. This can be interpreted as selling, on average, between 26 and 27 pairs of shoes per day.

6.2.6 Grouped data

For large data sets, it may be necessary to group data into smaller sets called **class intervals**. These class intervals must be the same size and must be set so that each value belongs to one interval only.

Class intervals are written differently for different types of data. Recall that **discrete data** are numerical data that can only take certain values (usually whole numbers), and **continuous data** are numerical data that can take any value within an interval.

A class interval for:

- discrete data is written showing the minimum and maximum values for the interval (e.g. 10–20 includes the values between and including 10 and 20, such as 12 or 16)
- continuous data is written like an inequation (e.g. for values of 10 up to values less than 20 $(10 \le x < 20)$, the class interval is written as $10- < 20$).

A frequency table shows the frequency of values appearing in each class interval. The class interval with the highest frequency is the mode of the grouped data and is called the **modal class**.

The **midpoint** of the class interval is the average of the maximum and minimum values for the class interval. The midpoint is used as the representative value for the class interval. It is assumed that half of the data values will be greater than the midpoint and half will be less than the midpoint.

The mean of grouped data is calculated using the midpoint of each class interval to represent the data values in the interval.

Calculating the mean from grouped data

$$\bar{x} = \frac{\text{sum of (midpoints} \times \text{frequency)}}{\text{sum of frequencies}} = \frac{\sum (x_{mid} \times f)}{n}$$

where:

x_{mid} is the midpoint of the class interval

f is the frequency of the class interval

n is the total number of data values

$\sum$ is the summation symbol.

tlvd-3716

WORKED EXAMPLE 4 Calculating the mean for grouped data

A takeaway food shop is trying to improve its ordering system. The manager takes a survey of the number of hamburgers made each day for a month and collects the data shown in the frequency table. Calculate the mean number of hamburgers made each day.

Class interval	Frequency (f)
0–4	6
5–9	8
10–14	2
15–19	3
20–24	5

THINK

1. To calculate the mean of grouped data, a single value is needed to represent the class interval.
   - To calculate the midpoint of the class intervals, calculate the mean of the maximum and minimum values for each class interval.
   - For the first class interval 0–4, $x_{mid} = \frac{0+4}{2} = 2$.
   - For the second class interval 5–9, $x_{mid} = \frac{5+9}{2} = 7$.

WRITE

Class interval	Frequency (f)	Midpoint (x_{mid})
0–4	6	②
5–9	8	⑦
10–14	2	12
15–19	3	17
20–24	5	22

2. The midpoint will be the representative value for the interval.
   - Add a new column to the table, $x_{mid} \times f$, and multiply the midpoint by the frequency of each class interval to calculate a value to represent the sum of all values in that class interval.
   - For the first class interval 0–4, $x_{mid} \times f = 2 \times 6 = 12$.
   - Calculate the total sum of all the data values by adding all values in the $x_{mid} \times f$ column. The total of the $x_{mid} \times f$ column is 253.
   - Calculate the total number of data values by adding all values in the frequency column. The total of the frequency column is 24.

Class interval	Frequency (f)	Midpoint (x_{mid})	$x_{mid} \times f$
0–4	6	2	**12**
5–9	8	7	56
10–14	2	12	24
15–19	3	17	51
20–24	5	22	110
Total	**24**		**253**

3. Substitute the total of the $x \times f$ column and the total of the frequency column into the formula for the mean of grouped data.

$$\bar{x} = \frac{\sum (x_{mid} \times f)}{n}$$
$$= \frac{253}{24}$$
$$\approx 10.54$$

4. Answer the question.

The mean number of hamburgers made per day is approximately 10.54. This means that, on average, between 10 and 11 burgers are made every day.

Resources

Video eLesson Mean and median (eles-1905)

Interactivities Measures of central tendency (int-4621)
Mean (int-3818)
Median (int-3819)
Mode (int-3820)

6.2 Exercise

Note: Where necessary, give answers correct to 1 decimal place.

1. **WE1** For each of the following data sets, calculate the:

   i. mean　　ii. median　　iii. mode.

   a. 3, 3, 4, 5, 5, 6, 6, 7, 8, 8, 8, 9
   b. 12, 18, 4, 17, 5, 12, 0, 10, 12
   c. 42, 29, 11, 28, 21
   d. 8, 2, 5, 6, 9, 9, 7, 3, 2, 9, 3, 7, 6, 8
   e. 5, 5, 6, 4, 8, 3, 4
   f. 3.7, 3.5, 3.8, 3.8, 3.5

2. For each of the following data sets, calculate the mean, median and mode.
   a. 10, 12, 21, 23, 23, 25, 44
   b. 7, 8, 10, 6, 9, 11, 4, 12, 2
   c. 50, 44, 50, 46, 50, 48
   d. 2.5, 1.4, 1.7, 2.1, 1.4, 1.8, 1.6, 1.7, 2.9
3. **MC** State which of the following are respectively the mean, median and mode of this data set.

   1024, 1032, 1067, 1112, 1112, 1178, 1236, 1269, 1290, 1301, 1345, 1357, 1365, 1377, 1400

   A. 1269, 1112, 1231
   B. 1231, 1269, 1112
   C. 1112, 1231, 1269
   D. 1231, 1112, 1269
4. The number of students standing in line at the school canteen 5 minutes after the start of lunch was recorded over a 2-week period. The results were as follows:

   52, 45, 41, 42, 53, 45, 47, 32, 52, 56

   Determine the mean number of students standing in line at this time for this 2-week period. Round your answer to the nearest whole number.

5. The police conducted a survey of the speed of cars down a highway. The lowest and highest speeds recorded were 91 km/h and 154 km/h respectively. The average speed was 104 km/h, and the police found that the speed most commonly recorded was 101 km/h. Half of the cars were also found to be travelling under 102 km/h. State the value of the mean, median and mode.

6. **WE2a** Calculate the mean of the data set shown.

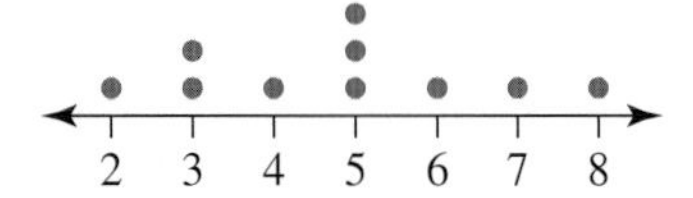

7. **WE2b** Calculate the mean of the data set shown.

Key: 1|6 = 16 years

Stem	*Leaf*
1	5 6 7 7 8 9 9
2	1 2 4 8 8
3	0 1 1 1 5
4	2 3
5	3

8. WE3 Calculate the mean, median and mode in each of the following frequency tables.

a.

Score (x)	Frequency (f)
1	12
2	10
3	8
4	7
5	2

b.

Score (x)	Frequency (f)
25	1
26	15
27	11
28	7
29	3

c.

Score (x)	Frequency (f)
1.5	2
2.0	9
2.5	7
3.0	11
3.5	4

9. WE4 Calculate the mean number of ebooks presented in this frequency table.

Number of ebooks (x)	Frequency (f)
1–15	3
16–30	9
31–45	8
46–60	11
61–75	10
76–90	14
91–105	15
106–120	18

10. Calculate the mean number of calls made on mobile phones in the month shown in the graph below.

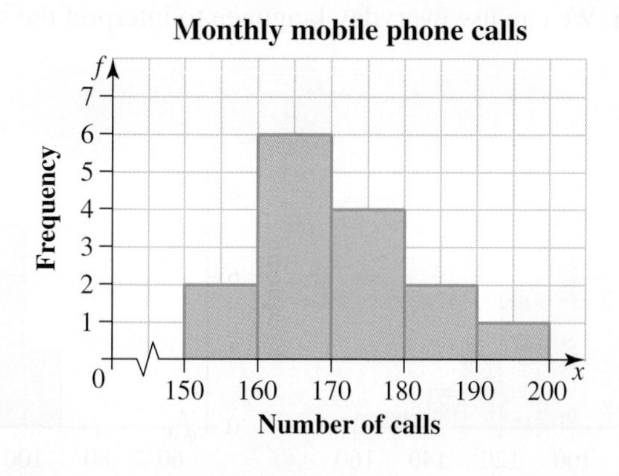

11. The number of goals a netballer scored in the 12 games of a season was as follows.

$$1, 1, 1, 1, 2, 2, 2, 3, 3, 3, 8, 12$$

A local newspaper reporter asked the netballer what their average was for the season.

a. State which measure of centre (mean or median) the netballer should give the reporter as their 'average' so that the value of the average is as high as possible.

b. State which measure of centre *you* would choose to best describe the 'average' number of goals the netballer scored each game. Explain why.

12. Create a data set that fits each of the following descriptions.
    a. Five data values with a mean of 3 and a mode of 3
    b. Five data values with a mean of 3 and a mode of 4
    c. Five data values with a mean of 3 and a median of 2
13. The mean length of three pieces of string is 145 cm. If they are joined together from end to end, determine their total length.
14. The median mark for a Science test was 45. No student actually achieved this result. Explain how this is possible.

6.3 Measures of spread

LEARNING INTENTION

At the end of this subtopic you should be able to:
- calculate quartiles, deciles and percentiles
- calculate the range and interquartile range (IQR)
- calculate the standard deviation.

6.3.1 Spread

The **spread** of a set of data indicates how far the data values are spread from the centre or from each other. This is also known as the distribution. There are many statistical measures of spread, including the range, interquartile range and the standard deviation. We can use everyday language to interpret the spread, such as 'spread out', 'dispersed' or 'tightly packed'.

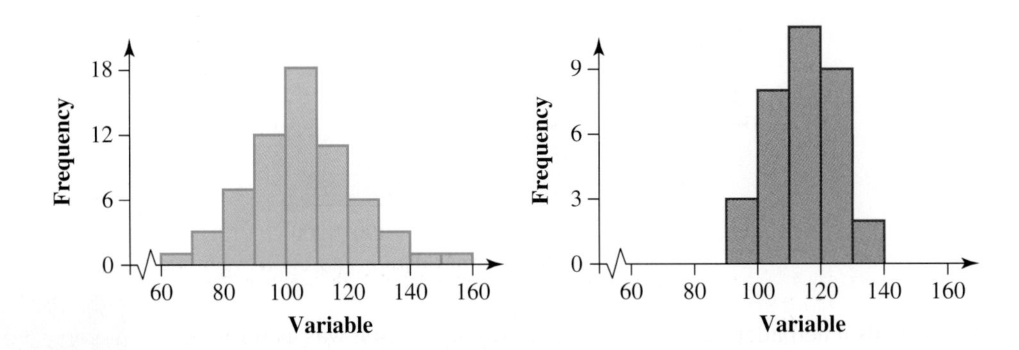

By comparing the two histograms, it can be seen that the histogram on the left has a larger spread of values compared to the histogram on the right.

6.3.2 Quartiles, deciles and percentiles

Data sets can be split up into any given number of equal parts called **quantiles**. Quantiles are named after the number of parts that the data is divided into. **Quartiles** divide the data into 4 equal-sized parts, **deciles** divide the data into 10 equal-sized parts and percentiles divide the data into 100 equal-sized parts.

Percentile	Quartile and symbol	Common name
25th percentile	First quartile, Q_1	Lower quartile
50th percentile	Second quartile, Q_2	Median
75th percentile	Third quartile, Q_3	Upper quartile
100th percentile	Fourth quartile, Q_4	Maximum

A percentile is named after the percentage of data that lies at or below that value. For example, 60% of the data values lie at or below the 60th percentile.

tlvd-3717

WORKED EXAMPLE 5 Calculating quartiles and percentiles

This cumulative frequency graph shows the cumulative frequency versus the number of skips with a skipping rope per minute for Year 11 students at a school.
Calculate:
a. **the median**
b. Q_1
c. Q_3
d. **the number of skips per minute for the 40th percentile.**

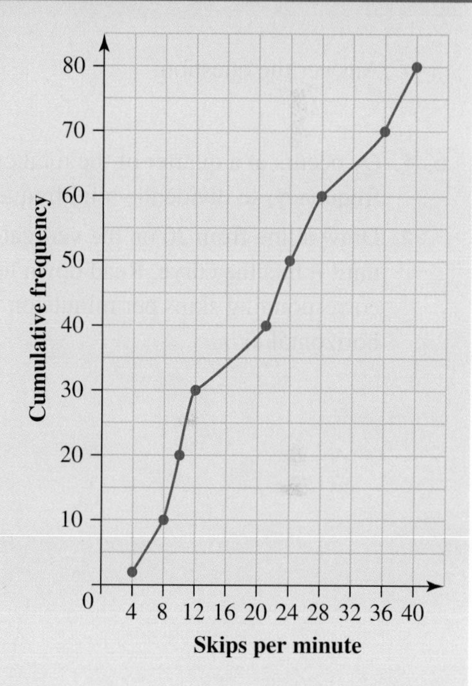

THINK

a. 1. Determine the total number of subjects. This is the highest vertical value on the cumulative frequency curve.

2. The median occurs at the middle of the total cumulative frequency, so divide the total frequency by 2.

WRITE

a. 80 students

The median occurs at the $\frac{80}{2} = 40$th person.

3. Draw a line from 40 on the vertical axis until it hits the curve. Read down to the corresponding skips per minute on the horizontal axis.

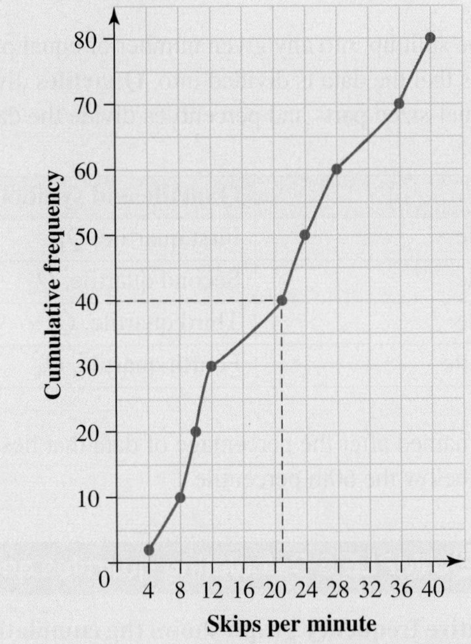

4. Answer the question.

The median number of skips per minute is approximately 21.

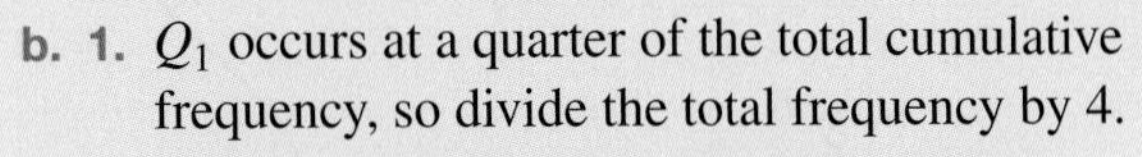

b. 1. Q_1 occurs at a quarter of the total cumulative frequency, so divide the total frequency by 4.

b. Q_1 occurs at the $\frac{80}{4} = 20$th person.

2. Draw a line from 20 on the vertical axis until it hits the curve. Read down to the corresponding skips per minute on the horizontal axis.

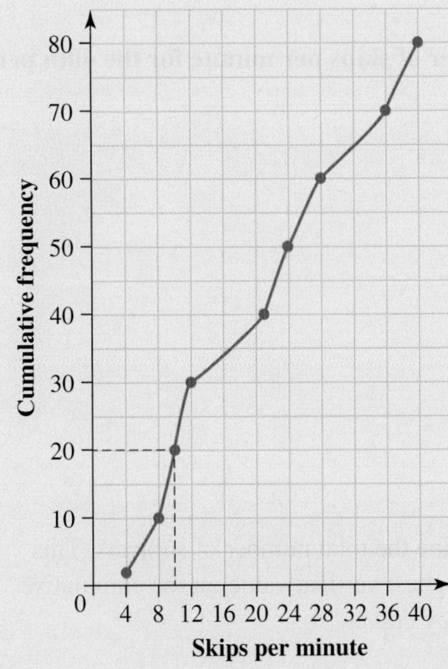

3. Answer the question.

Q_1 is at approximately 10 skips per minute.

c. 1. Q_3 occurs at three-quarters of the total cumulative frequency, so divide the total frequency by 4 and multiply the result by 3.

c. Q_3 occurs at the $\frac{80}{4} \times 3 = 60$th person.

2. Draw a line from 60 on the vertical axis until it hits the curve. Read down to the corresponding skips per minute on the horizontal axis.

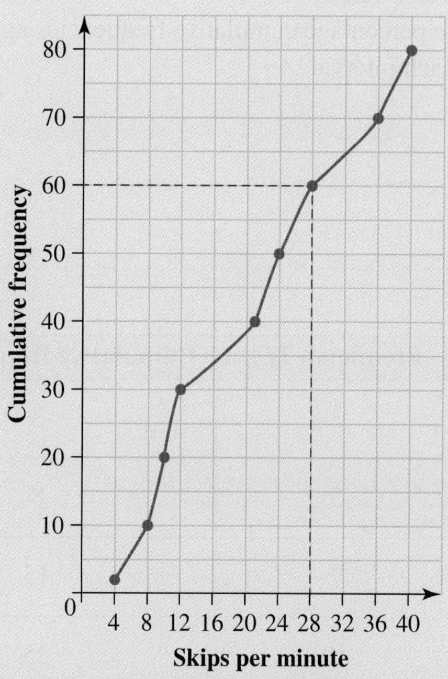

3. Answer the question.

Q_3 is at approximately 28 skips per minute.

d. 1. To determine the 40th percentile of the cumulative frequency total, multiply the total frequency by $\frac{40}{100}$.

d. $\frac{40}{100} \times 80 = 32$nd person

2. Draw a line from 32 on the vertical axis until it hits the curve. Read down to the corresponding skips per minute on the horizontal axis.

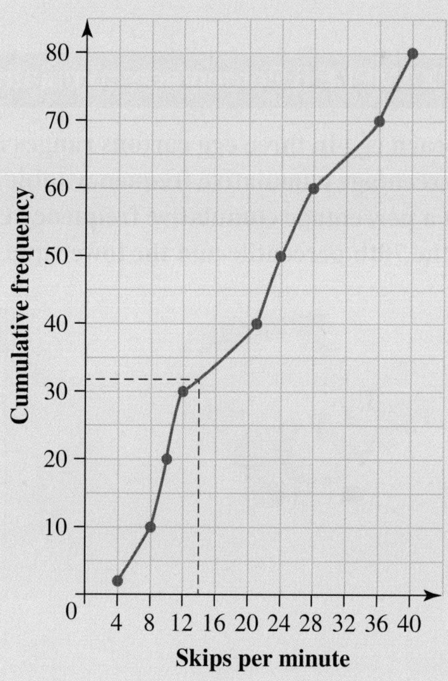

3. Answer the question.

The 40th percentile is at approximately 14 skips per minute.

Percentiles can be read off a percentage cumulative frequency curve.

A percentage cumulative frequency curve is created by:
- writing the cumulative frequencies as a percentage of the total number of data values
- plotting the percentage cumulative frequencies against the maximum value for each interval.

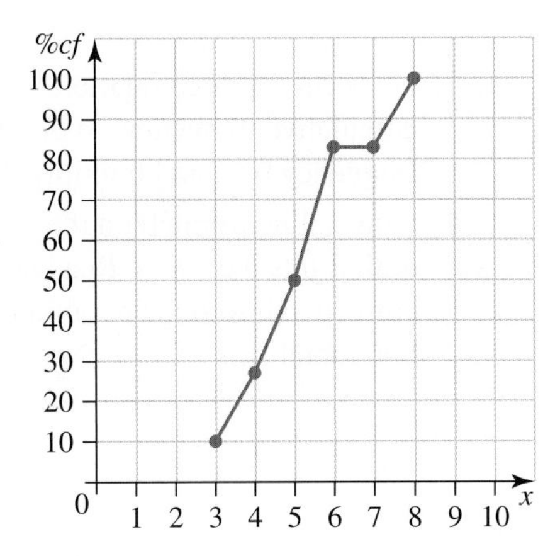

Data (x)	Frequency (f)	Cumulative frequency (cf)	Percentage cumulative frequency ($\%cf$)
3	3	3	$\frac{3}{30} \times \frac{100}{1}\% = 10\%$
4	5	8	$\frac{8}{30} \times \frac{100}{1}\% = 27\%$
5	7	15	$\frac{15}{30} \times \frac{100}{1}\% = 50\%$
6	10	25	$\frac{25}{30} \times \frac{100}{1}\% = 83\%$
7	0	25	$\frac{25}{30} \times \frac{100}{1}\% = 83\%$
8	5	30	$\frac{30}{30} \times \frac{100}{1}\% = 100\%$

tlvd-3718

WORKED EXAMPLE 6 Percentage cumulative frequency curves

The mass of each egg in three egg cartons ranges between 55 and 65 grams, as shown in the table.

a. Draw a percentage cumulative frequency table for the data.

b. Construct a percentage cumulative frequency curve (ogive).

c. Evaluate the 70th percentile and the lower quartile.

Mass (g)	Frequency
55– < 57	2
57– < 59	6
59– < 61	12
61– < 63	11
63– < 65	5

THINK	WRITE

a. 1. Construct the cumulative frequency table by calculating the cumulative frequency for each class interval, as shown in blue.

The total number of eggs in the three cartons is 36.

2. Calculate the percentage cumulative frequency for each interval by dividing the cumulative frequency for each interval by the total cumulative frequency, as shown in red.

a.

Mass (g)	Frequency (f)	Cumulative frequency (cf)	Percentage cumulative frequency ($\%cf$)
55– < 57	2	2	$\frac{2}{36} \times \frac{100}{1}\% \approx 6\%$
57– < 59	6	$2 + 6 = 8$	$\frac{8}{36} \times \frac{100}{1}\% \approx 22\%$
59– < 61	12	$8 + 12 = 20$	$\frac{20}{36} \times \frac{100}{1}\% \approx 56\%$
61– < 63	11	$20 + 11 = 31$	$\frac{31}{36} \times \frac{100}{1}\% \approx 86\%$
63– < 65	5	$31 + 5 = 36$	$\frac{36}{36} \times \frac{100}{1}\% = 100\%$

b. 1. Plot the percentage cumulative frequency curve. For the first interval (55– < 57), plot the minimum value for the interval (55) against 0%.

2. Plot the maximum value for each interval against the percentage cumulative frequency for the interval.

- For the first interval, plot 57 against 6%.
- For the second interval, plot 59 against 22%.

b.

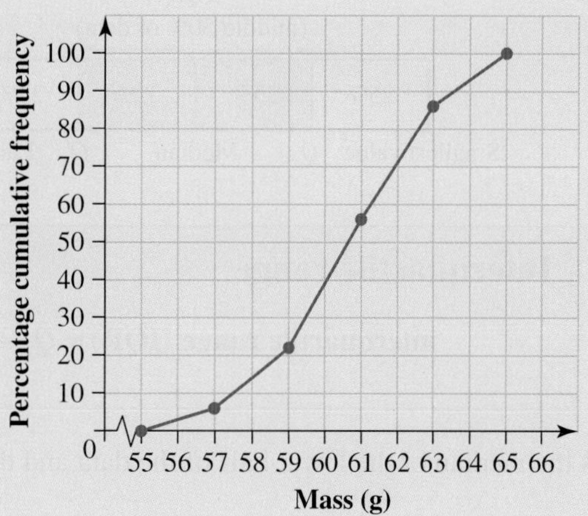

c. 1. The 70th percentile is read from the graph by following across from 70% on the vertical axis to the curve and then down to the horizontal axis, as shown in purple.

2. The lower quartile is the 25th percentile and is read by following across the graph from 25% on the vertical axis to the curve and then down to the horizontal axis, as shown in pink.

3. Answer the question.

c.

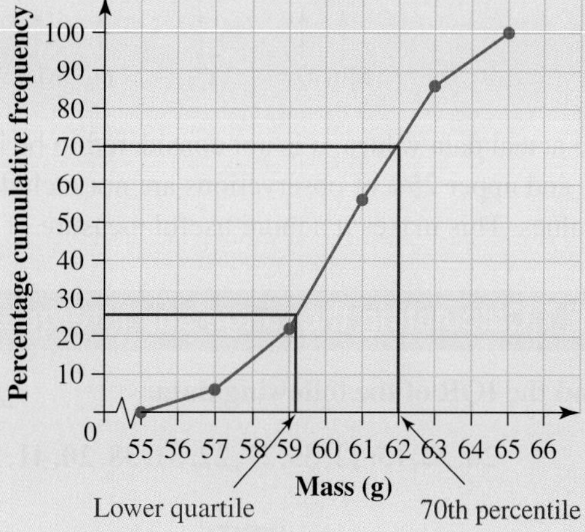

The 70th percentile is 62 grams. The lower quartile is approximately 59.2 grams.

6.3.3 The range and interquartile range

The **range** is the difference between the largest and smallest values of the data set.

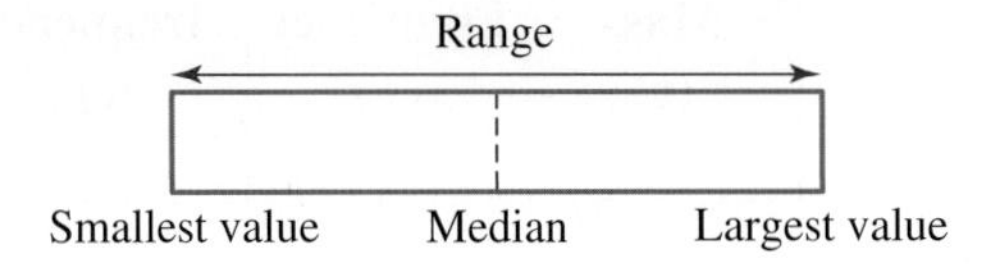

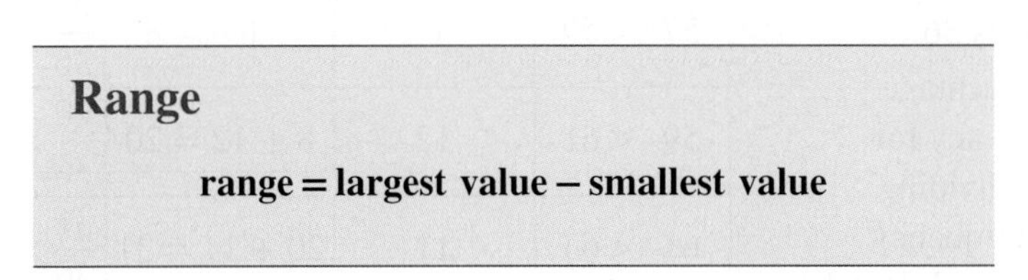

Range

range = largest value − smallest value

The **interquartile range (IQR)** is the range of the middle 50% of the data set. It measures the spread of the middle 50% of data.

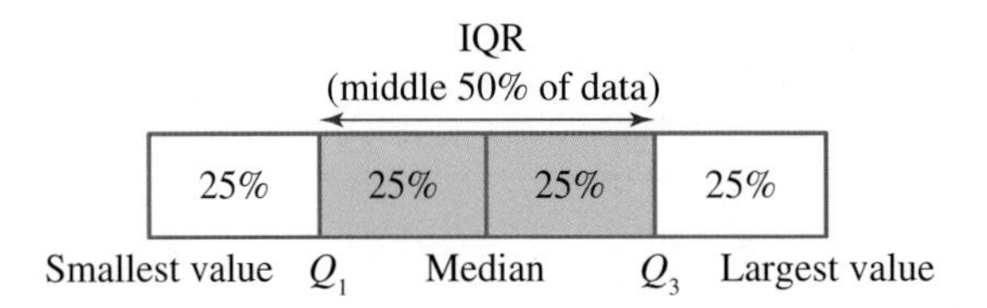

Interquartile range

interquartile range (IQR) = $Q_3 - Q_1$

The lower quartile (Q_1) is the median of the lower half of the data and the upper quartile (Q_3) is the median of the upper half of the data.

Lower half   Upper half

4, 9, 13, 13, (15), 17, 19, 29, 53

Min Q_1 Q_2 Q_3 Max

If the median is one of the actual data values, it is not considered to be in either the upper or lower half of the data. Since the lower 25% and upper 25% of observations are not included, the interquartile range is generally not affected by extreme values. This makes it a more useful measure of spread than the range.

tlvd-3719

WORKED EXAMPLE 7 Calculating the range and interquartile range

Calculate the range and the IQR of the following data.

$$26, 32, 15, 12, 35, 27, 22, 31, 38, 20, 41, 26, 17, 29$$

THINK	WRITE
1. Write the data in ascending order.	12, 15, 17, 20, 22, 26, 26, 27, 29, 31, 32, 35, 38, 41
2. The range is found by subtracting the smallest number from the largest number.	Range = 41 − 12 = 29

3. The median value will split the data into the top and bottom half. There are 14 values, so the median is at the $\frac{14+1}{2}=7.5$th value.	12, 15, 17, 20, 22, 26, 26, | 27, 29, 31, 32, 35, 38, 41 $\text{Median} = \frac{26+27}{2} = 26.5$
4. List the lower half of the data. Determine the median of the lower half of the data. This is Q_1. There are 7 values, so the median is at the $\frac{7+1}{2}=4$th value.	Lower half: 12, 15, 17, 20, 22, 26, 26 $Q_1 = 20$
5. List the upper half of the data. Determine the median of the upper half of the data. This is Q_3. There are 7 values, so the median is at the $\frac{7+1}{2}=4$th value.	Upper half: 27, 29, 31, 32, 35, 38, 41 $Q_3 = 32$
6. Calculate the IQR.	$\text{IQR} = Q_3 - Q_1 = 32 - 20 = 12$

6.3.4 The standard deviation

The **standard deviation**, like the IQR, is a measure of the spread of the data. The standard deviation measures the spread of data about the mean. The standard deviation of a population and the standard deviation of a sample are calculated differently:

Population standard deviation

$$\sigma = \sqrt{\frac{\sum(x-\mu)^2}{N}}$$

where:

σ = population standard deviation

x = a data value

μ = the population mean

N = the total number of population data values.

Sample standard deviation

$$s = \sqrt{\frac{\sum(x-\bar{x})^2}{n-1}}$$

where:

s = population standard deviation

x = a data value

$\bar{x}$ = the sample mean

n = the total number of population data values.

These formulas use the square of the distance between data points and the mean as part of the calculation. Therefore, a small standard deviation shows that the data is clustered closer to the mean, whereas a large standard deviation indicates that the distribution is more spread out about the mean. The sample standard deviation is more commonly used because questions usually involve samples instead of whole populations. When asked to calculate the standard deviation, assume it is referring to the sample standard deviation.

tlvd-3720

WORKED EXAMPLE 8 Calculating standard deviation

The number of lollies in a sample of 8 packets is:

11, 12, 13, 14, 16, 17, 18, 19

Calculate the standard deviation correct to 2 decimal places.

THINK

WRITE

1. Calculate the mean.

$$\bar{x} = \frac{11+12+13+14+16+17+18+19}{8}$$
$$= \frac{120}{8}$$
$$= 15$$

2. To calculate the deviations $(x-\bar{x})$, set up a table as shown and complete by subtracting the mean from each data value.

No. of lollies (x)	$(x-\bar{x})$
11	$11-15=-4$
12	-3
13	-2
14	-1
16	1
17	2
18	3
19	4
Total	

3. Add another column to the table to calculate the square of the deviations $(x-\bar{x})^2$ by squaring each value in the second column. Then sum the results: $\sum(x-\bar{x})^2$.

No. of lollies (x)	$(x-\bar{x})$	$(x-\bar{x})^2$
11	$11-15=-4$	16
12	-3	9
13	-2	4
14	-1	1
16	1	1
17	2	4
18	3	9
19	4	16
Total		$\sum(x-\bar{x})^2=60$

4. To calculate the standard deviation, divide the sum of the squares by one less than the number of data values, then take the square root of the result.

$$s=\sqrt{\frac{\sum(x-\bar{x})^2}{n-1}}$$

$$=\sqrt{\frac{60}{7}}$$

≈ 2.93 (correct to 2 decimal places)

5. Interpret the result.

The standard deviation is 2.93, which means that the number of lollies in each pack differs from the mean by an average of 2.93.

Calculations for standard deviation can be completed using different digital technologies, such as calculators, statistics programs or spreadsheets. This is generally the preferred method for calculating the standard deviation.

WORKED EXAMPLE 9 Calculating standard deviation using technology

The number of chocolate drops in a sample of different bags was recorded. These values were as follows.

271, 211, 221, 288, 209, 285, 230, 220, 296, 216

Calculate the standard deviation, correct to 1 decimal place, of the number of chocolate drops in a bag.

THINK

1. Enter the data into the digital technology of your choice. This example will be done with a scientific calculator.
On the calculator keypad, press **data** and enter the list of data represented in the dot plot into **L1**.

DISPLAY/WRITE

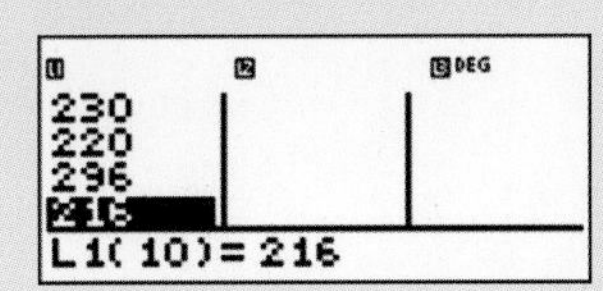

2. Press **2nd** and then **data**.
Select:
1: 1-Var Stats, then press ENTER.
Select:
DATA: **L1**
FRQ: **ONE**
Select:
CALC
Then press ENTER.

3. The answer appears on the screen.
As this is a sample only, S_x is the required standard deviation.

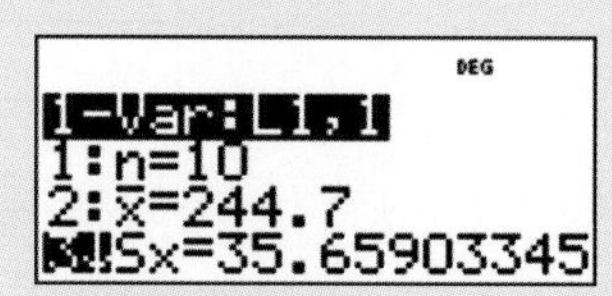

4. Write the answer.

$s_x = 35.7$ correct to 1 decimal place.

Resources

Interactivities Ogives (int-6174)
Range (int-3822)
The interquartile range (int-4813)
The standard deviation for a sample (int-4814)

6.3 Exercise

Students, these questions are even better in jacPLUS

Receive immediate feedback and access sample responses

Access additional questions

Track your results and progress

Find all this and MORE in jacPLUS

1. Put these expressions in order from the one with the smallest value to the one with the largest value: upper quartile; minimum; median; maximum; lower quartile.

2. MC For the following data, the values of Q_1 and Q_3 are:

$$7, 8, 4, 6, 2, 9, 13, 9, 11$$

A. $Q_1 = 4$ and $Q_3 = 9$
B. $Q_1 = 5$ and $Q_3 = 10$
C. $Q_1 = 4$ and $Q_3 = 10$
D. $Q_1 = 5$ and $Q_3 = 9$
E. $Q_1 = 4.5$ and $Q_3 = 9.5$

3. WE5 The number of steps per minute runners took at a local park run event was recorded. The cumulative frequency graph is as shown.

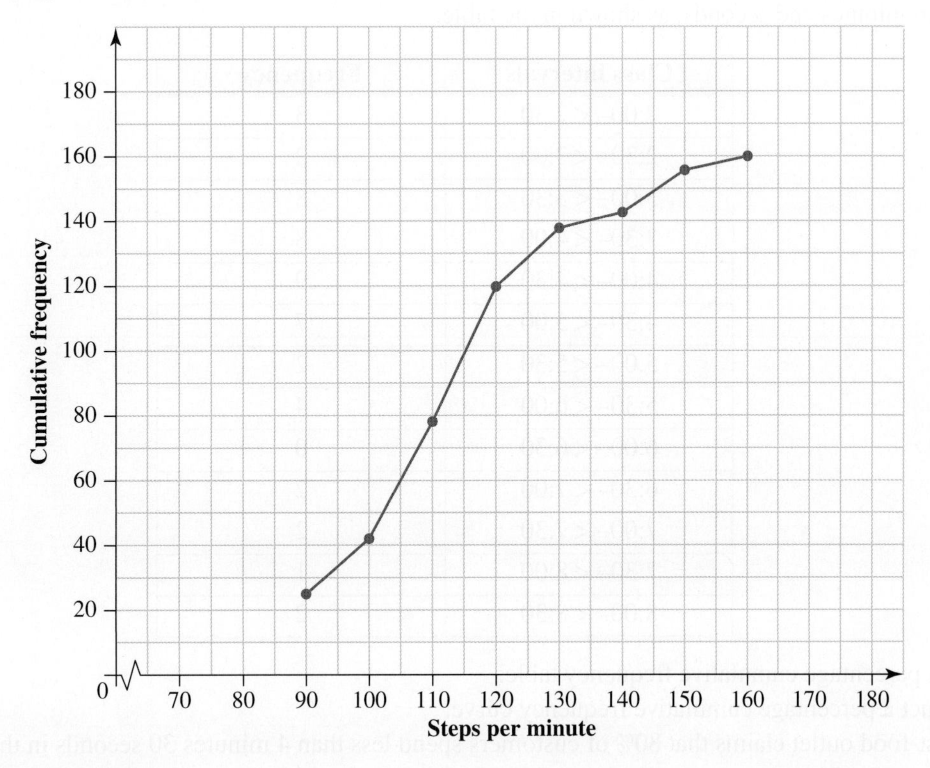

a. Calculate the median.
b. Determine Q_1.
c. Determine Q_3.
d. A local running group invited all people over the 65th percentile to join the group. Calculate how many steps per minute a runner would need to take to be invited to join the running group.

4. People who train for triathlons run a great distance every day. The distances (in km) that 60 athletes cover are shown in the frequency table.

Class interval	Frequency
10–15	3
15–20	7
20–25	12
25–30	18
30–35	14
35–40	6

a. Draw a cumulative frequency curve (ogive) for this data.
b. State how many of the distances run were less than 19 km.
c. Determine the median of the data.
d. Calculate the 30th percentile and the upper quartile.

5. **WE6** Sometimes when you go through a fast-food drive-through, they don't have your order ready, and you have to wait in the waiting bay for it to be brought out to you. The waiting times of 50 customers were recorded in minutes and seconds, as shown in the table.

Class intervals	Frequency
2:00– < 2:30	3
2:30– < 3:00	2
3:00– < 3:30	8
3:30– < 4:00	8
4:00– < 4:30	9
4:30– < 5:00	7
5:00– < 5:30	2
5:30– < 6:00	4
6:00– < 6:30	0
6:30– < 7:00	2
7:00– < 7:30	2
7:30– < 8:00	1
8:00– < 8:30	2

a. Draw a percentage cumulative frequency table.
b. Construct a percentage cumulative frequency curve.
c. The fast-food outlet claims that 80% of customers spend less than 4 minutes 30 seconds in the waiting bay. State whether you agree or disagree with this statement and justify your response.

6. Calculate the range of each of the following sets of data.

a. 4, 6, 8, 11, 15

b. 1.7, 1.9, 2.5, 0.5, 3.1, 1.9, 1.7, 1.6, 1.2

c.

Score (x)	Frequency (f)
110	12
111	9
112	18
113	27
114	5

7. **WE7** The speed of 20 cars (in km/h) is monitored along a stretch of road that is a designated 80 km/h zone. Calculate the range and IQR of the data.

80, 82, 77, 75, 80, 80, 81, 78, 79, 78,
80, 80, 85, 70, 79, 81, 81, 80, 80, 80

8. 30 pens are randomly selected off the conveyor belt at the factory and are tested to see how long they will last (with values given in hours). Determine the range and IQR of the data.

20, 32, 38, 22, 25, 34, 47, 31, 26, 29, 30, 36, 28, 40, 31,
26, 37, 38, 32, 36, 35, 25, 29, 30, 40, 35, 38, 39, 37, 30

***Use the following data set to answer questions* 9–11.**

171, 122, 182, 153, 167, 184, 171, 177,
189, 175, 128, 190, 135, 147, 171

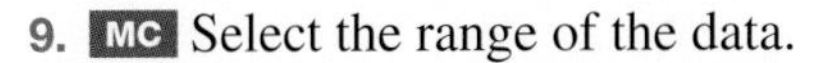

9. **MC** Select the range of the data.
   A. 67 B. 68 C. 69 D. 70

10. **MC** Select the IQR of the data.
   A. 32 B. 33 C. 34 D. 35

11. **WE8** **MC** Select the sample standard deviation of the data (correct to 1 decimal place).
   A. 19.5 B. 21.3 C. 22.0 D. 24.3

12. a. Calculate the maximum value of a data set if its minimum value is 23 and its range is 134.
   b. Calculate the minimum value of a data set if its range is 32 and its maximum value is 101.

13. Aptitude tests are often used by companies to help them decide who to employ. An employer gave 30 potential employees an aptitude test with a total of 90 marks. The scores achieved are shown below.

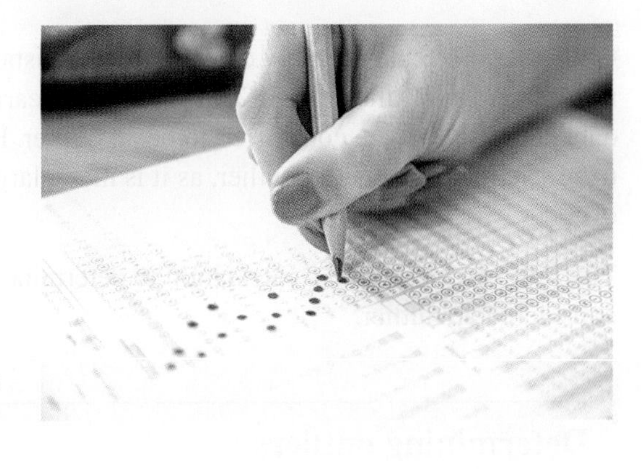

67, 67, 68, 68, 68, 69, 69, 72, 72, 73, 73, 74, 74, 75, 75,
77, 78, 78, 78, 79, 79, 79, 81, 81, 81, 82, 83, 83, 83, 86

Only applicants who score above the 80th percentile are invited to an interview. Use your knowledge of percentiles to work out how many interviews the employer will have to run.

14. **WE9** The ages of patients (in years) that came into a hospital emergency room during an afternoon were recorded. Using a digital technology of your choice, calculate the sample standard deviation of the data.

14, 1, 3, 87, 27, 42, 19, 91, 17, 73, 68, 83, 62, 29, 32, 2

6.4 Outliers

LEARNING INTENTION

At the end of this subtopic you should be able to:
- determine outliers and verify them using the interquartile range
- determine the effect an outlier has on the mean and the median.

6.4.1 Introduction

Outliers are extreme values on either end of a data set that appear very different from the rest of the data. They may be extremely large or extremely small compared to other data values in the set.

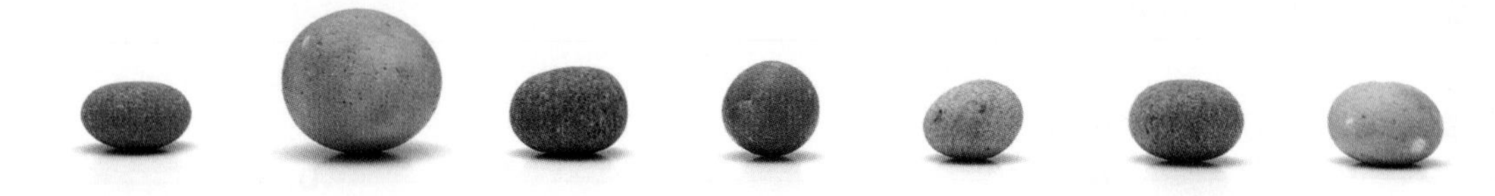

Outliers

An outlier is a value that is very far below or above the other values in the data set.

Outliers can be calculated by considering the distance a data point is from the mean or median compared to the rest of the data. If the value of a data point is clearly far away from the mean or median, as well as from other data points, it may be considered to be an outlier. For example, in the data set 2, 2, 4, 6, 7, 7, 100, it is reasonable to conclude that 100 is an outlier, as it is much larger than the rest of the data and is far away from the median of 5.

The IQR can be used in a calculation to determine if a data point is far enough away from the median to be considered an outlier.

Determining outliers

To determine if a data point is an outlier, first calculate the upper and lower fences:

$$\text{lower fence} = Q_1 - 1.5 \times \text{IQR}$$

$$\text{upper fence} = Q_3 + 1.5 \times \text{IQR}$$

where Q_1 is the lower quartile, Q_3 is the upper quartile and IQR is the interquartile range.

Any data point below the lower fence is an outlier.

Any data point above the upper fence is an outlier.

tlvd-3721

WORKED EXAMPLE 10 Determining outliers

Consider the following stem plot.
Calculate the IQR and use this to determine if there are any outliers.
***Note:* Technology can be used to determine the IQR.**

Key: 1|4 = 14

Stem	*Leaf*
0	1 3
1	0 0 1 3 4 4 4 6 8 9 9
2	2 2 5 5 6 7 8 8 8 9
3	4 5 7 7
4	0 1 1 2 9
5	
6	
7	1

THINK

WRITE

1. To calculate the IQR, you must first calculate Q_1 and Q_3. Start by locating the median. The median is the middle data value, so it occurs at the $\frac{33+1}{2} = 17\text{th}$ value.

Stem	*Leaf*
0	1 3
1	0 0 1 3 4 4 4 6 8 9 9
	Median
2	2 2 5 5 6 7 8 8 8 9
3	4 5 7 7
4	0 1 1 2 9
5	
6	
7	1

Median = 25

2. Locate the lower quartile. Q_1 is the middle value of the lower half of the data. There are 16 data values in the lower group, so Q_1 occurs at the $\frac{16+1}{2} = 8.5\text{th}$ value in the lower group.
Take the average of the 8th and 9th values.
$\frac{14+14}{2} = 14$

Stem	*Leaf*
0	1 3
	Q_1
1	0 0 1 3 4 4 | 4 6 8 9 9
2	2 2 5 5 6 7 8 8 8 9
3	4 5 7 7
4	0 1 1 2 9
5	
6	
7	1

$Q_1 = 14$

3. Locate the upper quartile. Q_3 is the middle value of the upper half of the data. There are 16 data values in the upper group, so Q_3 occurs at the $\frac{16+1}{2} = 8.5$th value in the upper group.
Take the average of the 8th and 9th values.
$\frac{35+37}{2} = 36$

Stem	Leaf
0	1 3
1	0 0 1 3 4 4 4 6 8 9 9
2	2 2 5 5 6 7 8 8 8 9
	Q_3
3	4 5 | 7 7
4	0 1 1 2 9
5	
6	
7	1

$Q_3 = 36$

4. Calculate the IQR.

$$\begin{aligned}\text{IQR} &= Q_3 - Q_1\\ &= 36 - 14\\ &= 22\end{aligned}$$

5. Calculate the lower fence.

$$\begin{aligned}\text{Lower fence} &= Q_1 - 1.5 \times \text{IQR}\\ &= 14 - 1.5 \times 22\\ &= -19\end{aligned}$$

Determine if there are any values below the lower fence.

No values are below this point.

6. Calculate the upper fence.

$$\begin{aligned}\text{Upper fence} &= Q_3 + 1.5 \times \text{IQR}\\ &= 36 + 1.5 \times 22\\ &= 69\end{aligned}$$

Determine if there are any values above the upper fence.

There is a value above this point; therefore 71 is an outlier.

6.4.2 The effect of outliers on the mean and median

The median is calculated using the number of data points, not the value of each data point. Therefore, outliers do not have much of an influence on the median. The mean involves a summation of the values of the data points, so extremely high or low values can greatly affect it. If outliers are present in a data set, the median is a better measure of central tendency, as it is less affected by outliers.

Outliers' effect on the mean and median

Outliers are extremely large or extremely small values that distort the mean to make it larger or smaller. Outliers have little effect on the median.

tlvd-3722

WORKED EXAMPLE 11 Outliers' effect on the mean and median

The number of days of sick leave per year taken by employees in a small company was recorded. The results were as follows.

5, 2, 4, 3, 26, 1, 1, 6, 4, 29, 0, 2, 1

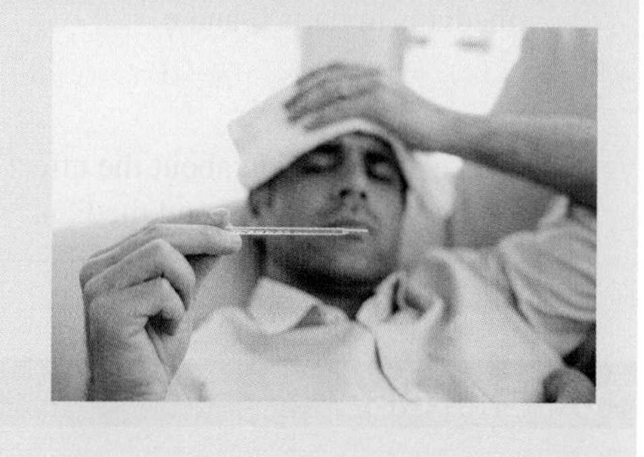

a. Using observation, identify any possible outliers.
b. For the data set, calculate:
  i. the mean **ii. the median.**
c. Exclude the outlier(s) in the data set and calculate:
  i. the mean **ii. the median.**
d. Comment on the difference in results between parts b and c.

THINK	WRITE
a. 1. Arrange the values in ascending order.	a. 0, 1, 1, 1, 2, 2, 3, 4, 4, 5, 6, 26, 29
2. Outliers are extreme values on either end of the data set.	0, 1, 1, 1, 2, 2, 3, 4, 4, 5, 6, 26, 29 26 and 29 are possible outliers.
b. i. 1. Calculate the sum of all the values.	b. i. $5+2+4+3+26+1+1+6+4+29+0+2+1=84$
2. Count the number of values.	Number of values = 13
3. The mean is the sum of all the values divided by the number of values.	$\bar{x}=\frac{84}{13}$ $=6.5$ The mean number of sick days employees took in a year was 6.5.
ii. 1. Arrange the values in ascending order. The median is the middle value.	ii. 0, 1, 1, 1, 2, 2, 3, 4, 4, 5, 6, 26, 29
2. There are 13 values, so the median is the $\frac{13+1}{2}=7$th value.	0, 1, 1, 1, 2, 2, 3, 4, 4, 5, 6, 26, 29 The median number of sick days employees took in a year was 3.
c. i. 1. Write out the data set, excluding the outliers. Calculate the sum of all the values.	c. i. 0, 1, 1, 1, 2, 2, 3, 4, 4, 5, 6 $5+2+4+3+1+1+6+4+0+2+1=29$
2. Count the number of values.	Number of values = 11
3. The mean is the sum of all the values divided by the number of values.	$\bar{x}=\frac{29}{11}$ $=2.6$ The mean number of sick days employees took in a year, when excluding outliers, was 2.6.
ii. 1. Arrange the values in ascending order. The median is the middle value.	ii. 0, 1, 1, 1, 2, 2, 3, 4, 4, 5, 6
2. There are 11 values, so the median is the $\frac{11+1}{2}=6$th value.	0, 1, 1, 1, 2, 2, 3, 4, 4, 5, 6 The median number of sick days employees took in a year, when excluding outliers, was 2.
d. 1. Comment on the difference between the means in parts b and c.	d. The mean sick days per year taken by employees reduced from 6.5 days to 2.6 days when outliers were excluded. Excluding the large outliers significantly decreased the mean.

2. Comment on the difference between the medians in parts b and c.	The median sick days per year taken by employees changed from 3 days to 2 days when outliers were excluded. Excluding the large outliers did not significantly change the median.
3. Make a conclusion about the effect of outliers on the mean and median.	The outliers had a greater effect on the mean than on the median.

6.4 Exercise

Students, these questions are even better in jacPLUS

Receive immediate feedback and access sample responses

Access additional questions

Track your results and progress

Find all this and MORE in jacPLUS

1. The following data represents car sales each month at a car yard.

   12, 9, 15, 10, 11, 23, 14, 8, 6, 11, 13, 15

   a. Calculate the IQR.
   b. Calculate any outliers.

2. WE10 Consider the following stem plot.

   Key: 1 | 6 = 16

Stem	*Leaf*
0	1
1	5
2	
3	
4	2
5	6 7
6	0 1 4 5 6
7	1 1 2 3 5 7 9
8	2 2 4 4 4 8 8
9	35

   Calculate the IQR and use it to determine if there are any outliers.

3. Calculate the lower and upper fence of the following data.

   34, 41, 53, 39, 48, 41, 33, 39, 40, 34, 44

4. Consider the following data:

   2.3, 3.1, 3.6, 1.8, 6.7, 4.4, 3.9, 2.8, 3.7, 4.0

   a. Calculate the lower and upper fence.
   b. Determine if there are any outliers in the data set.

5. The number of swimmers at a pool was recorded over a fortnight. The results are as follows.

56, 69, 59, 113, 9, 100, 80, 111, 94, 77, 57, 166, 101, 96

Determine if any outliers exist and justify your answer using a calculation.

6. **WE11** Below is the data collected during a survey of the number of pets people own.

2, 1, 0, 3, 1, 1, 2, 1, 2, 3, 3, 2, 0, 2, 1, 2, 3, 4,
1, 2, 1, 1, 3, 3, 1, 3, 2, 1, 2, 2, 2, 3, 21, 1, 2, 3

a. Determine if there are any possible outliers.
b. Calculate the mean and median.
c. Remove any outliers and recalculate the mean and median.
d. Comment on the difference in means and medians in parts **b** and **c**.

7. a. Select possible outliers in each of the following sets of numbers.
   i. 3, 5, 2, 6, 15, 1, 5, 4
   ii. 21, 33, 44, 34, 27, 3, 29, 30, 6, 31, 25, 36

   b. From parts **ai** and **aii**, state if the mean would increase or decrease with the outliers taken out of the data set.
   c. For parts **ai** and **aii**, calculate the mean before and after taking out the outliers.

8. a. Select possible outliers in each of the following sets of numbers.
   i. 0.03, 0.05, 0.07, 0.03, 0.9, 0.03, 0.04
   ii. −2.3, −3.6, −1.9, −2.1, 5.2, −3.9, −3.9, −2.7, 6.1, −1.7, −3.3, −2.4

   b. State if the median would increase or decrease with the outliers taken out of the data sets above.
   c. Calculate the median before and after taking out the outliers.

9. **MC** The mean of a data set was calculated to be 117. The median was calculated to be 107. The data set has two outliers of 178 and 190. If the outliers were replaced with values of 118 and 120:
   A. the median would increase, but the mean would stay the same.
   B. the mean would increase, but the median would remain the same.
   C. the median would decrease, but the mean would stay the same.
   D. the mean would decrease, but the median would stay the same.
   E. None of the above

10. A statistician investigated the monthly household sales in a suburb over a year. The Q_1, median and Q_3 were calculated to be 18, 27 and 35 respectively.

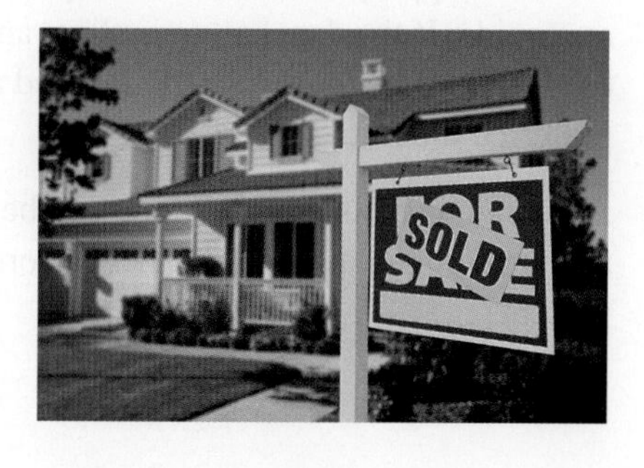

   a. Prove that 64 houses per month is an outlier.
   b. The statistician realised they made a mistake and the outlier of 64 was recorded incorrectly. The actual value was 46 houses per month. Using your answer from part **a** to help you, explain what would happen to the values of Q_1, Q_3 and the IQR if the data point was changed from 64 to 46 houses per month.

6.5 Applications of measures of central tendency

LEARNING INTENTION

At the end of this subtopic you should be able to:
- determine the most appropriate measure of centre and spread.

When analysing data or comparing data sets, it is common to discuss or compare measures of central tendency (mean or median) and the spread (range, IQR or standard deviation). Measures of central tendency give information about the centre of the distribution and the spread provides information on how spread out (widely spread or tightly packed) the data are.

For example, if the spread of one set of data was tightly packed compared to a set of data that was widely spread (dispersed), we could say that the data set that is tightly packed is more consistent than the set of data that is widely spread. This is because when a distribution's spread is tightly packed, the data collected would cluster around a certain value, implying that the data collected were consistent.

Given that there are several measures of centre and spread to select from when analysing data, it is often not necessary to calculate them all. Decisions need to be made about which measure of centre and which measure of spread to use when analysing and comparing data.
- The mean is calculated using every data value in the set. The median is the middle score of an ordered set of data, so it does not include every individual data value in its calculation.
- The mode is the most frequently occurring data value, so it also does not include every individual data value in its calculation.
- The range is calculated by determining the difference between the maximum and minimum data values, so it includes outliers. It provides only a rough idea about the spread of the data and insufficient detail for analysis. It is useful, however, when we are interested in extreme values such as high and low tides or maximum and minimum temperatures.
- The interquartile range is the difference between the upper and lower quartiles, so it does not include every data value in its calculation, but it will overcome the problem of outliers skewing data.
- The standard deviation is calculated using every data value in the set.

When to use mean or median

If the data has no outliers and is symmetric, either the mean or the median can be used as the measure of centre. If the data is clearly skewed and/or there are outliers, it is more appropriate to use the median as the measure of centre.

When to use range, IQR or standard deviation

The most appropriate measure of spread does depend on the type of data; however, the following can be used as a guide. If the data has no outliers and is symmetric, the standard deviation would be the preferred measure of spread. If the data is clearly skewed and/or there are outliers, it is more appropriate to use the IQR as the preferred measure of spread.

Although the above information can be applied in most cases, selecting the most appropriate measure of centre and spread may depend on other factors, such as wanting to include every data value in a set even if outliers are present.

WORKED EXAMPLE 12 Determining the most appropriate measure of centre and spread

The following are samples of scores achieved by two students in eight Mathematics tests throughout the year.

John: 45, 62, 64, 55, 58, 51, 59, 62

Penny: 84, 37, 45, 80, 74, 44, 46, 50

a. **Determine the most appropriate measure of centre and measure of spread to compare the performance of the students.**
b. **Determine which student had the better overall performance on the eight tests.**
c. **Determine which student was more consistent over the eight tests.**

THINK	WRITE
a. In order to include all data values in the calculation of measures of centre and spread, calculate the mean and standard deviation.	a. John: $\bar{x} = 57, s = 6.41$ Penny: $\bar{x} = 57.5, s = 18.62$
b. Compare the mean for each student. The student with the higher mean performed better overall.	b. Penny performed slightly better on average, as her mean mark was higher than John's.
c. Compare the standard deviation for each student. The student with the lower standard deviation performed more consistently.	c. John was the more consistent student because his standard deviation was much lower than Penny's. This means that his test results were closer to his mean score than Penny's were to hers.

6.5 Exercise

1. The number of M&M's in 20 packets was recorded by a group of students.

$$30, 32, 33, 35, 37, 37, 38, 38, 39, 39, 40, 40, 40, 41, 41, 41, 41, 41, 41, 50$$

   a. Calculate the mean number of M&M's per packet.
   b. Calculate the median number of M&M's per packet.
   c. Calculate the modal number of M&M's per packet.
   d. Based on your calculations, state how many M&M's you would expect to find in a packet.
   e. If you were advertising M&M's, determine which summary statistic (mean, median or mode) you would use. Explain.

2. Explain how the mean, median and mode for both ungrouped and grouped data are similar and different.

3. a. Calculate the mean and median of the following data sets.

Data set A: 20, 24, 29, 33, 37, 42, 51, 53, 96
Data set B: 20, 24, 29, 33, 37, 42, 51, 53, 66

b. Comment on the means and medians of the data sets and explain any similarities and differences you see.
c. Write a short statement to explain when to use the mean and when to use the median as a measure of central tendency.

4. **WE12** The Mathematics test results for two students over a year were recorded as follows.

Student A: 49, 52, 51, 50, 54, 49, 100
Student B: 65, 71, 64, 63, 60, 81, 0

a. Determine the most appropriate measure of centre and spread to compare the performance of the students.
b. Determine which student had the better overall performance over the year.
c. Determine which student was more consistent over the year.

5. Create a data set that fits the following descriptions.
a. 8 data values with a range of 43
b. 6 data values with a lower quartile of 5 and an upper quartile of 12
c. 9 data values with a lower quartile of 7 and an upper quartile of 13

6. A Mathematics teacher wanted to design a question for their class. They wanted the question to have 10 data points, a lower quartile of 14 and an IQR of 20. State a data set that could be used in the question.

7. The number of trees in several different parks was recorded, and the mean number of trees was calculated to be 15. The council wanted to determine whether the majority of parks contained close to 15 trees, or whether there was a large difference in the number of trees. Describe how the council could use a measure of spread to investigate this problem.

8. The following scores show the number of points scored by two AFL teams over the first 10 games of the season.

Sydney Swans	110	95	74	136	48	168	120	85	99	65
Brisbane Lions	125	112	89	111	96	113	85	90	87	92

a. Calculate the range of the scores for each team.
b. Based on the results shown, state which team you think is more consistent.

9. Two machines are used to put approximately 100 Smarties into boxes. A check is made on the operation of the two machines. Ten boxes filled by each machine have the number of Smarties in them counted. The results are shown below.

Machine A: 100, 99, 99, 101, 100, 101, 100, 100, 101, 108
Machine B: 98, 104, 96, 97, 103, 96, 102, 100, 97, 104

a. Calculate the range in the number of Smarties from the first machine.
b. Calculate the range in the number of Smarties from the second machine.
c. Ralph is the quality control officer and he argues that machine A is more consistent in its distribution of Smarties. Explain why.

10. The following frequency distribution gives the prices paid by a car wrecking yard for a sample of 40 car wrecks.

Price ($)	Frequency
0– < 500	2
500– < 1000	4
1000– < 1500	8
1500– < 2000	10
2000– < 2500	7
2500– < 3000	6
3000– < 3500	3

Determine the mean and standard deviation of the price paid for these wrecks.

11. The following stem plot represents the lifespan of different animals at an animal sanctuary. Determine which measure of centre is best to represent the data set.

Key: 1|2 = 12

Stem	*Leaf*
0	3 5 9
1	2 4 6 8
2	0 1 4 5 5 7 9
3	0 2 6
4	
5	
6	0 3

12. The following data set represents the salaries (in $1000s) of workers at a small business.

45, 50, 55, 55, 55, 60, 65, 65, 70, 70, 75, 80, 220

a. Calculate the mean of the salaries correct to 3 decimal places.
b. Calculate the median of the salaries.
c. When it comes to negotiating salaries, the workers want to use the mean to represent the data and the management want to use the median. Explain why this might be the case.

13. A sample of crime statistics over a two-year period are shown in the following table.

Crime	Year 1	Year 2
Theft from motor vehicle	46 700	42 900
Theft from shop	19 800	20 600
Theft of motor vehicle	15 650	14 670
Theft of bicycle	4200	4660
Theft (other)	50 965	50 650

a. Calculate the interquartile range and standard deviation (correct to 1 decimal place) for both years.
b. Recalculate the interquartile range and standard deviation for both years after removing the smallest category.
c. Comment on the effect of removing the smallest category on the interquartile ranges and standard deviations.

14. The following data shows the number of daylight hours in Alice Springs.

10.3, 9.8, 9.6, 9.5, 8.5, 8.4, 9.1, 9.8, 10.0, 10.0, 10.1, 10.0, 10.1, 10.1, 10.6, 8.7, 8.8, 9.0, 8.0, 8.5, 10.6, 10.8, 10.5, 10.9, 8.5, 9.5, 9.3, 9.0, 9.4, 10.6, 8.3, 9.3, 9.0, 10.3, 8.4, 8.9

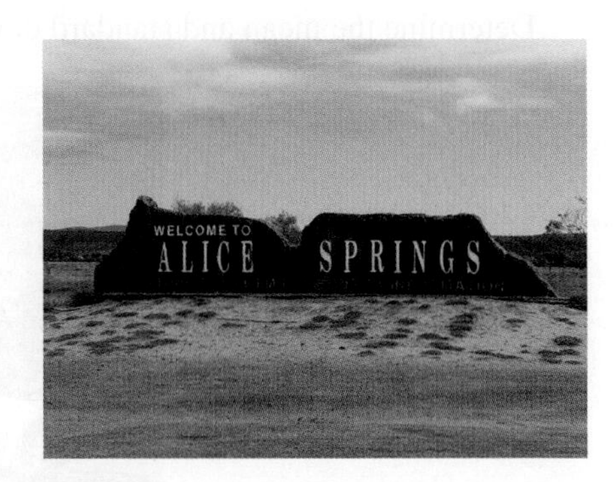

a. Calculate the range of the data.
b. Calculate the interquartile range of the data.
c. Comment on the difference between the two measures and what this indicates.

6.6 Review

6.6.1 Summary

doc-38014

Hey students! Now that it's time to revise this topic, go online to:

Review your results

Watch teacher-led videos

Practise questions with immediate feedback

Find all this and MORE in jacPLUS

6.6 Exercise

Multiple choice

1. **MC** The mean, median and mode of the following data set are:

$$11, 63, 24, 36, 25, 61, 29, 42$$

A. mean = 36.4; median = 32.5; no mode.
B. mean = 36; median = 32.5; mode = 63.
C. mean= 36.4; median = 32; mode = 52.
D. mean = 34.6; median = 32; no mode.
E. mean = 32.6; median = 32; no mode.

2. **MC** The following data was collected about the number of days people go away during the holidays.

$$2, 10, 5, 7, 9, 14, 2, 0, 6, 7, 0, 7, 14, 7, 8, 10, 12, 5, 2, 1, 16, 12, 10$$

The range, mean and median are:

A. range= 12; mean = 7; median = 7.
B. range = 16; mean = 7; median = 7.
C. range = 16; mean = 7.2; median = 7.
D. range = 16; mean = 7.2; median = 7.5.
E. range = 7.2; mean = 16; median = 7.5.

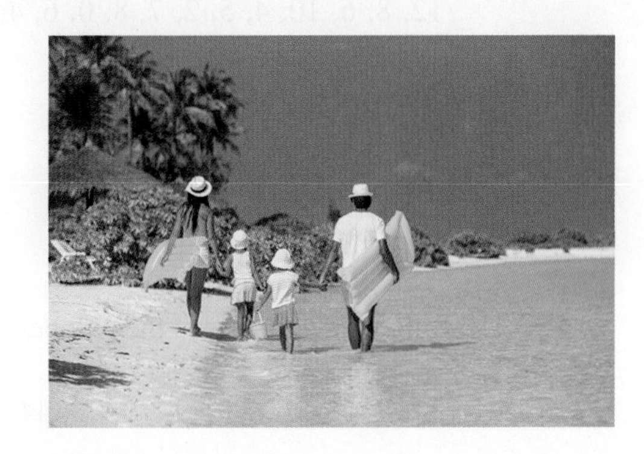

3. **MC** The screen time per day for a group of 0–14-year-olds is listed in the table. Calculate the mean number of minutes of screen time for 0–14-year-olds.

Age (years)	Screen time/day (minutes)
0–2	45
3–5	73
6–8	98
9–11	124
12–14	142

A. 98 minutes
B. 96.4 minutes
C. 94.6 minutes
D. 120 minutes
E. 135 minutes

4. **MC** A measure of spread that uses the distance from the mean in the calculation is called:

A. mean.
B. range.
C. median.
D. IQR.
E. standard deviation.

5. **MC** The number of books borrowed by 20 students over the period of a month is monitored. The IQR of the data is:

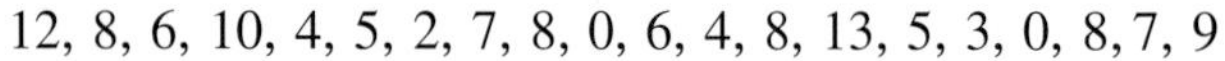
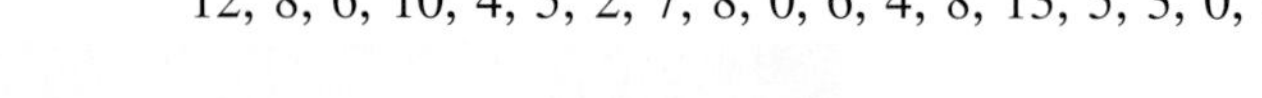

12, 8, 6, 10, 4, 5, 2, 7, 8, 0, 6, 4, 8, 13, 5, 3, 0, 8, 7, 9

A. 3
B. 4
C. 5
D. 6
E. 7

6. **MC** The following data shows the number of ice creams sold nightly over two weeks.

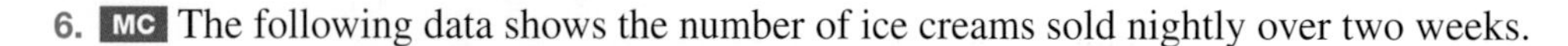

4, 6, 48, 29, 39, 48, 44, 45, 39, 47, 48, 32, 31, 51

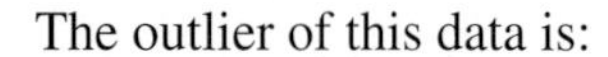

The outlier of this data is:

A. none.
B. 51
C. 4
D. 48
E. 29

The test marks for a class of students are shown in the histogram and frequency distribution table. Use this information to answer questions 7–10.

Class interval	Frequency (f)
0– < 5	2
5– < 10	2
10– < 15	5
15– < 20	5
20– < 25	6
25– < 30	3
Total	23

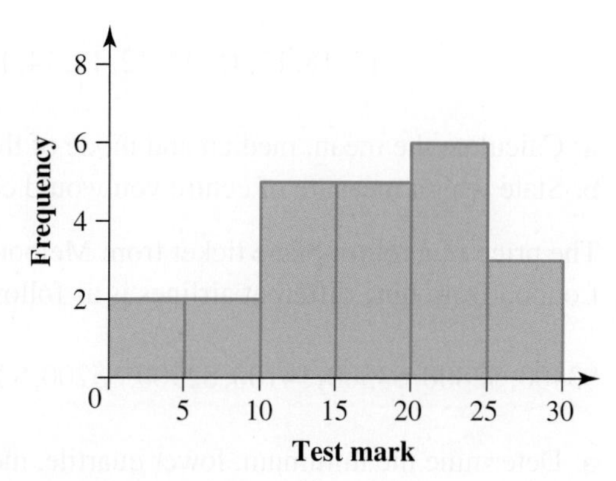

7. **MC** Select the category the median test mark for the class falls into.
   A. 5–10 B. 10–15 C. 15–20 D. 20–25 E. 25–30

8. **MC** Determine the mean of this data to 1 decimal place.
   A. 14.3 B. 16.8 C. 22.5 D. 19.3 E. 18.5

9. **MC** Determine the mode of this data.
   A. 5–10 B. 10–15 C. 15–20 D. 20–25 E. 25–30

10. **MC** Determine the range of this data.
    A. 5
    B. 15
    C. 20
    D. 30
    E. None of the above

Short answer

11. For the dot plot shown, calculate:
    a. the mean (to 2 decimal places)
    b. the median
    c. the IQR
    d. the standard deviation
    e. the range.

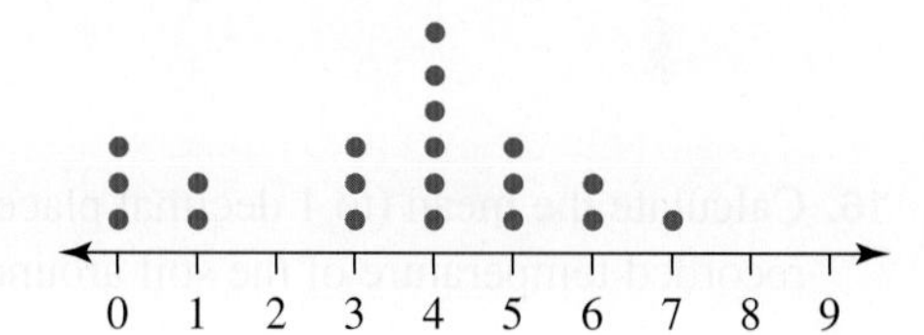

12. Consider the stem plot shown.
    Calculate:
    a. the mean
    b. the median
    c. the IQR
    d. the standard deviation.

Key: 11|2 = 112

Stem	*Leaf*
5	0 1 2
6	4 5 8 9
7	0 7
8	0 3 3
9	5 5 5 7 8
10	1 1 2 3 4 6 6
11	0 1 3 3 4 5 5 5 7 9
12	1 1 2 2 3 3 4 7
13	2 6 9

13. A survey of Year 12 students was conducted to determine the average travel time to school. The results (in minutes) were as follows.

$$12, 18, 17, 15, 15, 12, 14, 14, 14, 17, 10, 14, 11, 16, 14, 14, 11, 15, 10, 9, 13, 20$$

a. Calculate the mean, median and mode of this data set.
b. State which measure of centre you would choose as the average travel time. Explain why.

14. The price of a return plane ticket from Melbourne to London from nine different airlines is as follows:

$3400, $2800, $3500, $4100, $2900, $5200, $3900, $4000, $4575

a. Determine the minimum, lower quartile, median, upper quartile and maximum.
b. Calculate the interquartile range.

15. The following data set shows the number of doors people have in their houses.

18, 11, 9, 14, 11, 16, 14, 16, 6, 18, 19, 9, 10, 17, 36

a. State the minimum and maximum number of doors, and hence calculate the range.
b. Calculate the mean and median.
c. Determine if there are any outliers. If yes, remove these and recalculate the mean and median.
d. Compare your answers to parts a and c.

16. Calculate the mean (to 1 decimal place), median, mode, range and IQR of the following data, showing the recorded temperature of the soil around 25 germinating seedlings.

28.9, 27.4, 23.6, 25.6, 21.1, 22.9, 29.6, 25.7, 27.4, 23.6, 22.4, 24.6, 21.8,
26.4, 24.9, 25.0, 23.5, 26.1, 23.6, 25.3, 29.5, 23.5, 22.0, 27.9, 23.6

Extended response

17. A data set contains the numbers 5, 6, 6 and 10. A fifth number is added to the set. If this fifth number is a whole number, determine the possible values of this fifth number if:

a. the mean is now equal to the median
b. the mean is now greater than the mode
c. the mean is now greater than the range.

18. Kris and Loren were discussing the following histogram showing the number of magazines sold in a convenience store. Kris said that the best way to describe the central tendency for magazine sales would be to look at the median. Loren disagreed and said to use the mean. Determine who is correct and why.

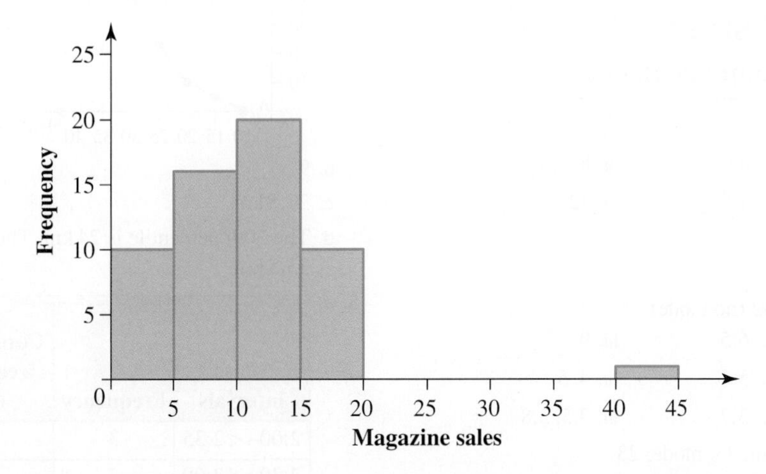

19. The data set used to generate the histogram relating to magazine sales in question **18** was as follows.

0, 1, 12, 2, 3, 3, 4, 4, 4, 5, 5, 5, 6, 6, 7, 7, 7, 7, 7, 8, 8, 8, 9, 9,
10, 10, 10, 10, 10, 11, 11, 11, 11, 11, 11, 11, 12, 12, 12, 12,
13, 13, 14, 14, 15, 15, 15, 15, 16, 17, 18, 18, 18, 19, 44

State whether the 44 magazine sales in a day are an outlier. Justify your answer using calculations.

20. Use the data shown to answer the following questions.

	Women who gave birth and Indigenous status by states and territories, 2009								
Status	**NSW**	**Vic.**	**Qld**	**WA**	**SA**	**Tas.**	**ACT**	**NT**	**Australia**
Indigenous	2904	838	3332	1738	607	284	107	1474	11 284
Non-Indigenous	91 958	70 328	57 665	29 022	18 994	5996	5601	2369	281 933

a. Calculate the mean births per state/territory of Australia in 2009 for both Indigenous and non-Indigenous groups. Give your answers correct to 1 decimal place.
b. Calculate the median births per state/territory of Australia in 2009 for both Indigenous and non-Indigenous groups.
c. Calculate the standard deviation (correct to 1 decimal place) and IQR for the data on births per state/territory of Australia in 2009 for both Indigenous and non-Indigenous groups.
d. Comment on the measures of centre and spread you have calculated for this data.

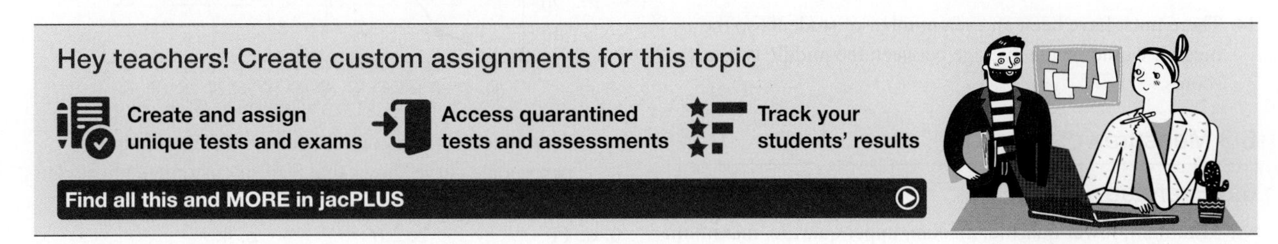

Answers

Topic 6 Measures of central tendency and spread

6.2 Measures of central tendency

6.2 Exercise

1. a. i. 6 ii. 6 iii. 8
   b. i. 10 ii. 12 iii. 12
   c. i. 26.2
      ii. 28
      iii. 11, 21, 28, 29, 42 (no mode)
   d. i. 6 ii. 6.5 iii. 9
   e. i. 5 ii. 5 iii. 4, 5
   f. i. 3.66 ii. 3.7 iii. 3.5, 3.8
2. a. Mean: 22.6; median: 23; mode: 23
   b. Mean: 7.7; median: 8;
      mode: 2, 4, 6, 7, 8, 9, 10, 11, 12 (no mode)
   c. Mean: 48; median: 49; mode: 50
   d. Mean: 1.9; median: 1.7; mode: 1.4, 1.7
3. B
4. 47
5. Mean: 104 km/h; median: 102 km/h; mode: 101 km/h
6. 4.8
7. 27
8. a. Mean: 2.4; median: 2; mode: 1
   b. Mean: 26.9; median: 27; mode: 26
   c. Mean: 2.6; median: 2.5; mode: 3.0
9. 73.5
10. 171
11. a. Mean
    b. Median. There are two possible outliers of 8 and 12 goals, which were not typical goal numbers for the netballer. The median is less affected than the mean by these two atypically high scores, so it is the best measure to use.
12. Sample responses can be found in the worked solutions in the online resources.
13. 435 cm
14. There must have been an even number of students so the median would be the average between the middle two test scores.

6.3 Measures of spread

6.3 Exercise

1. Minimum, lower quartile, median, upper quartile, maximum
2. B
3. a. 110
   b. 99
   c. 120
   d. > 117 steps/minute
4. a. (graph)
   b. 9
   c. 27.5 km
   d. The 30th percentile is 24 km. The upper quartile is 33.5 km.
5. a.

Class intervals	Frequency	Cumulative frequency (cf)	Percentage cumulative frequency ($\%cf$)
2:00–< 2:35	3	3	6%
2:30–< 3:00	2	5	10%
3:00–< 3:30	8	13	26%
3:30–< 4:00	8	21	42%
4:00–< 4:30	9	30	60%
4:30–< 5:00	7	37	74%
5:00–< 5:30	2	39	78%
5:30–< 6:00	4	43	86%
6:00–< 6:30	0	43	86%
6:30–< 7:00	2	45	90%
7:00–< 7:30	2	47	94%
7:30–< 8:00	1	48	96%
8:00–< 8:30	2	50	100%

   b.

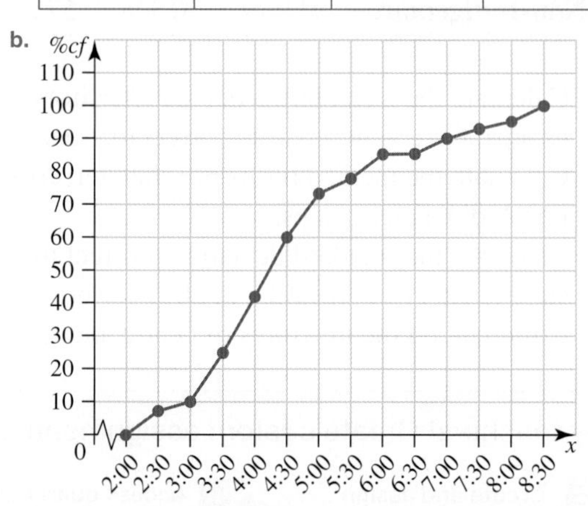

   c. This statement is incorrect because the cumulative percentage curve shows that 80% of customers must wait over five minutes for their food.
6. a. 11 b. 2.6 c. 4
7. Range = 15 km/h; IQR = 2 km/h
8. Range = 27 hours; IQR = 8 hours
9. B
10. D
11. C

12. a. 157 b. 69

13. 5

14. 32.0 years

6.4 Outliers

6.4 Exercise

1. a. 5 b. 23
2. IQR = 23; outliers are 1 and 15.
3. 19 and 59
4. a. 1 and 5.8 b. 6.7
5. 166 is an outlier. (It is above the upper fence of 164.)
6. a. 21
   b. Mean: 2.4; median: 2
   c. Mean: 1.9; median: 2
   d. Mean value lowered; median remained unchanged
7. a. i. 15 is larger than the rest of the data.
   ii. 3 and 6 are smaller than the rest of the data.
   b. Taking 15 out, which is larger than the rest of the data, would make the mean decrease.
   Taking out 3 and 6, which are smaller than the rest of the data, would make the mean increase.
   c. i. With: mean = 5.125
   Without: mean = 3.71
   ii. With: mean = 26.58
   Without: mean = 31
8. a. i. 0.9 is larger than the rest of the data.
   ii. 5.2 and 6.1 are larger than the rest of the data.
   b. Taking out 0.9, which is large, would therefore decrease the median.
   Taking out 5.2 and 6.1, which are large, would therefore decrease the median.
   c. With: median = 0.04
   Without: median = 0.035
   With: median = −2.3
   Without: median = −2.4
9. D
10. a. IQR = 35 − 18 = 17

$$\begin{aligned}\text{Upper fence} &= Q_3 + 1.5 \times \text{IQR}\\ &= 35 + 1.5 \times 17\\ &= 60.5\end{aligned}$$

Any value above 60.5 will be an outlier, so 64 is an outlier.

b. Q_1, Q_3 and IQR would remain unchanged.

6.5 Applications of measures of central tendency

6.5 Exercise

1. a. 39
   b. 40
   c. 41
   d. 40
   e. The mode, as it gives the highest number of M&M's in a packet
2. The mean of grouped data is calculated using the midpoint of the class interval, making it an approximation. The mean of ungrouped data will give an exact value that is not an approximation. Using both grouped and ungrouped data, a single value can be obtained for the mean.
   The median for grouped data will be a class interval, rather than an exact value as for ungrouped data.
   The mode is calculated the same way for grouped and ungrouped data: it is the value with the greatest frequency.
3. a. Data set A: mean: 42.8, median: 37
   Data set B: mean: 39.4, median: 37
   b. The mean of data set A is larger than the mean of data set B. This is due to the last value being 96 in data set A, compared to 66 in data set B. The medians of the two data sets are the same, as each data set has the same number of data values and all data values are the same, except the last value.
   c. The mean or median can be used as a good measure of central tendency when there are no outliers or extreme values. The median is best to use when there are outliers or extreme values in the data.
4. a. Student A: median = 51, IQR = 5;
   student B: median = 64, IQR = 11
   b. Student B
   c. Student A
5. Answers will vary. Example answers are shown.
   a. 2, 5, 17, 21, 29, 35, 39, 45. (Several different data sets are possible.)
   b. 1, **5**, 6, 8, **12**, 14. (Several different data sets are possible, but the numbers in bold must be present in the same position.)
   c. 5, 6, 8, 9, 10, 11, 12, 14, 15. (Several different data sets are possible.)
6. 5, 10, **14**, 21, 22, 23, 31, **34**, 35, 40. (Several different data sets are possible, but the numbers in bold must be present in the same position.)
7. Calculate the standard deviation. If the standard deviation is small, then the majority of parks will contain close to 15 trees.
8. a. Sydney = 120
   Brisbane = 40
   b. Brisbane is more consistent, since the scores have a lower range.
9. a. Machine A range: 9 Smarties
   b. Machine B range: 8 Smarties
   c. Only 2 boxes out of 10 sampled from machine A contain less than 100 Smarties. The box with 108 Smarties causes the range to be higher than normal.
10.

Price ($)	Class centre	Frequency
0–<500	250	2
500–<1000	750	4
1000–<1500	1250	8
1500–<2000	1750	10
2000–<2500	2250	7
2500–<3000	2750	6
3000–<3500	3250	3

Use the statistical function on a calculator. The sample standard deviation should be used.
Mean = $1825, sample SD = $797

11. The median, as the data set has two clear outliers
12. a. $74 231
    b. $65 000
    c. It would be in the workers' interest to use a higher figure when negotiating salaries, whereas it would be in the management's interest to use a lower figure.
13. a. Year 1: interquartile range = 38 907.5; standard deviation = 20 382.8
    Year 2: interquartile range = 37 110; standard deviation = 19 389.01
    b. Year 1: interquartile range = 31 107.5; standard deviation = 18 123.5
    Year 2: interquartile range = 29 140; standard deviation = 17 289.2
    c. Both values are reduced by a similar amount, but there is a bigger impact on the standard deviation than on the interquartile range.
14. a. 2.9
    b. 1.25
    c. The range is less than double the value of the interquartile range. This indicates that the data is quite tightly bunched with no outliers.

6.6 Review

6.6 Exercise

Multiple choice

1. A
2. C
3. B
4. E
5. B
6. C
7. C
8. B
9. D
10. D

Short answer

11. a. 3.45 b. 4
    c. 3 d. 2.1
    e. 7
12. a. 101.0 b. 106
    c. 37 d. 23.5
13. a. Mean: 13.9; median: 14; mode: 14
    b. 20 is an outlier. Any measure could be used in this case, as they all give similar results. However, since the data contains an outlier, the median is the more appropriate measure of centre.
14. a. 2800, 3150, 3900, 4337.5, 5200
    b. 1187.5
15. a. Minimum: 6; maximum: 36; range: 30
    b. Mean: 14.9; median: 14
    c. Possible outlier: 36; mean: 13.4; median: 14
    d. The median was unchanged by removing the outlier, whereas the mean decreased.
16. Mean = 25.0
    Median = 24.9
    Mode = 23.6
    Range = 8.5
    IQR = 3.4

Extended response

17. a. 3
    b. Any number greater than 3
    c. Any number from 4 to 12
18. Kris. There is a possible outlier at 40–45, so the median is a better measure of central tendency, as it is less affected by outliers than the mean.
19. 44 is above the upper fence of 23.5, so the data value is an outlier.
20. a. Indigenous mean = 1410.5
    Non-Indigenous mean = 35 241.6
    b. Indigenous median = 1156
    Non-Indigenous median = 24 008
    c. Indigenous: standard deviation = 1193.8, IQR = 1875.5
    Non-Indigenous: standard deviation = 33 949.03, IQR = 58 198
    d. The median and IQR are probably more appropriate due to the presence of potential extreme values in the data.

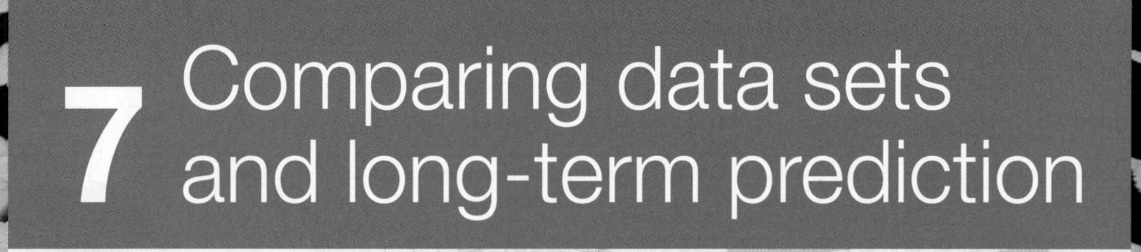

7 Comparing data sets and long-term prediction

LEARNING SEQUENCE

Fully worked solutions for this topic are available online.

7.1 Overview

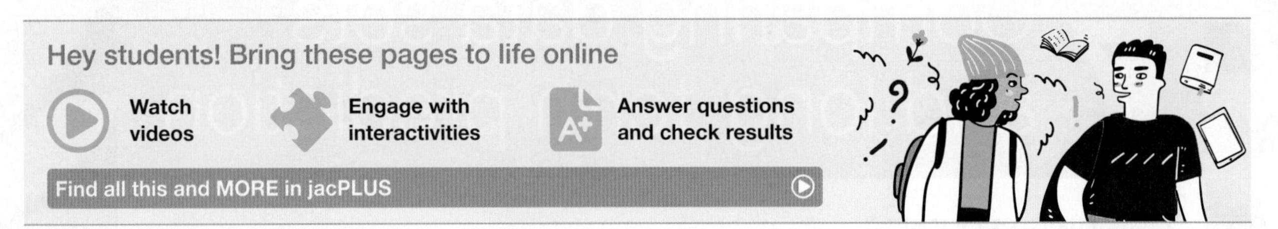

7.1.1 Introduction

A boxplot is a way of graphically representing groups of numerical data through their quartiles. The spaces between the different sections of the box indicate the degree of spread and skewness in the data. The boxplot gives a snapshot of a number of values, such as the interquartile range, maximum value, minimum value, range and medium of the data. It is also used to determine and show outliers.

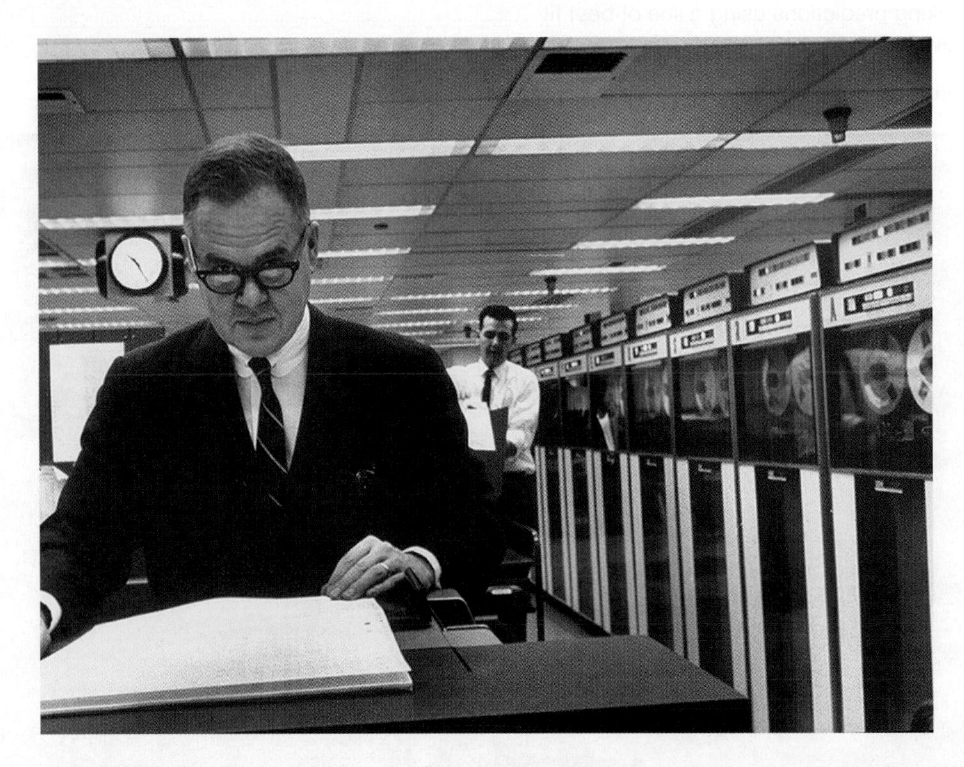

The boxplot was introduced by the mathematician John Tukey. He is regarded as one of the most influential statisticians of the past 50 years. Some of his work in modern statistics led to concepts that have played a central role in the creation of today's telecommunication technology. He is credited for the invention of the computer term *bit*.

It was in his 1977 book *Exploratory Data Analysis* that he introduced the boxplot.

KEY CONCEPTS

This topic covers the following key concepts from the VCE Mathematics Study Design:
- interpolation and extrapolation of data, predictions, limitations, inferences and conclusions comparing and interpreting data sets and graphs, including using measures of central tendency and spread (percentiles and standard deviation) and cumulative frequency.

Note: Concepts shown in grey are covered in other topics.

7.2 Constructing boxplots

LEARNING INTENTION

At the end of this subtopic you should be able to:
- use the five-number summary to construct boxplots
- interpret the spread of boxplots.

We will firstly revise the knowledge covered in topics 5 and 6 and then use this knowledge to make comparisons of different data sets.

7.2.1 Five-number summary

A **five-number summary** is a list consisting of the lowest score, lower quartile, median, upper quartile and highest score of a set of data. A five-number summary gives information about the centre and spread of a set of data. The convention is not to detail the numbers with labels but to present them in order; so, for example, the five-number summary 4, 15, 21, 23, 28 would be interpreted as lowest score 4, lower quartile 15, median 21, upper quartile 23 and highest score 28.

tlvd-3723

WORKED EXAMPLE 1 Using a five-number summary

From the following five-number summary, determine:

29, 37, 39, 44, 48

a. **the median** b. **the interquartile range** c. **the range.**

THINK	WRITE
The figures are presented in the order of lowest score, lower quartile, median, upper quartile, highest score.	Lowest score = 29, $Q_1 = 37$, median = 39, $Q_3 = 44$, highest score = 48
a. The median is the third number in the list.	a. Median = 39
b. The interquartile range is the difference between the upper and lower quartiles.	b. $\text{IQR} = Q_3 - Q_1$ $= 44 - 37$ $= 7$
c. The range is the difference between the highest score and the lowest score.	c. Range = highest score − lowest score $= 48 - 29$ $= 19$

7.2.2 Boxplots

A boxplot (or **box-and-whisker plot**) is a graphical representation of the five-number summary. It is a powerful way to show the centre and spread of data.
- Boxplots consist of a central divided box with attached 'whiskers'.
- The box spans the interquartile range.
- The median is marked by a vertical line inside the box.
- The whiskers indicate the minimum and maximum scores.

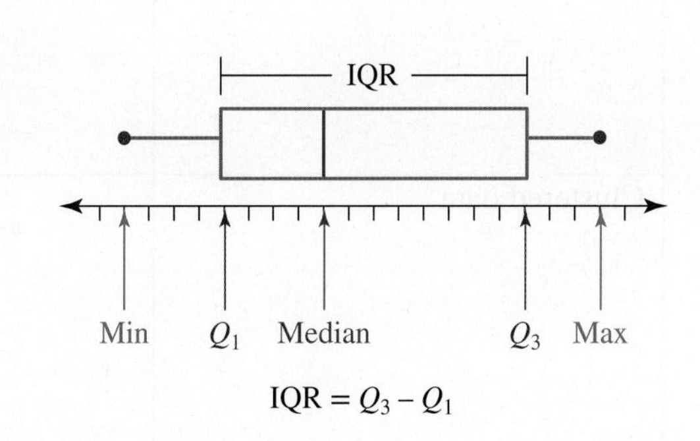

Boxplots are *always drawn to scale*. They are presented either with the five-number summary figures attached as labels (below left) or with a scale presented alongside the boxplot (below right).

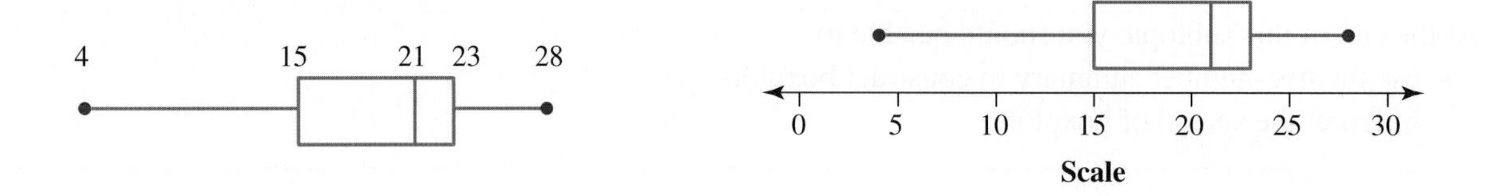

7.2.3 Interpreting a boxplot

The boxplot neatly divides the data into four sections. One-quarter of the scores lie between the lowest score and the lower quartile, one-quarter between the lower quartile and the median, one-quarter between the median and the upper quartile, and one-quarter between the upper quartile and the highest score.

It's easy to see where clustering of the data occurs. For example, a small box with relatively long whiskers would indicate that half of the data (from Q_1 to Q_3) would be confined to a small range, and the data could be described as clustered.

A wide box with relatively short whiskers would indicate that half of the data (from Q_1 to Q_3) would be spread over a wide range, and the data could be described as spread out.

Boxplots also clearly show how a set of data is distributed. Data that are evenly spaced around a central point can be described as **symmetrical**.

Distribution of boxplots

- **Positively skewed data has larger amounts of data at the lower end (the median is left of centre, closer to Q_1).**
- **Negatively skewed data has larger amounts of data at the higher end (the median is right of centre, closer to Q_3).**
- **A boxplot that is neither positively nor negatively skewed is described as symmetrical (the median is in the middle of the box).**

Consider the following boxplots with their matching histograms.

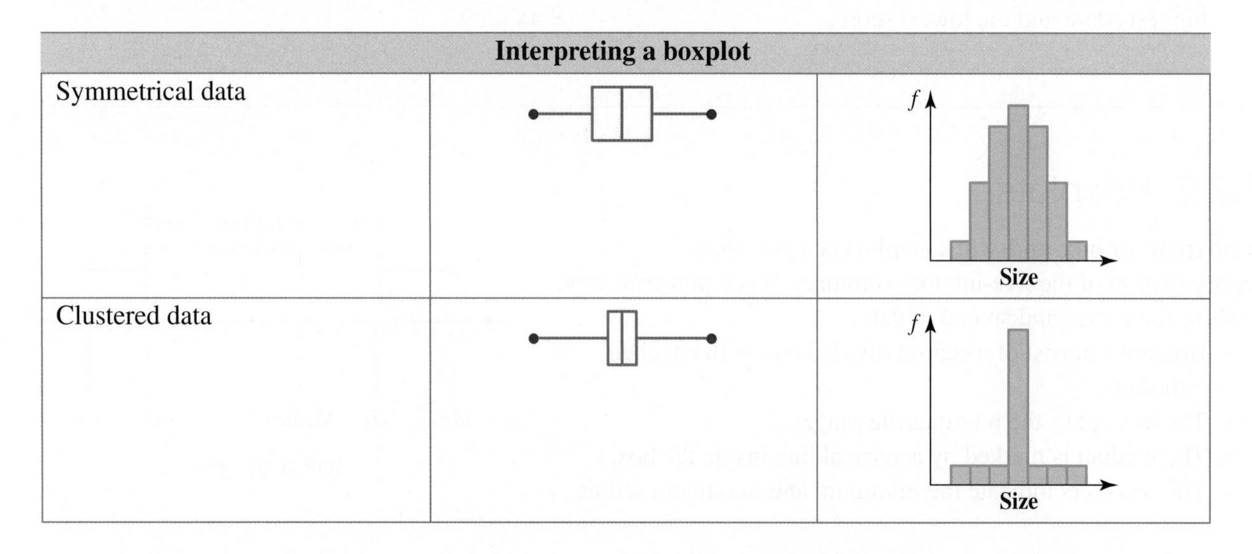

Interpreting a boxplot		
Symmetrical data		
Clustered data		

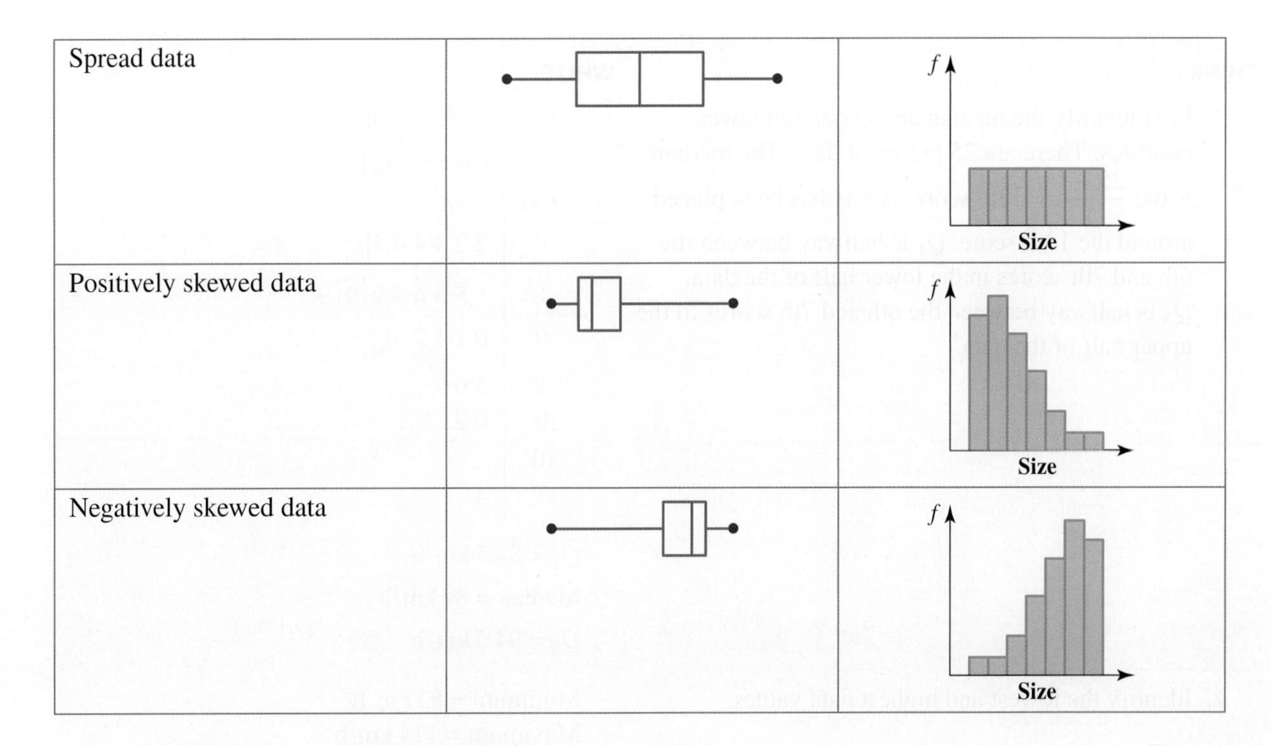

Spread data		
Positively skewed data		
Negatively skewed data		

7.2.4 Identification of outliers

Outliers often make the whiskers appear longer than they should and hence give the appearance that the data are spread over a much greater range than they really are.

If an outlier occurs in a set of data, it can be denoted by a small cross on the boxplot. The whisker is then shortened to the next largest (or smallest) value.

The boxplot below shows that the lowest score was 5. This was an outlier as the rest of the scores were located within the range of 15 to 42. The second lowest score was 15. This was not an outlier, so the left-hand whisker extends down to 15.

Recall from subtopic 6.4.1 that outliers can be identified by calculating the values of the lower and upper fences. Values that lie below the lower fence or above the upper fence are outliers.

0 5 10 15 20 25 30 35 40 45
Scale

$$\text{lower fence} = Q_1 - 1.5 \times \text{IQR}$$
$$\text{upper fence} = Q_3 + 1.5 \times \text{IQR}$$

tlvd-3724

WORKED EXAMPLE 2 Drawing and interpreting boxplots

The stem plot (stem-and-leaf plot) shown gives the speed of 25 cars caught by a roadside speed camera.

a. Prepare a five-number summary of the data.

b. Draw a boxplot of the data. (Identify any outliers.)

c. Describe the distribution of the data in terms of shape, centre and spread.

Key: 8|2 = 82 km/h
8*|6 = 86 km/h

Stem	*Leaf*
8	2 2 4 4 4 4
8*	5 5 6 6 7 9 9 9
9	0 1 1 2 4
9*	5 6 9
10	0 2
10*	
11	4

THINK	WRITE

a. 1. First identify the median and upper and lower quartiles. There are 25 pieces of data. The median is the $\frac{25+1}{2}$ = 13th score. A box has been placed around the 13th score. Q_1 is halfway between the 6th and 7th scores in the lower half of the data. Q_3 is halfway between the 6th and 7th scores in the upper half of the data.

a. Key: 8|2 = 82 km/h
8*|6 = 86 km/h

Stem	*Leaf*
8	2 2 4 4 4 4 $\|_{Q_1}$
8*	5 5 6 6 7 9 [9] 9
9	0 1 1 2 4 $\|_{Q_3}$
9*	5 6 9
10	0 2
10*	
11	4

$Q_1 = 84.5$ km/h
Median = 89 km/h
$Q_3 = 94.5$ km/h

2. Identify the lowest and highest data values.

Minimum = 82 km/h
Maximum = 114 km/h

3. Write the five-number summary: The lowest score is 82. The lower quartile is 84.5. The median is 89. The upper quartile is 94.5. The highest score is 114.

Five-number summary: 82, 84.5, 89, 94.5, 114

b. 1. Identify any outliers by calculating the upper and lower fences.

lower fence = $Q_1 - 1.5 \times \text{IQR}$

upper fence = $Q_3 + 1.5 \times \text{IQR}$

b. $\text{IQR} = 94.5 - 84.5$
$= 10$

Lower fence $= 84.5 - 1.5 \times 10$
$= 69.5$

Upper fence $= 94.5 + 1.5 \times 10$
$= 109.5$

114 is higher than the upper fence; hence it is an outlier.

2. To draw the boxplot, start by ruling a suitable scale. Remember to include the units of measurement. The box represents the interquartile range so it runs from 84.5 to 94.5. The median is a vertical line in the box at 89. The whiskers should extend to the lowest score (82) and the highest score that is not an outlier (102). The outlier at 114 should be indicated by a cross.

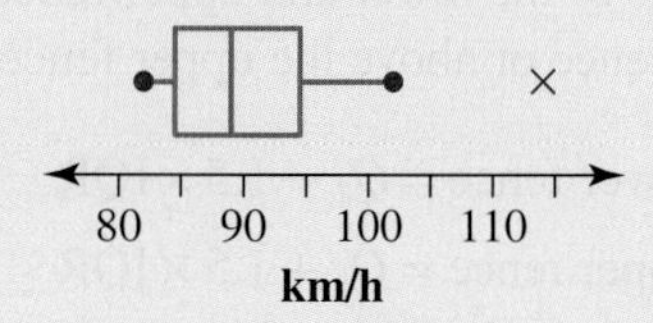

c. Because an outlier is present in this set of data, the most useful measure of centre is the median, and the best measure of spread is the interquartile range. The position of the outlier and the shape of the boxplot should also be commented on.

c. The distribution of car speeds is positively skewed with an outlier at 114 km/h. The average speed, as measured by the median, is 89 km/h. The spread of speeds, as measured by the interquartile range, is 10 km/h.

on Resources

Interactivities Box plots (int-6245)
Box-and-whisker plots (int-4623)

7.2 Exercise

Students, these questions are even better in jacPLUS

Receive immediate feedback and access sample responses

Access additional questions

Track your results and progress

Find all this and MORE in jacPLUS

1. **WE1** From the following five-number summary, determine:

6, 11, 12, 16, 32

a. the median **b.** the interquartile range **c.** the range.

2. From the following five-number summary, determine:

101, 119, 122, 125, 128

a. the median **b.** the interquartile range **c.** the range.

3. From the following five-number summary, determine:

39.2, 46.5, 49.0, 52.3, 57.8

a. the median **b.** the interquartile range **c.** the range.

4. The boxplot below shows the distribution of final points scored by a football team over a season's roster.

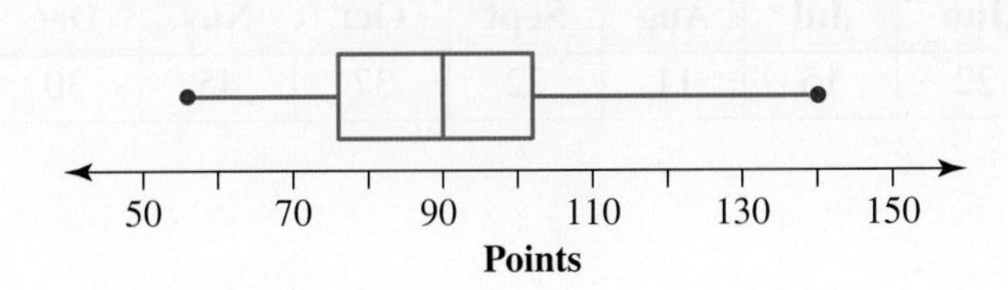

a. Determine the team's greatest score.
b. Determine the team's lowest score.
c. Determine the team's median score.
d. Determine the range of points scored.
e. Determine the interquartile range of points scored.

5. The boxplot shows the distribution of data formed by counting the number of honey bears in each of a large sample of packs.

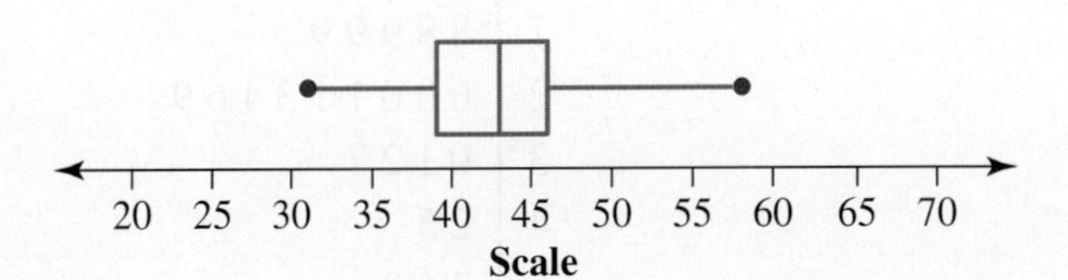

a. Determine the largest number of honey bears in any pack.
b. Determine the smallest number of honey bears in any pack.
c. Determine the median number of honey bears in any pack.
d. Determine the range of numbers of honey bears per pack.
e. Determine the interquartile range of honey bears per pack.

Questions 6–8 refer to the following boxplot.

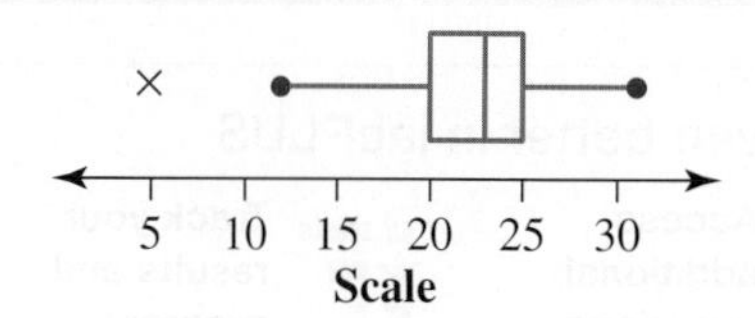

6. **MC** The median of the data is:

   A. 5 B. 20 C. 23 D. 25 E. 12

7. **MC** The interquartile range of the data is:

   A. 23 B. 26 C. 5 D. 20 to 25 E. 9

8. **MC** Determine which of the following is *not* true of the data represented by the boxplot.
   A. One-quarter of the scores are between 5 and 20.
   B. Half of the scores are between 20 and 25.
   C. The lowest quarter of the data is spread over a wide range.
   D. Most of the data are contained between the scores of 5 and 20.
   E. There is an outlier at 5.

9. The number of sales made each day by a salesperson is recorded over a 2-week period:

   25, 31, 28, 43, 37, 43, 22, 45, 48, 33

   a. Prepare a five-number summary of the data. (There is no need to draw a stem plot of the data.)
   b. Draw a boxplot of the data.

10. The data below show monthly rainfall in millimetres.

Jan	Feb	Mar	Apr	May	Jun	Jul	Aug	Sept	Oct	Nov	Dec
10	12	21	23	39	22	15	11	22	37	45	30

   a. Prepare a five-number summary of the data.
   b. Draw a boxplot of the data.

11. **WE2** The stem plot below details the ages of 25 offenders who were caught with a blood alcohol concentration over the limit during random breath testing.
   a. Prepare a five-number summary of the data.
   b. Draw a boxplot of the data.
   c. Describe the distribution of the data in terms of shape, centre and spread.

Key: 1|8 = 18 years

Stem	*Leaf*
1	8 8 9 9 9
2	0 0 0 1 1 3 4 6 9
3	0 1 2 7
4	2 5
5	3 6 8
6	6
7	4

12. The stem plot details the price at which 30 apartments in a particular suburb sold for.
   a. Prepare a five-number summary of the data.
   b. Draw a boxplot of the data.

Key: 32|4 = $324 000

Stem	Leaf
32	4 7 9
33	0 0 2 5 5
34	0 0 2 3 5 5 7 9 9
35	0 0 2 3 7 7 8
36	0 2 2 5 8
37	5

13. The following data detail the number of hamburgers sold by a fast-food outlet every day over a 4-week period.

Mon	Tue	Wed	Thu	Fri	Sat	Sun
125	144	132	148	187	172	181
134	157	152	126	155	183	188
131	121	165	129	143	182	181
152	163	150	148	152	179	181

   a. Prepare a stem plot of the data. (Use a class interval of size 10.)
   b. Draw a boxplot of the data.

14. The following data show the ages of 30 mothers upon the birth of their first baby.

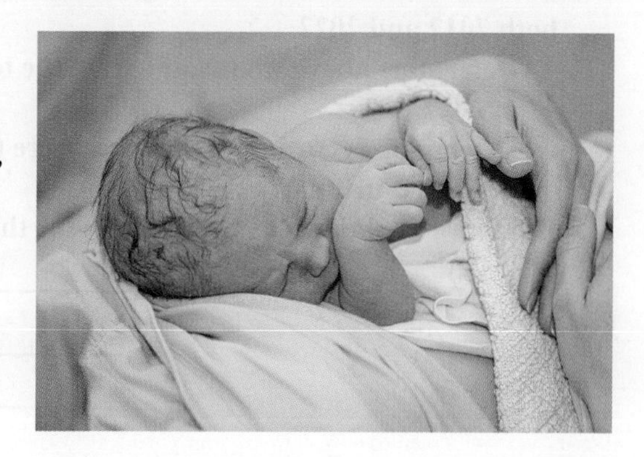

22, 18, 17, 22, 24, 25, 32, 19, 23, 28, 31, 19, 23, 25, 23,
21, 33, 23, 24, 20, 29, 18, 22, 24, 20, 22, 17, 48, 18, 20

   a. Prepare a stem plot of the data. (Use a class interval of size 5.)
   b. Draw a boxplot of the data. Indicate any extreme values appropriately.
   c. Describe the distribution in words. Explain what the distribution says about the age at which mothers have their first baby.

7.3 Comparing parallel boxplots

LEARNING INTENTION

At the end of this subtopic you should be able to:
- compare parallel boxplots using key features.

Parallel boxplots are drawn one above the other, using the same scale.

The set of parallel boxplots shown provides a means of comparing the results of four classes in a Maths test.

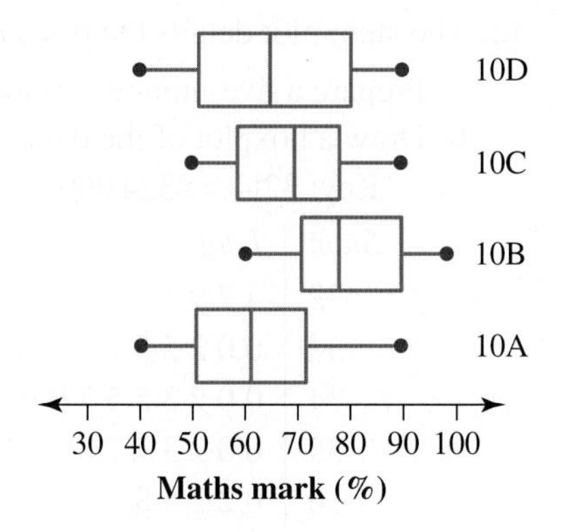

Reporting on a parallel boxplot involves comparing the individual boxplot features, including:

- central tendency (median)
- spread (both IQR and range)
- outliers
- shape of the distribution.

Comparisons made on parallel boxplots should always be in context. For example, to compare the parallel boxplots shown, comment on the median, spread, shape and outliers in terms of Maths marks for the different classes.

Comparing boxplots

Values of the median, IQR, range and/or outliers should always be included in a comparison of parallel boxplots.

tlvd-3725

WORKED EXAMPLE 3 Comparing boxplots

The following parallel boxplots show the number of hats produced in a factory, per hour, over a two-day period in 2012 and 2022.

a. State the five-number summary for the hourly hat production in both 2012 and 2022.

b. i. Determine above what value were the top 25% of hats produced in 2022.

ii. Determine between which values were the middle 50% of hats produced in 2012.

c. Compare the hourly hat production for the two different years.

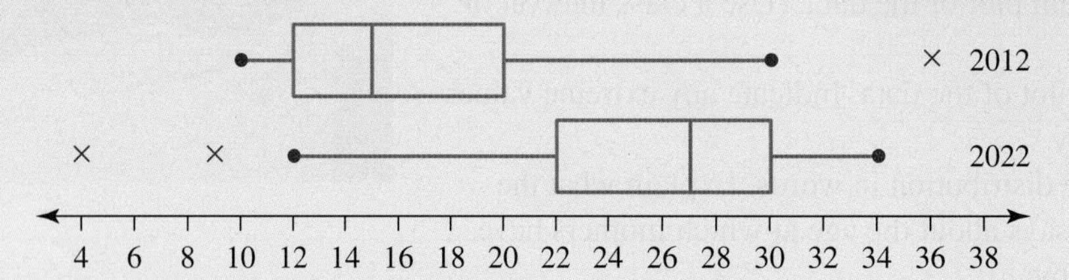

THINK	WRITE
a. 1. Look at the boxplot for 2012. Using the scale and the boxplot, determine the values for the five-number summary.	a. 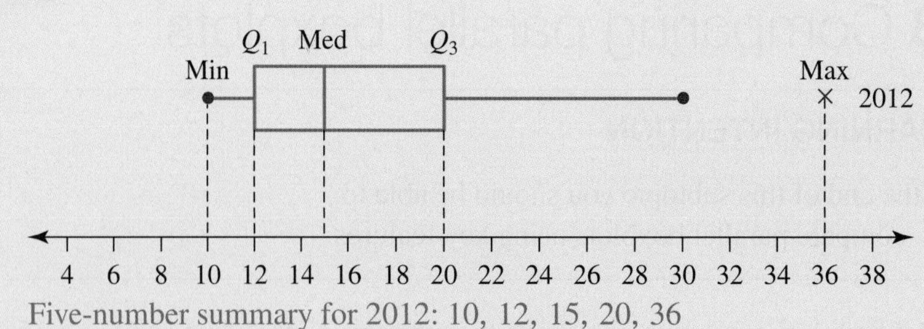 Five-number summary for 2012: 10, 12, 15, 20, 36

2. Look at the boxplot for 2022. Using the scale and the boxplot, determine the values for the five-number summary.

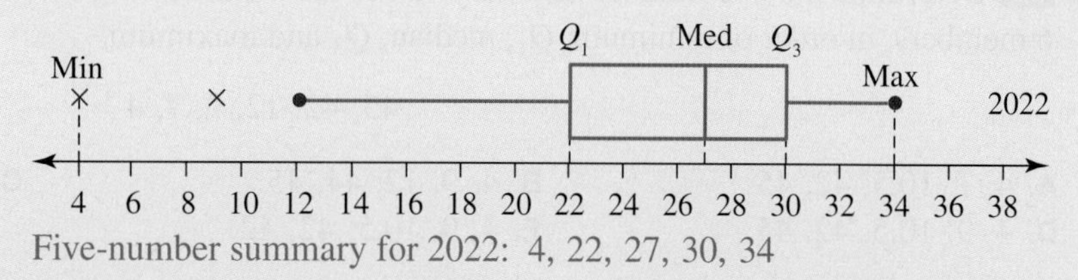

Five-number summary for 2022: 4, 22, 27, 30, 34

b. **i.** The top 25% on a boxplot occurs above the third quartile, Q_3.

b. **i.** The top 25% of hourly hats produced in 2022 occurred above 30 hats per hour.

ii. The middle 50% on a boxplot is represented by the IQR, or the length of the box.

ii. The middle 50% of hourly hats produced in 2012 occurred between 12 and 20 hats per hour.

c. **1.** Write a sentence, comparing the two boxplots for the median, IQR, range, shape and outliers.

c. The distribution of hat production in 2012 was positively skewed with an outlier at 36 hats per hour. The distribution of hat production in 2022 was negatively skewed with outliers at 4 and 9 hats per hour. The parallel boxplots show that median hourly hat production in 2012 (15 hats per hour) was lower than in 2022 (27 hats per hour). The IQR for both 2012 and 2022 was identical (8 hats per hour). The consistency of hat production was similar, as evidenced by the IQR.

2. Write a sentence summarising the comparison.

Overall, the hourly hat production over the two-day period in 2022 was generally greater than it was in 2012, as evidenced by the median. Hat production was slightly more consistent in 2012 due to the smaller IQR.

on Resources

Interactivity Parallel box plots (int-6248)

7.3 Exercise

1. A gardener recorded how long it took to mow each lawn on the weekly route (in minutes). The data set is shown.

32, 45, 56, 28, 19, 38, 26, 47, 54, 21, 33, 40, 17, 58, 21

State the five-number summary for the data set.

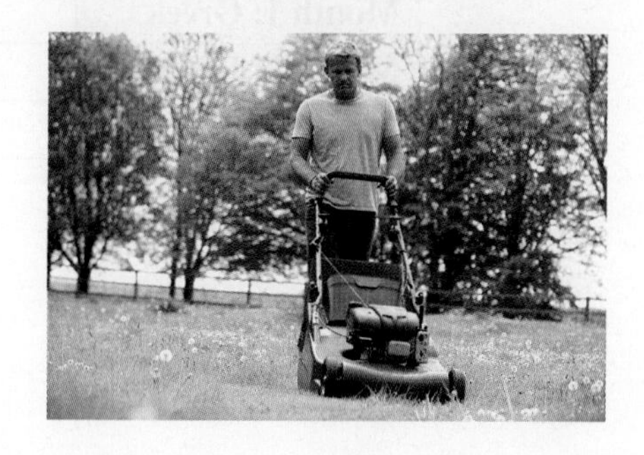

2. **MC** Determine the five-number summary values for the following data set of ages for a family with 6 members, in order of minimum, Q_1, median, Q_3 and maximum.

45, 42, 12, 9, 7, 4

A. 4, 7, 10.5, 42, 45
B. 4, 9, 12, 44, 45
C. 4, 7, 12, 42, 45
D. 4, 9, 10.5, 42, 45
E. 4, 9, 10.5, 42, 42

3. The weights of 14 boxes (in kilograms) being moved from one house to another are as follows.

3, 6, 13, 15, 15, 15, 17, 17, 18, 20, 20, 22, 26, 30

Draw a boxplot to display this data set.

4. The prices of 10 mobile phones are as follows.

\$349, \$469, \$265, \$497, \$159, \$52, \$999, \$489, \$599, \$577

Draw a boxplot to represent this data set.

5. The speeds of 15 cars on a school-zone road were as follows.

39, 51, 60, 42, 44, 38, 75, 45, 40, 52, 41, 42, 46, 41, 39

Draw a boxplot to represent this data set.

6. A restaurant trialled three different menus for a month each to work out which type of cuisine its patrons enjoy the most. The restaurant asked you for some help analysing data that they have collected.
   a. Draw a boxplot for each month of data on the same scale.
   b. Write a comparison of the boxplots to help the restaurant decide which menu will be the most successful.

	Week 1	Week 2	Week 3	Week 4
Month 1: Greek	75	62	48	50
Month 2: Italian	48	17	9	68
Month 3: Spanish	56	43	37	28

The following parallel boxplots, showing the ages at which boys and girls learn to ride a 2-wheel bike, should be used to answer questions 7–9.

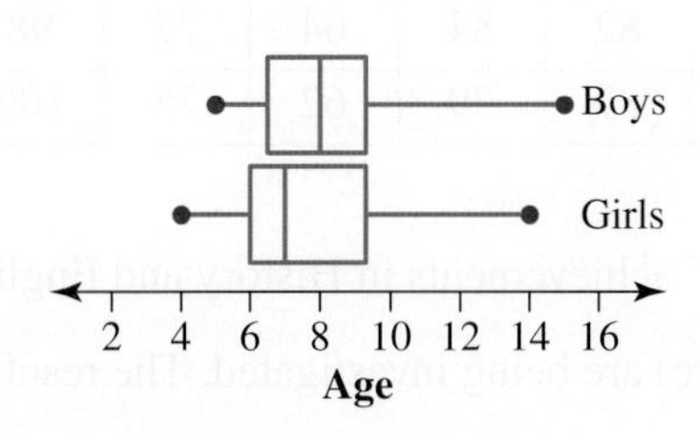

7. **MC** Looking at the parallel boxplots, the youngest girl and the oldest boy, respectively, learned to ride a two-wheel bike at the ages of:
   A. 15 and 4. **B.** 4 and 15. **C.** 5 and 14. **D.** 14 and 5. **E.** 5 and 15.
8. **MC** The value that is the same for both boxplots is:
   A. Q_1. **B.** Q_3. **C.** IQR. **D.** the range. **E.** the median.
9. **MC** The difference between the medians of the boxplots is:
   A. 1 year. **B.** 2 years. **C.** 3 years. **D.** 4 years. **E.** 5 years.
10. **a.** Explain what it means if the median on a boxplot is not placed exactly halfway between the upper and lower quartiles.
    b. Explain what it means if the median on a boxplot is closer to the upper quartile than the lower quartile.
    c. Write a statement explaining what the position of the median can tell you about a data set.
11. **WE3** The parallel boxplots below show the weekly sales for jars of peanut butter and Nutella in a shop.

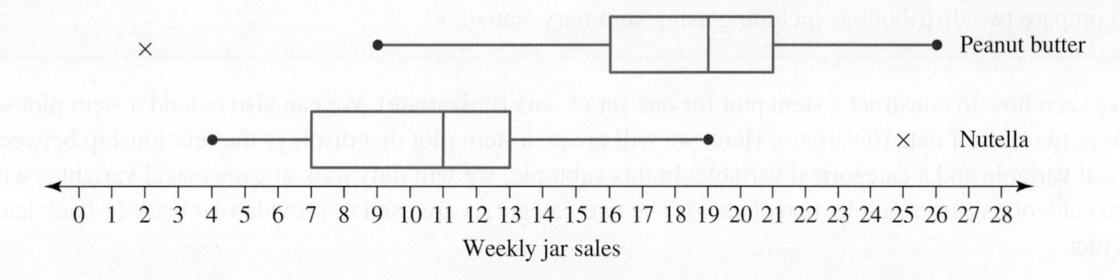

   a. State the five-number summary for the weekly sales of peanut butter and Nutella.
   b. i. Determine below which value were the lowest 25% of weekly peanut butter jar sales.
   ii. Determine between which values were the middle 50% of Nutella weekly jar sales.
   c. Compare the weekly sales for the two different products.
12. The following data was collected from a company that is comparing two different compounds for sunscreen to see how long (measured in minutes) they remain effective in water before they need to be reapplied.

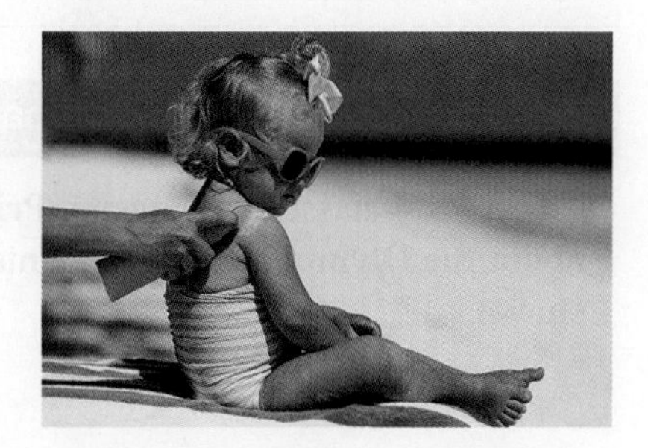

Compound 1	176	156	123	169	149	133	164	127	139	121	172	168
Compound 2	130	125	128	121	118	120	122	127	130	122	131	126

   a. Draw parallel boxplots for this data set.
   b. Compare the two compounds.

13. A comparison of Year 12 students' achievements (measured as a number out of 100) in History and English was recorded and the following results were obtained.

History	75	78	42	92	59	67	78	82	84	64	77	98
English	78	80	57	96	58	71	74	87	79	62	75	100

a. Draw parallel boxplots for this data set.
b. Explain what the parallel boxplots tell you about Year 12 students' achievements in History and English.

14. The heights of Year 11 and Year 12 students (to the nearest centimetre) are being investigated. The results of some sample data are shown.

Year 11	160	154	157	170	167	164	172	158	177	180	175	168	159	155	163	163	169	173	172	170
Year 12	160	172	185	163	177	190	183	181	176	188	168	167	166	177	173	172	179	175	174	180

a. Draw parallel boxplots for this data set.
b. Comment on what the plots tell you about the heights of Year 11 and Year 12 students.

7.4 Comparing back-to-back stem plots

LEARNING INTENTION

At the end of this subtopic you should be able to:
- construct back-to-back stem plots
- compare two distributions including using summary statistics.

We have seen how to construct a stem plot for one set of data (univariate). We can also extend a stem plot so that it displays two sets of data (bivariate). Here, we will create a stem plot that displays the relationship between a numerical variable and a categorical variable. In this subtopic, we will only look at categorical variables with just two categories (for example, two classes). The two categories are used to provide two back-to-back leaves of a stem plot.

Back-to-back stem plots

A back-to-back stem plot is used to display bivariate data involving a numerical variable and a categorical variable with 2 categories.

WORKED EXAMPLE 4 Comparing two sets of data using back-to-back stem plots

Two Year 4 classes at Kingston Primary School submitted projects about the Olympic Games. The marks they obtained out of 20 are shown.

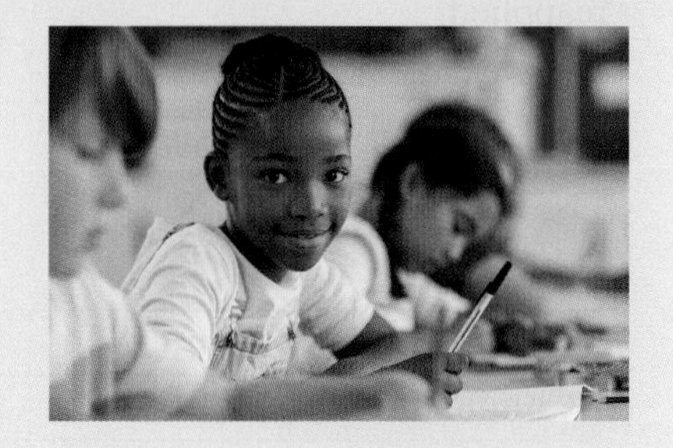

Class A	14	15	16	13	12	13	14	13	15	14
Class B	16	17	19	15	12	16	17	19	19	16

Display the data in the form of a back-to-back stem plot.

THINK

1. Identify the highest and lowest scores in order to decide on the stems.

WRITE

Highest score = 19
Lowest score = 12
Use a stem of 1, and divide into five class intervals. (Values of 12 and 13 will be included in the first class interval, values of 14 and 15 will be included in the next class interval, etc.)

2. Create an unordered stem plot first. Put class A scores on the left, and class B scores on the right.

Key: 1|2 = 12

Leaf: Class A	*Stem*	*Leaf: Class B*
3 2 3 3	1	2
4 5 4 5 4	1	5
6	1	6 7 6 7 6
	1	9 9 9

3. Now order the stem plot. The scores on the left should increase in value from right to left, while the scores on the right should increase in value from left to right.

Key: 1|2 = 12

Leaf: Class A	*Stem*	*Leaf: Class B*
3 3 3 2	1	2
5 5 4 4 4	1	5
6	1	6 6 6 7 7
	1	9 9 9

The back-to-back stem plot allows us to make some visual comparisons of two distributions. In Worked example 4:

- the centre of the distribution for class B is higher than the centre of the distribution for class A
- the spread of each of the distributions seems to be about the same
- for class A, the marks are grouped around 12–15; for class B, they are grouped around 16–19
- overall, we can conclude that class B obtained better marks than class A.

To get a more precise picture of the centre and spread of each of the distributions, we can use summary statistics. Specifically, we are interested in:

1. the mean and the median (to measure the centre of the distributions)
2. the interquartile range and the standard deviation (to measure the spread of the distributions).

If the distribution of either category is skewed and/or contains outliers, then the median and interquartile range are better measures of centre and spread than the mean and the standard deviation.

WORKED EXAMPLE 5 Using summary statistics to compare two distributions

The number of how-to-vote cards handed out by various Australian Labor Party and Liberal Party volunteers during the course of a polling day is shown below.

Labor	180	233	246	252	263	270	229	238	226	211
	193	202	210	222	257	247	234	226	214	204
Liberal	204	215	226	253	263	272	285	245	267	275
	287	273	266	233	244	250	261	272	280	279

a. Display the data using a back-to-back stem plot.

b. Use the stem plot, together with summary statistics, to compare the distributions of the number of cards handed out by the Labor and Liberal volunteers.

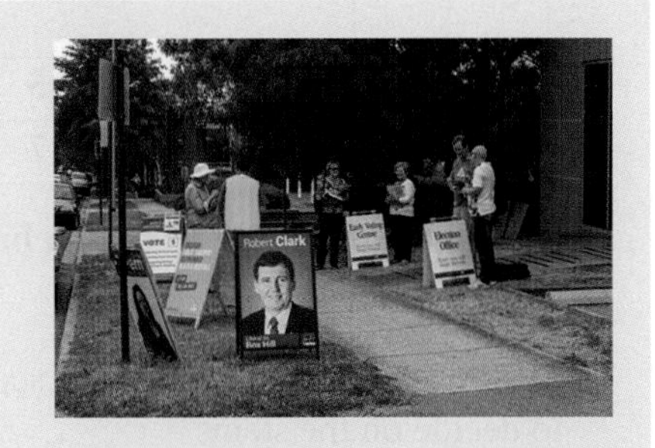

THINK

a. 1. Identify the highest and lowest scores in order to decide on the stems.

2. Construct the stem plot.

WRITE

a. The highest score is 287 and the lowest score is 180, so use stems 18 to 28.

Key: 18|0 = 180

Leaf: Labor	*Stem*	*Leaf: Liberal*
0	18	
3	19	
4 2	20	4
4 1 0	21	5
9 6 6 2	22	6
8 4 3	23	3
7 6	24	4 5
7 2	25	0 3
3	26	1 3 6 7
0	27	2 2 3 5 9
	28	0 5 7

THINK

b. 1. Use a form of technology to calculate the summary statistics for each party: the mean, the median, the standard deviation and the interquartile range. Enter each set of data as a separate list.

WRITE

b. For the Labor volunteers:

Mean = 227.9

Median = 227.5

Interquartile range = 36

Standard deviation = 23.9

For the Liberal volunteers:

Mean = 257.5

Median = 264.5

Interquartile range = 29.5

Standard deviation = 23.4

THINK

2. Comment on the relationship.

WRITE

From the stem plot we see that the Labor distribution is symmetric and therefore the mean and the median are very close, whereas the Liberal distribution is negatively skewed.

Since the distribution is skewed, the median is a better indicator of the centre of the distribution than the mean.

Therefore, comparing the medians, we have the median number of cards handed out for Labor at 228 and for Liberal at 265.

The standard deviations were similar, as were the interquartile ranges. There was not a lot of difference in the spread of the data.

Overall, the Liberal Party volunteers handed out a lot more how-to-vote cards than the Labor Party volunteers.

on Resources

Interactivities Back-to-back stem plots (int-6252)
Comparing data sets (int-4625)

7.4 Exercise

Students, these questions are even better in jacPLUS

Receive immediate feedback and access sample responses

Access additional questions

Track your results and progress

Find all this and MORE in jacPLUS

1. **WE4** The marks (out of 50), obtained for the end-of-term test by the students in German and French classes are given below. Display the data in the form of a back-to-back stem plot.

German	20	38	45	21	30	39	41	22	27	33	30	21	25	32	37	42	26	31	25	37
French	23	25	36	46	44	39	38	24	25	42	38	34	28	31	44	30	35	48	43	34

2. The birth masses of 20 newborns at two hospitals (in kilograms, to the nearest 100 grams) are recorded in the table below. Display the data using a back-to-back stem plot.

Hospital A	3.4	5.0	4.2	3.7	4.9	3.4	3.8	4.8	3.6	4.3
Hospital B	3.0	2.7	3.7	3.3	4.0	3.1	2.6	3.2	3.6	3.1

3. **WE5** The number of delivery trucks making deliveries to a supermarket each day over a 2-week period was recorded for two neighbouring supermarkets — supermarket A and supermarket B. The data are as follows.

A	11	15	20	25	12	16	21	27	16	17	17	22	23	24
B	10	15	20	25	30	35	16	31	32	21	23	26	28	29

a. Display the data using a back-to-back stem plot.
b. Use the stem plot, together with summary statistics, to compare the distributions of the number of trucks delivering to supermarkets A and B.

4. The marks out of 20 on a Science test for a Year 12 class are separated by where the students sit in the room, as shown.

Front	12	13	14	14	15	15	16	17
Back	10	12	13	14	14	15	17	19

a. Display the data using a back-to-back stem plot.
b. Use the stem plot, together with summary statistics, to compare the distributions of the marks of the students in the front compared to those in the back.

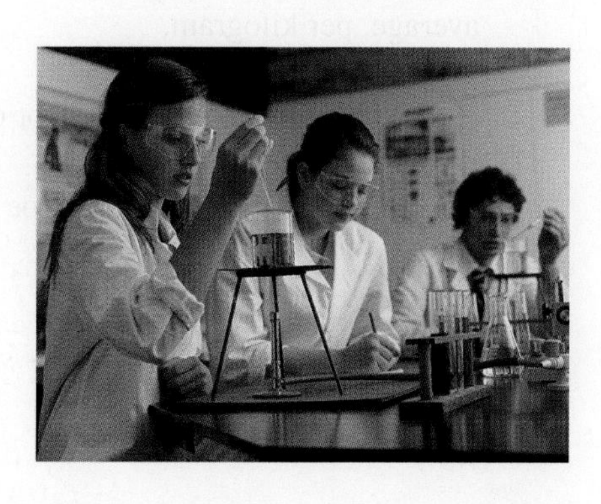

5. The end-of-year marks for 10 students in an English class were compared over 2 years. The marks for the first year and for the same students in the second year are shown.

First year	30	31	35	37	39	41	41	42	43	46
Second year	22	26	27	28	30	31	31	33	34	36

a. Display the data using a back-to-back stem plot.
b. Use the stem plot, together with summary statistics, to compare the distributions of the marks obtained by the students over the 2 years.

6. The age of two groups of people attending a fitness class at different times were recorded as follows.

7 am class	23	24	25	26	27	28	30	31
8 am class	22	25	30	31	36	37	42	46

a. Display the data using a back-to-back stem plot.
b. Use the stem plot, together with summary statistics, to compare the distributions of the ages of both fitness classes.

7. The scores on a board game are recorded for a group of kindergarten children and for a group of children in a preparatory school.

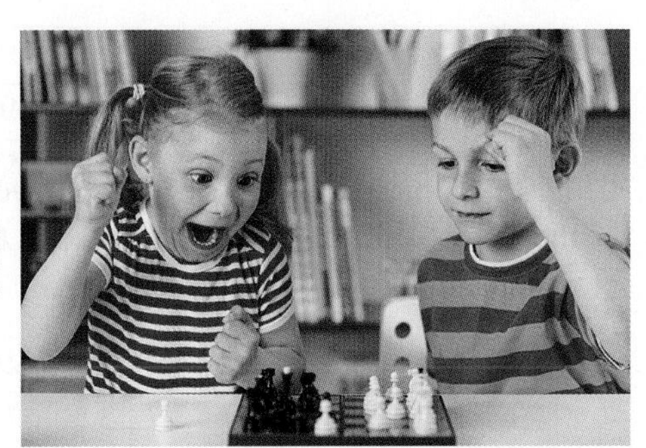

Kindergarten	3	13	14	25	28	32	36	41	47	50
Prep school	5	12	17	25	27	32	35	44	46	52

a. Display the data using a back-to-back stem plot.
b. Use the stem plot, together with summary statistics, to compare the distributions of the scores of the kindergarten children compared to the preparatory school children.

8. MC The pair of variables that could be displayed on a back-to-back stem plot are:
A. the height of a student and the number of people in the student's household.
B. the time put into completing an assignment and a pass or fail score on the assignment.
C. the weight of a businessperson and their age.
D. the religion of an adult and their head circumference.
E. the AFL team a person supports and their shoe size.

9. MC A back-to-back stem plot is a useful way of displaying the relationship between:

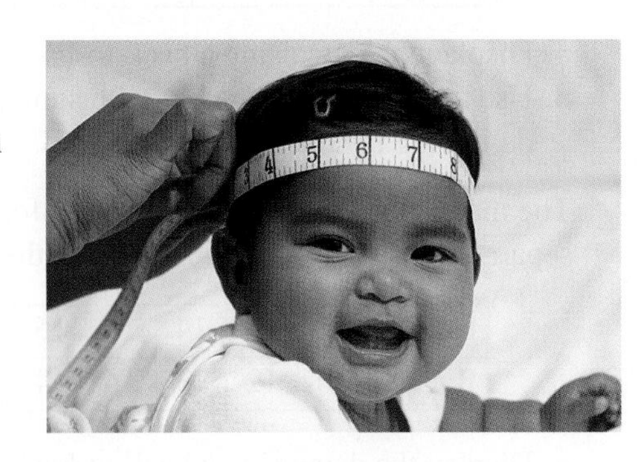

A. the proximity to markets (km) and the cost of fresh foods, on average, per kilogram.
B. height and head circumference.
C. age and attitude to gambling (for or against).
D. weight and age.
E. the brand of phone used and shoe size.

10. The two dot plots shown display the latest Maths test results for two Year 2 classes. The results show the marks out of 20.

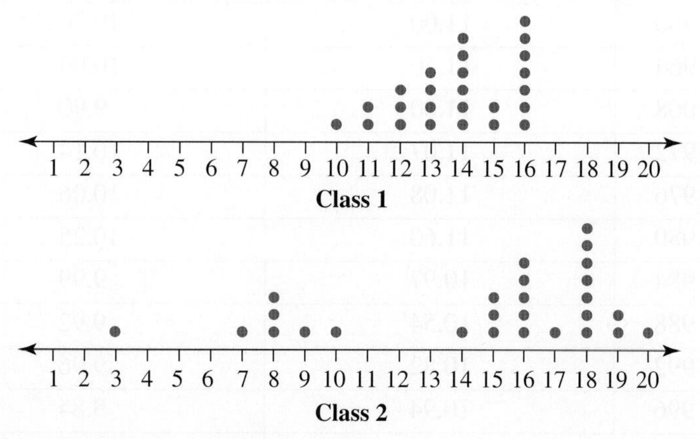

a. Determine the number of students in each class.
b. For each class, determine how many students scored 15 out of 20 on the test.
c. For each class, determine how many students scored more than 10 on the test.
d. Use the dot plots to describe the performance of each class on the test.

11. The following data show the ages of the morning and afternoon players at a ten-pin bowling centre. Draw a back-to-back stem plot of the data.

Morning:

20, 36, 16, 38, 32, 18, 19, 21, 25, 45, 29, 60, 31, 21,
16, 38, 52, 43, 17, 28, 23, 23, 43, 17, 22, 23, 32, 34

Afternoon:

21, 23, 30, 16, 31, 46, 15, 17, 22, 17, 50, 34, 65, 25, 27, 19,
15, 43, 22, 17, 22, 16, 48, 57, 54, 23, 16, 30, 18, 21, 28, 35

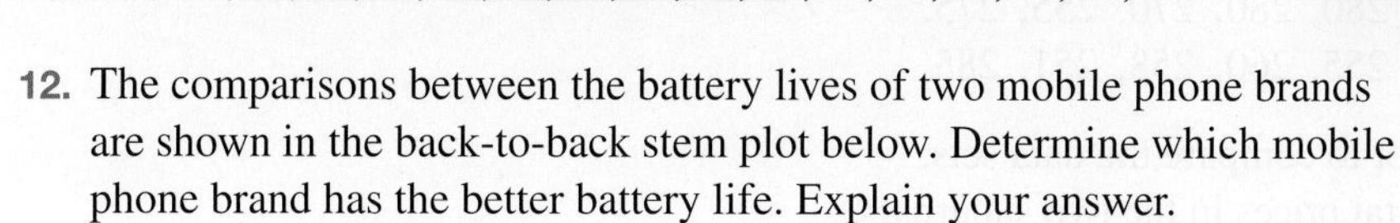

12. The comparisons between the battery lives of two mobile phone brands are shown in the back-to-back stem plot below. Determine which mobile phone brand has the better battery life. Explain your answer.

Key: 6|1 = 61 hours

Brand A	*Stem*	*Brand B*
8 8 7 5	0	7
9 7 4 1 0	1	0 5 5 5 7 9
2 2 2 1	2	0 2 2 6 7
8 6 4 2 0	3	0 2 4 6 8
	4	
	5	6
1	6	
	7	5

13. The winning times in seconds for the women's and men's 100-metre sprint in the Olympics are shown in the table.

Year	Women's 100-m sprint	Men's 100-m sprint
1928	12.20	10.80
1932	11.90	10.30
1936	11.50	10.30
1948	11.90	10.30
1952	11.50	10.40
1956	11.50	10.50

(continued)

(continued)

Year	Women's 100-m sprint	Men's 100-m sprint
1960	11.00	10.20
1964	11.40	10.00
1968	11.00	9.90
1972	11.07	10.14
1976	11.08	10.06
1980	11.60	10.25
1984	10.97	9.99
1988	10.54	9.92
1992	10.82	9.96
1996	10.94	9.84
2000	10.75	9.87
2004	10.93	9.85
2008	10.78	9.69

a. Display the winning times for women and men using a stem plot.
b. Determine if there is a large difference in winning times. Explain your answer.

14. The following data sets show the weekly rental prices (in $) of two-bedroom apartments in two different suburbs of Geelong.

Suburb A:

215, 225, 211, 235, 244, 210, 215, 210, 256, 207,
200, 200, 242, 225, 231, 205, 240, 205, 235, 280

Suburb B:

235, 245, 231, 232, 240, 280, 280, 270, 255, 275,
275, 285, 245, 265, 270, 255, 260, 258, 251, 285

a. Draw a back-to-back stem plot to compare the data sets.
b. Compare and contrast the rental prices in the two suburbs.

7.5 Comparing histograms

LEARNING INTENTION

At the end of this subtopic you should be able to:
- create and compare histograms.

7.5.1 Frequency histograms

A histogram is a useful way of displaying large, numerical data sets (e.g. with more than 50 observations). The vertical axis on the histogram displays the frequency and the horizontal axis displays class intervals of the variable (e.g. height or income). The frequency for each class interval determines the height of the column. When data are given in raw form — that is, just as a list of numbers in no particular order — it is helpful to first construct a frequency table.

tlvd-3726

WORKED EXAMPLE 6 Constructing a grouped frequency histogram

The data show the distribution of masses (in kilograms) of 60 students in Year 7 at Northwood State High School.

45.7, 34.2, 56.3, 38.7, 52.4, 45.7, 48.2, 52.1, 58.7, 62.3,
45.8, 52.4, 60.2, 48.5, 54.3, 39.8, 36.2, 54.3, 39.7, 46.3,
45.9, 52.3, 44.2, 49.6, 48.6, 42.5, 47.2, 51.3, 43.1, 52.4,
48.2, 51.8, 53.8, 56.9, 53.7, 42.9, 46.7, 51.9, 56.2, 61.2,
48.3, 45.7, 43.5, 43.8, 58.7, 59.2, 58.7, 54.6, 43.0, 48.2,
48.4, 56.8, 57.2, 58.3, 57.6, 53.2, 53.1, 58.7, 56.3, 58.3

a. **Complete a frequency distribution table for these data.**
b. **Construct a frequency histogram to display these data.**

THINK

a. 1. First construct a frequency table. The lowest data value is 34.2 and the highest is 62.3. Divide the data into class intervals. If we started the first class interval at, say, 30 kg and ended the last class interval at 65 kg, we would have a range of 35. If each interval was 5 kg, we would then have 7 intervals, which is a reasonable number of class intervals. While there are no set rules about how many intervals there should be, somewhere between about 5 and 15 class intervals is usual. So, in this example, we would have class intervals of 30–< 35 kg, 35–< 40 kg, 40–< 45 kg and so on. Count how many observations fall into each of the intervals and record these in a table.

2. Check that the frequency column totals 60.

b. A histogram can be constructed. Since we are dealing with grouped data, each column has the range of the values included in the class interval on either edge of the column. The column heights are determined by the frequency for each class interval.

WRITE

a.

Class interval	Frequency
30–< 35	1
35–< 40	4
40–< 45	7
45–< 50	16
50–< 55	15
55–< 60	14
60–< 65	3
Total	**60**

b.

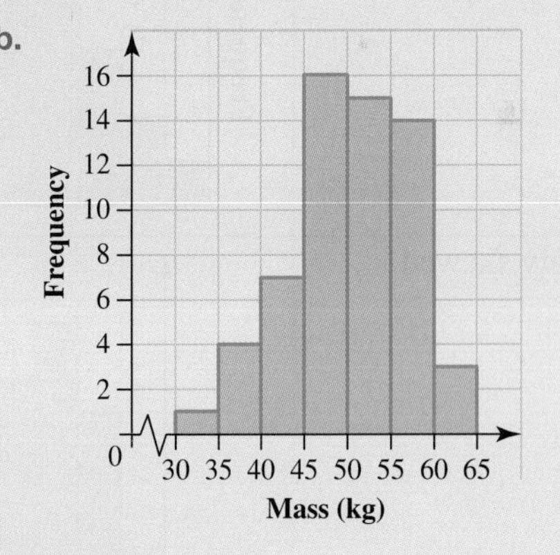

WORKED EXAMPLE 7 Constructing an ungrouped frequency histogram

The marks out of 20 received by 30 students for a book-review assignment are given in the frequency table.

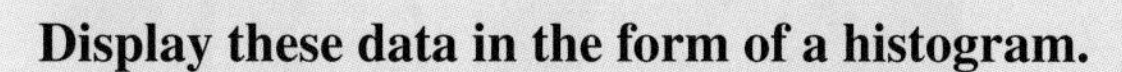

Mark	12	13	14	15	16	17	18	19	20
Frequency	2	7	6	5	4	2	3	0	1

Display these data in the form of a histogram.

THINK

In this case we are dealing with integer values, not grouped data. The integer values are written underneath the centre of each column. The column heights are determined by the frequency for each class interval.

WRITE

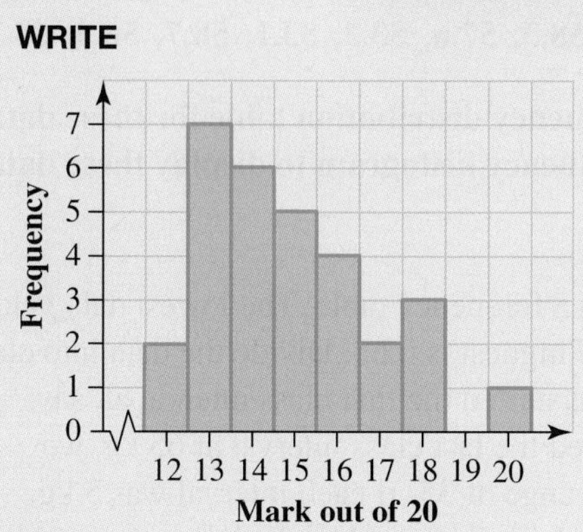

7.5.2 Describing the shape of a histogram

Each histogram below shows an example of a skewed distribution.

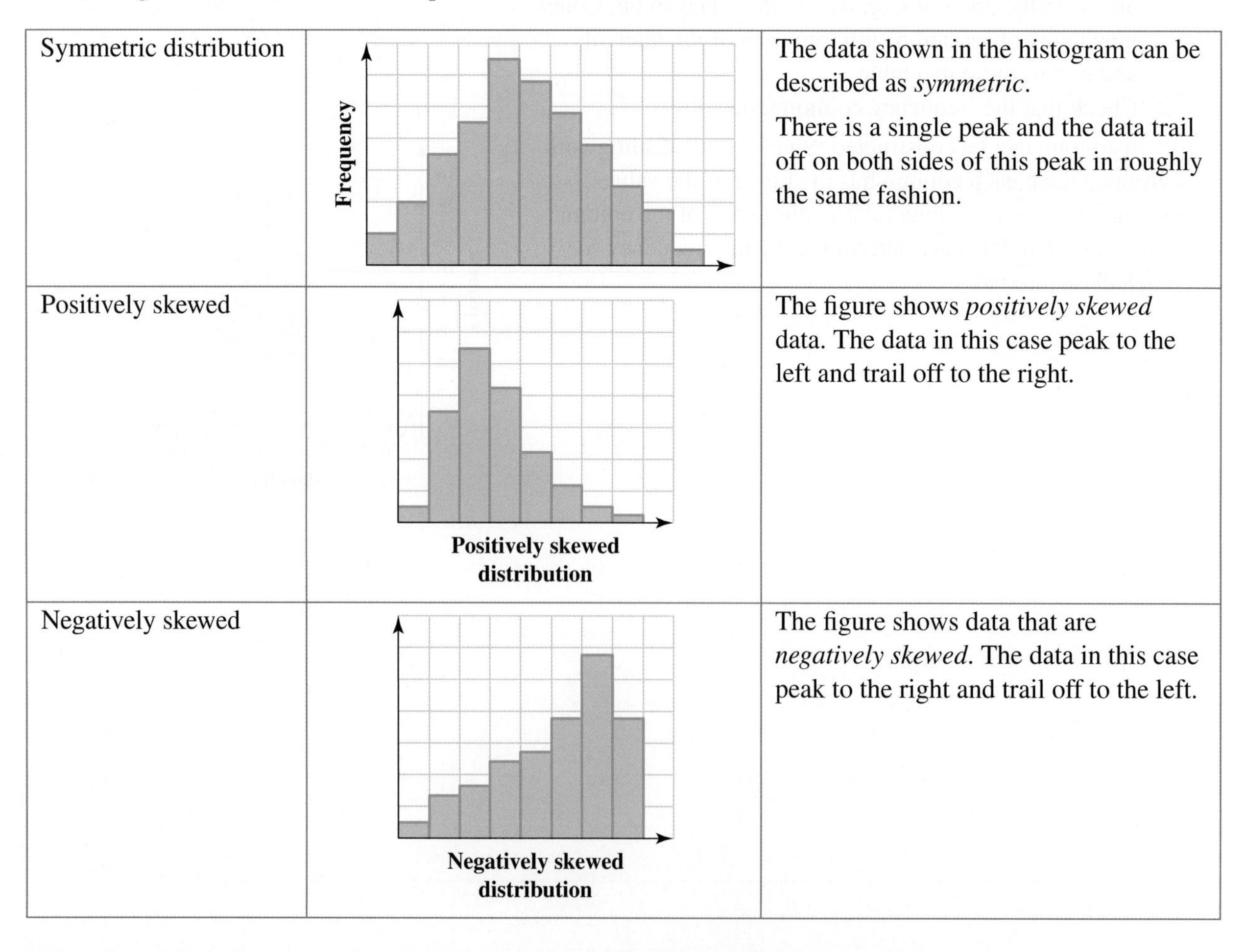

Symmetric distribution	Frequency	The data shown in the histogram can be described as *symmetric*. There is a single peak and the data trail off on both sides of this peak in roughly the same fashion.
Positively skewed	Positively skewed distribution	The figure shows *positively skewed* data. The data in this case peak to the left and trail off to the right.
Negatively skewed	Negatively skewed distribution	The figure shows data that are *negatively skewed*. The data in this case peak to the right and trail off to the left.

Outliers

Possible outliers can be identified as a column that lies well away from the main body of the histogram.

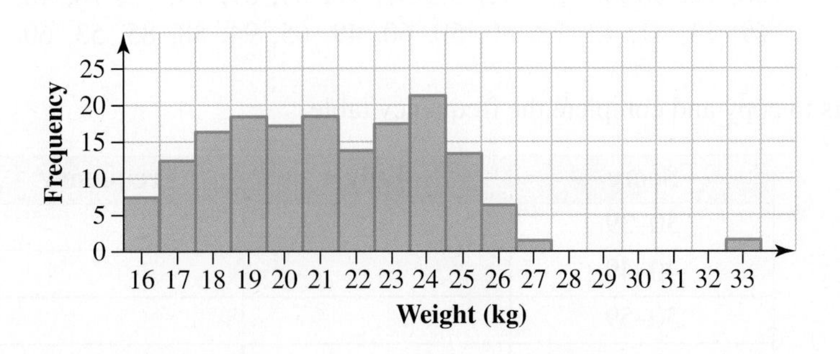

Bimodality

The mode of a distribution is the data value or class interval that has the highest frequency. This will be the longest column or row on display. When there is more than one mode, the data distribution is multimodal. This can indicate that there may be subgroups within the distribution that may require further investigation.

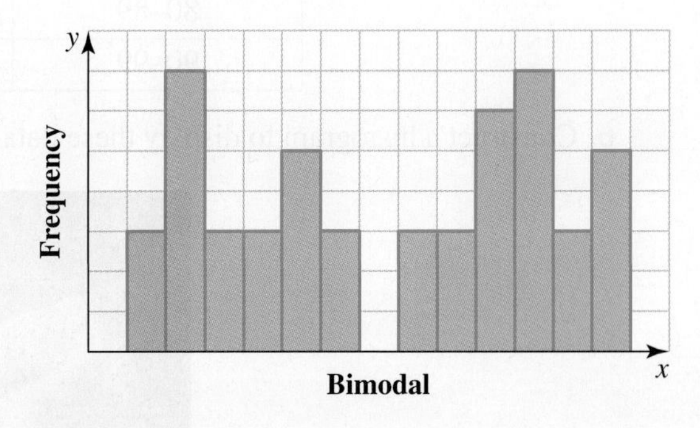

Bimodal distributions can occur when there are two distinct groups present, such as in data values that typically have clear differences between two categorical measurements.

7.5 Exercise

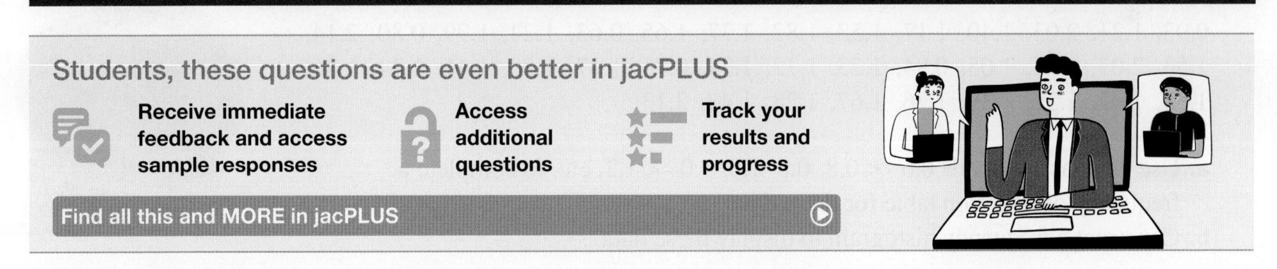

1. For each set of data:
   i. construct a frequency table
   ii. construct a histogram by hand
   iii. construct a histogram using technology. Compare with your hand-drawn histogram.
   a. 3, 4, 4, 5, 5, 6, 7, 7, 7, 8, 8, 9, 9, 10, 10, 12
   b. 4.3, 4.5, 4.7, 4.9, 5.1, 5.3, 5.5, 5.6, 5.2, 3.6, 2.5, 4.3, 2.5, 3.7, 4.5, 6.3, 1.3
   c. 11, 13, 15, 15, 16, 18, 20, 21, 22, 21, 18, 19, 20, 16, 18, 20, 16, 10, 23, 24, 25, 27, 28, 30, 35, 28, 27, 26, 29, 30, 31, 24, 28, 29, 20, 30, 32, 33, 29, 30, 31, 33, 34
   d. 0.4, 0.5, 0.7, 0.8, 0.8, 0.9, 1.0, 1.1, 1.2, 1.0, 1.3, 0.4, 0.3, 0.9, 0.6

2. A class of 30 students sat for a Mathematics test. Their results out of 100 are as follows.

68, 72, 58, 45, 69, 92, 38, 51, 70, 65, 69, 73, 52, 76, 48,
69, 73, 41, 42, 73, 80, 50, 60, 49, 65, 94, 88, 85, 53, 60

a. Use these results to copy and complete the frequency table.

Score	Tally	Frequency
30–39		
40–49		
50–59		
60–69		
70–79		
80–89		
90–99		

b. Construct a histogram to display these data.

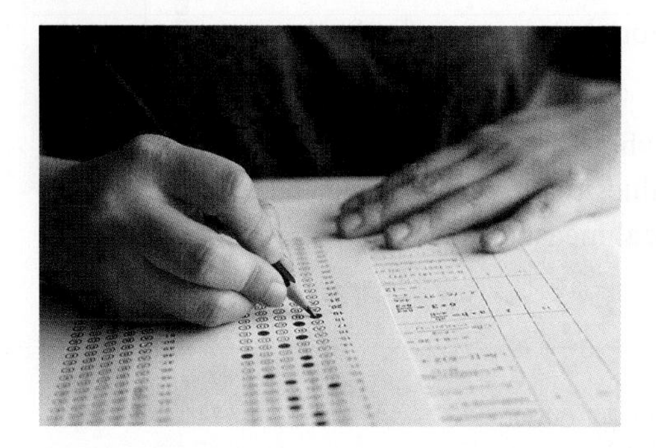

3. WE6 A farmer measures the heights of his tomato plants. The results, in metres, are as follows.

0.93, 1.21, 2.03, 1.40, 1.17, 1.53, 1.82, 1.77, 1.65, 0.63, 1.24, 1.99, 0.80, 2.14,
1.53, 2.07, 1.96, 1.05, 0.94, 1.23, 1.72, 1.34, 0.75, 1.17, 1.50, 1.41, 1.74, 1.86,
1.55, 1.42, 1.52, 1.39, 1.76, 1.67, 1.28, 1.43, 2.13

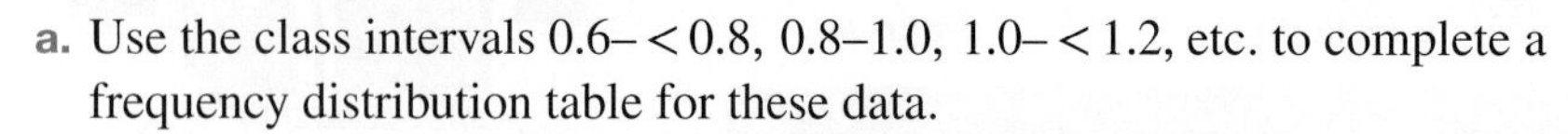

a. Use the class intervals 0.6– < 0.8, 0.8–1.0, 1.0– < 1.2, etc. to complete a frequency distribution table for these data.
b. Construct a frequency histogram to display these data.

4. The following data give the times (in seconds) taken by athletes to complete a 100-m sprint.

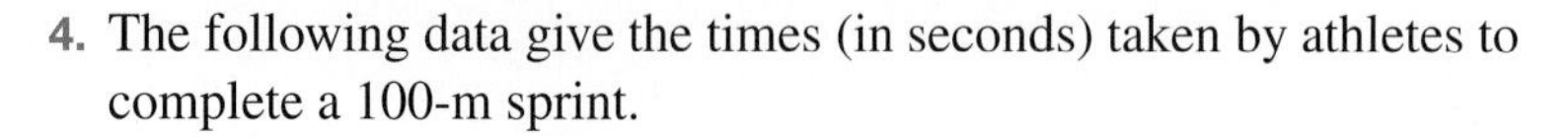

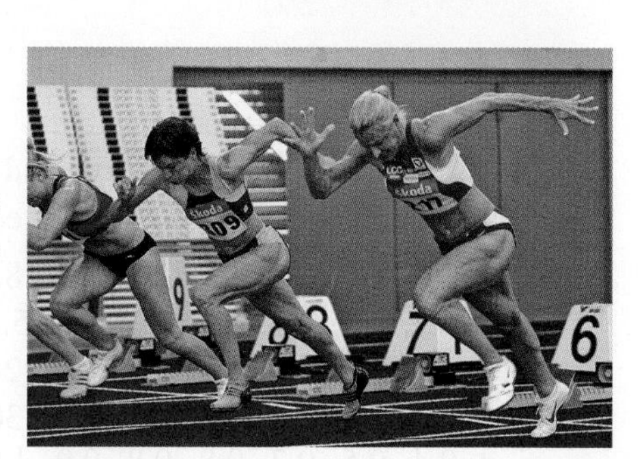

12.2, 12.0, 11.9, 12.0, 12.6, 11.7, 11.4, 11.0, 10.9, 11.7, 11.2, 11.8,
12.2, 12.0, 12.7, 12.9, 11.3, 11.2, 12.8, 12.4, 11.7, 10.8, 13.3, 11.7,
11.6, 11.7, 12.2, 12.7, 13.0, 12.2

a. Construct a frequency distribution table for the data. Use class intervals of 0.5 seconds.

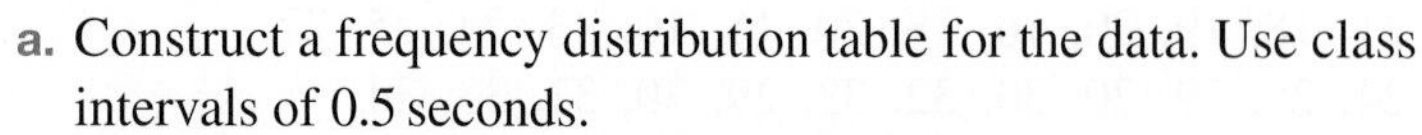

b. Construct a histogram to display these data.

5. **WE7** The marks out of 20 received by 26 students for a History assignment are given in the frequency table.

Score	Frequency
13	2
14	9
15	3
16	5
17	6
18	1

Display these data in the form of a histogram.

6. For each of the following histograms, describe the shape of the distribution of the data and comment on the existence of any outliers.

a. 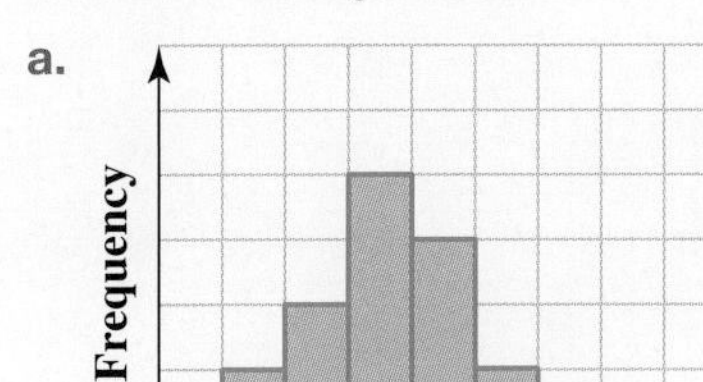

b. 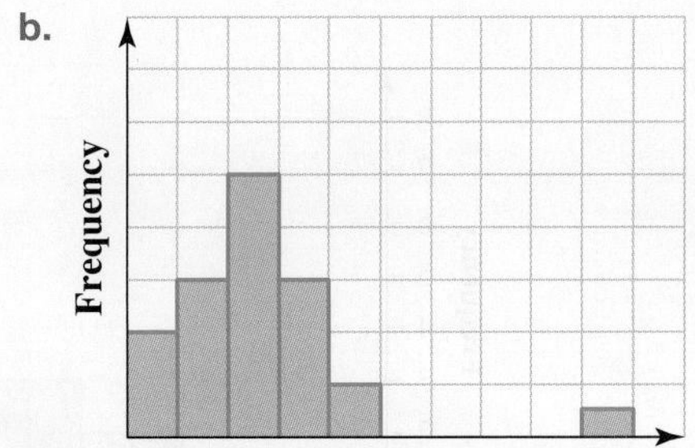

c. 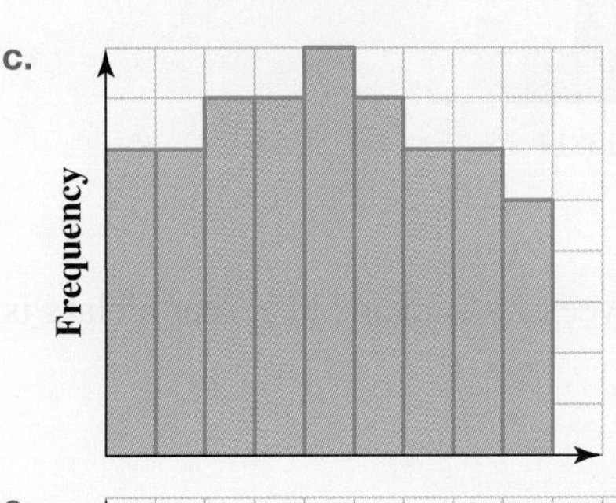

d. 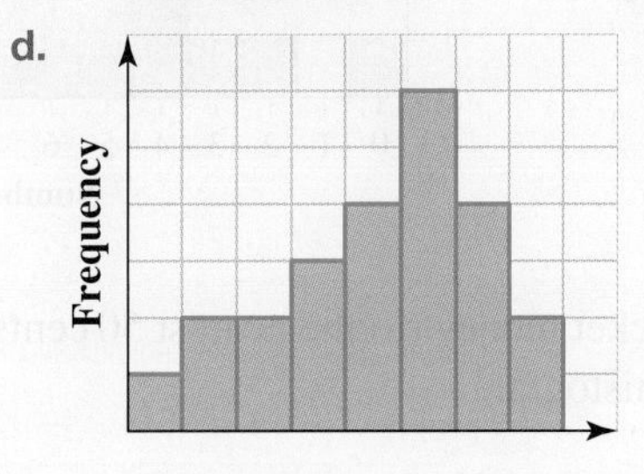

e. 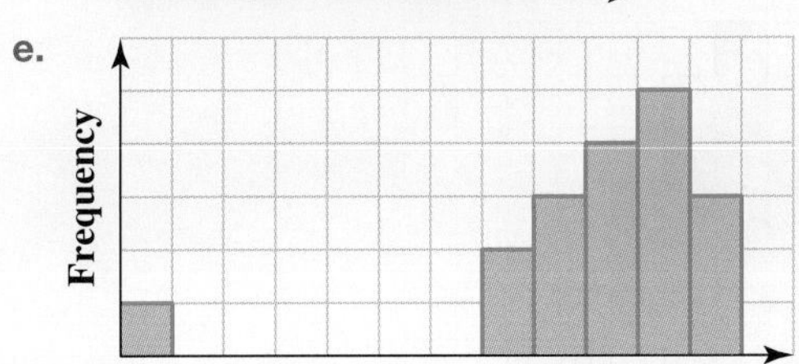

f. 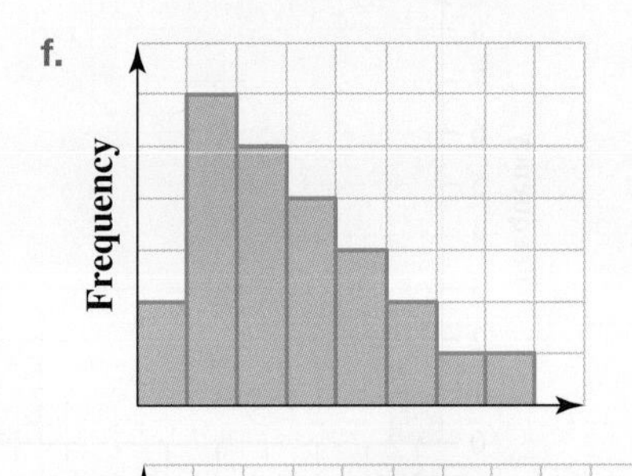

7. **MC** The distribution of the data shown in this histogram could be described as:

A. negatively skewed.
B. negatively skewed with outliers.
C. positively skewed.
D. positively skewed with outliers.
E. symmetrical.

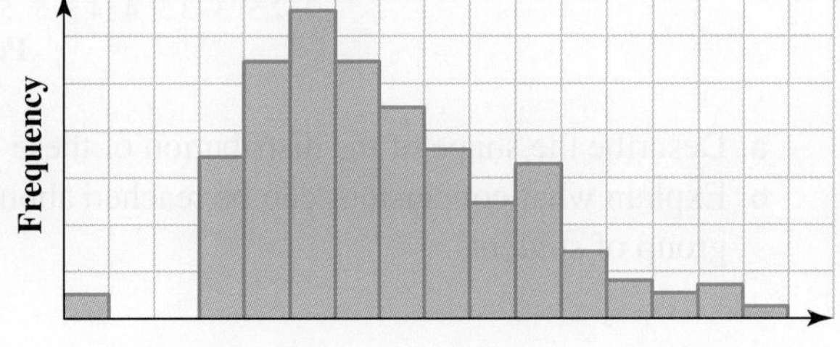

8. **MC** The histogram shown is:

A. unimodal and positively skewed.
B. bimodal and positively skewed.
C. unimodal and negatively skewed.
D. bimodal and negatively skewed.
E. bimodal and symmetrical.

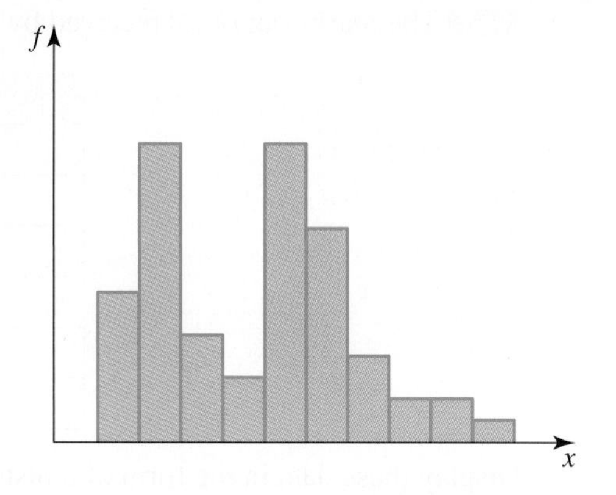

9. The average number of product enquiries per day received by a group of small businesses who advertised online is given below. Describe the shape of the distribution of these data and comment on the existence of any outliers.

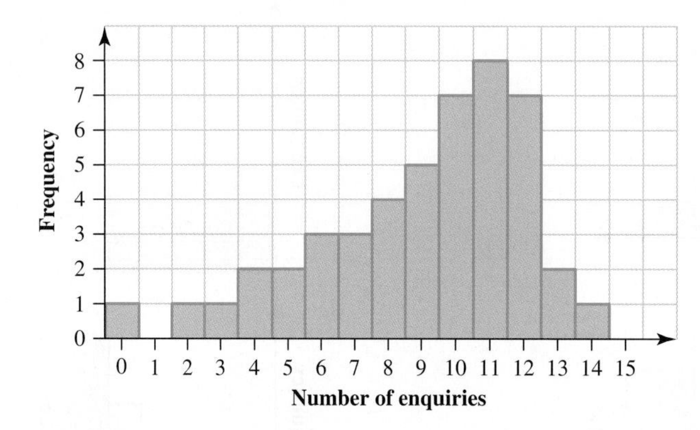

10. The amount of pocket money (to the nearest 50 cents) received each week by students in a Year 6 class is illustrated in this histogram.

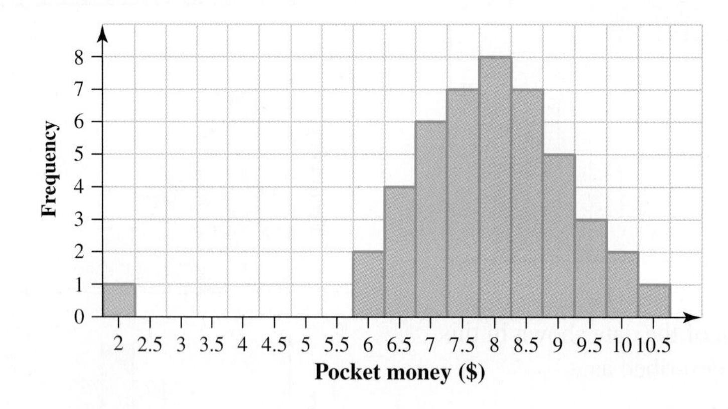

a. Describe the shape of the distribution of these data and comment on the existence of any outliers.
b. Explain what conclusions can be reached about the amount of pocket money received weekly by this group of students.

11. Ten workers were required to complete two tasks. Their supervisor observed them and gave them a score for the quality of their work on each task, where higher scores indicated better-quality work. The results are indicated in the following side-by-side bar chart.

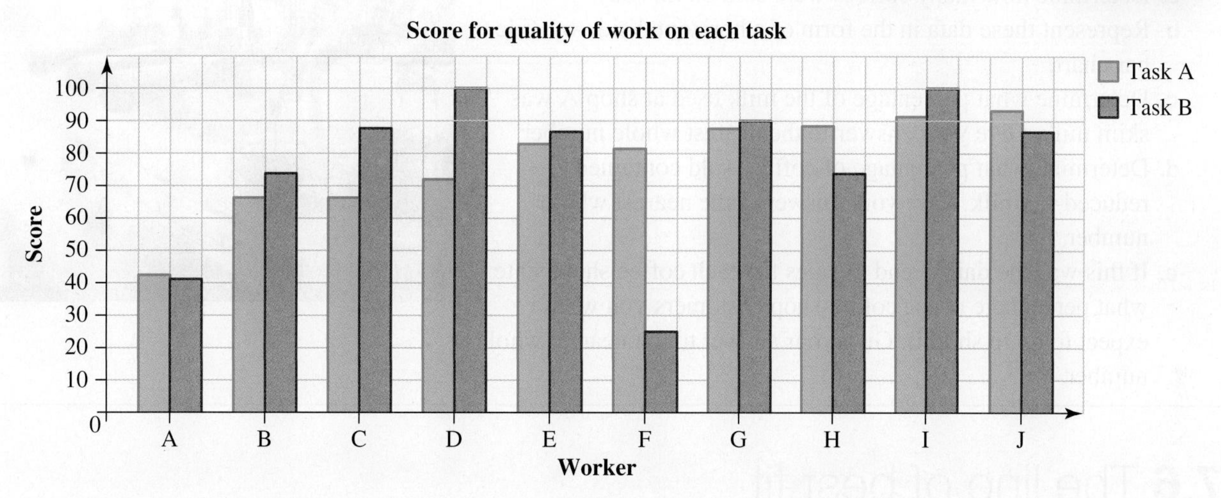

a. Determine which worker had the largest difference between scores for the two tasks.
b. Determine how many workers received a lower score for task B than for task A.

12. The side-by-side bar chart shows a monthly comparison of road fatalities from 2019 to 2021 in Australia.

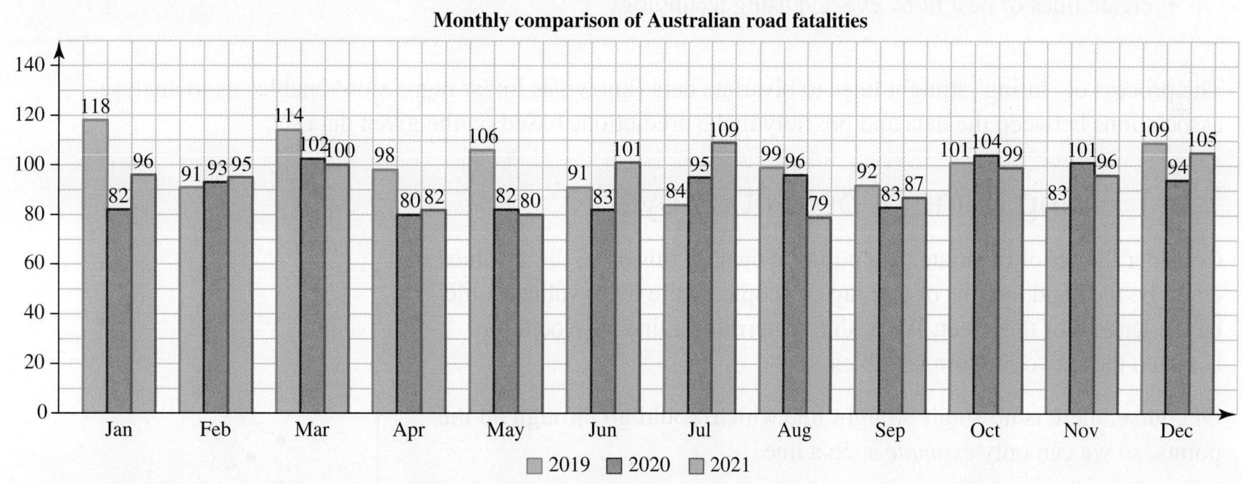

Source: Australian Government, Department of Infrastructure, Regional Development and Cities, Australian Road Deaths Database

a. Determine which year had the most fatalities on Australian roads.
b. Determine which year had the fewest fatalities from January to March.
c. Give a possible reason why the fatalities are lower in April–May in 2020 and 2021 compared to 2019.

13. Two coffee shops serve either skim, reduced-fat or whole milk in coffees. The coffees sold on a particular day are shown in the table below, sorted by the type of milk used at coffee shop A and shop B.

		Coffee shops	
		A	B
Type of milk	Skim	87	124
	Reduced fat	55	73
	Whole	112	49

a. Determine how many coffees were sold on this day.
b. Represent these data in the form of a horizontal side-by-side bar chart.
c. Determine what percentage of the milk used at shop A was skim milk. Give your answer to the nearest whole number.
d. Determine what percentage of coffees sold contained reduced-fat milk. Give your answer to the nearest whole number.
e. If this was the daily trend of sales for each coffee shop, state what percentage of the coffee-shop customers you would expect to go to shop B. Give your answer to the nearest whole number.

7.6 The line of best fit

LEARNING INTENTION

At the end of this subtopic you should be able to:
- create lines of best fit by eye and using technology.

The process of 'fitting' straight lines to bivariate data (known as linear regression) enables us to analyse associations between the data and possibly make predictions based on the given data set.

7.6.1 Fitting a line of best fit by eye

Consider the set of bivariate data points shown. In this case, the x-values could be the hand lengths of a group of people, while the y-values could be the lengths of their feet. We wish to determine a linear association between these two random variables.

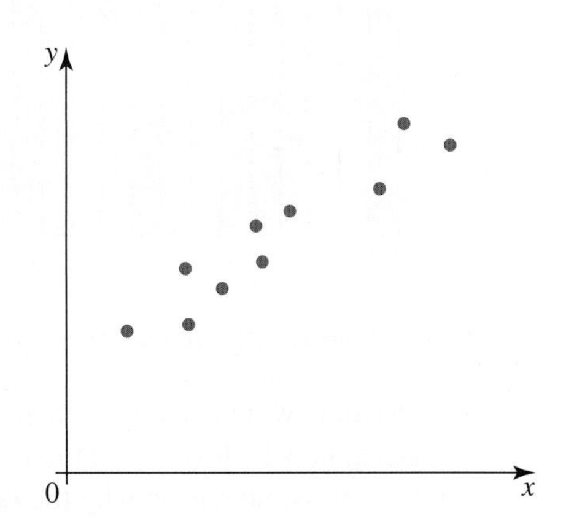

Of course, there is no single straight line which would go through all the points, so we can only *estimate* such a line.

Furthermore, the more closely the points appear to be on or near a straight line, the more confident we are that such a linear association may exist and the more accurate our fitted line should be.

Consider the estimate, drawn 'by eye' in the figure shown. The line extends through all data points, and most of the points are very close to this straight line, indicating a strong linear association between the two variables. Due to the strong linear association between foot length and hand length, we can conclude from this graph that there is evidence to suggest that a person's foot length is related to the length of their hand.

Regression analysis is concerned with finding these lines of best fit using various methods so that the number of points above and below the line is 'balanced'.

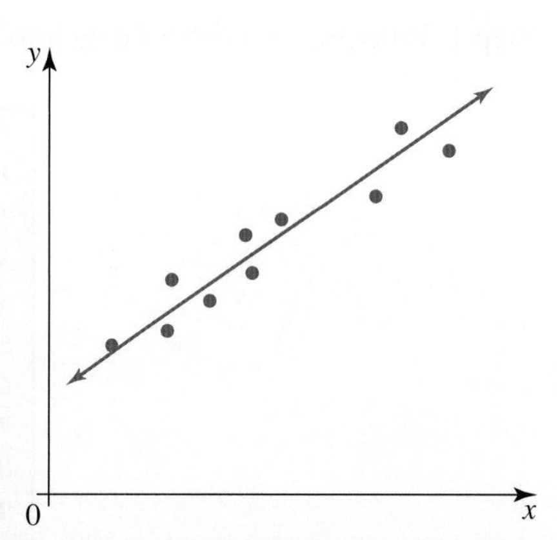

There are many different methods of fitting a straight line by eye. They may appear logical or even obvious, but fitting by eye involves a considerable margin of error. We are going to consider only one method: fitting the line by balancing the number of points. The technique of balancing the number of points involves fitting a line so that there is an *equal number of points* above and below the line. For example, if there are 12 points in the data set, 6 should be above the line and 6 below it.

tlvd-3727

WORKED EXAMPLE 8 Drawing lines of best fit

Draw the line of best fit for the data in the diagram using the equal-number-of-points method.

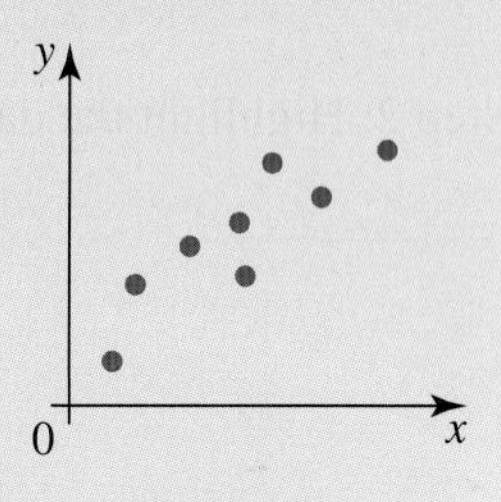

THINK

1. Note that the number of points (n) is 8.
2. Fit a line where 4 points (i.e. half n) are above the line. Using a clear plastic ruler, try to fit the best line. Make sure the line extends through all data points.
3. The first attempt shown has only 3 points below the line, where there should be 4. Make refinements.
4. The second attempt is an improvement, but the line is too close to the points above it.
5. Improve the position of the line until a better 'balance' between upper and lower points is achieved.

WRITE

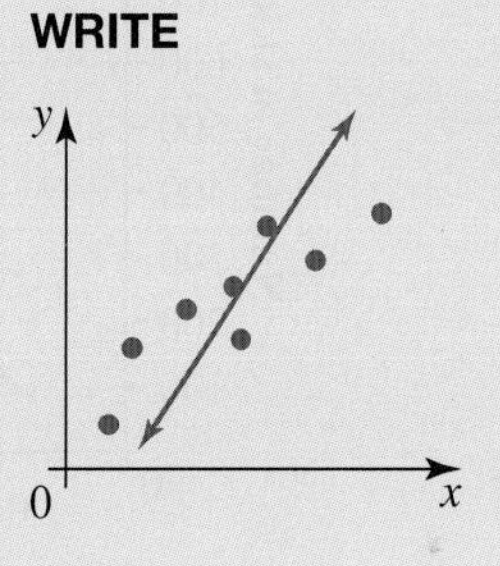

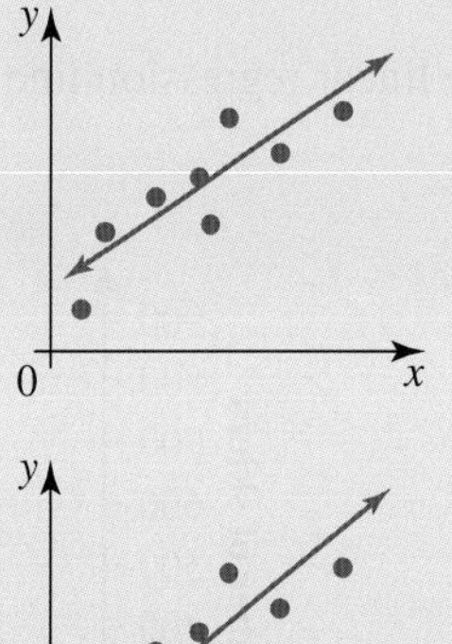

7.6.2 Determining the equation of a line of best fit

Using technology

Using a spreadsheet, the equation of the line of best fit, commonly referred to as a **regression line**, can readily be produced from two lists of data.

Consider the example used at the beginning of the chapter involving skiers and snow depth.

Step 1: Enter the two lists of data into two columns in the spreadsheet.

	A	B
1	Depth of snow (m)	Number of skiers
2	0.5	120
3	0.8	250
4	2.1	500
5	3.6	780
6	1.4	300
7	1.5	280
8	1.8	410
9	2.7	320
10	3.2	640
11	2.4	540
12	2.6	530
13	1.7	200

Step 2: Highlight the data and insert a scatterplot from the spreadsheet software's chart options.

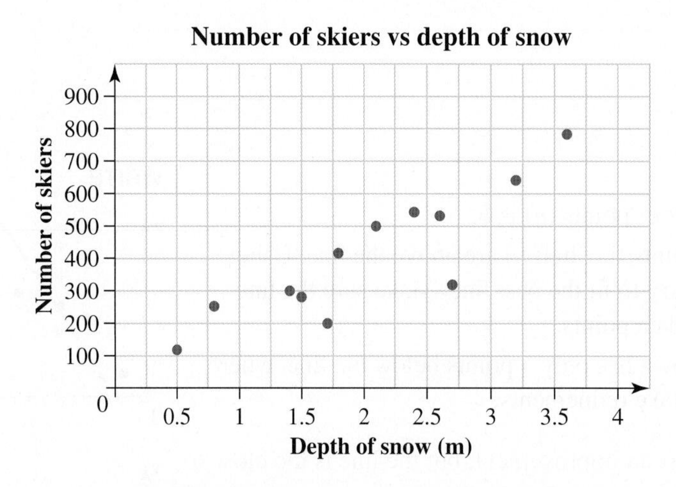

Step 3: Display the linear regression line (also known as a trendline) and its equation on the scatterplot.

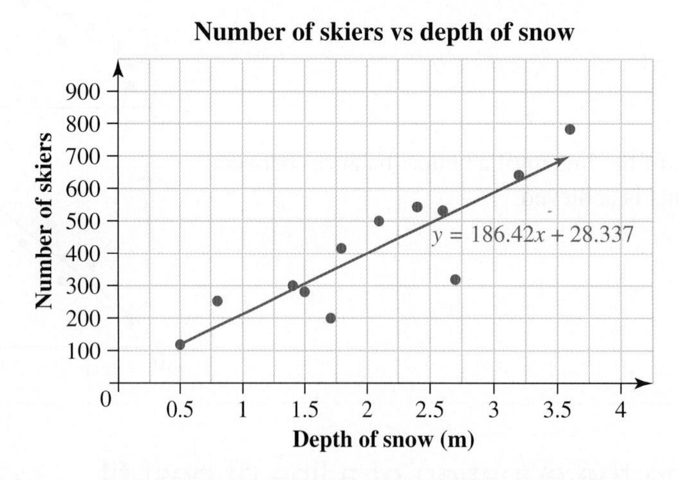

Step 4: Replace the variables x and y in the regression equation with the variable names:

$$\text{Number of skiers} = 186.42 \times \text{depth of snow} + 28.337$$

Manually

To determine the equation of a line of best fit by hand, we must determine the gradient and *y*-intercept of the drawn line. First, consider drawing the regression line as shown in Worked example 9.

tlvd-3728

WORKED EXAMPLE 9 Drawing regression lines

The table below shows the marks of 10 students in Physics and Chemistry tests. Using the variable Physics as the explanatory variable, draw a scatterplot showing the regression line.

Physics	2	5	6	7	7	8	8	9	9	10
Chemistry	4	7	5	8	6	6	9	7	10	9

THINK

1. Plot the points corresponding to each pair of marks.
2. Add a regression line, making sure that there is an equal number of points above and below the line.

WRITE

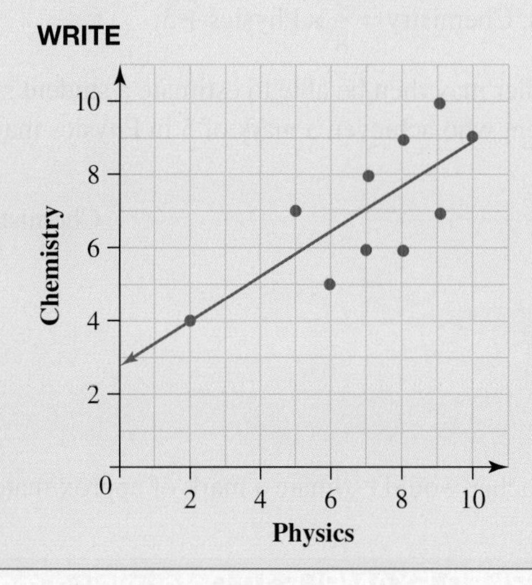

Once the regression line has been drawn, we can determine its equation. The equation of a straight line has the form $y = mx + c$, where m is the gradient and c is the *y*-intercept.

The gradient of a regression line is best found by calculating the rise and the run from two points on the line of regression. (Note that these two points are not necessarily given data points.) The gradient is then found using:

$$\text{gradient } (m) = \frac{\text{rise}}{\text{run}}$$

The *y*-intercept can then be seen by noting the point where the line crosses the *y*-axis. The *y*-intercept can only be determined from the graph if the scale on the horizontal axis starts at 0.

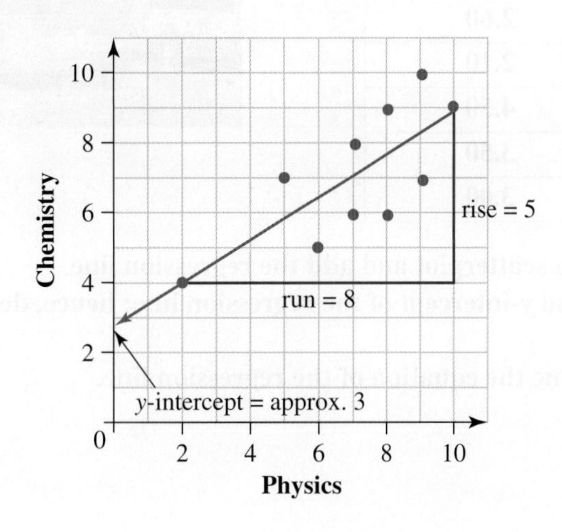

Consider Worked example 9. We can see that the points (2, 4) and (10, 9) lie on the line. The rise between these points is 5 and the run is 8, so the gradient of the line is:

$$\begin{aligned}\text{Gradient } (m) &= \frac{\text{rise}}{\text{run}} \\ &= \frac{5}{8}\end{aligned}$$

From the graph, we can see that the y-intercept is approximately (0, 3).

Substituting the value of the gradient and the y-coordinate of the y-intercept into the general form of a linear equation, we can determine that the equation of the regression line is approximately $y = \frac{5}{8}x + 3$ or $y = 0.625x + 3$.

That is, Chemistry $= \frac{5}{8} \times$ Physics $+ 3$.

A teacher may then be able to estimate a student's Chemistry mark by using their result in Physics. For example, a student who achieved a mark of 5 in Physics may get:

$$\begin{aligned}\text{Chemistry} &= \frac{5}{8} \times \text{Physics} + 3 \\ &= \frac{5}{8} \times 5 + 3 \\ &= 6\frac{1}{8} \text{ or } 6.125\end{aligned}$$

The teacher would estimate a mark of approximately 6 for this student in Chemistry.

WORKED EXAMPLE 10 Determining regression line equations by hand and using technology

The following table shows the bus fare charged by a bus company for different distances.

Distance (km)	Fare ($)
1.5	2.10
0.5	2.00
7.5	4.50
6	4.00
6	4.50
2.5	2.60
0.5	2.10
8	4.50
4	3.50
3	3.00

a. **Represent the data using a scatterplot and add the regression line.**
b. **Determine the gradient and y-intercept of the regression line; hence, determine its equation manually.**
c. **Use technology to determine the equation of the regression line.**

THINK

WRITE

a. 1. The fare would generally depend on the distance travelled. Show the distance on the x-axis and the fare on the y-axis.

2. Plot the points given by each pair.

3. Add the regression line to the graph.

a.

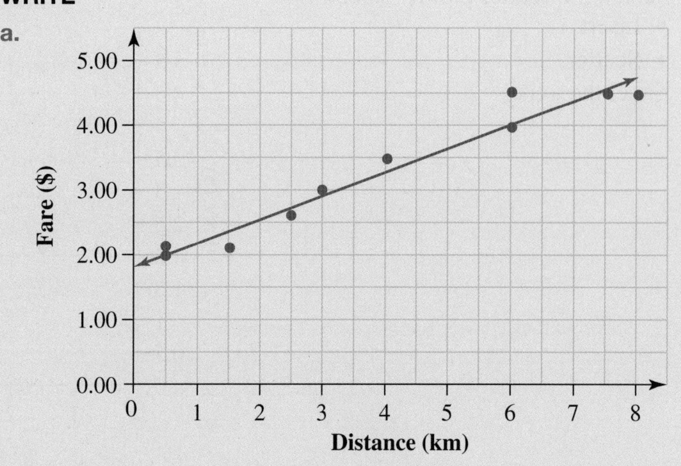

b. 1. Take two points on the regression line and determine the gradient between them by substituting in the formula $m = \frac{\text{rise}}{\text{run}}$ and simplifying. Use the points (0.5, 2) and (6, 4).

b.

Rise = 2; run = 5.5

$$m = \frac{2}{5.5}$$
$$\approx 0.36$$

2. Give an estimate of the y-intercept.

y-intercept ≈ 1.8

3. Substitute the gradient and y-intercept into the formula $y = mx + c$ and state the equation.

$y = 0.36x + 1.8$

c. 1. Enter the data into the spreadsheet.

c.

Distance (km)	Fare ($)
1.5	2.1
0.5	2
7.5	4.5
6	4
6	4.5
2.5	2.6
0.5	2.1
8	4.5
4	3.5
3	3

2. Sketch the scatter plot by selecting:
   - Insert
   - Scatter

   Label the axis.

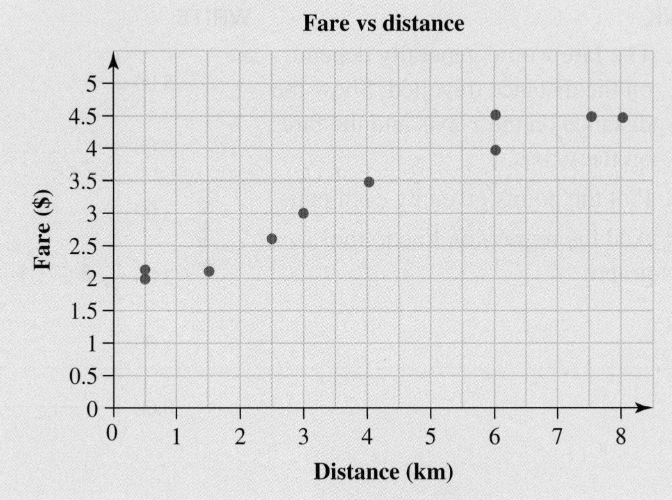

3. Complete a linear regression by selecting:
   - Chart Design
   - Quick Layout
   - Layout 9.

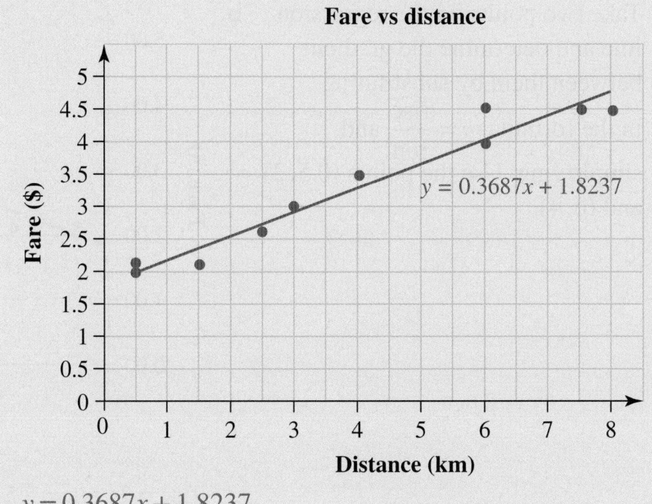

4. Write the answer.

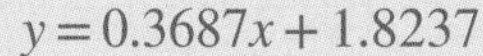

$y = 0.3687x + 1.8237$

Note that there may be some slight discrepancies in the calculated values when manually determining the regression line compared to using technology.

Resources

Interactivities Applying lines of best fit (int-2798)
Lines of best fit (int-6180)

7.6 Exercise

Students, these questions are even better in jacPLUS

Receive immediate feedback and access sample responses

Access additional questions

Track your results and progress

Find all this and MORE in jacPLUS

1. WE8 Draw the line of best fit for the scatterplot shown using the equal-number-of-points method.

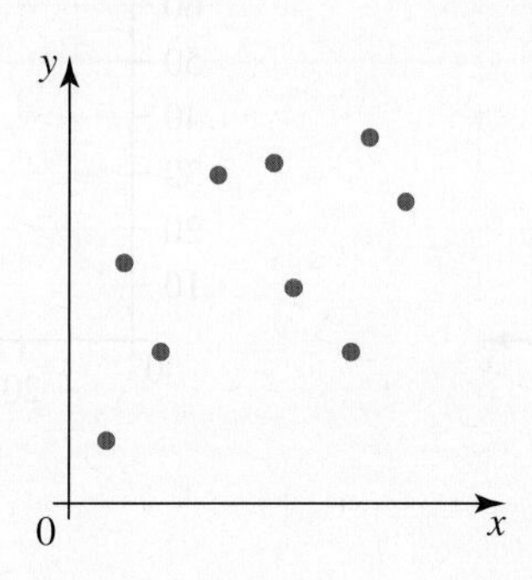

2. Draw the line of best fit for the scatterplot shown using the equal-number-of-points method.

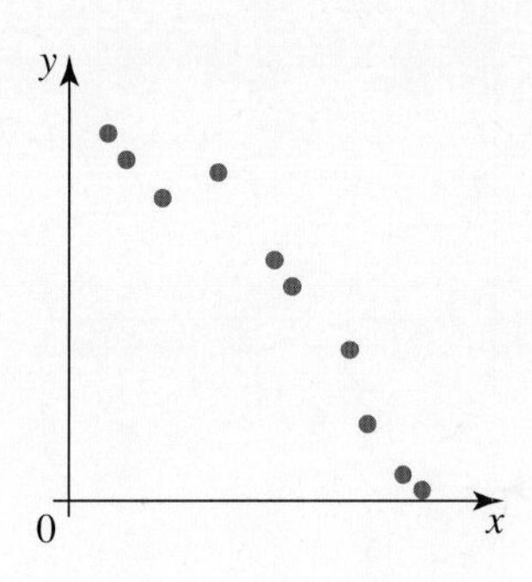

3. Fit a straight line to the data in the scatterplots using the equal-number-of-points method.

a.
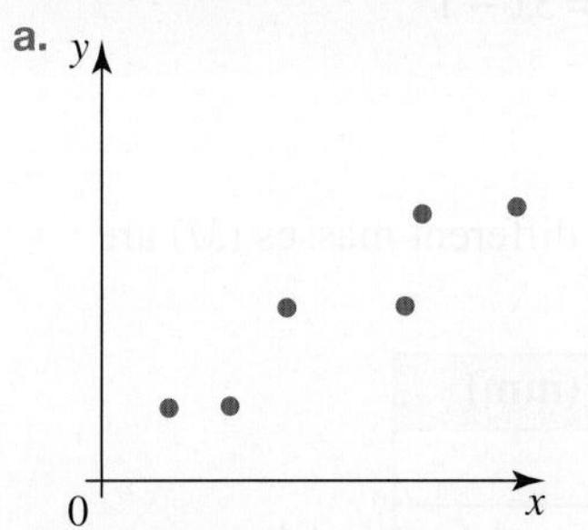

b.
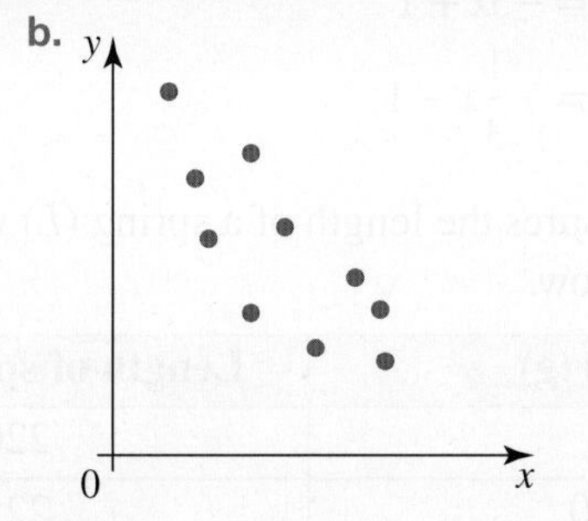

c.
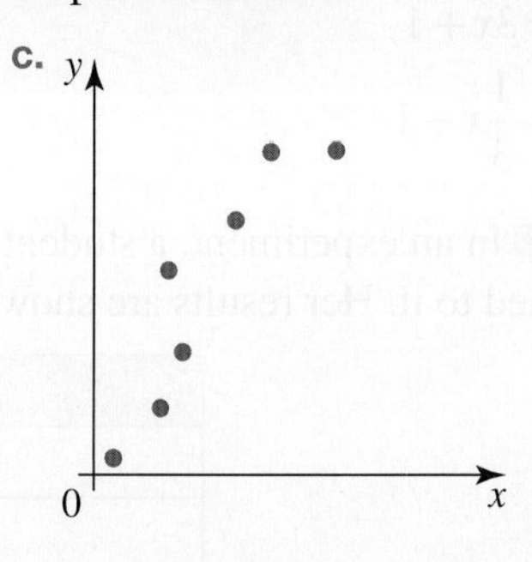

d.
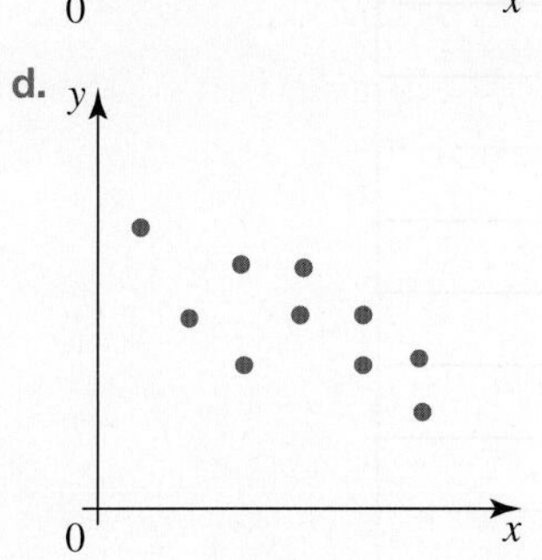

e.
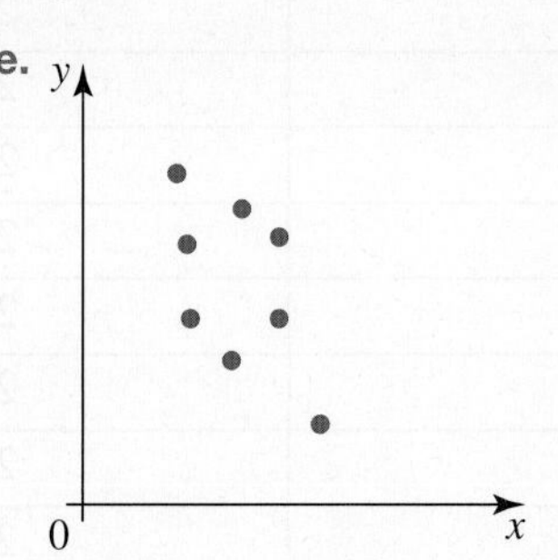

4. **WE9** The table below shows the marks achieved by a class of students in English and Maths.

English	64	75	81	63	32	56	47	59	73	64
Maths	76	62	89	56	49	57	53	72	80	50

Using English as the explanatory variable, draw a scatterplot showing the regression line.

5. Position the line of best fit through each of the following graphs and determine the equation of each.

a.

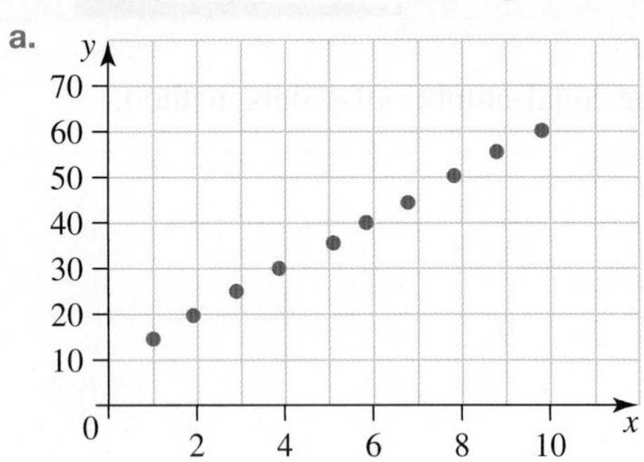

b.

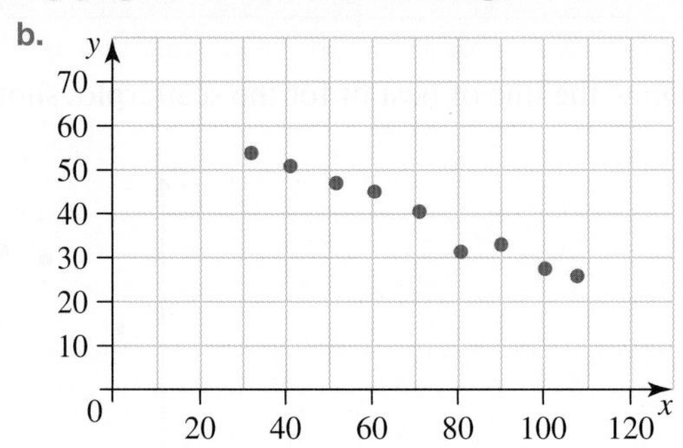

c.

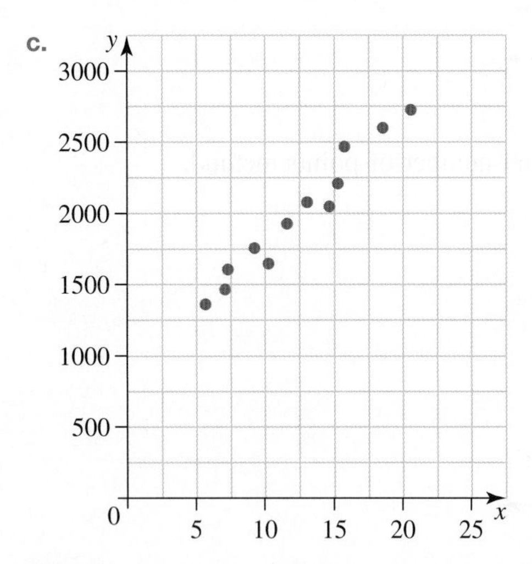

6. **MC** The equation of the line of best fit that passes through the points (3, 8) and (12, 35) is:

A. $y = 3x + 1$
B. $y = -3x + 1$
C. $y = 3x - 1$
D. $y = \frac{1}{3}x - 1$
E. $y = -\frac{1}{3}x - 1$

7. **WE10** In an experiment, a student measures the length of a spring (L) when different masses (M) are attached to it. Her results are shown below.

Mass (g)	Length of spring (mm)
0	220
100	225
200	231
300	235
400	242
500	246
600	250
700	254
800	259
900	264

a. Draw a scatterplot of the data showing the regression line.
b. Determine the gradient and y-intercept of the regression line and, hence, state the equation of the regression line. Write your equation in terms of the variables L and M.

8. A scientist who measures the volume of a gas at different temperatures provides the following table of values.

Temperature (°C)	Volume (L)
−40	1.2
−30	1.9
−20	2.4
0	3.1
10	3.6
20	4.1
30	4.8
40	5.3
50	6.1
60	6.7

a. Draw a scatterplot of the data and on it insert the regression line.
b. Determine the equation of the regression line. Write your equation in terms of the variables volume of gas, V, and its temperature, T.

9. A sports scientist is interested in the importance of muscle bulk to strength. He measures the biceps circumference of ten people and tests their strength by asking them to complete a lift test. His results are given in the following table.

Circumference of biceps (cm)	Lift test (kg)
25	50
25	52
27	58
28	51
30	60
30	62
31	53
33	62
34	61
36	66

a. Draw a scatterplot of the data and draw the regression line.
b. Determine a rule for predicting the ability of a person to complete a lift test, S, from the circumference of their biceps, B.

10. A sports scientist is looking at data comparing the heights of athletes and their performance in the high jump. The following table and scatterplot represent the data they have collected. A line of best fit by eye has been drawn on the scatterplot. Use technology to determine the equation of the line.

Height (cm)	168	173	155	182	170	193	177	185	163	190
High jump (cm)	172	180	163	193	184	208	188	199	174	186

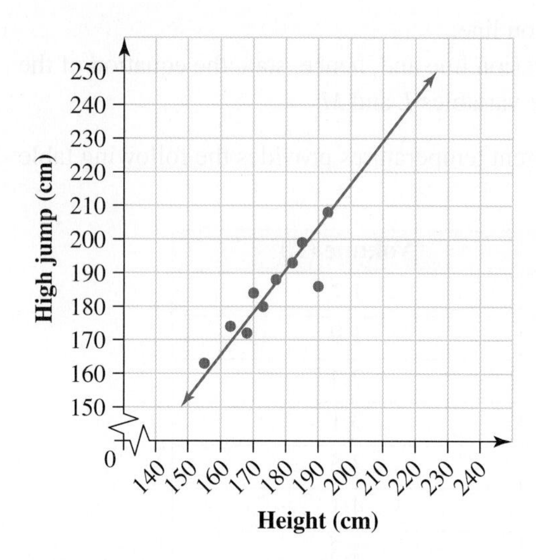

11. Nidya is analysing the data from question 10, but a clerical error means that she only has access to two points of data: (170, 184) and (177, 188).
    a. Determine Nidya's equation for the line of best fit, rounding all decimal numbers to 2 places.
    b. Add Nidya's line of best fit to the scatterplot of the data.
    c. Comment on the similarities and differences between the two lines of best fit.
12. The following table and scatterplot show the age and height of a field of sunflowers planted at different times throughout summer.

Age of sunflower (days)	63	71	15	33	80	22	55	47	26	39
Height of sunflower (cm)	237	253	41	101	264	65	218	182	82	140

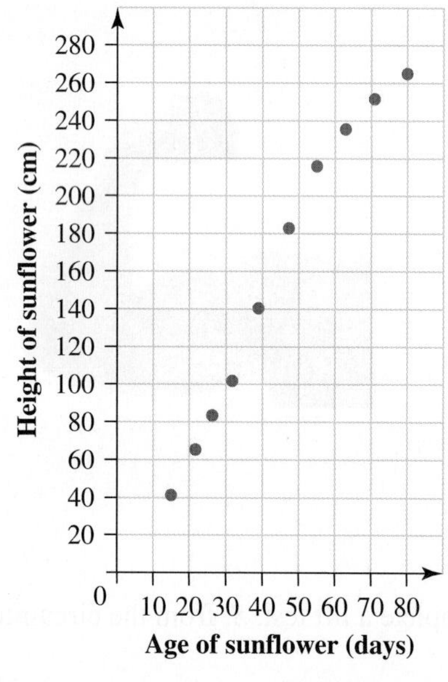

    a. Xavier draws a line of best fit by eye that goes through the points (10, 16) and (70, 280). Draw his line of best fit on the scatterplot and comment on his choice of line.
    b. Calculate the equation of the line of best fit using the two points that Xavier selected.
    c. Patricia draws a line of best fit by eye that goes through the points (10, 18) and (70, 258). Draw her line of best fit on the scatterplot and comment on her choice of line.
    d. Calculate the equation of the line of best fit using the two points that Patricia selected.
    e. State why the value of the y-intercept is not 0 in either equation.

13. Steve is looking at data comparing the sizes of different music venues across the country and the average ticket prices at these venues. After plotting his data in a scatterplot, he calculates a line of best fit for the data as $y = 0.04x + 15$, where y is the average ticket price in dollars and x is the capacity of the venue.
   a. Determine what the value of the gradient (m) represents in Steve's equation.
   b. Determine what the value of the y-intercept represents in Steve's equation.
   c. Determine if the y-intercept is a realistic value for this data.
14. A government department is analysing the population density and crime rate of different suburbs to see if there is a connection. The following table and scatterplot display the data that have been collected so far.

Population density (persons per km^2)	3525	2767	4931	3910	1572	2330	2894	4146	1968	5337
Crime rate (per 1000 people)	185	144	279	227	65	112	150	273	87	335

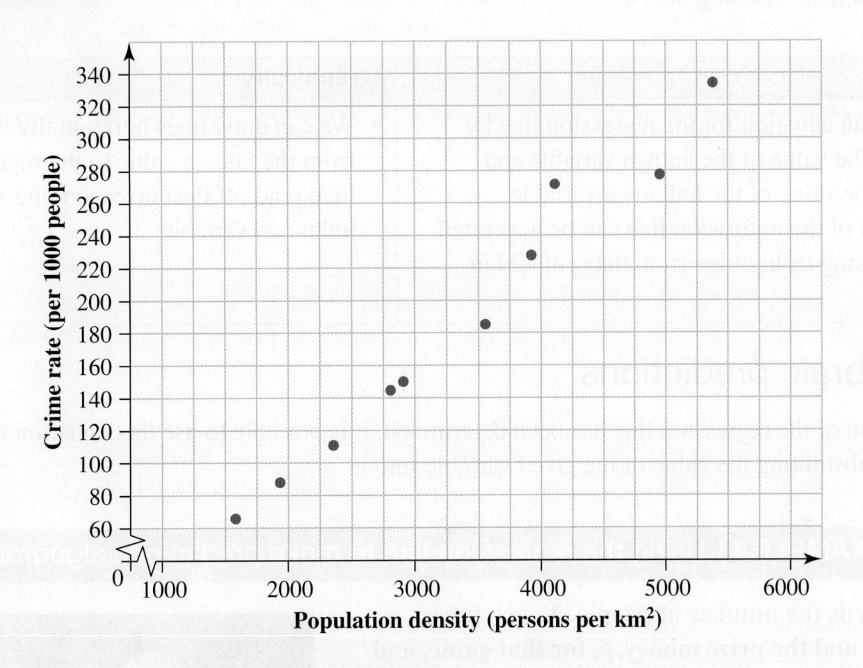

   a. Draw a line of best fit on the scatterplot of the data.
   b. Using technology or otherwise, determine the equation of the line of best fit.
   c. Explain what the value of the x-intercept means in terms of this problem.
   d. State if the x-intercept value is realistic. Explain your answer.

7.7 Making predictions using a line of best fit

LEARNING INTENTION

At the end of this subtopic you should be able to:
- create a line of best fit algebraically and graphically
- use the line of best fit to make interpolated and extrapolated predictions.

7.7.1 Methods of creating a line of best fit

If a linear association exists between a pair of variables, it is useful to be able to make predictions for one variable from the other. The line of best fit can be extended and then used to make predictions. When this is done, the line is called a regression line.

There are two methods of doing this:

Algebraically	Graphically
• We can find an equation for the regression line by substituting the value of the known variable and calculating the value of the unknown variable. • The equation of the regression line can be generated by hand or using technology from data entered in lists.	• We can draw lines horizontally or vertically from the known value to the regression line, then read off the corresponding value of the unknown variable.

7.7.2 Algebraic predictions

Once the equation of the regression line has been determined, it is possible to use this equation to make predictions by substituting the value of the given variable into it.

tlvd-3729

WORKED EXAMPLE 11 Using the regression line to make predictions algebraically

A casino records the number of people, N, playing a jackpot game and the prize money, p, for that game, and plots the results on a scatterplot. The regression line is found to have the equation $N = 0.07p + 220$.

a. Determine the number of people playing when the prize money is \$2500.

b. Calculate the likely prize on offer when there are 500 people playing.

THINK	WRITE
a. 1. Write the equation of the regression line.	a. $N = 0.07p + 220$
2. Substitute 2500 for p in the regression equation.	$N = 0.07 \times 2500 + 220$
3. Calculate N.	$= 395$
4. Give a written answer.	When the prize money is \$2500, it is estimated that 395 people will be playing.

b. 1. Write the equation of the regression line. | b. $N = 0.07p + 220$

2. Substitute 500 for N in the regression equation. | $500 = 0.07p + 220$

3. Solve the equation for p by subtracting 220 from both sides, then dividing the result by 0.07. | $280 = 0.07p$

$$p = \frac{280}{0.07}$$
$$= 4000$$

4. Give a written answer. | When 500 people are playing, the prize is estimated to be \$4000.

7.7.3 Graphical predictions

Once the regression line has been drawn, horizontal or vertical lines can be added from the axes to the regression line to read off the value of the unknown variable. Remember that the values determined are only estimates, and not exact values.

tlvd-3730

WORKED EXAMPLE 12 Using the regression line to make predictions graphically

Consider the graph drawn in Worked example 10, representing bus fares for travelling various distances. Use the graph to estimate:

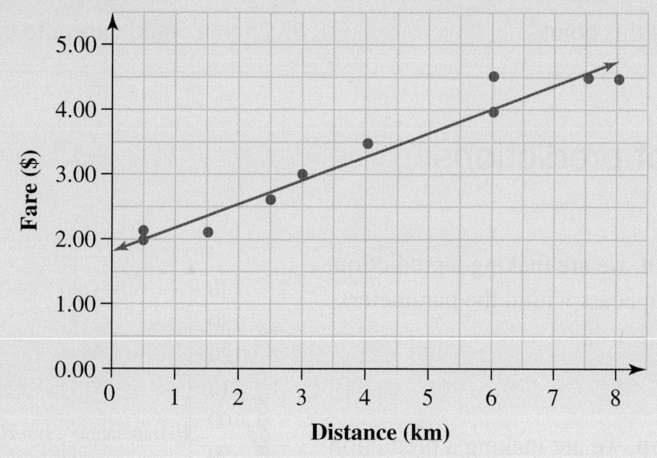

a. **the fare for a journey of 5 km**

b. **how far you could expect to travel for a cost of \$3.25.**

THINK	WRITE
a. 1. Locate the value of 5 km on the x-axis. **2.** Draw a vertical line from this point to meet the regression line. See the pink line. **3.** Draw a horizontal line from this point to the y-axis. See the pink line.	**a.** 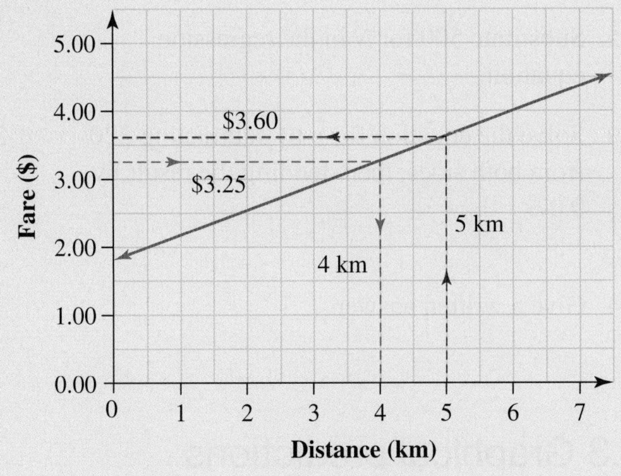
4. Read this value on the y-axis.	A distance of 5 km of bus travel is predicted to cost about \$3.60.
b. 1. Locate \$3.25 on the y-axis. **2.** Draw a horizontal line from this point to the regression line. See the green line. **3.** From this point, draw a vertical line to the x-axis. See the green line.	**b.** (Refer to the graph above.)
4. Read the x-value at this point.	For \$3.25 you could expect to travel about 4 km.

7.7.4 Reliability of predictions

Interpolation

When we use **interpolation**, we are making a prediction from a line of best fit that appears within the parameters of the original data set.

Extrapolation

When we use **extrapolation**, we are making a prediction from a line of best fit that appears outside the parameters of the original data set.

If we plot our line of best fit on the scatterplot of the given data, then extrapolation will occur before the first point or after the last point of the scatterplot. The more pieces of data there are in a set, the better the line of best fit you will be able to draw. More data points allow more reliable predictions.

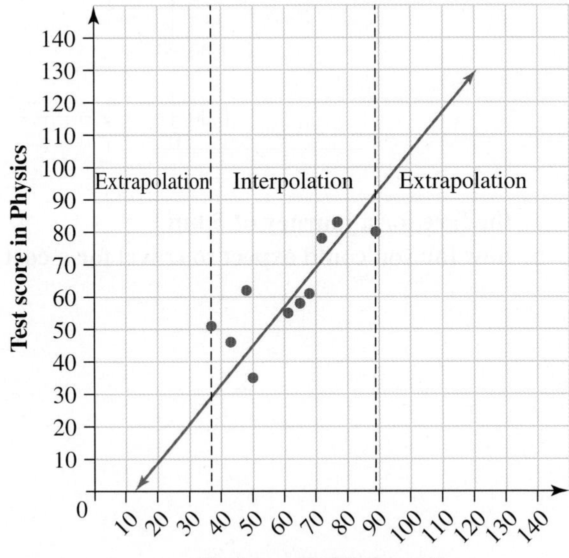

In general, interpolation is a far more reliable method of making predictions than extrapolation. However, there are other factors that should also be considered. Interpolation closer to the centre of the data set will be more reliable than interpolation closer to the edge of the data set. Extrapolation that appears closer to the data set will be much more reliable than extrapolation that appears further away from the data set.

tlvd-3731

WORKED EXAMPLE 13 Reliability of interpolation and extrapolation

The following data represent the air temperature (°C) and depth of snow (cm) at a popular ski resort.

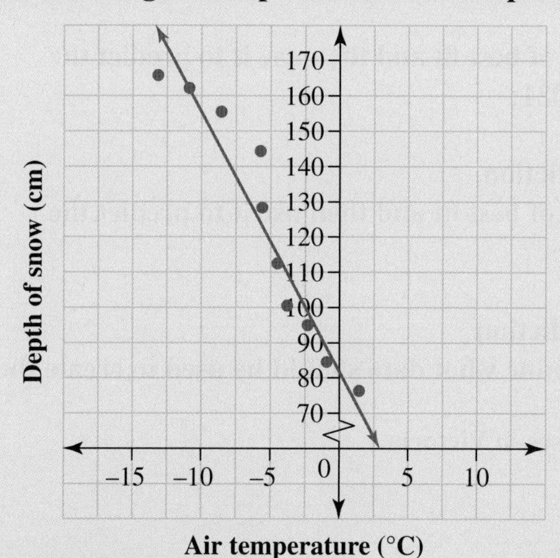

Air temperature (°C)	−4.5	−2.3	−8.9	−11.0	−13.3	−6.2	−0.4	1.5	−3.7	−5.4
Depth of snow (cm)	111.3	95.8	155.6	162.3	166.0	144.7	84.0	77.2	100.5	129.3

The equation of the line of best fit for this data set has been calculated as $y = -7.2x + 84$.

a. **Use the line of best fit to estimate the depth of snow if the air temperature is −6.5 °C.**

b. **Use the line of best fit to estimate the depth of snow if the air temperature is 25.2 °C.**

c. **Comment on the reliability of your estimations in parts a and b.**

THINK	WRITE
a. 1. Rewrite the regression equation in terms of the variables.	a. $y = -7.2x + 84$ Depth of snow $= -7.2 \times$ air temperature $+ 84$
2. Substitute −6.5 for the air temperature into the equation for the line of best fit.	Air temperature $= -6.5$ Depth of snow $= -7.2 \times$ air temperature $+ 84$
3. Evaluate the expression for the depth of snow.	$= -7.2 \times -6.5 + 84$ $= 130.8$
4. Write the answer.	When the air temperature is −6.5 °C, the depth of snow will be approximately 130.8 cm.
b. 1. Substitute 25.2 for the air temperature into the equation for the line of best fit.	b. Air temperature $= 25.2$ Depth of snow $= -7.2 \times$ air temperature $+ 84$
2. Evaluate the expression for the depth of snow.	$= -7.2 \times 25.2 + 84$ $= -97.4$ (to 1 decimal place)
3. Write the answer.	When the air temperature is 25.2 °C, the depth of snow will be approximately −97.4 cm.
c. Relate the answers back to the original data to check their reliability.	c. The estimate in part **a** was made using interpolation, with the point being comfortably located within the parameters of the original data. The estimate appears to be consistent with the given data and as such is reliable. The estimate in part **b** was made using extrapolation, with the point being located well outside the parameters of the original data. This estimate is clearly unreliable, as we cannot have a negative depth of snow.

tlvd-3732

WORKED EXAMPLE 14 Further applications of interpolation and extrapolation

The following data represent the daily number of COVID-19 cases in Victoria from December 2021 to January 2022.

a. If we use data from 9–14 December 2021 to form a line of best fit and then use it to predict the number of positive COVID-19 cases on 15 December 2021:
   i. state whether this is interpolation or extrapolation
   ii. explain whether this method would give a good prediction.

b. If we use data from 9–14 December 2021 to form a line of best fit and then use it to predict the number of positive COVID-19 cases on 3 January 2022:
   i. state whether this is interpolation or extrapolation
   ii. explain whether this method would give a good prediction
   iii. to get a better prediction for 3 January 2022, determine what data should be used to create the line of best fit.

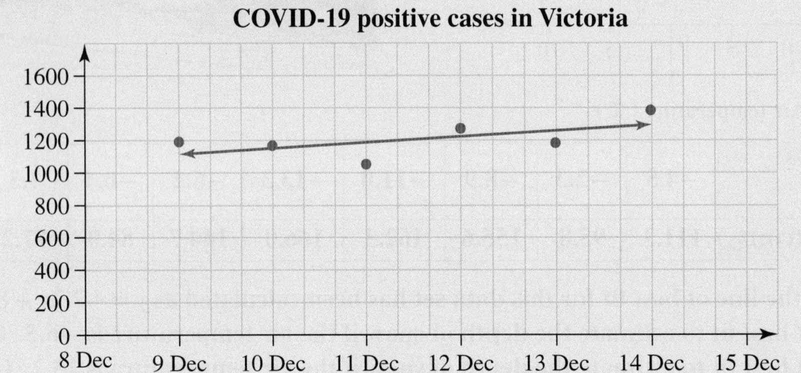

THINK	WRITE
a. i. Determine whether the predicted data is inside or outside the parameters.	a. i. Since the predicted data is outside the parameters, it is *extrapolation*.
ii. Since we are extrapolating, the further the prediction is from the most recent data, the less reliable the prediction is.	ii. 15 December 2021 is relatively close to the date of the most recent data used (14 December 2021) so without knowing the outcome, you would expect the prediction to be good. This is the case from the data provided since it follows a similar trend to the data used to create the line of best fit.
b. i. Determine whether the predicted data is inside or outside the parameters.	b. i. Since the predicted data is outside the parameters, it is *extrapolation*.
ii. Since we are extrapolating, the further the prediction is from the most recent data, the less reliable the prediction is.	ii. 3 January 2022 is further away from the date of the most recent data used (14 December 2021) so without knowing the outcome, you would expect the prediction to be less reliable.
iii. To make a better prediction, we need to use data that create a line of best fit that follows the recent trend.	iii. The line of best fit should be created using data from very late December and the first two days of January (the most recent data).

on Resources

Interactivities Extrapolation (int-1154)
Interpolation and extrapolation (int-6181)

7.7 Exercise

Students, these questions are even better in jacPLUS

Receive immediate feedback and access sample responses

Access additional questions

Track your results and progress

Find all this and MORE in jacPLUS

1. **WE11** A taxi company adjusts its meters so that the fare is charged according to the equation $F = 1.2d + 3$, where F is the fare, in dollars, and d is the distance travelled, in km.
   a. Determine the fare charged for a distance of 12 km.
   b. Determine the fare charged for a distance of 4.5 km.
   c. Determine the distance that could be covered on a fare of \$27.
   d. Determine the distance that could be covered on a fare of \$13.20.

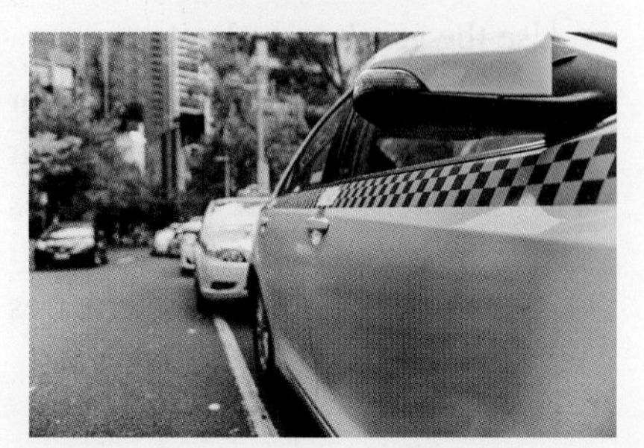

2. Detectives can use the equation $H = 6.1f - 5$ to estimate the height of a burglar who leaves footprints behind. (H is the height of the burglar, in cm, and f is the length of the footprint.)
   a. Determine the height of a burglar whose footprint is 27 cm in length.
   b. Determine the height of a burglar whose footprint is 30 cm in length.
   c. Determine the footprint length of a burglar of height 185 cm. (Give your answer correct to 2 decimal places.)
   d. Determine the footprint length of a burglar of height 152 cm. (Give your answer correct to 2 decimal places.)
3. A football match pie seller finds that the number of pies sold is related to the air temperature. The situation could be modelled by the equation $N = 870 - 23t$, where N is the number of pies sold and t is the air temperature in degrees Celsius.
   a. Determine the number of pies sold if the air temperature was 5 °C.
   b. Determine the number of pies sold if the air temperature was 25 °C.
   c. Determine the likely air temperature if 400 pies were sold.
   d. Determine how hot the day would have to be before the pie seller sold no pies at all.

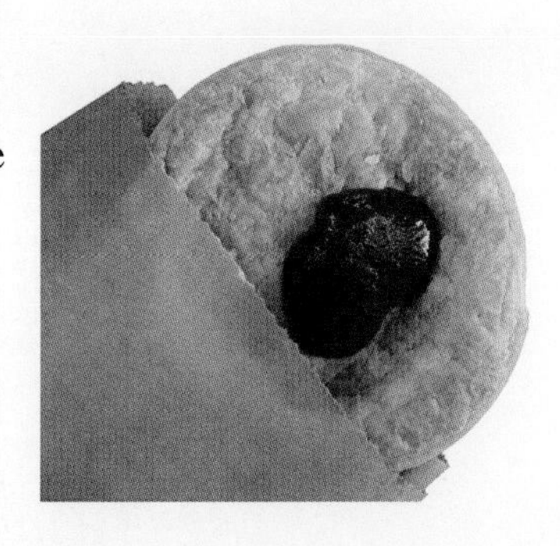

4. WE12 The following graph shows the average annual costs of running a car. It includes all fixed costs (registration, insurance, etc.) as well as running costs (petrol, repairs, etc.).

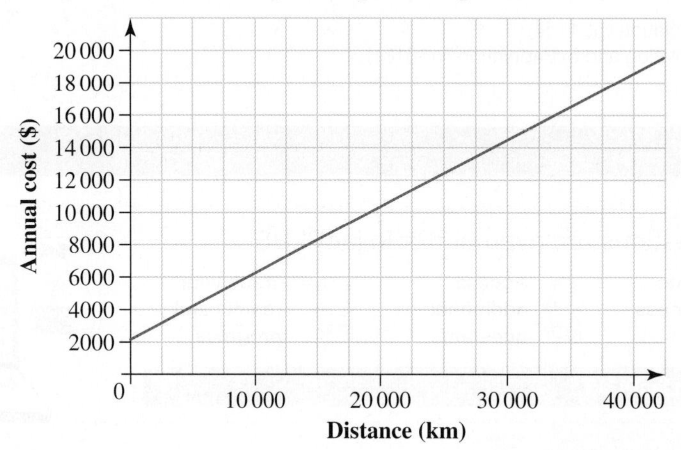

Use the graph to estimate:

a. the annual cost of running a car if it is driven 15 000 km
b. the annual cost of running a car if it is driven 1000 km
c. the likely number of kilometres driven if the annual costs were $8000
d. the likely number of kilometres driven if the annual costs were $16 000.

5. A market researcher finds that the number of people who would purchase Wise-up (the intelligent deodorant) is related to its price. He provides the following table of values.

Price ($)	Weekly sales (× 1000)
1.40	105
1.60	101
1.80	97
2.00	93
2.20	89
2.40	85
2.60	81
2.80	77
3.00	73
3.20	69
3.40	65

a. Draw a scatterplot of the data.
b. Draw the line of best fit.
c. Determine an equation that represents the association between the number of cans of Wise-up sold, N (in thousands), and its price, p.
d. Predict the number of cans sold each week if:

i. the price was $3.10 ii. the price was $4.60.

e. Determine the price Wise-up should be sold at if the manufacturers wish to sell 80 000 cans.
f. Given that the manufacturers of Wise-up can produce only 100 000 cans each week, determine the price it should be sold at to maximise production.

6. The following table gives the adult return airfares between some Australian cities.

City	Distance (km)	Price ($)
Melbourne–Sydney	713	150
Perth–Melbourne	2728	570
Adelaide–Sydney	1172	200
Brisbane–Melbourne	1370	240
Hobart–Melbourne	559	130
Hobart–Adelaide	1144	220
Adelaide–Melbourne	669	140

a. Draw a scatterplot of the data including the regression line.
b. Determine an equation that represents the association between the airfare, A, and the distance travelled, d.
c. Use the equation to predict the likely airfare (to the nearest dollar) from:
   i. Sydney to the Gold Coast (671 km)
   ii. Perth to Adelaide (2125 km)
   iii. Hobart to Sydney (1024 km)
   iv. Perth to Sydney (3295 km).

7. **WE13** The owner of an ice-cream parlour has collected data relating air temperature to ice-cream sales.

Air temperature (°C)	23.4	27.5	26.0	31.1	33.8	22.0	19.7	24.6	25.5	29.3
Ice-cream sales	135	170	165	212	204	124	86	144	151	188

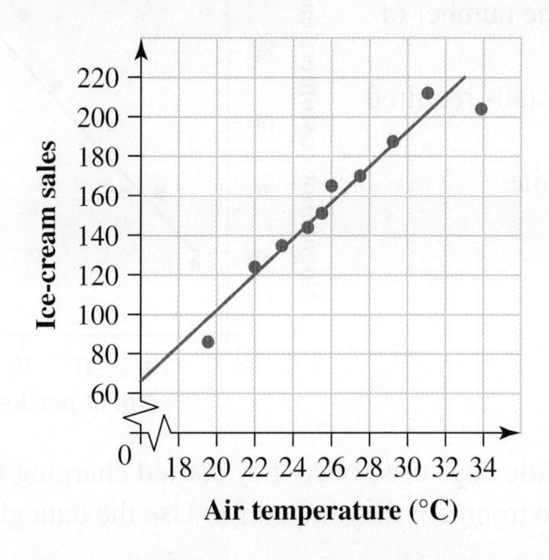

A line of best fit for this data has been calculated as ice-cream sales = 9 × air temperature −77.
a. Use the line of best fit to estimate ice-cream sales if the air temperature is 27.9 °C.
b. Use the line of best fit to estimate ice-cream sales if the air temperature is 15.2 °C.
c. Comment on the reliability of your answers to parts **a** and **b**.

8. Georgio is comparing the cost and distance of various long-distance flights, and after drawing a scatterplot he creates an equation for a line of best fit to represent his data. Georgio's line of best fit is $y = 0.08x + 55$, where y is the cost of the flight and x is the distance of the flight in kilometres.
   a. Estimate the cost of a flight between Melbourne and Sydney (713 km) using Georgio's equation.
   b. Estimate the cost of a flight between Melbourne and Broome (3121 km) using Georgio's equation.
   c. All of Georgio's data came from flights of distances between 400 km and 2000 km. Comment on the suitability of using Georgio's equation for shorter and longer flights than those he analysed. Explain what other factors might affect the cost of these flights.

9. Mariana is a scientist collecting data measuring lung capacity (in L) and time taken to swim 25 metres (in seconds). Unfortunately a spillage in her lab causes all of her data to be erased apart from the records of a person with a lung capacity of 3.5 L completing the 25 metres in 55.8 seconds and a person with a lung capacity of 4.8 L completing the 25 metres in 33.3 seconds.
   a. Use the remaining data to construct an equation for the line of best fit relating lung capacity (x) to the time taken to swim 25 metres (y). Give any numerical values correct to 2 decimal places.
   b. Determine the value that the gradient (m) represents in the equation.
   c. Use the equation to estimate the time it takes people with the following lung capacities to swim 25 metres.
      i. 3.2 litres
      ii. 4.4 litres
      iii. 5.3 litres
   d. Comment on the reliability of creating the equation from Mariana's two remaining data points.

10. WE14 The regression line shown on the scatterplot has the equation $c = 13.33 + 2.097m$, where c is the number of new customers each hour and m is the number of market stalls.
   a. Using the line of best fit, interpolate the data to find the number of new customers expected if there are 30 market stalls.
   b. Use the formula to extrapolate the number of market stalls required in order to expect 150 new customers.
   c. Explain why part a is an example of interpolation, while part b demonstrates extrapolation.

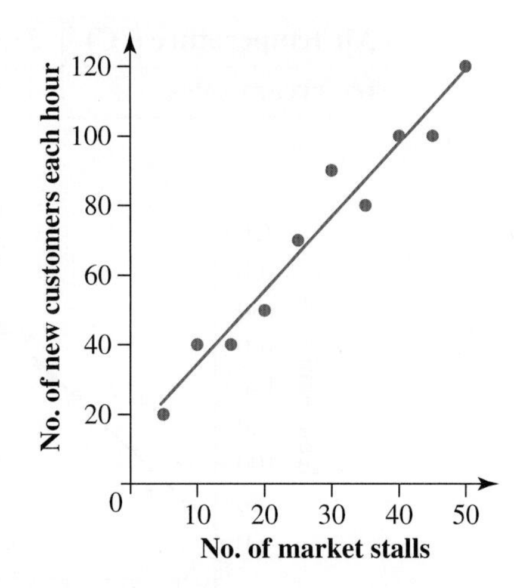

11. A supermarket was collecting data on the amount of plastic bags used once they started charging for plastic bags to see if it was having the positive impact on the environment they hoped for. Use the data given below to complete the following questions.

x (months)	10	11	12	13	14	15	16	17	18	19
y (plastic bags × 1000)	22	18	20	15	17	11	11	7	9	8

   a. Draw a scatterplot and a line of best fit by eye.
   b. Determine the equation of the line of best fit. Give values correct to 3 decimal places.
   c. Extrapolate the data to predict the value of y when $x = 23$.
   d. Explain what assumptions are made when extrapolating data.

12. While camping, a mathematician estimated that the number of mosquitoes around the fire $= 10.2 + 0.5 \times$ temperature of the fire (°C).

a. Determine the number of mosquitoes that would be expected if the temperature of the fire was 240 °C. Give your answer correct to the nearest whole number.
b. Determine the temperature of the fire if there were only 12 mosquitoes in the area.
c. Identify some factors that could affect the reliability of this equation.

13. Data about people's average monthly income and the amount of money they spend at restaurants was collected.

Average monthly income (×$1000)	**Money spent at restaurants per month ($)**
2.8	150
2.5	130
3.0	220
3.1	245
2.2	100
4.0	400
3.7	380
3.8	200
4.1	600
3.5	360
2.9	175
3.6	350
2.7	185
4.2	620
3.6	395

a. Draw a scatterplot of this data.
b. Find the equation of the line of best fit in terms of average monthly income in thousands of dollars (I) and money spent at restaurants in dollars (R). Give values correct to 1 decimal place.
c. Extrapolate the data to predict how much a person who earns $5000 a month might spend at restaurants each month.
d. Explain why part **c** is an example of extrapolation.
e. A person spent $265 eating out last month. Estimate their monthly income, giving your answer to the nearest $10. State if this is an example of interpolation or extrapolation.

7.8 Review

7.8.1 Summary

doc-38015

Hey students! Now that it's time to revise this topic, go online to:

Review your results

Watch teacher-led videos

Practise questions with immediate feedback

Find all this and MORE in jacPLUS

7.8 Exercise

Multiple choice

The test marks for a class of students are shown in the histogram below. Use this information to answer questions 1–3.

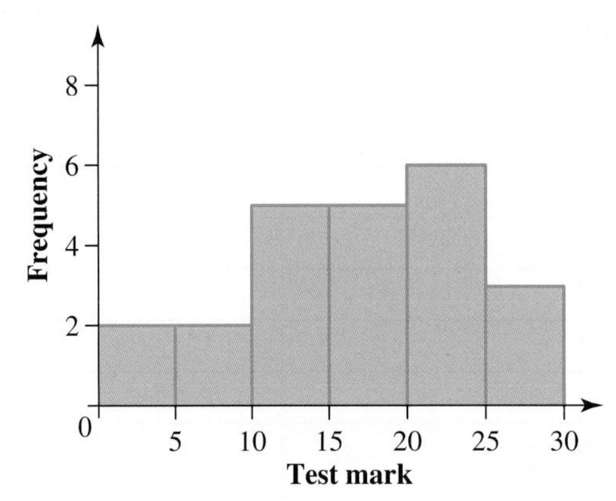

1. MC The mean of this histogram, correct to 1 decimal place, is:

   A. 14.3   B. 16.8   C. 22.5   D. 19.3   E. 21.7

2. MC The mode of this histogram is:

   A. 5–10   B. 10–15   C. 15–20   D. 20–25   E. 25–30

3. MC The range of this histogram is:

   A. 5   B. 10   C. 15   D. 20   E. 30

4. MC Match the boxplot with its most likely histogram.

A. f

Size

B. f

Size

C.

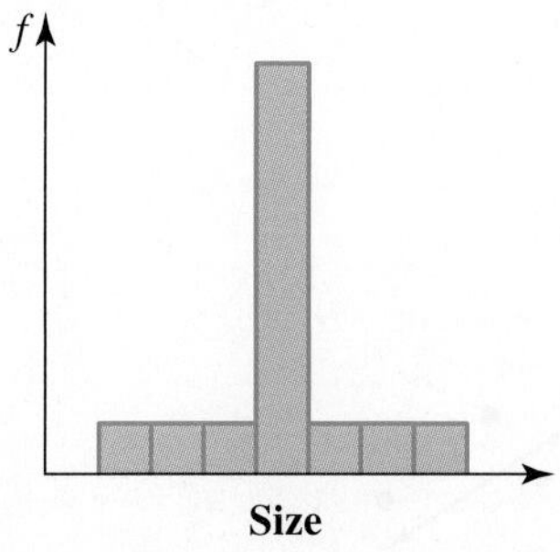

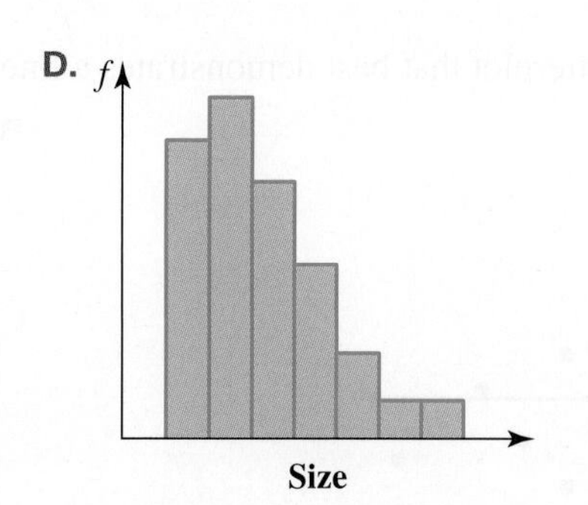

E.

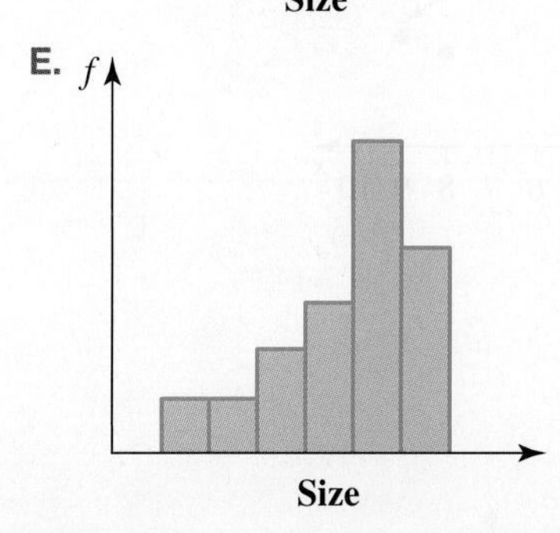

5. **MC** For the following sample data set, determine which of the following is an example of interpolating data.

x	1	5	15	25
y	10	16	18	22

A. Finding the value of x when $y = -7$
B. Finding the value of y when $x = 17$
C. Finding the value of x when $y = 27$
D. Finding the value of y when $x = 37$
E. Finding the value of y when $x = 0$

6. **MC** For the data set from question 5, the regression line equation is:

A. $y = 10 + x$
B. $y = 0.435 + 11.456x$
C. $y = 0.876 + 0.936x$
D. $y = 11.496 + 0.435x$
E. $y = 0.435 + 0.936x$

7. **MC** The regression line equation for the graph shown is closest to:

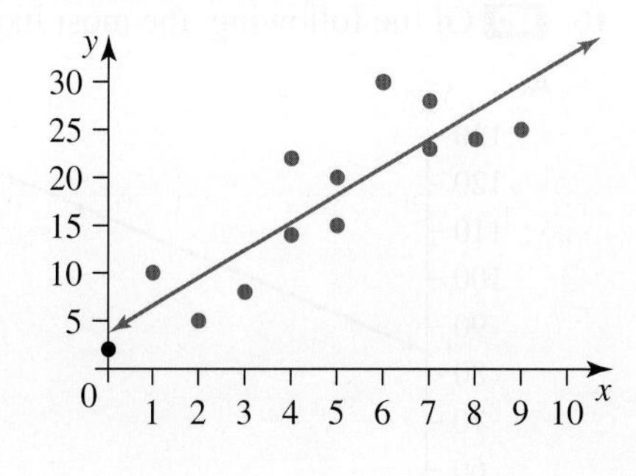

A. $y = 3.8 + 2.9x$
B. $y = -3.8 - 2.9x$
C. $y = -3.8 + 2.9x$
D. $y = 3.8 - 2.9x$
E. $y = 2.9 - 3.8x$

8. **MC** When $y = 0.54 + 15.87x$, the value of y when $x = 2.5$ is:

A. 18.91
B. 40.215
C. 39.135
D. 6.888
E. 24.782

9. MC Select the scatterplot that best demonstrates a line of best fit.

A.

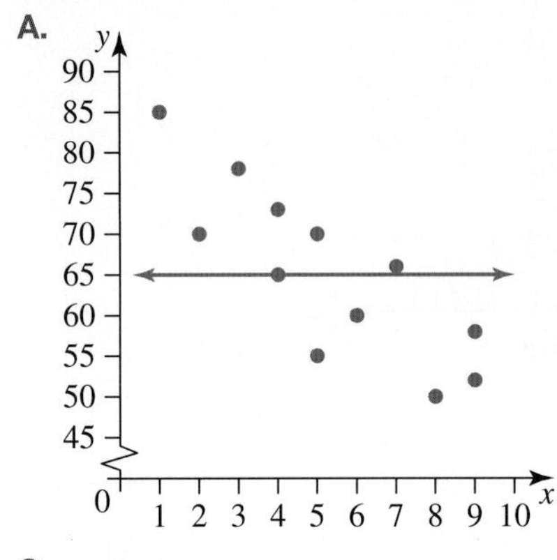

B.

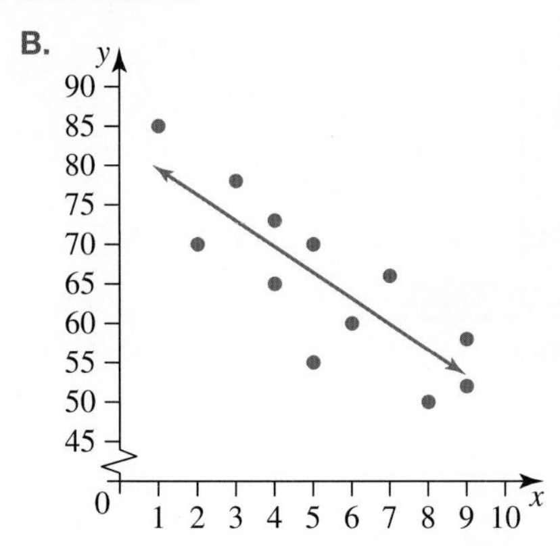

C.

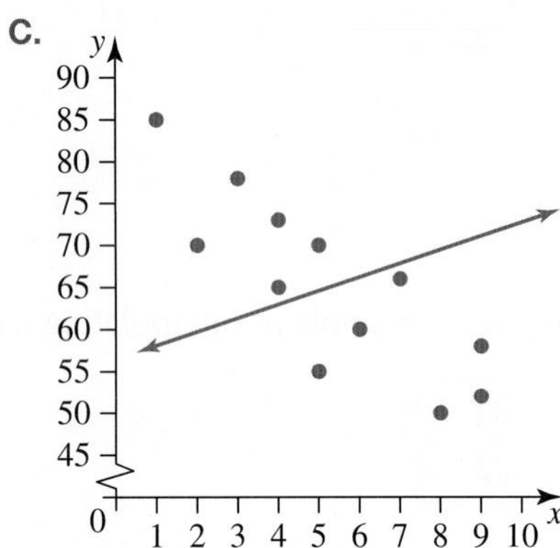

D.

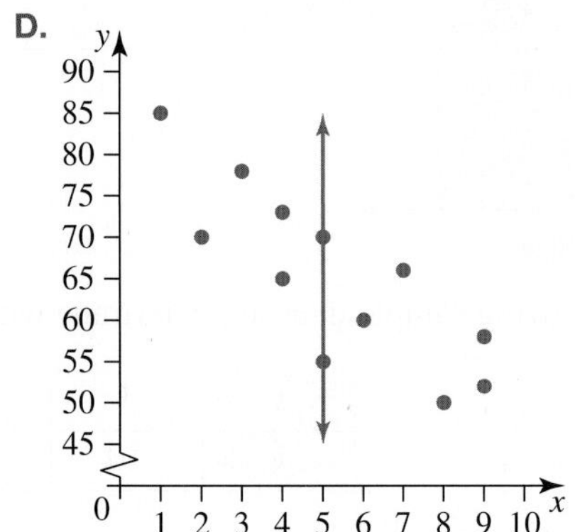

E.

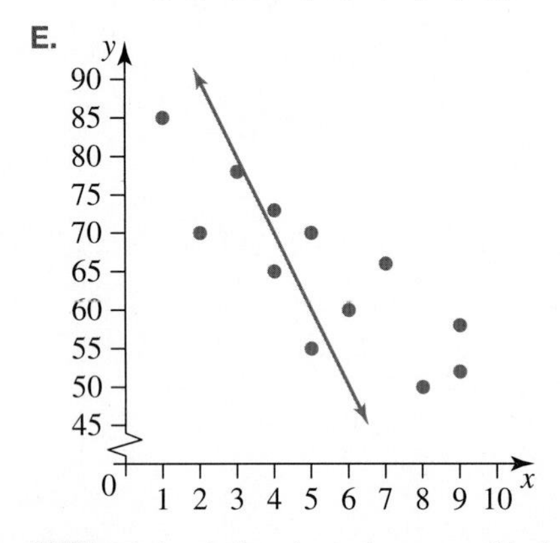

10. MC Of the following, the most likely graph for the regression line equation $y = 85 - 4x$ is:

A.

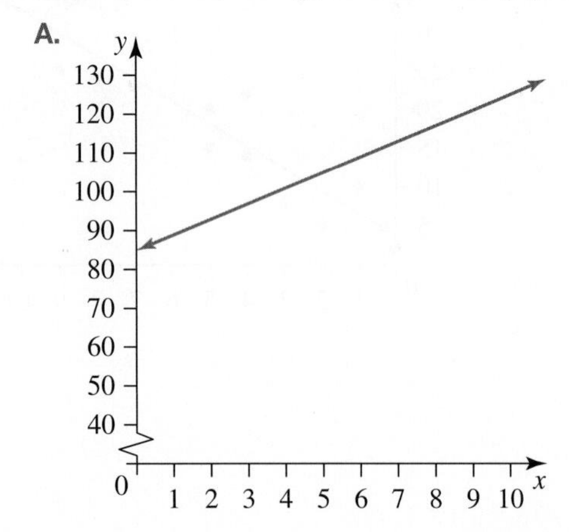

B.

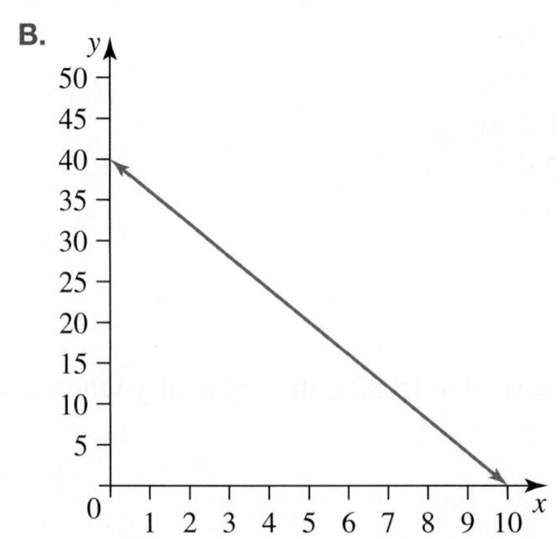

C.

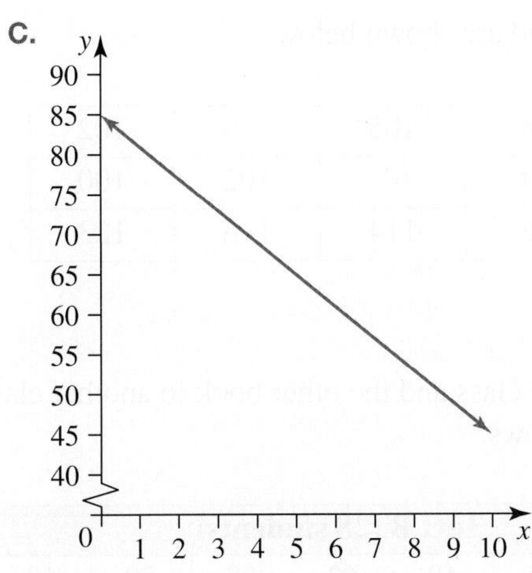

D.

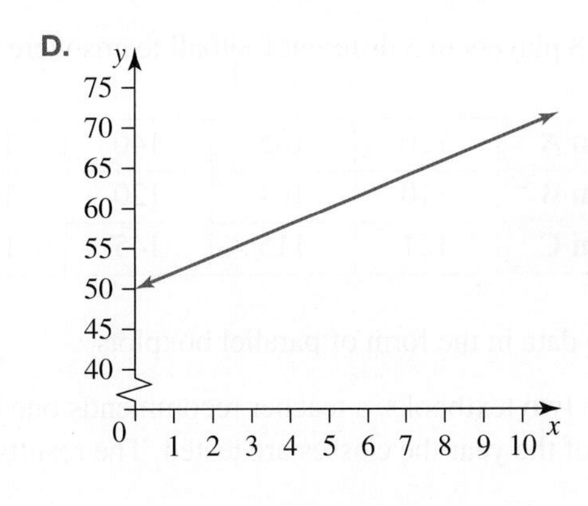

E.

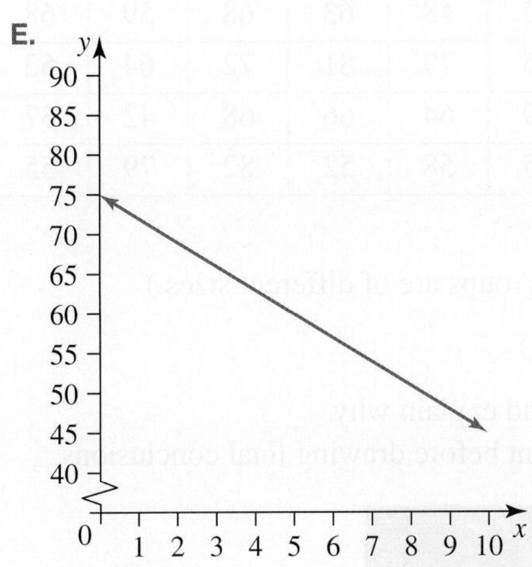

Short answer

11. The number of goals scored by two members of a basketball team during 10 matches is as follows.

Player 1	6	8	12	9	5	15	2	1	10	24
Player 2	6	8	7	8	7	5	8	6	7	7

a. Draw parallel boxplots for this data.
b. State which player you would rather have on your team and why.

12. The number of hours of counselling received by a group of 9 full-time firefighters and 9 volunteer firefighters after a serious bushfire is given in the table.

Full-time	2	4	3	5	2	4	6	1	3
Volunteer	8	10	11	11	12	13	13	14	15

a. Construct a back-to-back stem plot to display the data.
b. Comment on the distributions of the number of hours of counselling of the full-time firefighters and the volunteers.

13. The IQs of 8 players in 3 different football teams were recorded and are shown below.

Team A	120	105	140	116	98	105	130	102
Team B	110	104	120	109	106	95	102	100
Team C	121	115	145	130	120	114	116	123

Display the data in the form of parallel boxplots.

14. To compare two textbooks, a teacher recommends one book to one class and the other book to another class. At the end of the year the classes are tested. The results are as follows.

Text A (25 students)							Text B (28 students)						
44	52	95	76	13	94	83	65	72	48	63	68	59	68
72	55	81	22	25	64	72	62	75	79	81	72	64	53
35	48	56	59	84	98	84	58	59	64	66	68	42	37
21	35	69	28				39	55	58	52	82	79	55

a. Prepare a back-to-back stem plot of the data.
b. Prepare a five-number summary for each group. (Note that the groups are of different sizes.)
c. Prepare parallel boxplots of the data.
d. Compare the performance of each of the classes.
e. Determine which textbook students have better knowledge of and explain why.
f. Determine what other things you would need to take into account before drawing final conclusions.

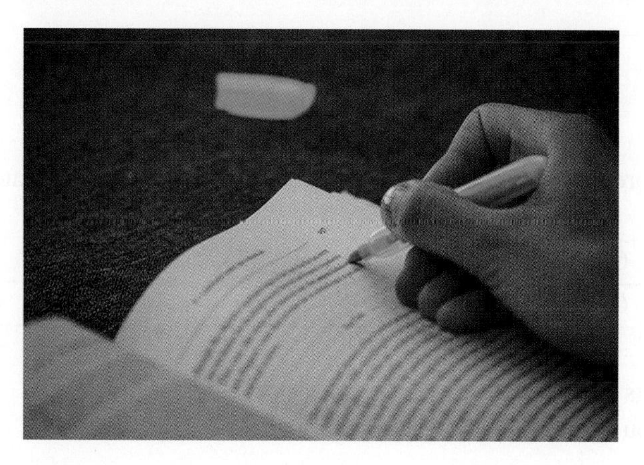

Extended response

15. A driver tracks how many kilometres (km) they travelled against the amount of fuel they used in litres (L). The data is shown in the table.

Distance travelled (km)	38	66	95	167	181	207	275	459
Fuel used (L)	3.2	5.6	7.2	12.8	14.1	16.9	21.4	37.5

a. State the response and explanatory variables.
b. Sketch a scatterplot of distance travelled against fuel used.
c. Draw a line of best fit.
d. State the equation of the line of best fit in terms of the variables. Give values correct to 3 decimal places.
e. Using your equation in part d, estimate the fuel used when travelling 50 km.
f. Explain the reliability of your answer in part e.

16. The weight of top-brand runners was tracked against the recommended retail price, and the results were recorded in the following scatterplot.

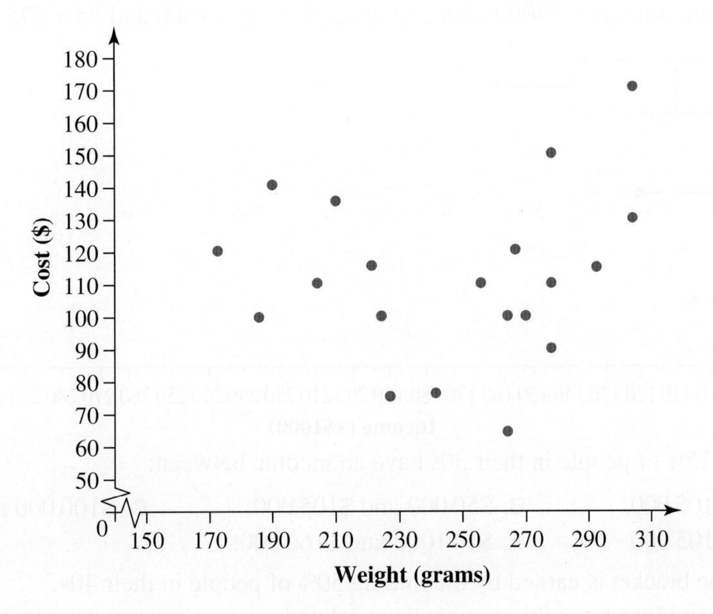

a. Identify the explanatory variable for this situation.
b. Identify two external factors that could explain the distribution of the data points.

17. The Bureau of Meteorology records data such as maximum temperatures and solar exposure on a daily and monthly basis. The following data table, for the Royal Botanic Gardens in Melbourne, shows the monthly average amount of solar energy that falls on a horizontal surface, and the monthly average maximum temperature. (*Note:* The data values have been rounded to the nearest whole number.)

Month	Jan	Feb	Mar	Apr	May	Jun	Jul	Aug	Sep	Oct	Nov	Dec
Average solar exposure (MJ)	25	21	17	11	8	6	7	10	13	18	21	24
Average max daily temperature (°C)	43	41	34	33	24	19	24	24	28	32	25	40

a. Identify the explanatory and response variables for this situation.
b. Using technology, display the data in the form of a scatterplot.
c. Draw the regression line for this data and write its equation in terms of the variables.
d. Using your equation, calculate the amount of solar exposure for a monthly maximum temperature of 37 °C.
e. Extrapolate the data to find the average maximum temperature expected for a month that recorded an average solar exposure of 3 MJ.
f. Explain why part **e** is an example of extrapolation.

18. A large company completed a survey of the incomes of their employees. The incomes were recorded for people in their 30s, 40s and 50s. The results are summarised in the parallel boxplots below.
Note: The incomes are written in $1000s; for example, 110 = $110 000 and 75 = $75 000.

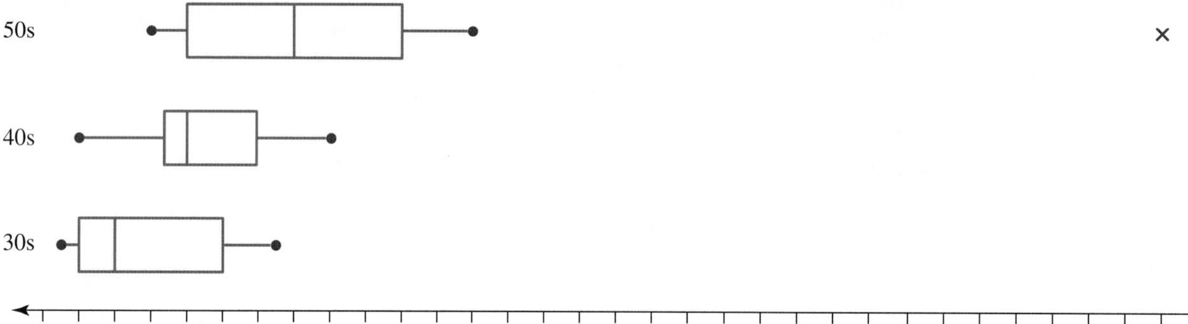

a. MC The highest 25% of people in their 30s have an income between:

A. $60 000 and $105 000.
B. $50 000 and $105 000.
C. $100 000 and $120 000.
D. $90 000 and $105 000.
E. $140 000 and $160 000.

b. State what income bracket is earned by the middle 50% of people in their 40s.
c. Explain how age and income at this company are related.
d. One of the employees surveyed at the company earns $46 000. Determine which age category they belong to.
e. The person in their 50s who earns $350 000 was demoted and now earns $190 000. Determine if that person's income is still an outlier. Justify your response.

19. The following table shows some key nutritional information about a sample of fruits and vegetables.

Food	Calcium (mg)	Serve weight (g)	Water (%)	Energy (kcal)	Protein (g)	Carbohydrates (g)
Avocado	19	173	73	305	4.0	12.0
Blackberries	46	144	86	74	1.0	18.4
Broccoli	205	180	90	53	5.3	10
Cantaloupe	29	267	90	94	2.4	22.3
Carrots	19	72	88	31	0.7	7.3
Cauliflower	17	62	92	15	1.2	2.9
Celery	14	40	95	6	0.3	1.4
Corn	2	77	70	83	2.6	19.4
Cucumber	4	28	96	4	0.2	0.8
Eggplant	10	160	92	45	1.3	10.6
Lettuce	52	163	96	21	2.1	3.8
Mango	21	207	82	135	1.1	35.2
Mushrooms	2	35	92	9	0.7	1.6
Nectarines	6	136	86	67	1.3	16.0
Peaches	4	87	88	37	0.6	9.6
Pears	19	166	84	98	0.7	25.1
Pineapple	11	155	86	76	0.6	19.2
Plums	10	95	84	55	0.5	14.4
Spinach	55	56	92	12	1.6	2.0
Strawberries	28	255	73	245	1.4	66.1

a. Use a form of technology to convert the water data into its equivalent weight in grams.
b. Compare the data for serve weight with your data for the weight of the water content using parallel boxplots.
c. Comment on the parallel boxplots from part b.
d. Use a form of technology to compare the data for protein and carbohydrates using parallel boxplots.
e. Comment on the parallel boxplots from part d.

20. The average maximum temperature (in °C) in Victoria for two 20-year time periods is shown in the following tables.
Time period (2002–2021):

Year	2002	2003	2004	2005	2006	2007	2008	2009	2010	2011
Temp.	22.3	22.6	21.8	22.1	22.4	22.7	22.1	22.7	23.1	23.1

Year	2012	2013	2014	2015	2016	2017	2018	2019	2020	2021
Temp.	22.7	23.4	23.4	23.1	22.7	22.1	22.9	22.6	22.6	22.7

Time period (1892–1911):

Year	1892	1893	1894	1895	1896	1897	1898	1899	1900	1901
Temp.	20.5	20.9	21.0	20.9	21.5	21.3	21.0	20.9	21.0	21.0

Year	1902	1903	1904	1905	1906	1907	1908	1909	1910	1911
Temp.	20.7	20.9	20.9	21.5	21.4	21.3	21.1	21.4	21.3	21.4

a. Use a form of technology to display the data:
  i. for the period 2002–2021 as a histogram using intervals of 0.4 °C commencing with the data value 20 °C
  ii. for the period 1892–1911 as a histogram using intervals of 0.4 °C commencing with the data value 20 °C.
b. Describe each display and comment on the differences between the two data sets.

Answers

Topic 7 Comparing data sets and long-term prediction

7.2 Constructing boxplots

7.2 Exercise

1. a. 12 b. 5 c. 26
2. a. 122 b. 6 c. 27
3. a. 49 b. 5.8 c. 18.6
4. a. 140 points b. 56 points c. 90 points
   d. 84 points e. 26 points
5. a. 58 b. 31 c. 43
   d. 27 e. 7
6. C
7. C
8. D
9. a. 22, 28, 35, 43, 48
   b.

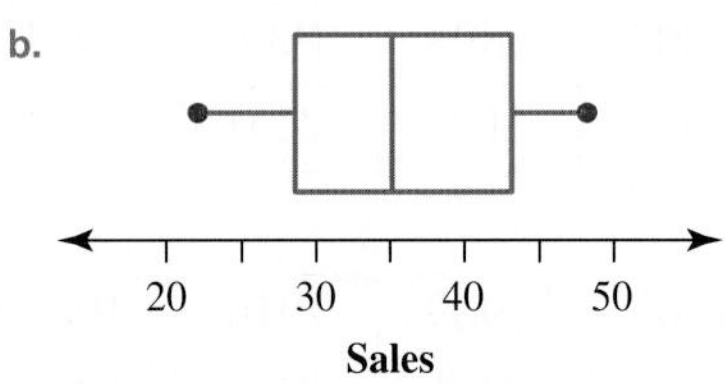

10. a. 10, 13.5, 22, 33.5, 45
    b.

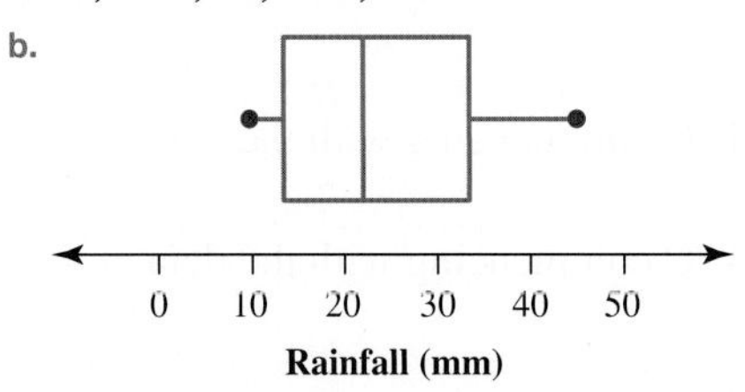

11. a. 18, 20, 26, 43.5, 74
    b.

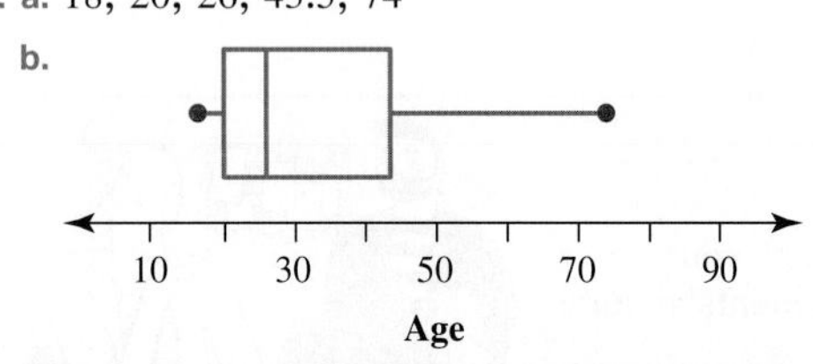

c. The data is positively skewed with no outliers. The average age of offenders, as measured by the median, is 26. The spread of ages, as measured by the range, is 56.

12. a. 324 000, 335 000, 348 000, 357 000, 375 000
    b.

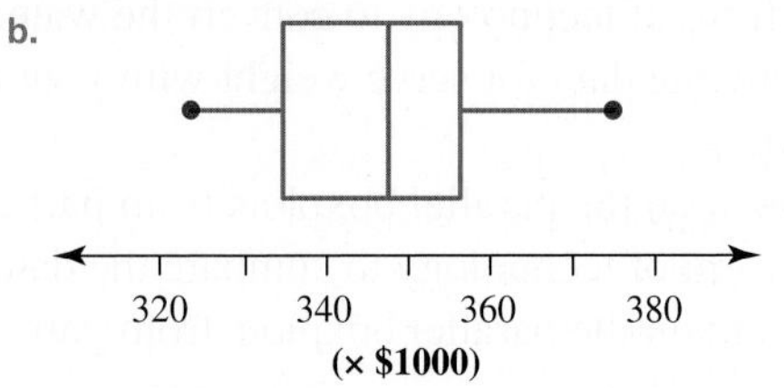

13. a. Key: 12|1 = 121

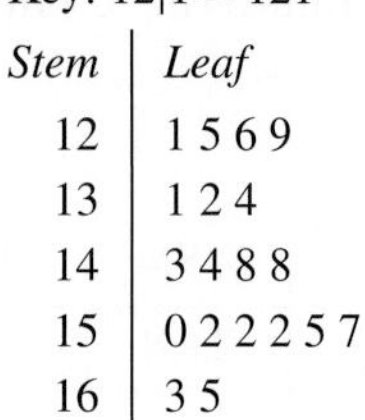

Stem	*Leaf*
12	1 5 6 9
13	1 2 4
14	3 4 8 8
15	0 2 2 2 5 7
16	3 5
17	2 9
18	1 1 1 2 3 7 8

b.

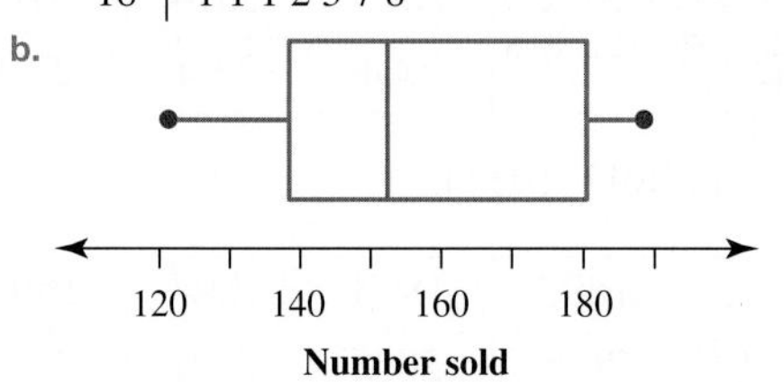

14. a. Key: 1*|7 = 17 years

Stem	*Leaf*
1*	7 7 8 8 8 9 9
2	0 0 0 1 2 2 2 2 3 3 3 3 4 4 4
2*	5 5 8 9
3	1 2 3
3*	
4	
4*	8

b.

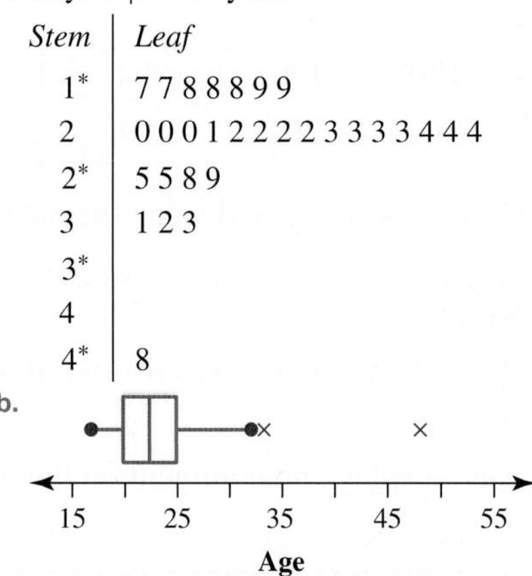

c. Data is positively skewed, with outliers at 33 and 48. The average age to have a first baby, as measured by the median, is 22.5 years of age. 50% of mothers have their first baby between the ages of 20 and 25.

7.3 Comparing parallel boxplots

7.3 Exercise

1. 17, 21, 33, 47, 58
2. A
3. See the figure at the bottom of the page.*

*3

Weight of boxes

0 1 2 3 4 5 6 7 8 9 10 11 12 13 14 15 16 17 18 19 20 21 22 23 24 25 26 27 28 29 30 31 32 33 34 35

Weight (kg)

4. See the figure at the bottom of the page.*
5. See the figure at the bottom of the page.*
6. a. See the figure at the bottom of the page.*
   b. The distribution of Greek food was positively skewed with no outliers. The distributions for Italian and Spanish food were approximately symmetric with no outliers. The parallel boxplots show that the median number of patrons who preferred Greek cuisine in month 1 (56) was higher than the median of those who preferred Italian in month 2 (32.5) and Spanish in month 3 (40).
   The IQR was the smallest in month 1 (14.5) and largest in month 2 (45). The range was also the smallest in month 1 (27) compared with the range in month 2 (59) and month 3 (28).
   Overall, patrons preferred Greek food over Italian and Spanish food, as month 1 (Greek) had a larger median and smaller measures of spread compared to the other two months. The Greek menu will therefore be the most successful.
7. B
8. B
9. A
10. a. The data set is likely to be asymmetric.
    b. The data are more clustered between the median and upper quartile than between the median and the lower quartile. The data is likely to be negatively skewed.
    c. If the median is halfway between the upper and lower quartiles, the data set has a symmetric distribution; if it is not, then the data set is asymmetric.
11. a. Peanut butter: 2, 16, 19, 21, 26
    Nutella: 4, 7, 11, 13, 25
    b. i. 16
       ii. 7 and 13
    c. The distribution of peanut butter sales was negatively skewed, with an outlier at 2 jars. The distribution of Nutella sales is also negatively skewed, with an outlier at 25 jars.
    The parallel boxplots show that the median weekly jar sales were much higher for peanut butter (19 jars) than for Nutella (11 jars). The IQR was approximately the same for both peanut butter and Nutella (5 and 6 jars, respectively), whereas the range was larger for peanut butter (24 jars) than for Nutella (21 jars). Overall, the average weekly jar sales for peanut butter were greater than for Nutella.
12. a.

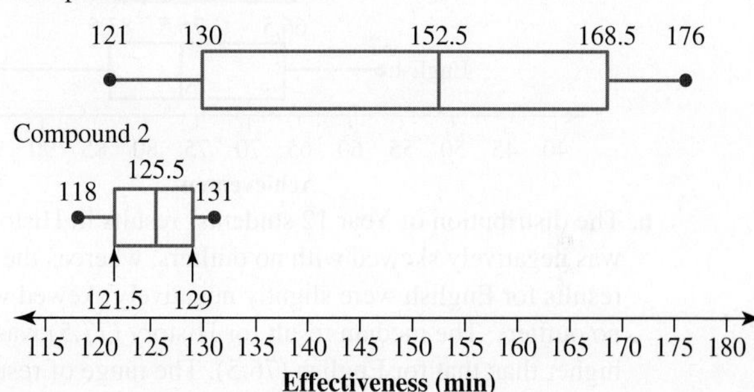

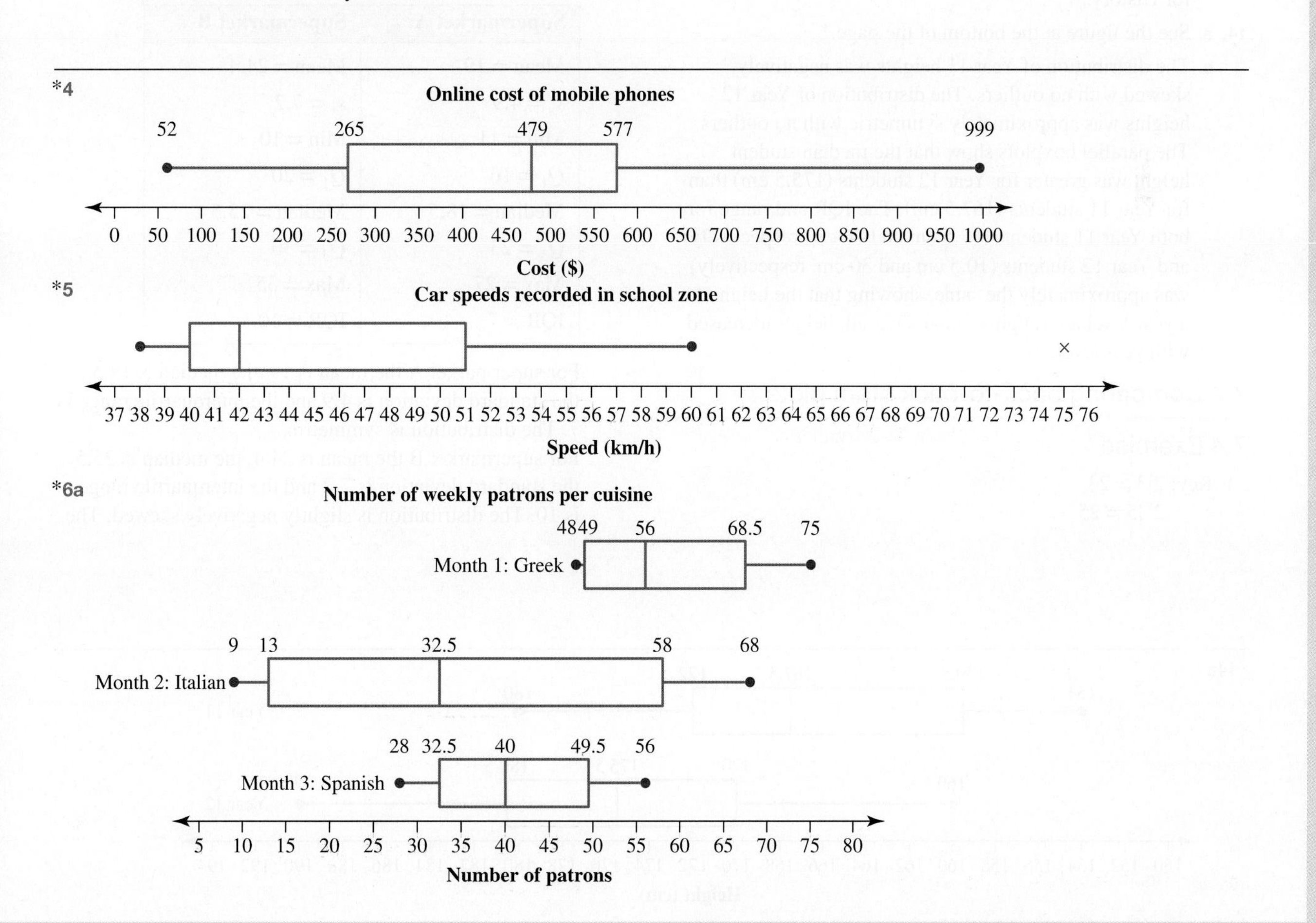

b. The distribution for compound 1 was slightly negatively skewed, with no outliers. Distribution for compound 2 was approximately symmetric, with no outliers. The parallel boxplots show that the median of minutes of effectiveness of the compounds in water was much higher for compound 1 (152.5 minutes) than for compound 2 (125.5 minutes). The IQR and ranges were also much higher for compound 1 (38.5 minutes and 55 minutes, respectively), showing more variation in minutes of effectiveness than compound 2 (7.5 minutes and 13 minutes, respectively). Overall, compound 1 was more effective in water before needing to be reapplied than compound 2.
However, the consistency of compound 2 gives the user a better indication of when they will have to reapply.

13. a.

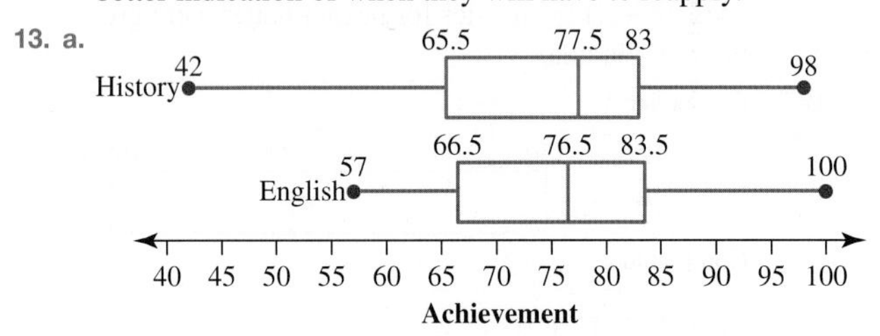

b. The distribution of Year 12 students' results in History was negatively skewed with no outliers, whereas the results for English were slightly negatively skewed with no outliers. The median result for History (77.5) was higher than that for English (76.5). The range of results was larger for History (56) than English (43), indicating the results for English were more consistent than those for History.

14. a. See the figure at the bottom of the page.*

b. The distribution of Year 11 heights was negatively skewed with no outliers. The distribution of Year 12 heights was approximately symmetric with no outliers. The parallel boxplots show that the median student height was greater for Year 12 students (175.5 cm) than for Year 11 students (167.5 cm). The IQR and range for both Year 11 students (12.5 cm and 26 cm, respectively) and Year 12 students (10.5 cm and 30 cm, respectively) was approximately the same, showing that the heights in a year level are roughly equal. Overall, height increased with year level.

7.4 Comparing back-to-back stem plots

7.4 Exercise

1. Key: $2|3 = 23$
$2^*|5 = 25$

German		*French*
2 1 1 0	2	3 4
7 6 5 5	2^*	5 5 8
3 2 1 0 0	3	0 1 4 4
9 8 7 7	3^*	5 6 8 8 9
2 1	4	2 3 4 4
5	4^*	6 8

2. Key: $2|3 = 2.3$ kg
$2^*|7 = 2.7$ kg

Hospital A		*Hospital B*
	2^*	6 7
4 4	3	0 1 1 2 3
8 7 6	3^*	6 7
3 2	4	0
9 8	4^*	
0	5	

3. a. Key: $2|3 = 23$ trucks
$2^*|5 = 25$ trucks

A		*B*
2 1	1	0
7 7 6 6 5	1^*	5 6
4 3 2 1 0	2	0 1 3
7 5	2^*	5 6 8 9
	3	0 1 2
	3^*	5

b. Statistical analysis

Supermarket A	Supermarket B
Mean = 19	Mean = 24.4
$s_x = 4.9$	$s_x = 7.2$
Min = 11	Min = 10
$Q_1 = 16$	$Q_1 = 20$
Median = 18.5	Median = 25.5
$Q_3 = 23$	$Q_3 = 30$
Max = 27	Max = 35
IQR = 7	IQR = 10

For supermarket A the mean is 19, the median is 18.5, the standard deviation is 4.9 and the interquartile range is 7. The distribution is symmetric.
For supermarket B the mean is 24.4, the median is 25.5, the standard deviation is 7.2 and the interquartile range is 10. The distribution is slightly negatively skewed. The

*14a

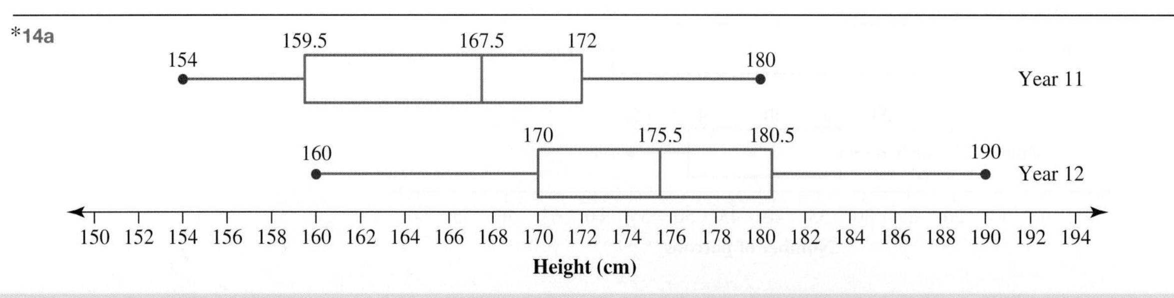

centre and spread of the distribution of supermarket B are higher than those of supermarket A. There is greater variation in the number of trucks arriving at supermarket B.

4. a. Key: 1|7 = 17 marks

Front		*Back*
	1	0
3 2	1	2 3
5 5 4 4	1	4 4 5
7 6	1	7
	1	9

b. Statistical analysis

Front	Back
Mean = 14.5	Mean = 14.25
$s_x = 1.6$	$s_x = 2.8$
Min = 12	Min = 10
$Q_1 = 13.5$	$Q_1 = 12.5$
Median = 14.5	Median = 14
$Q_3 = 15.5$	$Q_3 = 16$
Max = 17	Max = 19
IQR = 2	IQR = 3.5

For the marks of the front, the mean is 14.5, the median is 14.5, the standard deviation is 1.6 and the interquartile range is 2. The distribution is symmetric.
For the marks of the back, the mean is 14.25, the median is 14, the standard deviation is 2.8 and the interquartile range is 3.5. The distribution is symmetric.
The centre of each distribution is about the same.

5. a. Key: 2|4 = 24 marks
$2^*|6 = 26$ marks

First year		*Second year*
	2	2
	2*	6 7 8
1 0	3	0 1 1 3 4
9 7 5	3*	6
3 2 1 1	4	
6	4*	

b. Statistical analysis

First year	Second year
Mean = 38.5	Mean = 29.8
$s_x = 5.2$	$s_x = 4.2$
Min = 30	Min = 22
$Q_1 = 35$	$Q_1 = 27$
Median = 40	Median = 30.5
$Q_3 = 42$	$Q_3 = 33$
Max = 46	Max = 36
IQR = 7	IQR = 6

The distributions of marks for the first year and for the second year are symmetric.
For the first-year marks, the mean is 38.5, the median is 40, the standard deviation is 5.2 and the interquartile range is 7. The distribution is symmetric.
For the second-year marks, the mean is 29.8, the median is 30.5, the standard deviation is 4.2 and the interquartile range is 6.
The spread of each of the distributions is much the same but the centre of each distribution is quite different, with the centre of the second-year distribution quite a lot lower. The work may have become a lot harder!

6. a. Key: 2|3 = 23 years old
$2^*|5 = 25$ years old

7 am class		*8 am class*
4 3	2	2
8 7 6 5	2*	5
1 0	3	0 1
	3*	6 7
	4	2
	4*	6

b. Statistical analysis

7 am class	8 am class
Mean = 26.5	Mean = 33.6
$s_x = 2.8$	$s_x = 8.2$
Min = 23	Min = 22
$Q_1 = 24.5$	$Q_1 = 27.5$
Median = 26.5	Median = 33.5
$Q_3 = 29$	$Q_3 = 39.5$
Max = 31	Max = 46
IQR = 4.5	IQR = 12

For the distribution of the 7am class, the mean is 26.75, the median is 26.5, the standard deviation is 2.8 and the interquartile range is 4.5.
For the distribution of the 8am class, the mean is 33.6, the median is 33.5, the standard deviation is 8.2 and the interquartile range is 12.
The centre of the distributions is very different: it is much higher for the 8 am class. The spread of the 7 am class is very small but very large for the 8 am class. Older people are more likely to attend the later class than the early class.

7. a. Key: 3|2 = 32

Kindergarten		*Prep school*
3	0	5
4 3	1	2 7
8 5	2	5 7
6 2	3	2 5
7 1	4	4 6
0	5	2

b. Statistical analysis

Kindergarten	Prep school
Mean = 28.9	Mean = 29.5
$s_x = 15.4$	$s_x = 15.3$
Min = 3	Min = 5
$Q_1 = 14$	$Q_1 = 17$
Median = 30	Median = 29.5
$Q_3 = 41$	$Q_3 = 44$
Max = 50	Max = 52
IQR = 27	IQR = 27

For the distribution of scores of the kindergarten children, the mean is 28.9, the median is 30, the standard deviation is 15.4 and the interquartile range is 27.
For the distribution of scores for the prep children, the mean is 29.5, the median is 29.5, the standard deviation is 15.3 and the interquartile range is 27.
The distributions are very similar. There is not a lot of difference between the way the kindergarten children and the prep children scored.

8. B
9. C
10. a. Class 1: 25; class 2: 27
b. Class 1: 2; class 2: 3
c. Class 1: 24; class 2: 20
d. Class 1 results are clustered tightly with a range of 6 from 10 to 16 marks out of 20. Class 2 results are more widely spread, with a range of 17 from 3 to 20. 11 of 27 students scored higher than any member of class 1.
11. Key: 1|6 = 16 years

Morning		*Afternoon*
9 8 7 7 6 6	1	5 5 6 6 6 7 7 7 8 9
9 8 5 3 3 3 2 1 1 0	2	1 1 2 2 2 3 3 5 7 8
8 8 6 4 2 2 1	3	0 0 1 4 5
5 3 3	4	3 6 8
2	5	0 4 7
0	6	5

12. Both brand A and B battery life is clustered around 10–38 hours. Brand A had more batteries with a life of less than 10 hours (4 batteries) in comparison to brand B (1 battery). The median value of brand A battery life was 21 hours and for brand B it was 22 hours. From this comparison, brand B has the better battery life span.
13. a. Key: 9|69 = 9.69

Female		*Male*
	9	69
	9	84 85 87 90 92 96 99
	10	00 06 14 20 25 30 30 30 40
97 94 93 82 78 75 54	10	50 80
40 08 07 00 00	11	
90 90 60 50 50 50	11	
20	12	

b. There is not a large difference within each gender, but there is a large difference in time between the two genders.

14. a. Key: 23|1 = $231

Suburb A		*Suburb B*
7 5 5 0 0 0	20	
5 5 1 0 0	21	
5 5	22	
5 5 1	23	1 2 5
4 2 0	24	0 5 5
6	25	1 5 5 8
	26	0 5
	27	0 0 5 5
	28	0 0 5 5

b. Suburb A has a lower median rent ($215) compared to suburb B ($259). Suburb A has a positively skewed distribution for median rental price, whereas suburb B has an even distribution. The range of rental prices for suburb A is $200–256 compared to suburb B, which is $231–285.

7.5 Comparing histograms

7.5 Exercise

1. a. i.

Score	Frequency
3	1
4	2
5	2
6	1
7	3
8	2
9	2
10	2
11	0
12	1

ii.

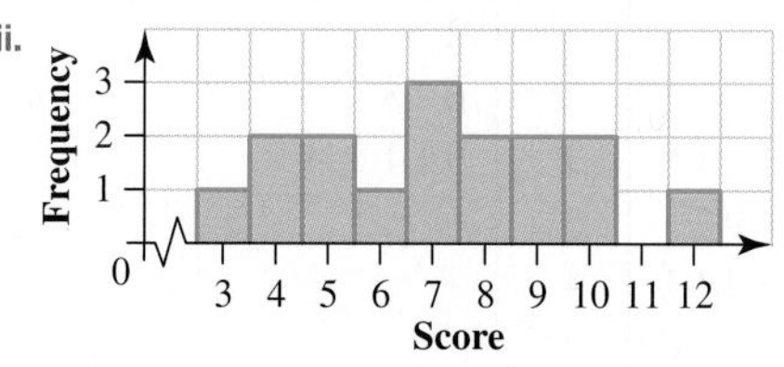

iii. Check your histograms against those shown in part ii solutions.

b. i.

Class interval	Frequency
1–<2	1
2–<3	2
3–<4	2
4–<5	6
5–<6	5
6–<7	1

ii. 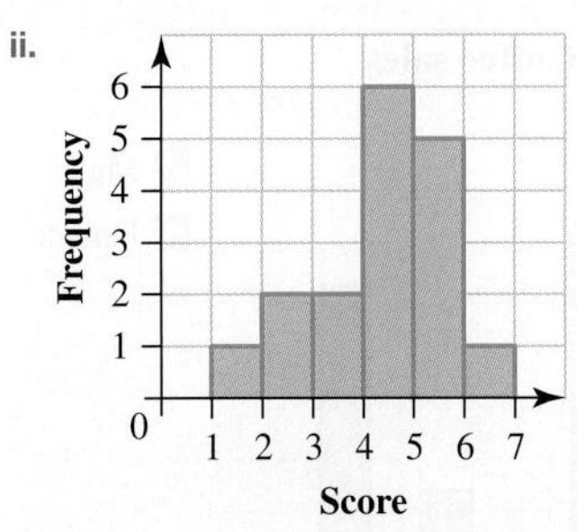

iii. Check your histograms against those shown in part ii solutions.

c. i.

Class interval	Frequency
10–< 15	3
15–< 20	9
20–< 25	10
25–< 30	10
30–< 35	10
35–< 40	1

ii. 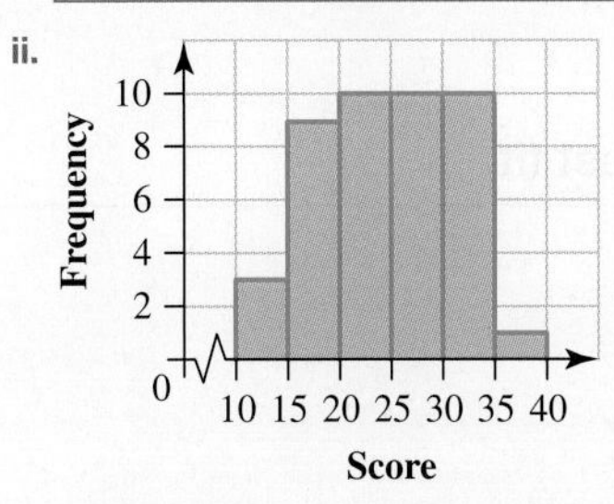

iii. Check your histograms against those shown in part ii solutions.

d. i.

Score	Frequency
0.3	1
0.4	2
0.5	1
0.6	1
0.7	1
0.8	2
0.9	2
1.0	2
1.1	1
1.2	1
1.3	1

ii. 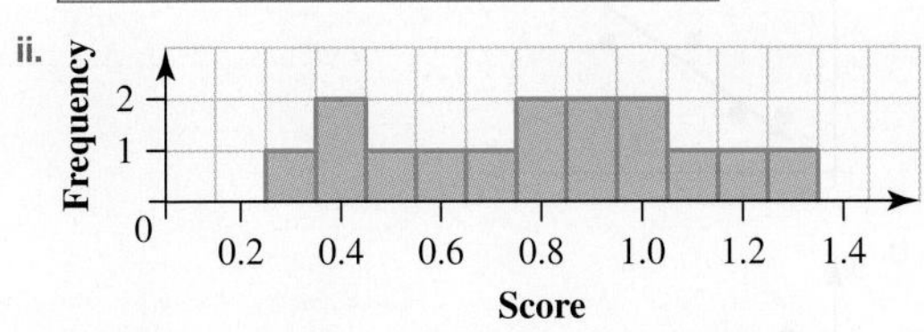

iii. Check your histograms against those shown in part ii solutions.

2. a.

Score	Tally	Frequency
30–39	|	1
40–49	~~||||~~	5
50–59	~~||||~~	5
60–69	~~||||~~ |||	8
70–79	~~||||~~ |	6
80–89	|||	3
90–99	||	2

b.

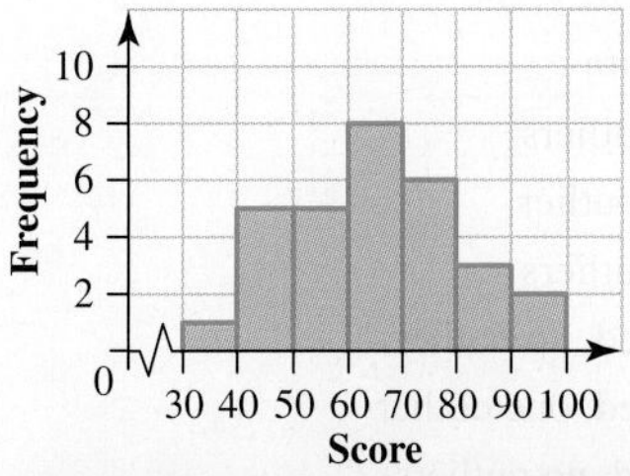

Frequency

Score

3. a.

Class	Class centre	Tally	Frequency
0.6–< 0.8	0.7	||	2
0.8–< 1.0	0.9	|||	3
1.0–< 1.2	1.1	|||	3
1.2–< 1.4	1.3	~~||||~~ |	6
1.4–< 1.6	1.5	~~||||~~ ||||	9
1.6–< 1.8	1.7	~~||||~~ |	6
1.8–< 2.0	1.9	||||	4
2.0–< 2.2	2.1	|||	3

b. 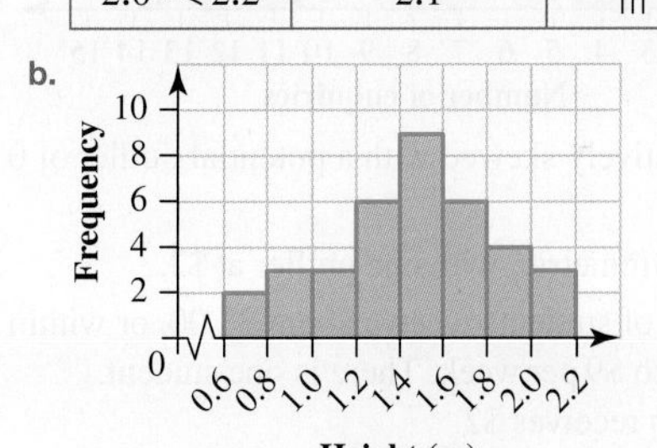

4. a.

Class	Class centre	Frequency
10.5–< 11	10.7	2
11.0–< 11.5	11.2	5
11.5–< 12	11.7	8
12.0–< 12.5	12.2	8
12.5–< 13	12.7	5
13.0–< 13.5	13.2	2

b. 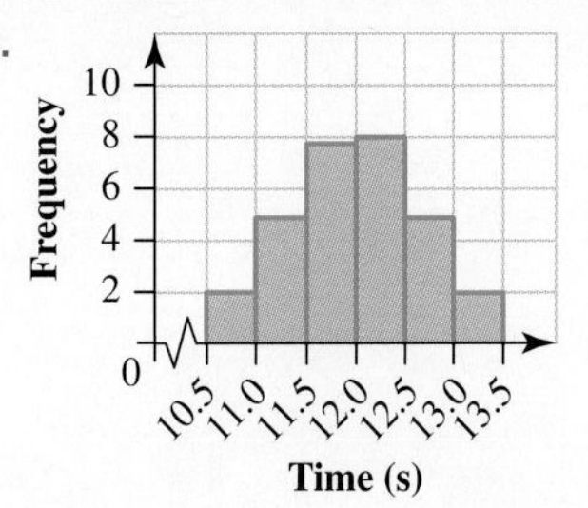

5.

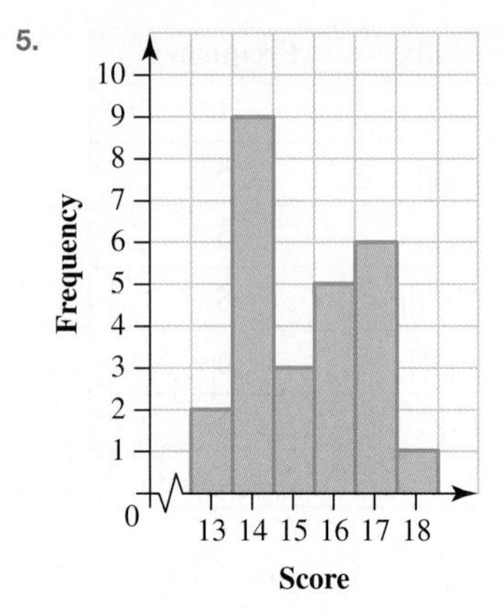

6. a. Symmetric, no outliers
   b. Symmetric, one outlier
   c. Symmetric, no outliers
   d. Negatively skewed, no outliers
   e. Negatively skewed, one outlier
   f. Positively skewed, no outliers
7. D
8. B
9.

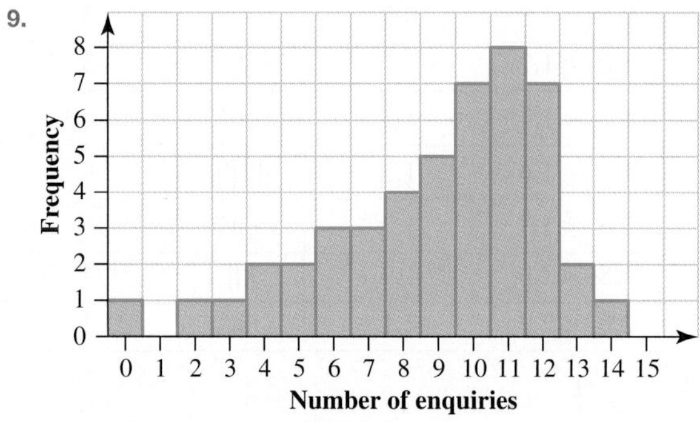

The data is negatively skewed with a potential outlier of 0 enquiries.

10. a. The data is symmetric, with one outlier at $2.
    b. The majority of students receive about $8.00, or within the range $7 to $9 per week. There is one student, however, who receives $2.
11. a. Worker F   b. 4
12. a. 2019
    b. 2020
    c. Sample responses can be found in the worked solutions in the online resources.
13. a. 500

b.

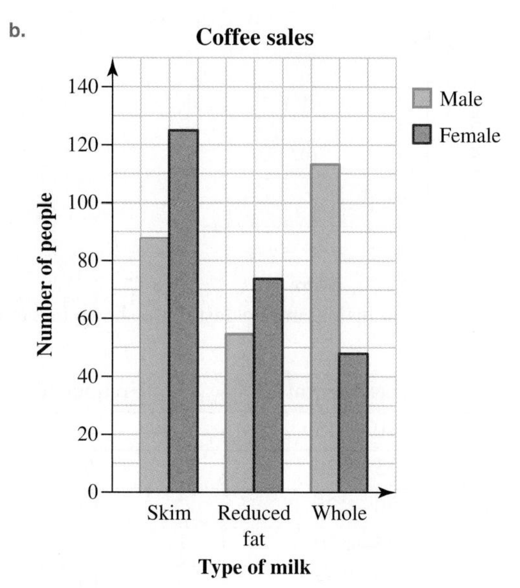

c. 34%
d. 26%
e. 49%

7.6 The line of best fit

7.6 Exercise

1.

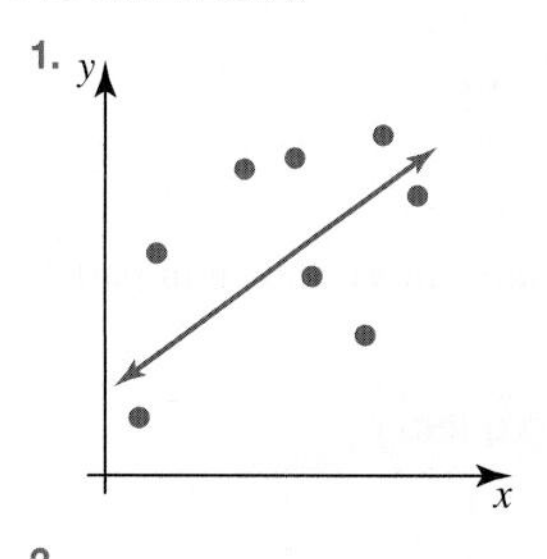

2.

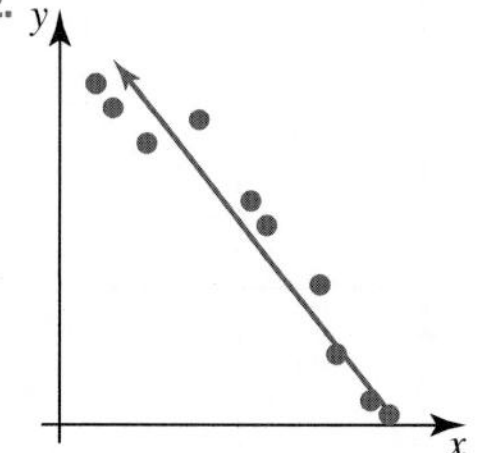

3. a.

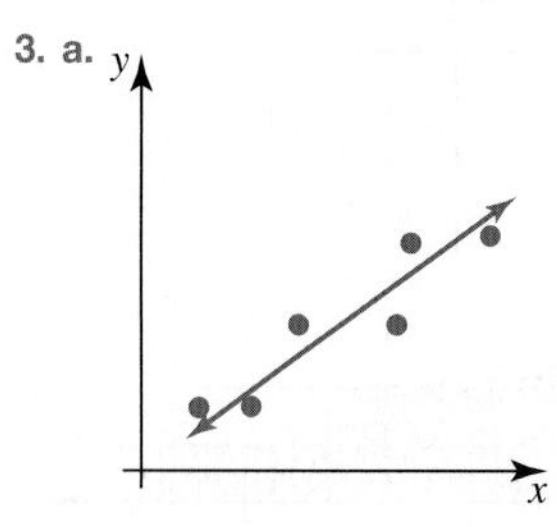

b.

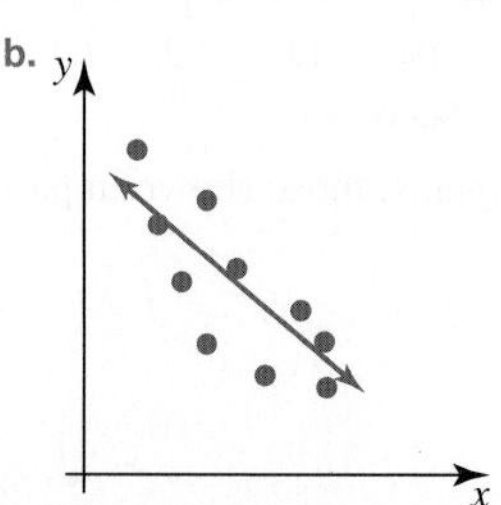

c.

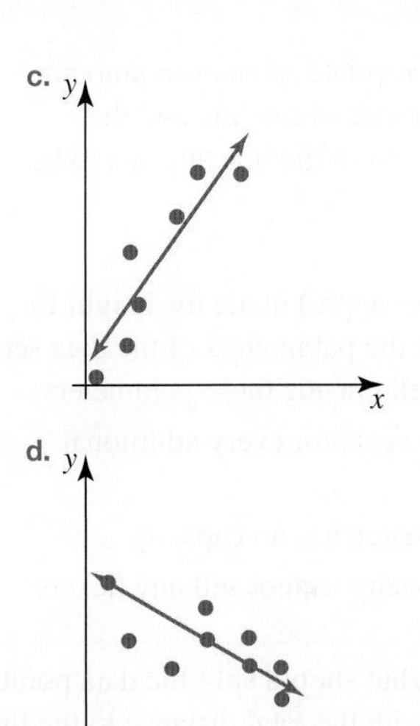

e.

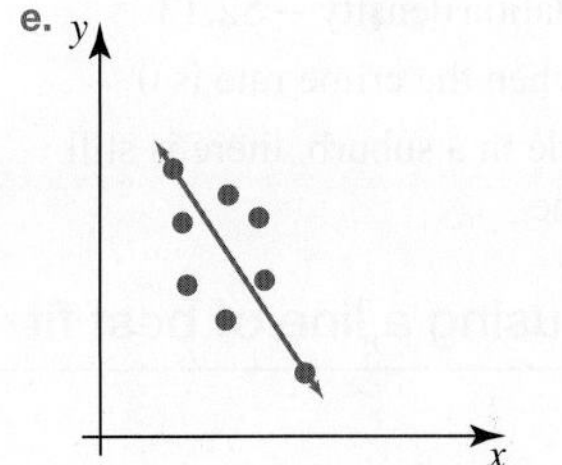

4.

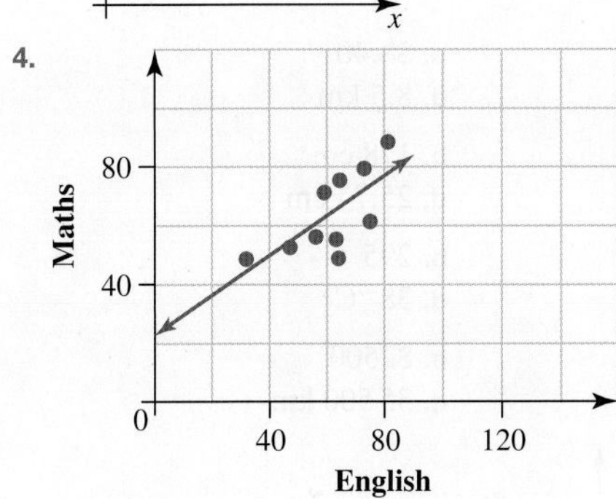

5. a.

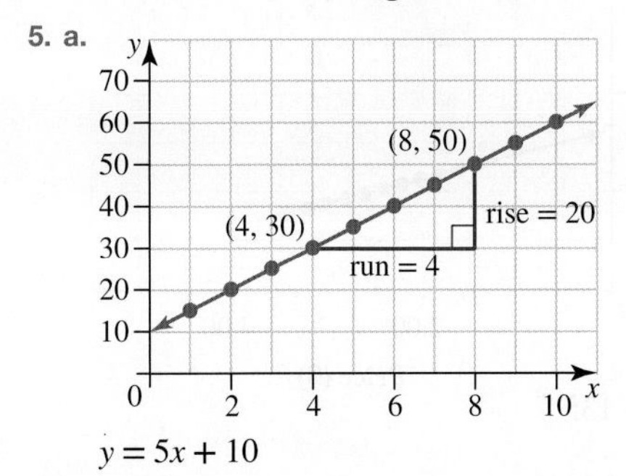

$y = 5x + 10$

b.

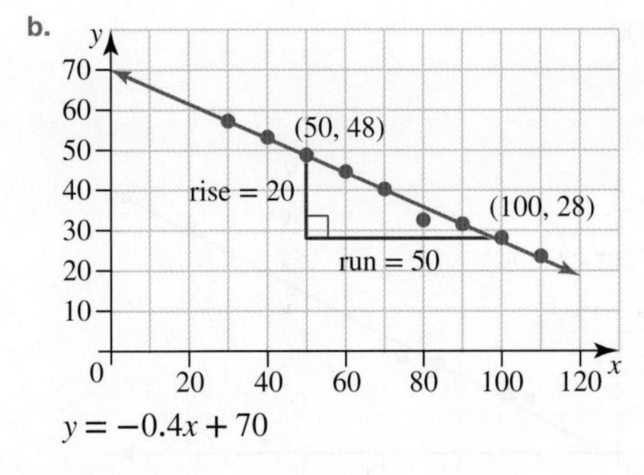

$y = -0.4x + 70$

c.

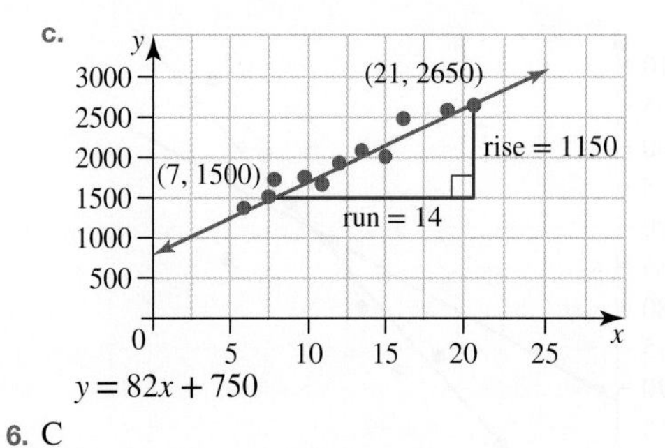

$y = 82x + 750$

6. C

7. a.

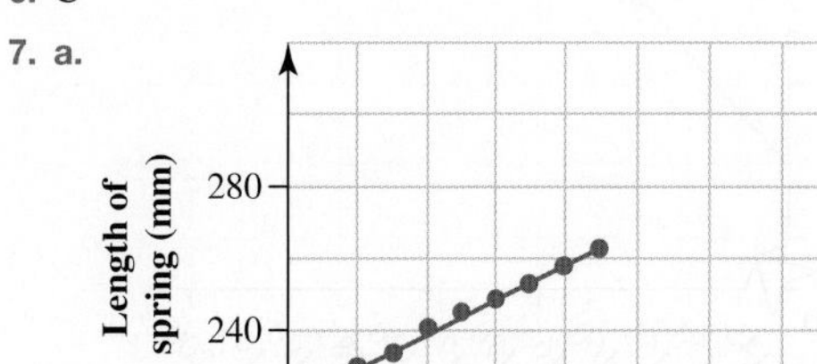

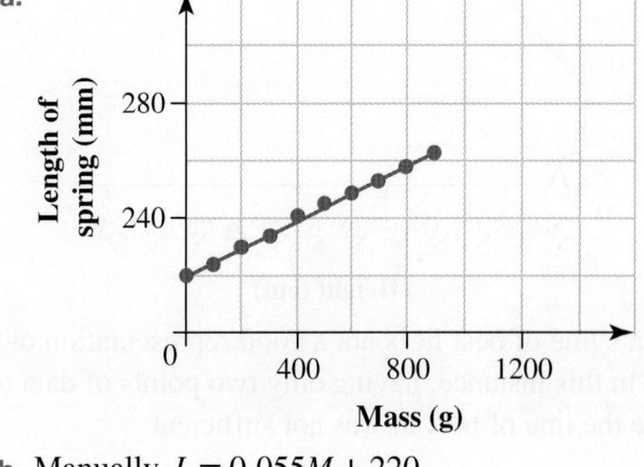

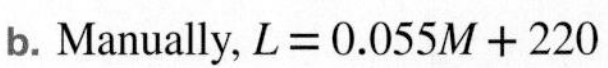

b. Manually, $L = 0.055M + 220$

8. a.

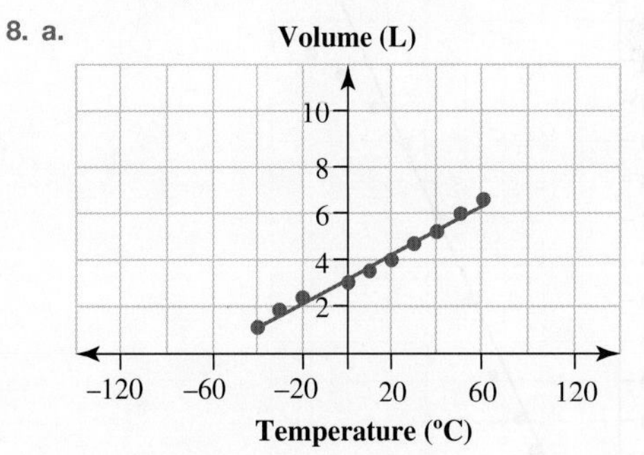

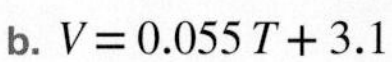

b. $V = 0.055\,T + 3.1$

9. a.

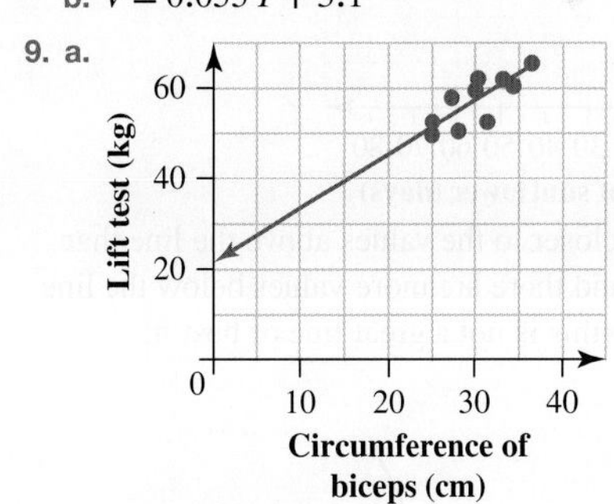

b. $S = 1.3B + 21$

10. $y = 1.25x - 33.75$

11. a. $y = 0.57x + 86.86$

b.

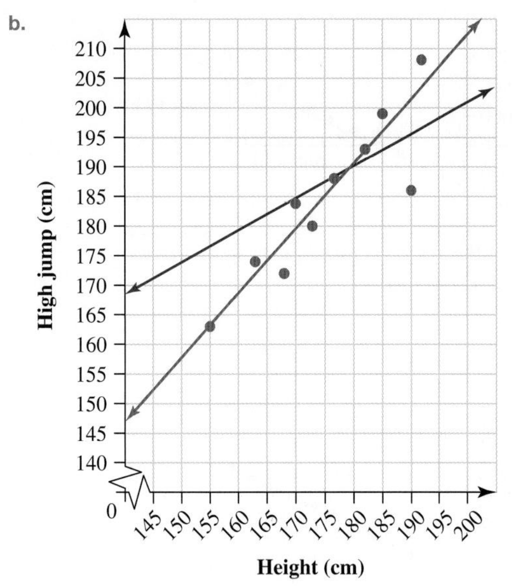

c. Nidya's line of best fit is not a good representation of the data. In this instance, having only two points of data to create the line of best fit was not sufficient.

12. a.

Xavier's line is closer to the values above the line than those below it, and there are more values below the line than above it, so this is not a great line of best fit.

b. $y = 4.4x - 28$

c.

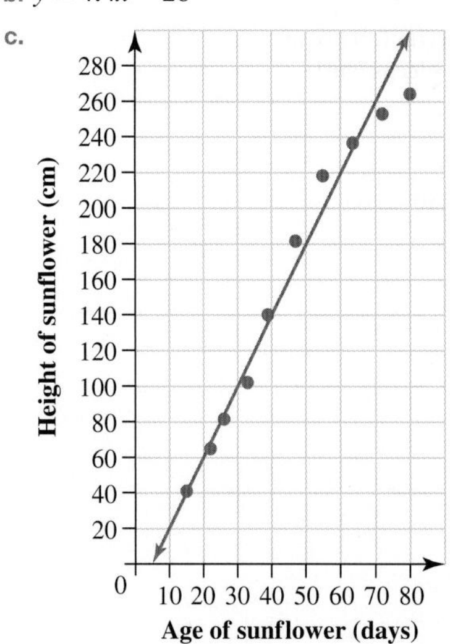

Patricia's line is more appropriate as an even amount of data points lie on either side of the line and the total distance of the points from the line appears to be minimal.

d. $y = 4x - 22$

e. The line of best fit does not approximate the height for values that appear outside the parameters of the data set and the y-intercept lies well outside these parameters.

13. a. The increase in price of 4 cents for every additional person the venue holds

b. The price of a ticket if a venue has no capacity

c. No, as a venue with 0 capacity cannot sell any tickets for $15.

14. a. Lines of best fit will vary, but should split the data points on either side of the line with the total distance to the line from the points minimised.

b. crime rate = 0.07 × population density − 52.13

c. The population density when the crime rate is 0

d. No; if there are 731 people in a suburb, there is still expected to be some crime.

7.7 Making predictions using a line of best fit

7.7 Exercise

1. a. $17.40 b. $8.40
c. 20 km d. 8.5 km

2. a. 159.7 cm b. 178 cm
c. 31.15 cm d. 25.74 cm

3. a. 755 b. 295
c. 20 °C d. 38 °C

4. a. $8300 b. $2500
c. 14 000 km d. 35 500 km

5. a–b.

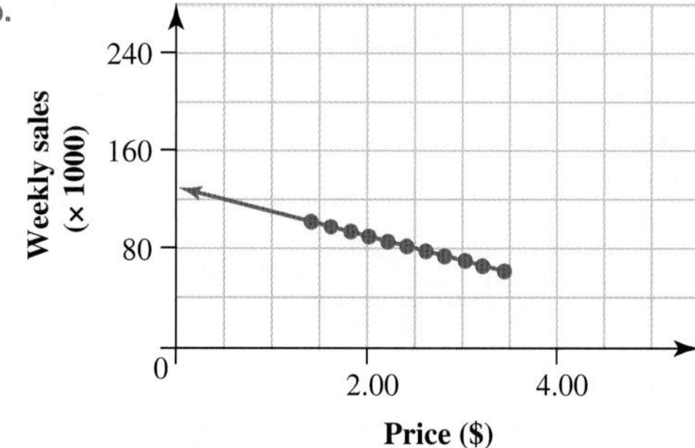

c. $N = -20p + 135$

d. i. 73 000
ii. 43 000

e. $2.75

f. $1.75

6. a.

b. $A = 0.19d + 8$

c. i. \$135 ii. \$412
iii. \$203 iv. \$634

7. a. 174
b. 60
c. The estimate in part a is more reliable as it was made using interpolation; it is located within the parameters of the original data set and appears consistent with the given data.
The estimate in part b is less reliable as it was made using extrapolation; it is located well outside the parameters of the original data set.

8. a. \$112
b. \$305
c. All estimates outside the parameters of Georgio's original data set (400 km to 2000 km) will be unreliable, with estimates further away from the data set more unreliable than those closer to the data set.
Other factors that might affect the cost of flights include air taxes, fluctuating exchange rates and the choice of airlines for various flight paths.

9. a. $y = -17.31x + 116.38$
b. For each increase in 1 L of lung capacity, the swimmers will take 17.31 seconds less to swim 25 m.
c. i. 61 seconds ii. 40.2 seconds
iii. 24.6 seconds
d. As Mariana has only two data points and we have no idea of how typical these are of the data set, the equation for the line of best fit and the estimates established from it are all very unreliable.

10. a. 76
b. 65
c. Part a looks at data within the parameters of original data set range, while part b asks to predict data outside the parameters of the original data set, with a range of 20–120 new customers each hour.

11. a.

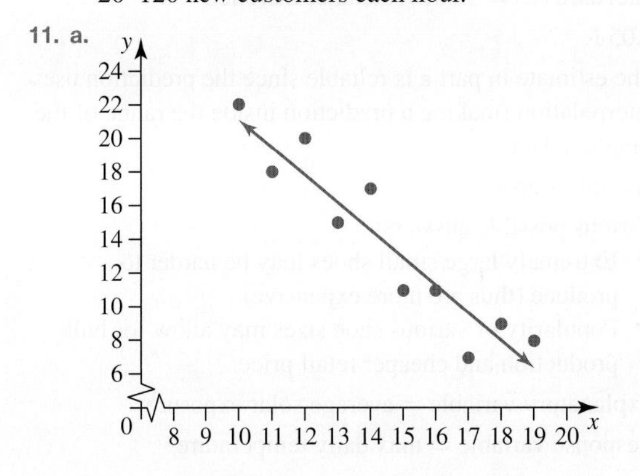

b. $y = 37.703 - 1.648x$
c. −0.201
d. It is assumed the data will continue to behave in the same manner as the data originally supplied.

12. a. 130
b. 3.6 °C
c. The location of the fire, air temperature, proximity to water, and so on

13. a.

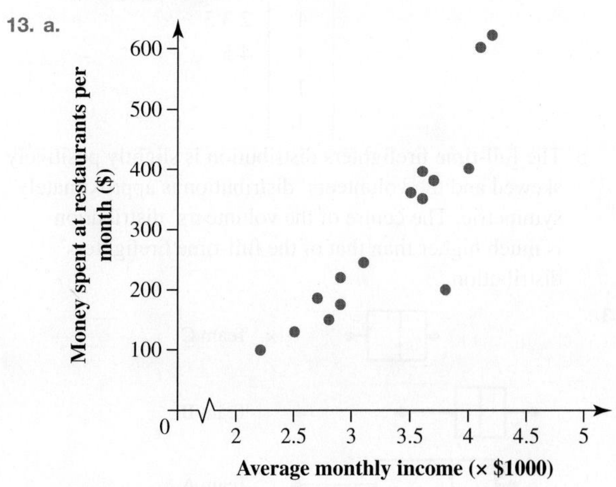

b. $R = -459.8 + 229.5I$
c. \$687.70
d. Part c is asking you to predict data outside of the original data set range.
e. To the nearest \$10, \$3160. This is an example of interpolation.

7.8 Review

7.8 Exercise

Multiple choice

1. B	2. D	3. E	4. C	5. B
6. D	7. A	8. B	9. B	10. C

Short answer

11. a. See the figure at the bottom of the page.*
b. Player 1. Higher median number of goals scored (8.5) than player 2 (7). Also, over 50% of the number of goals scored by player 1 across the ten matches lie above the maximum number of goals scored by player 2 across the 10 matches.

12. a. Highest – 15
Lowest – 1
Use stems of 1, divided into fifths.

*11a

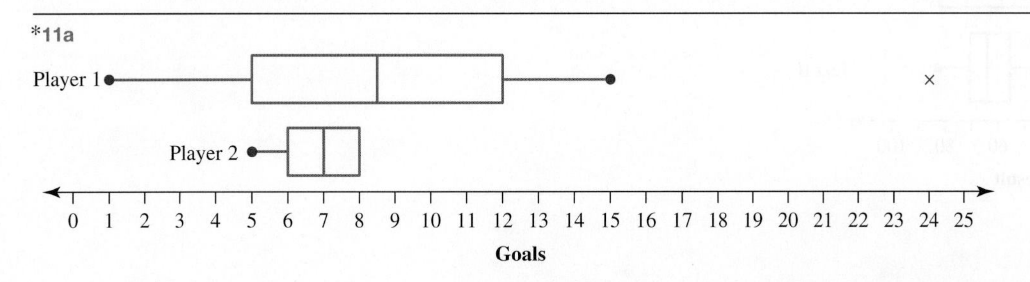

Key: 1|3 = 13 hours

Full-time		*Volunteer*
1	0	
3 3 2 2	0	
5 4 4	0	
6	0	
	0	8
	1	0 1 1
	1	2 3 3
	1	4 5
	1	
	1	

b. The full-time firefighters distribution is slightly positively skewed and the volunteers' distribution is approximately symmetric. The centre of the volunteers' distribution is much higher than that of the full-time firefighters' distribution.

13.

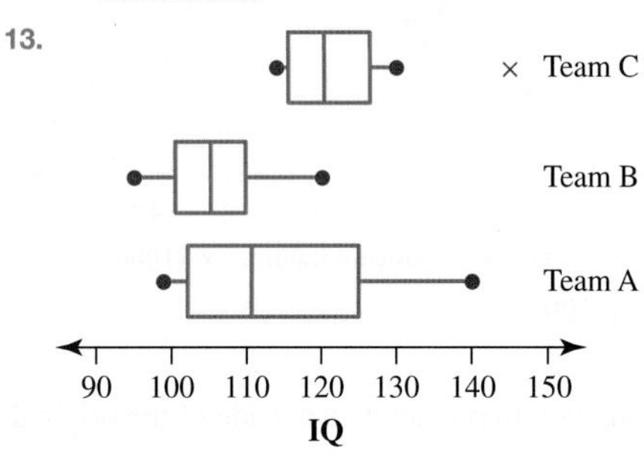

14. a. Key: 1|3 = 13

Text B		*Text A*
	1	3
	2	1 2 5 8
9 7	3	5 5
8 2	4	4 8
9 9 8 8 5 5 3 2	5	2 5 6 9
8 8 8 6 5 4 4 3 2	6	4 9
9 9 5 2 2	7	2 2 6
2 1	8	1 3 4 4
	9	4 5 8

b. Text A: $n = 25$

Min	Q_1	Median	Q_3	Max
13	35	59	82	98

Text B: $n = 28$

Min	Q_1	Median	Q_3	Max
37	55	63.5	70	82

c.

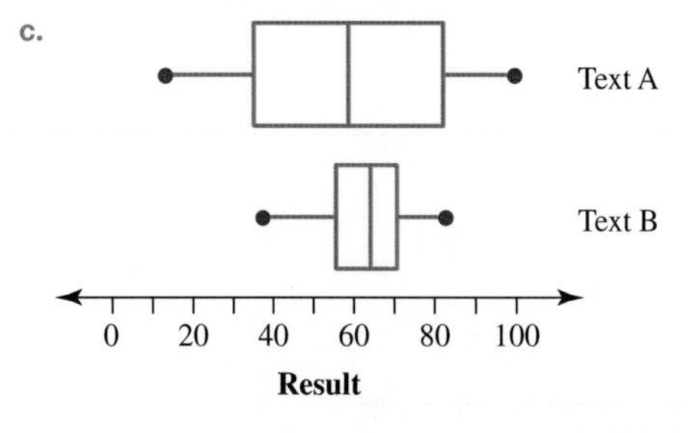

d. Performance on test. Students who used Text A had varied results while the students who used Text B were more consistent and they had a higher median score.
Text A: range = 85, IQR = 47, median = 59
Text B: range = 45, IQR = 15, median = 63.5

e–f. Text B had a higher median mark and more consistency/lower spread with a lower range and IQR. The other factors to take into account are the amount of time studying per class, the average mark using the same book per class, etc.

Extended response

15. a. Explanatory variable: distance travelled
Response variable: fuel used

b.

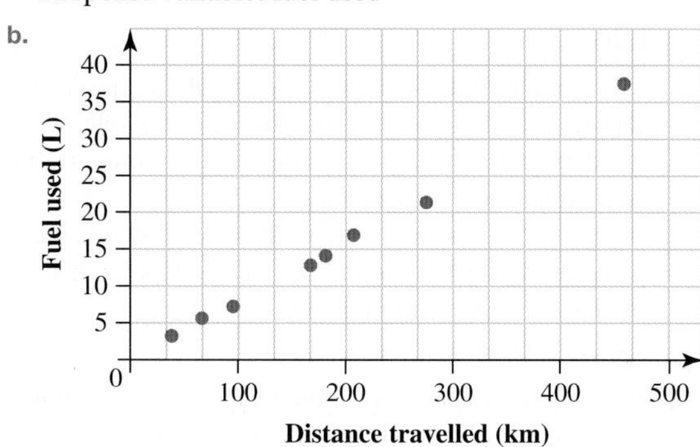

c.

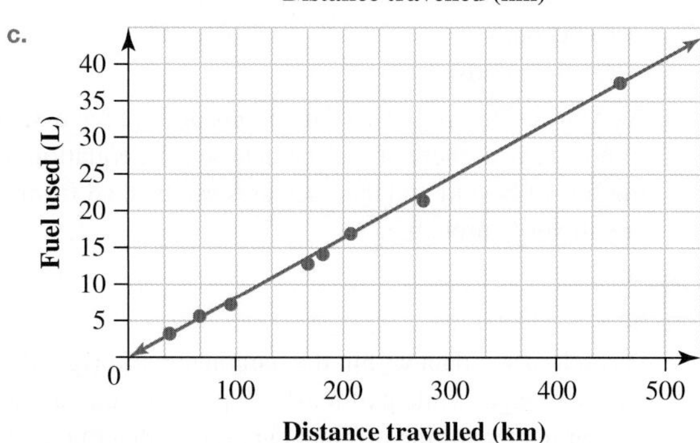

d. Fuel used (L) = 0.081 × distance travelled

e. 4.05 L

f. The estimate in part **e** is reliable since the prediction uses interpolation (making a prediction inside the range of the supplied data).

16. a. Weight (grams)

b. Various possible answers:
- Extremely large/small shoes may be harder to produce (thus are more expensive).
- Popularity of various shoe sizes may allow for bulk production and cheaper retail price.

17. a. Explanatory variable = average solar exposure
Response variable = max daily temperature

b.

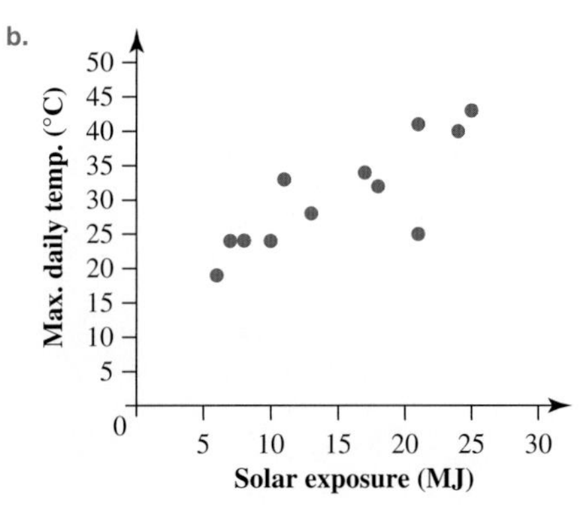

c.

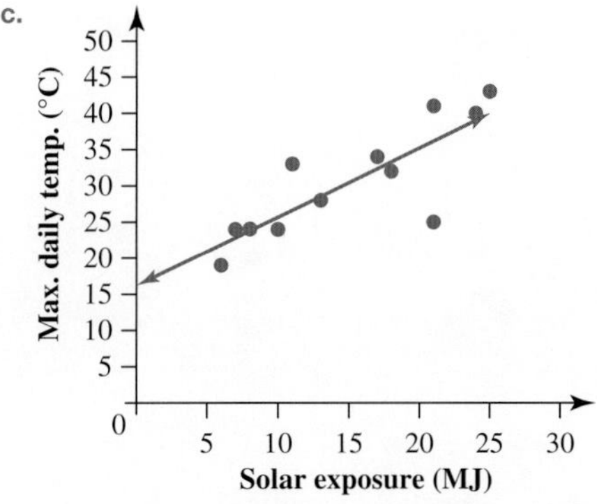

Max daily temperature = 16.232 + 0.9515 × average solar exposure

d. 22

e. 19

f. An x-value of 3 MJ is outside the parameters of the original data set.

18. a. D

b. Between $75 000 and $100 000

c. Generally, salary increases with age.

d. 30s

e. $190 000 is lower than the upper fence, so it is no longer an outlier.

19. a.

Food	Serve weight (g)	Water (%)	Water weight (g)
Avocado	173	73	126
Blackberries	144	86	124
Broccoli	180	90	162
Cantaloupe	267	90	240
Carrots	72	88	63
Cauliflower	62	92	57
Celery	40	95	38
Corn	77	70	54
Cucumber	28	96	27
Eggplant	160	92	147
Lettuce	163	96	156
Mango	207	82	170
Mushrooms	35	92	32
Nectarines	136	86	117
Peaches	87	88	77
Pears	166	84	139
Pineapple	155	86	133
Plums	95	84	80
Spinach	56	92	52
Strawberries	255	73	186

b.

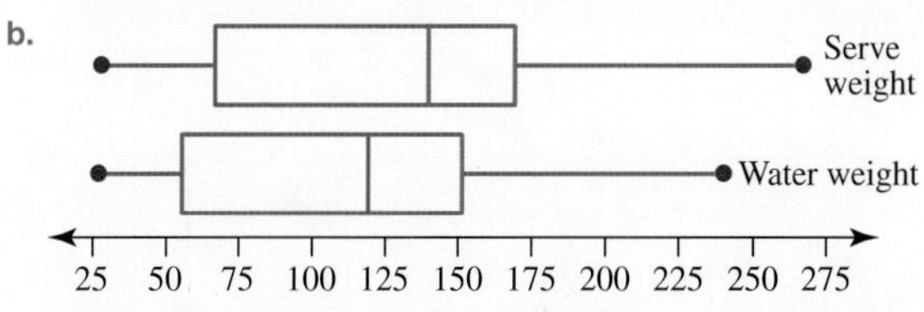

c. The boxplots appear to indicate that there are only slight differences between the serve weights and water weights of the samples. The distributions are very similar in shape, with the water weights being slightly smaller overall.

d. See the figure at the bottom of the page.*

e. Carbohydrate for this sample of foods is much greater but more variable than protein, as indicated by the larger range and IQR. The protein amounts are all less than the Q_1 for the carbohydrate amounts, with the exception of two upper outliers for protein. The distribution of protein amounts is slightly positively skewed compared to the distribution of carbohydrate amounts, which is approximately symmetrical.

*19d

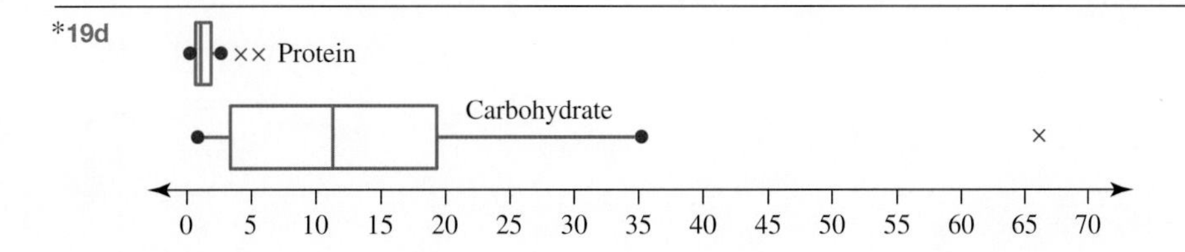

20. a. i.

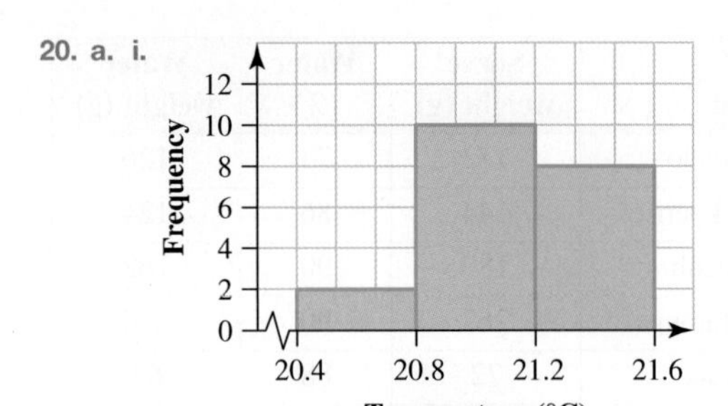

ii.

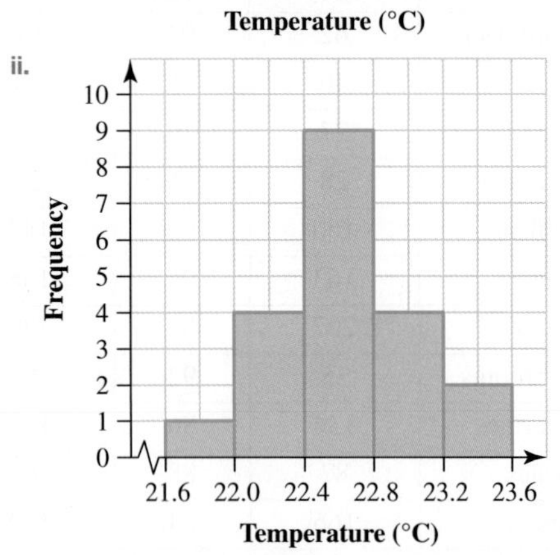

b. The distribution for the interval 2002–2021 is approximately symmetrical. The distribution for the interval 1892–1911 is negatively skewed, with a smaller overall range. Both distributions are unimodal.

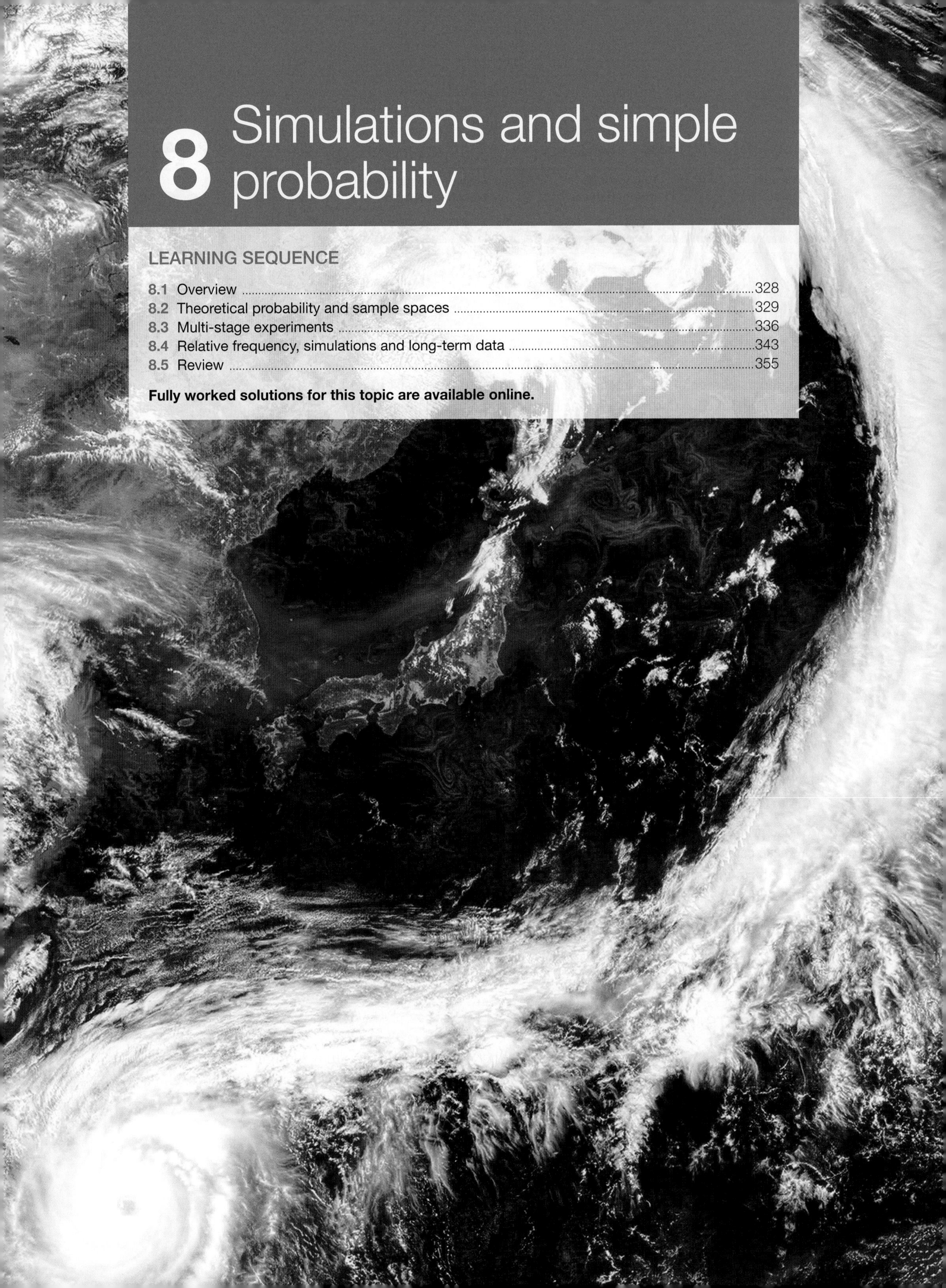

8 Simulations and simple probability

LEARNING SEQUENCE

Fully worked solutions for this topic are available online.

8.1 Overview

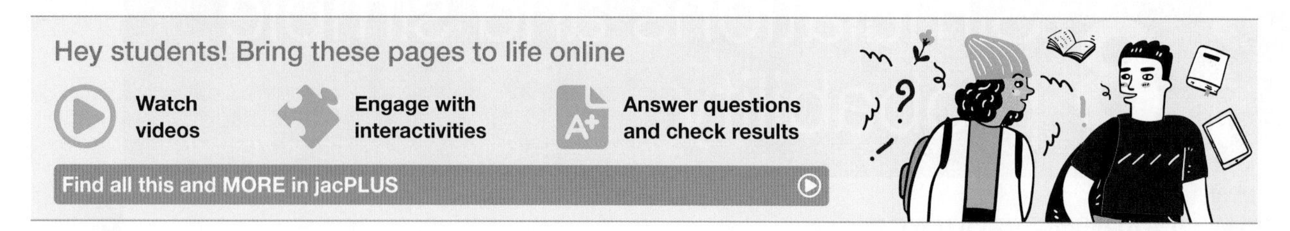

8.1.1 Introduction

Probability describes the chances of different events occurring. As in all branches of mathematics, probability has its own terminology and language. The ability to correctly interpret and apply these terms is essential for a deep understanding of the content and an important tool to assist in problem solving. We come across probabilities in the media all the time; some examples include 'There is a 50% chance of rain tomorrow', and 'Nick Kyrgios has an 85% chance of winning his next tennis match'.

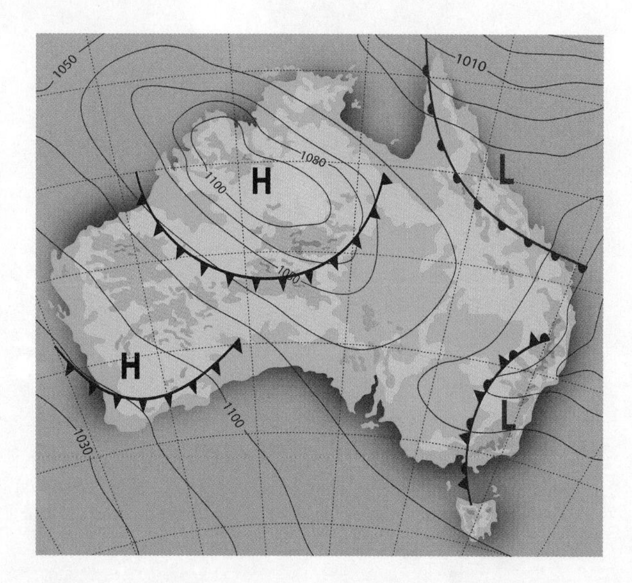

Lotteries use probability to determine the likelihood of a winning ticket being bought, and set the prize money based on these calculations. To win first prize in a particular lottery, you need to have selected the correct 6 balls from balls numbered 1 to 45. The chance of this happening is 1 in 8 145 060. This means that you have approximately a 1 in 8 million chance of winning the lottery!

Being able to calculate the probability of an event allows us to determine how likely it is to occur. Probabilities are used in weather forecasting, calculating prices of insurance premiums and determining the likelihood of winning games.

KEY CONCEPTS

This topic covers the following key concepts from the VCE Mathematics Study Design:

- long-term data and relative frequencies in practical situations such as in relation to epidemics, climate, environment, sport and marketing.

8.2 Theoretical probability and sample spaces

LEARNING INTENTION

At the end of this subtopic, you should be able to:
- evaluate events on the probability scale
- determine the sample space for an event
- calculate theoretical probability
- determine complementary events.

8.2.1 The probability scale

Probability is a measure of the likely occurrence of an event. The probability of an event occurring is measured with a scale ranging from and including 0 (**impossible** event) to 1 (**certain** event).

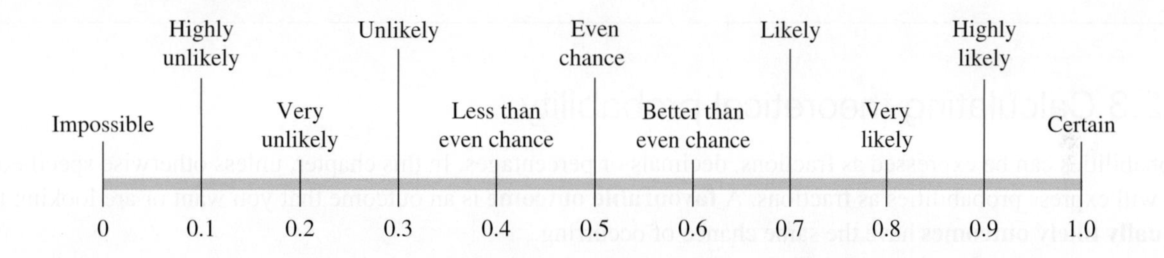

Probability can be written as a decimal number, fraction or percentage. For example, the probability of even chance may be written as 0.5, $\frac{1}{2}$ or 50%. The probability of an event (A) occurring can be denoted by Pr(A). This means that $0 \leq \Pr(A) \leq 1$, and $\Pr(A) = 0$ if A is an impossibility and $\Pr(A) = 1$ if A is a certainty.

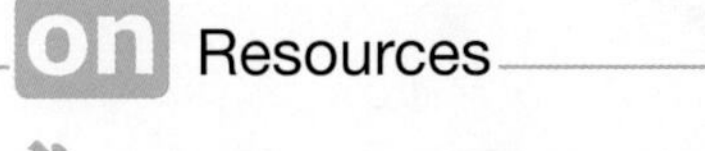

Interactivity Probability scale (int-3824)

8.2.2 Sample spaces

A sample space is often listed as a set of outcomes. For example, the sample space for the spinner shown is {pink, blue, green, orange}; the sample space includes all of the colours on the spinner.

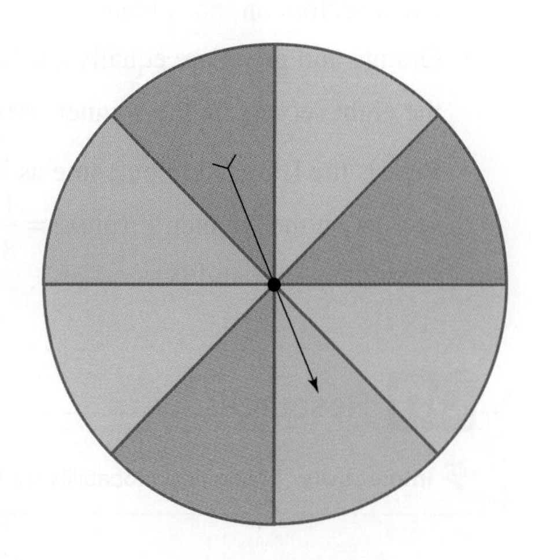

Outcome and sample space

An outcome is a particular result of an experiment.

A sample space is a list of all possible outcomes.

WORKED EXAMPLE 1 Sample space

For each of the following experiments, list the sample space to show the possible outcomes.
a. **A coin is flipped.**
b. **A die is rolled.**
c. **A circular spinner, with seven sectors labelled 1 to 7, is spun.**

THINK	WRITE
a. A coin has two sides. List each possible outcome.	a. When a coin is flipped, the two possible outcomes are Heads and Tails. Therefore, the sample space is {Heads, Tails}.
b. A regular die has 6 sides. List each possible outcome.	b. When a die is rolled, the six possible outcomes are 1, 2, 3, 4, 5 and 6.
c. There are seven different sectors on the spinner so there are seven possible outcomes. List each possible outcome.	c. When the circular spinner is spun, the seven possible outcomes are 1, 2, 3, 4, 5, 6 and 7. Therefore, the sample space is $\{1, 2, 3, 4, 5, 6, 7\}$.

8.2.3 Calculating theoretical probability

Probabilities can be expressed as fractions, decimals or percentages. In this chapter, unless otherwise specified, we will express probabilities as fractions. A **favourable outcome** is an outcome that you want or are looking for. **Equally likely outcomes** have the same chance of occurring.

Theoretical probability

The theoretical probability of an event can be calculated using the following probability formula.

$$\text{Pr(event)} = \frac{\text{number of favourable outcomes}}{\text{total number of outcomes}}$$

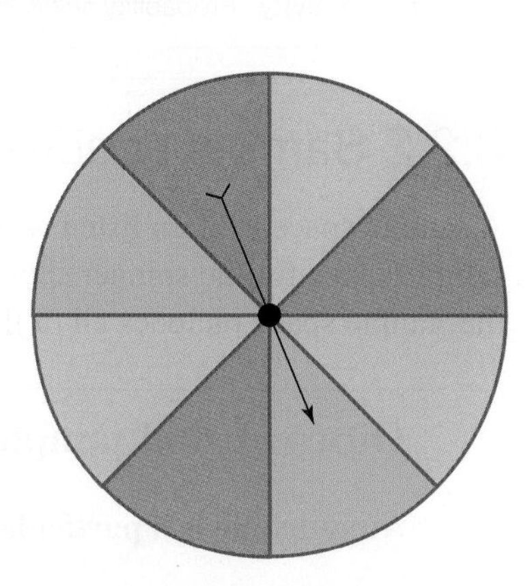

- For the spinner shown, $\text{Pr(blue)} = \frac{3}{8}$ because there are three favourable outcomes (blue sectors) out of eight possible outcomes (total number of equal sectors on the spinner).
- Orange and green are equally likely outcomes as they both occupy two of the eight sectors on the spinner. $\text{Pr(orange)} = \text{Pr(green)} = \frac{2}{8}$ or $\frac{1}{4}$.
- Pink is the least likely outcome as it occupies only one of the eight sectors on the spinner. $\text{Pr(pink)} = \frac{1}{8}$. Pink has the smallest value for its theoretical probability.

Resources

Interactivity Theoretical probability (int-6081)

tlvd-3946

WORKED EXAMPLE 2 Theoretical probability

A standard six-sided die is rolled. Calculate the probability of rolling:

a. a 3

b. an odd number.

THINK	WRITE
a. 1. The number 3 occurs once on a die.	a. Number of favourable outcomes = 1
2. The possible outcomes are 1, 2, 3, 4, 5 and 6.	Number of outcomes = 6
3. Calculate the probability of rolling a 3.	$\Pr(3) = \frac{\text{number of favourable outcomes}}{\text{total number of outcomes}}$ $= \frac{1}{6}$
b. 1. The odd numbers are 1, 3 and 5. There are six numbers on a die.	b. Number of favourable outcomes = 3 Number of outcomes = 6
2. Calculate the probability of rolling an odd number. Write the fraction in simplest form.	$\Pr(\text{odd number}) = \frac{\text{number of favourable outcomes}}{\text{total number of outcomes}}$ $= \frac{3}{6}$ The probability of rolling an odd number is $\frac{1}{2}$.

tlvd-3947

WORKED EXAMPLE 3 Theoretical probability 2

A normal pack of 52 playing cards is well shuffled and one card is drawn. Calculate the probability of drawing:

a. a club **b. an ace** **c. a red queen.**

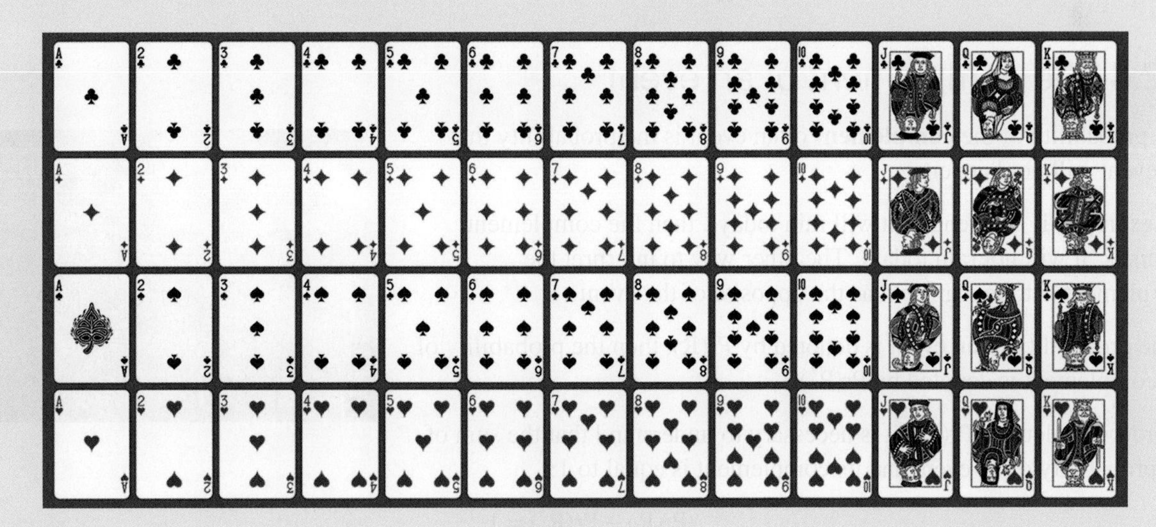

THINK	WRITE
a. 1. There are 13 clubs in a deck of cards.	a. Number of favourable outcomes = 13
2. There are 52 cards in the deck, making 52 possible outcomes.	Number of outcomes = 52

3. Calculate the probability of drawing a club, making sure to simplify the fraction.

$$\Pr(\text{club}) = \frac{\text{number of favourable outcomes}}{\text{total number of outcomes}} = \frac{13}{52} = \frac{1}{4}$$

b. 1. There are 4 aces in a deck of cards.

b. Number of favourable outcomes = 4

2. There are 52 cards in the deck, making 52 possible outcomes.

Number of outcomes = 52

3. Calculate the probability of drawing an ace. Write the fraction in simplest form.

$$\Pr(\text{ace}) = \frac{\text{number of favourable outcomes}}{\text{total number of outcomes}} = \frac{4}{52} = \frac{1}{13}$$

c. 1. There are 2 red queens in a deck of cards.

c. Number of favourable outcomes = 2

2. There are 52 cards in the deck, making 52 possible outcomes.

Number of outcomes = 52

3. Calculate the probability of drawing a red queen. Write the fraction in simplest form.

$$\Pr(\text{red queen}) = \frac{\text{number of favourable outcomes}}{\text{total number of outcomes}} = \frac{2}{52} = \frac{1}{26}$$

8.2.4 The complement of an event

The probability of the **complement** of an event is the probability that the event will not happen.

For example, if an event is 'it will rain today', then the complement of this is 'it will not rain today'. The other way to interpret the complement is to recognise it as the opposite of the event.

If the probability of an event is denoted by Pr(R), then the probability of the complement is denoted by Pr(R′).

In order to calculate Pr(R′) it is necessary to understand that the sum of the probability of an event and its complement is equal to 1:

$$\Pr(R) + \Pr(R') = 1$$

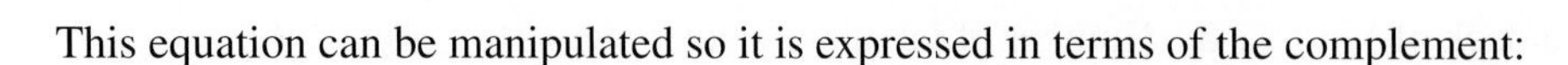

This equation can be manipulated so it is expressed in terms of the complement:

$$\Pr(R') = 1 - \Pr(R)$$

For example, if the probability of rain today is $\frac{7}{10}$, then:

$$\begin{aligned}\Pr(R') &= 1-\frac{7}{10}\\ &= \frac{3}{10}\end{aligned}$$

This means there is a 30% chance that it will not rain.

WORKED EXAMPLE 4 Complementary events

Niema has discovered that 35% of energy users have chosen green energy suppliers.

a. State the complement of this event.

b. Calculate the probability of the complement of this event. Give your answer as a percentage.

THINK	WRITE
a. The complement is the event not happening.	a. The complement of choosing green energy suppliers is not choosing green energy suppliers.
b. 1. Use the formula to calculate the complement.	b. Pr(choosing green energy suppliers) = $35\% = \frac{35}{100}$ Pr(not choosing green energy suppliers) $= 1 - \text{Pr(choosing green energy suppliers)}$ $= 1 - \frac{35}{100}$ $= \frac{65}{100}$
2. Convert the answer into a percentage.	$\frac{65}{100} = 65\%$

8.2 Exercise

Students, these questions are even better in jacPLUS

Receive immediate feedback and access sample responses

Access additional questions

Track your results and progress

Find all this and MORE in jacPLUS

1. Match the words with one of the numbers between 0 and 1 that are given below. Choose the number depending on what sort of chance the word indicates, between *impossible* and *certain*. You may use a number more than once.
   The numbers to choose from are 1, 0.75, 0.5, 0.25 and 0.
   a. Certain
   b. Likely
   c. Probable
   d. Improbable
   e. Definite
   f. Impossible
2. List two events that have a probability of:
   a. 0
   b. 0.5
   c. 1
   d. 0.75
   e. 0.25.

3. WE1 For each of the following experiments, list the sample space to show the possible outcomes.
   a. A standard 6-sided die is rolled.
   b. A marble is randomly selected from a bag containing 4 green, 2 yellow and 3 blue marbles.
   c. A letter is selected from all the letters of the alphabet.

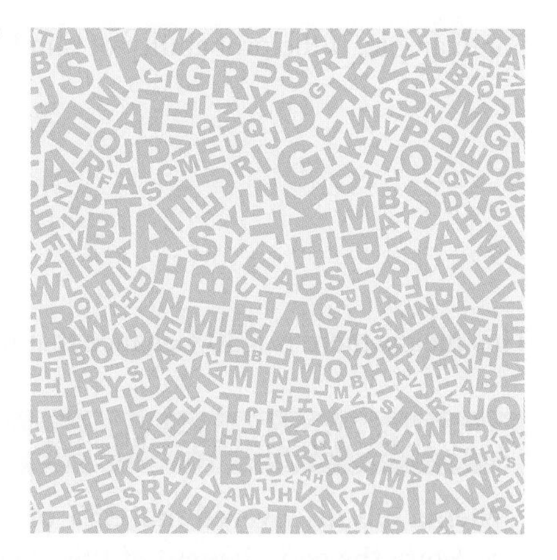

4. WE2 A standard 6-sided die is rolled. Calculate the probability of rolling:
   a. a 5
   b. a number less than 3
   c. an even number.
5. The letters of the word *MATHEMATICS* are each written on a small piece of cardboard and placed face down. If one card is selected at random, calculate the probability that it is:
   a. a vowel
   b. a consonant
   c. the letter M
   d. the letter C.
6. A spinner of 8 equally likely numbers is spun. The spinner is numbered 1–8.

   a. List the sample space.
   b. Calculate the probability, expressed as a percentage, of obtaining:
      i. an odd number
      ii. a number that is less than 6
      iii. the number 9
      iv. a number that is at most 8.

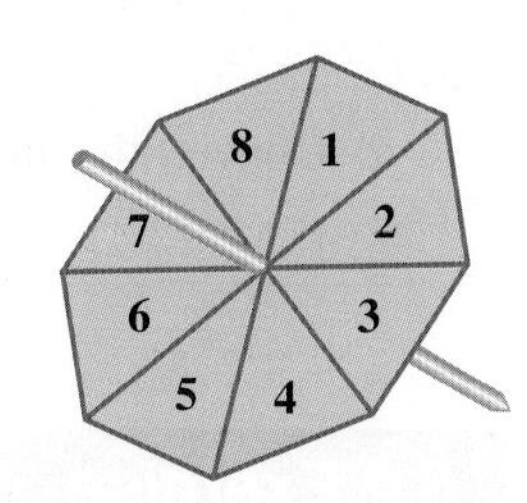

7. Answer these questions for each of the following spinners.
   i. Determine if there is an equal chance of landing on each colour. Explain your answer.
   ii. List all the possible outcomes.
   iii. Calculate the probability of each outcome.

a.
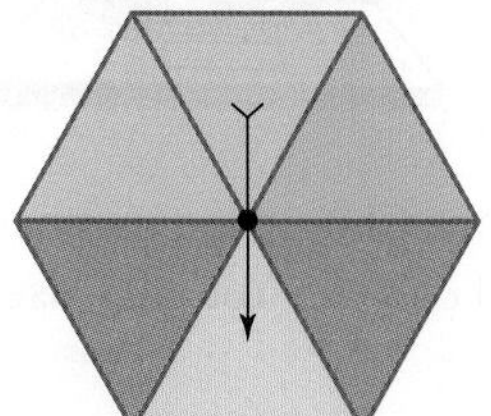
b.
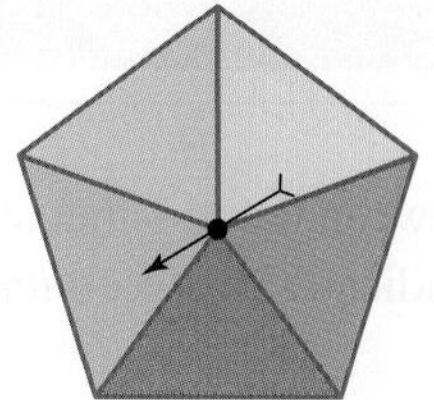
c.
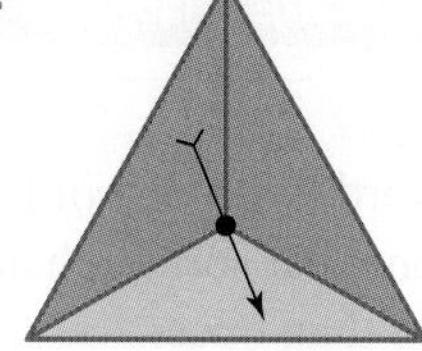

8. All the jelly beans shown are placed in a bag for a simple probability experiment.
   a. Determine which jelly bean colour is most likely to be selected from the bag. Explain your answer.
   b. Determine which jelly bean colour is least likely to be selected from the bag. Explain your answer.
   c. Calculate the probability of selecting each jelly bean colour from the bag.

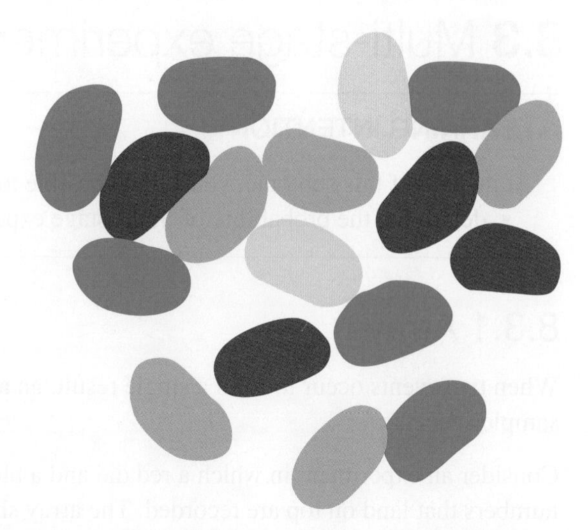

9. **MC** The probability of obtaining a number less than 5 on the spinner represents the chance of an advertising company being successful. This probability is:

   A. 1 B. 0 C. $\frac{1}{5}$
   D. $\frac{4}{5}$ E. $\frac{5}{4}$

10. **WE3** A normal pack of 52 playing cards is well shuffled and one card is drawn. Calculate the probability of drawing:
    a. a heart
    b. a red card
    c. an ace of diamonds
    d. a picture card (jack, queen or king).

11. Draw spinners with the following probabilities.
    a. $\Pr(\text{blue}) = \frac{1}{3}$ and $\Pr(\text{white}) = \frac{2}{3}$
    b. $\Pr(\text{blue}) = \frac{1}{2}$, $\Pr(\text{white}) = \frac{1}{4}$, $\Pr(\text{green}) = \frac{1}{8}$ and $\Pr(\text{red}) = \frac{1}{8}$
    c. $\Pr(\text{blue}) = 0.75$ and $\Pr(\text{white}) = 0.25$

12. State the complement of each of these events.
    a. Owning a dog
    b. Flipping a coin and getting a tail
    c. Sleeping
    d. Attending school
    e. Passing your learner permit knowledge test
    f. Owning a mobile phone

13. **WE4** Bol's doctor told him that he had a 20% chance of catching a very infectious virus in the next month based on the available data.
    a. State the complement of this event.
    b. Calculate the probability of the complement of this event. Give your answer as a percentage.

14. There are three different colours of flowers in a basket: red, yellow and purple. Assuming that you pick one flower from the basket, the complement of picking a red flower is picking a purple flower. State whether this is True or False and explain why.

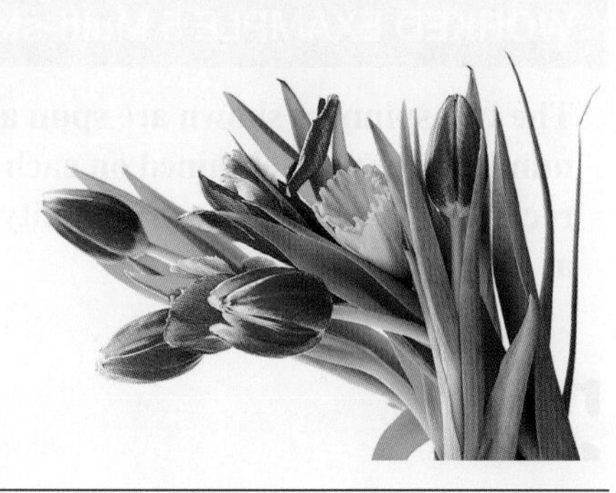

8.3 Multi-stage experiments

LEARNING INTENTION

At the end of this subtopic, you should be able to:
- determine the probability of multi-stage experiments using tables, tree diagrams and arrays.

8.3.1 Arrays

When two events occur to form a single result, an **array** can be used to display the sample space.

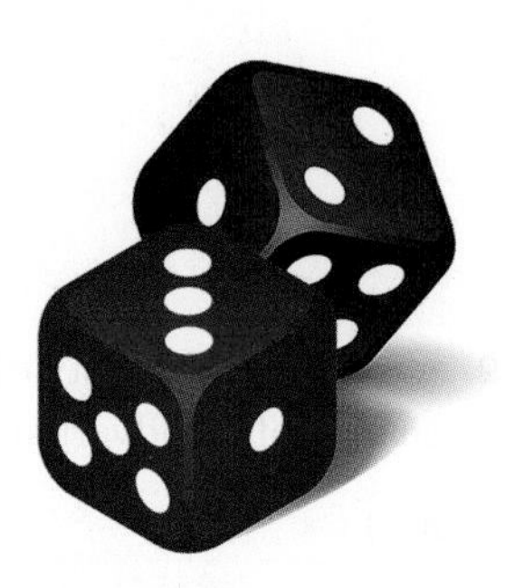

Consider an experiment in which a red die and a blue die are rolled, and the two numbers that land on top are recorded. The array shows the sample space for that experiment. There are 36 possible outcomes in this sample space.

		Blue die					
		1	2	3	4	5	6
Red die	1	(1, 1)	(1, 2)	(1, 3)	(1, 4)	(1, 5)	(1, 6)
	2	(2, 1)	(2, 2)	(2, 3)	(2, 4)	(2, 5)	(2, 6)
	3	(3, 1)	(3, 2)	(3, 3)	(3, 4)	(3, 5)	(3, 6)
	4	(4, 1)	(4, 2)	(4, 3)	(4, 4)	(4, 5)	(4, 6)
	5	(5, 1)	(5, 2)	(5, 3)	(5, 4)	(5, 5)	(5, 6)
	6	(6, 1)	(6, 2)	(6, 3)	(6, 4)	(6, 5)	(6, 6)

An array is also useful when there are two events, but a different number of outcomes for each event. For example, the following array shows the sample space for an experiment in which a coin is flipped and a die is rolled.

		Die					
		1	2	3	4	5	6
Coin	H	(H, 1)	(H, 2)	(H, 3)	(H, 4)	(H, 5)	(H, 6)
	T	(T, 1)	(T, 2)	(T, 3)	(T, 4)	(T, 5)	(T, 6)

tlvd-3948

WORKED EXAMPLE 5 Multi-stage experiments

The two spinners shown are spun and the numbers that are obtained on each spinner are recorded. Calculate the probability of spinning at least one 1.

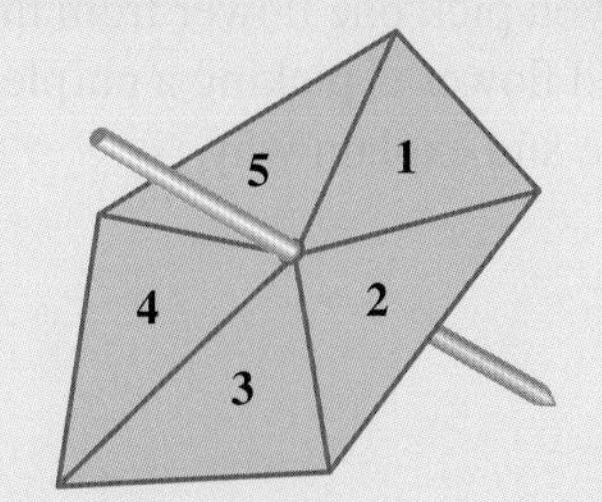

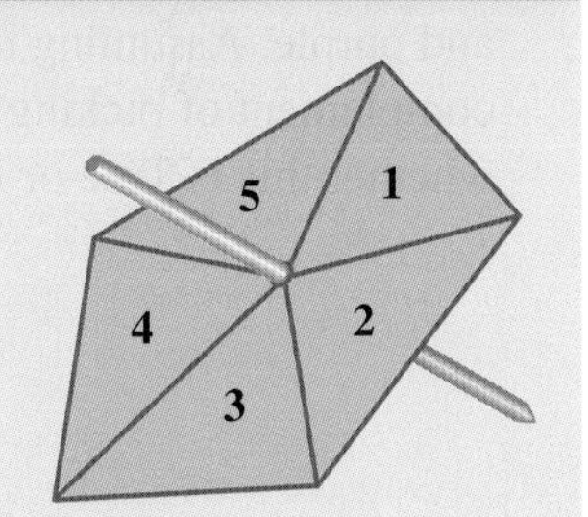

THINK

1. Draw the array for the sample space. Highlight or circle the outcomes that have at least one 1, as shown in pink.
There are 9 outcomes with at least one 1.

2. Calculate the probability using the formula.
There are 25 outcomes in the sample space.

WRITE

		Spinner 2				
		1	2	3	4	5
Spinner 1	1	(1, 1)	(1, 2)	(1, 3)	(1, 4)	(1, 5)
	2	(2, 1)	(2, 2)	(2, 3)	(2, 4)	(2, 5)
	3	(3, 1)	(3, 2)	(3, 3)	(3, 4)	(3, 5)
	4	(4, 1)	(4, 2)	(4, 3)	(4, 4)	(4, 5)
	5	(5, 1)	(5, 2)	(5, 3)	(5, 4)	(5, 5)

$$\Pr(\text{at least one } 1) = \frac{\text{number of favourable outcomes}}{\text{total number of outcomes}}$$
$$= \frac{9}{25}$$

8.3.2 Tree diagrams

Tree diagrams are used to list all possible outcomes of two or more events. The branches show the possible links between one outcome and the next. The tree diagram shown represents the sample space for flipping two coins. When all outcomes are equally likely, probabilities can be calculated from tree diagrams. For example, in the tree diagram shown, the probability of obtaining one Tail and one Head can be calculated:

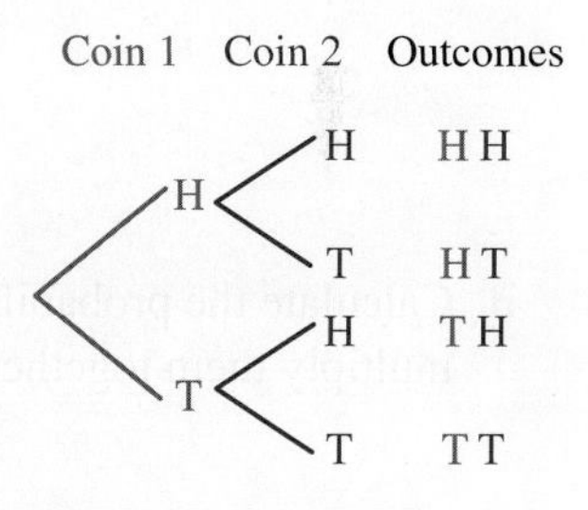

$$\Pr(\text{one H and one T}) = \frac{\text{number of outcomes with one H and one T}}{\text{total number of outcomes}}$$
$$= \frac{2}{4}$$
$$= \frac{1}{2}$$

Each event is treated as if it were a separate event, even if events occur at the same time. For example, a coin flipped twice and two coins flipped together once would give the same outcomes.

Resources

Video eLesson Tree diagrams (eles-1894)

tlvd-3949

WORKED EXAMPLE 6 Using tree diagrams to calculate probabilities

The uniform committee at school is deciding on a new colour combination for the school uniform. The two colour choices for the school pants, shirt and jumper are red and yellow.

a. **Use a tree diagram to show all the possible combinations.**
b. **Calculate the probability that the uniform will consist of only one colour.**
c. **Calculate the probability that a red jumper will be part of the uniform.**
d. **Calculate the probability that the uniform will be red pants, a yellow shirt and a red jumper.**

THINK

WRITE

a. The three events are the colour of the pants, the colour of the shirt and the colour of the jumper. There are two choices for each of these.

a.

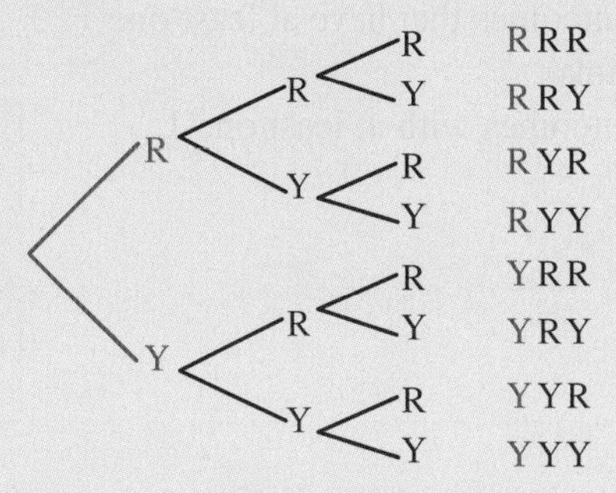

b. For the uniform to consist of only one colour, all pieces need to be red or all pieces need to be yellow. There is 1 outcome with all red and 1 outcome with all yellow. There are 8 outcomes altogether.

b. $\Pr(\text{all one colour}) = \frac{2}{8}$

$= \frac{1}{4}$

c. Four of the outcomes have a red jumper.

c. $\Pr(\text{red jumper}) = \frac{4}{8}$

$= \frac{1}{2}$

d. Calculate the probability of each option and multiply them together.

d. $\Pr(\text{red pants}) = \frac{1}{2}$

$\Pr(\text{yellow shirt}) = \frac{1}{2}$

$\Pr(\text{red jumper}) = \frac{1}{2}$

$\Pr(\text{RYR}) = \frac{1}{2} \times \frac{1}{2} \times \frac{1}{2}$

$= \frac{1}{8}$

8.3 Exercise

Note: Unless otherwise stated, express probabilities as fractions.

1. A spinner is divided into 5 sections (red, blue, green, yellow and orange). A six-sided die is also rolled. Draw the array for one spin of the spinner and one roll of the die.

2. **WE5** The two spinners shown are spun and the numbers that are obtained on each spinner are recorded. Calculate the probability of spinning exactly one 5.

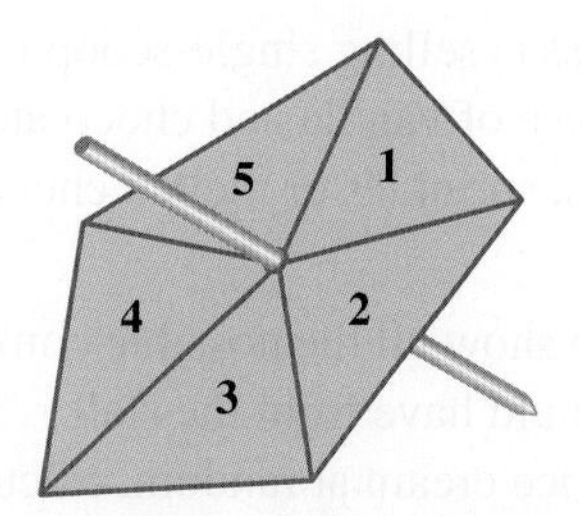

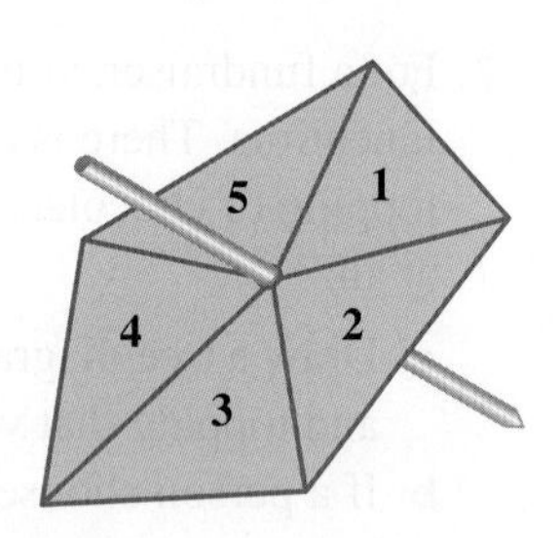

3. **MC** A six-sided die and a coin are tossed at the same time. The sample space is:
   A. {1, 2, 3, 4, 5, 6, 7, 8}
   B. {1, 2, 3, 4, 5, 6, H, T}
   C. {(1, H), (2, H), (3, H), (4, H), (5, H), (6, H)}
   D. {(1, H), (2, H), (3, H), (4, H), (5, H), (6, H), (1, T), (2, T), (3, T), (4, T), (5, T), (6, T)}
   E. {(1, H), (2, H), (3, H), (4, T), (5, T), (6, T)}

4. In a variation of the game Twister, the two spinners shown are spun and each contestant places the two body parts spun in contact and holds them there until the next two body parts are spun. If you fall over, you are out. The winner is the last person standing.
   Calculate the probability, expressed as a percentage, that when the spinners are spun, the contestants will need to place:

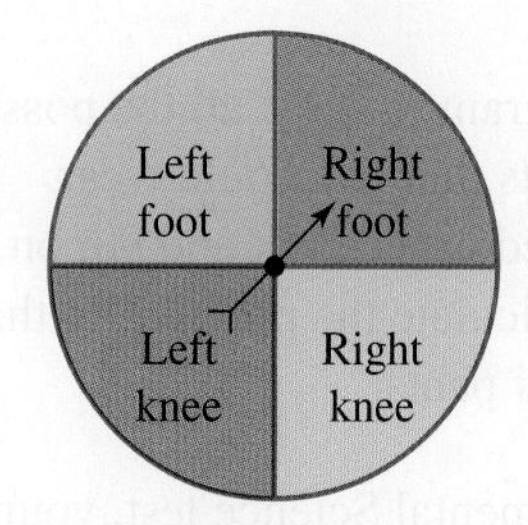

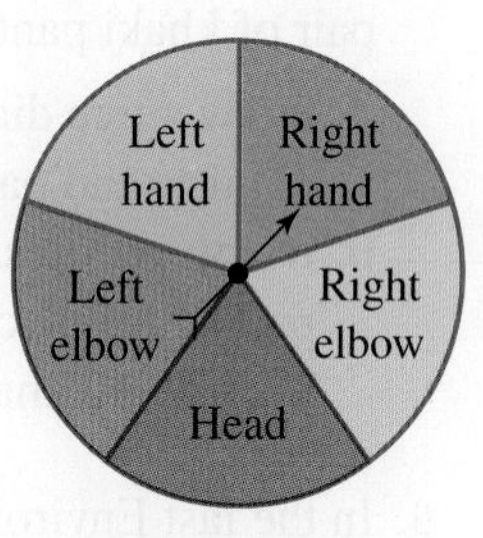

   a. their right foot on their head
   b. their left elbow on a knee
   c. their head on a foot
   d. their left hand on a knee or foot.

5. A restaurant has the menu shown.
   a. Use a tree diagram to show how many different combinations are possible for a meal with one selection at each course.
   b. If a person selects a dinner at random, calculate the probability that:
      i. they have soup for entrée
      ii. they have spring rolls and curry
      iii. they don't have ice cream for dessert
      iv. they have soup and pasta but not ice cream.

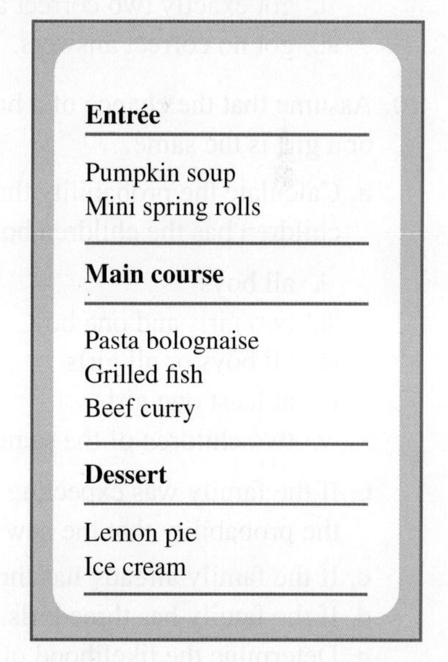

6. **WE6** The uniform committee at school is deciding on a new colour combination for the school uniform.
   The two colour choices for the school hat, bag and hoodie are blue and white.
   a. Use a tree diagram to show all the possible combinations.
   b. Calculate the probability that the uniform will consist of only one colour.
   c. Calculate the probability that a white hat will be part of the uniform.
   d. Calculate the probability that a white hat, blue bag and blue hoodie will be part of the uniform.

7. For a fundraiser, your class is selling single-scoop ice-cream cones at lunchtime. There is a choice of vanilla and chocolate ice cream, with a topping of chocolate curls, sprinkles or melted chocolate, or no topping at all.

   a. Draw a tree diagram to show all the possible combinations of ice cream and topping that you could have from the stall.
   b. If a person chooses an ice cream at random, calculate the probability that:
      i. both the ice cream and the topping are chocolate
      ii. it is a vanilla ice cream with sprinkles
      iii. there is no topping on the ice cream.

8. In your drawers at home, there are two white T-shirts, a green T-shirt and a red T-shirt. There is also a pair of black pants and a pair of khaki pants.

   a. Draw a tree diagram to show all the possible combinations of T-shirts and pants that you could wear.
   b. If you get dressed in the dark and put on one T-shirt and one pair of pants, calculate the probability that you put on the red T-shirt and khaki pants.

9. In the last Environmental Science test, your friend guessed the answers to three True/False questions.
   a. Use a tree diagram to show all the different answer combinations for the three questions.
   b. Calculate the probability that your friend:
      i. got all three answers correct
      ii. got exactly two correct answers
      iii. got no correct answers.

10. Assume that the chance of a baby being born a boy or a girl is the same.

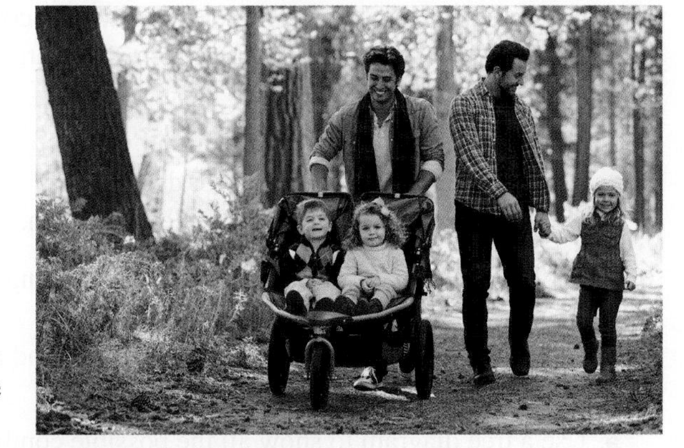

   a. Calculate the probability that a family with three children has the children born as:
      i. all boys
      ii. two girls and one boy
      iii. all boys or all girls
      iv. at least one girl
      v. two children of the same gender.
   b. If the family was expecting another baby, calculate the probability that the new baby will be a boy.
   c. If the family already has three boys, calculate the probability that the new baby will be a boy.
   d. If the family has three girls, calculate the probability that the new baby will be a boy.
   e. Determine the likelihood of the combination of children shown in the photo.

11. A fair coin is flipped 3 times. Calculate the probability of obtaining:

a. at least two Heads or at least two Tails
b. exactly two Tails.

12. **MC** A fair coin is flipped 3 times. The probability of obtaining at most 1 Tail is:

A. $\frac{1}{3}$ B. $\frac{1}{2}$ C. $\frac{3}{8}$ D. $\frac{1}{8}$ E. $\frac{1}{6}$

13. A spinner of 7 equally likely numbers (numbered 1–7) is spun twice and the two numbers are added. Calculate the probability of obtaining a total of:

a. 2 or 14 b. 9 c. at least 12.

14. Two dice are rolled and the product of the two numbers is found. Calculate the probability that the product of the two numbers is:

a. an odd number b. a prime number
c. more than 1 d. at most 36.

15. A fair die is rolled and a fair coin flipped. Calculate the probability of obtaining:

a. an even number and a Head b. a Tail from the coin
c. a prime number from the die d. a number less than 5 and a Head.

16. a. Use an array to display the sample space for an experiment where two dice are rolled and the sum of the two numbers appearing is found.
b. List any patterns you found in the sample space.
c. Copy and complete the following table.

Sum of two dice	2	3	4	5	6	7	8	9	10	11	12
Probability											

d. List any patterns that you notice in the probabilities you found in the table.

17. In a new game, two dice are rolled and the difference between the two dice is noted. If the difference is larger than 2, you win; otherwise you lose.

a. Use an array to determine the sample space for the game.
b. Calculate the probability of winning (having a difference greater than 2).
c. Calculate the probability of losing (having a difference of 2 or less).
d. A fair game is one in which the chances of winning are the same as the chances of losing. State if this game is fair.

18. The game rock, paper, scissors is sometimes used to make a decision between two people. It uses the three different hand signs shown in the photo. On the count of 3, each player displays one of the hand signs. If both players display the same hand sign, the game is a tie. The photos show all other possible results.

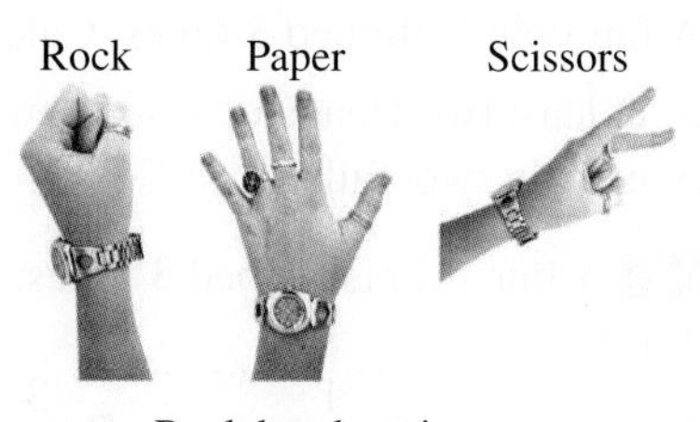

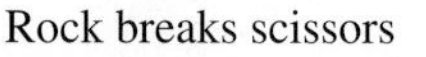
Rock breaks scissors

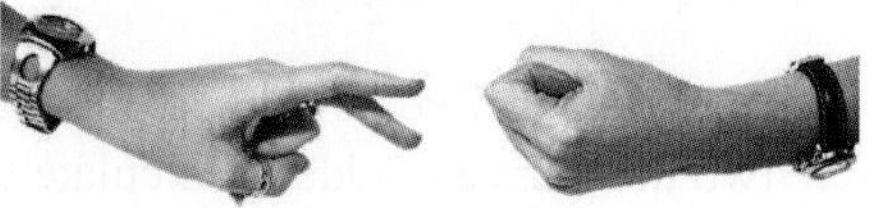
Rock wins

Paper covers rock

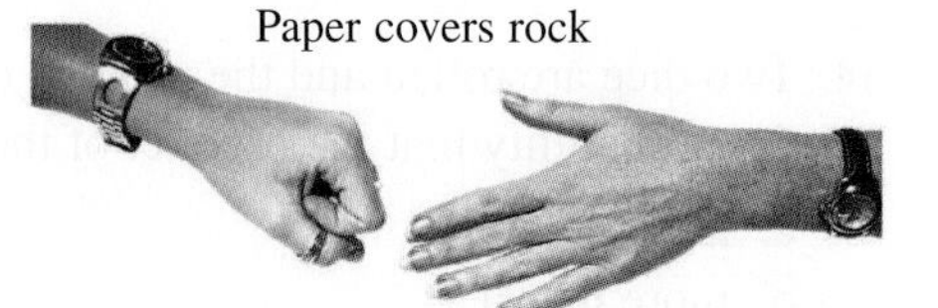
Paper wins

Scissors cut paper

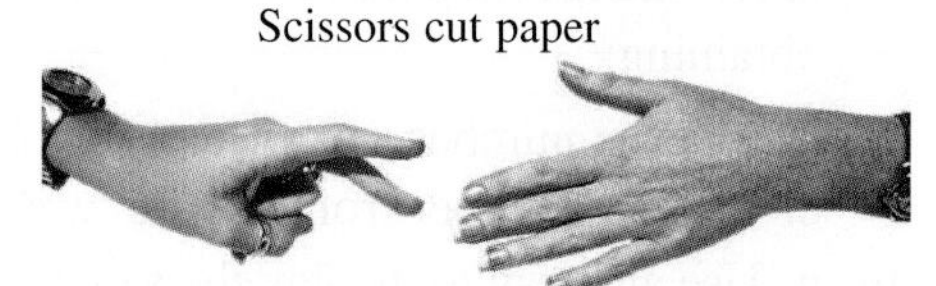
Scissors win

a. Draw an array to show all the possible combinations for a single game of rock, paper, scissors.

		Person 2		
		Rock	**Paper**	**Scissors**
Person 1	**Rock**			
	Paper			
	Scissors			

Calculate the probability that:

b. i. rock wins
ii. paper wins
iii. scissors win.

c. State if rock, paper, scissors is a fair game.

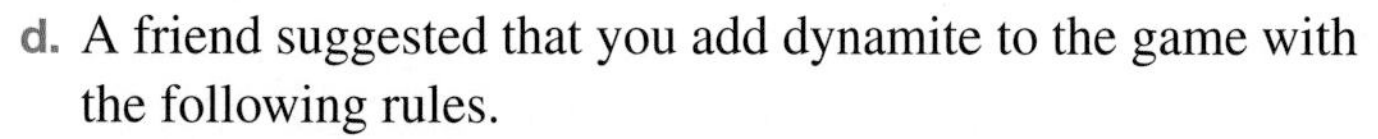

d. A friend suggested that you add dynamite to the game with the following rules.
- Dynamite blows up the rock and wins.
- Dynamite has its fuse cut by scissors and loses.
- Dynamite sets the paper on fire and wins.

i. Modify your array in part a to include dynamite.
ii. State if each hand sign has an equal chance of winning.
iii. Support your answer to part ii using the array and probabilities.

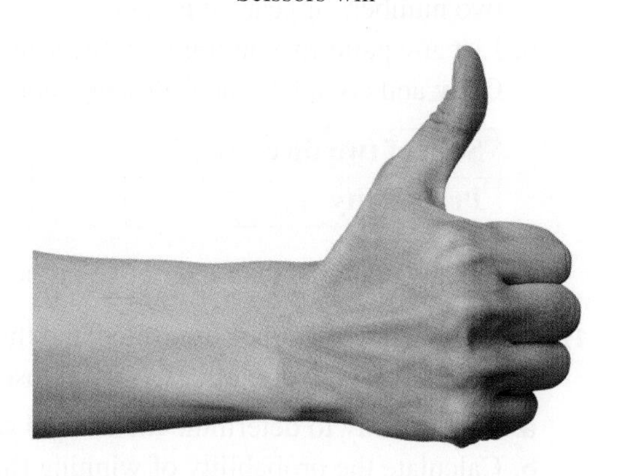

19. There are three different ways to go from school to the shops. There are two different ways to go from the shops to the library. There is only one way to go from the library to home.

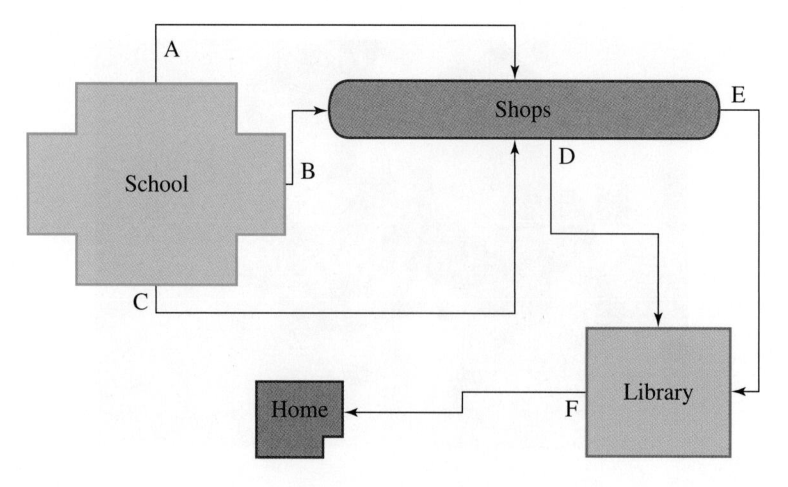

This afternoon you need to travel from school to home via the shops and the library.

a. Use a tree diagram to calculate the number of different routes you could use on your journey. (*Hint:* Use the letters A, B and C to represent the different routes from school to the shops and D and E to represent the different routes from the shops to the library. Use F to represent the route from the library to home.)

b. Consider if the number of outcomes would be different if we omitted the last leg of the journey from the library to home. Explain your answer.

20. A die is rolled and 2 coins are flipped. Display the sample space using an appropriate method.

8.4 Relative frequency, simulations and long-term data

LEARNING INTENTION

At the end of this subtopic, you should be able to:

- determine relative frequencies
- use technologies to conduct simulations and compare results
- determine limitations of simulations
- calculate successful trials.

8.4.1 Relative frequency

Often in real life it is not possible to calculate a theoretical probability for an event. In these situations, an experiment is carried out. An experiment that is repeated a number of times is called a **trial** and becomes part of a larger experiment. For example, a die is rolled and the number is recorded. This is an experiment. If the experiment is repeated another 49 times, the first roll of the die is now referred to as the first trial in an experiment where the die is rolled 50 times. A **successful trial** is a trial where the result is the outcome that you wanted.

Relative frequency

The relative frequency, sometimes called the empirical or experimental probability, of an event is calculated using the following formula:

$$\text{relative frequency} = \frac{\text{number of successful trials}}{\text{total number of trials}}$$

The relative frequency of an event for a very large number of trials gives an indication of the value of the theoretical probability. As the number of trials increases, the relative frequency of an event will gradually become closer in value to the theoretical probability. This is known as the 'law of large numbers'.

Resources

Interactivity Experimental probability (int-3825)

8.4.2 Simulations

Sometimes it is not possible to conduct trials of a real experiment because it is too expensive, too difficult or impracticable. In situations like this, outcomes of events can be modelled using devices such as spinners, dice or coins to simulate or represent what happens in real life. These simulated experiments are called **simulations**.

WORKED EXAMPLE 7 Simulations

To simulate whether a baby is born male or female, a coin is flipped. If the coin lands Heads up, the baby is a boy. If the coin lands Tails up, the baby is a girl. Student 1 flips a coin 10 times and obtains 3 Heads and 7 Tails, while student 2 flips a coin 100 times and obtains 43 Heads and 57 Tails.

a. **Determine the relative frequency of female babies in both cases. Give your answers as decimals.**

b. **Compare the results from the two simulations and determine which is a better estimate of the true probability.**

THINK	WRITE
a. 1. A baby born female is a successful trial. For student 1: in the simulation, females are Tails. There were 7 Tails, so 7 females were born and there were 7 successful trials out of 10 trials. Write the formula for relative frequency and substitute the results.	a. Student 1: $\text{Relative frequency} = \frac{\text{number of successful trials}}{\text{total number of trials}}$ $= \frac{7}{10}$ $= 0.7$
2. For student 2: there were 57 Tails, so 57 females were born and there were 57 successful trials out of 100 trials. Write the formula for relative frequency and substitute the results.	Student 2: $\text{Relative frequency} = \frac{\text{number of successful trials}}{\text{total number of trials}}$ $= \frac{57}{100}$ $= 0.57$
b. 1. Compare the results of the two simulations. Theoretically, there should be an equal number of males and females born, so we would expect the results to be close to 0.5.	b. Student / Relative frequency: 1 — 0.72; 2 — 0.57
2. Give your answer.	The true probability is likely to be close to 0.5, so student 2 achieved the better estimate of probability.

8.4.3 Using technology to perform simulations

Simulating experiments using manual devices such as dice and spinners can take a lot of time. A more efficient method of collecting results is to use a list of randomly generated numbers.

Random number generators can generate a series of numbers between two given values — for example, decimals between 0 and 0.9999.

The following table shows some Excel formulas that generate random numbers:

Formula	Output
= RAND()	A random decimal from 0 to 0.9999
= INT(6*RAND() + 1)	Integers between 1 and 6 that can be used to simulate rolling a die. Both of these formulas can be modified to generate the type of numbers that you require.
= RANDBETWEEN(1, 6)	Random numbers between 1 and 6

tlvd-3950

WORKED EXAMPLE 8 Using technology for simulation and to calculate probabilities

a. Use a random number generator to simulate the number of chocolate chips in 50 biscuits, with a maximum of 80 chocolate chips in each.

b. Calculate the probability that there will be more than 30 chocolate chips in a randomly chosen biscuit.

THINK

a. Use a random number generator to generate 50 numbers between 0 and 80.
Use either of the following formulas in Excel.
= INT(80 * RAND() + 1)
or
= RANDBETWEEN (1, 80)

WRITE

a.

8	55	8	62	80	50	42	80	57	39
15	7	64	73	47	12	74	74	16	42
41	22	50	33	68	72	64	16	6	72
70	72	52	48	14	22	59	48	65	34
67	62	72	59	10	30	13	7	40	18

b. 1. Count the number of biscuits with more than 30 chocolate chips. There are 34 in the example shown.

b.

8	55	8	62	80	50	42	80	57	39
15	7	64	73	47	12	74	74	16	42
41	22	50	33	68	72	64	16	6	34
70	72	52	48	14	22	59	48	65	34
67	62	72	59	10	30	13	7	40	18

2. The probability is the number of successful outcomes (34) divided by the number of trials (50).

$$\text{Pr(more than 30 chocolate chips)} = \frac{34}{50}$$
$$= \frac{17}{25}$$

Resources

Interactivity Generating random numbers on an Excel spreadsheet (int-4032)

8.4.4 Factors that could complicate simulations

Simulations are used to reproduce real-life situations when it is not feasible to implement the actual test. The reasons for not conducting an actual test can include financial or safety constraints, or just the overall complexity of running the test.

Most simulations are conducted using computers and various other mechanical devices. For example, airbag manufacturers run airbag tests using cars, but instead of using real humans, they use mannequins. These simulations are fairly accurate, but they have their limitations.

WORKED EXAMPLE 9 Limitations of simulations

Six runners are competing in a race.

a. Explain how simulation could be used to determine the winner.

b. Determine if there are any limitations in your answer.

THINK	WRITE
a. There are six runners, so six numbers are needed to represent them. A die is the obvious choice. Assign numbers 1 to 6 to each runner.	a. Each runner could be assigned a number on a die, which could be rolled to determine the winner of the race. When the die is rolled, it simulates a race. The number that lands uppermost is the winner.
b. In this simulation, each runner had an equally likely chance of winning each game.	b. The answer depends on each runner having an equally likely chance of winning the race. This is rarely the case.

8.4.5 Using relative frequencies: polls

Statisticians sometimes want to compare the attitudes of certain groups, with frequency tables used to visually show the statistical differences in the groups' attitudes. Frequency tables typically show results obtained from taking a random sample from a large population, or from gathering all the data from an entire population when that population is small.

When calculating relative frequencies, six steps are followed.

1. Gather the required statistics.
2. Determine the total number of people polled.
3. Determine each group's attitude in relation to the total number of people polled and write it in fractional form. For example:
   - Of the 100 people polled, 30 liked the caravan park $= \frac{30}{100}$.
   - Of the 100 people polled, 70 did not like the caravan park $= \frac{70}{100}$.
4. Convert the fraction to a decimal.

$$\frac{30}{100} = 0.3$$

5. Convert the decimal to a percentage.

$$\begin{aligned}\frac{30}{100} &= 0.3 \\ &= 30\%\end{aligned}$$

6. Check that all relative frequencies add to 100%.

Note: When the decimals or percentages are rounded off, the figure may be slightly below or above 100%.

WORKED EXAMPLE 10 Estimating probabilities using polls

In a recent election, there were 4 candidates. During a telephone poll conducted a week before the election, 500 randomly selected voters were asked to indicate their preferences for the 4 candidates. The results are shown in the table.

Candidate	1	2	3	4
Number of candidate's voters	115	168	145	72

Estimate each candidate's probability of winning the election, giving your answers as decimals.

THINK	WRITE
1. The probability of a candidate winning is the number of voters who prefer them divided by the total number of votes.	$\Pr(\text{candidate 1 wins}) = \dfrac{115}{500}$ $= 0.230$ $\Pr(\text{candidate 2 wins}) = \dfrac{168}{500}$ $= 0.336$ $\Pr(\text{candidate 3 wins}) = \dfrac{145}{500}$ $= 0.290$ $\Pr(\text{candidate 4 wins}) = \dfrac{72}{500}$ $= 0.144$
2. The closer the relative frequency is to 1, the higher the probability of success.	Candidate 2 is the favourite to win.

8.4.6 Calculating the number of successful trials

Sometimes, a relative frequency is known and may be used to calculate the number of individuals who responded a certain way, or the number of successful trials.

$$\text{number of successful trials} = \text{relative frequency} \times \text{total number of trials}$$

For example, you may have determined that the relative frequency of individuals who vaccinate their children is 88% or 0.88. In a population of 300 individuals, it can be assumed that approximately 264 individuals vaccinated their children ($0.88 \times 300 = 264$).

tlvd-3951

WORKED EXAMPLE 11 Calculating successful trials

In 2017, the Australian Marriage Law Postal Survey was conducted to determine the level of support within the population for legalising same-sex marriage in Australia.
The table shows the breakdown of votes per state as a decimal frequency.

		Yes	No
State/territory	New South Wales	0.578	0.422
	Victoria	0.649	0.351
	Queensland	0.607	0.393
	South Australia	0.625	0.375
	Western Australia	0.637	0.363
	Northern Territory	0.606	0.394
	Australian Capital Territory	0.740	0.260

a. 5000 people from Queensland were randomly selected. State how many answered 'yes'.
b. 3000 people from Western Australia were randomly selected. State how many answered 'no'.

THINK	WRITE
a. 1. Look on the chart and find the row with Queensland and those who answered 'yes'.	a. 0.607
2. Multiply 5000 (number of individuals) by the result from step 1. The answer represents the number of Queenslanders who answered 'yes'.	$0.607 \times 5000 = 3035$
3. State the answer.	Approximately 3035 Queenslanders answered 'yes'.
b. 1. Look on the chart and find the row with Western Australia and those who stated 'no'.	b. 0.363
2. Multiply 3000 (number of individuals) by the result from step 1. The answer represents the number of Western Australians who answered 'no'.	$0.363 \times 3000 = 1089$
3. State the answer.	Approximately 1089 Western Australians answered 'no'.

8.4.7 Long-term data

Unless we happen to 'know' the true probability of an event, a simulation as outlined previously provides us with a method of finding the 'experimental' probability. In fact, how do we know that the probability of getting a Tail when tossing a fair coin is really $\frac{1}{2}$? The answer to this lies in the belief that if we tossed a coin enough times, the proportion of Tails would be close to, if not exactly, $\frac{1}{2}$. However, there are no guarantees; different simulation experiments will yield different results, but it is the sum total of all coin-toss simulations that would be likely to yield a result close to 0.5.

An application of long-run proportion is to use experimental data to estimate the true probability of an event. Consider an application in the game of cricket.

WORKED EXAMPLE 12 Long-term data

A possible measure of a batsman's effectiveness in test cricket is the number of times he makes a run-scoring stroke as a proportion of balls faced. Consider this set of data from 10 innings over 5 test matches for a cricketer. Calculate the final long-run proportion of scoring strokes to balls faced.

Innings	1	2	3	4	5	6	7	8	9	10
Run-scoring strokes	34	1	67	38	12	15	47	69	43	18
Balls faced	62	6	107	87	29	19	75	119	67	31

THINK

1. Create 4 columns.
   a. Put the innings number in column 1.
   b. Put the cumulative (total) number of run-scoring strokes in column 2.
   c. Put the cumulative (total) number of balls faced in column 3.
2. Complete column 4.
   a. Divide the number in column 2 by the number in column 3.
   b. Enter this result in column 4. This is the long-run proportion for run-scoring strokes to balls faced.
   c. Round the answer to 3 decimal places.

WRITE

Column 1 (innings)	Column 2 (run-scoring strokes)	Column 3 (balls faced)	Column 4 (proportion)
1	34	62	$34 \div 62 = 0.548$
2	$34 + 1 = 35$	$62 + 6 = 68$	$35 \div 68 = 0.515$
3	$35 + 67 = 102$	$68 + 107 = 175$	$102 \div 175 = 0.583$
4	$102 + 38 = 140$	$175 + 87 = 262$	$140 \div 262 = 0.534$
5	$140 + 12 = 152$	$262 + 29 = 291$	$152 \div 291 = 0.522$
6	$152 + 15 = 167$	$291 + 19 = 310$	$167 \div 310 = 0.539$
7	$167 + 47 = 214$	$310 + 75 = 385$	$214 \div 385 = 0.556$
8	$214 + 69 = 283$	$385 + 119 = 504$	$283 \div 504 = 0.562$
9	$283 + 43 = 326$	$504 + 67 = 571$	$326 \div 571 = 0.571$
10	$326 + 18 = 344$	$571 + 31 = 602$	$344 \div 602 = \mathbf{0.571}$

This implies the batsman has a probability of making a run-scoring stroke of about 0.57. It is significant to note that this proportion didn't change much after the first 6 or 7 innings. Although this might not be the best way to measure a batsman's 'effectiveness' in cricket, it is similar to the method used in baseball to measure 'batting average', a method that has been used for over 100 years.

It is not necessary to calculate long-run proportion at all stages of an experiment. Often it is sufficient merely to take the data at the end of the experiment and perform the appropriate division.

WORKED EXAMPLE 13

A die that is suspected of being biased (unfair) is tossed 800 times and it is observed that 6 appeared 205 times. Calculate the long-run proportion and comment on the result.

THINK

1. Calculate the long-run proportion. In this case, divide the number of sixes by the total.
2. Compare with the 'theoretical' result.

WRITE

1. $\text{Long-run proportion} = \dfrac{205}{800}$

   $= 0.256$

2. We expect about $\dfrac{1}{6}$ of the results to be sixes, or about 133 out of 800. This proportion is 0.167. The experimental result indicates that the die is likely to be unfair, with 6 appearing too often.

It is only after a *very large* number of trials, say up to 100 000, that we could be confident the experimental probability was close to the theoretical probability.

8.4 Exercise

Students, these questions are even better in jacPLUS

Receive immediate feedback and access sample responses

Access additional questions

Track your results and progress

Find all this and MORE in jacPLUS

Note: Unless otherwise stated, express probabilities as fractions.

1. **WE7** To simulate whether or not the weather will be suitable for sailing, a coin is flipped. If the coin lands Heads up, the weather is perfect. If the coin lands Tails up, the weather is not suitable.
   Mandy flips a coin 20 times and obtains 13 Heads and 7 Tails. Sophia flips a coin 100 times and obtains 47 Heads and 53 Tails.
   a. Determine the relative frequency of perfect weather days in both cases.
   b. Compare the results from the two simulations and determine which is a better estimate of the true probability of the experiment.

2. A group of 50 students were tested for COVID and three were found to be positive.
   a. Calculate the relative frequency of testing positive for COVID.
   b. Determine how many students you would expect to be positive in the school population of 500 students.
3. The results of a class experiment that involved rolling a standard six-sided die 300 times are shown in the table.

Number on die	1	2	3	4	5	6
Number of times rolled	42	50	61	37	52	58

   Use the results to calculate the experimental probability of rolling:
   a. the number 5
   b. an odd number
   c. an even number.
4. In their last 20 games, a basketball player sank 17 free throws and missed 11. Estimate the probability that they will sink their next free throw.

5. **MC** A student rolls a pair of dice 10 times and records the number of times the total is 7. They repeat the experiment numerous times and their results are as follows.

$$1, 5, 3, 2, 1, 0, 2, 3, 4, 3, 2, 1, 0, 1, 2, 6, 2, 0, 1, 2$$

   The estimated probability of getting a total of 7 when 2 dice are tossed is:

   A. $\frac{7}{12}$ B. $\frac{41}{100}$ C. $\frac{41}{200}$ D. $\frac{41}{400}$ E. 0

6. a. **WE8** Use a random number generator to simulate the number of walnuts in 40 different carrot cakes, with a maximum of 60 in each cake.
   b. Calculate the probability that there will be more than 40 walnuts in a randomly chosen cake.

7. The gender of babies in a set of triplets is simulated by flipping 3 coins. If a coin lands Tails up, the baby is a boy. If a coin lands Heads up, the baby is a girl. In the simulation, the trial is repeated 40 times and the following results show the number of Heads obtained in each trial.

   0, 3, 2, 1, 1, 0, 1, 2, 1, 0, 1, 0, 2, 0, 1, 0, 1, 2, 3, 2,
   1, 3, 0, 2, 1, 2, 0, 3, 1, 3, 0, 1, 0, 1, 3, 2, 2, 1, 2, 1

   a. Calculate the probability, as a percentage, that exactly one of the babies in a set of triplets is female.
   b. Calculate the probability, as a percentage, that more than one of the babies in the set of triplets is female.

8. **WE9** There are 24 drivers lining up at the start of a Formula 1 race.
   a. Explain how simulation could be used to determine the winner.
   b. Consider if there are any limitations in your answer.

9. **WE10** A school canteen asked 150 random Year 12 students to indicate their favourite food before they asked the entire school to place their vote. The food with the most votes was going to be made half price for a week.
   The results from the Year 12 students are shown in the table.

Meal	Tally
Hamburger	45
Fish and chips	31
Macaroni and cheese	30
Lamb souvlaki	25
BBQ pork ribs	19

   Estimate each food's probability in the overall vote, giving your answers as decimals.

10. If a computer manufacturer wanted to simulate the probability of its laptops having a fault, determine if it would be appropriate to use a random number generator to select random computers on the assembly line. Explain your reasoning.

11. A random number generator was used to select how many students will receive a free lunch on any given day. The maximum number of students who can receive a free lunch is 10 and the minimum is 0.
   a. If the test is run for 30 days, calculate the theoretical probability that:
      i. 8 or more students are selected on any given day
      ii. 0 students are selected on any given day
      iii. 10 students are selected on any given day.
   b. Use a random number generator to simulate the free-lunch program over 30 days, and compare the results to those from part a.
   c. State if the results from the random number generator are the same as your predictions. Explain your answer.
12. Determine whether the following simulations are easy or difficult to undertake. If your response is 'difficult', explain why.
   a. Predicting how a helicopter pilot will react in an emergency
   b. Estimating how many times you will win when spinning a roulette wheel
   c. Predicting how many kilometres a new car will travel before any major mechanical problems
   d. Estimating how many car accidents you will have in your lifetime
   e. Estimating the number of defective mobile phones that one manufacturer will produce over 2 months
13. **WE11** In 2022, a survey was conducted to gauge support for changing the age of gaining a licence from a minimum of 18 years to a minimum of 17 years.
   The table shows the breakdown of votes by age as a decimal frequency.

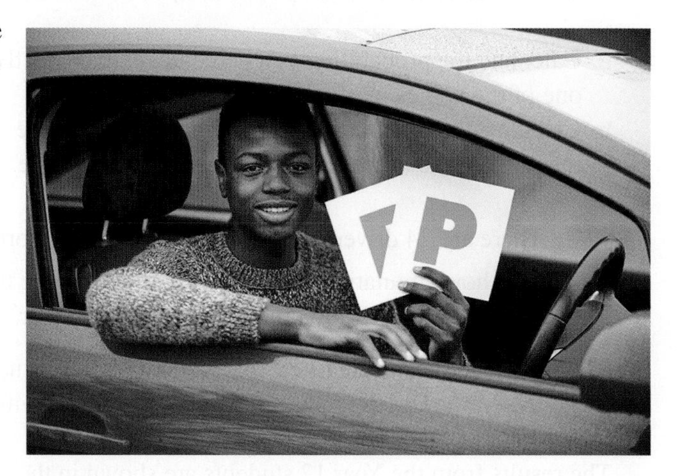

Ages of respondents	Yes	No
16–20	0.871	0.129
21–25	0.764	0.236
26–30	0.641	0.359
31–35	0.443	0.557
36–40	0.398	0.602
41–45	0.287	0.713
51–55	0.213	0.787
51–55	0.122	0.878

   2800 individuals aged 16–20 were randomly selected and asked for their vote.
   a. Calculate how many will state they answered 'yes'.
   b. Calculate how many will state they answered 'no'.
14. a. Five thousand new domestic university students are selected to take part in a study. The study shows the completion rates of higher education students in a 4-year cohort. Use the data in the chart to predict how many students did not come back after their first year of university for the 4-year period of 2021–2024.
   b. Based on the domestic trend for students not coming back after the first year, state whether your estimate in part a will be higher or lower than the actual percentage. Explain your answer.

Completion rates of higher education students — cohort analysis, 2011–2020					
National total (domestic students)	**Year**	**Completed (in any year)**	**Still enrolled at the end of the 4-year cohort period**	**Re-enrolled, but dropped out**	**Never came back after the first year**
	2011–2014	45.1%	34.5%	11.7%	8.7%
	2012–2015	44.2%	34.2%	12.1%	9.4%
	2013–2016	42.9%	34.3%	12.2%	10.6%
	2014–2017	42.0%	34.6%	12.5%	10.9%
	2015–2018	42.1%	34.5%	12.6%	10.9%
	2016–2019	43.0%	34.1%	12.6%	10.3%
	2017–2020	41.7%	35.9%	11.9%	10.5%

Source: Table extract from Department of Education and Training, *Completion Rates of Higher Education Students — cohort analysis*, 2011–2020; © Commonwealth of Australia.

15. The results of a coin-flipping experiment are shown.

Outcome	Frequency
Heads	38
Tails	62
Total	100

a. Calculate the relative frequencies, expressed as a percentage, of:

i. Heads
ii. Tails.

b. If the experiment was repeated, determine whether you would expect the same results. Explain your answer.

c. If the experiment was repeated 100 times, state what you expect to happen to the relative frequencies of the outcomes.

16. The youth of today were surveyed regarding their opinion on important environmental issues such as carbon emissions, electric cars and solar energy. The following results were found:

Environmental issue	Number surveyed	Number who agree
Agree their country will meet its carbon emission targets	535	145
Believe they will purchase an electric car in the next 10 years	386	106
Have or believe they will get solar panels in the next 10 years	424	275

From the data provided, answer the following.

a. Calculate the probability that:

i. the people surveyed agree their country will meet their carbon emission targets
ii. the people surveyed agree they will purchase an electric car in the next 10 years
iii. the people surveyed disagree they will get solar panels in the next 10 years.

b. Looking at the answers, predict the environmental issue that the people surveyed consider of most concern. Explain your reasoning.

c. Looking at the answers, predict the environmental issue that is of least concern. Explain your reasoning.

17. Research over the past 25 years shows that each November there is an average of two wet days on Sunnybank Island. Travelaround Tours offer one-day tours to Sunnybank Island at a cost of $150 each, with a money back guarantee against rain.
  a. Determine the relative frequency of wet November days as a percentage.
  b. If Travelaround Tours take 1200 bookings for tours in November, determine how many refunds they could expect to give.

18. An average of 200 robberies takes place each year in the town of Amiak. There are 10 000 homes in this town.
  a. Determine the relative frequency of robberies in Amiak.
  b. Each robbery results in an average insurance claim of $20 000. Determine the medium premium per home that the insurance company would need to charge to cover these claims.

19. A possible measure of a batsman's effectiveness in test cricket is the number of times he makes a run-scoring stroke as a proportion of balls faced. Consider the data from 12 innings over 6 test matches for a cricketer. Calculate the *final* long-run proportion of scoring shots to balls faced.

Innings	1	2	3	4	5	6	7	8	9	10	11	12
Scoring shots	23	26	45	9	23	34	56	37	18	23	31	46
Balls faced	57	57	89	19	46	72	100	68	31	50	66	89

20. A coin suspected of being biased is tossed 600 times and it is observed that a Head appeared 425 times. Calculate the long-run proportion and comment on the result.

21. A student answers 200 questions from a mathematics textbook and answers 156 of them correctly. Estimate the probability that the student will answer the next question correctly.

22. A cricketer wishes to measure her effectiveness against left- and right-handed bowlers and records the following data over 2 years.

	Innings caught	**Innings bowled**	**Innings not out**
Left-handers	6	4	4
Right-handers	24	26	6

  a. State whether she is more likely to be caught by left- or right-handers.
  b. State whether she is more likely to be caught or bowled.
  c. State if she is more likely to be not out against left- or right-handers.
  d. Use the results from a to c to comment on the cricketer's measure of effectiveness against left- and right-handed bowlers.

8.5 Review

8.5.1 Summary

doc-38016

Hey students! Now that it's time to revise this topic, go online to:

Review your results

Watch teacher-led videos

Practise questions with immediate feedback

Find all this and MORE in jacPLUS

8.5 Exercise

Multiple choice

1. MC The event that has the highest probability of occurring is:
   - **A.** a cyclone in Victoria.
   - **B.** the temperature rising above 27 degrees on a summer day in Melbourne.
   - **C.** snow in Cairns.
   - **D.** getting struck by lightning during a storm.
   - **E.** winning the lottery.

2. MC Two fair dice are rolled and the numbers added together. The most likely outcome is:
   - **A.** 12
   - **B.** 5
   - **C.** 7
   - **D.** 9
   - **E.** 2

3. MC Two fair dice are rolled and the numbers added together. The probability of getting a 9 is:
   - **A.** $\frac{1}{9}$
   - **B.** $\frac{1}{6}$
   - **C.** $\frac{1}{3}$
   - **D.** $\frac{2}{9}$
   - **E.** $\frac{1}{12}$

4. MC If the probability of an event is $\frac{5}{12}$, the probability of the complement occurring is:
   - **A.** $\frac{5}{12}$
   - **B.** $\frac{12}{5}$
   - **C.** $\frac{4}{12}$
   - **D.** $\frac{7}{12}$
   - **E.** $\frac{1}{2}$

The following table shows what mobile phone plan people use. This table is used for questions 5 and 6.

Telstra	Vodafone	Optus	Virgin
48%	27%	15%	8%

5. MC The percentage of people who use neither of the four mentioned providers is:
   A. 8% B. 27% C. 12% D. 15% E. 2%
6. MC The percentage of people who don't use Telstra or Vodafone is:
   A. 75% B. 23% C. 25% D. 27% E. 48%
7. MC The spinner shown is spun 21 times. The sample space is:
   A. $\frac{1}{3}$ B. 21 C. {red, blue, yellow}
   D. {blue, pink, green} E. 7

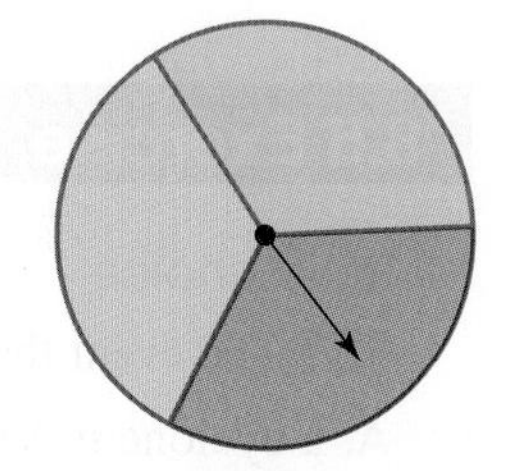

8. MC If 20% of VCE students have had COVID and the school has 350 VCE students, the number of VCE students who have had COVID at this school is:
   A. 20 B. 350 C. 70 D. 280 E. 80
9. MC A coin is tossed 3 times. The probability of getting three Tails in a row is:
   A. $\frac{1}{2}$ B. $\frac{1}{4}$ C. $\frac{1}{6}$ D. $\frac{1}{3}$ E. $\frac{1}{8}$
10. MC A die is rolled 200 times and the following results are recorded. From the data, the probability of rolling an even number is:

Number on die	1	2	3	4	5	6
Number of times rolled	32	38	27	31	33	39

   A. $\frac{27}{50}$ B. $\frac{1}{2}$ C. $\frac{11}{20}$ D. $\frac{23}{50}$ E. $\frac{13}{25}$

Short answer

11. Determine the sample space for the following scenarios.
    a. Rolling an 8-sided die numbered 1–8
    b. Drawing an 'a' from a bag containing the letters of the alphabet
    c. Drawing the winning ticket in a raffle with 100 tickets sold
12. Two fair dice are rolled and the product of the numbers is recorded.
    a. Construct a table to illustrate the sample space. List the sample space for the experiment.
    b. Determine the most likely outcome.
    c. Determine the least likely outcome.
13. a. Calculate the probability of rolling a 3 on a 6-sided die.
    b. Calculate the complement of the event in part a.
    c. Calculate the probability of the complement occurring.

14. The following table shows what people do with their old mobile phones in Australia.

Kept it	Shared it	Recycled it	Sold it	Lost it
48%	27%	15%	8%	2%

Source: Data from 2015 Deloitte Touche Tohmatsu, *Mobile Consumer Survey — The Australian Cut*

Calculate the probability of *not* selling or *not* losing your mobile phone in Australia. Give your answer as a percentage.

Extended response

15. In the game of backgammon, two dice are rolled together. The resulting numbers can be used separately in two separate moves, or their total can be used for one move.
   a. List the sample space of rolling two dice simultaneously.
   b. Calculate the probability of getting a 4 on a die or a total of 4 on both dice.
   c. Calculate the probability of *not* getting a 4 on a die or a total of 4 on both dice.
   d. Calculate the probability of getting a 4 or getting a 5 on a die or a total of 5 on both dice.

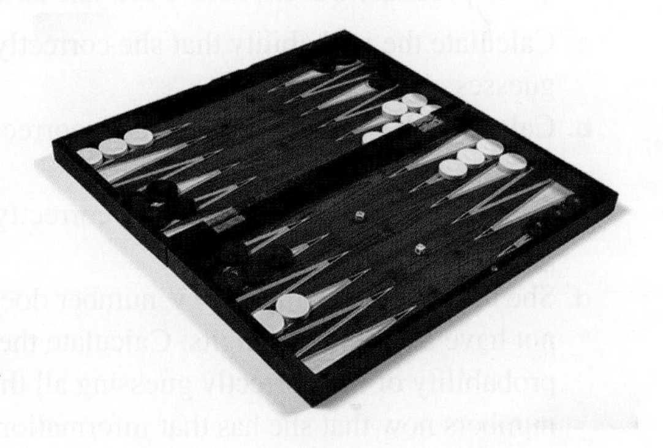

16. a. Calculate the number of times, to the nearest whole number, you could expect to roll a 6 if you rolled a fair die:
   i. 10 times   ii. 150 times   iii. 5000 times.
   b. If you actually did the experiments in part a, state whether you would get the exact number that you predicted. Explain your answer.

17. You are about to sit a Mathematics examination that contains 40 multiple-choice questions. You didn't have time to study so you are going to choose the answers completely randomly. There are four choices for each question.
Explain how you could use random numbers to select your answers for you.

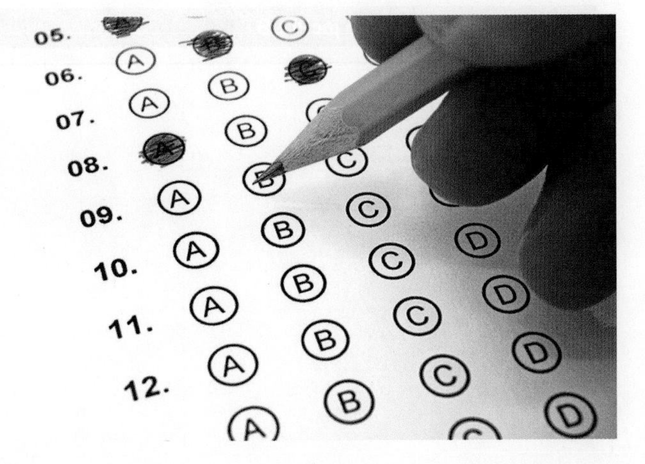

18. The spinner shown was spun 50 times and the outcome each time was recorded in the table.

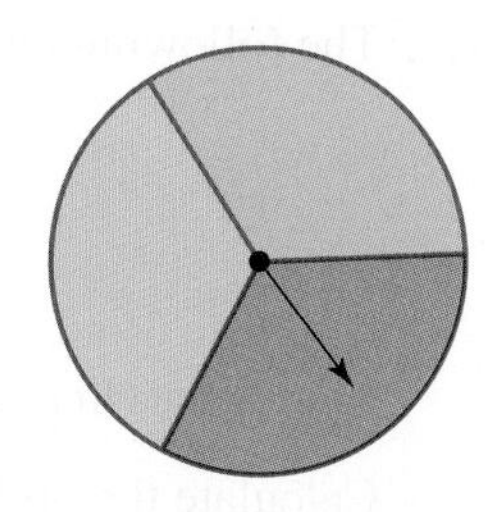

	Segment		
	Blue	Green	Pink
Tally	17	23	10

a. List the sample space.
b. Given the experimental results, determine the relative frequency for each segment.
c. The theoretical probability of the spinner landing on any particular segment with one spin is $\frac{1}{3}$. Explain how the experiment could be changed to give a better estimate of the true probabilities.

19. A friend offers you a chance to roll a single die three times, and if you roll a 3 on any one of the rolls, he will pay for your lunch. Otherwise, you will pay for his lunch. Decide if you should play his game. Explain your answer.

20. Sonya knows her credit card number for online shopping. However, she has forgotten the three-digit CVV number on the back of the card and wants to purchase a dress before the sale ends.

a. Calculate the probability that she correctly guesses all three digits.
b. Calculate the probability that she incorrectly guesses all three digits.
c. Calculate the probability that she correctly guesses at least one digit.
d. She remembers that her CVV number does not have any repeated digits. Calculate the probability of her correctly guessing all the numbers now that she has that information.

Answers

Topic 8 Simulations and simple probability

8.2 Theoretical probability and sample spaces

8.2 Exercise

1. a. 1 b. 0.75 c. 0.25
 d. 0.75 e. 0.25 f. 1
2. Answers will vary. Sample responses can be found in the worked solutions in the online resources.
3. a. $\{1, 2, 3, 4, 5, 6\}$
 b. {green, yellow, blue}
 c. {a, b, c, d, e, f, g, h, i, j, k, l, m, n, o, p, q, r, s, t, u, v, w, x, y, z}
4. a. $\frac{1}{6}$ b. $\frac{1}{3}$ c. $\frac{1}{2}$
5. a. $\frac{4}{11}$ b. $\frac{7}{11}$ c. $\frac{2}{11}$ d. $\frac{1}{11}$
6. a. $\{1, 2, 3, 4, 5, 6, 7, 8\}$
 b. i. 50% ii. 62.5%
 iii. 0% iv. 100%
7. a. i. No, the colours are not represented equally.
 ii. Blue, green, yellow, pink
 iii. $\Pr(\text{blue}) = \frac{1}{3}$, $\Pr(\text{green}) = \frac{1}{6}$, $\Pr(\text{pink}) = \frac{1}{3}$, $\Pr(\text{yellow}) = \frac{1}{6}$
 b. i. Yes, as there are equal numbers of the same colour.
 ii. Blue, green, yellow, purple, orange
 iii. $\Pr(\text{blue}) = \frac{1}{5}$, $\Pr(\text{green}) = \frac{1}{5}$, $\Pr(\text{purple}) = \frac{1}{5}$, $\Pr(\text{yellow}) = \frac{1}{5}$, $\Pr(\text{orange}) = \frac{1}{5}$
 c. i. No, the colours are not represented equally.
 ii. Pink, blue
 iii. $\Pr(\text{blue}) = \frac{1}{3}$, $\Pr(\text{pink}) = \frac{2}{3}$
8. a. Blue. There are more blue jelly beans (6) than beans of any other colour.
 b. Yellow. There are fewer yellow jelly beans (2) than beans of any other colour.
 c. $\Pr(\text{blue}) = \frac{6}{17}$, $\Pr(\text{green}) = \frac{5}{17}$, $\Pr(\text{red}) = \frac{4}{17}$, $\Pr(\text{yellow}) = \frac{2}{17}$
9. D
10. a. $\frac{1}{4}$ b. $\frac{1}{2}$ c. $\frac{1}{52}$ d. $\frac{3}{13}$
11. a.

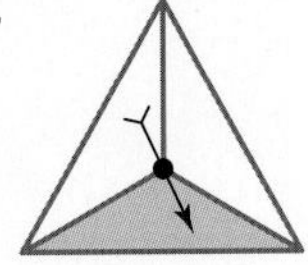

 b.

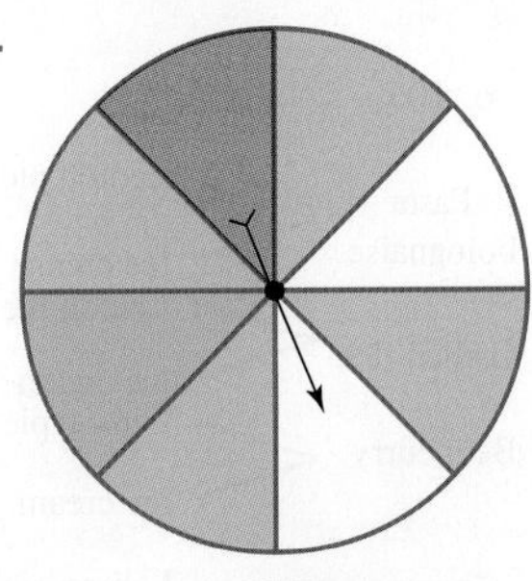

 c.

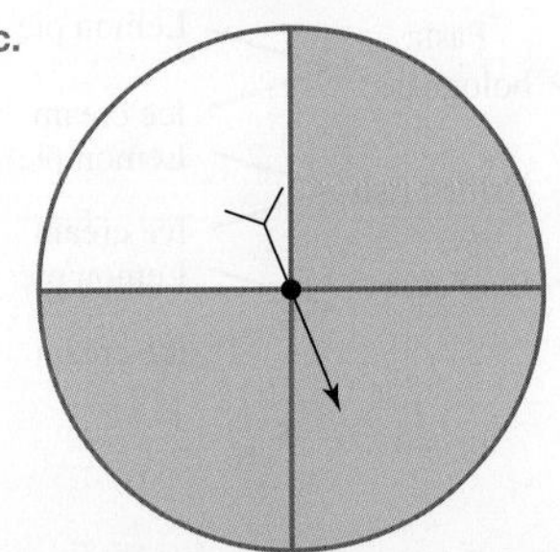

12. a. Not owning a dog
 b. Flipping a coin and getting a Head
 c. Being awake
 d. Truancy
 e. Failing your learner permit knowledge test
 f. Not owning a mobile phone
13. a. Failing the final exam
 b. 80%
14. False. The complement of picking a red flower is not picking a red flower; therefore, the complement of picking a red flower is picking a purple or yellow flower.

8.3 Multi-stage experiments

8.3 Exercise

1.

		Die					
		1	2	3	4	5	6
Spinner	Red	(R, 1)	(R, 2)	(R, 3)	(R, 4)	(R, 5)	(R, 6)
	Blue	(B, 1)	(B, 2)	(B, 3)	(B, 4)	(B, 5)	(B, 6)
	Green	(G, 1)	(G, 2)	(G, 3)	(G, 4)	(G, 5)	(G, 6)
	Yellow	(Y, 1)	(Y, 2)	(Y, 3)	(Y, 4)	(Y, 5)	(Y, 6)
	Orange	(O, 1)	(O, 2)	(O, 3)	(O, 4)	(O, 5)	(O, 6)

2.

		Spinner 2				
		1	2	3	4	5
Spinner 1	1	(1, 1)	(1, 2)	(1, 3)	(1, 4)	(1, 5)
	2	(2, 1)	(2, 2)	(2, 3)	(2, 4)	(2, 5)
	3	(3, 1)	(3, 2)	(3, 3)	(3, 4)	(3, 5)
	4	(4, 1)	(4, 2)	(4, 3)	(4, 4)	(4, 5)
	5	(5, 1)	(5, 2)	(5, 3)	(5, 4)	(5, 5)

$\Pr(\text{exactly one } 5) = \frac{8}{25}$

3. D

4. a. 5% b. 10% c. 10% d. 20%

5. a.

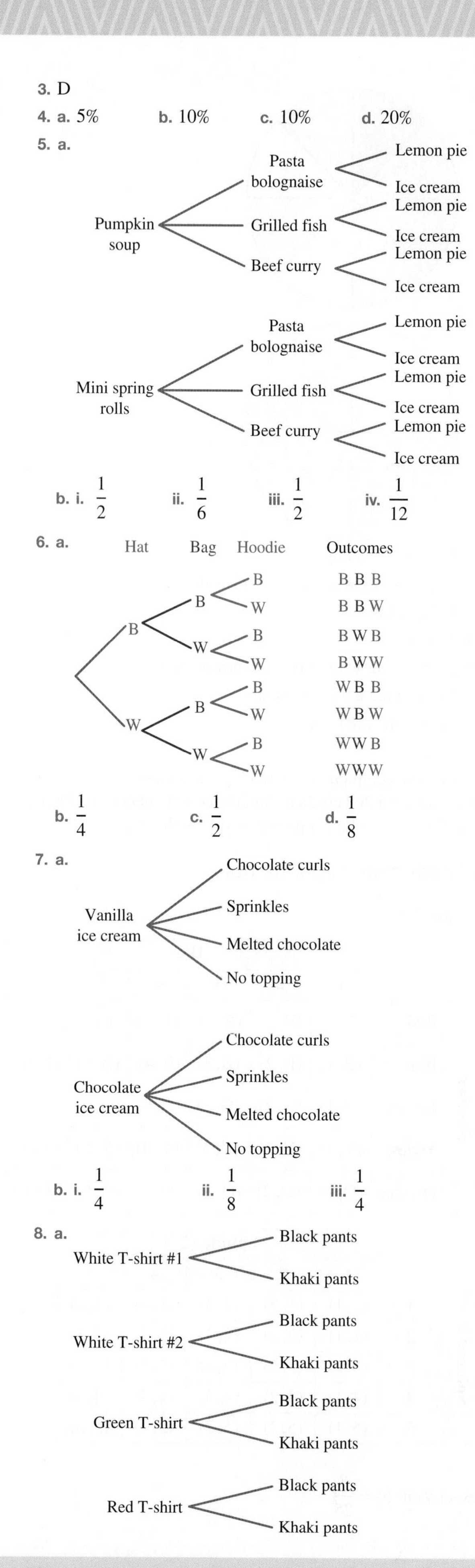

b. i. $\frac{1}{2}$ ii. $\frac{1}{6}$ iii. $\frac{1}{2}$ iv. $\frac{1}{12}$

6. a.

b. $\frac{1}{4}$ c. $\frac{1}{2}$ d. $\frac{1}{8}$

7. a.

b. i. $\frac{1}{4}$ ii. $\frac{1}{8}$ iii. $\frac{1}{4}$

8. a.

b. $\frac{1}{8}$

9. a.

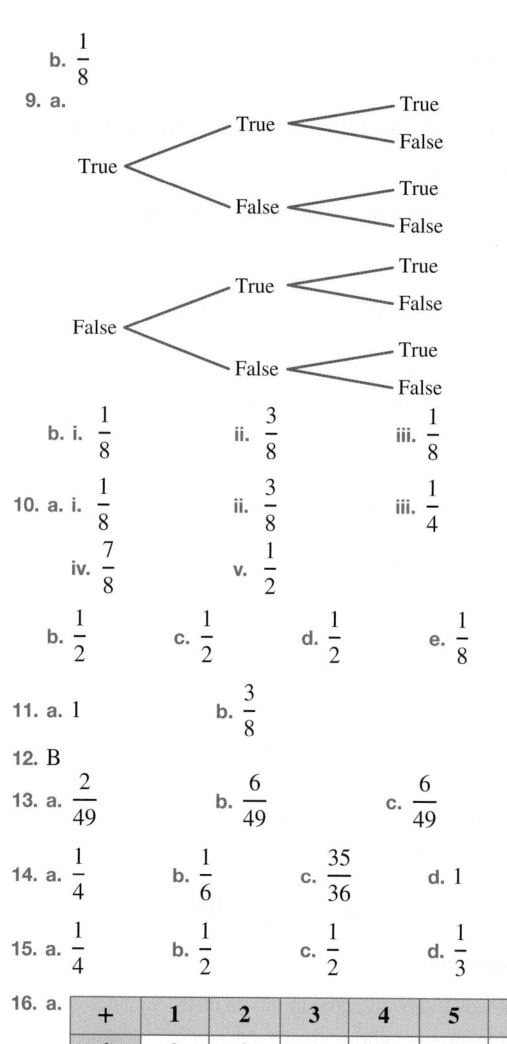

b. i. $\frac{1}{8}$ ii. $\frac{3}{8}$ iii. $\frac{1}{8}$

10. a. i. $\frac{1}{8}$ ii. $\frac{3}{8}$ iii. $\frac{1}{4}$ iv. $\frac{7}{8}$ v. $\frac{1}{2}$

b. $\frac{1}{2}$ c. $\frac{1}{2}$ d. $\frac{1}{2}$ e. $\frac{1}{8}$

11. a. 1 b. $\frac{3}{8}$

12. B

13. a. $\frac{2}{49}$ b. $\frac{6}{49}$ c. $\frac{6}{49}$

14. a. $\frac{1}{4}$ b. $\frac{1}{6}$ c. $\frac{35}{36}$ d. 1

15. a. $\frac{1}{4}$ b. $\frac{1}{2}$ c. $\frac{1}{2}$ d. $\frac{1}{3}$

16. a.

+	1	2	3	4	5	6
1	2	3	4	5	6	7
2	3	4	5	6	7	8
3	4	5	6	7	8	9
4	5	6	7	8	9	10
5	6	7	8	9	10	11
6	7	8	9	10	11	12

b. The diagonals in the sample space are repeating numbers. Also, 7 is the outcome that shows up the most, and each outcome further away from 7 has a reduced chance of occurring.

c. See the table at the bottom of the page.*

d. The probabilities on either side of the 7 decrease at the same rate.

17. a.

–	1	2	3	4	5	6
1	0	1	2	3	4	5
2	1	0	1	2	3	4
3	2	1	0	1	2	3
4	3	2	1	0	1	2
5	4	3	2	1	0	1
6	5	4	3	2	1	0

b. $\frac{1}{3}$ c. $\frac{2}{3}$ d. No

18. a.

		Person 2		
		Rock	**Paper**	**Scissors**
Person 1	**Rock**	RR	RP	RS
	Paper	PR	PP	PS
	Scissors	SR	SP	SS

b. i. $\frac{2}{9}$ ii. $\frac{2}{9}$ iii. $\frac{2}{9}$

c. Yes

d. i.

		Person 2			
		Rock	**Paper**	**Scissors**	**Dynamite**
Person 1	**Rock**	RR	RP	RS	RD
	Paper	PR	PP	PS	PD
	Scissors	SR	SP	SS	SD
	Dynamite	DR	DP	DS	DD

ii. No

iii. $\Pr(\text{rock wins}) = \frac{2}{16} = \frac{1}{8}$

$\Pr(\text{paper wins}) = \frac{2}{16} = \frac{1}{8}$

$\Pr(\text{scissors win}) = \frac{4}{16} = \frac{1}{4}$

$\Pr(\text{dynamite wins}) = \frac{4}{16} = \frac{1}{4}$

Scissors and dynamite are twice as likely to win as rock and paper.

19. a.

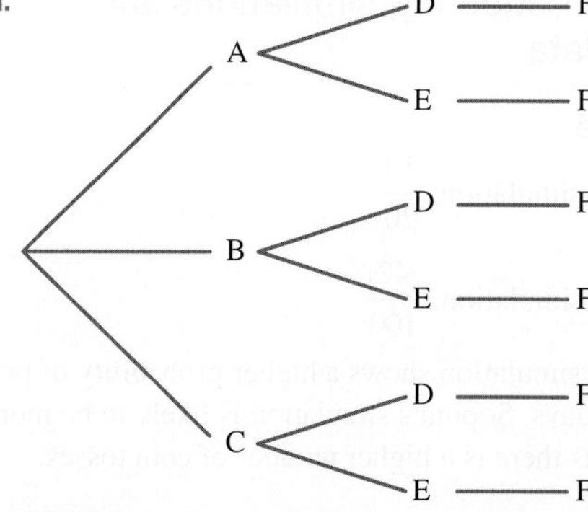

6 routes

b. No. There are still six paths, even if we omit home from the journey. Since there is only one choice to go from the library to home, the outcomes do not change.

20.

1
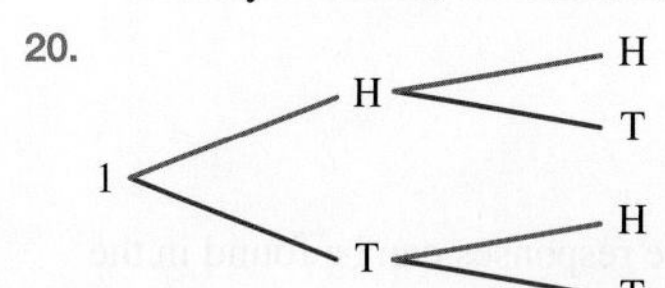

2
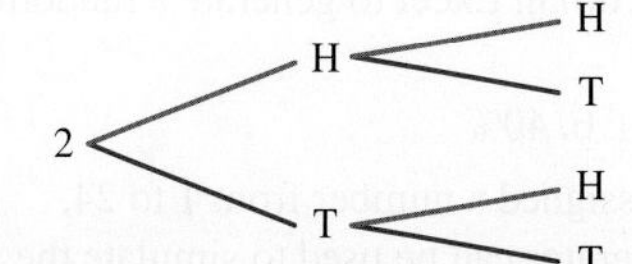

3
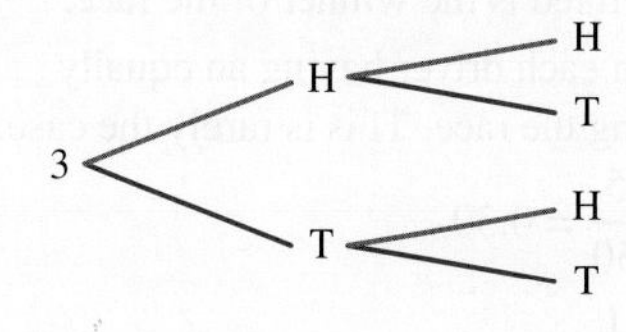

4
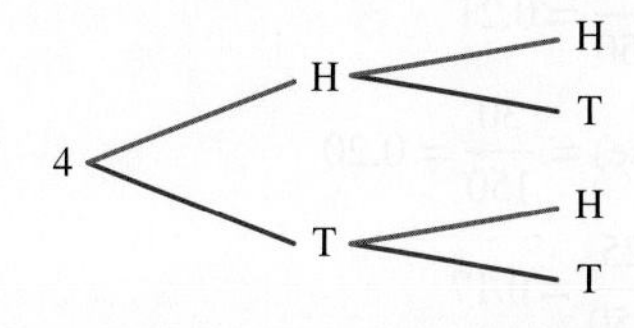

5
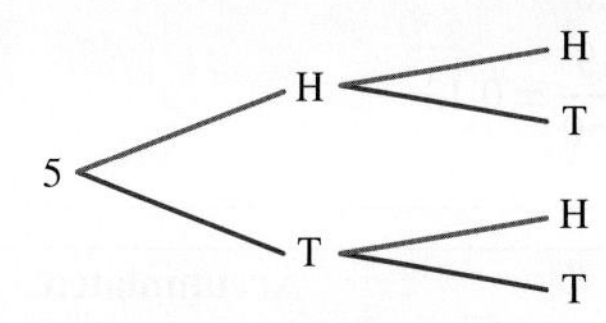

6
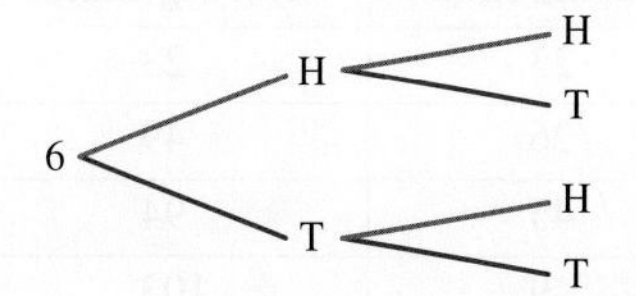

***16. c.**

Sum of two dice	2	3	4	5	6	7	8	9	10	11	12
Probability	$\frac{1}{36}$	$\frac{1}{18}$	$\frac{1}{12}$	$\frac{1}{9}$	$\frac{5}{36}$	$\frac{1}{6}$	$\frac{5}{36}$	$\frac{1}{9}$	$\frac{1}{12}$	$\frac{1}{18}$	$\frac{1}{36}$

8.4 Relative frequency, simulations and long-term data

8.4 Exercise

1. a. Mandy's simulation: $\frac{13}{20}$

   Sophia's simulation: $\frac{47}{100}$

   b. Mandy's simulation shows a higher probability of perfect weather days. Sophia's simulation is likely to be more reliable as there is a higher number of coin tosses.

2. a. $\frac{3}{50}$ b. 30 batteries

3. a. $\frac{13}{75}$ b. $\frac{31}{60}$ c. $\frac{29}{60}$

4. $\frac{17}{28}$

5. C

6. Answers will vary. Sample responses can be found in the worked solutions in the online resources. You can use =RANDBETWEEEN(1, 60) in Excel to generate a random number.

7. a. 35% b. 40%

8. a. Each driver could be assigned a number from 1 to 24. A random number generator can be used to simulate the race. The number generated is the winner of the race.

   b. The answer depends on each driver having an equally likely chance of winning the race. This is rarely the case.

9. a. $\Pr(\text{fish and chips}) = \frac{45}{150} = 0.30$

   b. $\Pr(\text{fish and chips}) = \frac{31}{150} = 0.21$

   c. $\Pr(\text{macaroni and cheese}) = \frac{30}{150} = 0.20$

   d. $\Pr(\text{lamb souvlaki}) = \frac{25}{150} = 0.17$

   e. $\Pr(\text{BBQ pork ribs}) = \frac{19}{150} = 0.13$

10. Yes. It is too expensive to check every computer for every possible fault. A random number generator will allow the manufacturer to gain insight into the probability of a computer having a fault.

11. a. i. $\frac{3}{11}$ ii. $\frac{1}{11}$ iii. $\frac{1}{11}$

    b. Answers will vary. Sample responses can be found in the worked solutions in the online resources.

    c. Answers will vary. Accuracy can be improved by using the random number generator to test more days. As more data is obtained, the results will be closer and closer to the theoretical probability.

12. a–e. Answers will vary. Sample responses can be found in the worked solutions in the online resources.

13. a. 2439 b. 361

14. a. 525 domestic students

    b. Answers may vary. However, based on the trend observed from 2011 to 2017, the number of students who never came back after their first year has steadily increased, peaking at 10.9% and remaining between 10 and 11%. If it remains in this bracket for the coming years, the estimate of 525 students for 2021−24 will be fairly accurate.

15. a. i. 38% ii. 62%

    b. No. If the experiment was repeated, it is unlikely that the results would be the same as or similar to the previous experiment.

    c. According to the law of large numbers, as the number of trials increases, the gap between the experimental probability and the theoretical probability decreases.

16. a. i. $\frac{29}{107}$ ii. $\frac{53}{193}$ iii. $\frac{149}{424}$

    b. The most important environmental issue was purchase of solar panels.

    c. The environmental issue of least concern was whether the country will meet its carbon emission targets.

17. a. 6.67% b. 80

18. a. 0.02 b. $400

19. See the table at the bottom of the page.*

*19.

Innings	Scoring shots	Accumulated scoring shots	Balls faced	Accumulated balls faced	Proportion
1	23	23	57	57	0.403 508 772
2	26	49	57	114	0.429 824 561
3	45	94	89	203	0.463 054 187
4	9	103	19	222	0.463 963 964
5	23	126	46	268	0.470 149 254
6	34	160	72	340	0.470 588 235
7	56	216	100	440	0.490 909 091
8	37	253	68	508	0.498 031 496
9	18	271	31	539	0.502 782 931
10	23	294	50	589	0.499 151 104
11	31	325	66	655	0.496 183 206
12	46	371	89	744	0.498 655 914

20. 0.708

Coment: One would expect the long-run proportion to be close to 0.5, so it seems the coin could be biased.

21. 0.78

22. a. Caught by left-handers = 0.429

Caught by right-handers = 0.429

Based on long-run proportions, there is no difference between being caught by a left-hander or a right-hander.

b. Bowled: $\frac{30}{70}$

Caught: $\frac{30}{70}$

There is no difference.

c. Not out (left) = 0.29
Not out (right) = 0.11
The value for not out is higher for left-handers than for right-handers.

d. The relatively small number of left-handed observations means comparisons are not very accurate. However, there seems little difference between left- and right-handed effectiveness.

8.5 Review

8.5 Exercise

Multiple choice

1. B
2. C
3. A
4. D
5. E
6. C
7. D
8. C
9. E
10. A

Short answer

11. a. $\{1, 2, 3, 4, 5, 6, 7, 8\}$

b. $\{a, b, c, d, e, f, g, h, i, j, k, l, m, n, o, p, q, r, s, t, u, v, w, x, y, z\}$

c. {winning ticket, losing ticket}

12. a.

×	1	2	3	4	5	6
1	1	2	3	4	5	6
2	2	4	6	8	10	12
3	3	6	9	12	15	18
4	4	8	12	16	20	24
5	5	10	15	20	25	30
6	6	12	18	24	30	36

Sample space is: $\{1, 2, 3, 4, 5, 6, 8, 9, 10, 12, 15, 16, 18, 20, 24, 25, 30, 36\}$

b. 6 or 12

c. $1, 9, 16, 25$ or 36

13. a. $\frac{1}{6}$

b. The complement is the set $\{1, 2, 4, 5, 6\}$ or not rolling a 3.

c. $\frac{5}{6}$

14. 90%

Extended response

15. a.

		Die 2					
		1	2	3	4	5	6
Die 1	1	(1, 1)	(1, 2)	(1, 3)	(1, 4)	(1, 5)	(1, 6)
	2	(2, 1)	(2, 2)	(2, 3)	(2, 4)	(2, 5)	(2, 6)
	3	(3, 1)	(3, 2)	(3, 3)	(3, 4)	(3, 5)	(3, 6)
	4	(4, 1)	(4, 2)	(4, 3)	(4, 4)	(4, 5)	(4, 6)
	5	(5, 1)	(5, 2)	(5, 3)	(5, 4)	(5, 5)	(5, 6)
	6	(6, 1)	(6, 2)	(6, 3)	(6, 4)	(6, 5)	(6, 6)

b. $\frac{7}{18}$ c. $\frac{11}{18}$ d. $\frac{2}{3}$

16. a. i. 1 ii. 25 iii. 833

b. No, these are estimated numbers calculated from the theoretical probabilities. As the experiment is carried out more times, the experimental probabilities will approach the theoretical probabilities. There is a slight chance in any experiment that you may obtain the theoretical probability, but it is unlikely.

17. You could run your random number generator using the numbers 1–4. These numbers would represent A, B, C and D respectively in the multiple-choice questions. You would run this simulation 40 times.

18. a. Blue, green, pink

b. $\Pr(\text{blue}) = \frac{17}{50}$

$\Pr(\text{green}) = \frac{23}{50}$

$\Pr(\text{pink}) = \frac{1}{5}$

c. Run the experiment more times (e.g. 10 000 times).

19. No. The theoretical probability of rolling a 3 in three rolls is only $\frac{91}{216}$.

20. a. $\frac{1}{1000}$ b. $\frac{999}{1000}$ c. $\frac{3}{10}$ d. $\frac{1}{720}$

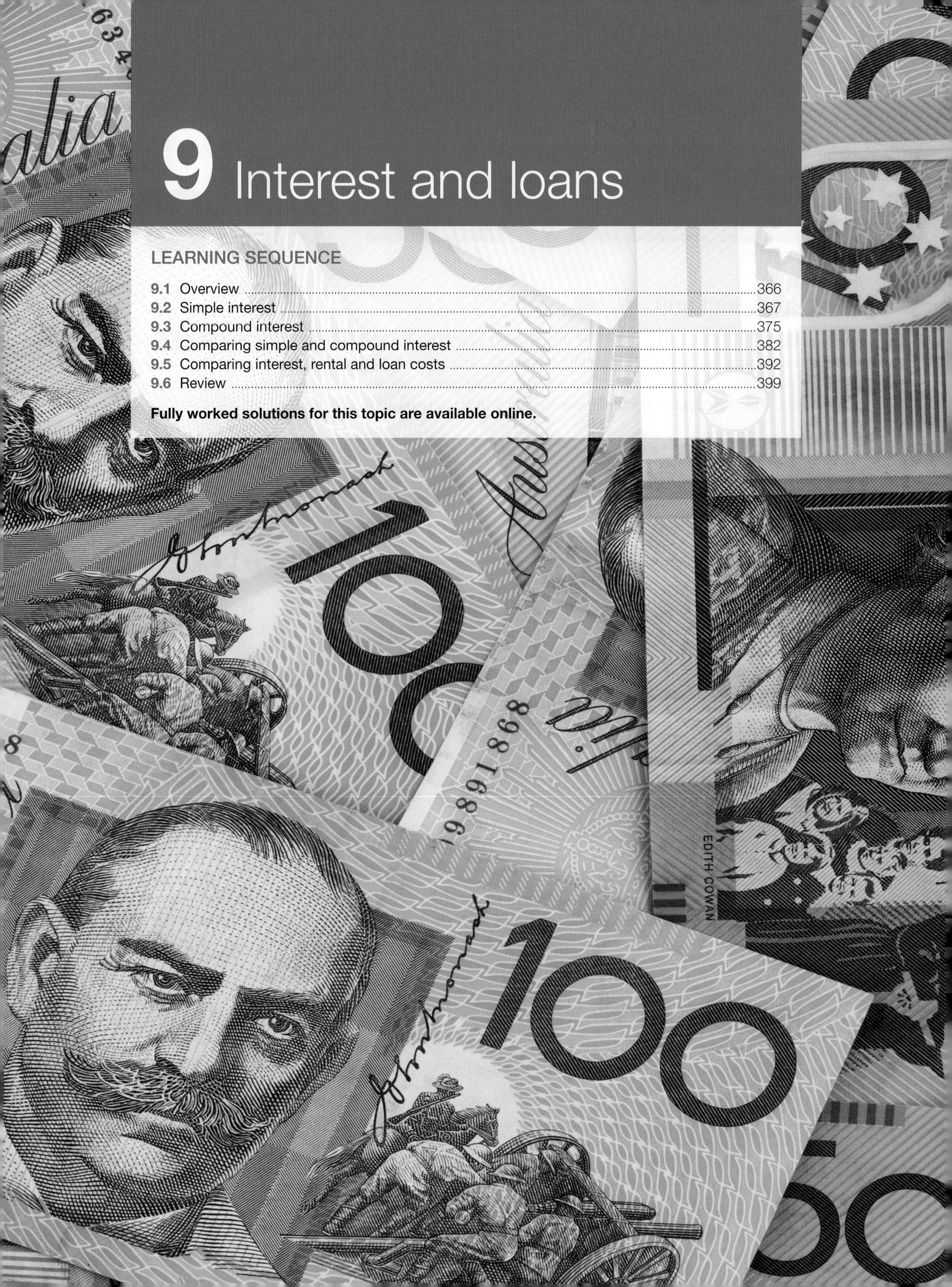

9 Interest and loans

LEARNING SEQUENCE

Fully worked solutions for this topic are available online.

9.1 Overview

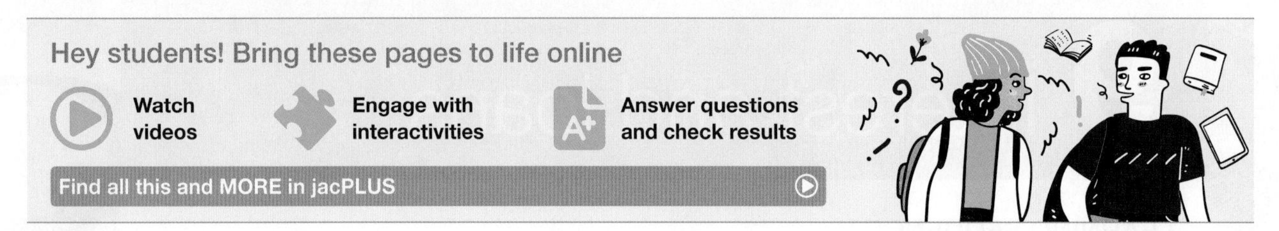

9.1.1 Introduction

Did you know that $1 today is worth more than $1 next year? Due to inflation, the value of the dollar gradually decreases each year, but this effect is usually more noticeable when looking at longer periods of time. However, there are ways to increase an amount of money. By putting money in a savings account, you will earn interest on your money, which means the amount in the account will steadily increase over time.

You may be able to save some money, but you might not be able to save enough quickly to afford more expensive items such as a house. We can invest money with a financial institution (such as a bank), or the institution can lend us money and charge interest. The rates of interest charged, or given, may vary and can fluctuate over time. In this topic, we will consider and compare different types of loans.

KEY CONCEPTS

This topic covers the following key concepts from the VCE Mathematics Study Design:

- money management including investments and loans, credit and debit, comparing mortgages versus rental costs and debt consolidation.

Source: VCE Mathematics Study Design (2023–2027) extracts © VCAA; reproduced by permission.

9.2 Simple interest

LEARNING INTENTION

At the end of this subtopic you should be able to:

- calculate simple interest, principal, interest rate and time period using the simple interest formula.

9.2.1 Introducing simple interest

When you lend money for a certain period of time (a **term deposit**) to a bank, building society or other financial institution, you expect to be rewarded by eventually getting your money back, plus an extra amount commonly known as **interest** (I).

Similarly, if you borrow money from any institution by taking out a loan or mortgage, you must pay back the original sum plus interest. The following examples deal with **simple interest** — that is, interest that is paid only on the original sum of money invested or borrowed.

The interest earned from an investment depends on the **interest rate**, the time period (**term**) during which it is invested, and the amount (**principal**) invested. This relationship can be expressed formally by a mathematical equation.

Simple interest

The formula used to calculate simple interest is given by:

$$I = \frac{Prn}{100}$$

where:

I = simple interest ($)

P = principal ($) — that is, the sum of money borrowed or invested

r = interest rate per time period

n = number of time periods — the period of time the sum of money is borrowed or invested. It must be expressed for the same time span as the rate (which will be in years).

The sum of the principal, P, and the interest, I, is called the *total amount* and is denoted by the symbol A.

Total amount

The formula used to calculate the total amount is given by:

$$A = P + I$$

where:

A = total amount at the end of the term ($)

P = principal ($)

I = simple interest ($).

tlvd-4115

WORKED EXAMPLE 1 Calculating simple interest and total amount

Calculate the amount of simple interest, I, earned and the total amount, A, at the end of the term, if:
a. $12 000 is invested for 5 years at 9.5% p.a.
b. $2500 is invested for 3 months at 4.5% p.a.

THINK	WRITE
a. 1. Write down the formula for simple interest.	a. $I = \frac{Prn}{100}$
2. Write down the known values of the variables.	$P = \$12\,000$ $r = 9.5$ $n = 5$ years
3. Substitute the values into the given formula.	$I = \frac{12\,000 \times 9.5 \times 5}{100}$
4. Evaluate.	$= 5700$
5. Answer the question and include the appropriate unit.	The amount of simple interest earned is $5700.
6. Write down the formula for the total amount.	$A = P + I$
7. Substitute the values for P and I.	$= 12\,000 + 5700$
8. Evaluate.	$= 17\,700$
9. Answer the question and include the appropriate unit.	The total amount at the end of the term is $17 700.
b. 1. Write down the formula for simple interest.	b. $I = \frac{Prn}{100}$
2. Write down the known values of the variables. *Note: n* must be expressed in years, so divide 3 months by 12.	$P = \$2500$ $r = 4.5$ $n = 3$ months $= \frac{3}{12}$ years $= 0.25$ years
3. Substitute the values into the given formula.	$I = \frac{2500 \times 4.5 \times 0.25}{100}$ $= 28.125$
4. Evaluate and round off the answer to 2 decimal places.	$I = 28.13$
5. Answer the question and include the appropriate unit.	The amount of simple interest earned is $28.13.
6. Write down the formula for the total amount.	$A = P + I$
7. Substitute the values for P and I.	$= 2500 + 28.13$
8. Evaluate.	$= 2528.13$
9. Answer the question and include the appropriate unit.	The total amount at the end of the term is $2528.13.

tlvd-4116

WORKED EXAMPLE 2 Calculating investment options using simple interest

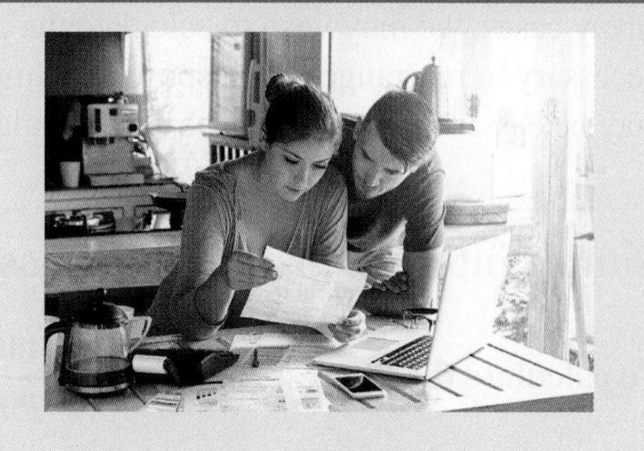

After comparing investment options from a variety of institutions, Lynda and Jason decided to invest their $18 000 in state government bonds at 7.75% p.a. The investment is for 5 years and the interest is paid biannually (twice per year). Calculate how much interest:

a. they receive in every payment

b. will be received in total.

THINK	WRITE
a. 1. Write down the formula for simple interest.	a. $I=\frac{Prn}{100}$
2. Write down the known values of the variables. *Note: n* must be expressed in years, so divide 6 months by 12.	$P=\$18\,000$ $r=7.75$ $n=6$ months $=\frac{6}{12}$ years $=0.5$ years
3. Substitute the values into the given formula.	$I=\frac{18\,000\times7.75\times0.5}{100}$
4. Evaluate.	$=697.5$
5. Answer the question and include the appropriate unit.	Lynda and Jason receive $697.50 in interest every 6 months.
b. **Method 1**	b.
1. Write down the formula for simple interest.	$I=\frac{Prn}{100}$
2. Write down the known values of the variables.	$P=\$\,18\,000$ $r=7.75$ $n=5$ years
3. Substitute the values into the given formula.	$I=\frac{18\,000\times7.75\times5}{100}$
4. Evaluate.	$=6975$
5. Answer the question and include the appropriate unit.	Lynda and Jason will receive a total of $6975 in interest.
Method 2	
1. Multiply the interest received in each 6-month period by the number of 6-month periods in 5 years; that is, multiply $697.50 by 10.	Interest obtained every 6 months $=\$697.50$ Number of payments to be received $=10$ Total interest received $=\$697.50\times10$ $=\$6975$
2. Answer the question.	Lynda and Jason will receive a total of $6975 in interest.

9.2.2 Rearranging the simple interest formula

In many cases, we may wish to calculate the principal, interest rate or time period of a loan. In these situations, it is necessary to rearrange or transpose the simple interest formula. This can be done before or after substitution. It may be easier to use the transposed formula when calculating r, n or P.

Simple interest formula transpositions

To calculate the interest rate, use $r = \dfrac{100I}{Pn}$.

To calculate the period of the loan or investment, use $n = \dfrac{100I}{Pr}$.

To calculate the principal, use $P = \dfrac{100I}{rn}$.

tlvd-4117

WORKED EXAMPLE 3 Calculating the interest rate using the simple interest formula

Calculate the interest rate offered when \$720 is invested for 36 months and earns \$205.20 simple interest. Express rates in % per annum.

THINK	WRITE
Method: transpose before substitution	
1. Write the transposed simple interest formula for r.	$r = \dfrac{100I}{Pn}$
2. List the values of P, I and n. n must be expressed in years.	$P = \$720$ $I = \$205.20$ $n = 36$ months $= 3$ years
3. Substitute into the formula.	$r = \dfrac{100 \times 205.20}{720 \times 3}$
4. Evaluate on a calculator. *Note:* The rate will be written as a decimal.	$= 9.5$
5. Write your answer.	The interest rate offered was 9.5% per annum.

tlvd-4118

WORKED EXAMPLE 4 Calculating the time period using the simple interest formula

Calculate the period of time for an investment of \$255 at simple interest of 8.5% p.a. to earn \$86.70 in interest.

THINK	WRITE
Method: transpose before substitution	
1. Write the transposed simple interest formula for n.	$n = \dfrac{100I}{Pr}$

2. List the values of P, I and i.	$P = \$255$ $I = \$86.70$ $r = 8.5$
3. Substitute into the formula.	$n = \dfrac{100 \times 86.70}{255 \times 8.5}$
4. Evaluate on a calculator.	$= 4$
5. Write your answer.	The period of the investment was 4 years.

tlvd-4119

WORKED EXAMPLE 5 Calculating the principal amount using the simple interest formula

Calculate the principal in an investment with simple interest of 9% p.a., earning \$215 interest over 4 years.

THINK	WRITE
Method: transpose after substitution	
1. Write the simple interest formula.	$I = \dfrac{Prn}{100}$
2. List the values of I, r and n.	$I = \$215$ $r = 9$ $n = 4$ years
3. Substitute into the formula.	$I = \dfrac{P \times r \times n}{100}$ $215 = \dfrac{P \times 9 \times 4}{100}$
4. Evaluate using a calculator to simplify the equation.	$215 = P \times 0.36$
5. Make P the subject by dividing both sides by 0.36.	$P = \dfrac{215}{0.36}$
6. Use a calculator to evaluate.	$= 597.22$
7. Write your answer.	The amount invested was \$597.22.

Resources

Interactivity Simple interest (int-6074)

9.2 Exercise

Students, these questions are even better in jacPLUS

Receive immediate feedback and access sample responses

Access additional questions

Track your results and progress

Find all this and MORE in jacPLUS

1. WE1 Calculate the amount of simple interest, I, earned and the total amount, A, at the end of the term for each of the following.
   a. \$680 for 4 years at 5% p.a.
   b. \$210 for 3 years at 9% p.a.
   c. \$415 for 5 years at 7% p.a.
   d. \$460 at 12% p.a. for 2 years
   e. \$1020 at $12\frac{1}{2}$% p.a. for 2 years
   f. \$713 at $6\frac{3}{4}$% p.a. for 7 years
2. WE2 Sue and Harry invested \$14 500 in state government bonds at 8.65% p.a. The investment is for 10 years and the interest is paid biannually (twice per year). Calculate how much interest:
   i. they receive with every payment
   ii. they will receive in total.
3. WE3 For each of the following, calculate the interest rate offered. Express rates in % per annum.
   a. Loan of \$10 000, with a \$2000 interest charge, for 2 years
   b. Investment of \$5000, earning \$1250 interest, for 4 years
   c. Loan of \$150, with a \$20 interest charge, for 2 months
   d. Investment of \$1400, earning \$178.50 interest, for 6 years
   e. Investment of \$6250, earning \$525 interest, for $2\frac{1}{2}$ years
4. WE4 For each of the following, calculate the period of time (to the nearest month) for which the principal was invested or borrowed.
   a. Investment of \$1000, at simple interest of 5% p.a., earning \$50 interest
   b. Loan of \$6000, at simple interest of 7% p.a., with an interest charge of \$630
   c. Loan of \$100, at simple interest of 24% p.a., with an interest charge of \$6
   d. Investment of \$23 000, at simple interest of $6\frac{1}{2}$% p.a., earning \$10 465 interest
   e. Loan of \$1 500 000, at simple interest of 1.5% p.a., with an interest charge of \$1875
5. WE5 For each of the following, calculate the principal invested.
   a. Simple interest of 5% p.a., earning \$307 interest over 2 years
   b. Simple interest of 7% p.a., earning \$1232 interest over 4 years
   c. Simple interest of 8% p.a., earning \$651 interest over 18 months
   d. Simple interest of $5\frac{1}{2}$% p.a., earning \$78 interest over 6 years
   e. Simple interest of 6.25% p.a., earning \$625 interest over 4 years
6. Determine the interest earned on the following investments.
   a. \$690 invested at 12% p.a. simple interest for 15 months
   b. \$7500 invested for 3 years at 12% per year simple interest

c. $25 000 invested for 13 weeks at 5.2% p.a. simple interest

d. $250 invested at 21% p.a. for $2\frac{1}{2}$ years

7. Determine the amount to which each investment has grown after the investment periods shown in the following examples.

a. $300 invested at 10% p.a. simple interest for 24 months
b. $750 invested for 3 years at 12% p.a. simple interest
c. $20 000 invested for 3 years and 6 months at 11% p.a. simple interest
d. $15 invested at $6\frac{3}{4}$% p.a. for 2 years and 8 months
e. $10.20 invested at $8\frac{1}{2}$% p.a. for 208 weeks

8. **MC** If John had $63 in his bank account and earned 9% p.a. over 3 years, the simple interest earned is:

A. $5.67 B. $17.01 C. $22.68 D. $80.01 E. $15.56

9. **MC** If $720 was invested in a fixed deposit account earning $6\frac{1}{2}$% p.a. for 5 years, the interest earned at the end of 5 years would be:

A. $23.40 B. $216.00 C. $234.00 D. $954.00 E. $46.80

10. **MC** Bodgee Bank advertised a special offer. If a person invests $150 for 2 years, the bank will pay 12% p.a. simple interest on the money. At the expiry date, the investor will have earned:

A. $36 B. $48 C. $186 D. $300 E. $84

11. **MC** Joanne asked Sally for a loan of $125 to buy new shoes. Sally agreed on the condition that Joanne paid it back in two years at 3% p.a. simple interest. At the end of the two years, the amount Joanne paid Sally is:

A. $7.50 B. $125 C. $130.50 D. $132.50 E. $137.50

12. **MC** Two banks pay simple interest on short-term deposits. Hales Bank pays 8% p.a over 5 years and Countrybank pays 10% p.a. over 4 years. If $2000 was invested in each account, the difference between the two banks' final payout figure is:

A. $0 B. $150 C. $800 D. $1200 E. $1000

13. **MC** Joanne's accountant found that for the past 2 years she had earned a total of $420 interest in an account paying 6% p.a. simple interest. The amount Joanne invested was:

A. $50.40
B. $350
C. $3500
D. $5040
E. $3850

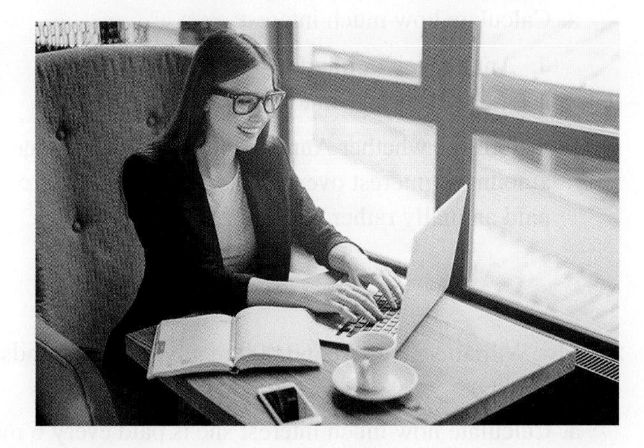

14. **MC** A loan of $1000 is taken over 5 years. The total amount repaid for this loan is $1800. The simple interest rate per year on this loan is:

A. 5% B. 8% C. 9% D. 16% E. 18%

15. **MC** Jarrod decides to buy a motorbike at no deposit and no repayments for 3 years. He takes out a loan of \$12 800 and is charged at 7.5% p.a. simple interest over the 3 years. The lump sum Jarrod has to pay in 3 years' time is:

A. \$2880 B. \$9920 C. \$13 760 D. \$15 680 E. \$16 640

16. Jill and John decide to borrow money to improve their yacht, but cannot agree which loan is the better value. They would like to borrow \$2550. Jill goes to Big 4 Bank and finds that they will lend her the money at $11\frac{1}{3}$% p.a. simple interest for 3 years. John finds that Friendly Building Society will lend the \$2550 to them at 1% per month simple interest for the 3 years.

a. Determine which institution offers the better rate over the 3 years.
b. Explain why.

17. Lennie Cavan earned \$576 in interest when she invested in a fund paying 9.5% p.a simple interest for 4 years. Calculate how much Lennie invested originally.

18. Lennie's sister Lisa also earned \$576 interest at 9% p.a. simple interest, but she only had to invest it for 3 years. Calculate Lisa's initial investment.

19. Kahn put some money away for 5 years in a bank account that paid $3\frac{3}{4}$% p.a. interest. He found from his bank statement that he had earned \$66. Calculate how much Kahn invested.

20. James needed to earn \$225. He invested \$2500 in an account earning simple interest at a rate of 4.5% p.a. Calculate how many months it will take James to achieve his aim.

21. Anna invested \$85 000 in Ski International shares. She earns 7.25% p.a., which is paid quarterly for one year.

a. Calculate how much interest:
   i. Anna receives quarterly
   ii. she will receive in total, over a year.
b. Determine whether Anna would receive the same amount of interest over a 3-year period if it were paid annually rather than quarterly.

22. Mrs Williams invested \$60 000 in government bonds at 7.49% p.a. with interest paid biannually (i.e. every 6 months).
a. Calculate how much interest she is paid every 6 months.
b. Calculate how much interest she is paid over $3\frac{1}{2}$ years.
c. Calculate how long the money would need to be invested to earn a total of \$33 705 in interest.

9.3 Compound interest

LEARNING INTENTION

At the end of this subtopic you should be able to:

- calculate compound interest using the compound interest formula
- calculate compound interest using technology.

9.3.1 Compound interest investment or loan

Compound interest is a different type of interest. Simple interest depends on the initial amount invested; however, it is more common for interest to be calculated on the changing value throughout the time period of a loan or investment. This is known as compounding.

In compounding, the interest is added to the balance, and the next interest calculation is made on the new value. This means that the amount of interest received changes over time. The final amount of a compound interest investment or loan depends on the rate of interest, the time period for which it is invested, and the amount invested. This relationship can be expressed formally by a mathematical equation.

Compound interest

The formula used to determine the value of a compound interest investment or loan is given by the rule:

$$A = P\left(1 + \frac{r}{100}\right)^n$$

where:

A = the future value of a compound interest investment or loan ($)

P = principal ($) – that is, the sum of money borrowed or invested

r = interest rate per time period

n = number of compounding time periods.

The compound interest formula gives the final amount (*total value*) of an investment, not just the interest earned as in the simple interest formula.

Another rule can be used to calculate the total interest.

Total interest compounded

The total interest compounded after n time periods, I, is given by the rule:

$$I = \text{future value} - \text{principal}$$
$$I = A - P$$

where:

P = principal ($)
A = future value ($)

Value after n time periods

If compound interest is used, the value of the investment at the end of each period grows by an increasing amount. Therefore, when plotted, the values of the investment at the end of each period form an exponential curve.

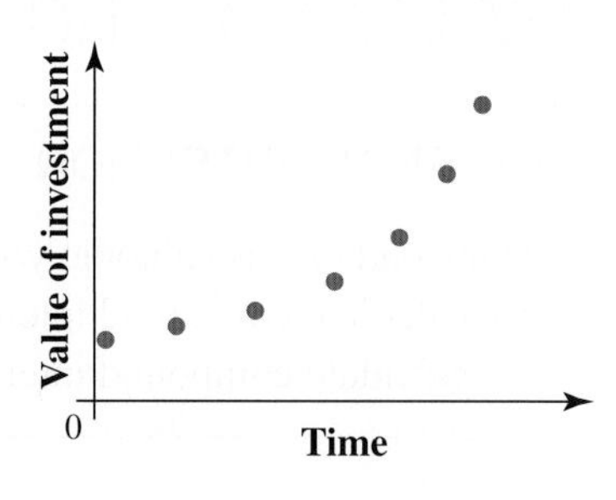

tlvd-4120

WORKED EXAMPLE 6 Calculating compound interest

\$5000 is invested for 4 years at 6.5% p.a. interest, compounded annually.

a. Generate the compound interest formula for this investment.

b. Calculate the amount in the balance after 4 years and the interest earned over this period.

THINK	WRITE
a. 1. Write the compound interest formula.	a. $A = P\left(1 + \frac{r}{100}\right)^n$
2. List the values of n, r and P.	$n = 4$ $r = 6.5$ $P = 5000$
3. Substitute the values into the formula.	$A = 5000\left(1 + \frac{6.5}{100}\right)^4$
4. Complete the addition inside the brackets to simplify.	$A = 5000(1.065)^4$
b. 1. Use your calculator to evaluate $5000(1.065)^4$ correct to 2 decimal places. This is the amount of the investment after 4 years.	b. $A = 5000(1.065)^4$ $= 6432.33$
2. To determine the amount of interest earned, subtract the principal from the balance.	I = final amount − principal $= 6432.33 - 5000$ $= \$1432.33$
3. Write your answer.	The amount of interest earned is \$1432.33 and the balance is \$6432.33.

9.3.2 Non-annual compounding

In Worked example 6, interest was compounded annually. However, in many cases, the interest is compounded more often than once a year — for example, quarterly (every 3 months), monthly, weekly or daily. In these situations, n and r still have their usual meaning and we calculate them as follows.

Adjusting the rate and number of periods for non-annual compounding

Number of interest periods, n = number of years × number of interest periods per year

$$\textbf{Interest rate per period, } r = \frac{\textbf{interest rate per annum}}{\textbf{number of interest periods per year}}$$

tlvd-4121

WORKED EXAMPLE 7 Compound interest compounded non-annually

If \$3200 is invested for 5 years at 6% p.a. interest compounded quarterly:

a. **determine the number of interest-bearing periods, n**

b. **calculate the interest rate per period, r**

c. **calculate the balance of the account after 5 years**

d. **graphically represent the balance at the end of each quarter for 5 years and describe the shape of the graph.**

THINK	WRITE
a. Calculate n.	a. $n = 5\text{ (years)} \times 4\text{ (quarters)}$ $= 20$
b. Convert % p.a. to % per quarter to match the time over which the interest is calculated. Divide r% p.a. by the number of compounding periods per year (4).	b. $r = \dfrac{6\%\text{ p.a.}}{4}$ $= 1.5\%$ per quarter
c. 1. Write the compound interest formula.	c. $A = P\left(1 + \dfrac{r}{100}\right)^n$
2. List the values of n, r and P	$n = 20$ $r = 1.5$ $P = 3200$
3. Substitute into the formula.	$A = 3200\left(1 + \dfrac{1.5}{100}\right)^{20}$
4. Simplify by completing the addition.	$= 3200(1.015)^{20}$
5. Use your calculator to evaluate $3200(1.015)^{20}$ correct to 2 decimal places.	$= \$4309.94$
6. Write your answer.	The balance of the account after 5 years is \$4309.94.
d. 1. Using a calculator, calculate the balance at the end of each quarter. This will involve using the formula $A = 3200(1.015)^n$ repeatedly for different values of n. 2. Plot these values on a set of axes. The first point is (0, 3200), which represents the principal.	d. \$3200, \$3284, \$3296.72, \$3346.17, ..., \$4309.94 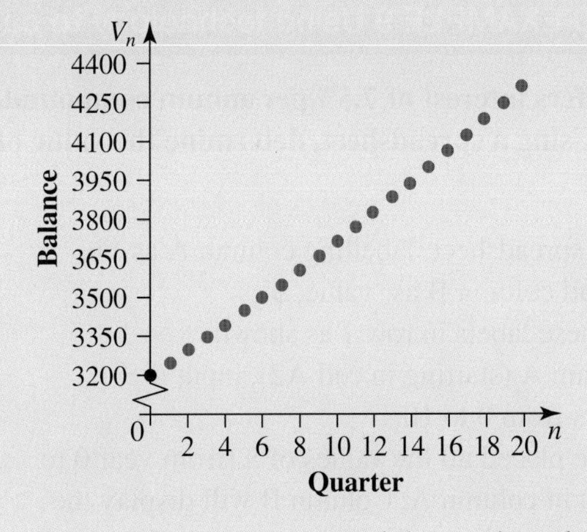
3. Comment on the shape of the graph.	The graph shows exponential growth as the interest is added at the end of each quarter and the following interest is calculated on the *new* balance.

The situation often arises where we require a certain amount of money by a future date. It may be to pay for a holiday or to finance the purchase of a car. It is then necessary to know what principal should be invested now so that it will increase in value to the desired final balance within the time available.

tlvd-4122

WORKED EXAMPLE 8 Determining the principal for future values

Determine the principal that will grow to $4000 in 6 years, if interest is compounded quarterly at 6.5% p.a.

THINK	WRITE
1. Write the compound interest formula.	$A = P\left(1 + \frac{r}{100}\right)^n$
2. Calculate n (there are 4 quarters in a year).	$n = 6 \times 4$ $= 24$
3. Calculate r.	$r = \frac{6.5\%}{4}$ $= 1.625\%$
4. State the value for A.	$A = 4000$
5. Substitute A, r and n into the rule and simplify.	$4000 = P\left(1 + \frac{1.625}{100}\right)^{24}$ $4000 = P(1.016\,25)^{24}$
6. Transpose to isolate P.	$P = \frac{4000}{(1.016\,25)^{24}}$
7. Use a calculator to evaluate correct to 2 decimal places.	$= 2716.73$
8. Write a summary statement.	The principal would need to be $2716.73.

9.3.3 Using a spreadsheet to calculate compound interest

Calculating compound interest over a long period of time requires numerous repetitive calculations. Spreadsheets are a useful tool to perform these calculations.

tlvd-4123

WORKED EXAMPLE 9 Using technology to calculate compound interest

A bank offers interest of 7.5% per annum compounded yearly on investments. $5000 is invested for 10 years. Using a spreadsheet, determine the value of the investment after the 10 years.

THINK

1. Open a spreadsheet, labelling column A as 'n, year' and column B as 'value, $'.
Place these labels in row 1 as shown.
In column A (starting in cell A2), input the numbers from 0 to 10.
We have placed all the values of n (from year 0 to year 10) in column A. Column B will display the value of the investment for its corresponding year number.

WRITE

	A	B
1	n, year	value, $
2	0	
3	1	
4	2	
5	3	
	...	
12	10	

2. Write down the formula for compound interest.

$$A = P\left(1 + \frac{r}{100}\right)^n$$

3. Calculate $1 + r$.

$$1 + \frac{r}{100} = 1 + \frac{7.5}{100}$$
$$= 1 + 0.075$$
$$= 1.075$$

4. Substitute this value and the value of P into the compound interest formula. We are now ready to input the compound interest formula into the spreadsheet.

$$A = 5000(1.075)^n$$

5. Use the cursor to click into cell B2. Type the following: =5000*(1.075)^A2 and press Enter. This should automatically input the value of 5000 in cell B2.

	A	B
1	n, year	value, $
2	0	5000
3	1	

6. Copy the calculation for the next 10 years by using the fill-down function (or drag the bottom right corner down the cells). Complete column B to show 10 years.

	A	B
1	n, year	value, $
2	0	5000
3	1	5375
4	2	5778.125
5	3	6211.484
6	4	6677.346
7	5	7178.147
8	6	7716.508
9	7	8295.246
10	8	8917.389
11	9	9586.193
12	10	10305.16

7. Answer the question.

After 10 years, the value of the investment is $10 305.16.

Resources

Digital document SpreadSHEET Compound interest (doc-9603)

9.3 Exercise

Students, these questions are even better in jacPLUS

Receive immediate feedback and access sample responses

Access additional questions

Track your results and progress

Find all this and MORE in jacPLUS

1. WE6 \$2500 is invested for 5 years at 7.5% p.a. compounding annually.
   a. Generate the compound interest formula for this investment.
   b. Calculate the amount in the balance after 5 years and the interest earned over this period.
2. Use the compound interest formula, $A = P(1 + r)^n$, to calculate the amount, A, for each of the following.
   a. $P = \$500$, $n = 2$, $r = 8\%$ p.a.
   b. $P = \$1000$, $n = 4$, $r = 13\%$ p.a.
   c. $P = \$3600$, $n = 3$, $r = 7.5\%$ p.a.
   d. $P = \$2915$, $n = 5$, $r = 5.25\%$ p.a.
3. Determine:
   i. the balance
   ii. the interest earned (interest compounded annually)
      a. if \$2000 is invested for 1 year at 7.5% p.a.
      b. if \$2000 is invested for 2 years at 7.5% p.a.
      c. if \$2000 is invested for 6 years at 7.5% p.a.
4. Determine the number of time periods, n, if interest is compounded:
   a. annually for 5 years
   b. quarterly for 5 years
   c. biannually for 4 years
   d. monthly for 6 years
   e. 6-monthly for $4\frac{1}{2}$ years
   f. quarterly for 3 years and 9 months.
5. Calculate the interest rate per period, r, if the annual rate is:
   a. 6% and interest is compounded quarterly
   b. 4% and interest is compounded half-yearly
   c. 18% and interest is compounded monthly
   d. 7% and interest is compounded quarterly.
6. Determine the amount in the account and interest earned after \$6750 is invested for 7 years at 5.25% p.a. interest compounded annually.
7. WE7 If \$4200 is invested for 3 years at 7% p.a. interest compounded quarterly:
   a. determine the number of interest-bearing periods, n
   b. calculate the interest rate per period, r
   c. calculate the balance of the account after 3 years
   d. graphically represent the balance at the end of each quarter for 3 years and describe the shape of the graph.

8. If \$7500 is invested for 2 years at 5.5% p.a. interest compounded monthly:
   a. determine the number of interest-bearing periods, n
   b. calculate the interest rate per period, r
   c. calculate the balance of the account after 2 years
   d. graphically represent the balance at the end of each month for 2 years and describe the shape of the graph.
9. WE8 Calculate the principal that will grow to \$5000 in 5 years, if interest is added quarterly at 7.5% p.a.
10. Calculate the principal that will grow to \$6300 in 7 years, if interest is added monthly at 5.5% p.a.
11. Assume \$1500 is invested for 2 years into an account paying 8% p.a.
    a. Calculate the balance if interest is compounded yearly.
    b. Calculate the balance if interest is compounded quarterly.
    c. Calculate the balance if interest is compounded monthly.
    d. Calculate the balance if interest is compounded daily.
    e. Compare your answers to parts **a–d**.
12. Use the compound interest formula to answer the following questions.
    a. If the balance in an account after 1 year is \$2612.50 at 4.5%, calculate the balance after 3 years.
    b. If the balance in an account after 2 years is \$4368.10 at 4.5%, calculate the balance after 5 years.
    c. If the balance in an account after 2 years is \$6552.15 at 4.5%, calculate the value of the initial investment.
13. Determine the amount that accrues in an account which pays compound interest at a nominal rate of:
    a. 7% p.a. if \$2600 is invested for 3 years (compounded monthly)
    b. 8% p.a. if \$3500 is invested for 4 years (compounded monthly)
    c. 11% p.a. if \$960 is invested for $5\frac{1}{2}$ years (compounded fortnightly)
    d. 7.3% p.a. if \$2370 is invested for 5 years (compounded weekly)
    e. 15.25% p.a. if \$4605 is invested for 2 years (compounded daily).
14. Use the compound interest formula to calculate the principal, P, for each of the following.
    a. $A = \$5000,\ r = 9\%,\ n = 4$
    b. $A = \$2600, r = 8.2\%, n = 3$
    c. $A = \$3550,\ r = 1.5\%,\ n = 12$
    d. $A = \$6661.15,\ r = 0.8\%,\ n = 36$

15. For each of the following, determine:
    i. the principal
    ii. the interest accrued.

    a. \$3000 in 4 years, if interest is compounded 6-monthly at 9.5% p.a.
    b. \$2000 in 3 years, if interest is compounded quarterly at 9% p.a.
    c. \$5600 in $5\frac{1}{4}$ years, if interest is compounded quarterly at 8.7% p.a.
    d. \$10 000 in $4\frac{1}{4}$ years, if interest is compounded monthly at 15% p.a.
16. WE9 A bank offers 2.90% per annum compounded yearly on investments. \$25 000 is invested for 5 years. Using a spreadsheet, determine the value of the investment after the 5 years.

9.4 Comparing simple and compound interest

LEARNING INTENTION

At the end of this subtopic you should be able to:
- compare various interest rates using simple and compound interest
- compare and contrast simple and compound interest.

9.4.1 Simple interest or compound interest?

Earlier in this chapter we have looked at simple interest (interest calculated on the principal) and compound interest (interest calculated on the principal and the interest). An example of the differences between simple and compound interest is shown over 5 years below.

Simple interest	Compound interest
Initial principal, $P = \$1000$	Initial principal, $P = \$1000$
Rate of interest, $r = 10\%$	Rate of interest, $r = 10\%$
Interest for year 1: 10% of \$1000, $I_1 = \$100$	Interest for year 1: 10% of \$1000, $I_1 = \$100$
Principal at the beginning of year 2: $P_2 = \$1000$	Principal at the beginning of year 2: $P_2 = \$1000 + \$100 = \$1100$
Interest for year 2: 10% of \$1000, $I_2 = \$100$	Interest for year 2: 10% of \$1100, $I_2 = \$110$
Principal at the beginning of year 3: $P_3 = \$1000$	Principal at the beginning of year 3: $P_3 = \$1100 + \$110 = \$1210$
Interest for year 3: 10% of \$1000, $I_3 = \$100$	Interest for year 3: 10% of \$1210, $I_3 = \$121$
Principal at the beginning of year 4: $P_4 = \$1000$	Principal at the beginning of year 4: $P_4 = \$1210 + \$121 = \$1331$
Interest for year 4: 10% of \$1000, $I_4 = \$100$	Interest for year 4: 10% of \$1331, $I_4 = \$133.10$
Principal at the beginning of year 5: $P_5 = \$1000$	Principal at the beginning of year 5: $P_5 = \$1331 + \$133.10 = \$1464.10$
Interest for year 5: 10% of \$1000, $I_5 = \$100$	Interest for year 5: 10% of \$1464.10, $I_5 = \$146.41$
The simple interest earned over a 5-year period is \$500.	The compound interest earned over a 5-year period is \$610.51.

The table above illustrates how, as the principal for compound interest increases periodically (i.e. $P = \$1000, \$1100, \$1210, \$1331, \$1464.10...$), so does the interest (i.e. $I = \$100, \$110, \$121, \$133.10, \$146.41...$), while for simple interest both the principal and the interest earned remain constant — that is, $P = \$1000$ and $I = \$100$. The difference of \$110.51 between the compound interest and simple interest earned in a 5-year period represents the interest earned on added interest.

If we were to place the set of data obtained in two separate tables and represent each set graphically, we would see that the simple interest investment grows at a constant rate while the compound interest investment grows exponentially.

<table>
<tr><th>Type of interest</th><th>Table of values</th><th>Graphic representation</th></tr>
<tr><td>Simple interest</td><td><table><tr><th>n</th><th>A</th></tr><tr><td>0</td><td>1000</td></tr><tr><td>1</td><td>1100</td></tr><tr><td>2</td><td>1200</td></tr><tr><td>3</td><td>1300</td></tr><tr><td>4</td><td>1400</td></tr><tr><td>5</td><td>1500</td></tr></table></td><td>The simple interest investment is represented by a straight line.</td></tr>
<tr><td>Compound interest</td><td><table><tr><th>n</th><th>A</th></tr><tr><td>0</td><td>1000</td></tr><tr><td>1</td><td>1100</td></tr><tr><td>2</td><td>1210</td></tr><tr><td>3</td><td>1331</td></tr><tr><td>4</td><td>1464.10</td></tr><tr><td>5</td><td>1610.51</td></tr></table></td><td>The compound interest investment is represented by a curve.</td></tr>
</table>

9.4.2 Graphing simple interest functions

Suppose that $10 000 is invested at 5% p.a. simple interest. The table below shows the amount of interest received after various lengths of time.

No. of years	1	2	3	4	5
Interest	$500	$1000	$1500	$2000	$2500

The amount of interest earned can be graphed by the **linear function** shown. Note that the gradient of this graph is 500, which is the amount of one year's interest, or 5% of the principal. This means that for every 1 unit increase on the x-axis (1 year in time) the y-axis (Interest) increases by 500 units ($500).

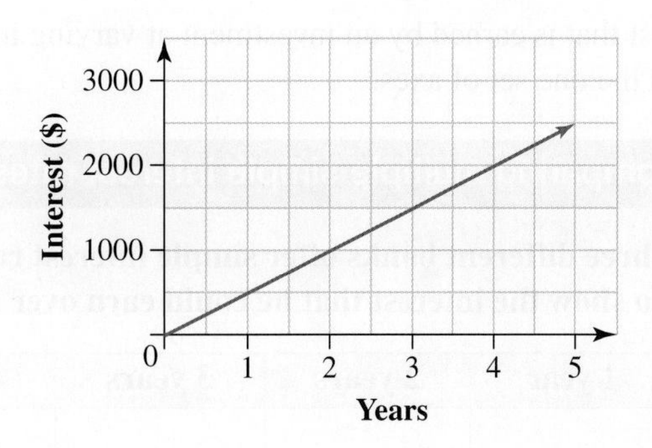

tlvd-4124

WORKED EXAMPLE 10 Graphing simple interest functions

Leilay invests $6000 into a savings account earning simple interest at a rate of 4% p.a.

a. Complete the table below to calculate the future value of the investment at the end of each year.

No. of years	1	2	3	4	5
Future value					

b. Draw a graph of the future value against the number of years the money is invested.

THINK	WRITE
a. 1. Calculate the interest earned after 1 year.	a. $I = \frac{Prn}{100}$ $= \frac{6000 \times 4 \times 1}{100}$ $= 240$
2. Since simple interest earns the same amount of interest each year ($240 in this case) we can continually add this value to the original principal to determine the future value.	Original principal = $6000 Future value after 1 year = $6000 + $240 = $6240 Future value after 2 years = $6240 + $240 = $6480 Future value after 3 years = $6480 + $240 = $6720
3. Complete the table.	(table below)

No. of years	1	2	3	4	5
Future value	$6240	$6480	$6720	$6960	$7200

b. Draw the graph with 'Years' on the horizontal axis and 'Future value' on the vertical axis.

b.

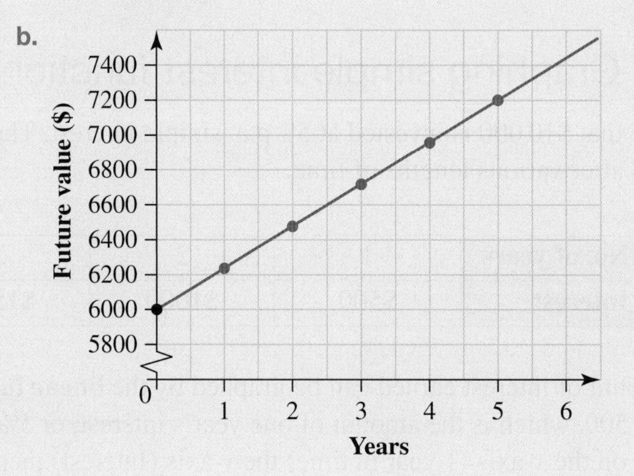

We are able to compare the interest that is earned by an investment at varying interest rates by graphing the interest earned at varying rates on the one set of axes.

tlvd-4125

WORKED EXAMPLE 11 Comparing multiple simple interest rates

Aston has $12 000 to invest. Three different banks offer simple interest rates of 4%, 5% and 6%.

a. Complete the table below to show the interest that he could earn over 5 years.

	1 year	2 years	3 years	4 years	5 years
Interest (4%)					
Interest (5%)					
Interest (6%)					

b. Show this information in graph form.

THINK	WRITE

a. 1. Use the simple interest formula to calculate the interest earned on $12 000 at 4% p.a. for 1, 2, 3, 4 and 5 years.

a. $I = \frac{Prn}{100}$

$I = \frac{12\,000 \times 4 \times 1}{100}$

$= 480$

$I = \frac{12\,000 \times 4 \times 2}{100}$

$= 960$

...

2. Use the simple interest formula to calculate the interest earned on $12 000 at 5% p.a. for 1, 2, 3, 4 and 5 years.

$I = \frac{Prn}{100}$

$I = \frac{12\,000 \times 5 \times 1}{100}$

$= 600$

$I = \frac{12\,000 \times 5 \times 2}{100}$

$= 1200$

...

3. Use the simple interest formula to calculate the interest earned on $12 000 at 6% p.a. for 1, 2, 3, 4 and 5 years to complete the table.

$I = \frac{Prn}{100}$

$I = \frac{12\,000 \times 6 \times 1}{100}$

$= 720$

$I = \frac{12\,000 \times 6 \times 2}{100}$

$= 1440$

...

	1 year	2 years	3 years	4 years	5 years
Interest (4%)	$480	$960	$1440	$1920	$2400
Interest (5%)	$600	$1200	$1800	$2400	$3000
Interest (6%)	$720	$1440	$2160	$2880	$3600

b. Draw a line graph for each investment.

b.

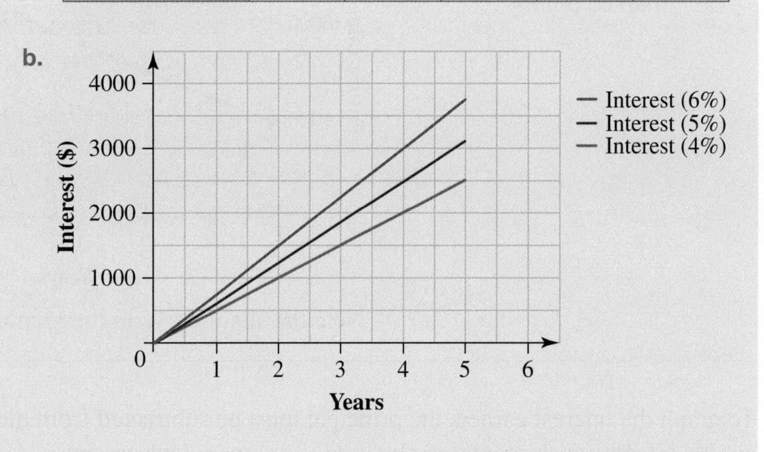

9.4.3 Graphing compound interest functions

With compound interest, the interest earned in each interest period increases, so when we graph the future value of the investment, the result is an exponential graph. The shape of the graph is a smooth curve that gets progressively steeper. We can use the compound interest formula to complete tables that will then allow us to graph a compound interest function.

tlvd-4126

WORKED EXAMPLE 12 Graphing compound interest

Olivia invests $5000 at 5% p.a. with interest compounded annually.

a. **Complete the table below to show the future value at the end of each year.**

No. of years	0	1	2	3	4	5
Future value						

b. **Draw a graph of the future value of the investment against the number of years.**

THINK

a. 1. Use the compound interest formula to calculate the final amount at the end of every year for 5 years.

WRITE

a. $A = P\left(1 + \dfrac{r}{100}\right)^n$

$A = 5000\left(1 + \dfrac{5}{100}\right)^1$

$= 5250$

$A = 5000\left(1 + \dfrac{5}{100}\right)^2$

$= 5512.50$

$A = 5000\left(1 + \dfrac{5}{100}\right)^3$

$= 5788.125$

...

2. Complete the table rounding all values correct to 2 decimal places.

No. of years	0	1	2	3	4	5
Future value	$5000	$5250	$5512.50	$5788.13	$6077.53	$6381.41

b. Draw a graph by drawing a smooth curve between the marked points.

b.

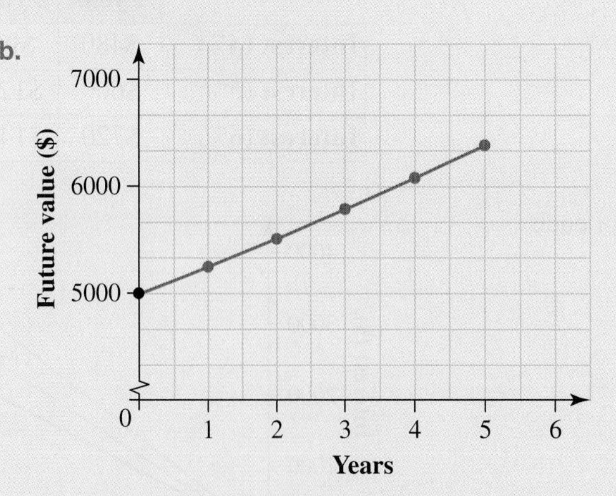

Note the slight curve in the graph.

To graph the interest earned, the principal must be subtracted from the future value of the investment. As with simple interest, such graphs can be used to compare investments.

tlvd-4127

WORKED EXAMPLE 13 Comparing compound interest graphically

Paul wants to invest \$2000 at 5% p.a., 6% p.a. and 7% p.a., compounded annually.

a. **Copy and complete the table below to calculate the amount of interest of each investment at the end of each year.**

	1 year	2 years	3 years	4 years	5 years
Interest (5%)					
Interest (6%)					
Interest (7%)					

b. **Draw a graph that will allow the amount of interest earned to be compared.**

THINK

a. 1. Use the compound interest formula $A = P\left(1 + \frac{r}{100}\right)^n$ to determine the value of each investment over a 5-year period. For example, to determine the value of the 5% p.a. investment after 1 year, evaluate $A = 2000\left(1 + \frac{5}{100}\right)^1$.

2. To calculate the amount of interest earned on the investments at the end of each year, subtract the principal (\$2000) from all values. We do this since the formula for calculating the amount of interest is $I = A - P$.

b. Draw each graph by joining the points with a smooth curve.

WRITE

a.

	1 year	2 years	3 years	4 years	5 years
Interest (5%)	\$2100	\$2205	\$2315.25	\$2431.01	\$2552.56
Interest (6%)	\$2120	2247.20	\$2382.03	\$2524.95	\$2676.45
Interest (7%)	\$2140	\$2289.80	\$2450.09	\$2621.59	\$2805.10

	1 year	2 years	3 years	4 years	5 years
Interest (5%)	\$100	\$205	\$315.25	\$431.01	\$552.56
Interest (6%)	\$120	\$247.20	\$382.03	\$524.95	\$676.45
Interest (7%)	\$140	\$289.80	\$450.09	\$621.59	\$805.10

b.

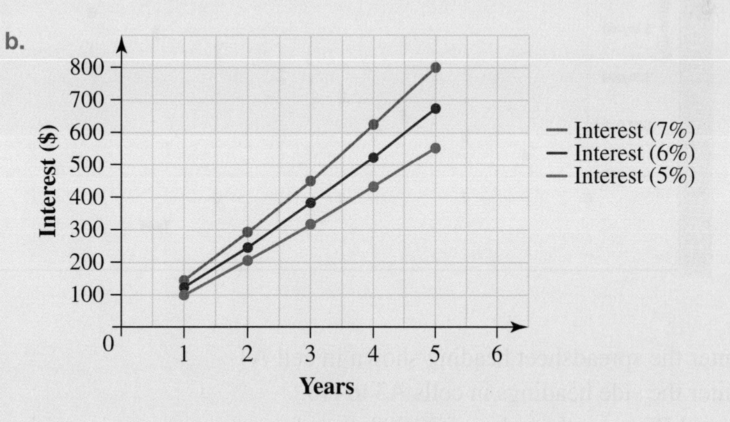

Comparing simple and compound interest functions using a spreadsheet

Your grandfather left you $20 000 in his will. You have no need to use the money at this stage, so you are looking at investing it for approximately 12 years. Your research has narrowed down your options to 4.25% p.a. simple interest or 3.6% p.a. interest compounding yearly. At this stage, you do not anticipate having to withdraw your money in the short term; however, it may become necessary to do so.

Use technology to determine which would be the better option if you were forced to withdraw your money at any period of time within 12 years.

The spreadsheet and graphs we are aiming to produce appear as follows.

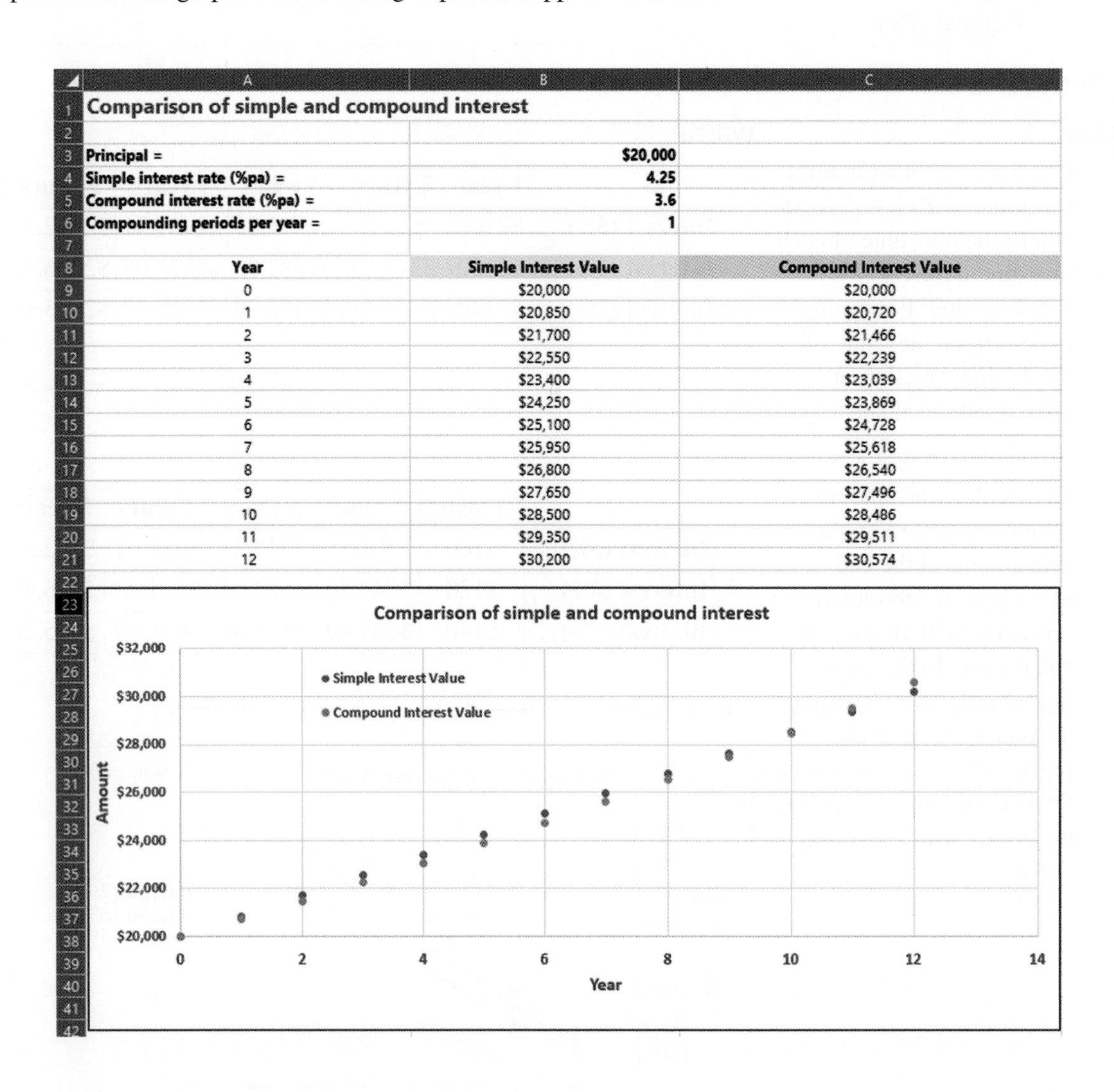

	A	B	C
1	Comparison of simple and compound interest		
2			
3	Principal =	$20,000	
4	Simple interest rate (%pa) =	4.25	
5	Compound interest rate (%pa) =	3.6	
6	Compounding periods per year =	1	
7			
8	Year	Simple Interest Value	Compound Interest Value
9	0	$20,000	$20,000
10	1	$20,850	$20,720
11	2	$21,700	$21,466
12	3	$22,550	$22,239
13	4	$23,400	$23,039
14	5	$24,250	$23,869
15	6	$25,100	$24,728
16	7	$25,950	$25,618
17	8	$26,800	$26,540
18	9	$27,650	$27,496
19	10	$28,500	$28,486
20	11	$29,350	$29,511
21	12	$30,200	$30,574

1. Enter the spreadsheet heading shown in cell A1.
2. Enter the side headings in cells A3 to A6.
3. In cell B3 enter the value of 20 000, then format it to currency with zero decimal places.
4. Enter the numeric values shown in cells B4, B5 and B6.
5. In row 8, enter the column headings shown.
6. In cell A9, enter the value 0.
7. In cell A10, enter the formula = A9 + 1. The value of 1 should appear. Copy this formula down from cell A11 to A21.

8. The formula for the simple interest value is:

$$\text{principal} + \text{interest} = \text{principal} + \text{principal} * \text{rate} * \text{time}/100$$

Enter the formula = B3 + B3 * B4 * A9/100 in cell position B9, then copy this formula down from B9 to B21. Format these cells to currency with zero decimal places. Check that the values that appear agree with those on the spreadsheet displayed.

9. The formula for the compound interest value for cell position C9 is:

= B3*(1 + B5/(100*B6))^A9

Enter this value, then copy it down from C9 to C21. Format these cells to currency with zero decimal places. Check the values with those on the spreadsheet displayed.

10. Use the graphing facility of your spreadsheet to produce a graph similar to the one shown.
11. From the table and the shape of the graphs it is obvious that a critical point occurs somewhere between the 10-year and 11-year marks. Modify your spreadsheet by inserting rows between these two years. Enter part-year values, such as 10.2, 10.5 and so on, in column A. Copy the formulas in columns B and C to complete the entries. Continue to investigate until you can determine a fairly exact value for the time when these two graphs cross.
12. Write a paragraph summarising the results of your spreadsheet and graphs. Describe which option would be the better one, considering that you may be forced to withdraw your money at any time within the 12 years. Support any conclusions by referring to your spreadsheet and graphs.

9.4 Exercise

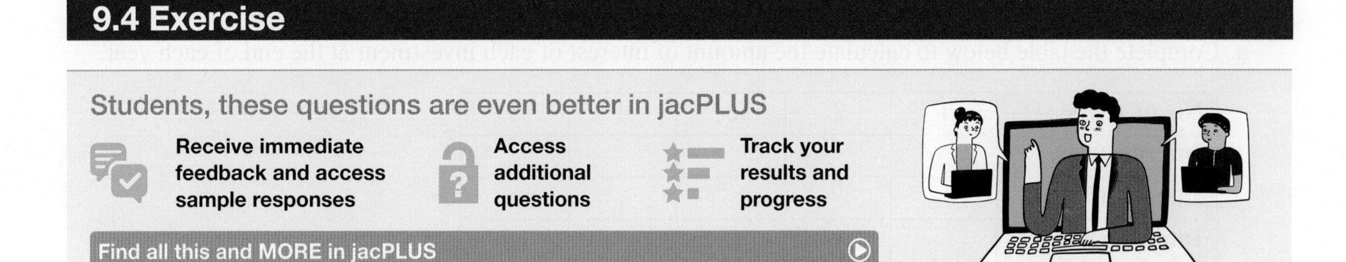

1. **WE10** An amount of $8000 is invested into a savings account earning simple interest at a rate of 5% p.a.
   a. Copy and complete the table below to calculate the future value of the investment at the end of each year.

No. of years	1	2	3	4	5
Future value					

   b. Draw a graph of the future value against the number of years the money is invested.

2. **WE11** Adam has $2500 to invest. Three different banks offer simple interest rates of 1.60%, 1.80% and 2% per annum.
   a. Complete the table below to show the interest that Adam could earn over 5 years.

No. of years	1	2	3	4	5
Interest (1.60%)					
Interest (1.80%)					
Interest (2%)					

   b. Show this information in graph form.

3. WE12 Monique invests $12 000 at 8% p.a., with interest compounded annually.

a. Copy and complete the table below to show the future value at the end of each year.

No. of years	1	2	3	4	5
Future value					

b. Draw a graph of the future value of the investment against the number of years.

4. Draw a graph to represent the future value of the following investments over the first five years.

a. $15 000 at 7% p.a. with interest compounded annually

b. $2000 at 10% p.a. with interest compounded annually

5. A graph is drawn to show the future value of an investment of $2000 at 6% p.a. with interest compounding six-monthly.

a. Complete the table below.

Years	0.5	1	1.5	2	2.5	3	3.5	4	4.5	5
Future value										

b. Use the table to draw the graph.

6. An amount of $1200 is invested at 4% p.a. with interest compounding quarterly.

a. Graph the future value of the investment at the end of each year for 10 years.

b. Graph the compound interest earned by the investment at the end of each year.

7. WE13 James has $8000 to invest at either 4% p.a., 6% p.a. or 8% p.a. compounding annually.

a. Complete the table below to calculate the amount of interest of each investment at the end of each year.

No. of years	1	2	3	4	5
Interest (4%)					
Interest (6%)					
Interest (8%)					

b. Draw a graph that will allow the amount of interest earned to be compared.

8. Petra has $4000 to invest at 6% p.a.

a. Complete the table below to show the future value of the investment at the end of each year, if interest is compounded:

i. annually

ii. six-monthly.

No. of years	1	2	3	4	5
Annually					
Six-monthly					

b. Show this information in graphical form.

9. Kerry invests $100 000 at 8% p.a. for a one-year term. For such large investments, interest is compounded daily.

a. Calculate the daily percentage interest rate, correct to 4 decimal places.

b. Calculate the compounded value of Kerry's investment on maturity.

c. Calculate the amount of interest earned on this investment.

d. Calculate the extra amount of interest earned, compared with the interest calculated at a simple interest rate.

10. Simon invests \$4000 for 3 years at 6% p.a. simple interest. Monica also invests \$4000 for 3 years but her interest rate is 5.6% p.a., with interest compounded quarterly.
 a. Calculate the value of Simon's investment on maturity.
 b. Show that the compounded value of Monica's investment is greater than Simon's investment.
 c. Explain why Monica's investment is worth more than Simon's, despite receiving a lower rate of interest.

11. a. A bank offers interest at 7% per annum, compounded yearly. If a customer puts \$900 in the bank, calculate how much there will be after 3 years.
 b. Another bank offers interest at 8% per annum, compounded yearly. If a customer puts \$14 000 in the bank, calculate how much there will be after 4 years.

12. A bank offers a term deposit for 3 years at an interest rate of 8% p.a. with a compounding period of 6 months. Calculate the end value of a \$5000 investment under these conditions.

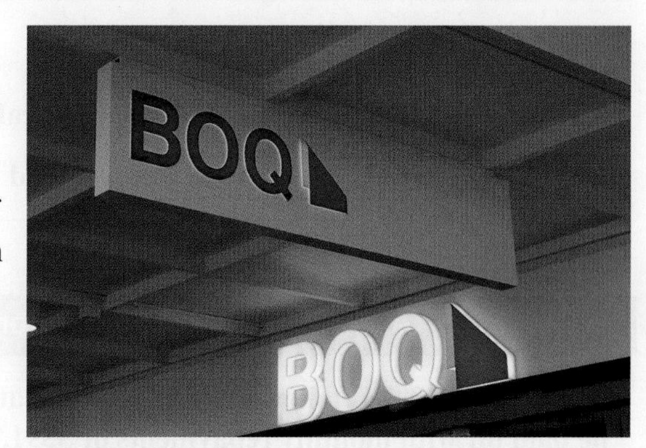

13. A bank offers compound interest on its savings account of 4% p.a. compounding yearly. A boy opens an account with \$1125.
 a. Using a spreadsheet or an online calculator, calculate the value of the boy's investment for the first four years.
 b. Using your values from part a, construct a compound interest graph.

14. A sum of \$1000 is invested at 10.5% p.a. interest. Using a spreadsheet or an online calculator, calculate the value of the investment for the following years and types of interest.

	Principal	Annual interest rate	Time (years)	Future value (simple interest)	Future value (compound interest with yearly compounding)
a.	\$1000	10.5%	3		
b.	\$1000	10.5%	6		
c.	\$1000	10.5%	9		
d.	\$1000	10.5%	12		

9.5 Comparing interest, rental and loan costs

LEARNING INTENTION

At the end of this subtopic you should be able to:
- calculate interest paid on loans
- compare rental and loan repayments.

9.5.1 Interest paid on loans

We will investigate the effect on the term of the loan, and on the total amount of interest charged, if the frequency of the repayments is changed. The value of the repayments will change but the total outlay will stay the same. For example, a $3000 quarterly (every three months) repayment will be compared to a $1000 monthly repayment. That is, the same amount is repaid during the same period of time in each case.

Interest paid

The calculation of the total interest paid is calculated as:

total interest paid = total payments − principal

WORKED EXAMPLE 14 Calculating interest paid on loans

A loan of $80 000 is taken out over 15 years at a rate of 4.0% p.a. (interest charged monthly) and is to be paid off with monthly repayments of $591.75. Determine the amount of interest paid after 15 years.

THINK	WRITE
1. Calculate the total amount paid back on the loan.	$12 \times 591.75 \times 15 = 106\,515$
2. Find the difference between what was paid back and the initial loan.	$106\,515 - 80\,000 = 26\,515$
3. Write the answer.	The amount of interest paid on the loan is $26 515.

WORKED EXAMPLE 15 Calculating interest paid on various repayment periods

Calculate the interest paid on a $33 000 loan, given:
a. quarterly repayments of $1180.05 for 10 years
b. monthly repayments of $393.35 for 119 months
c. fortnightly repayments of $181.55 for 257 fortnights.

THINK	WRITE
a. 1. Making quarterly payments over 10 years means that 40 (10×4) individual payments of $1180.05 are made. Use the formula: total interest = total payments − principal	a. For quarterly repayments: Total interest $= 1180.05 \times 40 - 33\,000$ $= \$14\,202$
2. Write the answer.	The total interest paid when repaying the loan quarterly is $14 202.

b. 1. Making monthly payments over 119 months means that 119 individual payments of \$393.35 are made. Use the formula:
total interest = total payments − principal

2. Write the answer.

b. $\text{Total interest} = 393.35 \times 119 - 33\,000$
$= \$13\,808.65$

The total interest paid when repaying the loan monthly is \$13 808.65.

c. 1. To calculate the total interest paid, use the formula:
total interest = total payments − principal

2. Write the answer.

c. $\text{Total interest} = 181.55 \times 257 - 33\,000$
$= \$13\,658.35$

The total interest paid when repaying the loan fortnightly is \$13 658.35.

9.5.2 Rent

Not everyone borrows money to purchase a house. Some people decide to rent. In this case, instead of paying off a loan, they pay rent to the owner of the house.

WORKED EXAMPLE 16 Determining yearly rent

Devon rents a property and pays \$420 per week in rent. Calculate how much rent Devon pays each year.

THINK	WRITE
1. There are 52 weeks in a year, so there will be 52 payments.	52 weekly payments per year
2. To calculate the yearly payment, multiply the weekly payment by 52.	$420 \times 52 = 21\,840$
3. Write the answer.	The amount of rent Devon pays each year is \$21 840.

tlvd-4129

WORKED EXAMPLE 17 Comparing rental options

A couple have two rental options:
- **Option 1: weekly rent of \$450**
- **Option 2: rent charged every three months (quarterly) at \$5950**

Assuming the properties are the same, determine which rental option is better.

THINK	WRITE
1. Option 1: There are 52 weeks in a year, so there will be 52 payments.	52 weekly payments per year
2. To calculate the yearly payment, multiply the weekly payment by 52.	$450 \times 52 = \$23\,400$
3. Write the answer.	The amount of rent the couple pay each year is \$23 400.
4. Option 2: There are 4 payments per year, since rent is paid quarterly.	4 payments per year

5. To calculate the yearly payment, multiply the quarterly payment by 4.	$5950 \times 4 = \$23\,800$
6. Write the answer.	The amount of rent paid each year is \$23 800.
7. Explain your response.	Option 2 cost \$23 800 per year in rent and option 1 cost \$23 400, so option 1 is better by \$400 per year.

9.5.3 Comparing rental and loan costs

tlvd-4130

WORKED EXAMPLE 18 Comparing rental and loan costs

Callum wants to live near the beach. He talked to a bank about getting a loan for a \$750 000 house and he was told his repayments would be \$776.70 per week to pay off the loan over 30 years. His other option is to rent a house nearby. The rental cost for this house is \$2340 per month.

a. **Calculate how much Callum would pay per year if he paid off the loan.**
b. **Calculate how much Callum would pay per year if he paid rent.**
c. **Calculate the difference in yearly costs of renting and paying off the loan.**

Callum wants to investigate his financial position after paying off his loan if it's assumed the value of the house increased by 325% over the time of the loan.

d. **Calculate the value of the purchased house after 30 years.**
e. **State whether Callum would be in a better position after 30 years if he rented or purchased the house. Explain.**

THINK	WRITE
a. 1. There are 52 weeks in a year, so there will be 52 payments.	a. 52 weekly payments per year
2. To calculate the yearly loan repayments, multiply the weekly loan repayments by 52.	$776.70 \times 52 = 40\,388.40$
3. Write the answer.	The amount of money paid each year to cover the loan is \$40 388.40.
b. 1. There are 12 months in a year, so there will be 12 payments.	b. 12 monthly payments per year
2. To calculate the yearly rental repayments, multiply the monthly rental repayments by 12.	$2340 \times 12 = 28\,080$
3. Write the answer.	The amount of money paid each year on rent is \$28 080.
c. 1. To find the difference, subtract the smaller value (rent) from the larger value (loan).	c. $\text{Difference} = 40\,388.40 - 28\,080 = 12\,308.40$
2. Write the answer.	The yearly loan amount is \$12 308.40 greater than the yearly rental amount.
d. 1. Determine 325% of the price of the house (\$750 000).	d. 325% of 750 000 $\frac{325}{100} \times 750\,000 = 2\,437\,500$
2. Write the answer.	After 30 years and an increase of 325%, the house is worth \$2 437 500.

e. 1. Calculate the total spent on the loan over 30 years.

e. Total loan repaid $= 30 \times 40\,388.40$
$= \$1\,211\,652$

2. Calculate the total rent paid over 30 years.

Total rent paid $= 30 \times 28\,080$
$= \$842\,400$

3. Compare and explain Callum's financial position.

Purchasing a house: Callum pays \$1 211 652, but after 30 years the house is worth \$2 437 500, so Callum would end up ahead by \$1 225 848.
Renting a house: Callum pays \$842 400 and ends up with no asset.
It seems obvious Callum would be better off taking out a loan to pay off the house over 30 years, but several factors need to be taken into account.

1. Can Callum get a loan from the bank with his current employment and income?
2. Can Callum afford to pay off the weekly repayments and still have enough money to live?
3. If Callum had a partner, they could look at taking out the loan together.

tlvd-4131

WORKED EXAMPLE 19 Comparing benefits of loans and rentals

James wants to see if he can afford to purchase a property or if he will decide to rent. James needs to borrow \$680 000 on top of his deposit to purchase a property at 3.75% p.a. over 30 years. If he chooses to rent, it will cost him \$495 per week, whereas if he takes out the loan, his weekly repayments are \$726.26.

a. Compare the difference between weekly loan costs to weekly rental costs.
b. Calculate how much more James would pay if he took out the loan over 30 years compared to paying rent over 30 years.
c. State what advice you would give James.

Looking into the future, James wonders what financial position he would be in if he were to sell the property after 10 years, assuming the property had increased in value by 22% over this time.

d. Calculate how much James would have paid in repayments over 10 years.
e. Calculate how much James would have paid in rent over 10 years.
f. Calculate how much the property is worth after 10 years.

THINK

a. 1. Weekly loan costs = \$726.26
Weekly rental costs = \$495

2. Write the answer.

WRITE

a. Difference $= 726.26 - 495$
$= 231.26$

The weekly loan repayments are \$231.26 greater than the weekly rental costs.

b. 1. Multiply the weekly difference found in part a by 52 to get the yearly difference, and then multiply by 30 to calculate the difference over 30 years.

$$\begin{aligned}\text{Difference} &= 231.26 \times 52 \times 30 \\ &= 360\,765.60\end{aligned}$$

2. Write the answer.

The difference over 30 years is \$360 765.60 more paid on the loan compared to paying rent.

c.

c. Even though James will pay \$360 765.60 more on the loan than if he were to rent over the 30-year period, at the end of the loan he will own the house; if he rents, he will not own the house. I would advise James that if he can afford to pay the loan off over the 30 years, he should do this so that he can then own the house and hope that its value goes up over the 30-year period. At the same time, James needs to budget so he has enough money for other living expenses.

d. Multiply the weekly loan repayments (\$726.26) by the number of weeks in 10 years.

d. $$\begin{aligned}\text{10 years of loan repayments} &= 10 \times 52 \times 726.26 \\ &= \$377\,655.52\end{aligned}$$

e. Multiply the weekly rental payments (\$495) by the number of weeks in 10 years.

e. $$\begin{aligned}\text{10 years of rent payments} &= 10 \times 52 \times 495 \\ &= \$257\,400\end{aligned}$$

f. 1. Calculate 122% of \$680 000, since the value of the property has increased by 22%.

f. $$\begin{aligned}\text{Property value after 10 years} &= \frac{122}{100} \times 680\,000 \\ &= \$829\,600\end{aligned}$$

2. Write the answer.

The property is worth \$829 600 after 10 years.

9.5 Exercise

Students, these questions are even better in jacPLUS

Receive immediate feedback and access sample responses

Access additional questions

Track your results and progress

Find all this and MORE in jacPLUS

1. WE14 A loan of \$75 000 is taken out over a 15-year period at a rate of 4.5% p.a. (interest charged monthly) and is to be paid back with monthly repayments of \$573.74. Determine the amount of interest paid after 15 years.
2. A loan of \$60 000 is taken out over a 20-year period at a rate of 6% p.a. (interest charged monthly) and is to be paid back with monthly repayments of \$429.86. Determine the amount of interest paid after 20 years.
3. Sonya has borrowed \$6000 to buy a ride-on mower. She agrees to pay the loan off over 3 years with equal monthly repayments of \$183.62. Determine the amount paid back after 3 years.
4. James borrowed \$14 000. Interest is charged at 7.4% p.a. (adjusted monthly) and he has been paying \$214.05 each month to pay off the loan. Calculate how much James pays over 5 years.

5. WE15 Calculate the interest paid on a $21 000 loan, given:
   a. Quarterly repayments of $708.39 for 10 years
   b. Monthly repayments of $236.13 for 119 months.
   c. Fortnightly repayments of $108.98 for 258 fortnights.
6. Peter's $17 000 loan at 6.9% p.a. gave the following three situations. Calculate the total interest paid by Peter in each case.
   a. Quarterly repayments of $871.05 for 6 years
   b. Monthly repayments of $290.35 for 71 months.
   c. Fortnightly repayments of $134.01 for 154 fortnights.
7. A loan of $13 000 is taken over 6 years. It is to be repaid with monthly repayments of $218.53 at a rate of 6.5% p.a. (debited monthly). Calculate the total interest paid.
8. A loan of $19 000 is taken over 7 years. It is to be repaid with monthly repayments of $288.62 at a rate of 7.2% p.a. (debited monthly). Calculate the total interest paid.
9. A loan of $18 000 attracts interest at 7.5% p.a. on the outstanding balance. The following three options are available. Calculate the interest paid on each of the three options.
   a. Half-yearly repayments of $1890.22 for 6 years
   b. Quarterly repayments of $945.11 for 5.75 years.
   c. Monthly repayments of $315.04 for 70 months.
10. WE 16 Layla decided to rent a property for $465 per week. Calculate how much rent Layla pays each year.
11. Lachlan found a property to rent at $990 per fortnight. Calculate how much rent Lachlan pays each year.
12. WE 17 A couple have two rental options:
   - Option 1: Weekly rent of $575
   - Option 2: Rent charged every three months (quarterly) at $7375

   Assuming the properties are the same, state which rental option is better.
13. A person wants to rent a property while they study at university. They have two rental options:
   - Option 1: Fortnightly rent of $890
   - Option 2: Monthly rent of $2000

   Assuming the properties are the same, state which rental option is better.
14. WE18 Kyle wants to find a house near his parents. After discussions with the bank, he found out that his repayments would be $923.50 per week in order to pay off the $994 500 loan over 30 years. His other option is to rent a house nearby; the rental cost for this house is $2562.75 per month.
   a. Calculate how much Kyle would pay per year if he paid off the loan.
   b. Calculate how much Kyle would pay per year if he paid rent.
   c. Calculate the difference in yearly costs of renting compared to yearly costs of paying off the loan.

   Kyle compares his financial position after paying off his loan. He assumes the value of the house will have increased by 245% over the time of the loan.
   d. Calculate the value of the purchased house after 30 years.
   e. Comparing renting with taking out a loan, state whether Kyle will be in a better position financially by renting or purchasing the house after the 30-year period. Explain your answer.
15. WE19 Noor is investigating the affordability of purchasing a property compared to that of renting. Noor needs to borrow $725 000 to purchase a property at 4.15% p.a. over 30 years, with weekly repayments of $812.74. If she chooses to rent, it will cost her $545 per week.
   a. Calculate the difference between the weekly cost of the loan and the weekly cost of renting.
   b. Calculate how much more Noor would pay if she took out the loan over 30 years compared to paying rent over 30 years.

Noor wants to understand her financial position if she were to sell the property after 8 years, assuming the property increased in value by 16% over this time.

c. Calculate how much Noor would pay in repayments over 8 years.
d. Calculate how much Noor would pay in rent over 8 years.
e. Calculate how much the property would be worth after 8 years.

16. A house is purchased at the start of 2023 for $568 500. The value of the property is predicted to increase by 21% over the next 5 years. Calculate the value of the house at the start of 2028.

17. A property was purchased for $675 000. It is anticipated that properties in this area will increase in value by an average of 2.2% p.a. Calculate the anticipated value of the property after 6 years.

18. A property was recently purchased for $722 000. The loan repayments are set at $789.60 per week over 25 years. If the value of the property grows by 285% over the time of the loan and the property is then sold, state whether a profit or loss will be made and the value of that profit or loss. Explain this profit/loss in real terms.

9.6 Review

9.6.1 Summary

doc-38017

Hey students! Now that it's time to revise this topic, go online to:

Review your results

Watch teacher-led videos

Practise questions with immediate feedback

Find all this and MORE in jacPLUS

9.6 Exercise

Multiple choice

1. **MC** An investment grew from \$2750 to \$3435 over 5 years. The interest earned over the 5 years is:
   A. \$3435 **B.** \$687 **C.** \$2750 **D.** \$550 **E.** \$685

2. **MC** Keira invested \$360 in a bank for 3 years at 8% simple interest each year. At the end of the 3 years, the total amount she will receive is:
   A. \$28.80 **B.** \$86.40 **C.** \$388.80 **D.** \$446.40 **E.** \$398.60

3. **MC** Philip borrowed \$7000 and intended to pay it back in 4 years. The terms of the loan indicated Philip was to pay 9% p.a. simple interest. The interest Philip paid on the loan is:
   A. \$630 **B.** \$2520 **C.** \$9520 **D.** \$9881 **E.** \$4250

4. **MC** A loan of \$5000 is taken over 5 years. The simple interest is calculated monthly. The interest bill on this loan is \$1125. The simple interest rate per year on this loan is:
   A. 3% **B.** 3.75% **C.** 4% **D.** $4\frac{1}{2}\%$ **E.** 4.75%

5. **MC** The principal invested in an investment bond that will accumulate \$2015 in simple interest after 6 months invested at $6\frac{1}{2}\%$ p.a. is:
   A. \$6000 **B.** \$50 000 **C.** \$60 000 **D.** \$62 000 **E.** \$65 000

6. **MC** A loan can be paid off quicker if the repayments are made:
   A. annually. **B.** monthly. **C.** quarterly. **D.** bi-annually. **E.** fortnightly.

7. **MC** Two banks pay simple interest on short-term deposits. Bank A pays 6% p.a. over 4 years and Bank B pays 6.5% p.a. over $3\frac{1}{2}$ years. If \$5000 was invested in each account, the difference between the two banks' final payout figures is:
   A. \$0 **B.** \$62.50 **C.** \$1137.50 **D.** \$1200 **E.** \$1317.50

8. **MC** Both Scott and Nia rent a one-bedroom apartment. Scott pays \$410 a week and Nia pays \$1755 a month. When comparing their rent for the year, the difference is:
   A. \$115 **B.** \$160 **C.** \$260 **D.** \$1380 **E.** \$1495

9. **MC** If \$12 000 is invested for $4\frac{1}{2}$ years at 6.75% p.a. compounded fortnightly, the amount of interest that would accrue would be closest to:
   A. \$3600 **B.** \$4253 **C.** \$5000 **D.** \$12 100 **E.** \$8996

10. MC Callie has borrowed \$50 000 to finance her new business venture. Callie makes fortnightly repayments of \$294 over 8 years. The amount of interest Callie pays on her loan is:

A. \$1394
B. \$7644
C. \$11 152
D. \$61 152
E. \$72 304

Short answer

11. Calculate the simple interest earned on each of the following investments.
   a. \$3600 at 9% p.a. for 4 years
   b. \$23 500 at 6% p.a. for 2 years
   c. \$840 at 2.5% p.a. for 2 years
   d. \$1350 at 0.2% p.a. for 18 months
   e. \$45 820 at 4.75% p.a. for $3\frac{1}{2}$ years

12. Bradley invests \$23 500 at 4.6% p.a. If he earned \$1351.25 in simple interest, calculate the length of time for which the money was invested.

13. An amount of \$7500 is to be invested at 6% p.a. simple interest.
   a. Copy and complete the table below to calculate the interest over 5 years.

No. of years	1	2	3	4	5
Interest					

   b. Draw a graph of the interest earned against the length of the investment.
   c. Determine the gradient of the linear graph drawn.
   d. Use your graph to determine the amount of interest that would have been earned after 10 years.

14. Maloo is currently renting a one-bedroom apartment for \$380 a week in Melbourne. Maloo wants to move into another rental that costs \$1600 a month. Calculate the difference in monthly rental price when Maloo moves.

15. Barry has an investment with a present value of \$4500. The investment is made at 6% p.a. with interest compounded biannually. Calculate the future value of the investment in 4 years.

16. Calculate the compounded value of each of the following investments.
   a. \$3000 at 7% p.a. for 4 years with interest compounded annually
   b. \$9400 at 10% p.a. for 3 years with interest compounded biannually
   c. \$11 400 at 8% p.a. for 3 years with interest compounded quarterly
   d. \$21 450 at 7.2% p.a. for 18 months with interest compounded biannually
   e. \$5000 at 2.6% p.a. for $2\frac{1}{2}$ years with interest compounded quarterly

17. Kim and Glenn each invest \$7500 for a period of 5 years.
   a. Kim invests her money at 9.9% p.a. with interest compounded annually. Calculate the compounded value of Kim's investment.
   b. Glenn invests his money at 9.6% p.a. with interest compounded quarterly. Calculate the compounded value of Glenn's investment.
   c. Explain why Glenn's investment has a greater compounded value than Kim's.

Extended response

18. $20 000 is to be invested at 4% p.a. with interest compounded annually.
   a. Copy and complete the table below to calculate the future value at the end of each year.

No. of years	1	2	3	4	5
Future value					

   b. Draw a graph of the interest earned against the length of the investment.
   c. Use your graph to determine the future value of the investment after 10 years.

19. A family recently purchased a 3-bedroom house for $845 000. The loan repayments were set at $732.50 a week over 30 years.
   a. Calculate how much the family pays per year.
   b. If the family wanted to change their payments to monthly, calculate their monthly repayments, rounding to the nearest cent.
   c. Calculate how much the family pays for the entirety of the loan.
   d. Determine how much interest is paid over the 30 years.
   e. If the property value is predicted to increase by 195% over the 30 years, determine whether a profit or loss is made, and what its value is.

20. Marco and Cameron want to buy a house. Their bank has offered a loan of $725 000 over 25 years, with loan repayments of $708 per week. Their other option is to rent a house nearby; the rental cost for this house is $2850 per month.
   a. Calculate how much they would pay per year if they took out the loan.
   b. Calculate how much they would pay over the entire loan period.
   c. Calculate how much they would pay per year if they rent.
   d. Calculate the difference in yearly costs of renting compared to yearly costs of paying off a loan.

   It is predicted that the purchased property will increase in value by 180% over the time of the loan.

   e. Calculate the value of the house after 25 years.
   f. Using your answers from parts **b** and **e**, explain Marco and Cameron's financial position after paying off the house.

Answers

Topic 9 Interest and loans

9.2 Simple interest

9.2 Exercise

1. a. \$136.00, \$816 b. \$56.70, \$266.70
   c. \$145.25, \$560.25 d. \$110.40, \$570.40
   e. \$255, \$1275 f. \$336.89, \$1049.89
2. a. \$627.13 b. \$12 542.50
3. a. 10% p.a. b. 6.25% p.a.
   c. 80% p.a. d. 2.125% p.a. or $2\frac{1}{8}$% p.a.
   e. 3.36% p.a.
4. a. 1 year b. 18 months
   c. 3 months d. 7 years
   e. 1 month
5. a. \$3070 b. \$4400 c. \$5425
   d. \$236.36 e. \$2500
6. a. \$103.50 b. \$2700
   c. \$325 d. 131.25
7. a. \$360 b. \$1020 c. \$27 700
   d. \$17.70 e. \$13.67
8. B
9. C
10. A
11. D
12. A
13. C
14. D
15. D
16. a. Big 4 Bank offers the better rate.
    b. Big 4 Bank charges $11\frac{1}{3}$% p.a. for a loan while Friendly Building Society charges 12% (12 × 1% per month).
17. \$1515.79
18. \$2133.33
19. \$352
20. 24 months
21. a. i. \$1540.63 ii. \$6162.50
    b. Yes
22. a. \$2247 b. \$15 729 c. $7\frac{1}{2}$ years

9.3 Compound interest

9.3 Exercise

1. a. $A = 2500(1.075)^5$
   b. $A = \$3589.07$, $I = \$1089.07$
2. a. \$583.20 b. \$1630.47 c. \$4472.27
   d. \$3764.86
3. a. i. \$2150 ii. \$150
   b. i. \$2311.25 ii. \$311.25
   c. i. \$3086.60 ii. \$1086.60
4. a. 5 b. 20 c. 8 d. 72 e. 9 f. 15
5. a. 1.5% b. 2% c. 1.5% d. 1.75%
6. $A = \$9657.36$
   $I = \$2907.36$
7. a. 12
   b. 1.75%
   c. \$5172.05
   d. 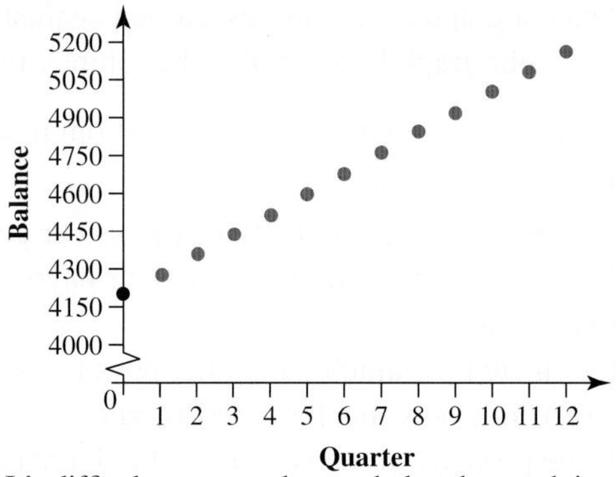

   It's difficult to see on the graph, but the graph is exponential.
8. a. 24
   b. 0.4583%
   c. \$8369.92
   d. 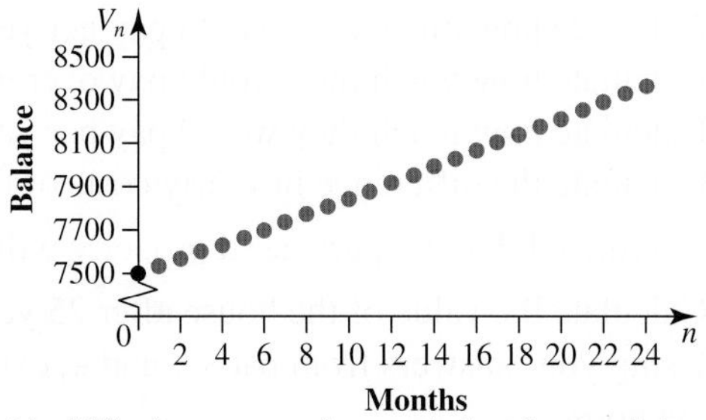

   It's difficult to see on the graph, but the graph is exponential.
9. \$3448.40
10. \$4290.73
11. a. \$1749.60
    b. \$1757.49
    c. \$1759.33
    d. \$1760.24
    e. The balance increases as the compounding periods become more frequent.
12. a. \$2852.92 b. \$4984.72 c. \$6000
13. a. \$605.60 b. \$1314.84 c. \$795.77
    d. \$1043.10 e. \$1641.82
14. a. \$3542.13 b. \$2052.54
    c. \$2969.18 d. \$5000
15. a. i. \$2069.61
       ii. \$930.39
    b. i. \$1531.33
       ii. \$468.67
    c. i. \$3564.10
       ii. \$2035.90

d. i. $5307.05

ii. $4692.95

16. $28 841.44

9.4 Comparing simple and compound interest

9.4 Exercise

1. a.

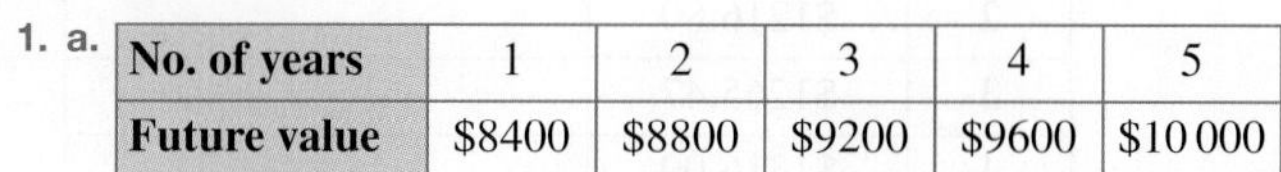

No. of years	1	2	3	4	5
Future value	$8400	$8800	$9200	$9600	$10 000

b.

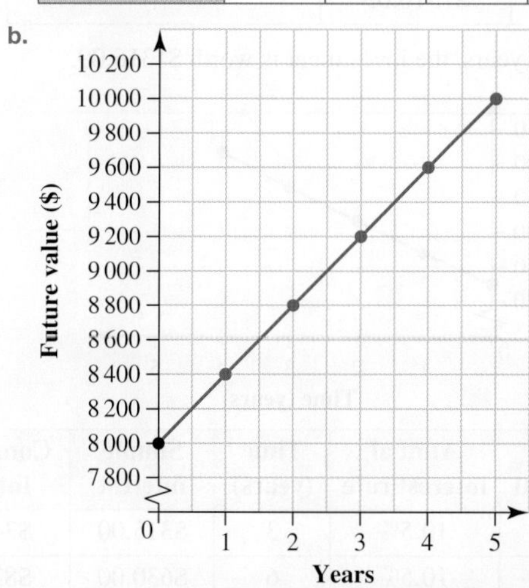

2. a.

	Number of years				
	1	2	3	4	5
Interest (1.60%)	$2540	$2580	$2620	$2660	$2700
Interest (1.80%)	$2545	$2590	$2635	$2680	$2725
Interest (2%)	$2550	$2600	$2650	$2700	$2750

b.

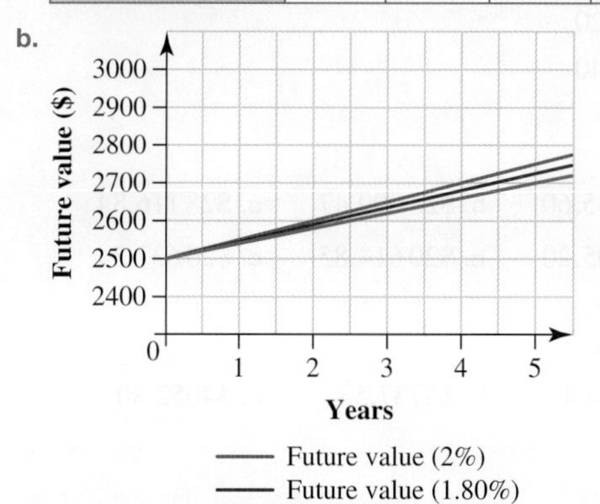

3. a.

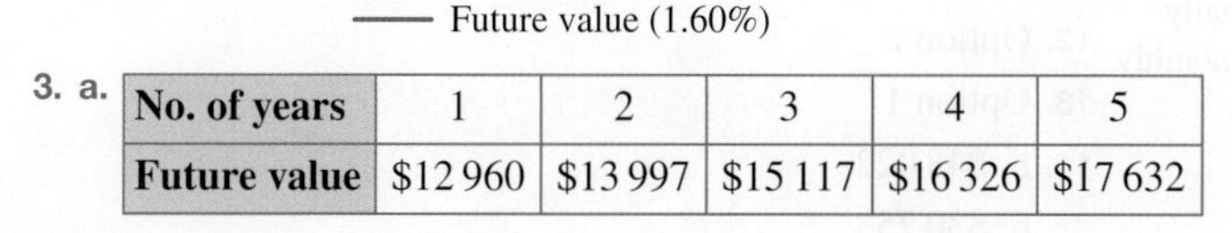

No. of years	1	2	3	4	5
Future value	$12 960	$13 997	$15 117	$16 326	$17 632

b.

4. a.

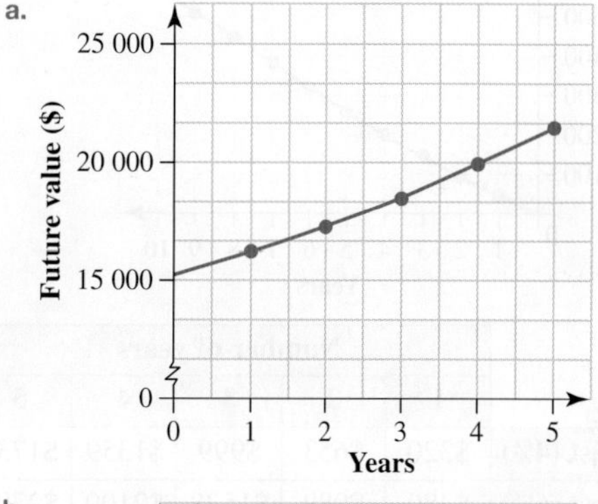

b.

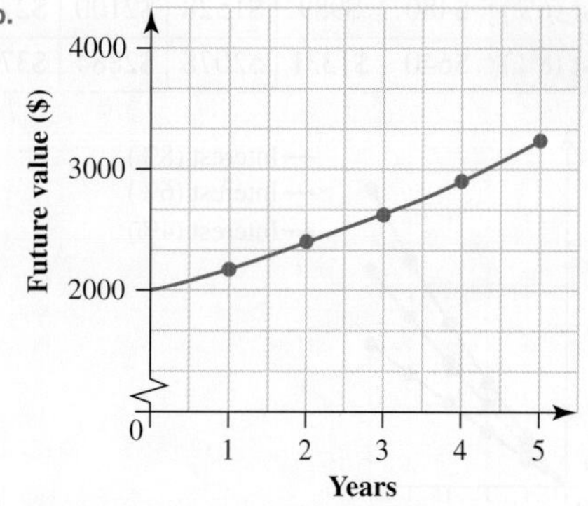

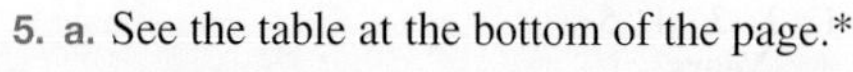

5. a. See the table at the bottom of the page.*

b.

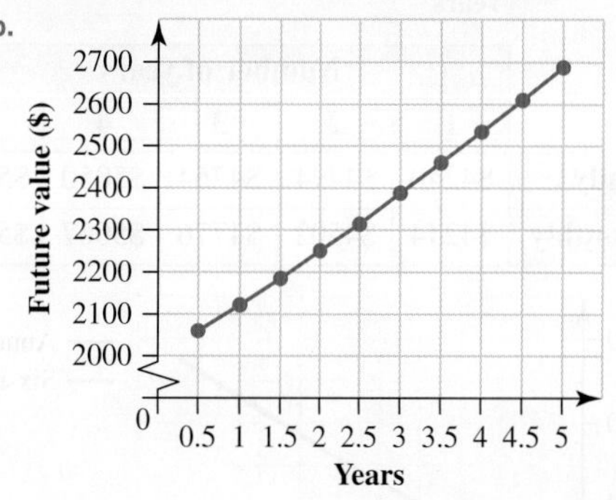

*5. a.

Years	0.5	1	1.5	2	2.5	3	3.5	4	4.5	5
FV	$2060	$2122	$2185	$2251	$2319	$2388	$2460	$2534	$2610	$2688

6. a.

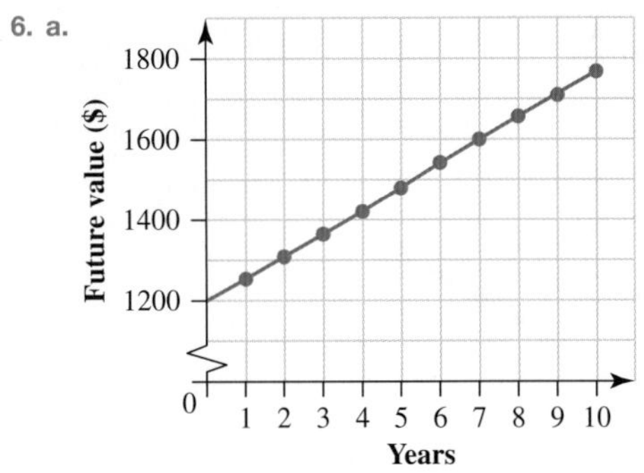

b.

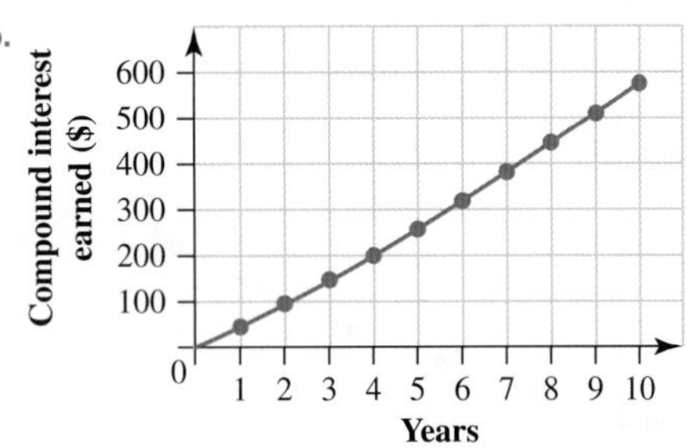

7. a.

	Number of years				
	1	2	3	4	5
Interest (4%)	$320	$653	$999	$1359	$1733
Interest (6%)	$480	$989	$1528	$2100	$2706
Interest (8%)	$640	$1331	$2078	$2884	$3755

b.

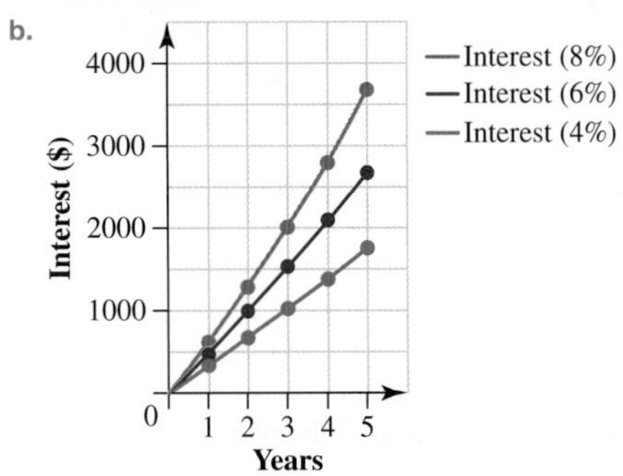

8. a.

	Number of years				
	1	2	3	4	5
Annually	$4240	$4494	$4764	$5050	$5353
Six-monthly	$4244	$4502	$4776	$5067	$5376

b.

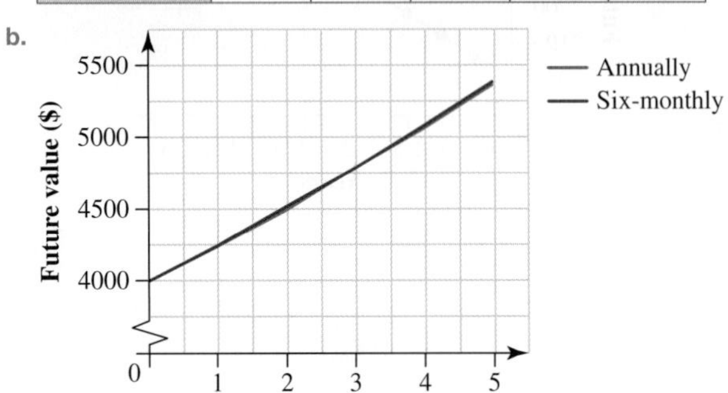

9. a. 0.0219%   b. $108 320.72
   c. $8320.72   d. $320.72

10. a. $4720   b. $4726.24
    c. Compounding interest

11. a. $1102.54   b. $19 046.85

12. $6326.59

13. a.

t	Value	Formula
0	$1125.00	
1	$1170.00	= (B2 ∗ 0.04) + B2
2	$1216.80	
3	$1265.47	
4	$1316.09	

After 4 years, the investment is worth $1316.09.

b.

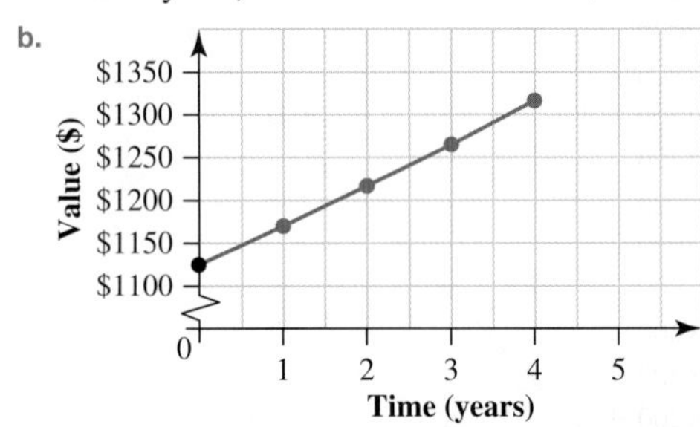

14.

Principal	Annual interest rate	Time (years)	Simple interest	Compound interest
$1000	10.5%	3	$315.00	$349.23
$1000	10.5%	6	$630.00	$820.43
$1000	10.5%	9	$945.00	$1456.18
$1000	10.5%	12	$1260.00	$2313.96

9.5 Comparing interest, rental and loan costs

9.5 Exercise

1. $28 273.20
2. $43 166.40
3. $610.32
4. $12 603
5. a. $28 335.60   b. $28 099.47   c. $28 116.84
6. a. $20 905.20   b. $20 614.85   c. $20 637.54
7. $2734.16
8. $5244.08
9. a. $4682.64   b. $3737.53   c. $4052.80
10. $24 180
11. $25 740
12. Option 2
13. Option 1
14. a. $48 022
    b. $30 753
    c. $17 269
    d. $2 436 525
    e. Loan = $1 440 660
    Rent = $922 590
    From the calculated values, it seems clear that Kyle would be better off taking out a loan to pay off the house over 30 years. However, a number of factors need to be taken into account.

- Can Kyle get a loan from the bank with his current employment and income?
- Can Kyle afford to pay off the weekly repayments and still have enough money to live?
- If Kyle had a partner, they could look at taking out the loan together.
- The impact of inflation on Kyle's profit needs to be considered.

15. a. $267.74 b. $417 674.40 c. $1 267 874.40
d. $850 200 e. $841 000

16. $687 885

17. $769 146.64

18. Profit = $1 031 220
This might be a profit value but, due to inflation, the value of $1 when the house was originally purchased would not be the same 25 years later. It would be lower.

9.6 Review

9.6 Exercise

Multiple choice

1. E
2. D
3. B
4. D
5. D
6. E
7. B
8. C
9. B
10. C

Short answer

11. a. $1296 b. $2820 c. $42
d. $4.05 e. $7617.58

12. 15 months

13. a.

No. of years	1	2	3	4	5
Interest	$450	$900	$1350	$1800	$2250

b.

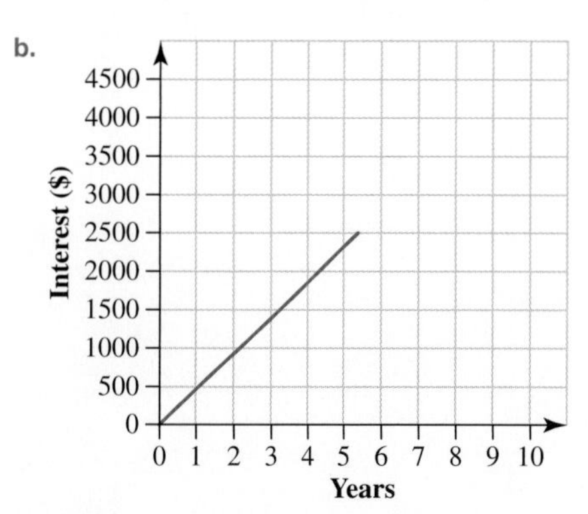

c. 450
d. $4500

14. $46.76 a month

15. $5700.47

16. a. $3932.39 b. $12 596.90 c. $14 457.96
d. $23 851 e. $5334.67

17. a. $12 024.02
b. $12 052.04
c. Compounding interest

Extended response

18. a.

No. of years	1	2	3	4	5
Future value	$20 800	$21 632	$22 497	$23 397	$24 333

b. See the figure at the bottom of the page.*
c. $29 500

19. a. $38 090
b. $3174.17
c. $1 142 700
d. $297 700
e. A profit of $1 350 050 is made.

20. a. $36 816
b. $920 400
c. $34 200
d. $2616
e. $2 030 000
f. Marco and Cameron will have paid $920 400 but will have a house worth $2 030 000.

*18. c.

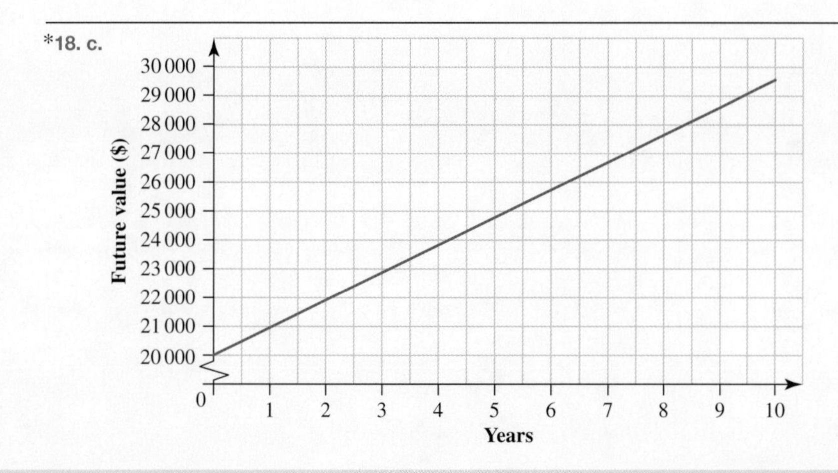

10 Financial and consumer mathematics

LEARNING SEQUENCE

Fully worked solutions for this topic are available online.

10.1 Overview

Hey students! Bring these pages to life online

Watch videos

Engage with interactivities

Answer questions and check results

Find all this and MORE in jacPLUS

10.1.1 Introduction

Money is needed to pay for all the essentials for us to live and provides options for living a comfortable and enjoyable life. Money can be difficult to earn, but easy to spend, so most of us must be conscious of where our money comes from and where it goes to. Therefore, understanding the basic principles of finance is very helpful for managing everyday life.

Every branch of industry and business, whether large or small, international or domestic, will have to pay their employees a salary or wage and contribute to their superannuation funds to assist in saving for retirement. Businesses also need to deduct taxation from employees' earnings to provide revenue for government, and complete a business activity statement if collecting the goods and services tax.

People use the money they earn to pay for services such as council rates, mobile phones, insurance and memberships. We use these services all the time, so we owe it to ourselves to understand how they work so that we can make the best use of our money.

KEY CONCEPTS

This topic covers the following key concepts from the VCE Mathematics Study Design:

- taxation systems at the personal and business level
- income and expenditure calculations such as GST, invoicing and BAS
- comparison of financial products and services such as insurance
- informal consideration of financial risk at the national and global level (short, medium and long term)
- analysis and interpretation of financial information and data sets, trends and economic indicators and their impact (at the personal, community, national or global level) such as gender pay gap, career trends and interruption, currency fluctuations and inflation, stock market movements and recessions.

Source: VCE Mathematics Study Design (2023–2027) extracts © VCAA; reproduced by permission.

10.2 Taxation and superannuation

LEARNING INTENTION

At the end of this subtopic you should be able to:
- calculate council rates
- calculate the amount of superannuation
- calculate income tax and tax deductions
- calculate the Medicare levy.

10.2.1 Council rates

Councils provide services and infrastructure to people living in communities, such as parklands, libraries, rubbish collection and road maintenance. To support councils in providing these services, home owners pay **council rates.**

Council rates are calculated as a percentage on the capital improved value of the land and buildings within the council municipality. The rate at which rates are charged — the **rate in the dollar** — is determined by calculating how much revenue the council plans to raise and dividing this by the amount of capital improvement of land and buildings within the council municipality.

Calculating the rate in the dollar and council rates

$$\textbf{rate in the dollar} = \frac{\textbf{revenue raised}}{\textbf{capital improved value in municipality}}$$

$$\textbf{council rates} = \textbf{rate in the dollar} \times \textbf{capital improved value of property}$$

tlvd-4665

WORKED EXAMPLE 1 Calculating council rates

A council municipality plans to raise \$15 million. There is \$3.25 billion in capital improved value in the municipality. Calculate the payable council rates for a property with \$275 000 capital improved value. Write your answer correct to the nearest cent.

THINK	WRITE
1. Calculate the rate in the dollar. Divide the revenue to be raised by the capital improved value.	$\frac{15\,000\,000}{3\,250\,000\,000} = \frac{15}{3250}$ $= 0.004\,62...$
2. Multiply the rate in the dollar by the capital improved value.	$0.004\,62... \times 275\,000 = \$1269.23$
3. Write the answer as a sentence.	The council rates payable are \$1269.23.

10.2.2 Superannuation

All employees in Australia have money set aside by their employer to help save for their retirement. This money is known as **superannuation**. From July 2022, the law requires employers to pay an additional 10.5% of annual salary into a recognised superannuation fund. This is known as the **super guarantee percentage**, and it is planned to increase on a sliding scale until it reaches 12% in 2025.

Some workers choose to contribute additional funds from their wages to increase their superannuation. There are tax incentives, such as paying lower tax rates on superannuation lump sums, to encourage employees to contribute and save for their retirement.

tlvd-4666

WORKED EXAMPLE 2 Calculating fortnightly superannuation contributions

An employee's hourly rate is $28.75 and they work 38 hours each week. They are paid fortnightly and will receive superannuation at a rate of 10.5%. Calculate the amount of superannuation the employer pays on their behalf each fortnight.

THINK	WRITE
1. Calculate the fortnightly wage.	$28.75 \times 38 \times 2 = \$2185$
2. Calculate 10.5% of the fortnightly wage. Remember to convert 10.5% to a decimal by dividing by 100.	$10.5\% \text{ of } 2185 = \frac{10.5}{100} \times 2185$ ≈ 229.43
3. Write the answer in a sentence.	The employer must pay $229.43 into the superannuation fund each fortnight.

10.2.3 Income tax

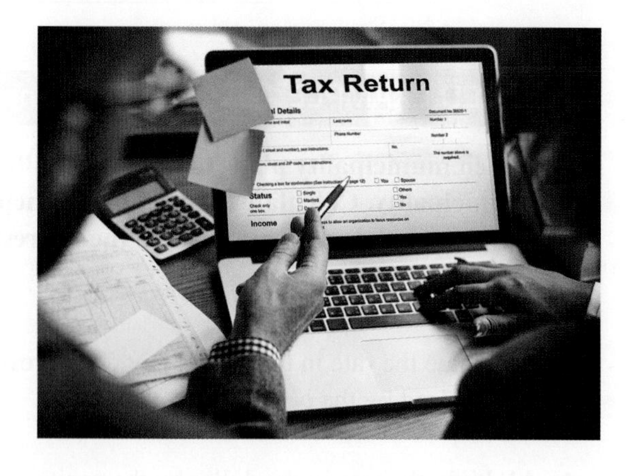

State and federal governments raise revenue for public services, welfare and community needs by imposing charges on citizens, organisations and businesses. These charges are known as taxation and the money raised can be put toward education, health, roads, defence, unemployment benefits, public transport and much more.

One of the key revenue sources for the federal government is income tax.

Calculating taxable income

The income tax is a tax levied on the taxable income of people and can be calculated as:

taxable income = assessable income − allowable deductions

***Note:* You can only claim deductions when you incur the cost and weren't reimbursed. You would need to show evidence of your expenses.**

Assessable income is income that you can pay tax on, if you earn enough to exceed the tax-free threshold. Deductions that you apply reduce the amount of income you pay tax on. You do not deduct them directly from your tax amount. Most people will need to lodge a tax return with the Australian Taxation Office (ATO) at the end of each financial year. This can be done with the assistance of a tax agent or accountant, or you can do it yourself using the ATO's online services.

10.2.4 Tax deductions

If a person spends their own money on work-related expenses, they are entitled to claim the amount spent as a tax deduction when completing their annual tax return. Tax deductions are deducted from total income before tax is calculated and will reduce the taxable income, which means it also reduces the amount of tax to be paid.

What can be claimed as a tax deduction is determined by the ATO and evidence, such as receipts, must be provided. Common tax deductions include:

- purchasing materials
- using your car to travel to a work-related event
- depreciation of equipment used in performing your job
- charitable donations
- expenses associated with maintaining a home office.

WORKED EXAMPLE 3 Calculating total deductions

A large company employs Ken as a plumber. Ken claims deductions of \$1400 for tools, \$25 for gumboots, \$200 for two pairs of work overalls, \$5 per week for dry-cleaning the overalls and \$1.50 per week for work-related telephone calls. Calculate Ken's total deductions for the year.

THINK	WRITE
1. Calculate Ken's total dry-cleaning and telephone deductions.	Dry-cleaning $= \$5 \times 52$ $= \$260$ Telephone $= \$1.50 \times 52$ $= \$78$
2. Add all of Ken's deductions.	Deductions $= \$1400 + \$25 + \$200 + \$260 + \$78$ $= \$1963$
3. Write the answer as a sentence.	Ken's total tax deductions are \$1963.

WORKED EXAMPLE 4 Calculating travel deductions

Raylene is a computer programmer. As part of her job, she uses her own car to visit clients and to attend training seminars. In the 2021–22 financial year, Raylene is allowed a deduction of 72 cents per kilometre. Calculate the amount of the tax deduction if she travelled 2547 km for work-related matters.

THINK	WRITE
1. Multiply the number of kilometres (2547) by the rate per kilometre (0.72). Be sure to convert the rate in cents to dollars.	Travel deduction $= 2547 \times 0.72$ $= \$1833.84$
2. Write the answer as a sentence.	The tax deduction for Raylene's travel expenses is \$1833.84.

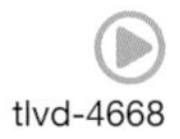
tlvd-4668

WORKED EXAMPLE 5 Calculating a depreciating tax deduction

Trevor is an accountant who works from home. He owns a personal computer he uses for work. Trevor bought a new computer on 1 July 2020 for \$3200. Each year he is allowed a 33% deduction for the depreciation of the computer. Calculate the tax deduction that Trevor was allowed in:

a. the 2020–21 financial year

b. the 2021–22 financial year.

THINK	WRITE
a. 1. The depreciation was 33% of the purchase price, so find 33% of \$3200.	a. Tax deduction $= 33\%$ of \$3200 $= 0.33 \times \$3200$ $= \$1056$
2. Write the answer as a sentence.	Trevor is allowed a \$1056 deduction in the 2020–21 financial year.
b. 1. Calculate the value of the computer at the beginning of 2021–22, by subtracting the depreciation in the previous financial year from the purchase price.	b. Computer value $= \$3200 - \$1056$ $= \$2144$
2. Calculate 33% of the value of the computer at the end of the last financial year to find the depreciation.	Tax deduction $= 33\%$ of \$2144 $= 0.33 \times \$2144$ $= \$707.52$
3. Write the answer as a sentence.	Trevor is allowed a \$707.52 deduction in the 2021–22 financial year.

10.2.5 Medicare levy

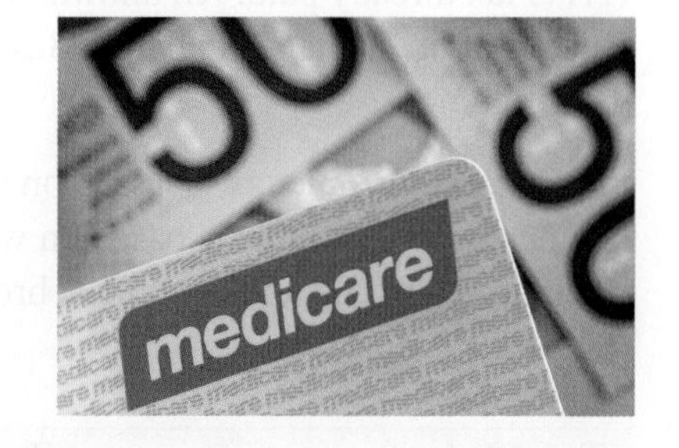

Australian residents have access to healthcare through Medicare, which is partly funded by taxpayers through the payment of the **Medicare levy**. The current Medicare levy is 2% of taxable income, but may be reduced if taxable income is below a certain amount. In the 2020–2021 financial year, single income earners with a taxable income of $29 033 or less were entitled to a reduction, and those with a taxable income equal to or less than $23 226 did not need to pay the Medicare levy.

In addition to the Medicare levy, residents who do not have private health insurance may also be required to pay a Medicare levy surcharge.

WORKED EXAMPLE 6 Calculating the Medicare levy

A plumber's taxable income for the financial year is $65 850. They claim tax deductions of $5680 in work-related expenses and also have private health insurance. Determine the amount they have to pay for the Medicare levy.

THINK	WRITE
1. Calculate the taxable income.	Taxable income $= \$65\,850 - \$5680$ $= \$60\,170$
2. Calculate 2% of taxable income. Remember to convert 2% to a decimal.	$2\% \times 60\,170 = 0.02 \times 60\,170$ $= \$1203.40$
3. Write the answer as a sentence.	They will pay $1203.40 in Medicare levy.

10.2.6 Calculating tax

The ATO administers the Pay As You Go (PAYG) withholding tax system whereby employers deduct tax from each pay and send it to the ATO on their employees' behalf.

This amount is withheld from the employees' regular earnings (gross pay) and contributes towards the tax they have to pay at the end of the financial year.

Calculating net pay

net pay = gross pay − tax withheld

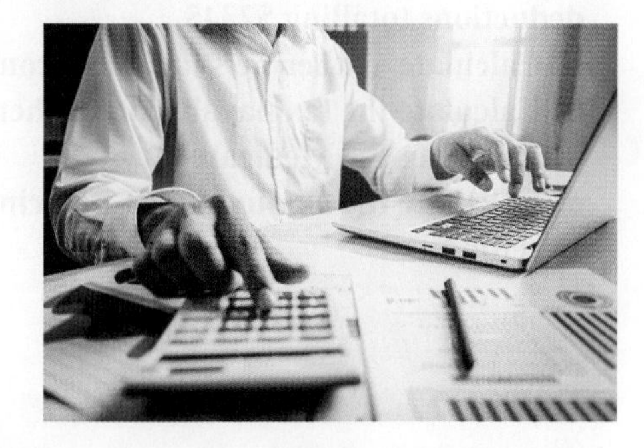

PAYG requires employers to calculate the amount of income tax to withhold from employees based on the pay being the employee's only source of income, and without considering tax deductions. In most cases, this means that the amount of tax paid by the end of the financial year will not be correct. For this reason, every taxpayer must complete a tax return so this amount can be adjusted.

Before completing a tax return, taxpayers must collect a **PAYG payment summary statement** (formerly known as a group certificate) from each of their employers, or employers will complete this online and it can be accessed via the ATO portal of the myGov website. A payment summary statement is a statement of gross earnings and the amount of PAYG tax that has been deducted from those earnings.

In a tax return, all payment summary statements are collected to determine the total gross income and total PAYG tax already paid. All allowable deductions are then subtracted to calculate taxable income. The correct amount of tax is then calculated. Based on this calculation, the taxpayer will then either receive a refund or pay the difference.

In Australia, income tax is levied on a sliding scale, which means that as you earn more money, the rate of tax increases. The table is based on whole dollar amounts, so any cents earned are ignored for the purposes of calculating tax. Taxable income is broken into five income tax brackets, as shown in the 2021–22 income tax table.

Taxable income	Tax on this income
0–\$18 200	Nil
\$18 201–\$45 000	19 cents for each \$1 over \$18 200
\$45 001–\$120 000	\$5092 plus 32.5 cents for each \$1 over \$45 000
\$120 001–\$180 000	\$29 467 plus 37 cents for each \$1 over \$120 000
\$180 001 and over	\$51 667 plus 45 cents for each \$1 over \$180 000

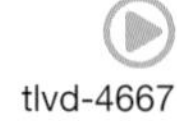
tlvd-4667

WORKED EXAMPLE 7 Calculating income tax payable

Malcolm has a taxable income of \$48 000. Calculate the annual tax payable on his income.

THINK	WRITE
1. Determine the income tax bracket for an income of \$48 000 from the taxable income table.	\$48 000 is in the \$45 001–120 000 tax bracket so the tax payable is \$5092 plus 32.5 cents for each \$1 over \$45 000.
2. Calculate the amount over \$45 000 by subtracting \$45 000 from \$48 000.	\$48 000 − \$45 000 = \$3000
3. Calculate the total tax payable by applying the rule \$5092 plus 32.5 cents for each \$1 over \$45 000.	Tax payable = 5092 + 0.325 × 3000 = \$6067
4. Write the answer as a sentence.	Malcolm needs to pay \$6067 in income tax.

WORKED EXAMPLE 8 Calculating tax refunds and debts

Catherine's gross annual salary as a veterinarian's assistant is \$49 500. She has paid \$7000 in PAYG tax. Catherine has also earned \$560.40 in interest from an investment and has tax deductions totalling \$2345.

a. Calculate Catherine's taxable income.

b. Calculate the tax payable on Catherine's taxable income, including the Medicare levy.

c. Calculate the amount that Catherine should receive as a tax refund.

THINK	WRITE
a. 1. Calculate taxable income by adding all incomes and subtracting any tax deductions.	a. Taxable income = \$49 500 + \$560.40 − \$2345 = \$47 715.40

2. Ignore cents when stating the taxable income.	Taxable income = $47 715
3. Write the answer as a sentence.	Catherine's taxable income is $47 715.
b. 1. Determine the income tax bracket for an income of $47 715 from the taxable income table.	**b.** $47 715 is in the $45 001–$120 000 tax bracket, so the tax payable is $5092 plus 32.5 cents for each $1 over $45 000.
2. Calculate the amount over $45 000 by subtracting $45 000 from $47 715.	$47 715 − $45 000 = $2715
3. Calculate the total tax payable by applying the rule $5092 plus 32.5 cents for each $1 over $45 000.	$\text{Income tax} = 5092 + 0.325 \times 2715$ $= 5974.38$
4. Calculate the Medicare levy.	Medicare levy = 2% of $47 715 = $954.30
5. Calculate the total tax payable by adding the income tax and the Medicare levy.	$\text{Total tax payable} = 5974.38 + 954.30$ $= 6928.68$
6. Write the answer as a sentence.	Catherine's total tax payable is $6928.68.
c. 1. Catherine has paid more tax than she needed to, so she gets a refund.	**c.** $7000 - 6928.68 = 71.32$
2. Calculate the size of the refund by subtracting the amount she should pay ($6928.68) from the amount that she paid ($7000).	
3. Write the answer as a sentence.	Catherine should receive a refund of $71.32.

10.2 Exercise

1. **WE1** A council municipality plans to raise $12 million. There is $1.8 billion in capital improved value in the municipality. Calculate the payable council rates for a property with $275 000 capital improved value.
2. The council rates for a home that has a capital improved value of $325 000 are $1950.
   a. Calculate the rate in the dollar charged by the council.
   b. Calculate the amount of revenue the council intends to collect if the total amount of capital improved value for the municipality is $2.67 billion.
3. **WE2** A worker's hourly rate is $29.45 and she works 25.75 hours each week. She is paid fortnightly. Calculate the amount of superannuation the employer pays on her behalf each fortnight, assuming a 10.5% contribution.
4. A school principal is on an annual salary of $155 750.
   a. Calculate the amount they earn per month.
   b. Calculate their superannuation fund payment if the fund is paid 10.5% of their annual salary.

5. A salary earner makes $62 000 per year.
   a. Calculate the amount they earn each month.
   b. Calculate their superannuation fund payment each month if they receive 10.5% superannuation.
   c. Calculate the total amount deposited into the fund in a year.
6. A graduate teacher makes $72 100 per year.
   a. Calculate the amount they get paid each fortnight.
   b. Calculate the amount of money that is paid into their superannuation fund (assuming a 10.5% contribution) each fortnight.

7. An architect makes a salary of $85 500 per year and is given 11% extra in the form of superannuation.
   a. Calculate the amount of superannuation they get each year.
   b. In the following year, the architect is not given a pay increase, but instead their superannuation increases to 14% of their salary. Calculate the amount of the increase in superannuation.
8. **WE3** Darren is a pest exterminator. Darren is allowed tax deductions for three sets of protective clothing at $167.50 each, two pairs of goggles at $34 each and four face masks at $13.60 each. Darren also uses a spray tank costing $269 and pays $5 per week to have his clothing professionally cleaned. Calculate Darren's total annual tax deductions.
9. Vladimir works as a waiter. Vladimir must wear a uniform comprising a white shirt with black pants, belt and bow tie. Vladimir buys three shirts at $45.00 each, two pairs of pants at $76.90 each, a belt for $15 and a bow tie for $14.90. Vladimir's uniform must be dry-cleaned each week at a cost of $5.70. Vladimir has other tax deductions: $345 for union fees, $60 for having his tax return prepared by an accountant, and $50 in charity donations. Calculate Vladimir's total tax deductions.

10. **WE4** Rajid uses his car as part of his job as an insurance assessor. In the 2017–18 financial year he was allowed a deduction of 66 cents per kilometre and he travelled 3176 km for work. Calculate Rajid's tax deduction.
11. The table below shows the rate per kilometre allowed as a tax deduction for travel in a private vehicle across various financial years.

Financial year	Allowable deduction
2017–18	66 cents per km
2018–19	68 cents per km
2019–20	68 cents per km
2020–21	71 cents per km
2021–22	72 cents per km

    Calculate the total tax deduction allowed for a person who claimed 2645 km in:
    a. 2021–22
    b. 2019–20
    c. 2017–18.
12. **WE5** Selina is a teacher with a home computer purchased for $2500. If a 40% tax deduction is allowed for depreciation, calculate the tax deduction that Selina is allowed in:
    a. the first financial year
    b. the second financial year
    c. the third financial year.
13. Jess is a builder. At the end of the financial year, Jess's building equipment was valued at $12 350. If Jess is allowed a tax deduction of 25% for depreciation of their equipment, calculate their deduction.

14. Kate is a motor mechanic who runs her own garage. Kate has $85 000 in capital equipment that she depreciates at a rate of 27.5% p.a. She travels 2750 km on work-related trips in her van, for which she is allowed a deduction of 68 cents per km. Calculate Kate's total tax deductions for this financial year.

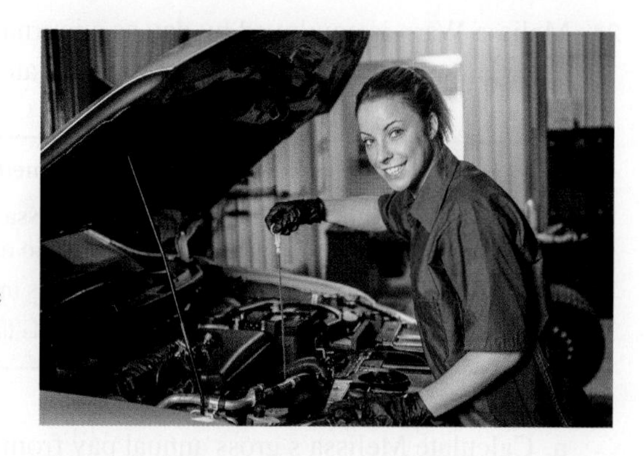

15. **WE6** Calculate the Medicare levy for the following taxable incomes.
    a. $60 400
    b. $77 300
    c. $89 400
    d. $108 423

16. Alvin has a gross weekly wage of $451.75.
    a. Calculate Alvin's gross annual wage.
    b. Calculate the amount of Medicare levy that Alvin pays annually.

17. **WE7** Use the income tax table provided in this subtopic to calculate the income tax payable on an annual taxable income of $55 450.

18. Mira receives a gross pay of $327.68 per week.
    a. Calculate Mira's gross annual pay.
    b. Calculate the annual amount of income tax that Mira must pay, based on this amount.

19. Thor earns a gross pay of $963.80 per fortnight. Calculate the annual amount of income tax that Thor must pay, based on this amount.

20. Hiro earns a gross weekly pay of $1623.60. Calculate the amount of PAYG tax deducted each week by Hiro's employer.

21. Sun earns a gross fortnightly pay of $5600. Calculate the amount of PAYG tax that Sun's employer should deduct each fortnight.

22. **WE8** The payment summary statement for Wendell Hancock is shown. Wendell has also earned $372 in interest from an investment and has tax deductions totalling $1298.
    a. Calculate Wendell's taxable income.
    b. Calculate the tax payable on Wendell's taxable income, including the Medicare levy.
    c. Calculate the amount that Wendell should receive as a tax refund.

Payment Summary Statement	
Wendell Hancock	
Gross income	$39 600.00
PAYG tax deducted:	$8054.00

23. Kamala earns a gross weekly pay of $1748.90.
    a. Calculate Kamala's gross annual pay.
    b. During the year Kamala earned $45.15 in bank interest, and had tax deductions of $1296 in total. Calculate the amount of tax that Kamala should pay for the year, based on their annual taxable income, including the Medicare levy.
    c. Calculate Kamala's refund or tax debt if Kamala had $350 PAYG tax deducted each week.

24. Melissa Warn is employed by day as a journalist and by night as a radio announcer. Her payment summary statements are shown.

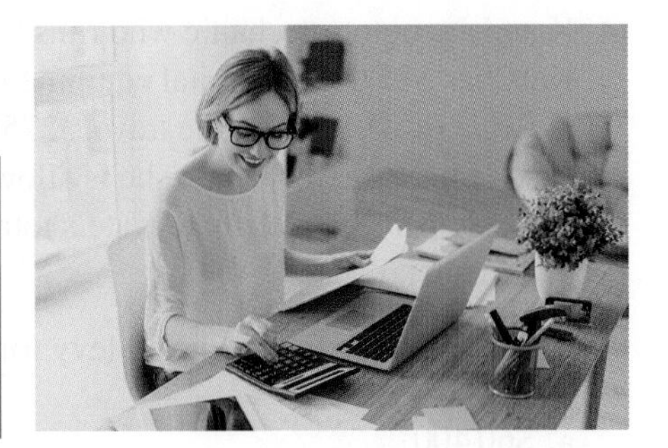

Job 1

Payment Summary Statement	
Melissa Warn	
Journalist	
Gross income	$35 000.00
PAYG tax deducted:	$7405.00

Job 2

Payment Summary Statement	
Melissa Warn	
Radio announcer	
Gross income	$9605.00
PAYG tax deducted:	$4535.50

a. Calculate Melissa's gross annual pay from both jobs and the total amount of PAYG tax that she has paid.
b. Melissa earned $184.40 in interest from bank accounts and had $3276 worth of tax deductions for the year. Calculate her taxable income.
c. Calculate the amount of tax that Melissa *should* have paid throughout the year, including the Medicare levy.
d. Calculate the tax refund that Melissa is owed.

25. An accountant pays $1719.40 in Medicare levy.
If they have private health insurance, calculate their taxable income for the year.

26. A mechanical engineer pays $14 812 in tax.
a. Determine the tax bracket their taxable income falls within.
b. Determine their taxable income, to the nearest dollar.
c. After the financial year they realise that they forgot to claim $985 in tax deductions. Determine their taxable income and the tax refund they should expect.

10.3 Business income and expenditure

LEARNING INTENTION

At the end of this subtopic you should be able to:
- calculate the total price and GST for items
- complete a tax invoice
- understand business activity statements (BAS)
- calculate leave loading
- prepare wage sheets.

10.3.1 Goods and services tax (GST)

In Australia there is a 10% tax that is charged on most purchases, known as a **goods and services tax (GST)**. Some essential items, such as medicine, education and certain types of food, are exempt from GST, but for all other goods GST is added to the cost of items bought or services paid for.

Businesses need to collect GST from their customers and pay this money to the Australian Tax Office (ATO).

Calculating GST

To calculate the amount of GST for an item, multiply the price without GST by the decimal equivalent of 10% (0.1) or divide by 10. If the price is quoted inclusive of GST, the amount of GST can be evaluated by dividing the price by 11.

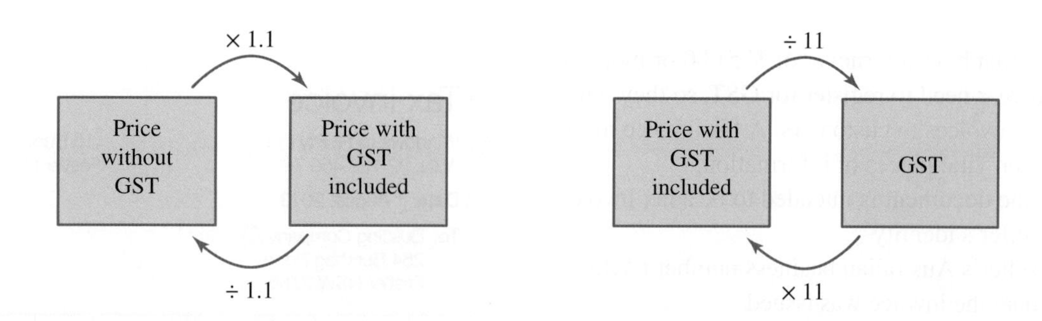

Calculating price

To calculate the total price of an item that does not have GST included, multiply the price without GST by 1.1. If the price is quoted inclusive of GST, price without GST can be evaluated by dividing the price by 1.1.

WORKED EXAMPLE 9 Calculating GST payable

A cricket bat has a pre-GST price of $127.50. Calculate the GST payable on the purchase of the bat.

THINK	WRITE
1. Calculate 10% of $127.50.	GST payable $= 10\%$ of 127.50 $= 0.1 \times 127.50$ $= \$12.75$
2. Write the answer as a sentence.	The GST payable on the cricket bat is $12.75.

tlvd-4669

WORKED EXAMPLE 10 Calculating GST included in a sale

Calculate the amount of GST included in an item purchased for a total of $280.50.

THINK	WRITE
1. Determine whether the price already includes the GST.	Yes, GST is included.
2. If GST is not included, calculate 10% of the value. If GST is included, divide the value by 11.	GST is included, so divide $280.50 by 11. $280.50 \div 11 = 25.5$
3. Write the answer as a sentence.	The amount of GST is $25.50.

10.3.2 Invoices

An invoice is a document that outlines a sale transaction and is provided by a seller to a buyer in order to collect payment.

Businesses that have a turnover of $75 000 or more in a financial year need to register for GST, so they must provide tax invoices to customers. A tax invoice must include seven vital pieces of information:

1. That the document is intended to be a tax invoice
2. The seller's identity
3. The seller's Australian business number (ABN)
4. The date the invoice was issued
5. A brief description of the items sold, including the quantity (if applicable) and the price
6. The GST amount (if any) payable — this can be shown separately or, if the GST amount is exactly one-eleventh of the total price, a statement that states 'Total price includes GST'
7. The extent to which each sale on the invoice is a taxable sale

1 Tax invoice

2 Windows to Fit Pty Ltd
ABN: 32 123 456 789 3

15 Burshag Road
Festler NSW 2755

4 Date: 1 August 2013

To: Building Company 8
254 Burshag Road
Festler NSW 2755

Qty	Description of supply	Unit price	GST	Total
5 50	Window frames	$150	$15	$8,250
10	Deadlocks	$40	$4	$440
			6	
TOTAL AMOUNT PAYABLE				$8,690

The total price includes GST 7

Business activity statements (BAS)

If a business has collected GST, it must complete a business activity statement (BAS) to report and pay the collected GST to the Australian Taxation Office (ATO). The ATO will automatically send a BAS through when it is time to lodge if a business has an ABN and is registered for GST. This could be annually, quarterly or monthly, depending on the turnover of the business.

As well as reporting on GST, a BAS helps a business provide information about other taxes, including:

- pay as you go (PAYG) instalments
- PAYG withholding tax
- fringe benefits tax (FBT) instalments
- luxury car tax
- wine equalisation tax.

The BAS can be completed online, or with the assistance of a registered agent.

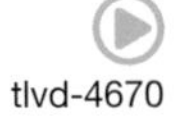
tlvd-4670

WORKED EXAMPLE 11 Completing a tax invoice

Complete the following tax invoice by calculating the unknown amounts A–F.

Office Supplies

TAX INVOICE

INVOICE # 100
DATE: 18TH JANUARY 2022

Melbourne
Australia
Phone: 04xx xxx xxx
ABN:

PURCHASED BY:
Customer name
Street address
City, State/region, Postal code

QUANTITY	DESCRIPTION	UNIT PRICE	GST	TOTAL
1	Laser Printer	270.91	27.09	A
1	A4 copy paper	24.50	B	C
1	Ink cartridge	D	E	165.00
			TOTAL DUE	F

THINK	WRITE
Amount A is the total of the unit price and GST for the laser printer.	$A = 270.91 + 27.09$ $= \$298$
Amount B is the 10% GST on the copy paper unit price of \$24.50.	$B = 0.1 \times 24.50$ $= \$2.45$
Amount C is the total of the unit price and GST for the copy paper.	$C = 24.50 + 2.45$ $= \$26.95$
Amount D is the unit price for the ink cartridge and can be found by dividing the total amount by 1.1.	$D = 165 \div 1.1$ $= \$150$
Amount E is the GST for the ink cartridge and can be found by dividing the total price by 11.	$E = 165 \div 11$ $= \$15$
The total due, amount F, can be calculated by adding up the three amounts in the total column.	$F = 298 + 26.95 + 165.00$ $= \$489.95$

EXAMPLE: Business activity statement – front

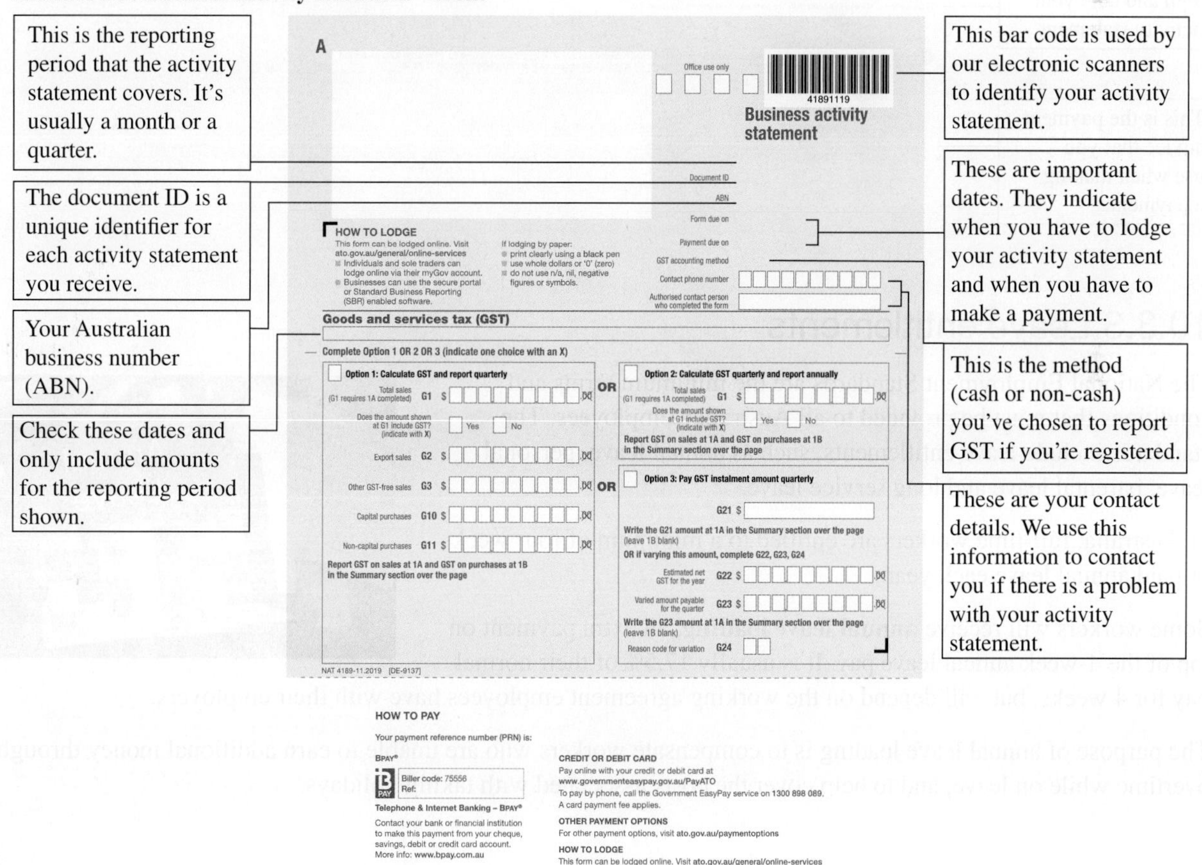

A

Office use only

41891119

Business activity statement

Document ID

ABN

Form due on

Payment due on

GST accounting method

Contact phone number

Authorised contact person who completed the form

HOW TO LODGE

This form can be lodged online. Visit **ato.gov.au/general/online-services**
- Individuals and sole traders can lodge online via their myGov account.
- Businesses can use the secure portal or Standard Business Reporting (SBR) enabled software.

If lodging by paper:
- print clearly using a **black** pen
- use whole dollars or '0' (zero)
- do not use n/a, nil, negative figures or symbols.

Goods and services tax (GST)

Complete Option 1 OR 2 OR 3 (indicate one choice with an X)

Option 1: Calculate GST and report quarterly

Total sales (G1 requires 1A completed) **G1** $ .00

Does the amount shown at G1 include GST? (indicate with X) Yes No

Export sales **G2** $ .00

Other GST-free sales **G3** $ .00

Capital purchases **G10** $ .00

Non-capital purchases **G11** $ .00

Report GST on sales at 1A and GST on purchases at 1B in the Summary section over the page

OR

Option 2: Calculate GST quarterly and report annually

Total sales (G1 requires 1A completed) **G1** $ .00

Does the amount shown at G1 include GST? (indicate with X) Yes No

Report GST on sales at 1A and GST on purchases at 1B in the Summary section over the page

OR

Option 3: Pay GST instalment amount quarterly

G21 $

Write the G21 amount at 1A in the Summary section over the page (leave 1B blank)

OR if varying this amount, complete G22, G23, G24

Estimated net GST for the year **G22** $ .00

Varied amount payable for the quarter **G23** $ .00

Write the G23 amount at 1A in the Summary section over the page (leave 1B blank)

Reason code for variation **G24**

NAT 4189-11.2019 [DE-9137]

HOW TO PAY

Your payment reference number (PRN) is:

BPAY®

Biller code: 75556
Ref:

Telephone & Internet Banking – BPAY®

Contact your bank or financial institution to make this payment from your cheque, savings, debit or credit card account. More info: **www.bpay.com.au**

CREDIT OR DEBIT CARD

Pay online with your credit or debit card at **www.governmenteasypay.gov.au/PayATO**

To pay by phone, call the Government EasyPay service on **1300 898 089**.

A card payment fee applies.

OTHER PAYMENT OPTIONS

For other payment options, visit **ato.gov.au/paymentoptions**

HOW TO LODGE

This form can be lodged online. Visit **ato.gov.au/general/online-services**

EXAMPLE: Business activity statement – rear

These dates will show whether you are reporting monthly or quarterly for PAYG withholding.

This is the section you need to complete if you withhold from payments to others.

PAYG tax withheld

Total salary, wages and other payments W1 $ .00

Amount withheld from payments shown at W1 W2 $ .00

Amount withheld where no ABN is quoted W4 $ .00

Other amounts withheld (excluding any amount shown at W2 or W4) W3 $ .00

Total amounts withheld (W2 + W4 + W3) W5 $ .00

Write the W5 amount at 4 in the Summary section below

Reason for varying (G24 & T4)	Code	Obligation
Change in investments	21	PAYG only
Current business structure not continuing	22	GST & PAYG
Significant change in trading conditions	23	GST & PAYG
Internal business restructure	24	GST & PAYG
Change in legislation or product mix	25	GST & PAYG
Financial market changes	26	GST & PAYG
Use of income tax losses	27	PAYG only

PAYG income tax instalment

Complete Option 1 OR 2 (indicate one choice with X)

Option 1: Pay a PAYG instalment amount quarterly

T7 $

Write the T7 amount at 5A in the Summary section below
OR if varying this amount, complete T8, T9, T4

Estimated tax for the year T8 $ .00

Varied amount payable for the quarter T9 $ .00

Write the T9 amount at 5A in the Summary section below

Reason code for variation T4

OR

Option 2: Calculate PAYG instalment using income times rate

PAYG instalment income T1 $ .00

T2 %

OR New varied rate T3 . %

T1 x T2 (or x T3) T11 $ .00

Write the T11 amount at 5A in the Summary section below

Reason code for variation T4

Summary

Amounts you owe the ATO

GST on sales or GST instalment 1A $ .00

PAYG tax withheld 4 $ .00

PAYG income tax instalment 5A $ .00

Deferred company/fund instalment 7 $ .00

1A + 4 + 5A + 7 8A $ .00

Amounts the ATO owes you

GST on purchases 1B $ .00

Do not complete 1B if using GST instalment amount (Option 3)

Credit from PAYG income tax instalment variation 5B $ .00

1B + 5B 8B $ .00

This is the section where you summarise your reporting obligations and calculate whether you are due a refund or you need to pay.

Payment or refund?

Is 8A more than 8B? (indicate with X)

Yes, then write the result of 8A minus 8B at 9. This amount is payable to the ATO.

No, then write the result of 8B minus 8A at 9. This amount is refundable to you (or offset against any other tax debt you have).

Your payment or refund amount 9 $ .00

Do not use symbols such as +, −, /, $

This is where you sign and date your activity statement.

Declaration I declare that the information given on this form is true and correct, and that I am authorised to make this declaration. The tax invoice requirements have been met.

Signature Date / /

Return this completed form to

Taxation laws authorise the ATO to collect information including personal information about individuals who may complete this form. For information about privacy and personal information go to ato.gov.au/privacy. Activity statement instructions are available from ato.gov.au or can be ordered by phoning 13 28 66.

Australian Government
Australian Taxation Office

This is the payment advice that you use when making a payment.

10.3.3 Leave entitlements

The National Employment Standards are the minimum terms and conditions that must be provided to all Australian employees. The standards include leave entitlements, such as annual leave, personal leave, parental leave and long service leave.

In Australia, full-time workers are entitled to a minimum of four weeks of paid annual leave each year.

Some workers will receive **annual leave loading**, an extra payment on top of the 4-week annual leave pay. It is usually 17.5% of their normal pay for 4 weeks, but will depend on the working agreement employees have with their employers.

The purpose of annual leave loading is to compensate workers who are unable to earn additional money through overtime while on leave, and to help cover the costs associated with taking holidays.

WORKED EXAMPLE 12 Calculating leave loading

A worker's annual salary is $58 056. Calculate the amount, in dollars, the employer pays for annual leave loading before tax.

THINK	WRITE
1. Calculate the weekly salary by dividing the annual salary by 52.	$\frac{58\,056}{52} \approx \$1116.46$
2. Calculate the wage for 4 weeks.	$\$1116.46 \times 4 = 4465.84$ 17.5% of 4465.84
3. Calculate 17.5% of the 4-week wage. Remember to convert 17.5% to a decimal by dividing by 100.	$= \frac{17.5}{100} \times 4465.84$ $= 781.52$
4. Write the answer as a sentence.	The amount paid for annual leave loading, before tax, is $781.52.

10.3.4 Preparing a wage sheet

A wage sheet shows a list of employees and details of their earnings. These include net wages, pay rates, tax withheld and allowances. A wage sheet may look similar to the one shown here.

Employee	Pay rate ($)	Normal hours worked	Overtime	Penalty rate		Allowance ($)	Gross pay ($)	Tax withheld ($)	Net pay ($)
			1.5	1.5	2				
Dave	19.50	38	2	5		25.5	945.75	163.00	782.75
Jose	21.80	32	4	3.5	5		1253.50	271.00	982.50
Neeve	25.70	25		4		15	1002.30	183.00	819.30
Nihe	22.45	38				10	853.10	131.00	722.10
Yen	29.85	36			6		1074.60	208.00	866.60

A pay sheet (or pay slip) shows individual employee details, including the hours worked, pay rate, net wage, tax deduction, leave days and superannuation paid.

A wage sheet can be set up in a spreadsheet containing formulas that perform the necessary calculations.

tlvd-4671

WORKED EXAMPLE 13 Preparing a wage sheet

Using a spreadsheet, calculate the gross pay and net pay for each employee in the wage sheet.

Employee	Pay rate ($)	Normal hours	Overtime	Penalty rate		Allowance ($)	Gross pay ($)	Tax withheld ($)	Net pay ($)
			1.5	1.5	2				
Honura	**38.95**	**38**	**4**		**6**	**15**		**620**	
Skye	**28.40**	**38**	**3**	**5**				**289**	
Beau	**19.15**	**35**		**5**				**234**	

THINK | **WRITE**

1. Set up a spreadsheet.

	A	B	C	D	E	F	G	H	I	J
1			Normal						Tax	
2		Pay	hours	Overtime	Penalty rate		Allowance	Gross	with-	Net
3	Employee	rate ($)	worked	1.5	1.5	2	($)	pay ($)	held ($)	pay ($)
4	Honura	38.95	38	4		6	15		620	
5	Skye	28.40	38	3	5				289	
6	Beau	19.15	35		5				234	

2. Determine gross pay for hours worked.
   Pay for normal hours:
   (e.g. 38.95 * 38)

   B4 * C4

   Overtime pay calculation:
   (e.g. 4 * 1.5 * 38.95)

   D4 * D3 * B4

   Pay for working on penalty rates:
   (e.g. 0 * 1.5 * 38.95 + 6 * 2 * 38.95)

   E4 * E3 * B4 + F4 * F3 * B4

3. Calculate gross pay by adding all the amounts determined above, plus allowances.

   In cell H4 enter the formula:
   H4 = B4 * C4 + D4 * D3 * B4 + E4 * E3 * B4 + F4 * F3 * B4 + G4

4. Complete the table by filling the formula in column H down for all employees.

	A	B	C	D	E	F	G	H	I	J
1			Normal						Tax	
2		Pay	hours	Overtime	Penalty rate		Allowance	Gross	with-	Net
3	Employee	rate ($)	worked	1.5	1.5	2	($)	pay ($)	held ($)	pay ($)
4	Honura	38.95	38	4		6	15	2196.20	620	
5	Skye	28.40	38	3	5			1420	289	
6	Beau	19.15	35		5			813.88	234	

5. Calculate the net pay by deducting tax withheld from the gross pay.

   in cell J4, type: = H4 – I4

6. Complete the wage sheet.

	A	B	C	D	E	F	G	H	I	J
1–3	Employee	Pay rate ($)	Normal hours worked	Overtime	Penalty rate		Allowance ($)	Gross pay ($)	Tax withheld ($)	Net pay ($)
				1.5	1.5	2				
4	Honura	38.95	38	4		6	15	2196.20	620	1576.20
5	Skye	28.40	38	3	5			1420	289	1131
6	Beau	19.15	35		5			813.88	234	579.88

10.3 Exercise

1. **WE9** A book has a pre-GST price of $35.60. Calculate the GST payable on the purchase of the book.

2. Calculate the GST payable on each of the following items (prices given are pre-tax):
   a. A bottle of dishwashing liquid at $2.30
   b. A basketball at $68.90
   c. A pair of cargo pants at $98.50

3. A pair of runners that cost $112.50 have 10% GST added to their cost. Calculate the total cost of the runners.

4. **WE10** Calculate the amount of GST included in an item purchased for a total of:
   a. $34.98
   b. $586.85
   c. $56 367.85
   d. $2.31.

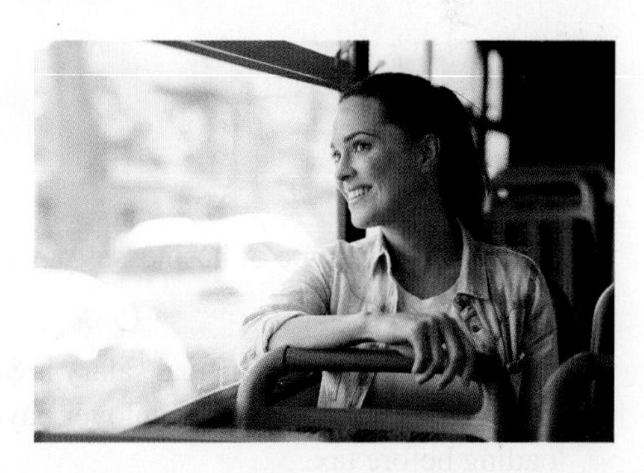

5. A restaurant bill totals $108.35, including 10% GST. Calculate the actual price of the meal before the GST was added.

6. A bus fare was $7.60, including the 10% GST. Calculate:
   a. the bus fare without the GST
   b. how much GST was paid.

7. WE11 Complete the following tax invoice by calculating the unknown amounts A–E.

Sporting Goods

TAX INVOICE

9 Winners Boulevard
Melbourne
Australia
Phone: 04xx xxx xxx
ABN:

INVOICE # 101
DATE: 19TH JANUARY 2022

PURCHASED BY:
Customer name
Street address
City, State/region, Postal code

QUANTITY	DESCRIPTION	UNIT PRICE	GST	TOTAL
1	Tennis racquet	A	B	349.00
1	Tennis balls	34.50	C	D
			TOTAL DUE	E

8. A tax invoice from a clothing order is given below.
   a. Calculate the total amount of GST payable.
   b. Calculate the total amount due.

Clothing

TAX INVOICE

2 Shoppers Place
Melbourne
Australia
Phone: 04xx xxx xxx
ABN:

INVOICE # 102
DATE: 21ST AUGUST 2022

PURCHASED BY:
Customer name
Street address
City, State/region, Postal code

QUANTITY	DESCRIPTION	UNIT PRICE	GST	TOTAL
1	Shorts			77.00
1	Shoes	128.00		
			TOTAL DUE	

9. Identify the due date for a business to lodge a BAS if it is lodging:
   a. annually
   b. quarterly
   c. monthly.

10. Research which type of businesses pay the luxury car tax and at what rate.

11. WE12 An employee's annual salary is $85 980. Calculate the amount, in dollars, the employer pays for annual leave loading before tax.

12. An employer is proposing a new working agreement: the removal of the 17.5% annual leave loading, an increase of 5% in annual wages and 11.5% superannuation contributions. Advise the employees whether to accept the agreement. Justify your answer using calculations.

13. **WE13** Using a spreadsheet, calculate the gross pay and net pay for each employee in the wage sheet.

Employee	Pay rate ($)	Normal hours worked	Overtime	Penalty rate		Allowance ($)	Gross pay ($)	Tax withheld ($)	Net pay ($)
			1.5	1.5	2				
Rex	15.85	28		7	7			124	
Tank	22.15	35		5		15		167	
Gert	30.10	32	1.5	5				367	

14. A pay sheet for an individual employee is shown. Complete the pay sheet using a spreadsheet, and hence state the employee's net weekly pay and amount of superannuation paid into her superannuation fund.

Entitlements	Unit	Rate	Total
Wages for ordinary hours worked	30 hours	$35.05	
TOTAL ORDINARY HOURS = 30 hours			
Penalty (double time)	5 hours	$70.10	
		Gross payment	
Deductions			
Taxation			$325
		Total deductions	
		Net payment	
Employer superannuation contribution			
Contribution			

15. Jules is shopping for groceries and buys the following items.

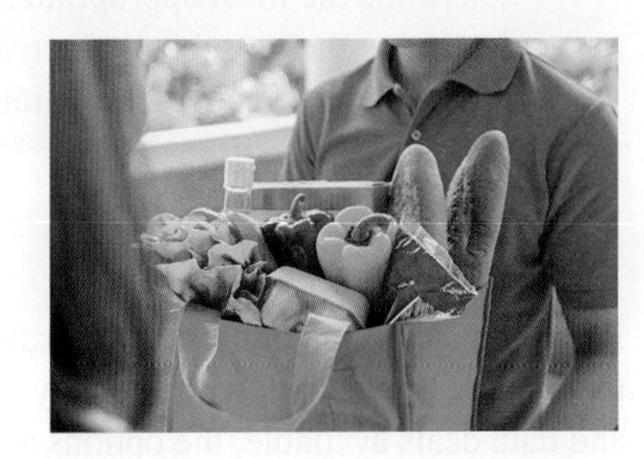

Bread — $3.30*
Fruit juice — $5.50*
Meat pies — $5.80
Ice cream — $6.90
Breakfast cereal — $5.00*
Biscuits — $2.90
All prices are listed before GST has been added on.

a. The items marked with an asterisk (*) are exempt from GST.
Calculate the total amount Jules has to pay for the shopping.
b. Calculate the additional amount Jules would have to pay if all of the items were eligible for GST.

16. Chris is a builder who works for Creative Constructions. A customer has asked Chris to provide a quote for an extension to a house. The company gives Chris the following guidelines for quoting on jobs such as this.

1. Calculate labour at $95 per hour (this figure includes the use of all tools and equipment).
2. Estimate the cost of materials delivered.
3. Calculate the total cost.
4. Add 15% as the company's margin.
5. Add 10% GST.

Chris makes the following notes when creating a quote for this work.

Labour	
Remove existing wall	— 2 builders @ 2 hours
Foundations	— 2 builders @ 14 hours
Framework	— 2 builders @ 24 hours
Roof	— 2 builders @ 8 hours
Interior fitting and finishing	— 4 builders @ 16 hours
Materials	
Concrete	— \$600
Timber	— \$4500
Tiles	— \$2300
Windows/doors	— \$1200
Mouldings, etc.	— \$800
Electricals	— \$1100

a. Create a Creative Constructions quote for this work.
b. Calculate the amount of GST that Creative Constructions should give the government when completing their BAS.

10.4 Financial products

LEARNING INTENTION

At the end of this subtopic you should be able to:
- determine the most appropriate, cost-effective mobile phone and internet plans
- determine the most cost-effective car insurance and club membership contracts
- calculate the effective rate of interest.

10.4.1 Mobile phones

Approximately nine out of ten Australian teenagers now have access to a mobile phone and, with so many different mobile phone plans and data deals available, the options should be carefully considered. Most mobile phone plans include unlimited texts and calls to Australian numbers, but it is important to check this, and using a comparison website can be helpful when doing so. Such websites can also help to compare charges for data, international calls and texts, and additional entertainment packages.

The first big decision to consider when entering into a mobile phone contract is whether you can continue to use your current phone or will need a new one. If you use your current phone, you can access a prepaid plan, which might have cost benefits. If instead you need a new phone, then the options include buying the phone outright by paying its full cost upfront, or buying the phone on a plan by paying a nominated amount each month.

WORKED EXAMPLE 14 Determining the most appropriate mobile phone package

Consider the three mobile phone package options shown below.

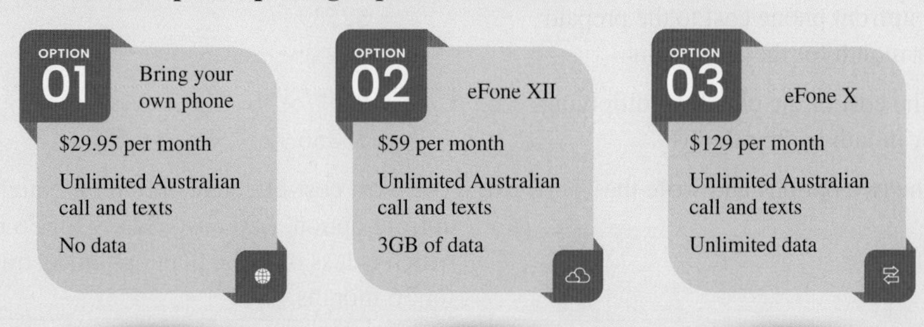

Determine the most appropriate option for the following situations. Explain your choice.

a. **Tom likes to have the latest phone. He uses his phone for casual web browsing and keeping in contact with friends and family.**
b. **Darcy mainly uses his phone to play games and stream movies. He would like to update his phone.**
c. **Bianca has a phone gifted by her parents for Christmas and uses it to text her friends and call her dad when she needs picking up from netball training.**

THINK	WRITE
Look at the three options and decide which choice is appropriate for which person.	a. Tom likes to have the latest phone, which can be achieved with option 2; with unlimited calls and texts, he can keep in contact with friends and family, and will have enough data for casual web browsing.
	b. Darcy should choose option 3 as he will be able to update his phone and have unlimited data for his games and movies.
	c. The most appropriate plan for Bianca is option 1, as she already has a phone and this plan provides unlimited calls and texts for keeping in contact with her friends and calling her dad.

WORKED EXAMPLE 15 Determining the most cost-effective mobile phone plan

Natalya needs a new mobile phone and is considering the following options.

Outright	Plan
Buying the phone upfront for $1099 and then utilising a prepaid service for $40 per month	**Buying the phone on a contract for $90.47 per month across 36 months**

a. **Calculate the cost of the outright option over 36 months.**
b. **Calculate the cost of the plan option over 36 months.**
c. **Determine the most cost-effective option for Natalya.**

THINK	WRITE
a. Calculate the cost of the outright option by adding the upfront phone cost to the prepaid amount per month for the 36 months.	a. $\text{Cost} = 1099 + 40 \times 36$ $= \$2539$
b. Calculate the cost of the plan by multiplying the cost per month by 36 months.	b. $\text{Cost} = 90.47 \times 36$ $= \$3256.92$
c. Compare the two options and write the answer.	c. The most cost-effective option for Natalya is the upfront option, as it costs \$2539 for 36 months, which is less than the plan option costing \$3256.92 for 36 months.

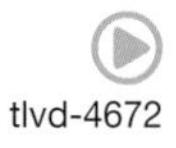
tlvd-4672

WORKED EXAMPLE 16 Determining the break-even point

Telephone company A charges its business customers \$60 per month plus \$0.28 per local call. Telephone company B charges \$90 per month plus \$0.16 per local call.

a. Determine the better deal for a company that makes 500 local calls per month.

b. Calculate the number of local calls at the break-even point, where the monthly charge is the same.

THINK	WRITE
a. 1. Find the monthly cost charged by company A for 500 local calls.	a. $60 + 0.28 \times 500 = \$200$
2. Find the monthly cost charged by company B for 500 local calls.	$90 + 0.16 \times 500 = \$170$
3. Write the answer as a sentence.	Company B is the better deal when 500 calls are made, as it offers a lower monthly cost (\$170) than company A (\$200).
b. 1. Write an equation that gives the monthly cost for each company. Let C = monthly cost and n = the number of local calls made per month.	b. company A: $C_A = 60 + 0.28n$ $C_B = 90 + 0.16n$
2. Solve the equations simultaneously by letting $C_A = C_B$.	$60 + 0.28n = 90 + 0.16n$ $60 + 0.12n = 90$ $0.12n = 30$ $n = 250$
3. Write the answer as a sentence.	The number of local calls at the break-even point is 250.

Club membership contracts

Some clubs, including health and fitness clubs, require people to sign a contract prior to becoming a member. The contract may include a joining fee, the frequency and cost of the membership payments, terms of cancellation and a cooling-off period for long-term contracts.

Before signing a contract, do the maths to work out how much it will cost per week, how many times per week you will use it, and whether the cost is worth the benefits of the membership.

10.4.2 Insurance

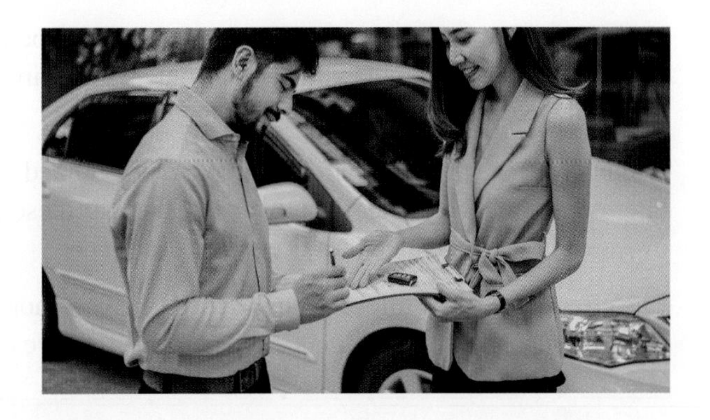

Insurance is a very important financial planning product, as it helps to cover the cost of unexpected life events, which are often expensive. The financial risk of these unexpected events can be transferred to an insurer if appropriate insurance is taken out. Payments, known as premiums, are made to an insurer in exchange for a guarantee that compensation will be provided if the unexpected occurs. Insurance premiums will differ depending on the type and size of the risk and how likely it is that a claim will be made. Most insurers will offer a reduction in premiums by increasing the amount of excess that needs to be paid when making a claim.

There are many types of insurance offered by insurance companies, covering things such as cars, health, life, home, home contents and travel.

tlvd-4673

WORKED EXAMPLE 17 Determining the most cost-effective car insurance

A 24-year-old driver has three options to purchase comprehensive car insurance for their car. All options include replacing the car, and costs for injuries for all passengers.

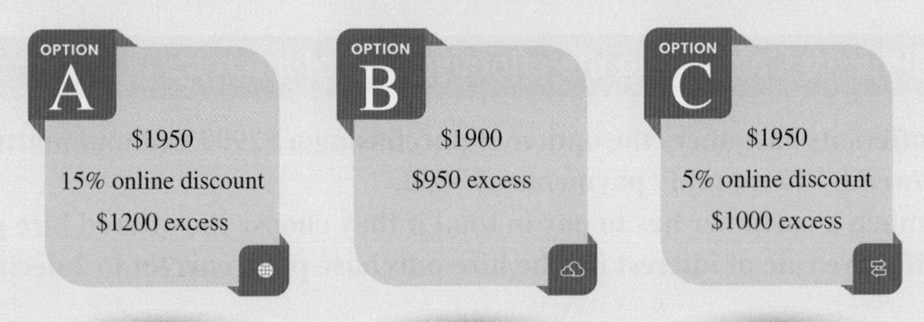

The excess fee is the amount that is to be paid to the insurer if a claim is made.
If the driver had an accident in the first year of the policy, determine which option is the most cost-effective.

THINK	WRITE
1. Calculate the policy amount with the 15% discount for option A.	Option A with 15% discount: $\frac{15}{100} \times 1950 = \$292.50$ Policy amount: $1950 - 292.50 = \$1657.50$
2. Add the excess to the policy amount for option A to determine the total yearly cost.	Option A with excess: $1657.50 + 1200 = \$2857.50$
3. Calculate the total policy amount for option B.	Option B with excess: $1900 + 950 = \$2850$
4. Calculate the policy amount with the 5% discount for option C.	Option C with 5% discount: $\frac{5}{100} \times 1950 = \$97.50$ Policy amount: $1950 - 97.50 = \$1852.50$
5. Add the excess to the policy amount for option C to determine the total yearly cost.	Option C with excess: $1852.50 + 1000 = \$2852.50$
6. Write the answer.	Option B is the least expensive.

10.4.3 Hire purchase

A hire purchase is a type of instalment plan that can be used when a customer wants to make a large purchase but does not have the money to pay upfront. With a hire purchase, the customer pays an initial amount upfront, and then pays weekly or monthly instalments.

The interest rate of a hire purchase may be determined by using the simple interest formula. However, the actual interest rate will be higher than that calculated, as these calculations don't take into account the reducing balance owing after each payment has been made.

The **effective rate of interest** can be used to give a more accurate picture of how much interest is actually charged on a hire purchase plan. To determine this, we can use the following formula:

The effective rate of interest

$$R_{ef} = \frac{2400I}{P(m+1)}$$

where R_{ef} is the effective rate of interest, I is the total amount of interest paid, P is the principal (the cash price minus the deposit) and m is the number of monthly payments.

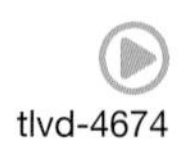
tlvd-4674

WORKED EXAMPLE 18 Calculating the effective rate of interest

A furniture store offers its customers the option of purchasing a $2999 bed and mattress by paying $500 upfront, followed by 12 monthly payments of $230.

a. Calculate how much a customer has to pay in total if they choose the offered hire purchase plan.

b. Calculate the effective rate of interest for the hire purchase plan, correct to 2 decimal places.

THINK	WRITE
a. 1. Determine the total amount to be paid under the hire purchase plan.	a. Total payment $= 500 + 12 \times 230$ $= 500 + 2760$ $= 3260$
2. State the answer.	The total amount paid under the hire purchase plan is $3260.
b. 1. Calculate the total amount of interest paid.	b. $I = 3260 - 2999$ $= 261$
2. Calculate the principal (the cash price minus the deposit).	$P = 2999 - 500$ $= 2499$
3. Identify the components of the formula for the effective rate of interest.	$I = 261$ $P = 2499$ $m = 12$

4. Substitute the information into the formula and determine the effective rate of interest.	$R_{\text{ef}} = \dfrac{2400I}{2499(m+1)}$ $= \dfrac{2400 \times 261}{2499(12+1)}$ $= 19.28\%$ (to 2 decimal places)
5. State the answer.	The effective rate of interest for the hire purchase plan is 19.28% per annum.

10.4 Exercise

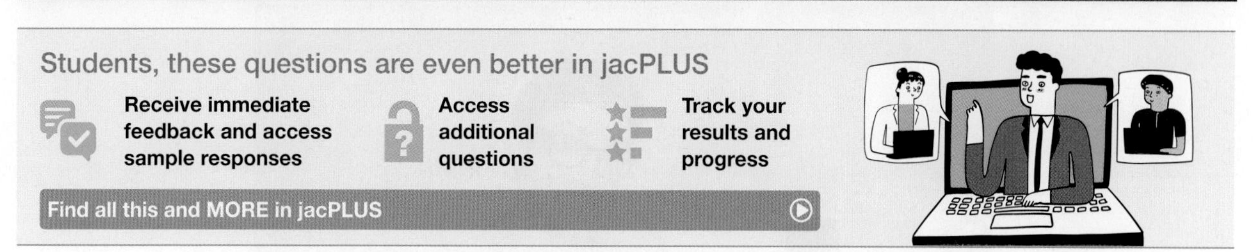

1. WE14 Consider the three mobile phone package options shown.

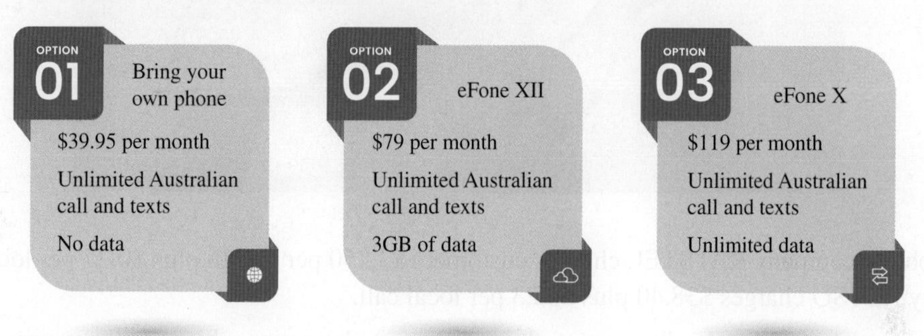

Determine the most appropriate option for the following situations. Explain your choice.

a. Bruce would like to update his old phone to the latest model. He uses his phone to regularly check work emails and social media.
b. Hussein has purchased a phone for $1299. He is always connected to Wi-Fi and just needs access to text and calls to ensure he is contactable for his job.
c. Kang would like a new phone, which he would use for casual web browsing and keeping in contact with friends and family.
d. Determine the overall cost for Kang over:
   i. 12 months
   ii. 36 months
   iii. 5 years.

2. WE15 Novak needs a new mobile phone and is considering the following options.

Outright	Plan
Buying the phone upfront for $999 and then using a prepaid service at $45 per month	Buying the phone on a contract for $89 per month across 24 months

a. Calculate the cost of the outright option over 24 months.
b. Calculate the cost of the plan option over 24 months.
c. Determine the more cost-effective option for Novak.

3. Santo has the option of purchasing the latest phone for $799 and accessing a prepaid plan for $39 per month, or entering into a three-year contract at $60 per month for his phone, calls and data. Determine the more cost-effective option for Santo.

4. WE16 Telephone company RATSTEL charges customers $32.50 per month plus $0.27 per local call. The company PUTSO charges $38.40 plus $0.23 per local call.
   a. If a customer makes 61 local calls in a month (on average), determine which company provides the better deal.
   b. Determine how many calls would have to be made for the PUTSO charges to be cheaper.

5. The telephone company Davophone charges home customers $42.50 per month plus $0.24 per local call. The company Four charges $36.00 plus $0.30 per local call.
   a. If a customer makes 51 local calls in a month (on average), determine which company has the better deal.
   b. Determine how many calls would have to be made for the Davophone charges to be cheaper.

6. When deciding which mobile phone to use, a customer has the following choices:
   Choice 1: a purchase price of $100 plus $20 per month
   Choice 2: $24 per month with no purchase price
   a. Over a 2-year period, determine which deal is better.
   b. Calculate the number of months at which the two plans would be equal.

7. Car rental agencies offer a usage plan similar to that of telephone companies. Rental company Saiv charges $45 per day plus $0.42 per kilometre travelled. Rental company Budgie charges $50 per day plus $0.35 per kilometre.
   a. A customer wishes to rent a car for 3 days and travel 1300 km. Determine which company they should use.
   b. Determine the break-even point for a 3-day trip.
   c. Determine the break-even point for a 5-day trip, to the nearest kilometre.

Questions 8 and 9 refer to the telephone bill shown below.

Previous Account	Payments and Adjustments	GST Adjustments	Balance Forward
71.79	**71.79CR**	**0.00**	**0.00**

MR BILL SAMPLE
41364 TELCO ST
TUSVILLE NSW 1234

New charges	**$71.20**
New charges due	**18 Apr**
Total amount due	**$71.20**

SUMMARY OF CHARGES

Recurring Charges	62.73
Other charges and credits	2.00
Usage charges	4.96
Discounts	4.96CR
GST	6.47
New charges (including GST)	71.20
Balance forward	0.00
Total Amount Due (including GST)	**$71.20**

Issue Date
04 Apr
Account period
04 Mar to 03 Apr

ACCOUNT DETAILS

Payments and Account Adjustments

Date	*Details*	*Amount*
21 Mar	(4) BPAY Pmt Rec'd Thank you	$71.79CR
		$71.79CR

ACCOUNT LEVEL CHARGES **Total Cost** **$2.00**

Other Charges and Credits

Date	*Details*	*Quantity*	*Rate*	*Amount*
03 Apr	Acct Processing Fee	1	2.00	2.00
				$2.00

SERVICE LEVEL SUMMARY

Line Rental / Network Access	62.73
Local Calls	2.52
13/1300 Calls	0.75
National Calls	0.81
Calls to Mobile	0.88
Discounts	4.96CR

SERVICE DETAILS

Optus Telephone Service: 089XXXXXXX **Total Cost** **$62.73**

Discounts

Date	*Details*	*Amount*
04 Apr	Unlimited free calls	0.88CR
04 Apr	Unlimited free calls	4.08CR
		$4.96CR

Line Rental / Network Access

From	*To*	*Details*	*Quantity*	*Rate*	*Amount*
04 Apr	03 May	Call Return	0	0.00	0.00
04 Apr	03 May	Three Way Call	0	0.00	0.00
04 Apr	03 May	Calling Number Display Sending	0	0.00	0.00
04 Apr	03 May	Optus Fusion $69	1	62.73	62.73
					$62.73

Local Calls

Date	*Description*	*Quantity*	*Rate*	*Amount*
03 Apr	13 calls	13	0.1800	2.3400
03 Apr	1 calls	1	0.1818	0.1818
Total for Local Calls				$2.52

13/1300 Calls

Date	*Description*	*Quantity*	*Rate*	*Amount*
03 Apr	3 calls	3	0.2500	0.7500
Total for 13/1300 Calls				$0.75

National Calls

Date	*Time*	*Destination*	*Tel No*	*Min:sec*	*Rate*	*Amount*
16 Mar	11:56am	Sydney	028XXXXXX	0:00:16	National	0.42
30 Mar	10:59am	Wee Waa	028XXXXXX	0:00:05	National	0.39
2 calls		**Total duration:**		**0:00:21**		$0.81

Calls to Mobile

Date	*Time*	*Destination*	*Tel No*	*Min:sec*	*Rate*	*Amount*
15 Mar	07:14pm	Mobile	0414XXXXXX	0:00:05	Mob	0.41
15 Mar	07:14pm	Mobile	0414XXXXXX	0:00:17	Mob	0.47
2 calls		**Total duration:**		**0:00:22**		$0.88

8. a. Determine how many local calls were made during this period.
   b. Determine how many calls to mobile phones were made during this period.
   c. If an extra 42 local calls were made during the period, calculate how much this would have added to the total bill.
9. a. In the summary of charges section, $2 was included under 'other charges and credits'. Explain why this amount was charged.
   b. The line rental/network access cost of the $69 plan is $62.73. Explain why it is not $69.
   c. The phone plan is called the $69 plan and it costs $71.20. Explain why it is not $69.
10. The yoga studio Yuan goes to is offering 15 classes for $285 or 20 classes for $340. Determine the most cost-effective option for Yuan.
11. The local golf club has an annual membership fee of $414. Dale plays golf 8 times each month. Determine the cost for each round of golf during the year.

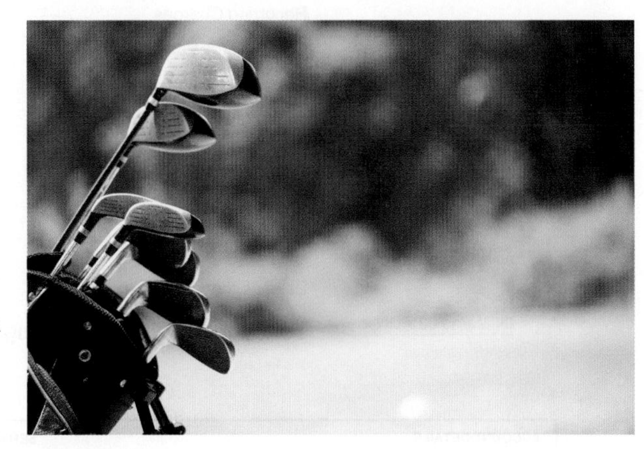

12. The Southern Sharks offer an annual membership of $774 for supporters to attend all 18 games of the season. For supporters without a membership, the cost per game is $55. Determine the number of games a supporter would need to attend in a season to make the membership cost-effective.
13. **WE17** Max has three options to purchase comprehensive car insurance for his car. All options include replacing the car, and costs for injuries for all passengers.
    Option A: $2700, 8% online discount, $1200 excess
    Option B: $3000, $800 excess
    Option C: $2400, 5% online discount, $1300 excess
    The excess fee is the amount that is to be paid to the insurer if a claim is made.
    If Max had an accident in the first year of his policy, determine which option is the most cost-effective.
14. Brook is 21 years old and has been driving since they were 18 years old. They just had a car accident in which they were at fault. The following excesses apply to claims for at-fault motor vehicle accidents for their comprehensive car insurance:
    - Basic excess of $650 for each claim
    - An additional age excess of $1400 for drivers under 25 years of age
    - An additional age excess of $300 for drivers 25 years of age or over, with no more than 2 years' driving experience

    Determine how much excess Brook is required to pay the insurance company.

15. HRU Hospital + Extras offers Armelle private health insurance to cover hospital costs for $175 per month.
    a. Calculate the annual cost for Armelle.
    b. Armelle doesn't join HRU until she turns 37. Determine the cost per month for Armelle, taking into account the Lifetime Health Cover Loading, which charges an extra 2% of your insurance fee for every year after you turn 30 if you have not joined a private health fund prior to turning 30.

Questions 16 to 19 refer to a city council rates notice shown below.

Rate Account

Property Location: **1 TEST ST**
TESTVILLE

Issue Date 02 Jan

The rates and charges set out in this notice are levied by the service of this notice and are due and payable within 30 days of the issue date. ***Full payment by the Due Date includes $15.00 Discount (rounded).***

Nett Amount Payable

$244.05

Due Date

01 February

Summary of Charges

Opening Balance	0.00
City Council Rates & Charges	505.22
City Council Remissions	229.15 CR
State Government Charges	35.00
State Government Subsidy	52.00 CR
Gross Amount	**259.07**
Discount (rounded) allowed if received by Due Date	**15.02 CR**
Nett Amount Payable	**244.05**

Page 1 of 4

If mailing your payment please tear off this slip and return with payment.
Please do not staple this slip.
See reverse for payment methods.

Pay in person at participating newsagents

***000058 5000 1999 9999 999**

Pay in person at any Post Office

*** 439 5000 1999 9999 999**

MR I M RATEPAYER
1 TEST ST
TESTVILLE

Due Date

01 Feb

Phone Pay
Transaction no.
Date

Biller Code : 78550
Ref. 5000 1999 9999 999

50

Gross Amount
$259.07

Nett Amount
$244.05

Property Details

Owner	MR I M RATEPAYER	
Property Location	1 TEST ST TESTVILLE	
Valuation effective from	01 Jul	$280,000
	01 Jul	$280,000
	01 Jul	$280,000
Averaged Rateable Valuation (A R V)		$280,000

Account Details Account number: 5000 0000 1234 567

Opening Balance

Closing Balance Of Last Bill	242.80
Payment Received - 04-Oct	227.80 CR
Discount Allowed	15.00 CR
Total	**0.00**

Period : 01 Jan - 31 Mar

City Council Rates & Charges

General Rates - Category 1(Annually 0.374 Cents In The A R V $)	261.80
Sewerage Flat Rate	94.10
Water Access Charge	35.00
Waste Mgt - Mobile Bin 240 Ltr Charge - 1 Service(S) @ $53.88 Qtr	53.88
Bushland Preservation Levy	8.29
Environmental Mgt And Compliance Levy - Cat 1(Annually 0.0184 Cents In The A R V $)	12.88
Water Meter Consumption from 16/07/ to 15/10/	
Meter AZ024756 Readings 2178 to 2211	
Total Consumption 33 kL	
Average Daily Consumption 0.36 kL 16/07/ to 15/10/	
Tier 1 Consumption 33 kL @ $1.19 Per kL	39.27
Total	**505.22**

City Council Remissions

Pensioner Remission (Full)	165.00 CR
Owner Occupier Remission	64.15 CR
Total	**229.15 CR**

State Government Charges

State Government Fire Service Levy - Group B	35.00
Total	**35.00**

State Government Subsidy

State Government Subsidy - Fire Service Levy	7.00 CR
State Government Subsidy - Rates	45.00 CR
Total	**52.00 CR**

Bill Number
5000 1999 9999 999
Page 3 of 4

16. a. Determine what amount is to be paid if the bill is paid by 1 February.
 b. Determine what amount is to be paid if the bill is paid after 1 February.
 c. Explain why Mr Ratepayer's bill is subject to a remission.
 d. Calculate the bill to be paid if the subsidies and remissions were not given.
 e. Estimate Mr Ratepayer's yearly rates bill.
 f. If Mr Ratepayer is planning a monthly budget, state how much he should allocate for rates.

17. a. Calculate how much water was used during the measurement period.
 b. State the cost of water consumption.
 c. State the value of the water access charge.
 d. Calculate the total water charges for the period.
 e. Estimate how much water would be used in a year.

18. If Mr Ratepayer's water readings were 2178 to 2245 instead, calculate his water bill for the period. Include the water access charge.

19. If the average rateable value (ARV) increased from $280 000 to $310 000, determine the increase for:
 a. the general rates charges
 b. the environmental management and compliance levy
 c. the total rates bill.

20. **WE18** A car dealership offers its customers the option of purchasing a $13 500 car by paying $2500 upfront, followed by 36 monthly payments of $360.
 a. Determine how much a customer pays in total if they choose the hire purchase plan.
 b. Determine the effective rate of interest for the hire purchase plan.

21. Georgie is comparing purchasing plans for the latest 4K TV. The recommended retail price of the TV is $3500. Georgie goes to three stores and they offer her the following hire purchase plans.
 - Store 1: $250 upfront + 12 monthly payments of $300
 - Store 2: 24 monthly payments of $165
 - Store 3: $500 upfront + 6 monthly payments of $540

 a. Calculate the total amount payable for each purchase plan.
 b. State which purchase plan has the lowest effective rate of interest.

22. Michelle uses all of the $12 000 in her savings account to buy a new car worth $25 000 on a hire purchase plan. The purchase plan also requires 24 monthly payments of $750.

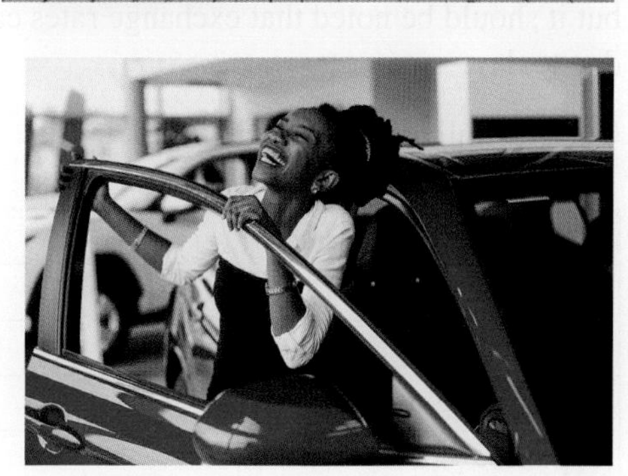

 a. State how much Michelle pays in total for the car.
 b. Determine the effective rate of interest for the hire purchase plan.

 Michelle gets a credit card to help with her cash flow during this 24-month period, and over this time her credit card balance averages $215 per month. The credit card has an interest rate of 23.75% p.a.
 c. State how much interest Michelle pays on her credit card over this period.
 d. In another 18 months Michelle could have saved the additional $13 000 she needed to buy the car outright. State how much she would have saved by choosing to save this money first.

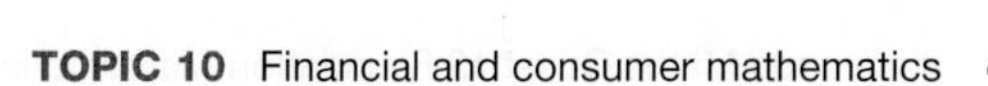

23. Javier purchases a new kitchen on a hire purchase plan. The kitchen usually retails for $24 500, but instead Javier pays an upfront fee of $5000 plus 30 monthly instalments of $820.
   a. Determine how much Javier pays in total.
   b. Determine the effective rate of interest of the hire purchase plan.

   If Javier paid an upfront fee of $10 000, he would only have to make 24 monthly instalments of $710.
   c. Determine how much Javier would save by going for the second plan.
   d. Calculate the effective rate of interest of the second plan.

10.5 Financial and economic data

LEARNING INTENTION

At the end of this subtopic you should be able to:
- convert one currency to another by applying the exchange rate
- determine future prices with inflation
- analyse data trends and make careful choices to make future predictions.

10.5.1 The exchange rate

When visiting other countries, Australian currency cannot be used, so it needs to be exchanged to the currency of the country being visited. The rate at which money can be exchanged varies according to the **exchange rate**.

The selling rate is the rate at which the Australian dollar is sold to a consumer, whilst the buying rate is the exchange rate at which the money is being bought from the consumer.

An example of exchange rates is given in the table below, but it should be noted that exchange rates can change from day to day.

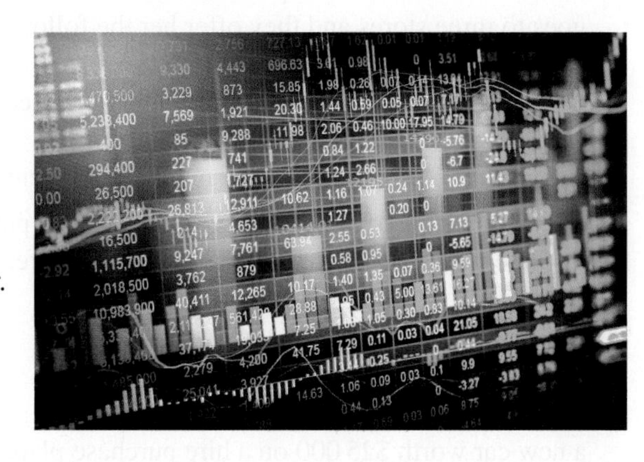

One Australian dollar is equivalent to:

Currency	Buying	Selling
Canadian dollar	0.9224	0.8361
EU euro	0.6422	0.5782
Hong Kong dollar	7.3802	6.6104
Japanese yen	108.5900	95.1300
New Zealand dollar	1.2163	1.1027
South African rand	7.3170	5.7666
Thai baht	29.5520	23.2350
UK pound sterling	0.4624	0.4315
United States dollar	0.9101	0.8610

WORKED EXAMPLE 19 Exchanging money from Australian dollars

If Karla exchanges A$400 for NZ$, calculate the amount she will receive.

THINK	WRITE
1. Use the selling price in the table.	The bank will sell NZ$1.1027 for A$1.
2. Multiply by 400.	A$400 is worth NZ$$1.1027 \times 400$.
3. State the answer.	Karla will receive NZ$441.08.

WORKED EXAMPLE 20 Exchanging money to Australian dollars

If Karla exchanges NZ$350 for A$ when she returns, calculate the amount she will receive.

THINK	WRITE
1. Use the buying price in the table.	The bank will buy NZ$1.2163 for A$1.
2. Divide 350 by 1.2163.	NZ$350 is worth A$$350 \div 1.2163$.
3. State the answer.	Karla will receive A$287.76.

10.5.2 Inflation

One of the measures of how an economy is performing is the rate of inflation. Inflation is the rise in prices within an economy and is generally measured as a percentage. In Australia this percentage is called the Consumer Price Index (CPI). Inflation needs to be taken into account when analysing profits and losses over a period of time. It can be analysed using the compound interest formula.

Calculating inflation using the compound interest formula

$$A = P\left(1 + \frac{r}{100}\right)^n$$

where:

A = the final amount of a compound interest investment or loan ($)

P = principal ($), the sum of money borrowed or invested

r = interest rate per time period

n = the number of compounding time periods.

tlvd-4675

WORKED EXAMPLE 21 Calculating the future price with inflation

The cost of an LED TV is $800. If the average inflation rate is 4% per annum, estimate the cost of the TV after 5 years.

THINK	WRITE
1. Write down the values of P, i and n.	$P = 800$, $r = 4$, $n = 5$
2. Substitute the values into the compound interest formula.	$A = P\left(1 + \frac{r}{100}\right)^n$ $A = 800 \times \left(1 + \frac{4}{100}\right)^5$ $= \$973.32$
3. State the answer.	After five years, the TV will cost \$973.32.

WORKED EXAMPLE 22 Determining whether an investment was profitable by taking inflation into account

An investment property is purchased for \$300 000 and is sold 3 years later for \$320 000. If the average annual inflation is 2.5% p.a., explain whether this has been a profitable investment.

THINK	WRITE
1. Recall that inflation is an application of compound interest and identify the components of the formula.	$P = 300\,000$ $r = 2.5$ $n = 3$
2. Substitute the values into the formula and evaluate the amount.	$A = P\left(1 + \frac{r}{100}\right)^n$ $= 300\,000\left(1 + \frac{2.5}{100}\right)^3$ $= 323\,067.19$ (to 2 decimal places)
3. Compare the inflated amount to the selling price.	Inflated amount: \$323 067.19 Selling price: \$320 000
4. State the answer.	This has not been a profitable investment, as the selling price is less than the inflated purchase price.

10.5.3 Data trends

To benefit from monetary opportunities and maximise our own financial wellbeing, we should have knowledge of local, community and national financial and economic data trends. By examining data trends and making careful choices, we may be able to make future predictions that will build us wealth over time.

tlvd-4676

WORKED EXAMPLE 23 Analysing trends to estimate future price

The graph shows the share price of a company over a 3-month period.

a. Draw a line of best fit on the graph.

b. Use your line of best fit to estimate the share price after another three months.

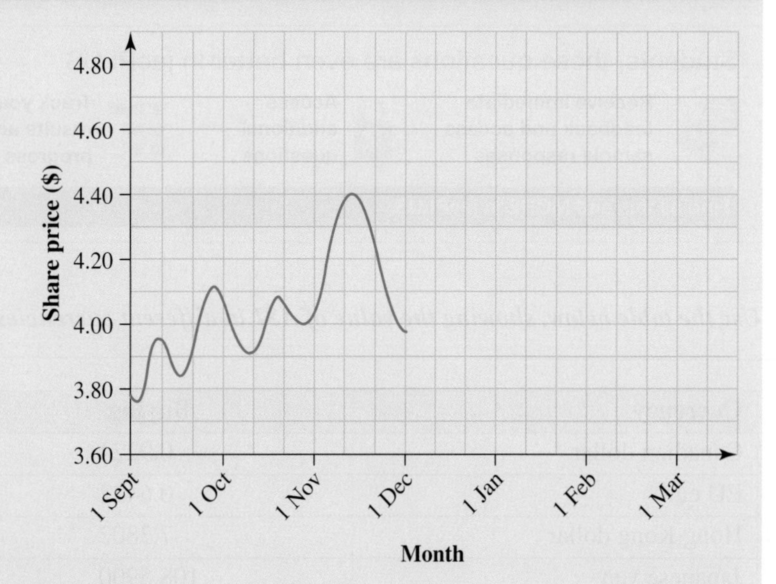

THINK	WRITE
a. Draw a line on the graph that fits best between the points marked.	a. 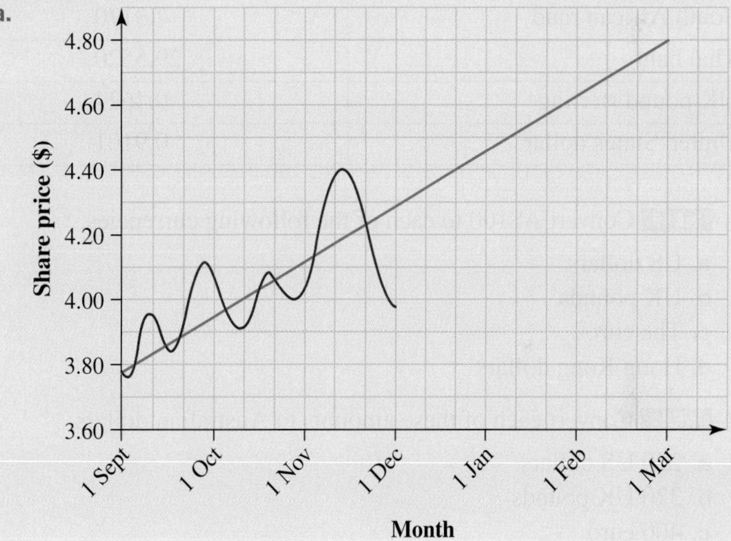
b. Extend the line of best fit for three months and read the predicted share price.	b. The predicted share price is $4.80.

10.5 Exercise

Students, these questions are even better in jacPLUS

Receive immediate feedback and access sample responses

Access additional questions

Track your results and progress

Find all this and MORE in jacPLUS

Use the table below, showing the value of A$1 in different currencies, to help you answer questions 1 to 4.

Currency	Buying	Selling
Canadian dollar	0.9224	0.8361
EU euro	0.6422	0.5782
Hong Kong dollar	7.3802	6.6104
Japanese yen	108.5900	95.1300
New Zealand dollar	1.2163	1.1027
South African rand	7.3170	5.7666
Thai baht	29.5520	23.2350
UK pound sterling	0.4624	0.4315
United States dollar	0.9101	0.8610

1. **WE19** Convert A$100 to each of the following currencies.
   a. US dollars
   b. UK pounds
   c. The euro
   d. Hong Kong dollars
2. **WE20** Convert each of these amounts to Australian dollars.
   a. 220 US dollars
   b. 320 UK pounds
   c. 400 euro
   d. 20 000 Japanese yen
3. Angie plans to visit Tokyo on business. She changes A$800 into Japanese yen.
   a. Determine how much she receives in yen.
   b. If the trip is suddenly cancelled and she changes the yen she has back to A$, determine how much she will have.
   c. Determine how much money she has lost because of this 'double' exchange.

4. Holly travels to Germany. She changes A$660 into euro.
   a. Determine the amount she has in euros.
   b. When in Germany, she spends 220 euros. Determine how many euros she has left.
   c. If she changes these back to Australian dollars, determine how much she will have.

5. **WE21** The cost of a skateboard is \$550. If the average inflation rate is predicted to be 3% p.a., estimate the cost of a new skateboard in 4 years' time.

6. The cost of a litre of milk is \$1.70. If the inflation rate is an average 4% p.a., estimate the cost of a litre of milk after 10 years.
7. A daily newspaper costs \$2.00. With an average inflation rate of 3.4% p.a., estimate the cost of the newspaper after 5 years (to the nearest 5c).
8. If a basket of groceries cost \$98.50 in 2009, determine the estimated cost of the groceries in 2016 if the average inflation rate for that period was 3.2% p.a.
9. **WE22** An investment property is purchased for \$325 000 and is sold 5 years later for \$370 000. If the average annual inflation rate is 2.73% p.a., explain whether this has been a profitable investment.

10. A business is purchased for \$180 000 and is sold 2 years later for \$200 000. If the average annual inflation rate is 1.8% p.a., determine whether a real profit has been made.
11. During an economic crisis in 1998, Indonesia experienced severe inflation. In one week, on Monday, A\$1 would have bought 9500 rupiah, whereas on Thursday A\$1 would have bought 10 900 rupiah. On holidays in Indonesia at this time, Joel exchanged A\$120 and paid for a camera on Monday. Determine how much he would have saved if he had waited to make the transaction on Thursday (assuming the marked price did not change).
12. The 1968 Australian 2c piece is very rare. If a coin collector purchased one in 2009 for \$400 and the value of the coin increases by 15% p.a., calculate its value in 2022 (to the nearest \$10).
13. **WE23** The graph shows the movement in a share price over a 3-month period.
    a. Copy the graph into your book and on it draw a line of best fit.
    b. Use your line of best fit to estimate the share price after 6 months.

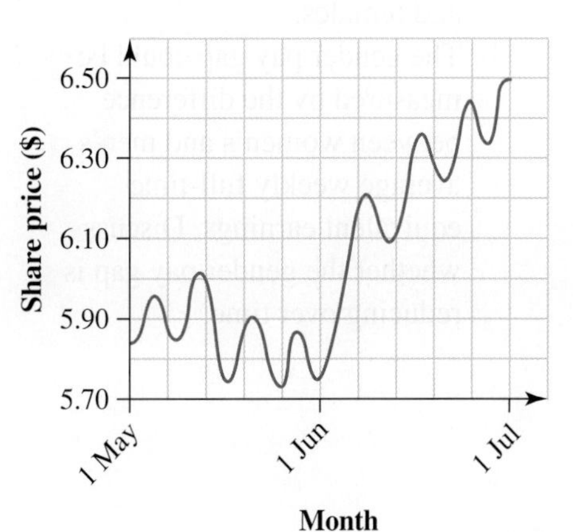

14. The graph shows the movement in a share price over a 9-month period.

    a. Copy the graph into your book and draw a line of best fit on the graph.
    b. Use your graph to predict the value of the share after a further 12 months.

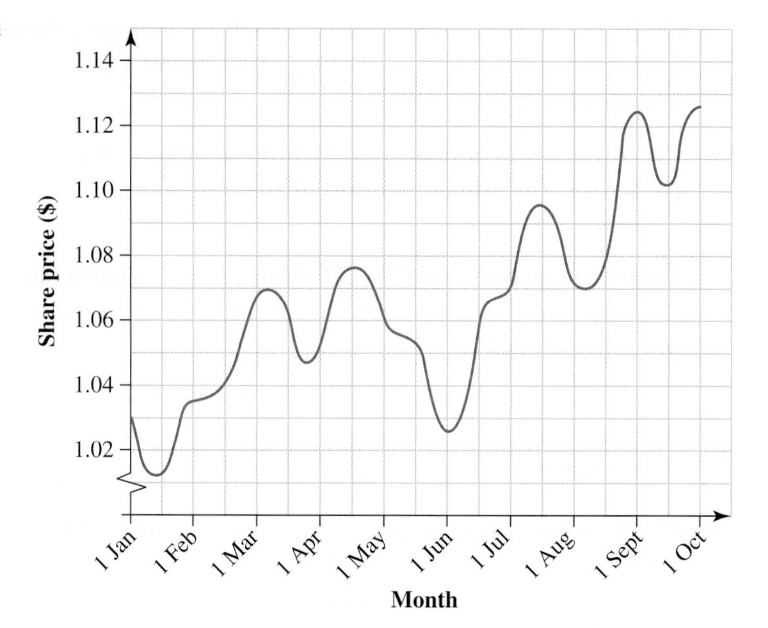

15. The table shows the share price of BigCorp Productions Ltd over a period of one year.

Month	Share price	Month	Share price
January	$12.40	July	$13.17
February	$12.82	August	$13.62
March	$12.67	September	$13.41
April	$13.05	October	$13.30
May	$13.06	November	$13.46
June	$12.89	December	$13.20

    a. Graph the share price for each month and show a line of best lit.
    b. Use your line of best fit to predict the share price after a further year.

16. The graph shows the average weekly earnings of Australians over a period of 10 years.

    a. Discuss the trend in average weekly earnings for both males and females.
    b. The gender pay gap could be measured by the difference between women's and men's average weekly full-time equivalent earnings. Discuss whether the gender pay gap is reducing over time.

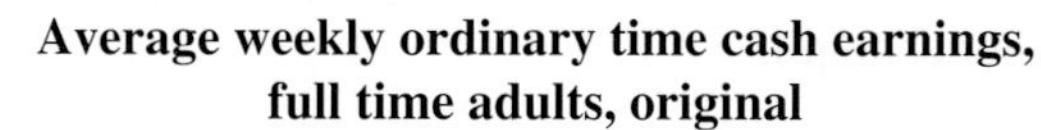

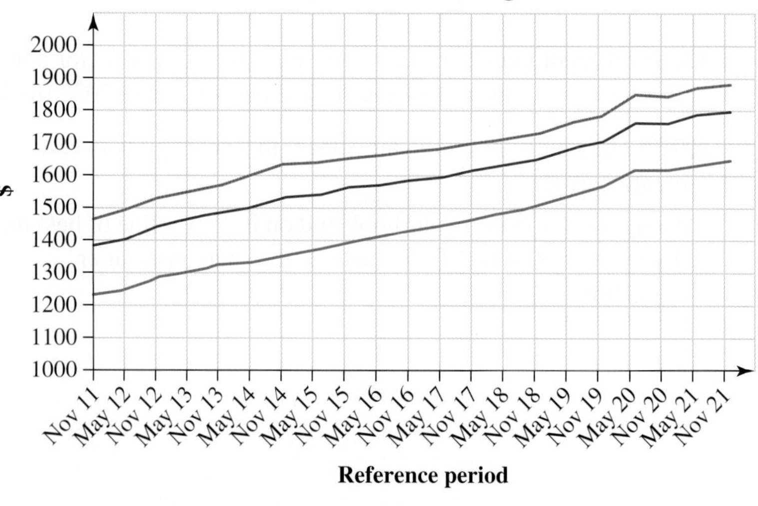

10.6 Review

10.6.1 Summary

doc-38018

Hey students! Now that it's time to revise this topic, go online to:

Review your results

Watch teacher-led videos

Practise questions with immediate feedback

Find all this and MORE in jacPLUS

10.6 Exercise

Multiple choice

1. **MC** An employee's annual salary is \$84 750. The amount of superannuation, at 10.5%, an employer is required to pay to the employee's superannuation fund is:
   - **A.** \$1050.00
   - **B.** \$8898.75
   - **C.** \$8051.25
   - **D.** \$8475.00
   - **E.** \$8071.43

2. **MC** In its annual budget, the local council plans to raise \$25 million in revenue to repair local infrastructure. The capital improvement for the area is valued at \$1.5 billion. The council rate quarterly instalment for a ratepayer whose house has a capital improved value of \$475 000 is:
   - **A.** \$712.50
   - **B.** \$791.67
   - **C.** \$1979.17
   - **D.** \$2850.00
   - **E.** \$7916.67

3. **MC** A worker receives 17.5% leave loading. If the worker's annual salary is \$57 840, the amount of leave loading received will be closest to:
   - **A.** \$195
   - **B.** \$409
   - **C.** \$779
   - **D.** \$844
   - **E.** \$10 122

4. **MC** A manager's annual salary is \$65 800. He receives \$450 in bank interest and a travel allowance of \$9900. He claims \$10 800 in deductions. The tax payable at the end of the financial year is closest to:
   - **A.** \$11 706
   - **B.** \$21 239
   - **C.** \$26 331
   - **D.** \$6614
   - **E.** \$5092

5. **MC** A bottle of soft drink costs \$2.50. If the inflation rate is predicted to average 2% p.a. for the next five years, the cost of the soft drink in five years will be:

A. \$2.60 B. \$2.65 C. 2.70
D. 2.75 E. 2.76

6. **MC** The GST payable on a house which sold for \$342 650 exclusive of GST is:

A. \$34 265 B. \$31 150 C. \$308 385
D. \$10 E. \$3426.50

7. **MC** A new outdoor furniture set normally priced at \$1599 is sold for an upfront fee of \$300 plus 6 monthly instalments of \$240. The effective rate of interest is:

A. 27.78% B. 30.23% C. 31.51%
D. 37.22% E. 43.42%

8. **MC** An annual salary of \$56 200 results in a fortnightly pay of:

A. \$4683.33 B. \$2161.54 C. \$2341.67
D. \$1080.77 E. \$2161.53

9. **MC** A delivery driver earns \$19.85 per hour. They work 70 hours in a fortnight. The amount of Medicare levy, at a rate of 2%, they will be expected to pay at the end of the financial year is closest to:

A. \$28 B. \$723 C. \$722
D. \$728 E. \$1445

10. **MC** The formula $A = P \times 1.011^n$ is used to determine the future inflated price of goods. From the formula, the annual inflation, % p.a., is:

A. 0.011 B. 0.11 C. 1.10
D. 1.011 E. 1.1

Short answer

11. The following table shows the fluctuations in a share's price over a period of 1 year.

Month	Share price	Month	Share price
January	\$15.76	July	\$16.60
February	\$16.04	August	\$16.77
March	\$16.27	September	\$16.51
April	\$16.12	October	\$16.71
May	\$16.49	November	\$16.69
June	\$16.39	December	\$16.98

a. On a set of axes plot the share price for each month.
b. Draw a line of best fit on your graph and use your line to predict the share price after a further one year.

12. Sophie bought an investment property for \$250 000, and 4 years later she sold it for \$275 000. If the average annual inflation was 2.82% per annum, determine whether this has been a profitable investment for Sophie.

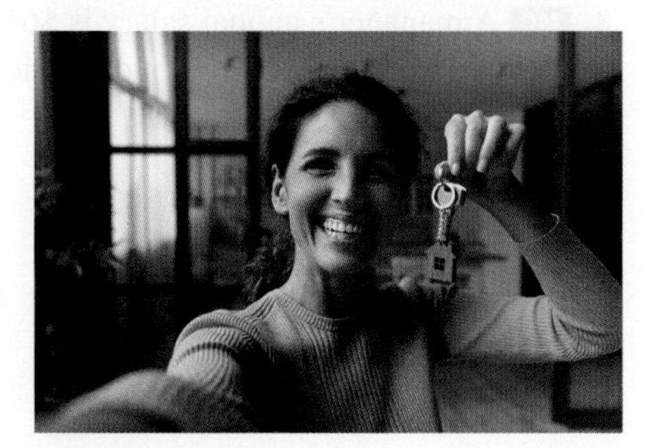

13. If a \$5000 computer can be depreciated for tax purposes at a rate of 33% p.a., calculate the tax deduction in the second year after the computer was purchased.

14. A librarian pays \$1137.60 in Medicare levy. Determine their annual salary.

15. Ben needs to convert US$100 to UK pounds and the only way this can be done is to convert the US$ to A$ and then change the A$ to UK pounds. Calculate the amount Ben will have in UK pounds using the exchange rates given in subtopic 10.5.1.

Extended response

16. A council raises its council rates by 5% to meet the growing costs associated with improving infrastructure. Last year the council raised $19 million in revenue, with total capital improved value of $1.6 billion.
    a. Calculate the rate in the dollar for last year's revenue.
    b. Determine the amount of revenue the council plans to raise this year based on the 5% increase.
    c. A ratepayer's council rates last year were $4135.25. Calculate this year's council rates for this ratepayer after the 5% increase, assuming the total capital improvement value is the same as last year.

    Over the year, the capital improved value for properties in the municipality increased by 2.5% due to the property market.

    d. Show that the rate in the dollar charged by the council for this year is approximately 0.012.
    e. Explain why the overall increase in the rate in the dollar is closer to 2.5% than the 5% stated by the council. Use calculations to support your explanation.

17. A physiotherapist's annual salary is $63 821. Her annual travel claim is 950 km and she is allowed to claim 66c for each km travelled as a tax deduction. She is paid monthly and her employer withholds $900 in tax.

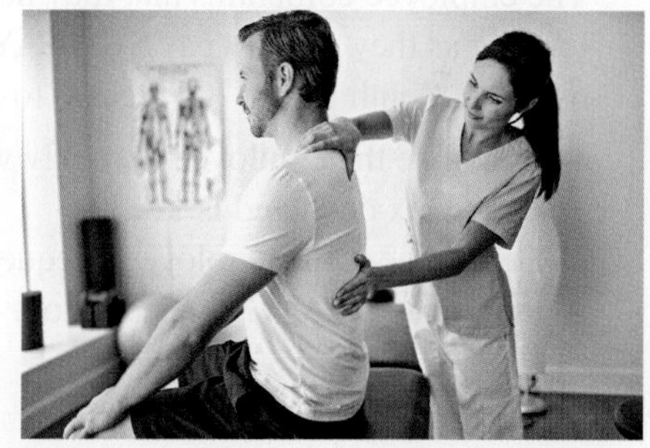

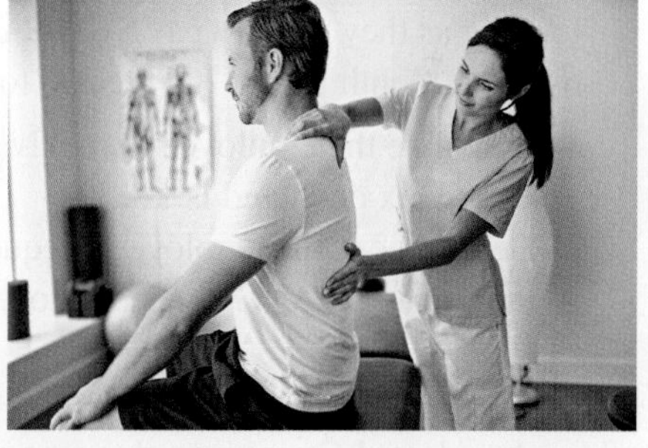

    a. Calculate her taxable income, tax payable, Medicare levy and employer superannuation contribution based on 10.5% contributions.
    b. Explain why she will have to pay tax. Justify your answer using calculations.

18. A plumber quotes their clients the cost of any parts required plus $74.50 per hour for labour, and then adds on the required GST.
    a. Calculate how much they quote for a job that requires $250 in parts (excluding GST) and should take 4 hours to complete.
    b. If the job ends up being faster than first thought, and they end up charging the client for only 3 hours of labour, determine the percentage discount on the original quote this represents. Write the percentage correct to the nearest whole number.

19. Lorna is purchasing $1500 worth of furniture for her new house. She can make the purchase using the store's hire purchase plan with a deposit of $400 and 12 monthly repayments of $105.
    a. Determine how much Lorna will end up paying for the furniture.
    b. Determine the effective rate of interest for the hire purchase plan.

    Alternatively, Lorna could put the purchase on her credit card with an interest rate of 15.66% p.a. and no interest-free period. She will repay $450 per month.

    c. Determine how long it would take her to repay the full amount.
    d. Based on the amount of interest payable, determine which payment option Lorna should select.

20. An employee receives 17.5% annual loading just before their four-week annual leave. They are paid an hourly rate of $20.50 for working a 38-hour week. They do not receive overtime or penalty rates.

a. Calculate the amount of annual loading they will receive for the four weeks of annual leave.

b. Complete the payslip for the week they receive their annual leave loading.

Entitlements	Unit	Rate	Total
Wages for ordinary hours worked	38 hours	$20.50	$
Annual leave loading (17.5%)			$
		Gross payment	**$**

Deductions	
Taxation	$295
Total deductions	
Net payment	

The employee complains that they have paid too much tax in the week that they receive their annual leave loading as they usually pay $106 PAYG tax per week. The employee asks to have their leave loading paid over a 6-month period (26 weeks) to avoid paying too much tax.

c. Calculate the employee's weekly wage if their annual leave loading is paid over 26 weeks, and the weekly PAYG tax deducted is $113.

d. Determine if the employee's request to receive their annual leave loading over 6 months helps to decrease their annual PAYG tax deduction.

Answers

Topic 10 Financial and consumer mathematics

10.2 Taxation and superannuation

10.2 Exercise

1. $1833.33
2. a. 0.006 b. $16 020 000
3. $159.25
4. a. $12 979.17 b. $1362.81
5. a. $5166.67 b. $542.50 c. $6510
6. a. $2773.08 b. $291.17
7. a. $9405 b. $2565
8. $1153.90
9. $1070.10
10. $2096.16
11. a. $1904.40 b. $1798.60 c. $1745.70
12. a. $1000 b. $600 c. $360
13. $3087.50
14. $25 245
15. a. $1208 b. $1546
    c. $1788 d. $2168.46
16. a. $23 491 b. $469.82
17. $8488.25
18. a. $17 039.36
    b. Mira's gross annual pay is below $18 200, so she doesn't have any income tax payable.
19. $1303
20. $344.34
21. $1497.65
22. a. $38 674
    b. $4663.54
    c. Tax refund of $3390.46
23. a. $90 942.80
    b. $21 410.40
    c. A tax debt of $3210.40 needs to be paid.
24. a. Gross annual pay = $44 605
    PAYG tax paid = $11 940.50
    b. $41 513.40
    c. $5259.82
    d. Tax refund of $6680.68
25. $85 970
26. a. Since they paid more than $5092 and less than $29 467, their taxable income is within the third tax bracket.
    b. $74 908
    c. Taxable income = $73 923
    Tax refund = $320.02

10.3 Business income and expenditure

10.3 Exercise

1. $3.56
2. a. $0.23 b. $6.89 c. $9.85
3. $123.75
4. a. $3.18 b. $53.35
   c. $5124.35 d. $0.21
5. $98.50
6. a. $6.91 b. $0.69
7. A = $317.27
   B = $31.72
   C = $3.45
   D = $37.95
   E = $386.95
8. a. $8.28 b. $206.28
9. a. 31 October each year
   b.

Quarter	Due date
1. July, August and September	28 October
2. October, November and December	28 February
3. January, February and March	28 April
4. April, May and June	28 July

   c. The 21st day of the following month
10. The luxury car tax is paid by businesses that sell or import luxury cars (dealers), and by individuals who import luxury cars. It is imposed at the rate of 33% on the amount above the luxury car threshold.
    https://www.ato.gov.au/business/luxury-car-tax/
11. $1157.42
12. Assume an annual salary of $100 000.
    Current agreement: 17.5% annual leave loading and 10.5% superannuation

$$100\,000 + \frac{10.5}{100} \times 100\,000 + \frac{17.5}{100} \times \frac{100\,000}{52} \times 4$$
$$= 100\,000 + 10\,500 + 1346.15$$
$$= \$111\,846.15$$

New agreement: 5% pay rise and 11.5% superannuation

$$100\,000 \times 1.05 + \frac{11.5}{100} \times 100\,000 = 105\,000 + 11\,500$$
$$= \$116\,500$$

Under the new agreement workers will receive more money, so they should choose the new agreement.

13. Gross = pay rate × normal hours + pay rate × overtime hours × penalty rate (1.5) + pay rate × overtime hours penalty rate (2) + allowance
Net pay = gross pay − tax withheld
See the table at the bottom of the page.*

14. See the table at the bottom of the page.**

15. a. $30.96 b. $1.38

16. a. $32 510.50 b. $2955.50

10.4 Financial products

10.4 Exercise

1. a. Bruce needs a large amount of data and a new phone, so option 3 is the most appropriate.
   b. Hussein already has a phone, so option 1 would be the most appropriate as it is the least expensive.
   c. Kang would get a new phone and enough data for his needs with option 2.
   d. i. $948 ii. $2844 iii. $4740
2. a. $2079 b. $2136
   c. The outright option costs less over 24 months so it is the most cost-effective.
3. Prepaid = $2203
   Contract = $2160
   Over three years it would be more cost-effective for Santo to take out a contract for his phone.
4. a. RATSTEL provides the better deal.
   b. 148 or more
5. a. Four b. 109 or more
6. a. Choice 2 is the better option.
   b. 25 months
7. a. Budgie b. 214 km c. 357 km
8. a. 14 b. 2
   c. $0, as the plan has unlimited local calls
9. a. Account-processing fee
   b. $62.73 + GST of $6.27 = $69
   c. The extra $2.20 is the $2 account-processing fee plus GST on this fee.
10. The most cost-effective option is the package of 20 classes.
11. $51.75
12. A supporter would need to attend more than 14 games for the membership to be cost-effective.
13. Option C is the most cost-effective.
14. $2050
15. a. $2100 b. $199.50
16. a. $244.05
    b. $259.05
    c. Pensioner and owner-occupier
    d. $525.20
    e. $976.20
    f. $81.35
17. a. 33 kL b. $1.19 per kL c. $35
    d. $74.27 e. 132 kL
18. $114.73
19. a. $28.05 per quarter b. $1.38 per quarter
    c. $29.43 extra per quarter

*13

Employee	Pay rate ($)	Normal hours worked	Overtime	Penalty rate		Allowance ($)	Gross pay ($)	Tax withheld ($)	Net pay ($)
			1.5	1.5	2				
Rex	15.85	28		7	7		**832.13**	124	**708.13**
Tank	22.15	35		5		15	**956.38**	167	**789.38**
Gert	30.10	32	1.5	5			**$1256.68**	367	**$889.68**

**14

Entitlements	Unit	Rate	Total
Wages for ordinary hours worked	30 hours	$35.05	$30 \times 35.05 = \$1051.50$
TOTAL ORDINARY HOURS = 30 hours			
Penalty (double time)	5 hours	$70.10	$5 \times 70.10 = \$350.50$
		Gross payment	$1051.50 + 350.50 = \$1402.00$
Deductions			
Taxation			$325
		Total deductions	**$325**
		Net payment	$1402.00 - 325 = \$1077.00$
Employer superannuation contribution			
Contribution			$\frac{10.5}{100} \times 1402 = \$147.21$

20. a. $15 460 b. 11.56%

21. a. Store 1 = $3850
Store 2 = $3960
Store 3 = $3740

b. Store 2

22. a. $30 000 b. 26.67% p.a.
c. $102.13 d. $5102.13

23. a. $29 600 b. 20.25%
c. $2560 d. 16.82%

10.5 Financial and economic data

10.5 Exercise

1. a. US$86.10 b. 43.15 UK pounds
c. 57.82 euro d. HK$661.04

2. a. $241.73 b. $692.04
c. $622.86 d. $184.18

3. a. 76 104 yen b. $700.84 c. $99.16

4. a. 381.61 euros b. 161.61 euros c. $251.65

5. $619.03

6. $2.52

7. $2.35

8. $122.80

9. The inflated value is $371 851.73, so it is not profitable.

10. The inflated value is $186 538.32, so it is profitable.

11. $15.41

12. $2461.12

13. a.

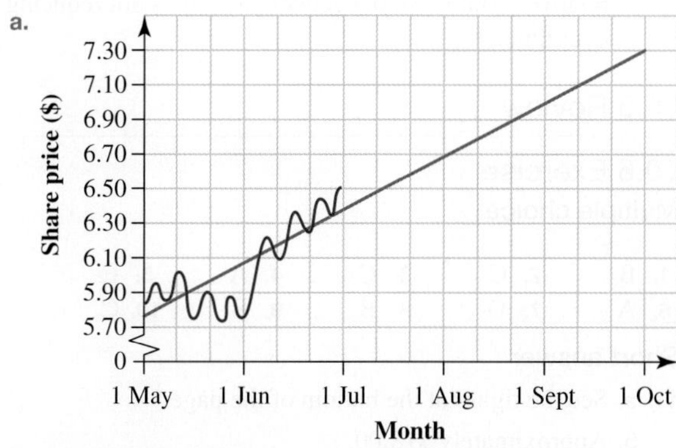

b. Approximately $7.00

14. a. See the figure at the bottom of the page.*
b. Approximately $1.20

15. a. See the figure at the bottom of the page.**
b. Approximately $14.50

*14. a.

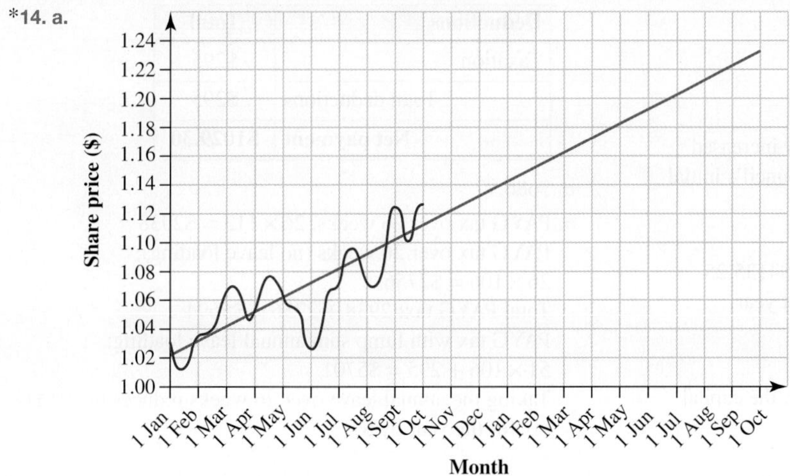

**15. a.

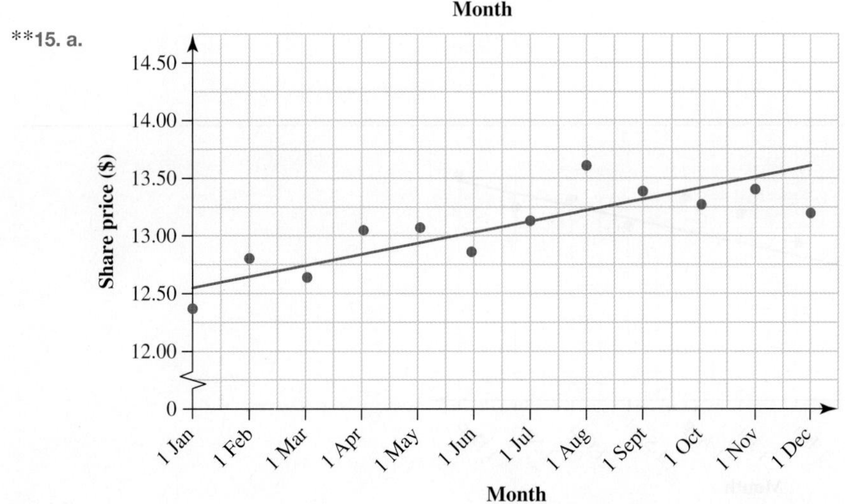

16. a. The average weekly earnings for both males and females have an increasing trend.

b. From the graph it can be seen that the difference between male and female average weekly earnings remains relatively stable, so the gender pay gap is not reducing over time.

10.6 Review

10.6 Exercise

Multiple choice

1. B	2. C	3. C	4. A	5. E
6. A	7. D	8. B	9. B	10. C

Short answer

11. a. See the figure at the bottom of the page.*

b. Approximately $18.00

12. Taking into account inflation, the value of the house when it sold was $279 415.44. Sophie sold it for $4415.44 less than this, so it was not a profitable investment.

13. $1105.50

14. $56 880

15. 47.41 UK pounds

Extended response

16. a. 0.011 875

b. $19.95 million

c. $4342.01

d. 1.025×1.6 billion $= \$1.64$ billion

Rate in the dollar: $\frac{19\,950\,000}{1\,640\,000\,000} = 0.012$

e. The value of properties in the municipality increased by 2.5%, which was not included in the council's initial calculations.

$5 - 2.5 = 2.5\%$

For example, property owner's rate fee = $4135.25.

Capital improved value of the property last year:

$\frac{4135.25}{0.011\,875} = \$348\,232$

With a 2.5% increase in value for this year, the capital improved value of the property is:

$359\,937 \times 0.011\,81 = \$4247.26$

The rates payable based on last year's property value with a 5% increase are $4342.01 (from part c).

Percentage increase: $\frac{4342.01 - 4247.26}{4247.26} \times 100 = 2.23\%$

The rate in the dollar has increased by 2.23% when the increase in property values is taken into account.

17. a. Taxable income: $63 194

Tax payable: $11 005.05

Medicare levy: $1263.88

Superannuation: $6635.37

b. The physiotherapist has paid $10 800 in tax across the year but is required to pay $11 005.05, so will need to pay an extra $205.05.

18. a. $602.80 b. 14%

19. a. $1660 b. 26.85%

c. 4 months d. Credit card

20. a. $545.30

b.

Entitlements	Unit	Rate	Total
Wages for ordinary hours worked	38 hours	$20.50	$779.00
Annual leave loading (17.5%)			$545.30
TOTAL ORDINARY HOURS 38 hours			
		Gross payment	**$1324.30**

Deductions	Total
Taxation	$295
Total deductions	$295
Net payment	**$1029.30**

c. $686.97

d. PAYG tax over 26 weeks: $26 \times 113 = \$2938$

PAYG tax over 26 weeks (no leave loading):

$26 \times 106 = \$2756$

Total PAYG tax: $2938 + 2756 = \$5694$

PAYG tax with lump sum annual leave loading:

$51 \times 106 + 295 = \$5701$

Taking the annual leave over 26 weeks reduces his PAYG tax for the year by $7.

*11. a.

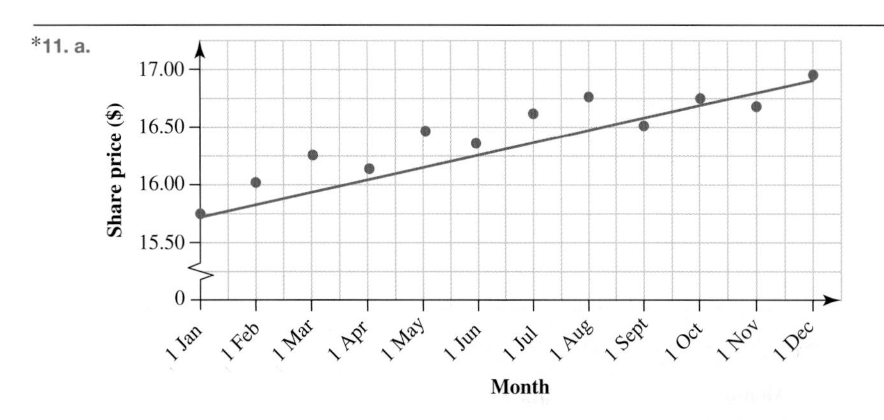

11 Scales, plans and models

LEARNING SEQUENCE

Fully worked solutions for this topic are available online.

11.1 Overview

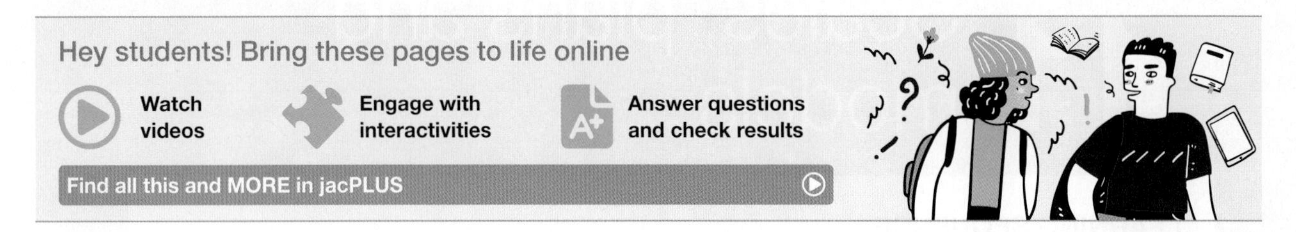

11.1.1 Introduction

Every day in the city you will see cranes in the sky, working on new office or apartment buildings. As you drive in and out of the city, you will come across roadworks. These projects have undergone extensive planning, and a detailed set of plans has been created to make sure the construction company knows how to complete them. Since such projects are large-scale projects, the plans need to be scaled down so they can fit on the paper. This is the case for a lot of construction and manufacturing work, such as building a new car or a new house. It is important to understand scale so you can understand how to read these plans and have an idea of size and space related to the plans.

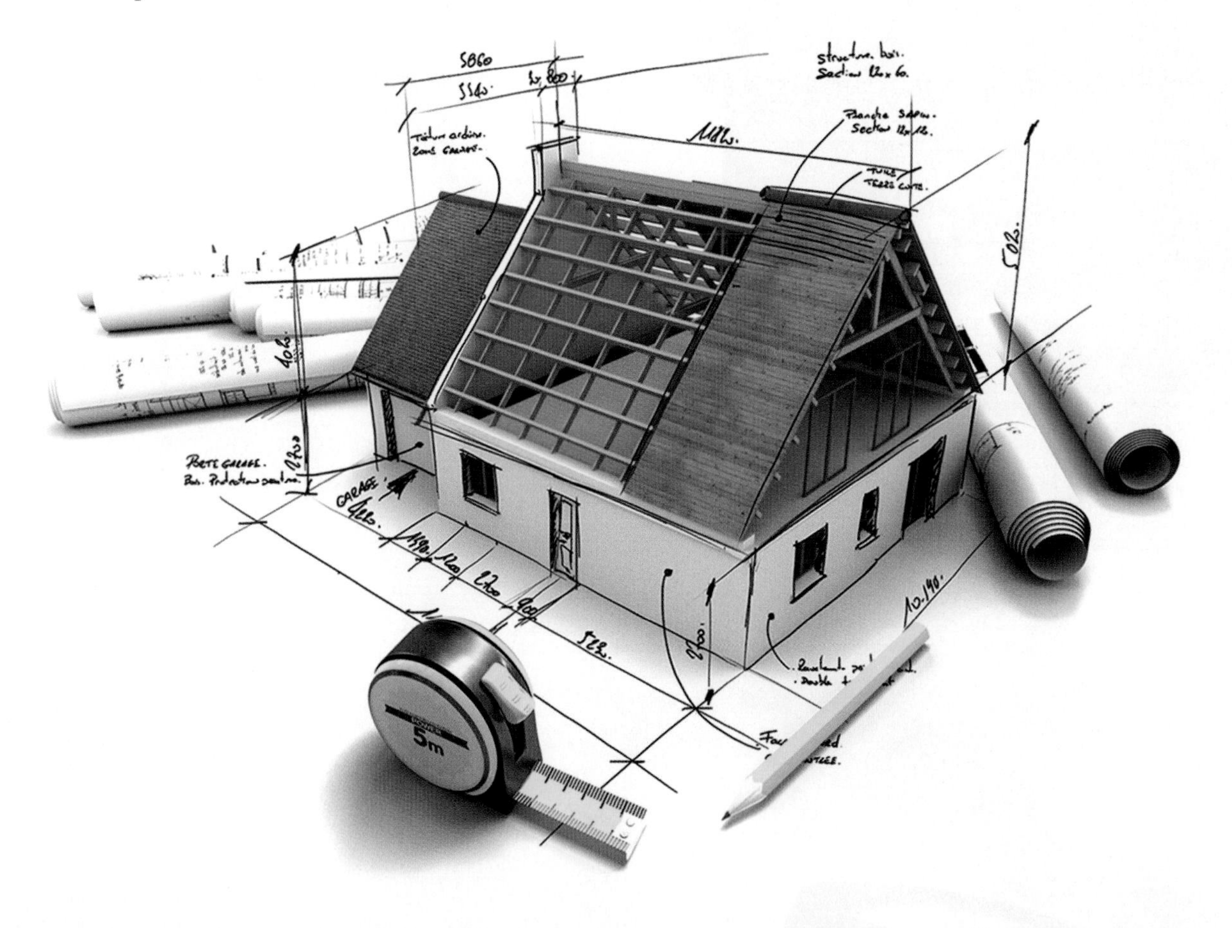

KEY CONCEPTS

This topic covers the following key concepts from the VCE Mathematics Study Design:

- calculations of enlargement and reduction using scaling techniques for two-dimensional and three-dimensional plans, diagrams and models.

11.2 Reading and interpreting scale drawings

LEARNING INTENTION

At the end of this subtopic you should be able to:
- determine scale and scale factors
- use scales to solve real-world problems.

11.2.1 Scale and scale factor

A **scale** is a ratio of the length on a drawing to the actual length.

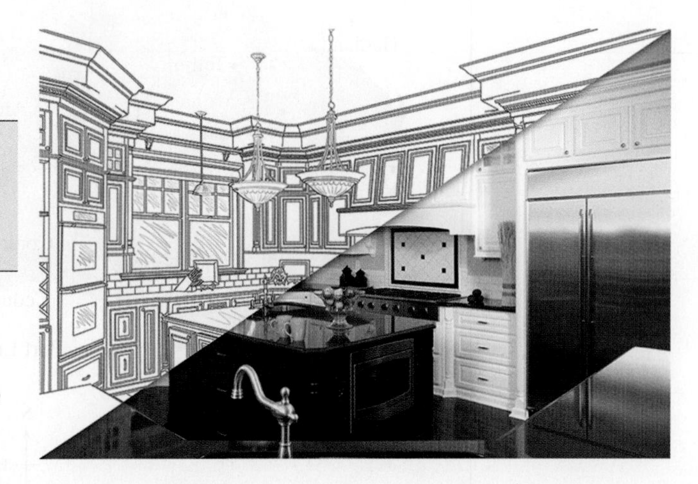

Scale

scale = length of drawing : actual length

The symbol : is read as 'to'.

Scales are usually written with no units. If a scale is given in two different units, the larger unit has to be converted into the smaller unit.

A **scale factor** is the ratio of two corresponding lengths in two similar shapes. The scale factor of $\frac{1}{2}$ or a scale (ratio) of 1 : 2 means that 1 unit on the drawing represents 2 units in actual size. The unit can be mm, cm, m or km.

Scale factor

Scale factor is a ratio of the same units to enlarge or reduce the size of any shape.

tlvd-3733

WORKED EXAMPLE 1 Scale and scale factor

Determine the scale and the scale factor of a drawing where 8 cm on the diagram represents 4 km in reality.

THINK	WRITE
1. Convert the larger unit into the smaller unit.	$4\text{ km} = 4 \times 1000 \times 100$ $= 400\,000\text{ cm}$
2. Write the scale of the drawing with the same unit.	$8\text{ cm} : 400\,000\text{ cm}$ $\Rightarrow 8 : 400\,000$
3. Simplify the scale by dividing both sides of the ratio by the highest common factor. Divide both sides of the ratio by 8, as 8 is the highest common factor of both 8 and 400 000. State the scale.	$1 : 50\,000$
4. State the scale factor of the drawing as a fraction.	The scale factor is $\frac{1}{50\,000}$.

11.2.2 Maps and scales

Maps are always drawn at a smaller scale. A map has its scale written or drawn using a diagram.

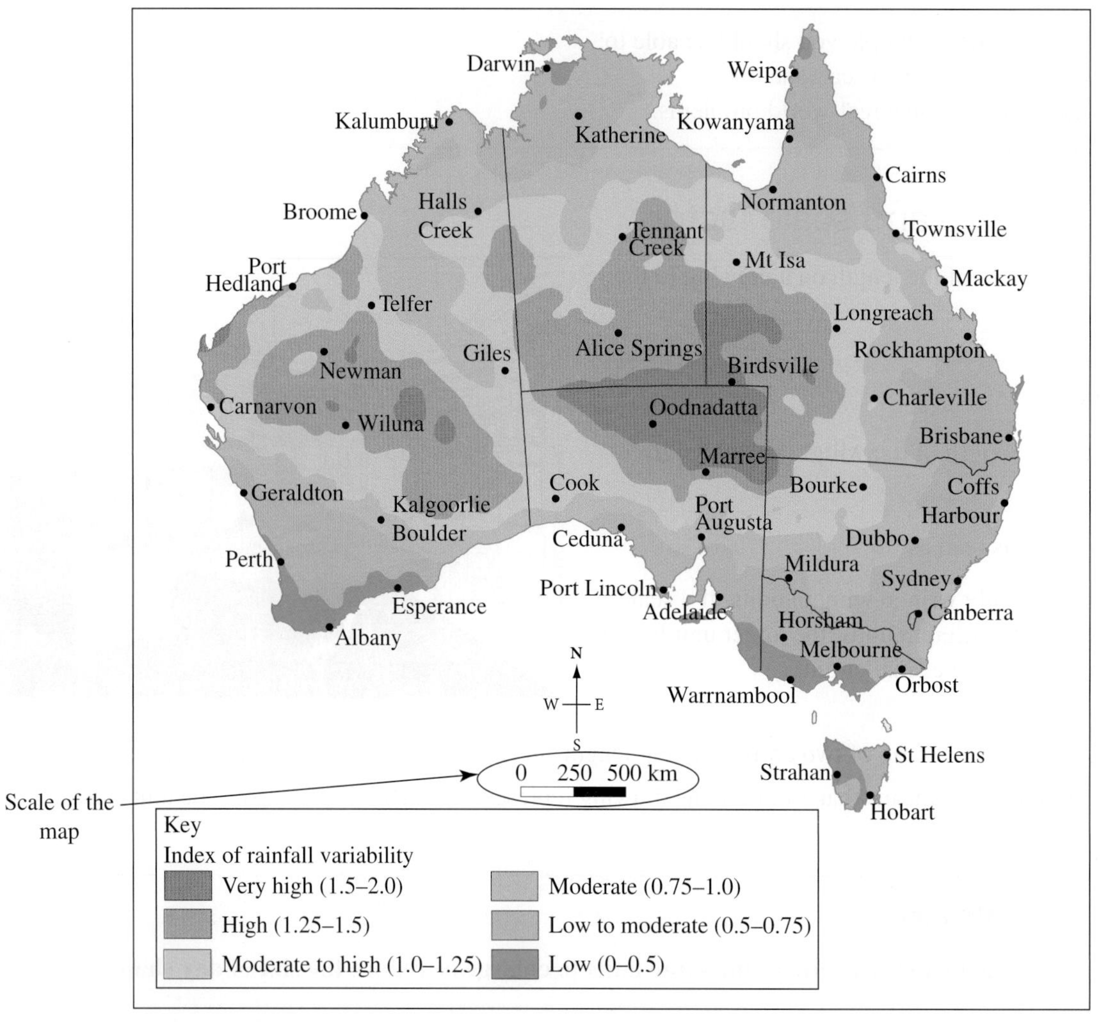

Source: MAP graphics Pty Ltd, Brisbane

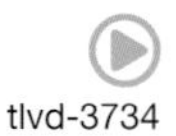
tlvd-3734

WORKED EXAMPLE 2 Scales of maps

State the scale of a map as a ratio using the graphical scale shown, where the length of each partition is 1 cm.

0 500 1000 km

THINK	WRITE
1. Measure the length of each partition in the diagram.	Each partition is 1 cm.
2. Determine the length that each partition represents in reality.	Each partition represents 500 km.
3. Write the scale as a ratio.	1 cm : 500 km
4. Convert the larger unit into the smaller unit.	$500\text{ km} = 500 \times 1000 \times 100$ $= 50\,000\,000\text{ cm}$

5. Write the scale as a ratio in the same unit. — 1 cm : 50 000 000 cm
6. State the scale of the map in ratio form. — 1 : 50 000 000
   1 cm on the map represents 50 000 000 cm in actual distance.

11.2.3 Scale factor

Calculating scale factor

$$\textbf{scale factor} = \frac{\textbf{dimension on the drawing}}{\textbf{actual dimension}}$$

We can rearrange the scale factor formula to determine the actual dimension.

Actual dimension

$$\text{actual dimension} = \frac{\text{dimension on the drawing}}{\text{scale factor}}$$

To calculate the actual dimensions, we measure the dimensions on the diagram and then divide these by the scale factor.

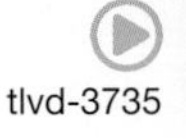
tlvd-3735

WORKED EXAMPLE 3 Using scale to determine actual dimensions

A scale of 1 : 200 was used for the diagram of the house shown. Calculate both the length and the width of the master bedroom, given the dimensions on the plan are length = 2.5 cm and width = 1.5 cm.

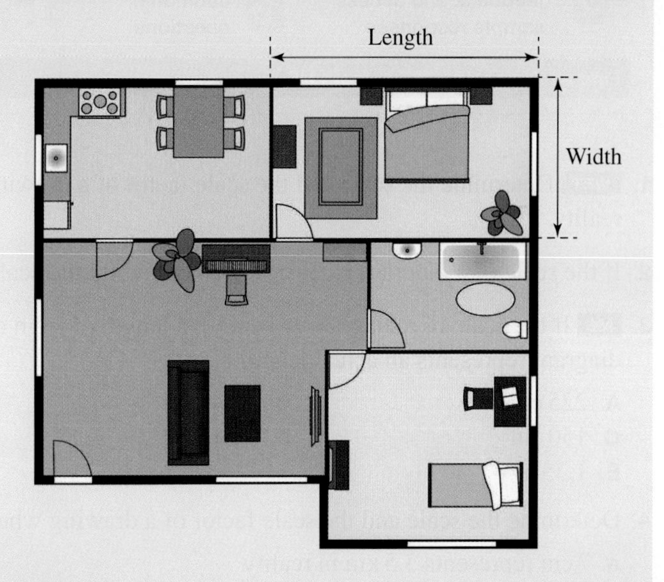

THINK	WRITE
1. State the length and the width of the room in the diagram.	Length = 2.5 cm Width = 1.5 cm
2. Write the scale of the drawing and state the scale factor.	1 : 200 This means that every 1 cm on the drawing represents 200 cm of the actual dimension. The scale factor is $\frac{1}{200}$.

3. Divide both dimensions by the scale factor.	Length of the bedroom $= 2.5 \div \frac{1}{200}$ $= 2.5 \times 200$ $= 500$ cm Width of the bedroom $= 1.5 \div \frac{1}{200}$ $= 1.5 \times 200$ $= 300$ cm
4. State the answer in reasonable units.	The length of the bedroom is 5 metres and the width is 3 metres.

Resources

Interactivities Scales (int-4146)
House plans (int-4147)
Scales and house plans (int-4391)

11.2 Exercise

Students, these questions are even better in jacPLUS

Receive immediate feedback and access sample responses

Access additional questions

Track your results and progress

Find all this and MORE in jacPLUS

1. WE1 Determine the scale and the scale factor of a drawing where 150 mm on the diagram represents 45 m in reality.
2. If the scale of a diagram is 3.9 cm : 520 km, write the scale using the same unit.
3. MC If the scale of a diagram is 1 : 450, a length of 3 cm on the diagram represents an actual length of:
   A. 225 cm
   B. 1350 cm
   C. 150 cm
   D. 1.5 m
   E. 1.35 m

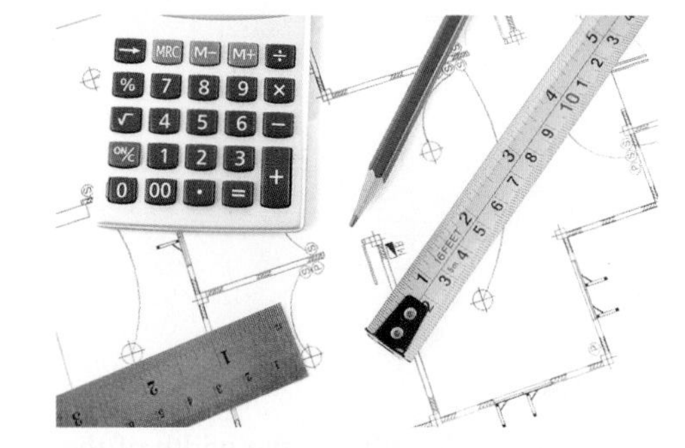

4. Determine the scale and the scale factor of a drawing where:
   a. 7 cm represents 3.5 km in reality
   b. 16 mm represents 6.4 m in reality.
5. Calculate the scale and scale factor, given:
   a. 5 cm represents 185 m in reality
   b. 12 mm represents 19.2 m in reality.
6. WE2 State the scale of a map as a ratio using the graphical scale shown, given each partition is 1 cm long.

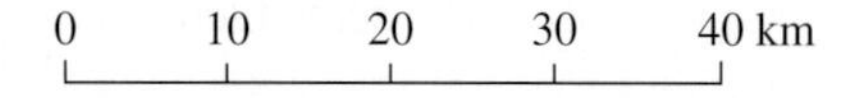

7. State the scale of a map as a ratio using the graphical scale shown, where the length of each partition is 1 cm.

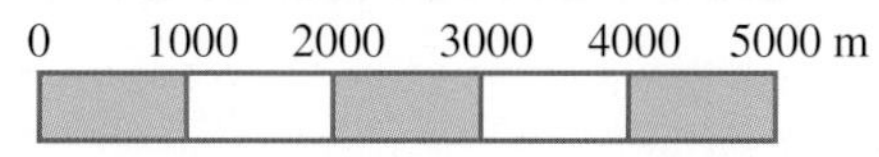

8. State the scale of a map as a ratio using the graphical scale shown, given:

a. each partition is 1.5 cm

0 150 300 km

b. each partition is 1 cm.

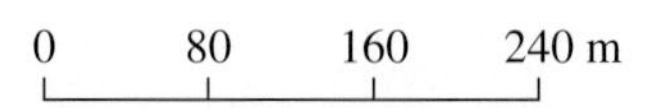

9. State the scale of a map as a ratio using the graphical scale shown, given each partition is 2 cm.

0 10 20 30 40 50 km

10. A house plan is drawn to a scale of 1 : 1000.

a. Calculate the actual length of the house if it is represented by 3.2 cm on the plan.

b. The width of the house is 17 m. Calculate the width of the house on the plan.

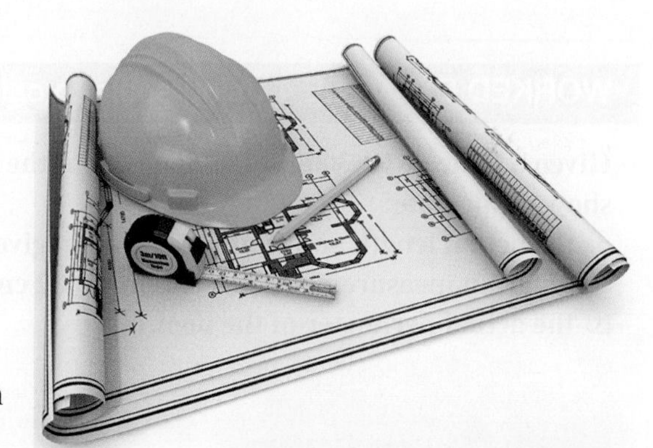

11. A house plan is drawn to a scale of 1 : 500.

a. Calculate the actual length of the lounge room if it is represented by 0.6 cm on the plan.

b. The width of the bedroom is 5.2 m. Calculate the width of the bedroom on the house plan.

12. Calculate the dimensions of the carpet shown, given that its diagram was drawn using a scale of 1 : 75 and its length and width on the diagram are 3 cm and 2 cm, respectively.

13. WE3 A scale of 1 : 300 was used for the diagram of the house shown. Calculate both the height and width, in metres, of the garage door, given the dimensions on the plan are height = 0.8 cm and width = 2 cm.

14. MC A 15 m long fence is represented by a straight line 4.5 cm long on a drawing. Calculate the scale of the drawing.

A. 1000 : 3
B. 4.5 : 1.5
C. 45 : 150
D. 3 : 1000
E. 9 : 30

11.3 Calculating measurements from scale drawings

LEARNING INTENTION

At the end of this subtopic you should be able to:
- use scale drawings to calculate lengths, perimeters and areas
- use scale drawings to solve real-world problems.

All diagrams, maps or plans are drawn using a given scale. This scale is used to convert the lengths in the diagram to the actual lengths.

11.3.1 Perimeters

Calculating the actual perimeters of shapes from scaled drawings involves first calculating the lengths required. Once the lengths of the sides of the shape are known, the perimeter can be calculated by adding up all the sides of the shape.

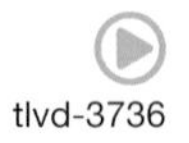
tlvd-3736

WORKED EXAMPLE 4 Using scale to calculate lengths and perimeters

Given a scale of 1 : 300 for the diagram of the backyard shown, calculate:

a. the actual length and width of the pool, given the diagram measurements are 2.4 cm by 1.2 cm

b. the actual perimeter of the pool.

THINK	WRITE
a. 1. Write the length and width of the pool on the diagram.	a. Length = 2.4 cm Width = 1.2 cm
2. Write the scale of the drawing and state the scale factor.	1 : 300 This means that every 1 cm on the drawing represents 300 cm of the actual dimension. The scale factor is $\frac{1}{300}$.

3. Divide both dimensions by the scale factor.	Length of the pool $= 2.4 \div \frac{1}{300}$ $= 2.4 \times 300$ $= 720$ cm $= 7.2$ m Width of the pool $= 1.2 \div \frac{1}{300}$ $= 1.2 \times 300$ $= 360$ cm $= 3.6$ m
b. Calculate the perimeter of the pool.	b. Perimeter of the pool $= 2 \times \text{length} + 2 \times \text{width}$ $= 2 \times 7.2 + 2 \times 3.6$ $= 14.4 + 7.2$ $= 21.6$ m

11.3.2 Areas

Areas of surfaces from scaled drawings can be calculated in a similar way to the perimeter. The dimensions of the shape have to be calculated first. Once the lengths of the shape are known, the area can be calculated using an appropriate formula.

tlvd-3737

WORKED EXAMPLE 5 Calculating actual areas from scale drawings

Given a scale of 1 : 15 for the diagram of the tile shown, calculate:

a. the actual length and width of the tile, given the diagram measurements are 2 cm by 2 cm

b. the actual area of the tile

c. the actual area covered by 20 tiles.

THINK	WRITE
a. 1. Write the length and the width of the shape.	a. Length = 2 cm Width = 2 cm
2. Write the scale of the drawing and state the scale factor.	1 : 15 This means that 1 cm on the drawing represents 15 cm of the actual dimension. The scale factor is $\frac{1}{15}$.
3. Since both dimensions are the same, divide the dimension by the scale factor.	Length/width of the tile $= 2 \div \frac{1}{15}$ $= 2 \times 15$ $= 30$ cm The length and width of the tile are 30 cm by 30 cm.

b. Calculate the area of the tile using the formula for the area of a square. Use $A = l^2$.	**b.** Area of one tile = length^2 $= 30^2$ $= 900\text{ cm}^2$
c. Calculate the area required by multiplying the area of one tile by 20. *Note:* Recall the conversion from cm^2 is converted to m^2 by dividing by 100^2.	**c.** Total area = area of one tile $\times$ 20 $= 900 \times 20$ $= 18\,000\text{ cm}^2$ $= 18\,000 \div 100^2$ $= 1.8\text{ m}^2$

11.3.3 Problem-solving with scale drawings

Scaled diagrams are used in many areas of work in order to estimate elements such as production costs and quantities of materials required.

Packaging

There are many products that are usually placed in a box when purchased or delivered. Boxes are cut out of cardboard or other materials using a template.

The dimensions of the package are determined by using a net of the three-dimensional shape, as shown in the diagram.

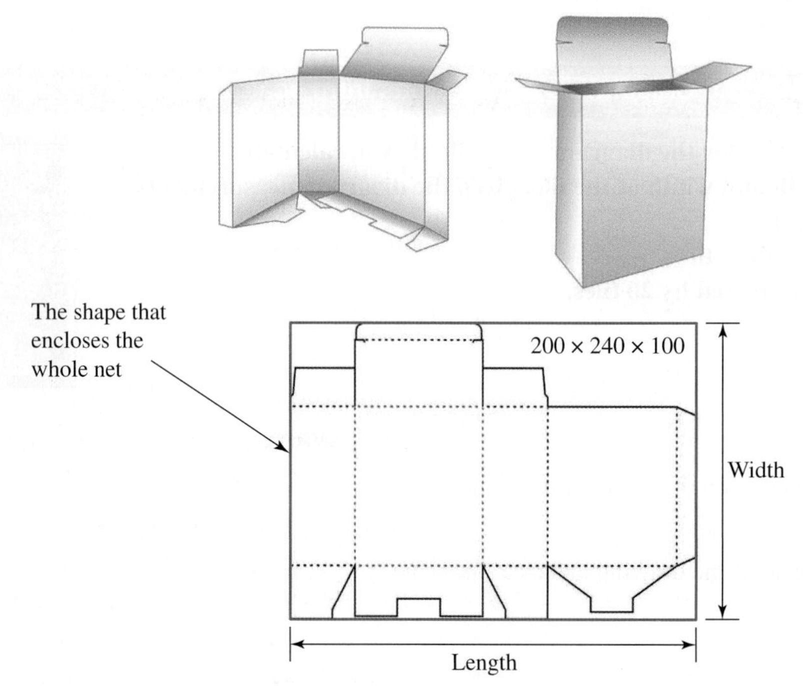

The required area can be estimated by calculating the area of the rectangle that encloses the whole net.

Costs

To calculate the cost of the materials used, we use the cost per m or m^2, known as the cost per unit.

Cost of materials

$$\text{cost required} = \frac{\text{area of material}}{\text{area per squared unit}} \times \text{cost per unit}$$

If the area of the packaging is 520 cm^2 and the cost of the material is \$1.20 per square metre, then the cost for packaging is as follows.

$$\begin{aligned}\text{Cost required} &= \frac{\text{area of material}}{\text{area per squared unit}} \times \text{cost per unit} \\ &= \frac{520\text{ cm}^2}{1\text{ m}^2} \times 1.20\end{aligned}$$

(*Note:* The two area units have to be the same: 1 m^2 = 10 000 cm^2.)

$$\begin{aligned}&= \frac{520}{10\,000} \times 1.20 \\ &= 0.624\ldots \\ &\approx \$0.62 \text{ per package}\end{aligned}$$

tlvd-3738

WORKED EXAMPLE 6 Solving problems using scales

Consider the gift box shown and the template required to make it. The scale of the diagram is 1 : 15, and the template measurements are 6 cm by 4 cm.

a. Calculate the width and the height of the packaging template.

b. Calculate the total area of material required.

c. If the cost of materials is \$0.25 per square metre, calculate how much it would cost to make the template for the gift box.

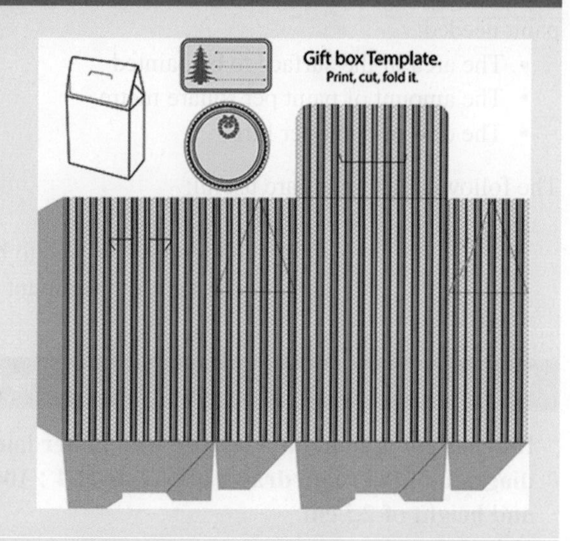

THINK	WRITE
a. 1. State the width and the height of the template.	**a.** Width = 6 cm Height = 4 cm
2. Write the scale of the drawing and state the scale factor.	1 : 15 This means that every 1 cm on the drawing represents 15 cm of the actual dimension. The scale factor is $\frac{1}{15}$.
3. Divide both dimensions by the scale factor.	$\text{Width of template} = 6 \div \frac{1}{15} = 6 \times 15 = 90\text{ cm}$ $\text{Height of template} = 4 \div \frac{1}{15} = 4 \times 15 = 60\text{ cm}$

b. Calculate the area of the box template using the formula for the area of a rectangle.

b. Area of template = width × height

$= 90 \times 60$

$= 5400\ \text{cm}^2$

c. 1. Calculate the cost of the material.

c. Cost is \$0.25 per m^2 or 10 000 cm^2.

$$\text{Cost for the area required} = \frac{\text{area of material}}{\text{area per squared unit}} \times \text{cost per unit}$$

$$= \frac{5400}{10\,000} \times 0.25$$

$$= \$0.135$$

$$\approx 0.14$$

2. State the answer.

It will cost \$0.14 or 14 cents to make the template for the gift box.

Painting

Painting a wall requires some calculations. Three pieces of information are required to calculate the amount of paint needed:

- The area of the surface to be painted
- The amount of paint per square metre
- The cost of paint per litre

The following formulas are useful:

$$\text{amount of paint required} = \text{area to be painted} \times \text{amount of paint per square metre}$$
$$\text{total cost} = \text{amount of paint required} \times \text{cost of paint per litre}$$

tlvd-3739

WORKED EXAMPLE 7 Solving problems involving area

Indra is going to paint a feature wall in her lounge room. She has a diagram of the room drawn at a scale of 1 : 100, with a length of 4 cm and height of 2.5 cm.

Calculate:

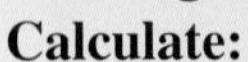

a. the length and the height of the wall
b. the area of the wall
c. the amount of paint Indra has to buy if the amount of paint needed is 3 L per m^2
d. the total cost of the paint if the price is \$7.60 per litre.

THINK / **WRITE**

a. 1. State the length and the height of the shape.

a. Length = 4 cm
Height = 2.5 cm

2. Write the scale of the drawing and state the scale factor.

1 : 100
This means that every 1 cm on the drawing represents 100 cm of the actual dimension.

The scale factor is $\frac{1}{100}$.

3. Divide both dimensions by the scale factor.

$$\text{Length of the wall} = 4 \div \frac{1}{100}$$
$$= 4 \times 100$$
$$= 400\text{ cm}$$
$$= 4\text{ m}$$

$$\text{Height of the wall} = 2.5 \div \frac{1}{100}$$
$$= 2.5 \times 100$$
$$= 250\text{ cm}$$
$$= 2.5\text{ m}$$

b. Calculate the area of the wall using the formula for the area of a rectangle.

b. Area wall = length × width (or height)
$= 4 \times 2.5$
$= 10\text{ m}^2$

c. Calculate the amount of paint required.

c. Amount of paint required
= area to be painted × amount of paint per square metre
$= 10 \times 3\text{ L}$
$= 30\text{ L}$

d. Calculate the cost of the paint.

d. Total cost = amount of paint required × cost of paint per litre
$= 30 \times 7.60$
$= \$228$

Bricklaying

Bricks are frequently used as a basic product in the construction of houses or fences.

The dimensions of a standard brick are shown in the diagram using a scale of 1 : 10.

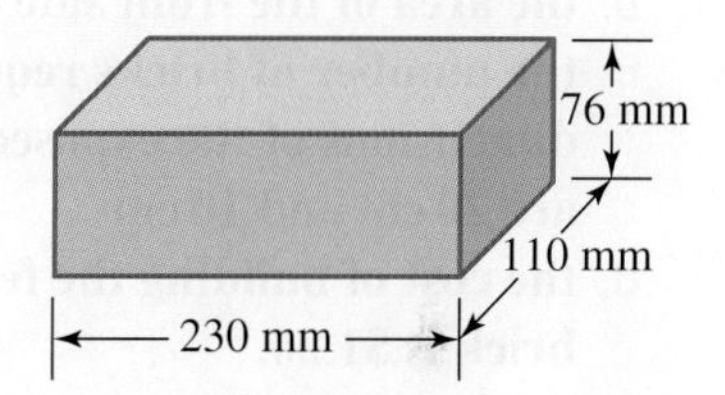

Bricklayers have to calculate the number of bricks required to build a wall by considering the length, width and height of the brick, and the dimensions of the wall.

To give the wall strength, bricks have to be laid so that their ends sit in the middle of the layer underneath, as shown in the diagram.

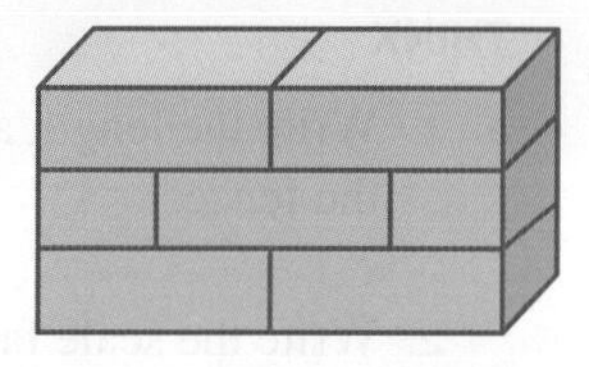

Calculating the number of bricks

Imagine that you want to calculate the number of bricks required to build a 2 m^2 wall.

$$\text{number of bricks} = \frac{\text{area of the wall}}{\text{area of the exposed side of the brick}}$$

The area of the brick is the area of the side that shows on the wall. In the diagram of the wall, using the standard brick dimensions, the exposed side of the brick has length 230 mm and height 76 mm.

The area of the exposed side of the brick is $230 \times 76 = 17\,480\text{ mm}^2$.

The two areas have to be written in the same unit.

$$\begin{aligned}\text{Area of the wall} &= 2\,\text{m}^2\\ &= 2\,000\,000\,\text{mm}^2\\ \text{Number of bricks} &= \frac{\text{area of the wall}}{\text{area of the exposed side of the brick}}\\ &= \frac{2\,000\,000}{17\,480}\\ &= 114.4\text{ bricks}\\ &= 115\text{ bricks}\end{aligned}$$

Note: Notice that although the answer is 114.4 bricks, we round up because we need 115 bricks so we can cut 0.4 of a brick to finish the whole wall.

This answer is an estimated value, as we did not take into account the mortar that has to be added to the structure of the wall.

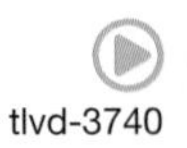
tlvd-3740

WORKED EXAMPLE 8 Solving problems using scales 2

Francis is going to build a brick fence with the dimensions shown in the diagram. The scale factor of the drawing is $\frac{1}{50}$.

Calculate:

a. **the length and the height of the fence, given that it measures 7 cm by 1 cm on the diagram**

b. **the area of the front side of the fence**

c. **the number of bricks required if the dimensions of the exposed side of one brick are 20 cm and 10 cm**

d. **the cost of building the fence if the price of one brick is $1.20.**

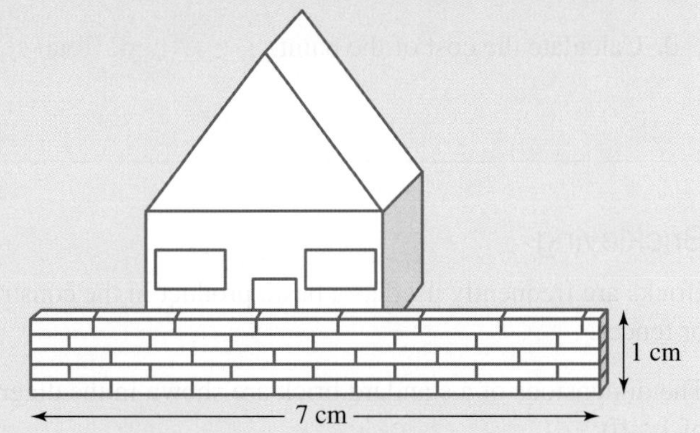

THINK	WRITE
a. 1. Write the length and the height of the fence.	a. Length = 7 cm Height = 1 cm
2. Write the scale factor.	The scale factor is $\frac{1}{50}$.
3. Divide both dimensions by the scale factor.	$\text{Length of the fence} = 7 \div \frac{1}{50} = 7 \times 50 = 350\text{ cm}$ $\text{Height of the fence} = 1 \div \frac{1}{50} = 1 \times 50 = 50\text{ cm}$

b. Calculate the area of the front face of the fence using the formula for the area of a rectangle.

b. Area of the front face of the fence = length × width (or height)

$$= 350 \times 50$$
$$= 17\,500 \text{ cm}^2$$

c. 1. Calculate the area of the exposed face of one brick.

c. Area of the exposed face of the brick = 20×10

$$= 200 \text{ cm}^2$$

2. Calculate the number of bricks required.

$$\text{Number of bricks} = \frac{17\,500}{200}$$
$$= 87.5 \text{ bricks}$$
$$= 88 \text{ bricks}$$

d. Calculate the cost.

d. Cost = number of bricks × cost of one brick

$$= 88 \times 1.20$$
$$= \$105.60$$

11.3 Exercise

Students, these questions are even better in jacPLUS

Receive immediate feedback and access sample responses

Access additional questions

Track your results and progress

Find all this and MORE in jacPLUS

1. **WE4** Given a scale of 1 : 250 for the floor plan of the apartment shown, calculate:
   a. the actual length and width of the apartment, given the diagram measurements are 5 cm by 3 cm
   b. the actual perimeter of the apartment.

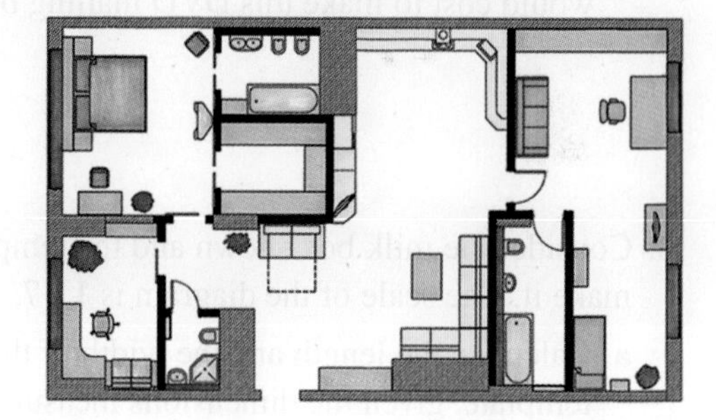

2. If the scale of the diagram shown is 1 : 800, calculate, to the nearest metre, the radius and the circumference of the Ferris wheel shown, given the radius measures 1.5 cm on the diagram. (*Note:* circumference = $2\pi r$)

3. WE5 Given a scale of 1 : 40 for the window shown, calculate:

   a. the actual width and height of each piece of glass needed to cover the window, given that the large piece of glass measures 1.5 cm by 2.5 cm on the diagram, and the width of the small piece of glass measures 0.75 cm
   b. the actual areas of the two glass pieces
   c. the total area covered by glass.

4. If the scale of the diagram shown is 1 : 20, calculate, to the nearest centimetre, the radius and the area of the clock face shown, given the radius measures 1.5 cm on the diagram. (*Note:* area = πr^2)

5. WE6 Consider the DVD mailing box shown and the template required to make it. The scale of the diagram is 1 : 8, and the diagram measurements are 3.5 cm by 1.5 cm.

   a. Calculate the length and the width of the packaging template.
   b. Estimate the total area of material required.
   c. If the cost of materials is \$0.85 per square metre, calculate how much it would cost to make this DVD mailing box.

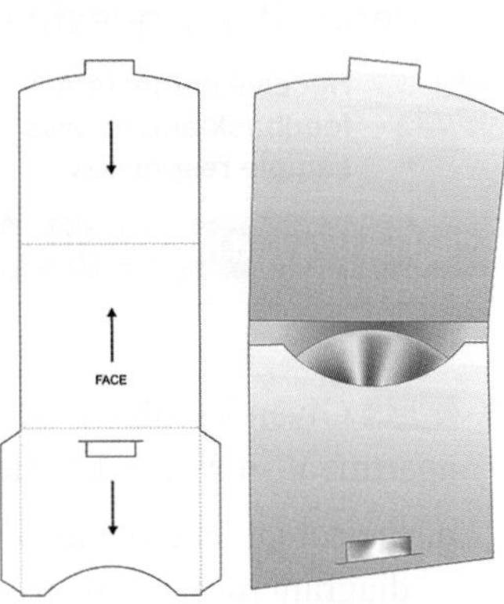

6. Consider the milk box shown and the template required to make it. The scale of the diagram is 1 : 7.

   a. Calculate the length and the width of the packaging template, given the dimensions measure 3 cm by 3 cm on the diagram.
   b. Estimate the total area of material required.
   c. If the cost of materials is \$0.45 per square metre, calculate how much it would cost to make this milk box.

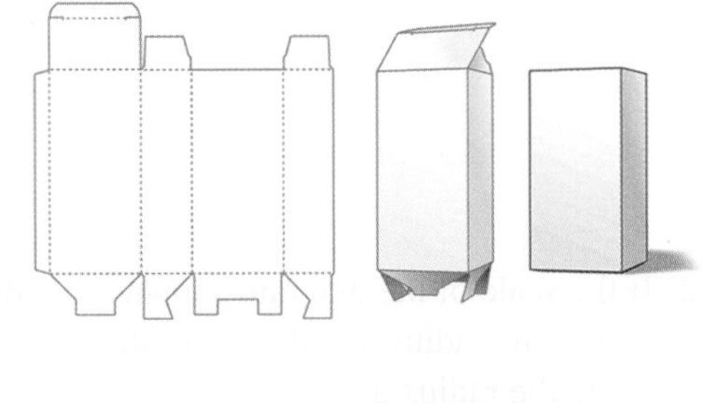

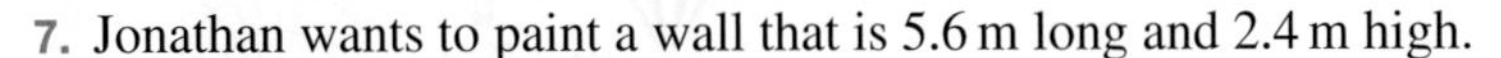

7. Jonathan wants to paint a wall that is 5.6 m long and 2.4 m high.

   a. Calculate how many litres of paint are required if 1.5 litres of paint cover 1 m^2.
   b. Calculate the cost of the paint if the price per litre is \$15.70.

8. **WE7** Ilia was asked by a friend to paint a bedroom wall. The friend gave Ilia a diagram of the bedroom drawn at a scale of 1 : 100. If the length and height of the wall are 1.5 cm and 2.5 cm respectively, calculate:
   a. the length and the height of the wall
   b. the area of the wall
   c. the amount of paint Ilia has to buy if the amount of paint needed is 2.6 L per m^2
   d. the total cost of the paint if the price is $12.80 per litre.

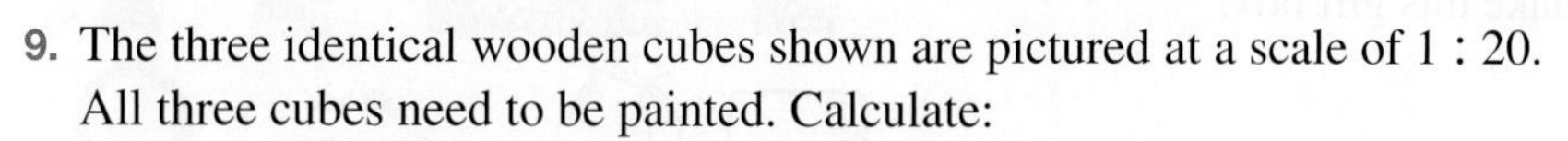

9. The three identical wooden cubes shown are pictured at a scale of 1 : 20. All three cubes need to be painted. Calculate:
   a. the length of their sides, given each side measures 1.3 cm on the diagram
   b. the area of each face
   c. the total area of the three cubes to be painted (state the answer in m^2 correct to 1 decimal place)
   d. the amount of paint required if 2L of paint are needed per m^2
   e. the total cost of the paint if the price is $9.25 per litre.

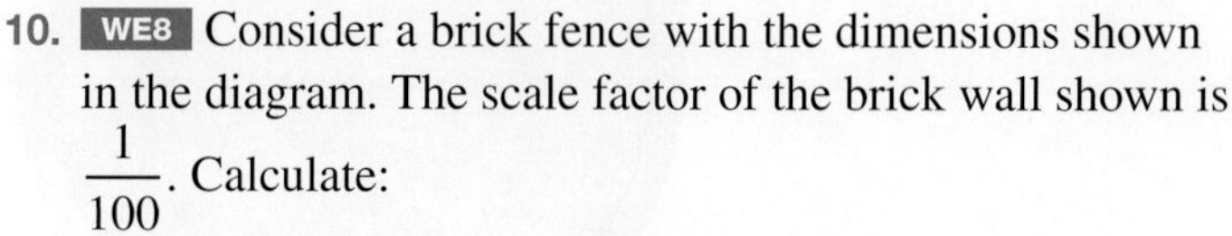

10. **WE8** Consider a brick fence with the dimensions shown in the diagram. The scale factor of the brick wall shown is $\frac{1}{100}$. Calculate:
   a. the length and the height of the fence, given that it measures 6 cm by 2.5 cm on the diagram
   b. the area of the front side of the fence
   c. the number of bricks required if the dimensions of the exposed side of one brick are 20 cm by 5 cm
   d. the cost of the fence if the price of one brick is $0.75.

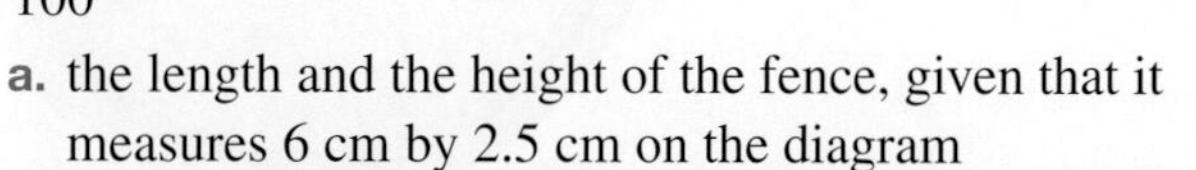

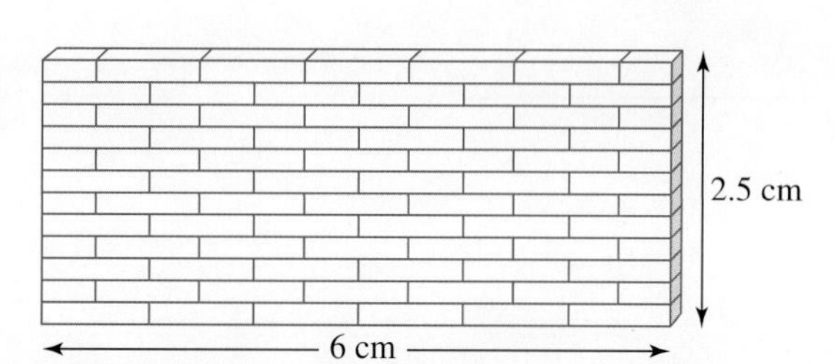

11. Sky wants to pave a garden path with bricks. The garden path is shown at a scale of 1 : 150. Calculate:
   a. the actual dimensions of the garden path, given the diagram measures 8.5 cm by 0.75 cm
   b. the area of the garden path
   c. the number of bricks required if the bricks used are squares with side length 24 cm
   d. the cost of paving the garden path if the price of one brick is $1.65.

12. The scale of the house plan shown is 1 : 200. Calculate:
   a. the length of the bathroom, given that it measures 1.2 cm by 1 cm on the plan
   b. the width of the hallway, given that it is 0.6 cm wide on the plan
   c. the perimeter of the kitchen, given that it measures 1 cm by 2 cm on the plan
   d. the floor area of the house, given that it measures 4.5 cm by 3.5 cm on the plan.

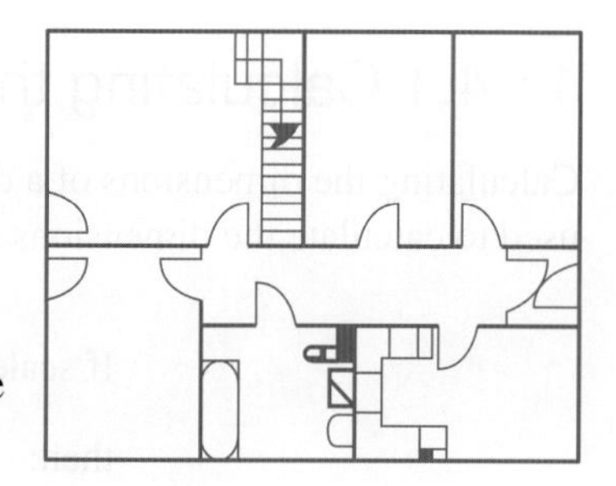

13. A brick has length 215 mm and height 65 mm. A bricklayer is building a wall 6 m long and 7.5 m high. Calculate:
   a. the number of bricks required
   b. the total cost of building the wall if one brick costs $1.95.

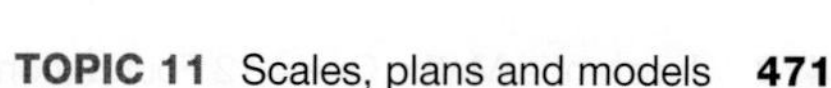

14. Consider the gift box shown and the template required to make it. The scale of the diagram is 1 : 8.
   a. Calculate the actual length and the width of the packaging template, given that it measures 8 cm by 5 cm on the diagram.
   b. Estimate the total area of material required.
   c. If the cost of materials is \$2.15 per square metre, calculate how much it would cost to make this gift box.

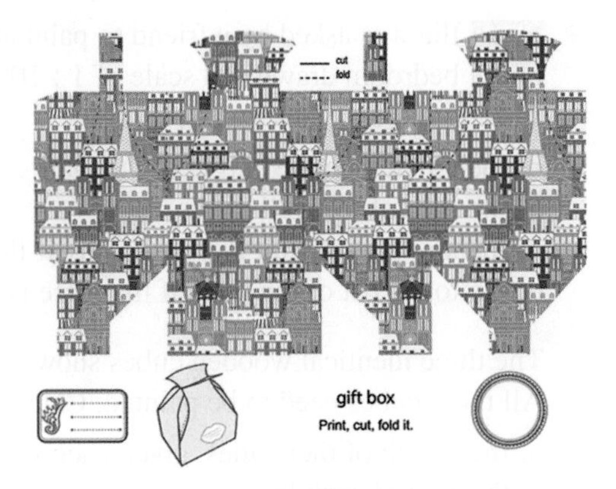

15. Calculate, to the nearest centimetre, the radius, the circumference and the area of the circular chopping board shown, given the radius measures 1.5 cm on the diagram and the scale is 1 : 12.

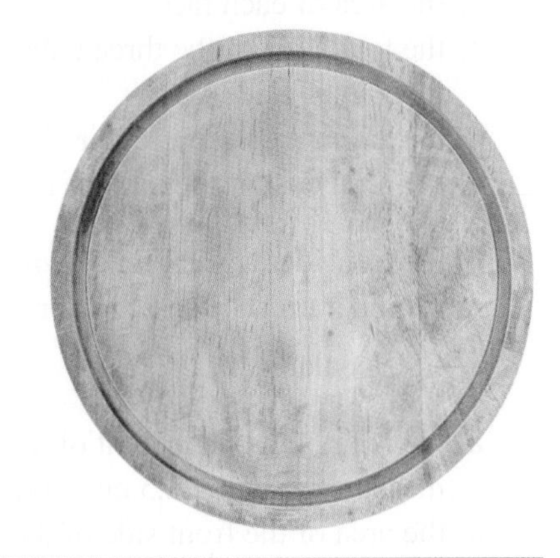

11.4 Creating scale drawings

LEARNING INTENTION

At the end of this subtopic you should be able to:
- calculate the dimensions for a scale drawing
- construct and label scale drawings.

Scale drawings are drawings on paper of real-life objects. All dimensions of the drawn object are kept in the same ratio to the actual dimensions of the object using a scale.

11.4.1 Calculating the dimensions of the drawing

Calculating the dimensions of a drawing always requires a scale or a scale factor. The formula for the scale is used to calculate the dimensions of the drawing.

$$\text{If scale factor} = \frac{\text{dimension on the drawing}}{\text{actual dimension}},$$

then:

dimension on the drawing = scale factor × actual dimension.

tlvd-3741

WORKED EXAMPLE 9 Calculating dimensions

Calculate the dimensions required to create a scale drawing of a shipping container 12 m long, 2.5 m wide and 3 m high using a scale of 1 : 100.

THINK	WRITE
1. Convert all dimensions to an appropriate drawing unit.	The dimensions of the diagram are going to be smaller than the actual dimensions of the shipping container. Convert m to cm. 12 m = 1200 cm 2.5 m = 250 cm 3 m = 300 cm
2. Write the scale as a scale factor.	$1:100=\frac{1}{100}$
3. Calculate the length of the object on the scale drawing.	$\frac{\text{dimension on the drawing}}{\text{actual dimension}}=\frac{1}{100}$ Length on the drawing = scale factor × actual dimension $=\frac{1}{100}\times 1200$ $=\frac{1200}{100}$ $=12\text{ cm}$
4. Calculate the width of the object on the scale drawing.	Width on the drawing = scale factor × actual dimension $=\frac{1}{100}\times 250$ $=\frac{250}{100}$ $=2.5\text{ cm}$
5. Calculate the height of the object on the scale drawing.	Height on the drawing = scale factor × actual dimension $=\frac{1}{100}\times 300$ $=\frac{300}{100}$ $=3\text{ cm}$

11.4.2 Constructing scale drawings

Scale diagrams are drawn on graph paper or plain paper using a pencil and a ruler. Accuracy is very important when constructing these drawings.

WORKED EXAMPLE 10 Constructing scale drawings

Construct a plan-view, front-view and side-view scale drawing of a shipping container with length 12 cm, width 2.5 cm and height 3 cm.

THINK	WRITE/DRAW
1. Write the drawing dimensions of the plan view of the object.	Drawing dimensions of the plan view of the shipping container are length = 12 cm and width = 2.5 cm.
2. Draw the plan view of the object. Recall that a plan view is a diagram of the object from above.	To draw the length of the object, construct a horizontal line 12 cm long.

Construct two 90° angles on both sides of the line segment. Ensure the two vertical lines are 2.5 cm long.

Connect the bottom ends of the vertical lines to complete the rectangle.

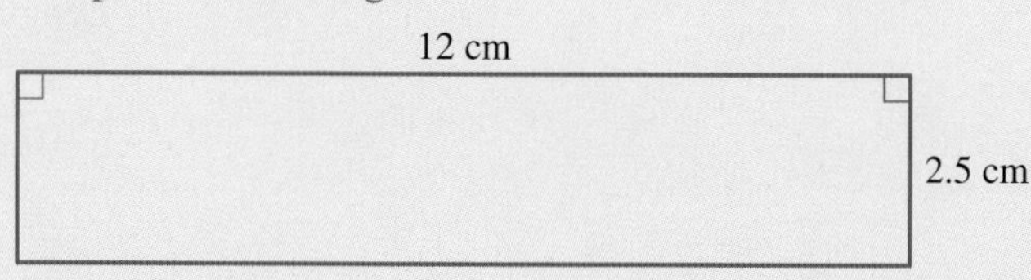

THINK	WRITE/DRAW
3. State the drawing dimensions of the front view of the object.	The drawing dimensions of the front view of the shipping container are length = 12 cm and height = 3 cm.
4. Draw the front view of the object. Recall that the front view is a diagram of the object looking straight at the object from the front.	To draw the length of the object, construct a horizontal line 12 cm long.

Construct two 90° angles on both sides of the line segment. Ensure the two vertical lines are 3 cm long.

Connect the bottom ends of the vertical lines to complete the rectangle.

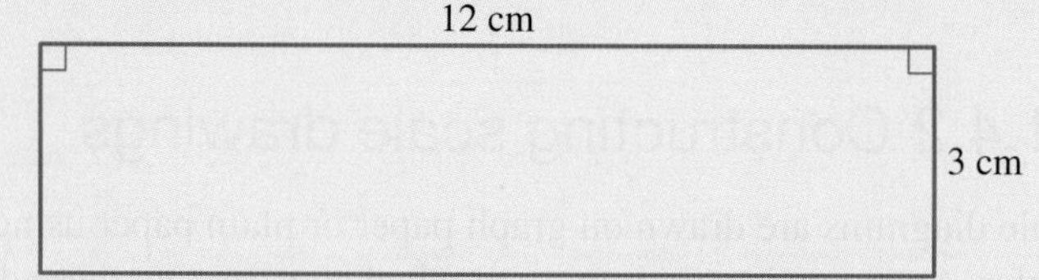

THINK	WRITE/DRAW
5. State the drawing dimensions of the side view of the object.	The drawing dimensions of the side view of the shipping container are width = 2.5 cm and height = 3 cm.

6. Draw the side view of the object. Recall that the side view is a diagram of the object looking at the object from one side.	Draw the length of the object, which is the width of the shipping container. Draw a horizontal line 2.5 cm long. Construct two 90° angles on both sides of the line segment. Ensure the two vertical lines are 3 cm long. 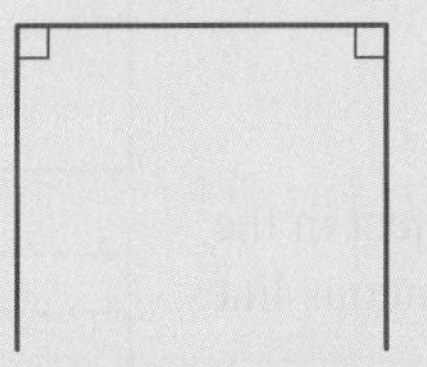 Connect the bottom ends of the vertical lines to complete the rectangle. 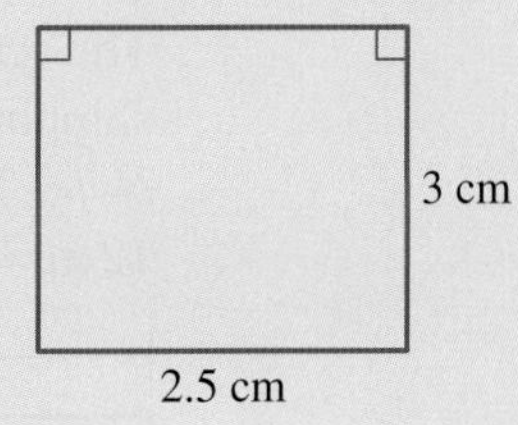

11.4.3 Labelling scale drawings

Scale diagrams are labelled using the actual dimensions of the object. The labels of a scale drawing have to be clear and easily read.

Conventions and line styles

The line style and its thickness indicate different details of the drawing.

- A *thick continuous line* represents a visible line — a contour of a shape or an object.
- A *thick dashed line* represents a hidden line — a hidden contour of a shape or an object.
- *Thin continuous lines* are used to mark the dimensions of the shape or object.

Dimension lines are thin continuous lines showing the dimension of a line. They have arrowheads at both ends.

Projection lines are thin continuous lines drawn perpendicular to the measurement shown. These lines do not touch the object.

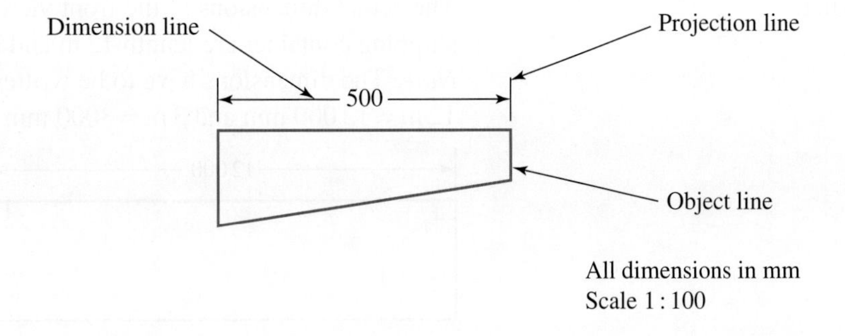

All dimensions on a drawing must be given in millimetres; however, 'mm' is not to be written on the drawing. This can be written near the scale, as shown in the diagram above.

Dimensions must be written in the middle of the dimension line.

WORKED EXAMPLE 11 Labelling scale drawings

Label the three scale drawings of the shipping container from Worked example 10.

THINK | **WRITE/DRAW**

1. Draw the projection lines of the object in the plan-view drawing using thin continuous lines.
2. Draw the dimension lines of the object in the plan-view drawing using thin continuous lines with arrowheads at both ends.

3. Label the drawing.

The actual dimensions of the plan view of the shipping container are length 12 m and width 2.5 m.
Note: The dimensions have to be written in mm.
12 m = 12 000 mm and 2.5 m = 2500 mm

4. Draw the projection lines of the object in the front-view drawing using thin continuous lines.
5. Draw the dimension lines of the object in the front-view drawing using thin continuous lines with arrowheads at both ends.

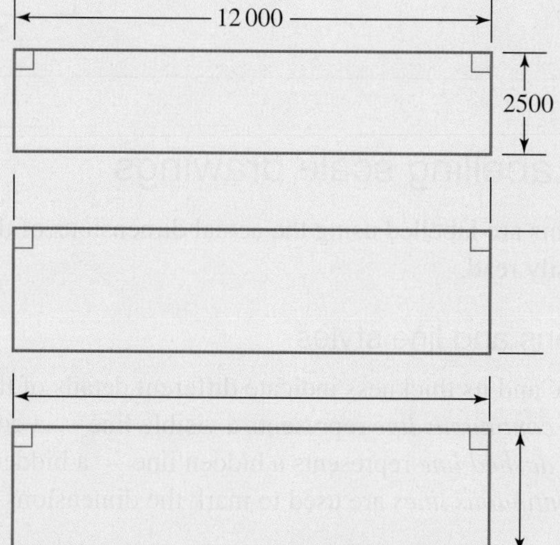

Note: The arrowheads of the dimension line for the height are directed from the outside of the projection line. This happens because the space between the projection lines is too small.

6. Label the drawing.

The actual dimensions of the front view of the shipping container are length 12 m and height 3 m.
Note: The dimensions have to be written in mm.
12 m = 12 000 mm and 3 m = 3000 mm

7. Draw the projection lines of the object in the side-view drawing using thin continuous lines.

8. Draw the dimension lines of the object in the side-view drawing using thin continuous lines with arrowheads at both ends.

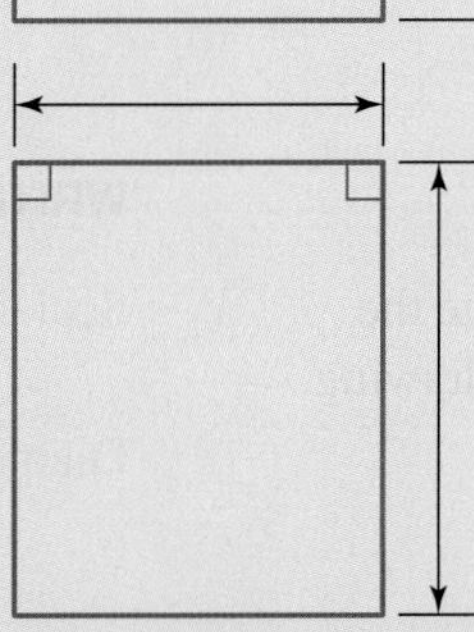

9. Label the drawing.

The actual dimensions of the side view of the shipping container are 2.5 m wide and 3 m high.
Note: The dimensions have to be written in mm.
2.5 m = 2500 mm and 3 m = 3000 mm

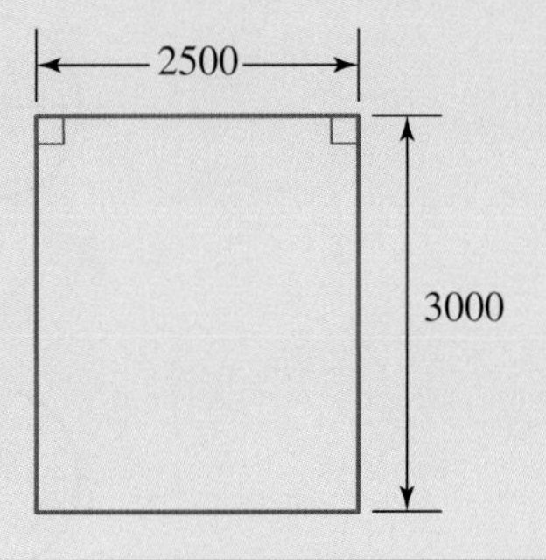

Labelling circles

A circle on a scale diagram is labelled using either the Greek letter Φ (phi) for diameter or R for radius. The diagram shows three correct notations for a 2-cm diameter. *Note:* The dimensions must be written in millimetres.

WORKED EXAMPLE 12 Labelling circle scale drawings

The food can shown in the diagram has a diameter of 10 cm. Construct a scale drawing of the plan view of the food can using a scale of 1 : 4.

THINK	WRITE/DRAW
1. State the scale factor and calculate the dimensions of the object on the drawing.	Scale factor $= \frac{1}{4}$ Diameter on the diagram $= \frac{1}{4} \times 10$ $= \frac{10}{4}$ $= 2.5$ cm $= 25$ mm
2. Draw the plan view of the object.	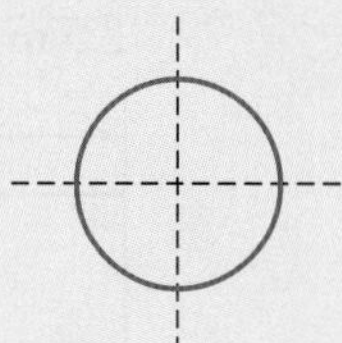
3. Label the drawing.	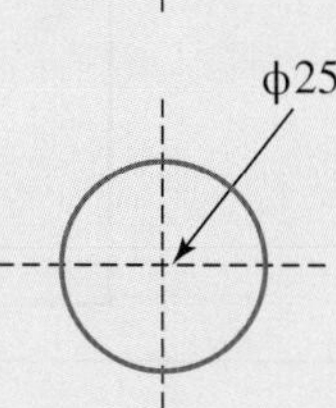 *Note:* The dimensions are written in mm.

11.4 Exercise

1. MC A projection line is drawn using:

   A. a thin continuous line with two arrowheads.
   B. a thick dashed line.
   C. a thin continuous line.
   D. a thick continuous line.
   E. a thin dashed line.

2. WE9 Calculate the dimensions required to create a scale drawing of a classroom 17 m long and 12 m wide to a scale of 1 : 100.

3. a. Calculate the dimensions required to create a scale drawing of a door 1.5 m wide, 2.4 m tall and 9 cm thick to a scale of 1 : 30.
   b. State the measurements that are to be written on a scale drawing.
   c. Explain whether the projection lines are parallel or perpendicular to the measurement shown.
   d. Write the symbol used to label the diameter of a circle on a drawing.

4. The drawings described are constructed to a scale of 1 : 50.
   a. Calculate the height, on the drawing, of a tree 3.6 m tall.
   b. Calculate the dimensions, on the drawing, of a carpet 2.8 m by 1.4 m.
   c. Calculate the height, on the drawing, of a student 1.8 m tall.

5. a. WE10 Construct a plan-view, front-view and side-view scale drawing of a skip bin with drawing dimensions of length 5 cm, width 2 cm and depth 1.5 cm.
   b. WE11 Label the three scale drawings of the skip from part a, given that the actual dimensions of the skip are length 2.5 m, width 1 m and depth 0.75 m.

6. Construct a plan-view, front-view and side-view scale drawing of a rectangular lunch box with drawing dimensions of length 5.3 cm, width 2.7 cm and depth 0.9 cm.

7. Calculate the dimensions to construct a scale drawing of the following shapes to the scale stated in brackets.
   a. A square of side length 1.7 cm (2 : 1)
   b. A rectangular pool table with dimensions 4.8 m by 4.0 m (1 : 80)
   c. A circular table with diameter 1.4 m (1 : 200)

8. Draw a house floor plan with actual dimensions 27.6 m by 18.3 m drawn to a scale of 1 : 600.

9. MC The dimension on a drawing is calculated using the formula:
   A. scale factor $\times$ actual length
   B. scale factor + actual length
   C. $\dfrac{\text{scale factor}}{\text{actual length}}$
   D. actual length − scale factor
   E. $\dfrac{\text{actual length}}{\text{scale factor}}$

10. Construct and label a scale drawing of the floor of a bedroom 3.5 m long and 2.5 m wide drawn to a scale of 1 : 100.

11. Construct and label a scale drawing of a window frame 1.8 m wide and 1.4 m tall drawn to a scale of 1 : 40.
12. Draw and label the projection and dimension lines on the following diagrams.
    a. Square with side length 6.1 m

    b. Rectangle 430 cm long and 210 cm wide

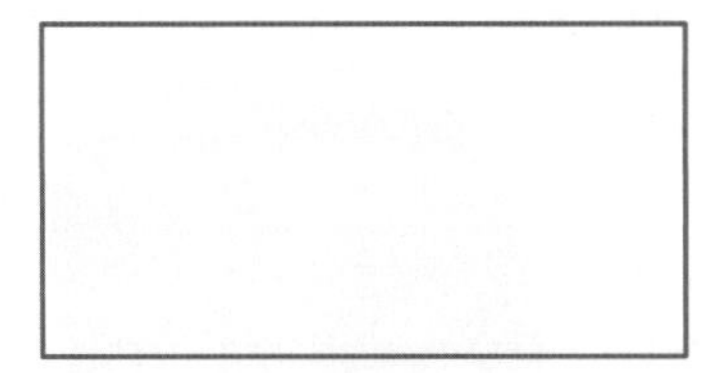

    c. Equilateral triangle with side length 49.7 cm and height 43.0 cm

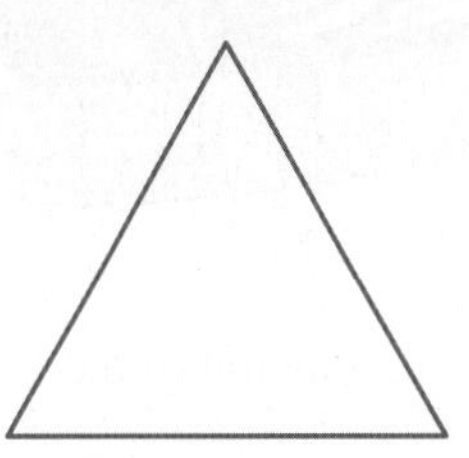

    d. Circle with diameter 7.6 m

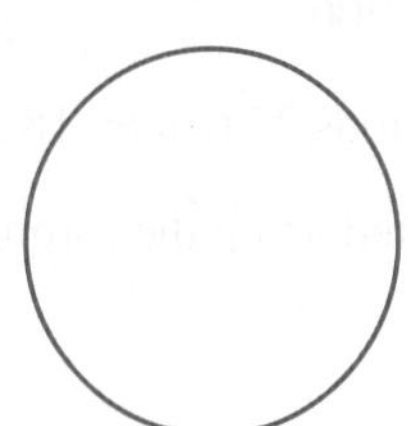

13. WE12 The pipe shown has a diameter of 28 cm. Construct a scale drawing of the plan view of the pipe drawn to a scale of 1 : 10.

14. The hat box shown in the diagram has a diameter of 36 cm. Construct a scale drawing of the plan view of the box drawn to a scale of 1 : 6.

15. A sculptor plans a sculpture by first creating a scale drawing. Determine the height of the sculpture on a drawing if its actual height is intended to be 5.2 m and the scale factor is $\frac{1}{26}$.

16. Measure the dimensions of your bedroom, bed and desk and construct a scale drawing using these measurements and an appropriate scale.

11.5 Review

11.5.1 Summary

eles-38019

Hey students! Now that it's time to revise this topic, go online to:

Review your results

Watch teacher-led videos

Practise questions with immediate feedback

Find all this and MORE in jacPLUS

11.5 Exercise

Multiple choice

1. MC If the scale factor of a diagram is 1 : 250, a length of 7.5 cm on the diagram represents an actual length of:
   A. 33.33 m
   B. 1875 cm
   C. 7500 cm
   D. 1875 m
   E. 75 m

2. MC The distance from Frankston to Melbourne is 45 km. If Google Maps is using a scale of 1 : 100 000, then the distance on Google Maps would measure:
   A. 4.5 m B. 4.5 cm C. 45 cm D. 54 cm E. 5.4 cm
3. MC A front fence is 24 m long and 1.2 m high. It is to be made with bricks 40 cm long and 12 cm high. An estimate on the number of bricks required to build the fence is:
   A. 240 B. 300 C. 400 D. 450 E. 600
4. MC A house plan uses the scale 1 : 250. If the main wall in the kitchen is 3.8 cm long on the plan, the actual length of the wall is:
   A. 950 mm B. 2.5 m C. 3.8 m D. 9.5 m E. 2500 cm
5. MC A plan of a new car that is 2.6 m long is drawn on a plan that uses the scale 1 : 20. The length of the car on the plan is:
   A. 26 cm B. 13 cm C. 52 cm D. 1.3 cm E. 2.6 cm
6. MC Kirrily wants to paint a feature wall in her lounge room. The wall is 6.2 m by 2.8 m and the paint she wants to use requires 2 L per m^2 and costs $9.85 per litre. The cost to put two coats of paint on the wall is closest to:
   A. $683.98 B. $341.99 C. $586.58 D. $872.50 E. $720.25
7. MC On a map with a scale 1 : 2 500 000, the distance between two towns is 3.9 cm. The actual distance between the two towns is:
   A. 3.9 km
   B. 97.5 km
   C. 64.1 km
   D. 15.6 km
   E. 36.6 km

8. MC The total cost of the paint required to paint an area of 68 m^2 at \$16.90 per litre if 2 litres of paint are required per m^2 is:

A. \$1149.20
B. \$2298.40
C. \$574.60
D. \$136
E. \$778.50

9. MC The drawing that uses the correct conventions is:

A.
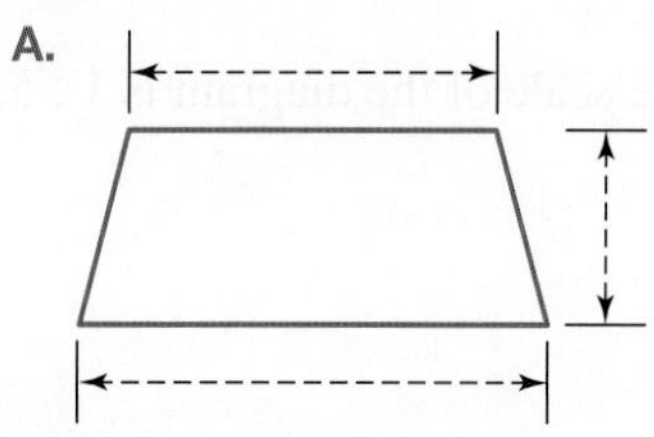

B.
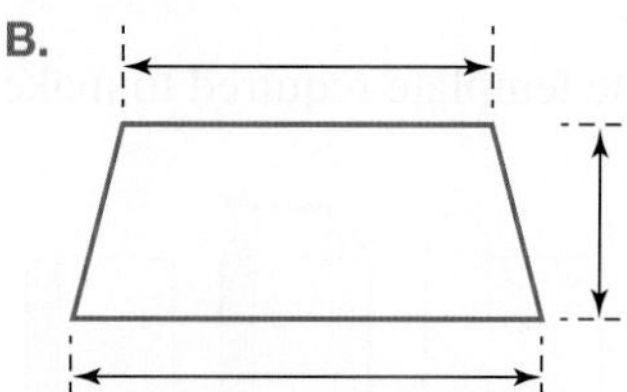

C.
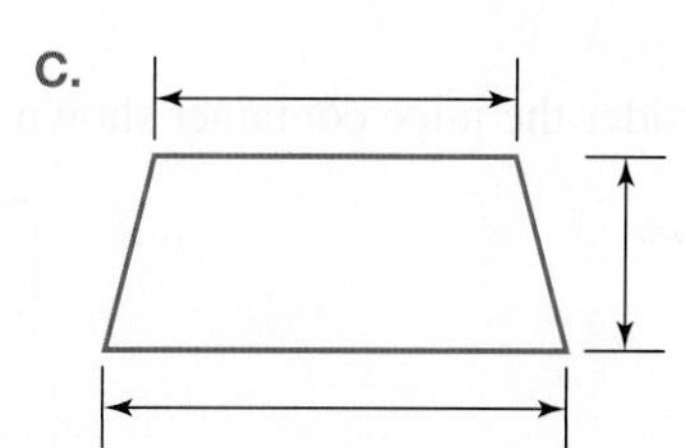

D.
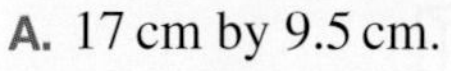

E.
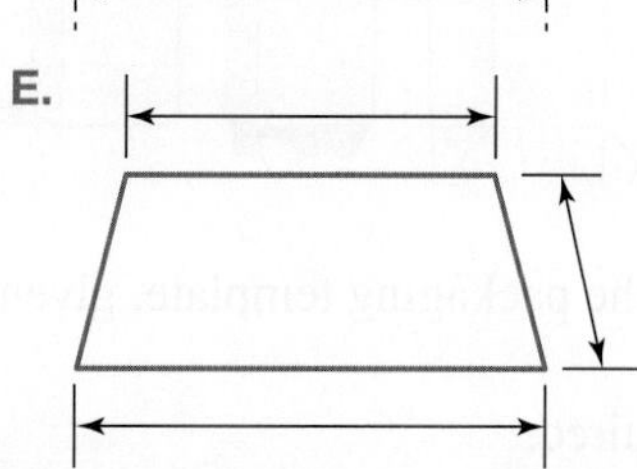

10. MC A shoe box is 34 cm long, 19 cm wide and 15 cm high, as shown in the diagram. If a scale of 1 : 2 is used, a scale diagram of the side view should be drawn with dimensions:

A. 17 cm by 9.5 cm.
B. 17 cm by 7.5 cm.
C. 9.5 cm by 7.5 cm.
D. 19 cm by 15 cm.
E. 19 cm by 17 cm.

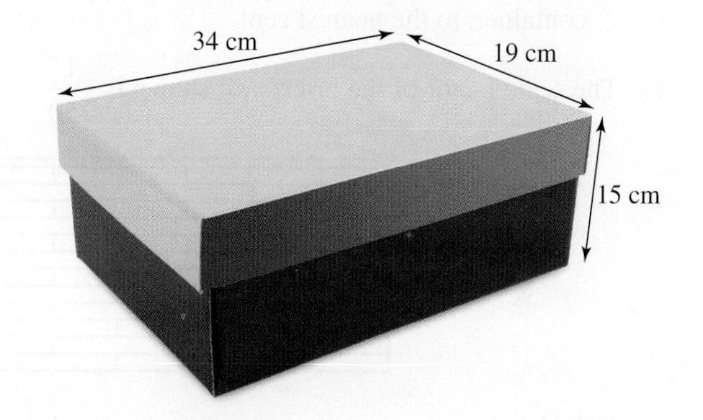

Short answer

11. If the scale of a diagram is 5 cm : 100 km, convert the scale using the same unit.

12. The graphical scale on a map is shown below. State the scale of the map as a ratio, if each partition measures 2.5 cm.

0 150 300 km

13. Given a scale of 1 : 300 for the floor plan of the apartment shown, calculate:

a. the length and the width of the apartment, given that it measures 6 cm by 4 cm on the diagram
b. the perimeter of the apartment.

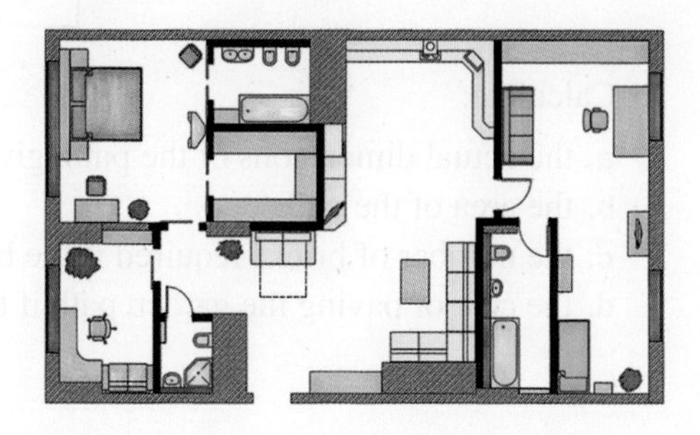

14. A sculptor plans a sculpture by first creating a scale drawing. Determine the height of the sculpture on a drawing if its actual height is intended to be 6.3 m and the scale factor is $\frac{1}{20}$.

Extended response

15. A brick has length 210 mm and height 55 mm. A bricklayer is building a wall 10 m long and 8 m high. Calculate:
   a. the number of bricks required
   b. the total cost of building the wall if one brick costs $2.15.

16. Consider the juice container shown and the template required to make it. The scale of the diagram is 1 : 5.

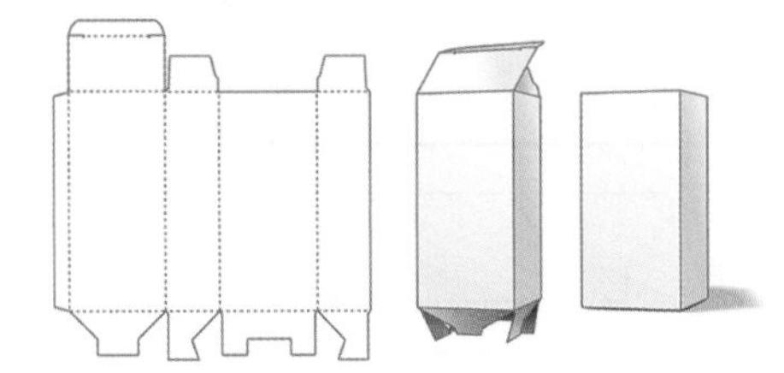

   a. Calculate the length and the width of the packaging template, given they both measure 4 cm on the diagram.
   b. Estimate the total area of material required.
   c. If the cost of materials is $0.55 per square metre, calculate how much it would cost to make this juice container, to the nearest cent.

17. The scale factor of the brick wall shown is $\frac{1}{250}$.

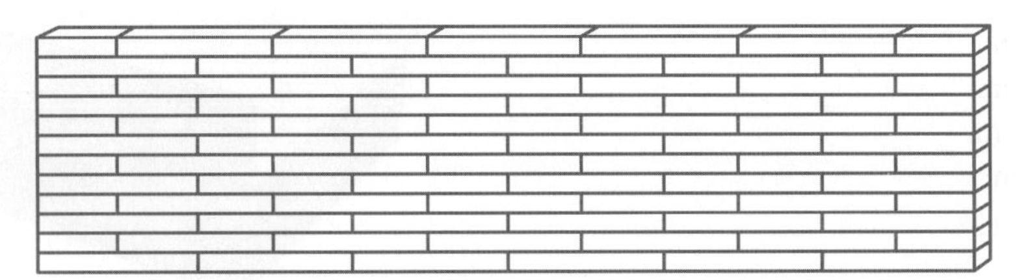

   Calculate:
   a. the length and the width of the wall, given that it measures 10 cm by 2.5 cm on the diagram
   b. the area of the front side of the wall
   c. the number of bricks required if the dimensions of the exposed side of one brick are 20 cm by 5 cm
   d. the cost of the wall if the price of one brick is $0.65.

18. Shae wants to pave a path towards her front door with bricks. The path is shown at a scale of 1 : 100.

   Calculate:
   a. the actual dimensions of the path, given that it measures 10.5 cm by 0.75 cm
   b. the area of the path
   c. the number of bricks required if the bricks used are squares with side length 25 cm
   d. the cost of paving the garden path if the price of one brick is $1.85.

19. The scale of the house plan shown is 1 : 250.

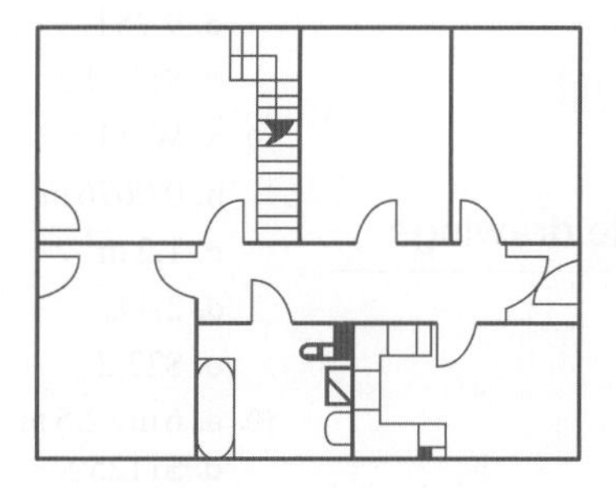

Calculate:

a. the dimensions of the top left room, given that it measures 1.8 cm by 1.7 cm on the plan
b. the width of the hallway, given that it is 0.6 cm wide on the plan
c. the perimeter of the kitchen, given that it measures 1 cm by 1 cm on the plan
d. the floor area of the house, given that it measures 4.5 cm by 3.5 cm on the plan.

20. The objects in the diagram shown are drawing instruments.

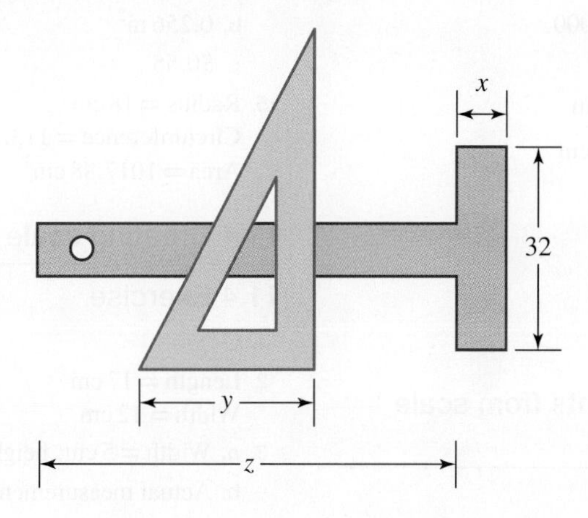

a. Calculate the scale factor of the drawing, given the length that is 32 mm measures 24 mm on the scale drawing.
b. Calculate the actual measurements of the unknown lengths x, y and z, given they measure 6 mm, 21 mm and 51 mm, respectively, on a scale drawing.
c. Determine the lengths of x, y and z on a drawing with a scale of 5 : 4.
d. Construct a scale drawing using the new drawing dimensions from part **c**. Clearly label all measurements.

Answers

Topic 11 Scales, plans and models

11.2 Reading and interpreting scale drawings

11.2 Exercise

1. $1 : 300, \frac{1}{300}$
2. $3 : 40\,000\,000$
3. B
4. a. $1 : 50\,000, \frac{1}{50\,000}$ b. $1 : 400, \frac{1}{400}$
5. a. $1 : 3700, \frac{1}{3700}$ b. $1 : 1600, \frac{1}{1600}$
6. $1 : 1\,000\,000$
7. $1 : 100\,000$
8. a. $1 : 10\,000\,000$ b. $1 : 8000$
9. $1 : 500\,000$
10. a. 32 m b. 1.7 cm
11. a. 3 m b. 1.04 cm
12. Length = 2.25 m
    Width = 1.5 m
13. Height = 2.4 m
    Width = 6 m
14. D

11.3 Calculating measurements from scale drawings

11.3 Exercise

1. a. Width = 7.5 m
      Length = 12.5 m
   b. 40 m
2. Radius = 12 m
   Circumference = 75 m
3. a. Height = 1 m
      Big glass width = 0.6 m
      Small glass width = 0.3 m
   b. Area of big glass = 0.6 m^2
      Area of small glass = 0.3 m^2
   c. 0.9 m^2
4. Radius = 30 cm and area = 2827 cm^2
5. a. Length = 28 cm
      Width = 12 cm
   b. 336 cm^2
   c. 3 cents each
6. a. Length = width = 21 cm
   b. 441 cm^2
   c. 2 cents each
7. a. 20.16 L
   b. $316.51
8. a. Length = 1.5 m
      Height = 2.5 m
   b. 3.75 m^2
   c. 9.75 L
   d. $124.80
9. a. Width = length = 0.26 m
   b. 0.0676 m^2
   c. 1.2 m^2
   d. 2.4 L
   e. $22.2
10. a. 6 m, 2.5 m b. 15 m^2 c. 1500 bricks
    d. $1125
11. a. 12.75 m, 1.125 m b. 14.34 m^2
    c. 249 bricks d. $410.85
12. a. Length = 2.4 m b. Width = 1.2 m
    c. Perimeter = 12 m d. Area = 63 m^2
13. a. 3221 bricks b. $6280.95
14. a. Length = 64 cm
       Width = 40 cm
    b. 0.256 m^2
    c. $0.55
15. Radius = 18 cm
    Circumference = 113.10 cm
    Area = 1017.88 cm^2

11.4 Creating scale drawings

11.4 Exercise

1. C
2. Length = 17 cm
   Width = 12 cm
3. a. Width = 5 cm; height = 8 cm; depth = 0.3 cm
   b. Actual measurements
   c. Perpendicular to the measurement shown
   d. Φ (the Greek letter phi)
4. a. Height = 7.2 cm b. 5.6 cm; 2.8 cm
   c. Height = 3.6 cm
5. a.

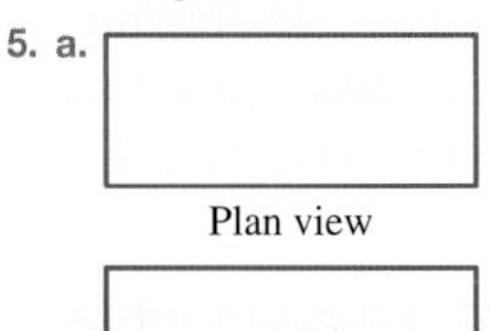

Plan view

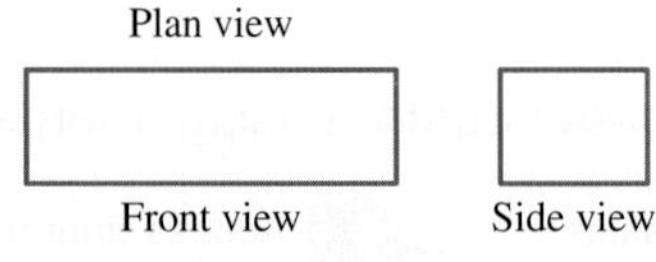

Front view Side view

b.

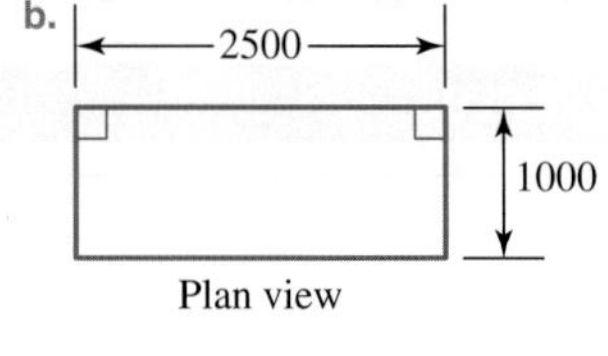

Plan view

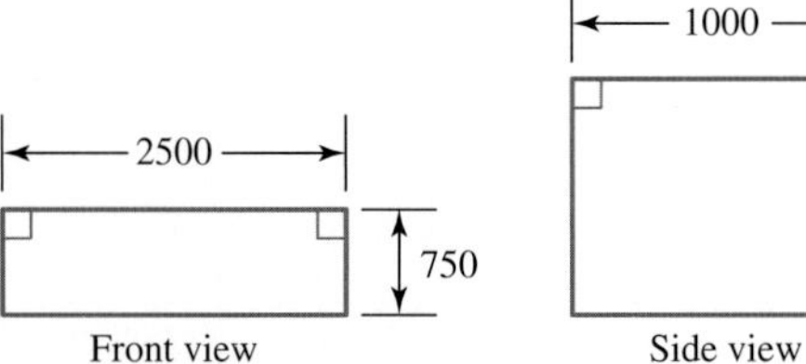

Front view Side view

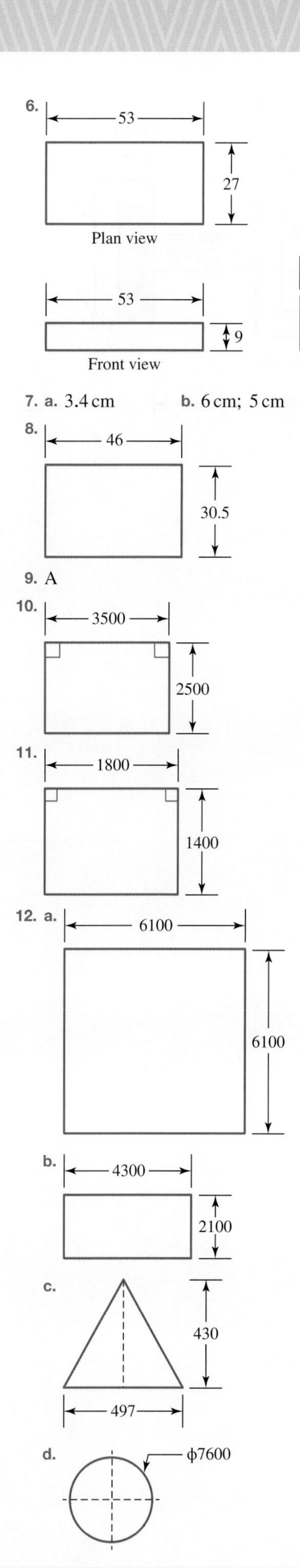

7. a. 3.4 cm b. 6 cm; 5 cm c. 0.7 cm

9. A

13.

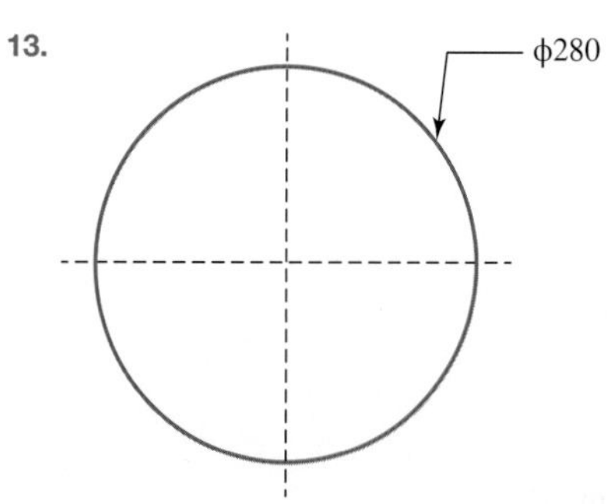

14.

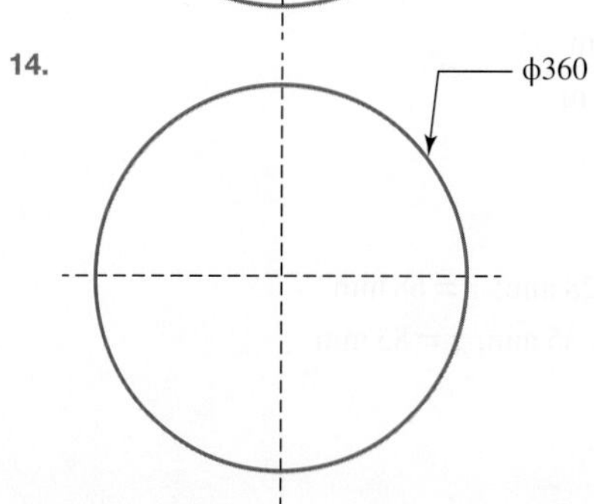

15. 20 cm

16. Sample responses can be found in the worked solutions in the online resources.

11.5 Review

11.5 Exercise

Multiple choice

1. B
2. C
3. E
4. D
5. B
6. A
7. B
8. B
9. C
10. C

Short answer

11. 1 : 2 000 000
12. 1 : 6 000 000
13. a. 18 m; 12 m b. 60 m
14. 31.5 cm

Extended response

15. a. 6927 bricks b. $14 893.05
16. a. Length = width = 20 cm
    b. 400 cm^2
    c. 2 cents each
17. a. 25 m, 6.25 m
    b. 156.25 m^2
    c. 15 625 bricks
    d. $10 156.25

18. a. 10.5 m, 0.75 m

b. 7.875 m^2

c. 126 bricks

d. $233.10

19. a. Length = 4.5 m
Width = 4.25 m

b. 1.5 m

c. Length = 2.5 m
Width = 2.5 m
Perimeter = 10 m

d. Length = 11.25 m
Width = 8.75 m
Area = 98.44 m^2

20. a. 3 : 4

b. $x = 8$ mm; $y = 28$ mm; $z = 68$ mm

c. $x = 10$ mm; $y = 35$ mm; $z = 85$ mm

d.

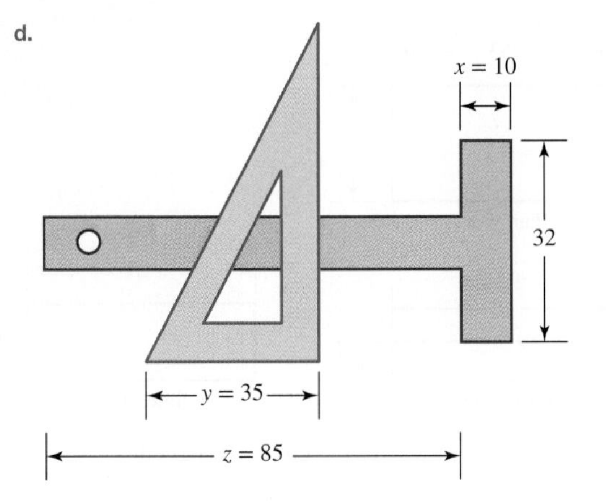

12 Right-angled triangles

LEARNING SEQUENCE

Fully worked solutions for this topic are available online.

12.1 Overview

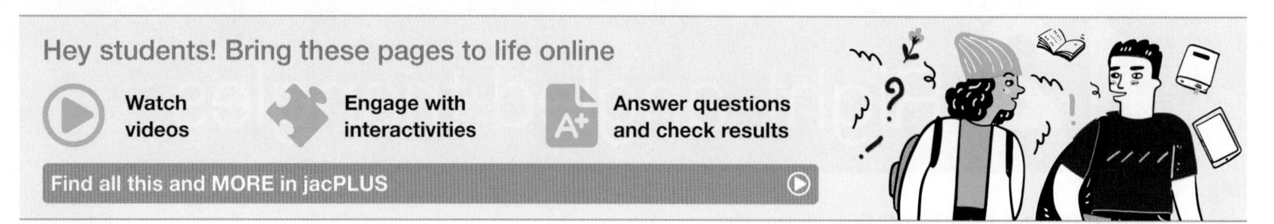

12.1.1 Introduction

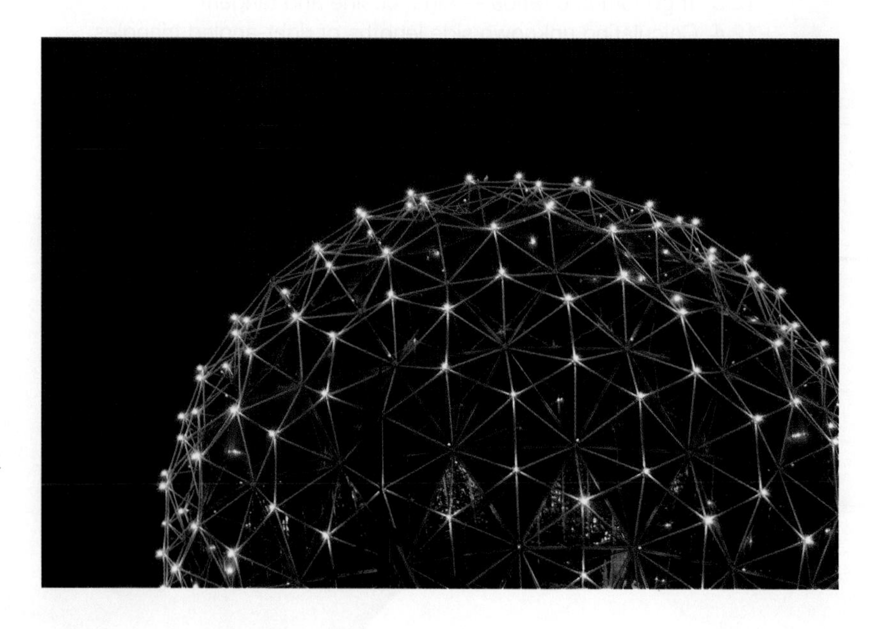

Thousands of years before the common era, ancient civilisations were able to build enormous structures – like Stonehenge in England and the pyramids in Egypt. How did they do that? Human ingenuity helps us achieve many things that at first seem impossible. How do we determine the height of a tall object, like a column, without physically scaling it? Sometimes, even tasks that appear to be very simple (e.g. planning and designing a staircase) present us with unexpected problems. How, in fact, do we go about the practical task of planning and designing a staircase?

None of the structures we build would be possible without our understanding of geometry and trigonometry. Engineers apply the principles of geometry and trigonometry regularly to make sure that buildings are strong, stable and capable of withstanding extreme conditions. Triangles are particularly useful to engineers and architects because they are the strongest shape. Any forces applied to a triangular frame will be distributed equally to all of the frame's sides and joins. This fact has been known for thousands of years — triangular building frames were used as far back as the sixth century BCE. A truss is an example of a structure that relies on the strength of triangles. Trusses are often used to hold up the roofs of houses and to keep bridges from falling down. Triangular frames can even be applied to curved shapes.

Geodesic domes, like the one shown here, are rounded structures that are made up of many small triangular frames connected together. This use of triangular frames makes geodesic domes very strong, but also very light and easy to build.

The theorem of Pythagoras and the geometry of right-angled triangles (trigonometry) will be explored in this topic, and we'll learn how to use them to solve real-world problems.

KEY CONCEPTS

This topic covers the following key concepts from the VCE Mathematics Study Design:

- spatial and geometric constructions including transformations, similarity, symmetry and projections.

12.2 Pythagoras' theorem

LEARNING INTENTION

At the end of this subtopic you should be able to:
- determine the hypotenuse of triangles
- use Pythagoras' theorem to find unknown lengths of a right-angled triangle
- use Pythagoras' theorem to solve problems with right-angled triangles.

12.2.1 Introduction

Pythagoras' theorem allows us to calculate the length of a side of a right-angled triangle if we know the lengths of the other two sides. Consider triangle ABC shown. AB is the **hypotenuse** (the longest side). It is always opposite the right angle.

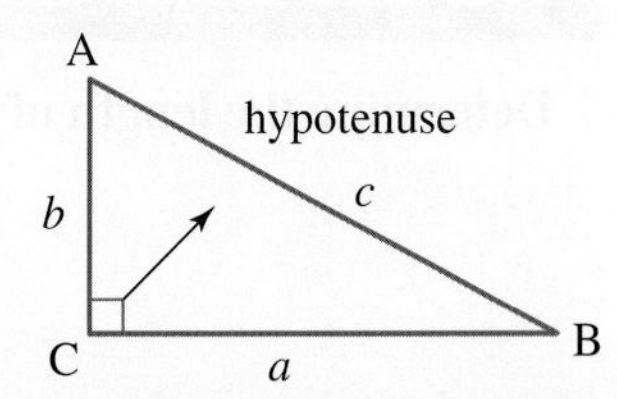

Note that the sides of a triangle can be named in one of two ways.
1. A side can be named by the two capital letters given to the vertices at each end. This is what has been done in the figure shown to name the hypotenuse AB.
2. We can also name a side by using the lower-case letter of the opposite vertex. In the figure shown, we could also have named the hypotenuse 'c'.

WORKED EXAMPLE 1 Labelling the hypotenuse

Name the hypotenuse in the triangle shown.

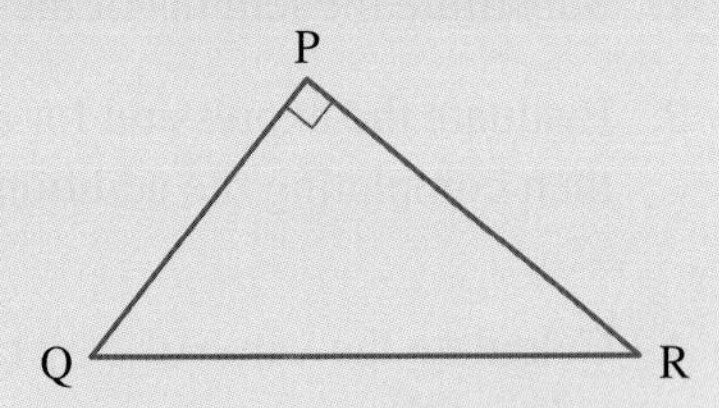

THINK	WRITE
1. Identify the hypotenuse by locating the longest side.	The longest side is opposite the right angle, between vertices Q and R, and opposite vertex P.
2. Name the hypotenuse.	The hypotenuse is QR or p.

Consider the right-angled triangle ABC with sides 3 cm, 4 cm and 5 cm. Squares have been constructed on each of the sides. The area of each square has been calculated by squaring the side length, and indicated inside the square.

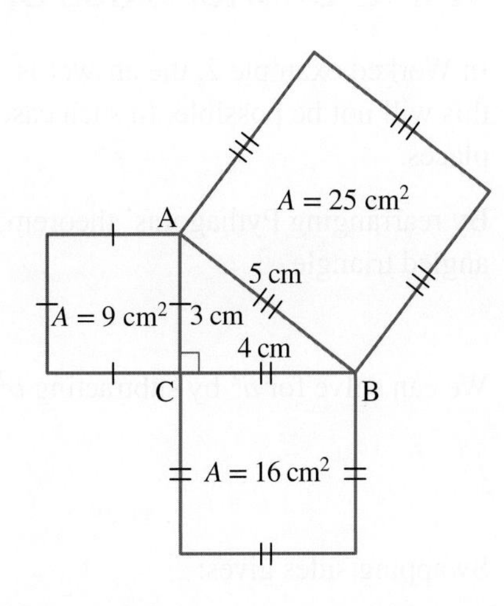

$$25\text{ cm}^2 = 16\text{ cm}^2 + 9\text{ cm}^2$$

$$\text{Alternatively: } (5\text{ cm})^2 = (4\text{ cm})^2 + (3\text{ cm})^2$$

$$\text{This means: hypotenuse}^2 = \text{base}^2 + \text{height}^2$$

$$c^2 = a^2 + b^2$$

This result is known as **Pythagoras' theorem**.

Note the area of the square on the hypotenuse is equal to the sum of the areas of the squares on the other two sides.

Pythagoras' theorem

In any right-angled triangle, the square of the length of the hypotenuse is equal to the sum of the squares of the lengths of the two shorter sides. That is:

$$c^2 = a^2 + b^2$$

where a and b are the two shorter sides and c is the hypotenuse.

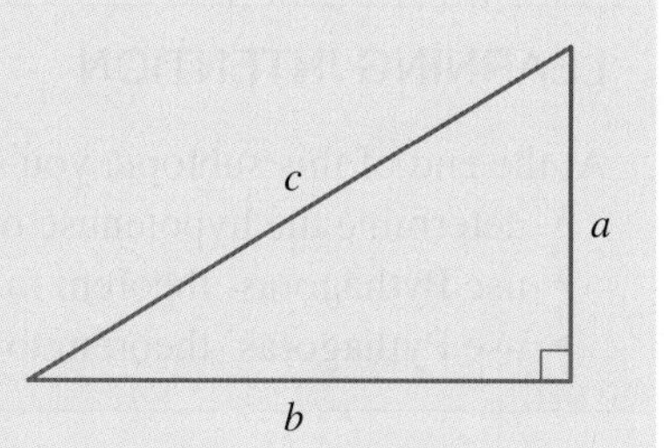

WORKED EXAMPLE 2 Calculating the length of the hypotenuse

Determine the length of the hypotenuse in the triangle shown.

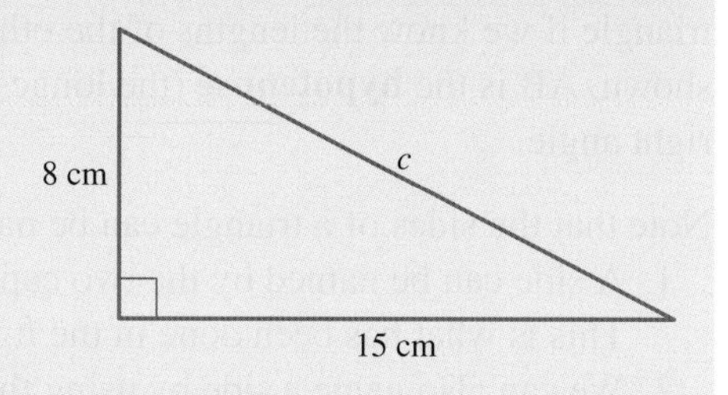

THINK	WRITE
1. This is a right-angled triangle, so we can use Pythagoras' theorem. Write the formula for Pythagoras' theorem.	$c^2 = a^2 + b^2$
2. Substitute the lengths of the shorter sides.	$c^2 = 15^2 + 8^2$
3. Evaluate the expression for c^2 by squaring 15 and 8, and then completing the addition.	$= 225 + 64$ $= 289$
4. Calculate the value of c by taking the square root of both sides of the equation.	$\sqrt{c^2} = \sqrt{289}$ $c = \sqrt{289}$ $= 17$ cm
5. State the answer.	The length of the hypotenuse is 17 cm.

12.2.2 Shorter sides of right-angled triangles

In Worked example 2, the answer is a whole number because we can calculate $\sqrt{289}$ exactly. In most examples, this will not be possible. In such cases, we are asked to write the answer correct to a given number of decimal places.

By rearranging Pythagoras' theorem, we can write the formula to calculate the length of a shorter side of a right-angled triangle.

$$c^2 = a^2 + b^2$$

We can solve for a^2 by subtracting b^2 on both sides of the equation.

$$c^2 - b^2 = a^2 + b^2 - b^2$$
$$c^2 - b^2 = a^2$$

Swapping sides gives:

$$a^2 = c^2 - b^2$$

Following a similar method, we can rearrange for b^2.

$$b^2 = c^2 - a^2$$

Pythagoras' theorem — hypotenuse and short sides

Calculating the hypotenuse:

$$c^2 = a^2 + b^2$$

Calculating the shorter side:

$$a^2 = c^2 - b^2$$

or

$$b^2 = c^2 - a^2$$

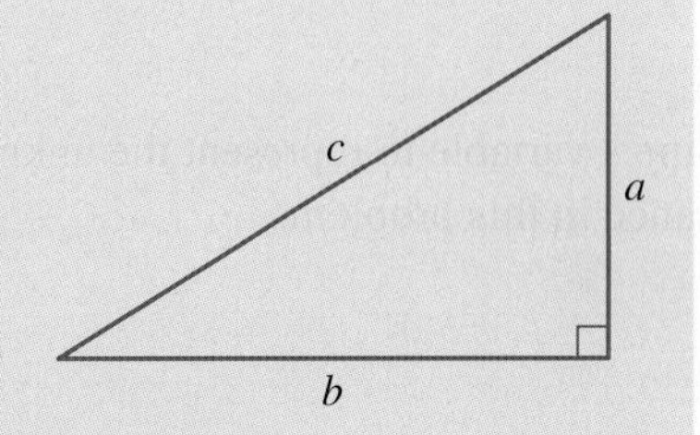

WORKED EXAMPLE 3 Calculating the length of a shorter side

Calculate the length of side PQ in triangle PQR, correct to 1 decimal place.

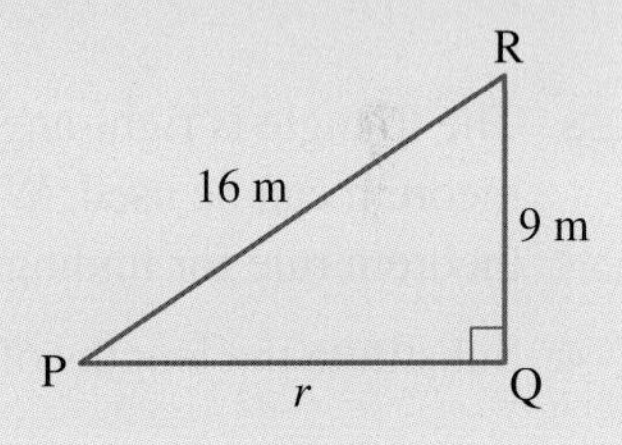

THINK	WRITE
1. This is a right-angled triangle, so we can use Pythagoras' theorem to calculate the length of a shorter side. Write the formula for Pythagoras' theorem.	$b^2 = c^2 - a^2$
2. Substitute the lengths of the known sides. Place the unknown variable on the left-hand side of the equation.	$r^2 = 16^2 - 9^2$
3. Evaluate the expression for r^2.	$= 256 - 81$ $= 175$
4. Calculate the answer by taking the square root of both sides of the equation and rounding the result to 1 decimal place.	$\sqrt{r^2} = \sqrt{175}$ $r = \sqrt{175}$ ≈ 13.2 m
5. State the answer.	The length of PQ is 13.2 m.

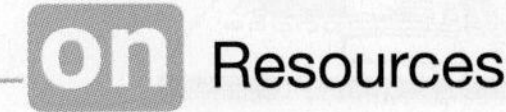

Resources

Interactivities Finding the hypotenuse (int-3844)
Finding the shorter side (int-3845)

12.2.3 Using Pythagoras' theorem to solve real-world problems

Pythagoras' theorem can be used to solve more practical problems. In these cases, it is necessary to draw a diagram to help you decide the appropriate method for finding a solution.

tlvd-3647

WORKED EXAMPLE 4 Using Pythagoras' theorem to solve practical problems

The fire brigade attends a blaze in a tall building. They need to rescue a person from the 6th floor of the building, which is 30 metres above ground level. Their ladder is 32 metres long and must be placed at least 10 metres from the foot of the building. Determine if the ladder can be used to reach the people needing rescue.

THINK	WRITE
1. Define a variable to represent the unknown distance in this problem.	For the fire brigade's ladder to be long enough, the straight-line distance from a point on the ground 10 m from the base of the building to the window 30 m above the ground must be less than 32 m. Let c represent the unknown distance.
2. Draw a diagram showing all important information.	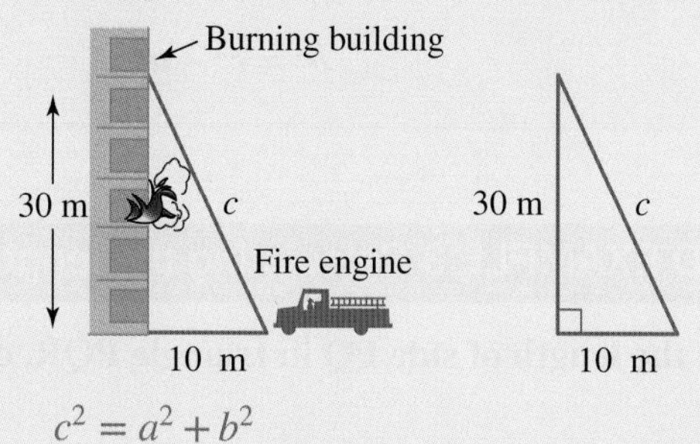
3. The triangle is right-angled, so Pythagoras' theorem can be used. Write the Pythagoras' theorem rule for finding the hypotenuse.	$c^2 = a^2 + b^2$
4. Substitute the lengths of the known sides.	$c^2 = 10^2 + 30^2$
5. Evaluate the expression on the right-hand side of the equation.	$= 100 + 900$ $= 1000$
6. Calculate the answer by taking the square root of both sides of the equation.	$\sqrt{c^2} = \sqrt{1000}$ $c = \sqrt{1000}$ $\approx 31.62\text{ m}$
7. Answer the question.	The ladder will be long enough to make the rescue, since it is 32 m long.

12.2 Exercise

1. WE1 Name the hypotenuse in each of the following triangles.

a. 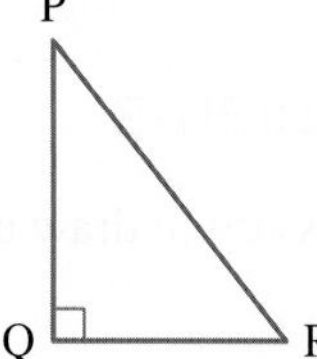

b. 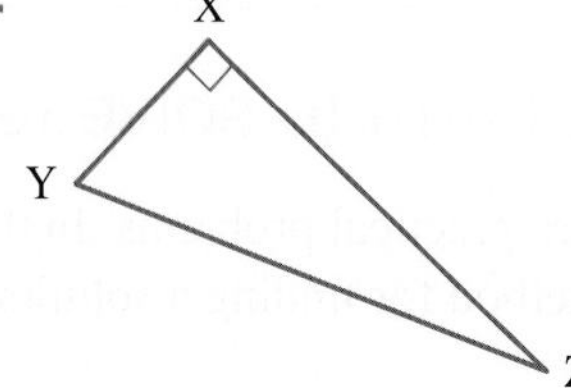

c. 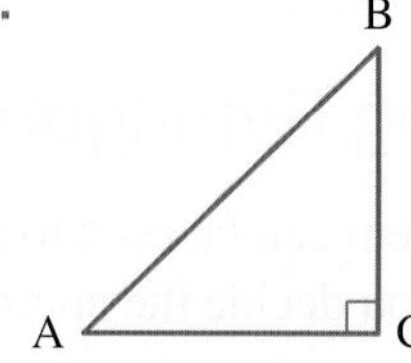

2. WE2 Determine the length of the hypotenuse in each of the following triangles.

a.

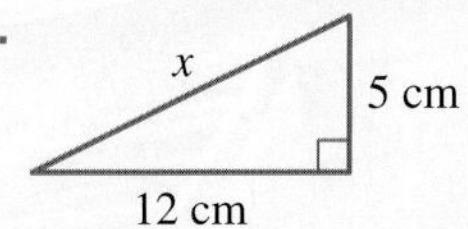

b.

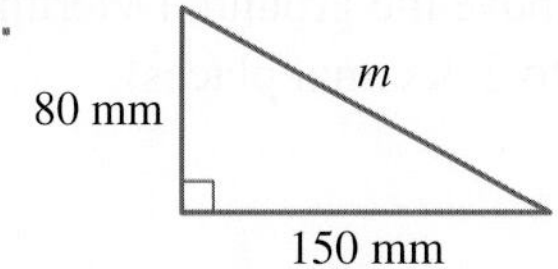

c.

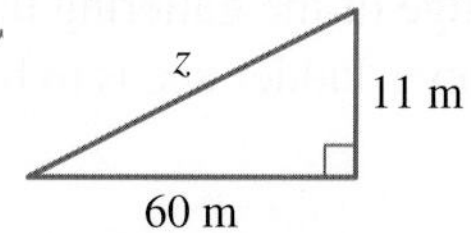

3. In each of the following, determine the length of the hypotenuse, correct to 2 decimal places.

a.

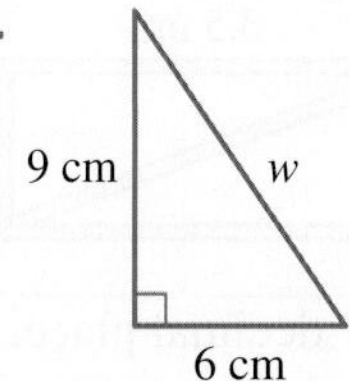

b.

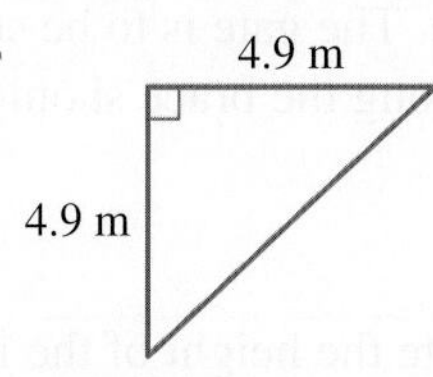

c.

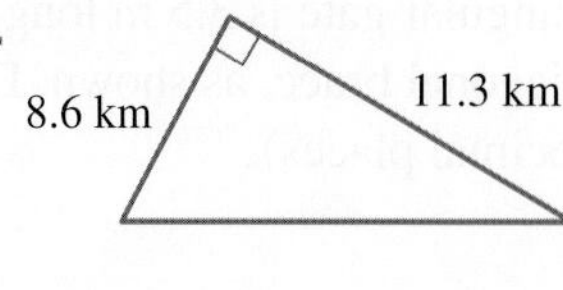

4. WE3 Calculate the length of the unknown shorter side in each right-angled triangle, correct to 1 decimal place.

a.

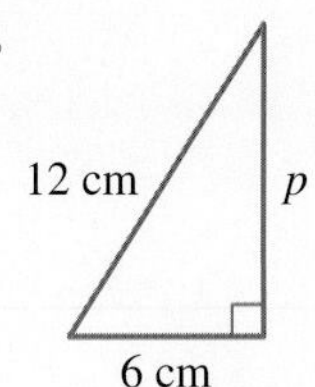

b.

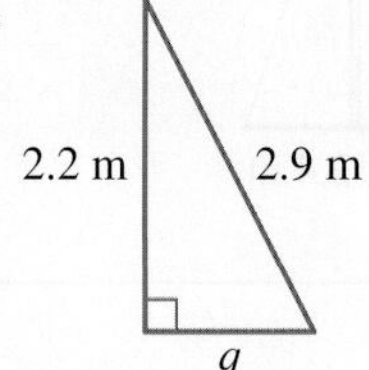

c.

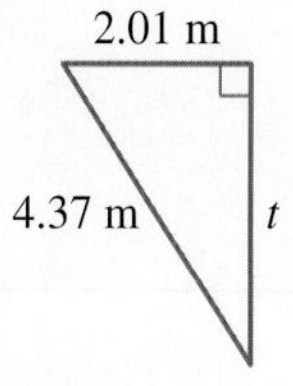

5. In each of the following right-angled triangles, determine the length of the side marked with a variable, correct to 1 decimal place.

a.

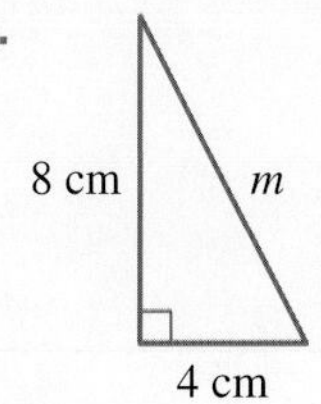

b.

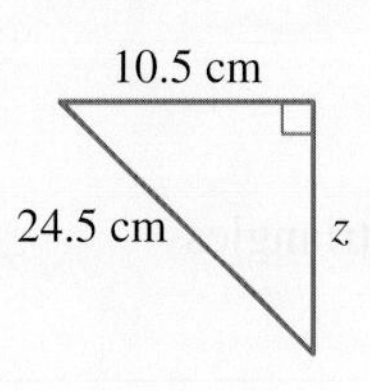

c.

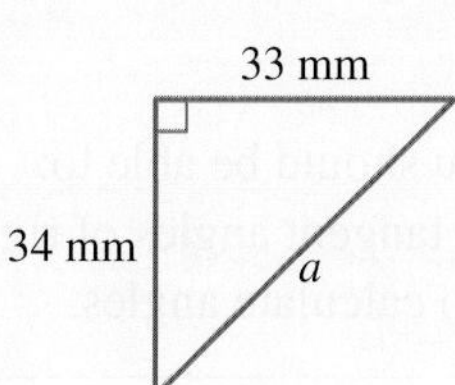

d.

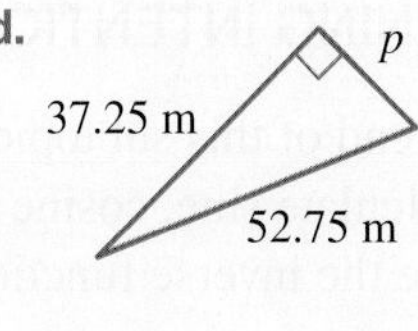

6. MC The hypotenuse in triangle WXY shown is:

A. WX
B. XY
C. YZ
D. ZW
E. ZX

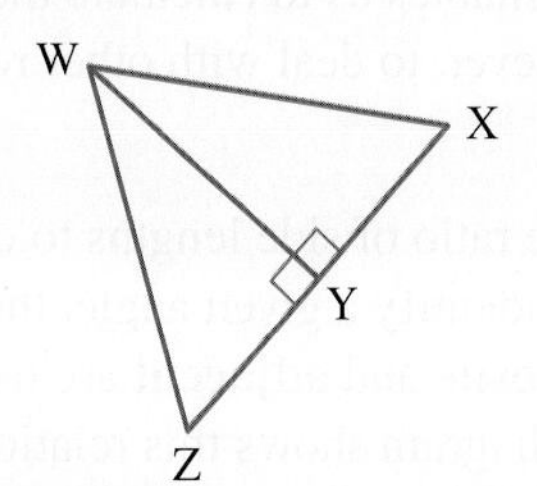

7. WE4 A television antenna is 12 m high. To support it, wires are attached from the top to the ground 5 m from the bottom of the antenna. Determine the length of the wires.

8. Charley needs to clean the guttering on their roof. They place a ladder 1.2 m from the edge of the guttering that is 3 m above the ground. Determine how long Charley's ladder needs to be (correct to 2 decimal places).

9. A rectangular gate is 3.5 m long and 1.3 m wide. The gate is to be strengthened by a diagonal brace, as shown. Determine how long the brace should be (correct to 2 decimal places).

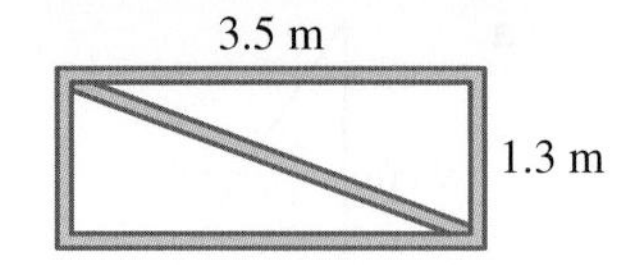

10. Use the measurements in the diagram to calculate the height of the flagpole, correct to 1 decimal place.

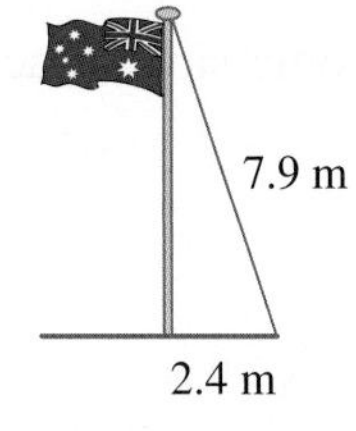

12.3 Trigonometric ratios — sine, cosine and tangent

LEARNING INTENTION

At the end of this subtopic you should be able to:

- calculate sine, cosine and tangent angles of right-angled triangles
- use the inverse function to calculate angles.

Pythagoras' theorem enables us to calculate the length of one side of a right-angled triangle given the lengths of the other two. However, to deal with other relationships in right-angled triangles, we need to turn to **trigonometry**.

Trigonometry uses the ratio of side lengths to calculate the lengths of sides and the size of angles. To identify a given angle, the shorter sides need individual names. The sides **opposite** and **adjacent** are used to show the relationship to the given angle. The diagram shows this relationship between the sides and the angle, θ. The opposite side is always opposite the angle and the adjacent side is always next to the angle (not the hypotenuse).

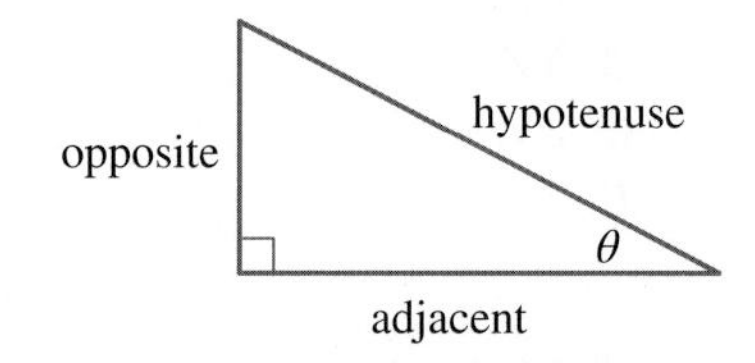

12.3.1 The sine ratio

For all right-angled triangles that contain the same angles, the ratio of the length of the opposite side to the length of the hypotenuse will remain the same, regardless of the size of the triangle.

Sine ratio

The formula for the sine ratio (abbreviated 'sin') is:

$$\sin\theta = \frac{\text{opposite side}}{\text{hypotenuse}}$$

WORKED EXAMPLE 5 The sine ratio

Write an expression for the sine ratio of θ for the triangle shown.

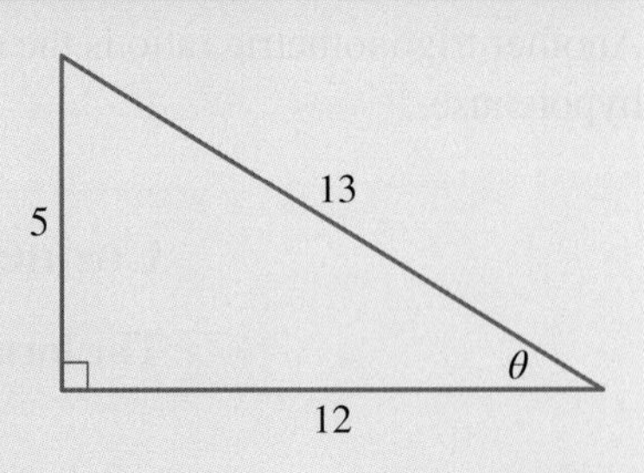

THINK	 WRITE
1. Label the sides of the triangle as opposite, adjacent and hypotenuse according to the position of the angle θ.	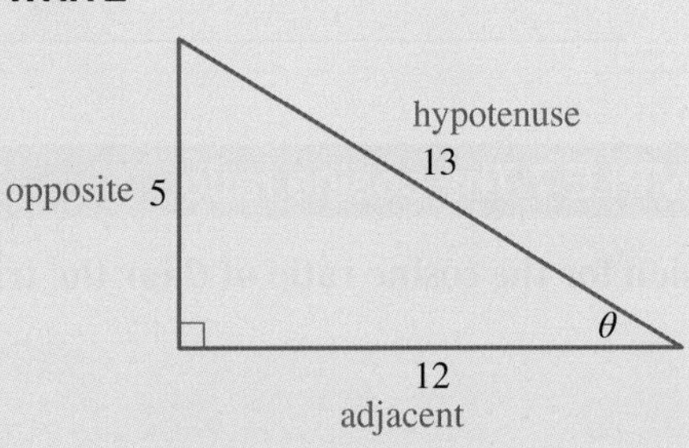
2. Write the general form of the sine ratio.	$\sin\theta = \dfrac{\text{opposite side}}{\text{hypotenuse}}$
3. Substitute the lengths of the opposite side and the hypotenuse.	$\sin\theta = \dfrac{5}{13}$

For all calculations in trigonometry you will need to make sure that your calculator is in degrees mode. Check the set-up on your calculator to ensure that this is the case.

The value of the sine ratio for any angle can be found using the sin function on the calculator.

WORKED EXAMPLE 6 Calculating the value of sine angles

Using your calculator, determine the value of the following, correct to 3 decimal places:

a. $\sin(57°)$ **b.** $9\sin(45°)$ **c.** $\dfrac{18}{\sin(44°)}$ **d.** $9.6\sin(26°12')$

THINK	DISPLAY/WRITE
a. On a scientific calculator, press (sin) and enter 57, then press (=).	**a.** $\sin(57°) = 0.839$
b. Enter 9, press (×) and (sin), enter 45, then press (=).	**b.** $9\sin(45°) = 6.364$
c. Enter 18, press (÷) and (sin), enter 44, then press (=).	**c.** $\dfrac{18}{\sin(44°)} = 25.912$
d. Enter 9.6, press (×) and (sin), enter 26°12′, then press (=).	**d.** $9.6\sin(26°12') = 4.238$

Note: **Check the sequence of button presses required by your calculator.**

Note: The symbol ′ represents minutes, which is a measure of part of a degree.

12.3.2 The cosine ratio

Another trigonometric ratio is the cosine ratio. This ratio compares the length of the adjacent side and the hypotenuse.

Cosine ratio

The formula for the cosine ratio (abbreviated 'cos') is:

$$\cos\theta = \frac{\text{adjacent side}}{\text{hypotenuse}}$$

WORKED EXAMPLE 7 The cosine ratio

Write an expression for the cosine ratio of θ for the triangle shown.

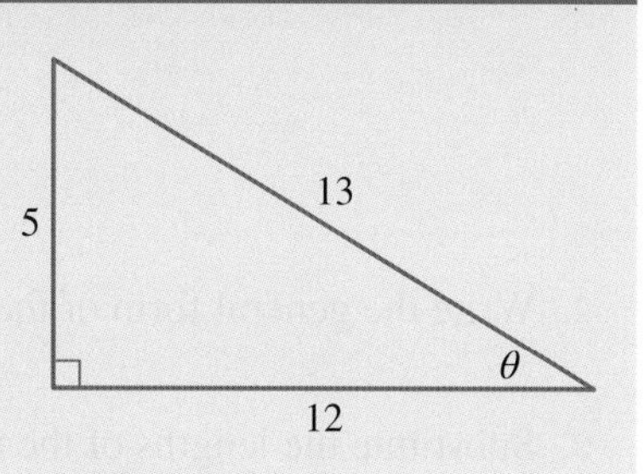

THINK	WRITE
1. Label the sides of the triangle as opposite, adjacent and hypotenuse according to the position of the angle θ.	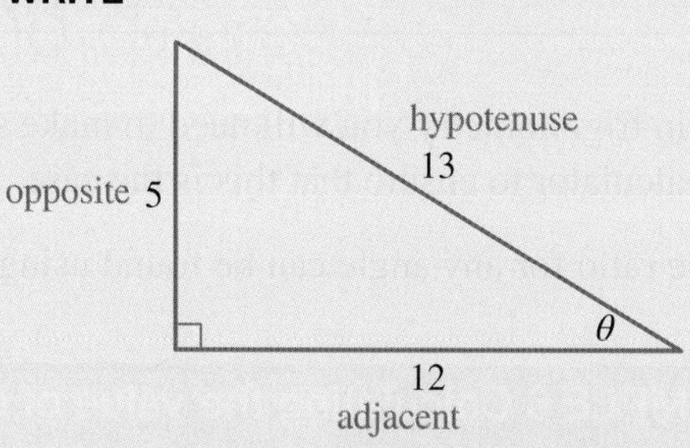
2. Write the general form of the cosine ratio.	$\cos\theta = \frac{\text{adjacent side}}{\text{hypotenuse}}$
3. Substitute the lengths of the adjacent side and the hypotenuse.	$\cos\theta = \frac{12}{13}$

The value of the cosine ratio for any angle can be found using the cos function on a calculator.

WORKED EXAMPLE 8 Calculating the value of cosine angles

Using your calculator, determine the value of each of the following, correct to 3 decimal places.

a. $\cos(27°)$
b. $6\cos(55°)$
c. $\frac{21.3}{\cos(74°)}$
d. $\frac{4.5}{\cos(82°46')}$

THINK	DISPLAY/WRITE
a. On a scientific calculator, press (cos) and enter 27, then press (=).	a. $\cos(27°) = 0.891$

b. Enter 6, press (×) and (cos), enter 55, then press (=). **b.** $6\cos(55°) = 3.441$

c. Enter 21.3, press (÷) and (cos), enter 74, then press (=). **c.** $\dfrac{21.3}{\cos(74°)} = 77.275$

d. Enter 4.5, press (÷) and (cos), enter 82°46′, then press (=). **d.** $\dfrac{4.5}{\cos(82°46')} = 35.740$

Note: **Check the sequence of button presses required by your calculator.**

12.3.3 The tangent ratio

The ratio of the opposite side to the adjacent side is called the tangent ratio (abbreviated 'tan'). This ratio is fixed for any particular angle.

The tangent ratio

The formula for the tangent ratio is:

$$\tan\theta = \frac{\text{opposite side}}{\text{adjacent side}}$$

WORKED EXAMPLE 9 The tangent ratio

Write an expression for the tangent ratio of θ for the triangle shown.

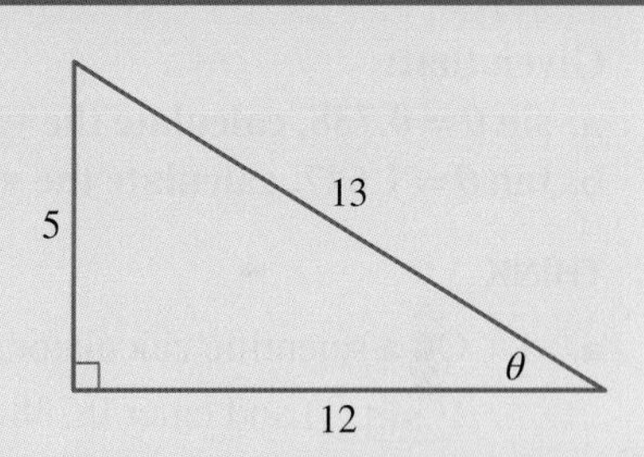

THINK	WRITE
1. Label the sides of the triangle as opposite, adjacent and hypotenuse according to the position of the angle θ.	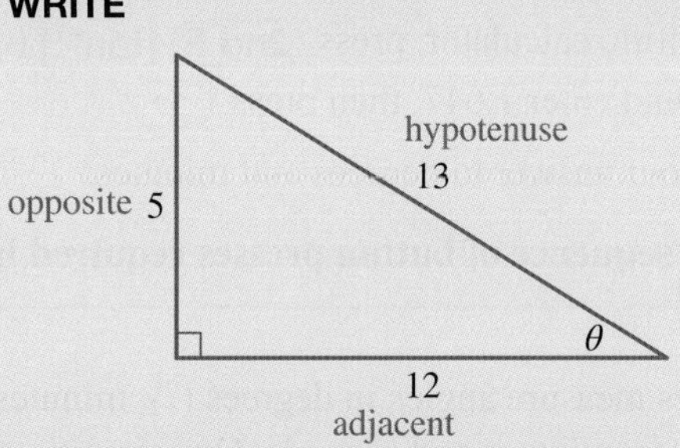
2. Write the general form of the tangent ratio.	$\tan\theta = \dfrac{\text{opposite side}}{\text{adjacent side}}$
3. Substitute the lengths of the opposite and adjacent sides.	$\tan\theta = \dfrac{5}{12}$

The value of the tangent ratio for any angle can be found using the tan function on a calculator.

WORKED EXAMPLE 10 Calculating the value of tangent angles

Using your calculator, determine the value of the following, correct to 3 decimal places.

a. $\tan(60°)$ b. $15\tan(75°)$ c. $\dfrac{8}{\tan(69°)}$ d. $\tan(49°32')$

THINK	DISPLAY/WRITE
a. On a scientific calculator, press (tan) and enter 60, then press (=).	a. $\tan(60°) = 1.732$
b. Enter 15, press (×) and (tan), enter 75, then press (=).	b. $15\tan(75°) = 55.981$
c. Enter 8, press (÷) and (tan), enter 69, then press (=).	c. $\dfrac{8}{\tan(69°)} = 3.071$
d. Press (tan), enter 49°32′, then press (=).	d. $\tan(49°32') = 1.172$

12.3.4 Inverse trigonometric functions

If we are given the sine, cosine or tangent of an angle, we can calculate the size of that angle using a calculator. We do this using the inverse trigonometric functions. On most calculators these are the second function of the sin, cos and tan functions and are denoted $\sin^{-1}$, $\cos^{-1}$ and $\tan^{-1}$.

WORKED EXAMPLE 11 Calculating angles using inverse trigonometric functions

Given that:
a. **$\sin\theta = 0.738$, calculate the value of θ, correct to the nearest degree**
b. **$\tan\theta = 1.647$, calculate the value of θ correct to the nearest minute.**

THINK	DISPLAY/WRITE
a. 1. On a scientific calculator, press (2nd F)[$\sin^{-1}$] (or press SHIFT (sin)) and enter 0.738, then press (=).	a. $\sin^{-1}(0.738) = 47.56...$
2. Round your answer to the nearest degree.	$\theta = 48°$
b. 1. On a scientific calculator, press (2nd F) [$\tan^{-1}$] (or press SHIFT (tan)) and enter 1.647, then press (=).	b. $\tan^{-1}(1.647) = 58.73...$
2. Convert your answer to degrees and minutes.	$\theta = 58°44'$

Note: **Check the sequence of button presses required by your calculator.**

Problems sometimes measure angles in degrees (°), minutes (′) and seconds (″), or require answers to be written in the form of degrees, minutes and seconds. On scientific calculators, you will use the DMS (Degrees, Minutes, Seconds) function or the (° ′ ″) function. To round an angle to the nearest minute, you need to look at the number of seconds. If there are 30 or more seconds, you will need to round the number of minutes up by 1.

on Resources

Interactivity Trigonometric ratios (int-2577)

12.3 Exercise

1. Label the sides of the following right-angled triangles using the words hypotenuse, adjacent and opposite.

a. 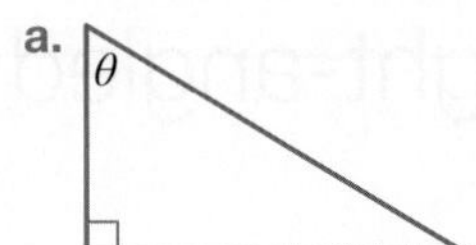

b. 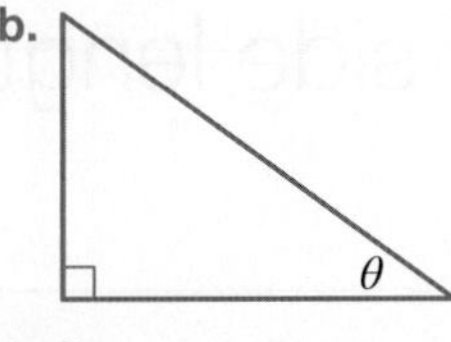

c. 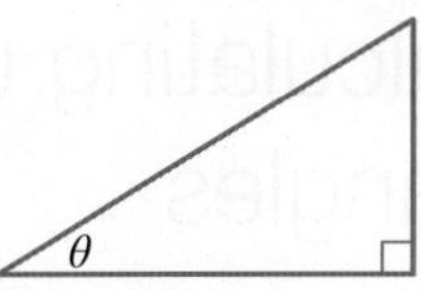

d. 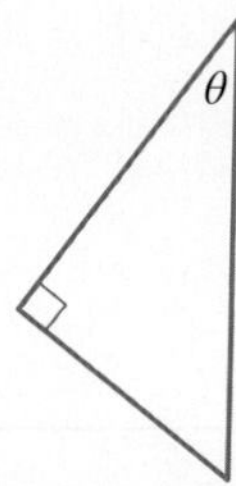

e. 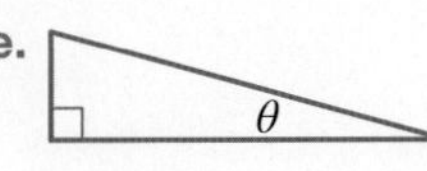

f. 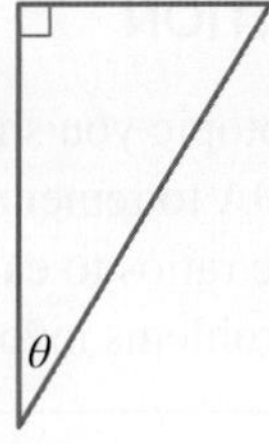

2. Write an expression for each of the following ratios for the angle θ shown.

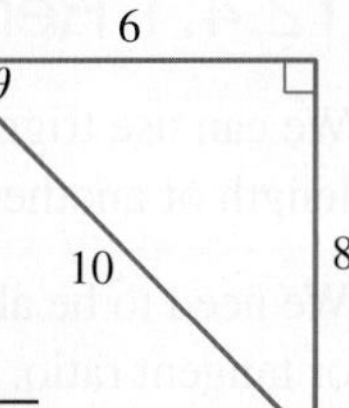

a.  WE5 sine
b. WE7 cosine
c. WE9 tangent

3. WE6 Using your calculator, determine the value of the following, correct to 3 decimal places:

a. $\sin(37°)$
b. $9.3\sin(13°)$
c. $\dfrac{14.5}{\sin(72°)}$
d. $\dfrac{48}{\sin(67°40')}$

4. WE8 Using your calculator, determine the value of the following, correct to 3 decimal places:

a. $\cos(45°)$
b. $0.25\cos(9°)$
c. $\dfrac{6}{\cos(24°)}$
d. $5.9\cos(2°3')$

5. WE10 Using your calculator, determine the value of the following, correct to 3 decimal places:

a. $\tan(57°)$
b. $9\tan(63°)$
c. $\dfrac{8.6}{\tan(12°)}$
d. $\tan(33°19')$

6. Calculate the value of each of the following, correct to 3 decimal places, if necessary.

a. $\sin(30°)$
b. $\cos(15°)$
c. $\tan(45°)$
d. $48\tan(85°)$
e. $128\cos(60°)$
f. $9.35\sin(8°)$

7. Calculate the value of each of the following, correct to 3 decimal places, if necessary.

a. $\dfrac{4.5}{\cos(32°)}$
b. $\dfrac{0.5}{\tan(20°)}$
c. $\dfrac{15}{\sin(72°)}$

8. Calculate the value of each of the following, correct to 2 decimal places.

a. $\sin(24°38')$
b. $\tan(57°21')$
c. $\cos(84°40')$
d. $9\cos(55°30')$
e. $4.9\sin(35°50')$
f. $2.39\tan(8°59')$

9. Calculate the value of each of the following, correct to 2 decimal places.

a. $\dfrac{19}{\tan(67°45')}$ b. $\dfrac{49.6}{\cos(47°25')}$ c. $\dfrac{0.84}{\sin(75°5')}$

10. WE11a Calculate the value of θ, correct to the nearest degree, given that $\sin\theta = 0.167$.

11. WE11b Given that $\tan\theta = 2.26$, calculate the value of θ correct to the nearest minute.

12. Determine θ, correct to the nearest degree, given that:

a. $\sin\theta = 0.698$ b. $\cos\theta = 0.173$ c. $\tan\theta = 1.517$.

12.4 Calculating unknown side lengths of right-angled triangles

LEARNING INTENTION

At the end of this subtopic you should be able to:
- use SOHCAHTOA to remember trigonometric ratios
- use trigonometric ratios to calculate unknown side lengths
- solve practical problems involving right-angled triangles.

12.4.1 Remembering trigonometric ratios

We can use trigonometric ratios to calculate the length of one side of a right-angled triangle if we know the length of another side and an angle.

We need to be able to look at a problem and then decide if the solution can be determined using the sine, cosine or tangent ratio. To do this, we need to examine the three formulas.

$$\sin\theta = \frac{\text{opposite side}}{\text{hypotenuse}} \qquad \cos\theta = \frac{\text{adjacent side}}{\text{hypotenuse}} \qquad \tan\theta = \frac{\text{opposite side}}{\text{adjacent side}}$$

Remembering trigonometric ratios

To remember the trigonometric ratios more easily, we can use this acronym:

SOHCAHTOA

The initials of the acronym represent the three trigonometric formulas.

$$\begin{pmatrix}S\\O\\H\end{pmatrix} \sin\theta = \frac{\text{opposite}}{\text{hypotenuse}} \longrightarrow \begin{pmatrix}C\\A\\H\end{pmatrix} \cos\theta = \frac{\text{adjacent}}{\text{hypotenuse}} \longrightarrow \begin{pmatrix}T\\O\\A\end{pmatrix} \tan\theta = \frac{\text{opposite}}{\text{adjacent}}$$

Consider the triangle shown on the next page.

In this triangle, we are asked to determine the length of the opposite side and have been given the length of the adjacent side and the value of the angle.

We know that the tangent ratio uses the opposite and adjacent sides:

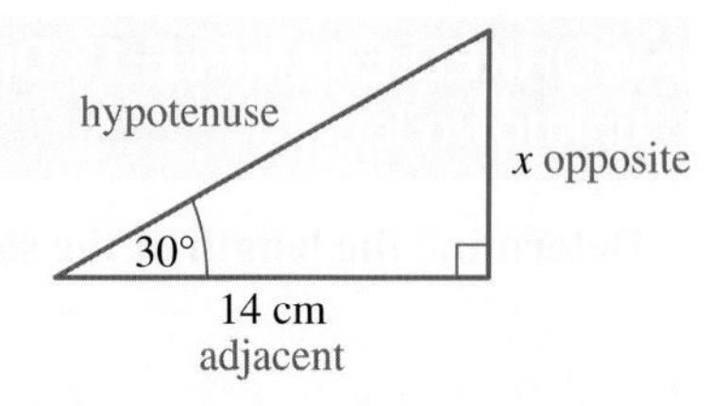

$\tan\theta = \dfrac{\text{opposite}}{\text{adjacent}}$. In this example, $\tan(30°) = \dfrac{x}{14}$.

We can use this equation to determine the value of x.

$$\tan\theta = \frac{\text{opposite}}{\text{adjacent}}$$

$$\tan(30°) = \frac{x}{14}$$

$$14 \times \tan(30°) = \frac{\cancel{14}x}{\cancel{14}} \quad \text{Multiply both sides by 14.}$$

$$x = 14\tan(30°) \quad \text{Enter } 14\tan(30°) \text{ into your calculator.}$$

$$= 8.083 \text{ cm}$$

tlvd-3649

WORKED EXAMPLE 12 Using the tangent ratio to determine a length

Use the tangent ratio to determine the value of h in the triangle shown, correct to 2 decimal places.

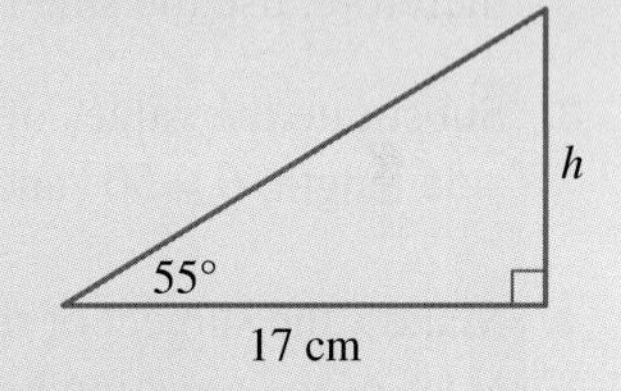

THINK	WRITE
1. Label the sides of the triangle as opposite, adjacent and hypotenuse.	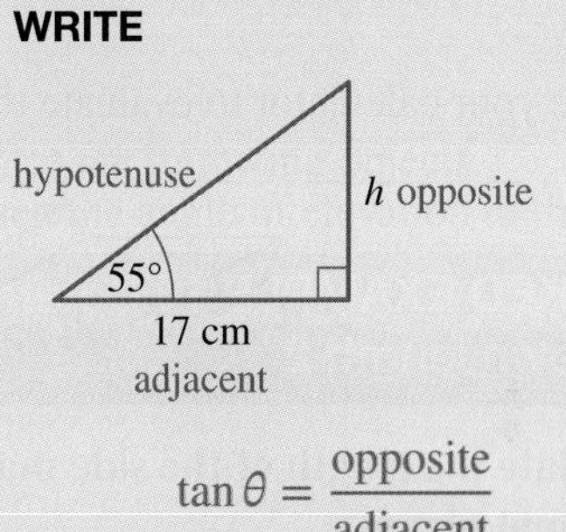
2. Write the tangent ratio.	$\tan\theta = \dfrac{\text{opposite}}{\text{adjacent}}$
3. Substitute the values of the known angle and the known side length: $\theta = 55°$ and adjacent = 17.	$\tan(55°) = \dfrac{h}{17}$
4. Make h the subject of the equation by multiplying both sides of the equation by 17.	$17 \times \tan(55°) = \dfrac{\cancel{17}h}{\cancel{17}}$ $h = 17\tan(55°)$
5. Use your calculator to evaluate the expression for h.	$= 24.28$ cm

12.4.2 Determining trigonometric ratios

When determining which trigonometric ratio to use, it is important to label the sides of the right-angled triangle to see what information is given. Once you have labelled the diagram, you can use the given information to determine the missing side.

tlvd-3650

WORKED EXAMPLE 13 Determining which trigonometric ratio to use to calculate an unknown length

Determine the length of the side marked x, correct to 2 decimal places.

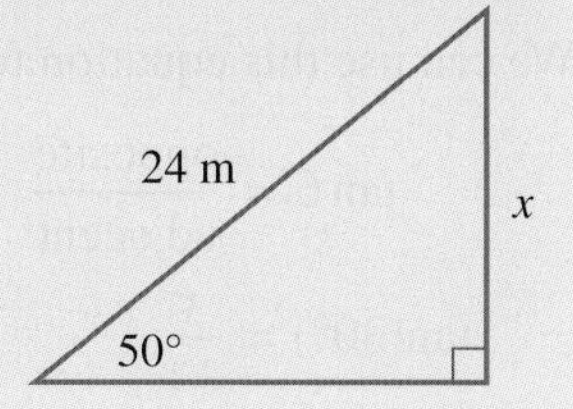

THINK | **WRITE**

1. Label the sides of the triangle as opposite, adjacent and hypotenuse.

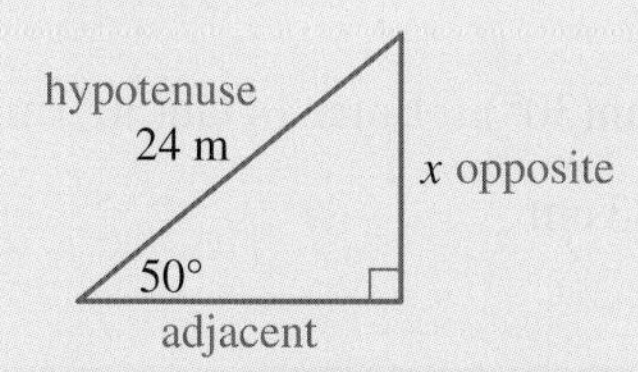

2. x is the opposite side and 24 m is the hypotenuse; therefore, use the sine ratio.

$$\sin\theta = \frac{\text{opposite}}{\text{hypotenuse}}$$

3. Substitute the values of the known angle and the known side length: $\theta = 50°$ and hypotenuse $= 24$.

$$\sin(50°) = \frac{x}{24}$$

4. Make x the subject of the equation by multiplying both sides of the equation by 24.

$$24 \times \sin(50°) = \frac{\cancel{24}x}{\cancel{24}}$$

$$x = 24\sin(50°)$$

5. Use your calculator to evaluate the expression for x.

$$= 18.39 \text{ m}$$

WORKED EXAMPLE 14 Determining which trigonometric ratio to use to calculate an unknown length 2

Calculate the length of the side marked z in the triangle shown, correct to 2 decimal places.

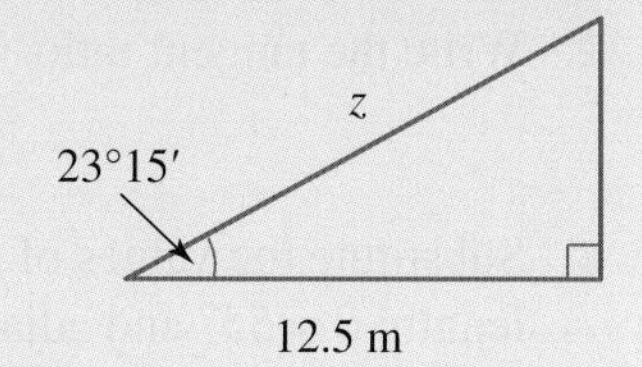

THINK | **WRITE**

1. Label the sides of the triangle as opposite, adjacent and hypotenuse.

hypotenuse
z
23°15′
opposite
12.5 m
adjacent

2. We are finding the hypotenuse and have been given the adjacent side, so choose the cosine ratio to solve this problem. Write the cosine ratio.

$$\cos\theta = \frac{\text{adjacent}}{\text{hypotenuse}}$$

3. Substitute the values of the known angle and the known side length: $\theta = 23°15'$ and adjacent $= 12.5$.

$$\cos(23°15') = \frac{12.5}{z}$$

4. Multiply both sides of the equation by z to remove z from the denominator.	$z \times \cos(23°15') = \frac{\cancel{z} \times 12.5}{\cancel{z}}$ $z\cos(23°15') = 12.5$
5. Divide both sides of the equation by $\cos(23°15')$ to make z the subject.	$\frac{z\cancel{\cos(23°15')}}{\cancel{\cos(23°15')}} = \frac{12.5}{\cos(23°15')}$ $z = \frac{12.5}{\cos(23°15')}$
6. Use your calculator to evaluate the expression for z.	$z = 13.60$ m

12.4.3 Solving practical problems

Trigonometry can be used to solve many practical problems. In these cases, it is necessary to draw a diagram to represent the problem and then identify which trigonometric ratio to use to solve the problem.

tlvd-3651

WORKED EXAMPLE 15 Solving practical problems with trigonometric ratios

A flying fox is used in an army training camp. The flying fox is supported by a cable that runs from the top of a cliff face to a point on the ground 100 m from the base of the cliff. The cable makes a 15° angle with the horizontal. Determine the length of the cable used to support the flying fox, correct to 1 decimal place.

THINK	WRITE
1. Draw a diagram and show all important information.	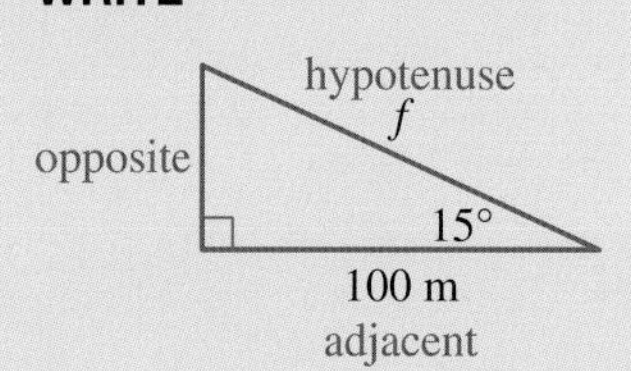
2. Label the sides of the triangle as opposite, adjacent and hypotenuse.	
3. Choose the cosine ratio to solve this problem, because we are finding the hypotenuse and have been given the adjacent side.	
4. Write the cosine ratio.	$\cos\theta = \frac{\text{adjacent}}{\text{hypotenuse}}$
5. Substitute the values of the known angle and the known side length: $\theta = 15°$ and adjacent $= 100$.	$\cos(15°) = \frac{100}{f}$
6. Make f the subject of the equation by multiplying both sides by f and then dividing both sides by $\cos(15°)$.	$f \times \cos(15)° = \frac{100\cancel{f}}{\cancel{f}}$ $f\cos(15)° = 100$ $\frac{f\cancel{\cos(15)°}}{\cancel{\cos(15)°}} = \frac{100}{\cos(15)°}$ $f = \frac{100}{\cos(15)°}$
7. Use your calculator to evaluate the expression for f.	$f = 103.5$ m
8. Answer the question.	The cable is approximately 103.5 m long.

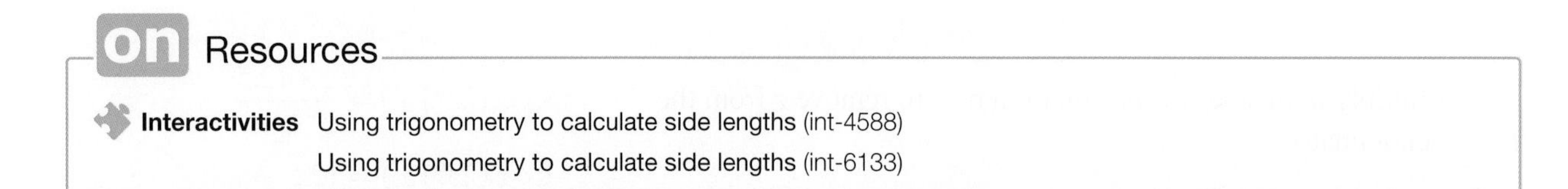

12.4 Exercise

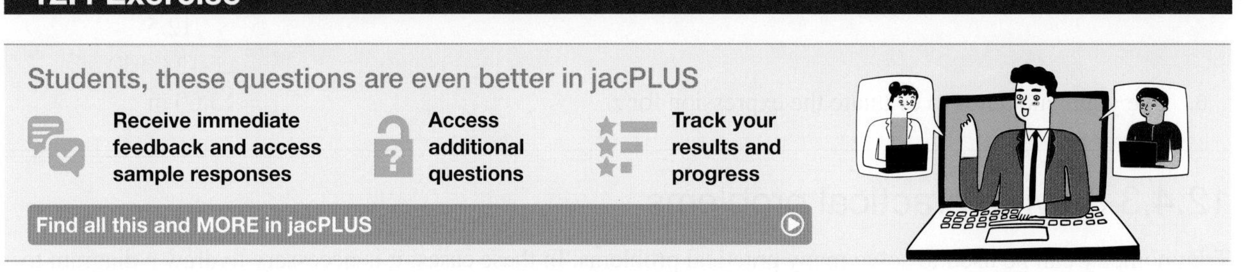

1. Label the sides of each of the following triangles with respect to the angle marked with the variable.

a. 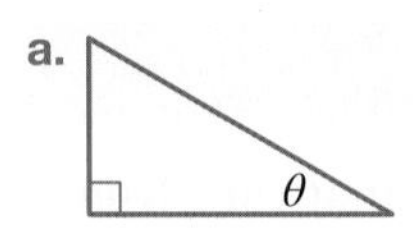

b. 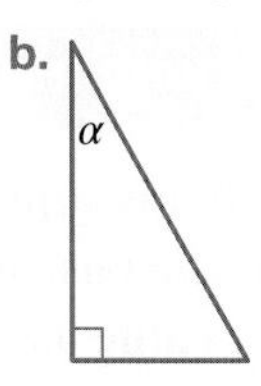

c. 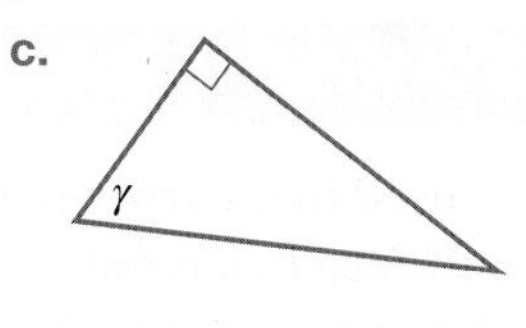

2. **WE12** Use the tangent ratio to determine the length of the side marked x (correct to 1 decimal place).

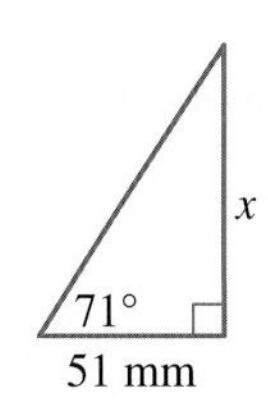

3. Use the sine ratio to determine the length of the side marked a (correct to 2 decimal places).

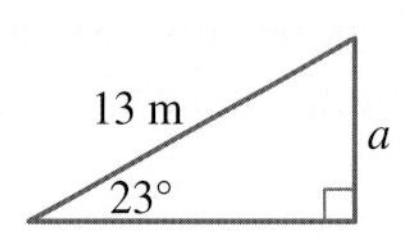

4. Use the cosine ratio to determine the length of the side marked d (correct to the nearest whole number).

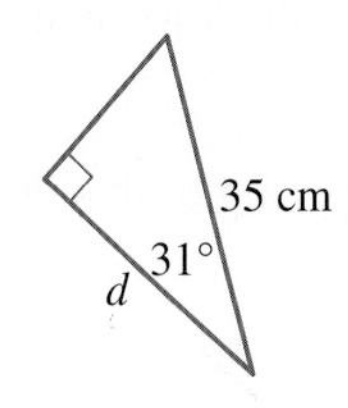

5. **WE13** Determine the length of the side marked with the variable, correct to 1 decimal place.

a. 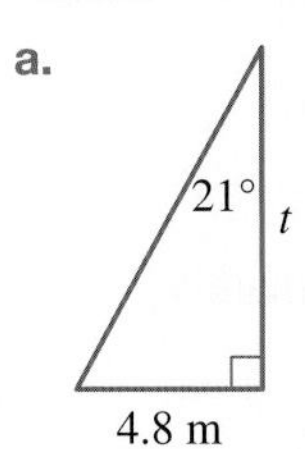

b. 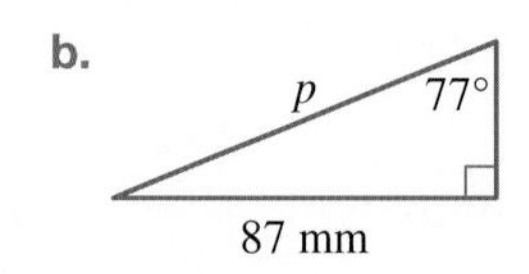

c. 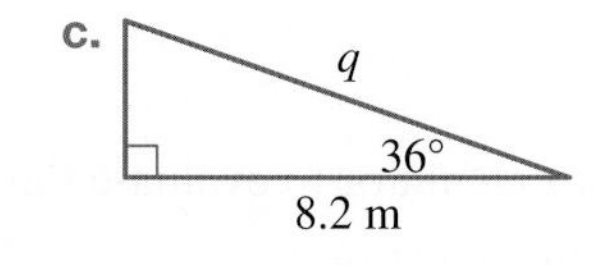

6. **WE14** Calculate the length of the side marked with the variable in each of the following (correct to 1 decimal place).

a.

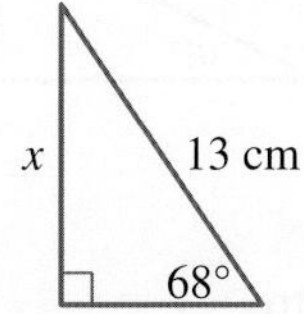

b.

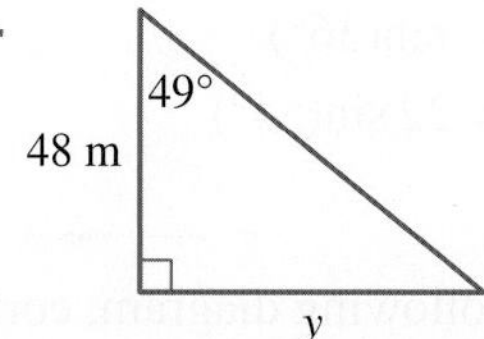

c.

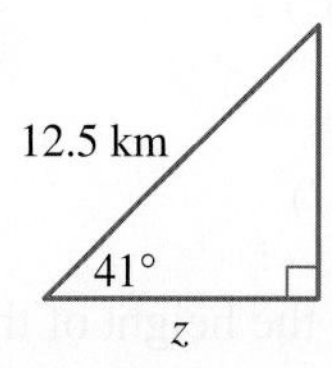

7. Calculate the length of the side marked with the variable in each of the following (correct to 1 decimal place).

a.

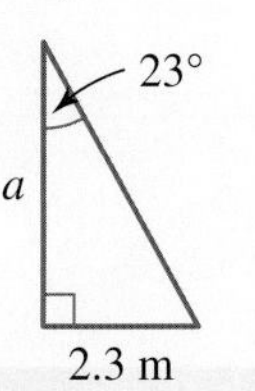

b.

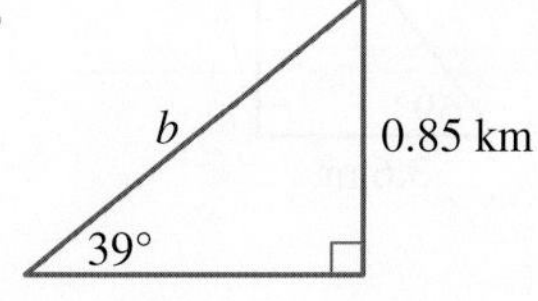

c.

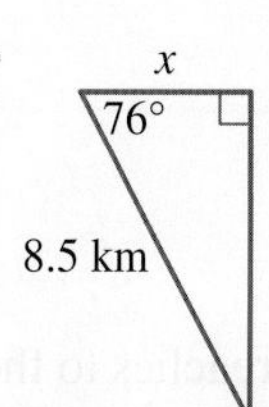

d.

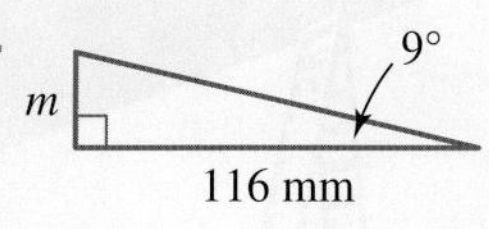

e.

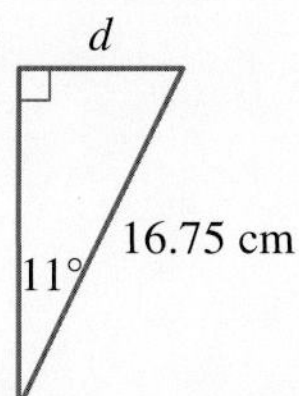

f.

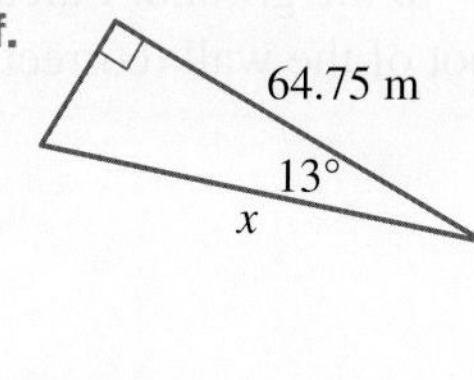

8. Calculate the length of the side marked with the variable in each of the following (correct to 1 decimal place).

a.

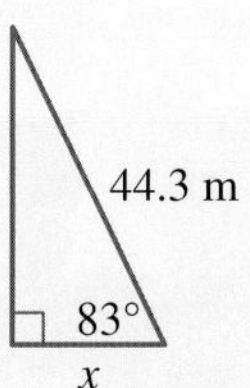

b.

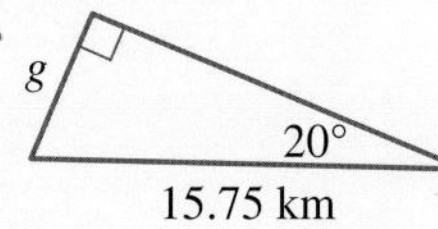

c.

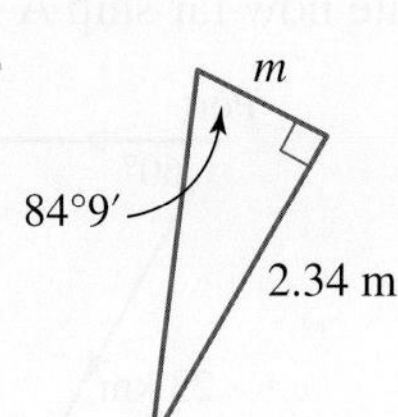

d.

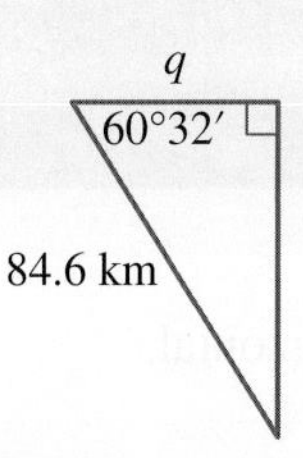

e.

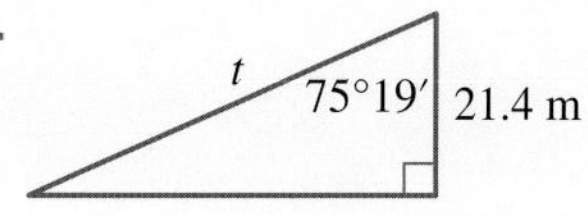

f.

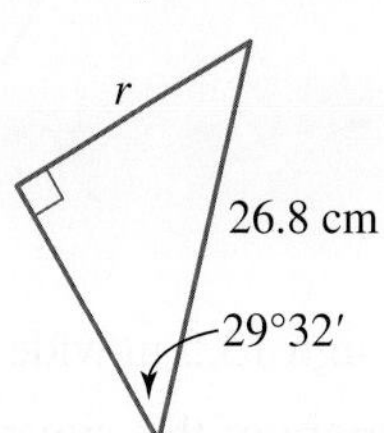

9. **MC** Look at the diagram and state which of the following is correct.

A. $x = 9.2\sin(69°)$ **B.** $x = \dfrac{9.2}{\sin(69°)}$ **C.** $x = 9.2\cos(69°)$

D. $x = \dfrac{9.2}{\cos(69°)}$ **E.** None of the above

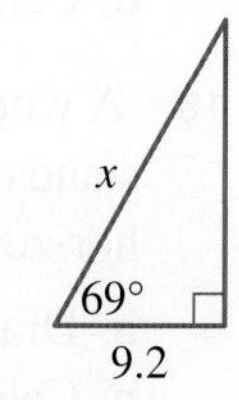

10. **MC** Study the triangle and state which of the following is correct.

A. $\tan\phi = \dfrac{8}{15}$ **B.** $\tan\phi = \dfrac{15}{8}$ **C.** $\sin\phi = \dfrac{15}{17}$

D. $\cos\phi = \dfrac{17}{15}$ **E.** $\sin\phi = \dfrac{17}{8}$

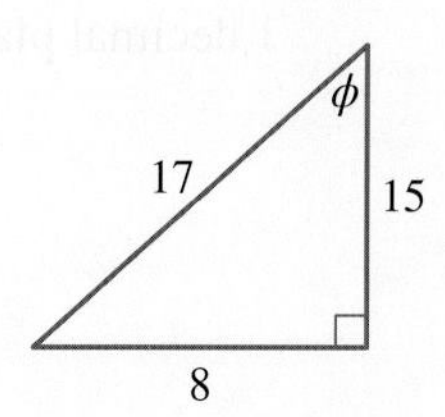

11. **MC** Study the diagram and state which of the following is correct.

A. $w = 22\cos(36°)$

B. $w = \dfrac{22}{\sin(36°)}$

C. $w = 22\cos 54°$

D. $w = 22\sin(54°)$

E. $w = 22\tan(54°)$

12. **WE15** Calculate the height of the tree in the following diagram, correct to the nearest metre.

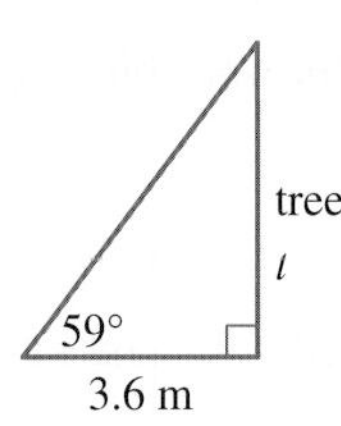

13. A 10-m ladder just reaches to the top of a wall when it is leaning at an angle of 65° to the ground. Calculate the distance of the base of the ladder from the foot of the wall (correct to 1 decimal place).

14. The diagram shows the paths of two ships, A and B, after they have left the port. If ship B sends a distress signal, calculate how far ship A must sail to give assistance (to the nearest kilometre).

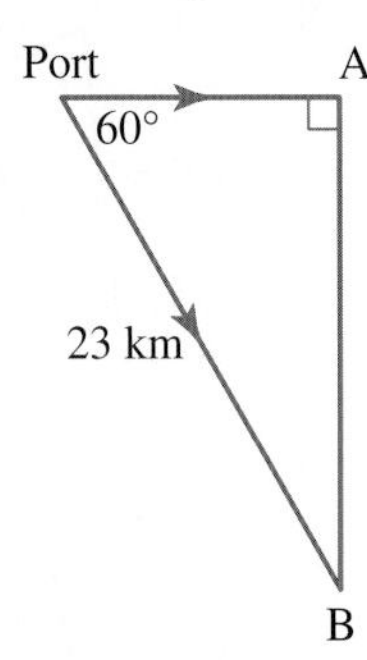

15. A rectangular sign 13.5 m wide has a diagonal brace that makes a 24° angle with the horizontal.

a. Draw a diagram of this situation.

b. Calculate the height of the sign, correct to the nearest metre.

16. A wooden gate has a diagonal brace built in for support. The gate stands 1.4 m high and the diagonal makes a 60° angle with the horizontal.

a. Draw a diagram of the gate.

b. Calculate the length that the diagonal brace needs to be (correct to 1 decimal place).

17. The wire support for a flagpole makes a 70° angle with the ground. If the support is 3.3 m from the base of the flagpole, calculate the length of the wire support (correct to 2 decimal places).

18. A ship drops anchor vertically with an anchor line 60 m long. After one hour the anchor line makes a 15° angle with the vertical.
    a. Draw a diagram of this situation.
    b. Calculate the depth of water, correct to the nearest metre.
    c. Calculate the distance that the ship has drifted, correct to 1 decimal place.

12.5 Calculating unknown angles in right-angled triangles

LEARNING INTENTION

At the end of this subtopic you should be able to:
- calculate unknown angles using inverse trigonometric ratios
- solve practical problems involving right-angled triangles.

12.5.1 Introduction

We can use trigonometry to calculate the sizes of angles when we have been given two side lengths. We can use the inverse trigonometric ratios $\sin^{-1}$, $\cos^{-1}$ and $\tan^{-1}$ to calculate the size of an unknown angle in a right-angled triangle.

Inverse trigonometric ratios

If $\sin\theta = a$, then $\sin^{-1}(a) = \theta$.

If $\cos\theta = a$, then $\cos^{-1}(a) = \theta$.

If $\tan\theta = a$, then $\tan^{-1}(a) = \theta$.

Consider the right-angled triangle shown.

We want to calculate the size of the angle marked θ.

Using the formula $\sin\theta = \dfrac{\text{opposite}}{\text{hypotenuse}}$, we know that in this triangle:

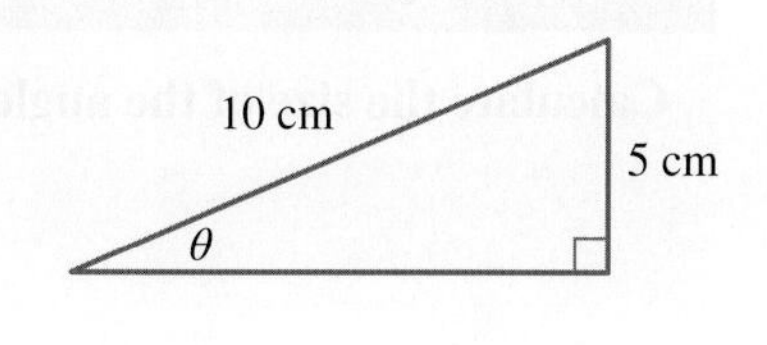

$$\begin{aligned}\sin\theta &= \frac{5}{10}\\ &= \frac{1}{2}\\ &= 0.5\end{aligned}$$

We then take the inverse sin of both sides of the equation:

$$\theta = \sin^{-1}(0.5)$$

and calculate:

$$\theta = 30°$$

As with all trigonometry, it is important that you have your calculator set to degrees mode.

tlvd-3652

WORKED EXAMPLE 16 Calculating unknown angles to the nearest degree

Determine the size of angle θ in the triangle, correct to the nearest degree.

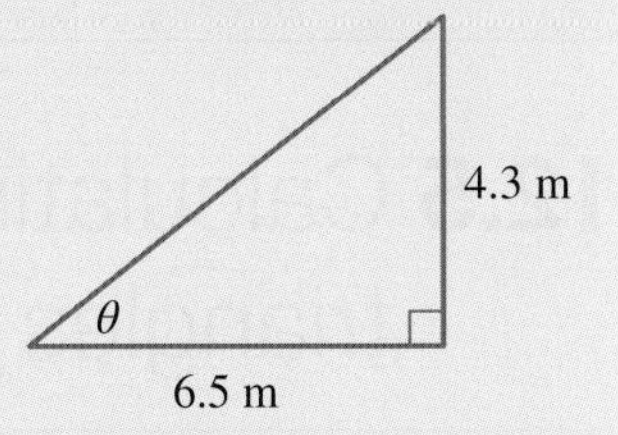

THINK	WRITE
1. Label the sides of the triangle.	hypotenuse; 4.3 m opposite; θ; 6.5 m adjacent
2. We are given the opposite and adjacent side lengths, so choose the tangent ratio to solve the problem. Write the tangent ratio.	$\tan\theta = \dfrac{\text{opposite}}{\text{adjacent}}$
3. Substitute the values of the known side lengths: opposite $= 4.3$ and adjacent $= 6.5$.	$\tan\theta = \dfrac{4.3}{6.5}$
4. Make θ the subject of the equation by taking the inverse tan of both sides.	$\theta = \tan^{-1}\left(\dfrac{4.3}{6.5}\right)$
5. Use your calculator to evaluate the expression for θ.	$\theta = 33°$

In many cases, we will need to calculate the size of an angle correct to the nearest minute. The same method for finding the solution is used; however, you will need to use your calculator to convert to degrees and minutes.

WORKED EXAMPLE 17 Calculating unknown angles to the nearest minute

Calculate the size of the angle θ in the triangle, correct to the nearest minute.

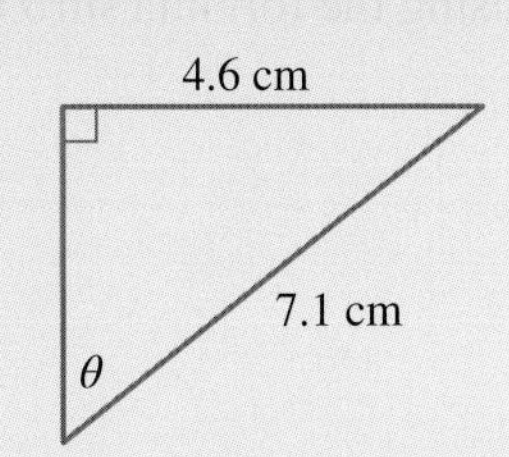

THINK

WRITE

1. Label the sides of the triangle.

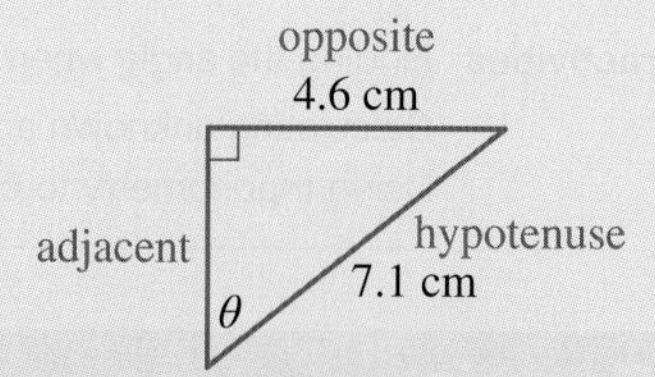

2. We are given the opposite side and the hypotenuse, so choose the sine ratio to solve this problem.
Write the sine ratio.

$\sin\theta = \dfrac{\text{opposite}}{\text{hypotenuse}}$

3. Substitute the values of the known side lengths: opposite = 4.6 and hypotenuse = 7.1.

$\sin\theta = \dfrac{4.6}{7.1}$

4. Make θ the subject of the equation by taking the inverse sin of both sides.

$\theta = \sin^{-1}\left(\dfrac{4.6}{7.1}\right)$

5. Use your calculator to evaluate the expression for θ and convert your answer to degrees and minutes.

$\theta = 40°23'$

12.5.2 Solving practical problems

The same methods can be used to solve problems requiring an unknown angle to be found. As with finding sides, we set the question up by drawing a diagram of the situation.

tlvd-3653

WORKED EXAMPLE 18 Solving practical problems using trigonometric ratios

A ladder is leant against a wall. The foot of the ladder is 4 m from the base of the wall and the ladder reaches 10 m up the wall. Calculate the angle that the ladder makes with the ground, correct to the nearest minute.

THINK

WRITE

1. Draw a diagram showing all important information and label the sides.

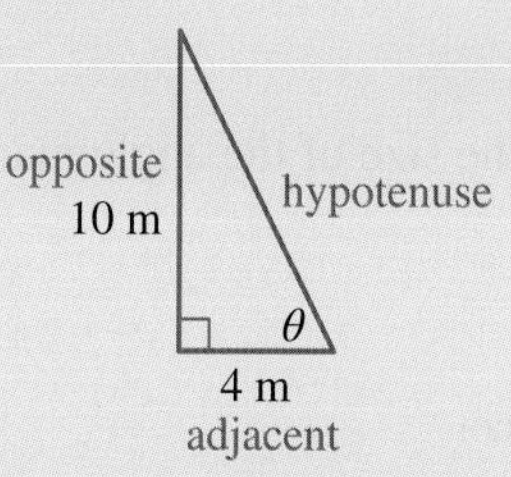

2. We are given the opposite and adjacent sides, so choose the tangent ratio to solve this problem.
Write the tangent ratio.

$\tan\theta = \dfrac{\text{opposite}}{\text{adjacent}}$

3. Substitute the values of the known side lengths: opposite = 10 and adjacent = 4.

$\tan\theta = \dfrac{10}{4}$

4. Make θ the subject of the equation by taking the inverse tan of both sides.

$\theta = \tan^{-1}\left(\dfrac{10}{4}\right)$

5. Use your calculator to evaluate the expression for θ.

$= 68°12'$

6. Answer the question.

The ladder makes an angle of 68°12′ with the ground.

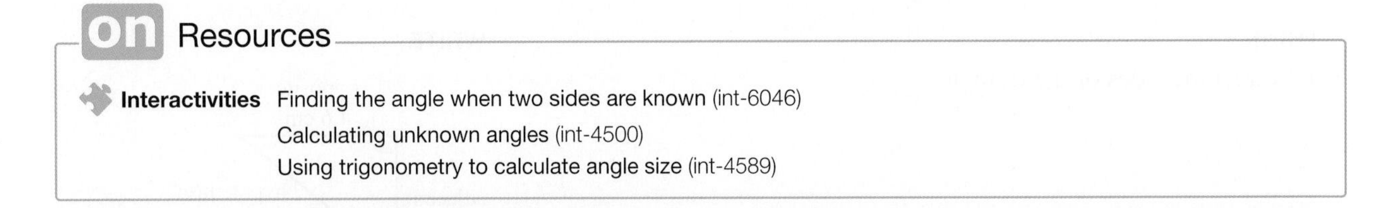

12.5 Exercise

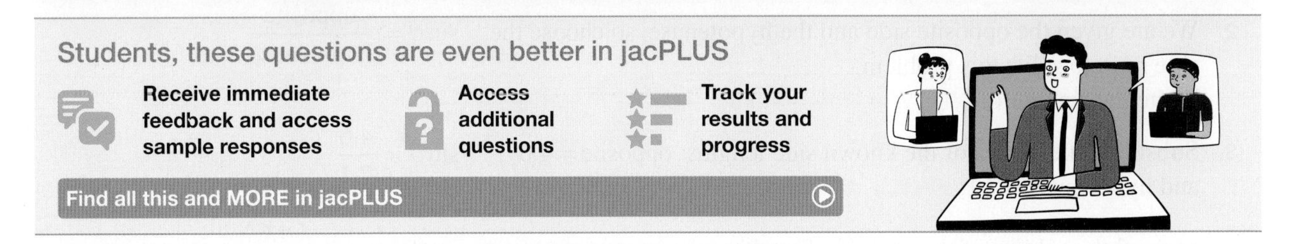

1. **WE16** Determine the size of the angle marked with the variable in each of the following, correct to the nearest degree.

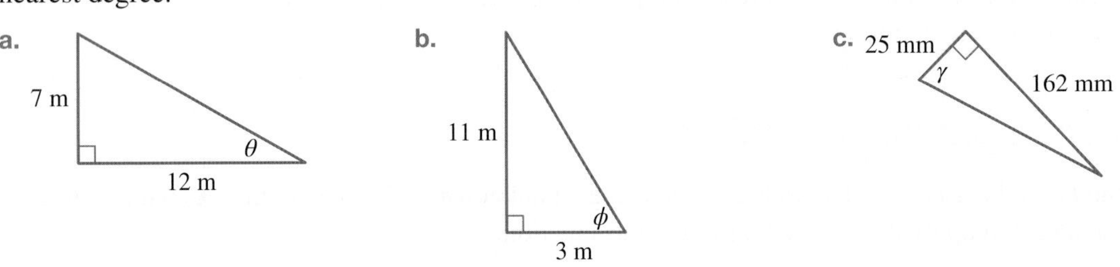

2. **WE17** Determine the size of the angle marked with the variable in each of the following, correct to the nearest minute.

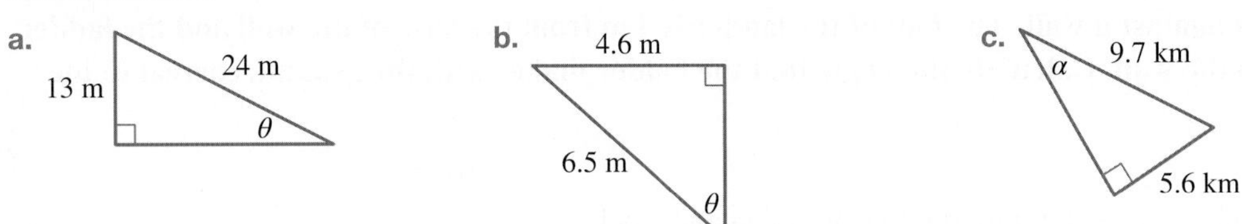

3. Determine the size of the angle marked with the variable in each of the following, correct to the nearest minute.

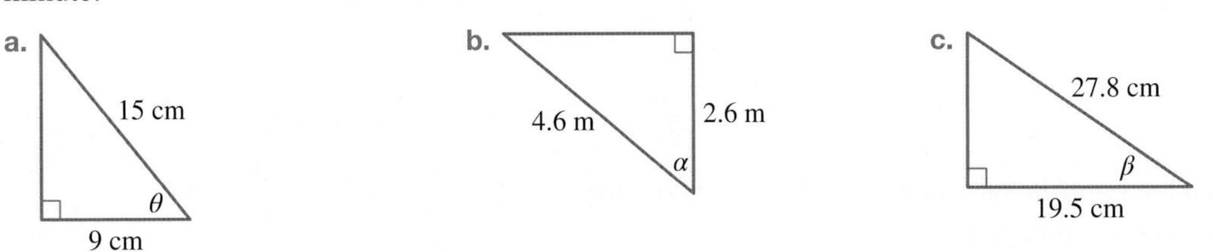

4. Calculate the size of the angle marked θ in the following triangles, correct to the nearest degree.

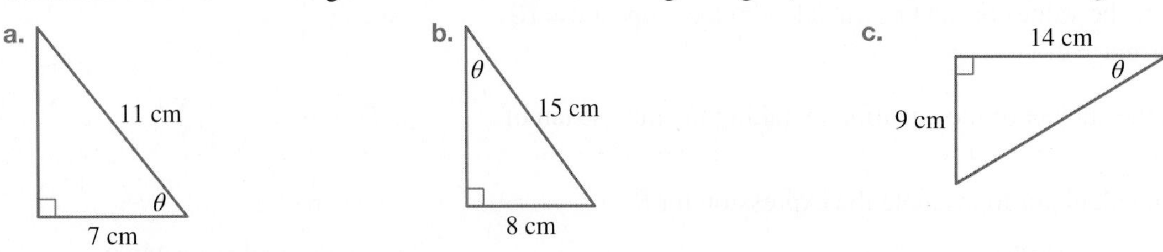

d.

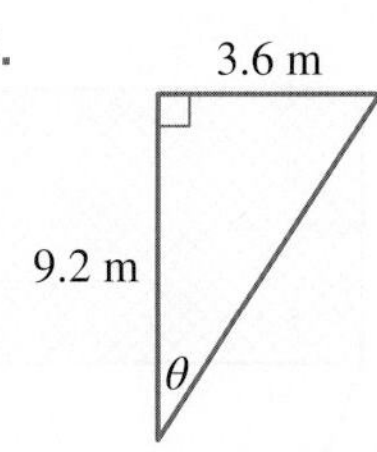

e.

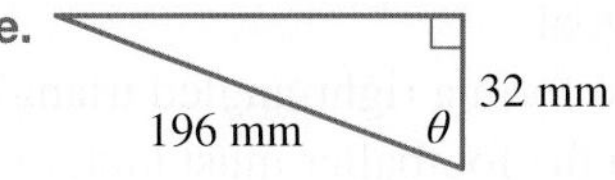

f.

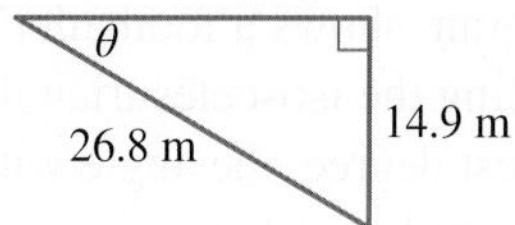

5. In each of the following calculate the size of the angle marked θ, correct to the nearest minute.

a.

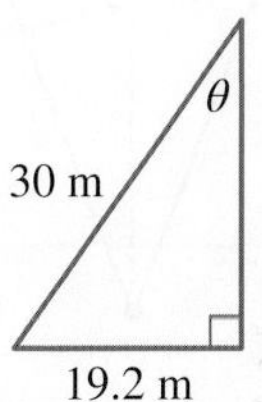

b.

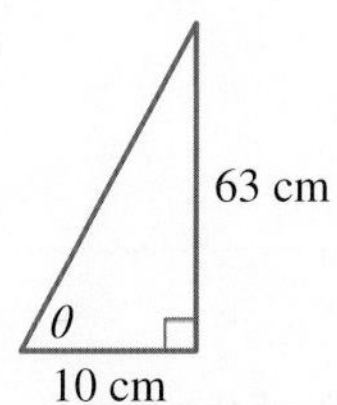

c.

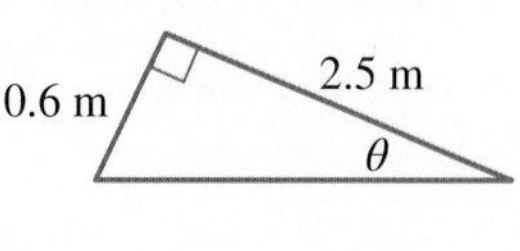

d.

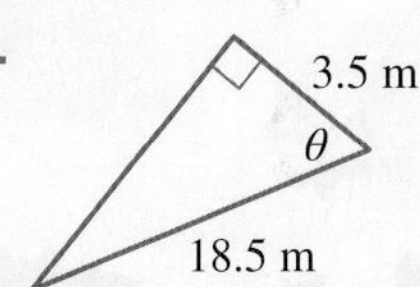

e.

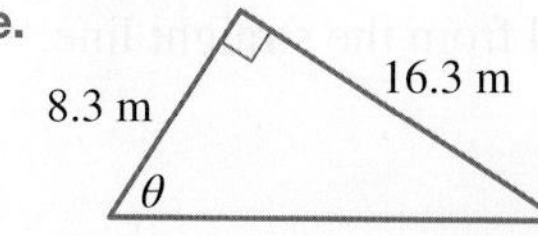

f.

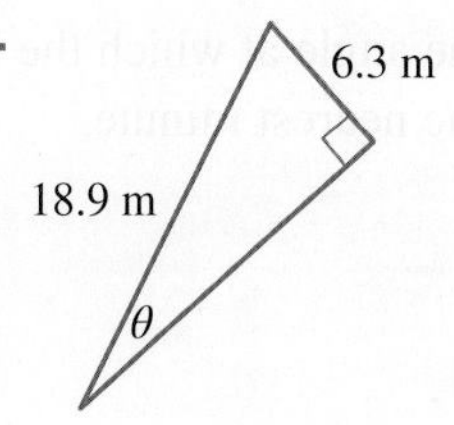

6. **MC** Look at the triangle. State which of the statements below is correct.

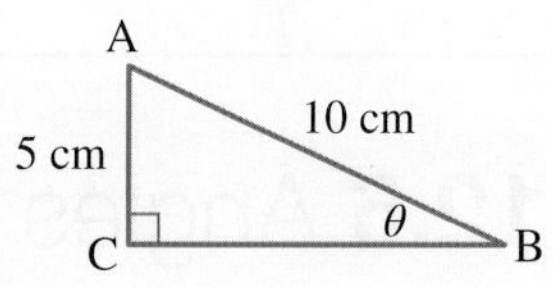

A. $\angle ABC = 30°$ B. $\angle CAB = 30°$ C. $\angle ABC = 45°$
D. $\angle CAB = 45°$ E. $\angle BAC = 30°$

7. **MC** Consider the triangle. The correct value of θ is:

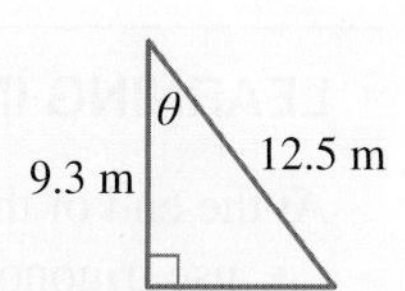

A. 36°39′ B. 41°55′ C. 41°56′ D. 48°4′ E. 44°44′

8. **WE18** A 10-m ladder leans against a wall 6 m high. Calculate the angle that the ladder makes with the horizontal, correct to the nearest degree.

9. A kite is flying on a 40-m string. The kite is flying 10 m away from the vertical, as shown in the figure. Calculate the angle the string makes with the horizontal, correct to the nearest minute.

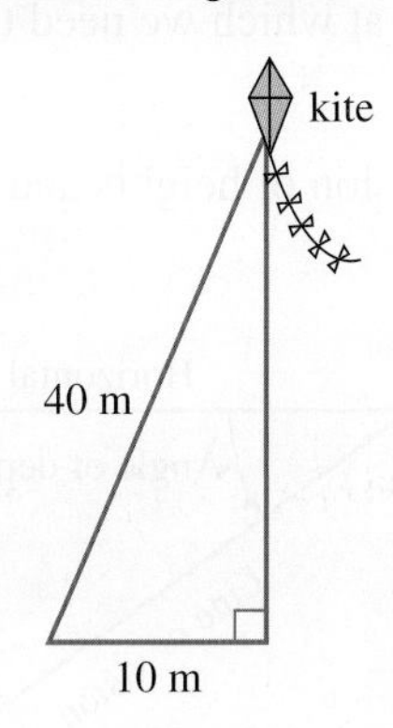

10. A ship's compass shows a course due east of the port from which it sails. After sailing 10 nautical miles (nmi), it is found that the ship is 1.5 nmi off course, as shown in the figure.
Determine the error in the compass reading, correct to the nearest minute.

port
10 nmi
1.5 nmi

11. The diagram shows a footballer's shot at goal.
By dividing the isosceles triangle in half to form a right-angled triangle, calculate, to the nearest degree, the angle within which the footballer must kick to get the ball to go between the posts.

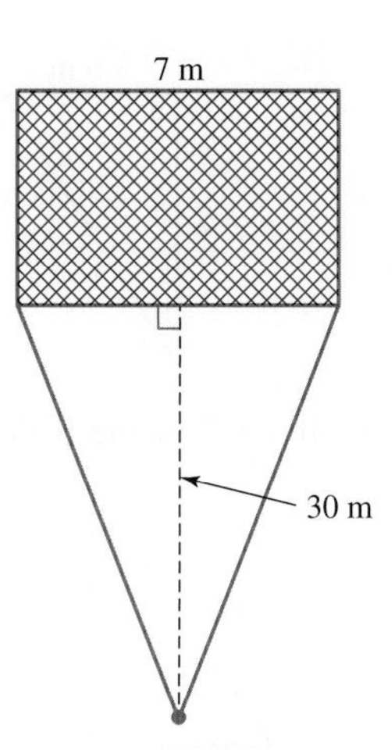

12. A golfer hits a ball 250 m, but it lands 20 m to the left of the target.
Calculate the angle at which the ball deviated from the straight line, correct to the nearest minute.

12.6 Angles of elevation and depression

LEARNING INTENTION

At the end of this subtopic you should be able to:
- use trigonometric ratios to solve problems involving angles of elevation and depression.

12.6.1 Solving problems involving angles of elevation and depression

The **angle of elevation** is measured upwards from the **horizontal** and refers to the angle at which we need to look up to see an object. Similarly, the **angle of depression** is the angle at which we need to look down from the horizontal to see an object.

We can use trigonometry to calculate the angles of elevation and depression of heights and distances of objects that would otherwise be difficult to measure.

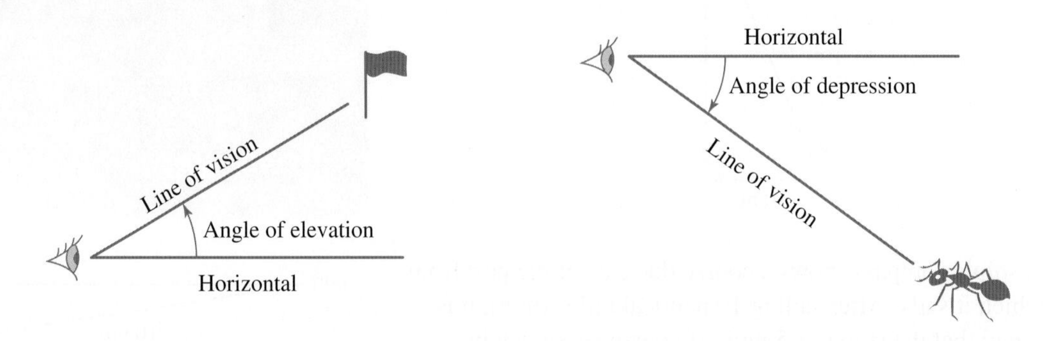

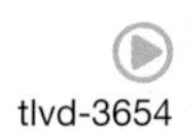
tlvd-3654

WORKED EXAMPLE 19 Solving problems involving angles of elevation

Bryan measures the angle of elevation to the top of a tree as 64°, from a point 10 m from the foot of the tree. If the height of Bryan's eyes is 1.6 m above the ground, calculate the height of the tree, correct to 1 decimal place.

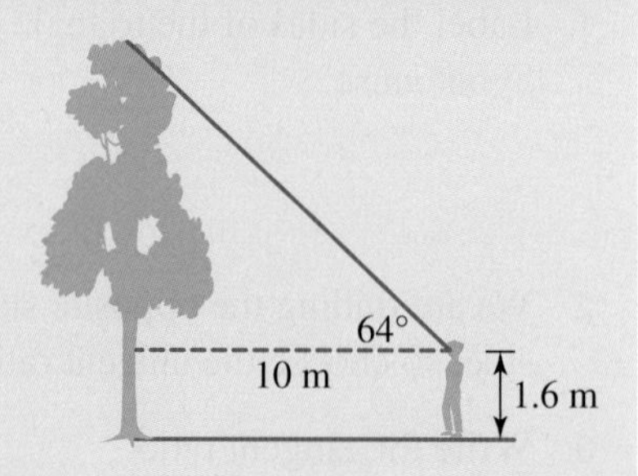

THINK | **WRITE**

1. Label the sides opposite, adjacent and hypotenuse.

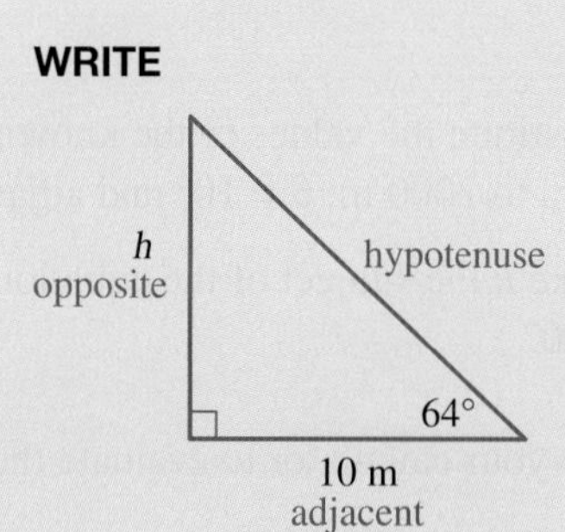

2. We are finding the opposite side and have been given the adjacent side, so choose the tangent ratio.

3. Write the tangent ratio.
$$\tan\theta = \frac{\text{opposite}}{\text{adjacent}}$$

4. Substitute the values of the known side and angle: $\theta = 64°$ and adjacent $= 10$.
$$\tan(64°) = \frac{h}{10}$$

5. Make h the subject of the equation by multiplying both sides by 10.
$$10 \times \tan(64°) = \frac{\cancel{10}h}{\cancel{10}}$$
$$h = 10\tan(64°)$$

6. Use your calculator to evaluate the expression for h.
$$h = 20.5 \text{ m}$$

7. The height of the tree is equal to the sum of h and the eye height.
$$\text{Height of tree} = 20.5 + 1.6 = 22.1$$

8. Answer the question.
The height of the tree is approximately 22.1 m.

A similar method for finding the solution is used for problems that involve an angle of depression.

WORKED EXAMPLE 20 Solving problems involving angles of depression

When an aeroplane in flight is 2 km from a runway, the angle of depression to the runway is 10°. Calculate the altitude of the aeroplane correct to the nearest metre.

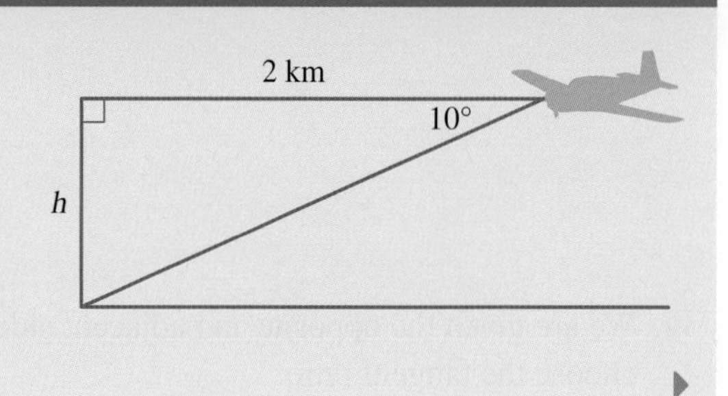

THINK | **WRITE**

1. Label the sides of the triangle as opposite, adjacent and hypotenuse.

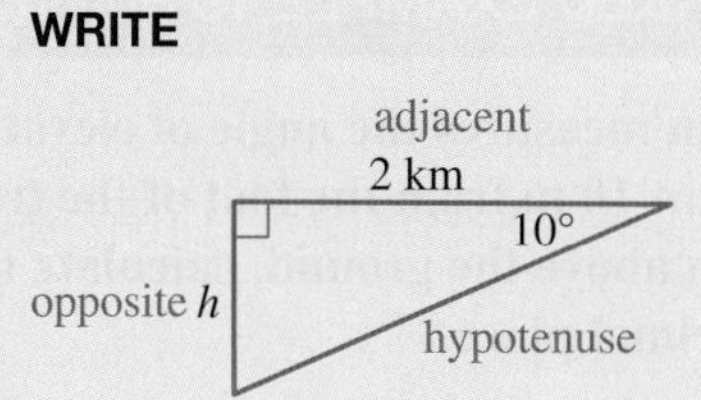

2. We are finding the opposite side and have been given the adjacent side, so choose the tangent ratio.

3. Write the tangent ratio.

$$\tan\theta = \frac{\text{opposite}}{\text{adjacent}}$$

4. Substitute the values of the known side and angle, converting 2 km to 2000 m: $\theta = 10°$ and adjacent $= 2000$.

$$\tan(10°) = \frac{h}{2000}$$

5. Make h the subject of the equation by multiplying both sides by 2000.

$$2000 \times \tan(10°) = \frac{\cancel{2000}h}{\cancel{2000}}$$
$$h = 2000\tan(10°)$$

6. Use your calculator to evaluate the expression for h.

$$h = 353 \text{ m}$$

7. Answer the question.

The altitude of the aeroplane is approximately 353 m.

12.6.2 Calculating angles of elevation and depression

Angles of elevation and depression can also be calculated by using known measurements. This is done by drawing a right-angled triangle to represent a situation.

tlvd-3655

WORKED EXAMPLE 21 Determining missing angles

A 5.2-m building casts a 3.6-m shadow. Calculate the angle of elevation of the sun, correct to the nearest degree.

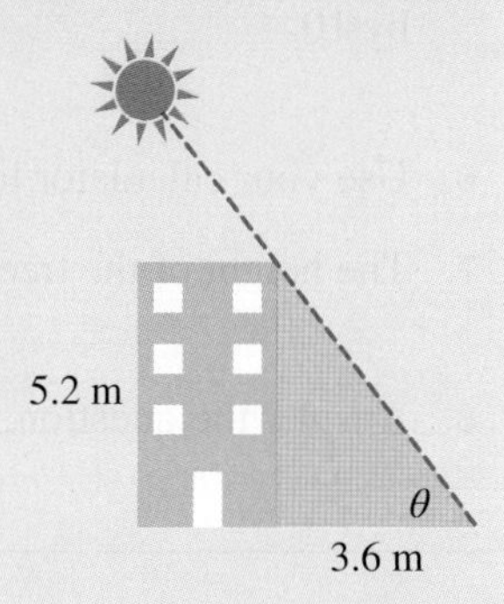

THINK | **WRITE**

1. Label the sides opposite, adjacent and hypotenuse.

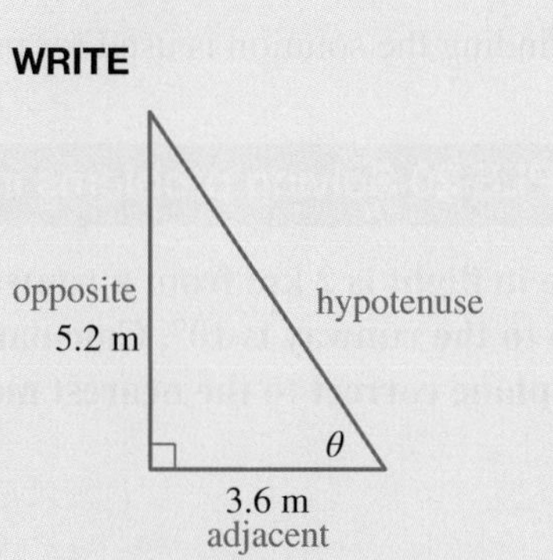

2. We are given the opposite and adjacent sides, so choose the tangent ratio.

3. Write the tangent. $\tan\theta = \frac{\text{opposite}}{\text{adjacent}}$

4. Substitute the values of the known side lengths: opposite = 5.2 and adjacent = 3.6. $\tan\theta = \frac{5.2}{3.6}$

5. Make θ the subject of the equation by taking the inverse tan of both sides. $\theta = \tan^{-1}\left(\frac{5.2}{3.6}\right)$

6. Use your calculator to evaluate the expression for θ. $\theta = 55°$

7. Answer the question. The angle of elevation of the sun is approximately 55°.

Resources

Interactivities Finding the angle of elevation and angle of depression (int-6047)
Angles of elevation and depression (int-4501)

12.6 Exercise

Students, these questions are even better in jacPLUS

Receive immediate feedback and access sample responses

Access additional questions

Track your results and progress

Find all this and MORE in jacPLUS

1. **WE19** From a point 50 m from the foot of a building, Rod sights the top of the building at an angle of elevation of 37°. Given that Rod's eyes are 1.5 m above the ground, calculate the height of the building, correct to 1 decimal place.

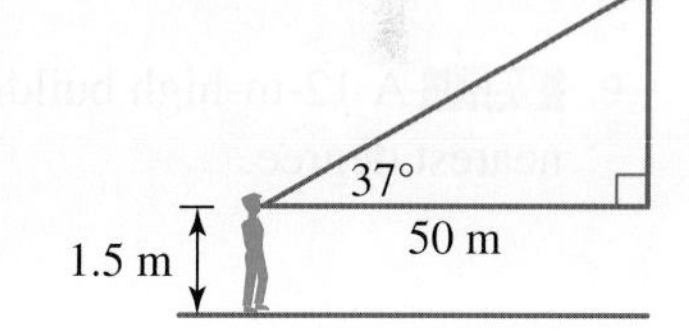

2. From a ship, the angle of elevation to an aeroplane is 60°. The aeroplane is located at a horizontal distance of 2300 m away from the ship. Calculate the altitude of the aeroplane, correct to the nearest metre.

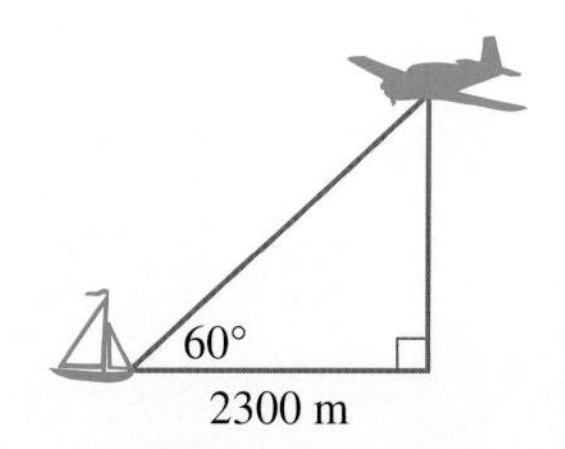

3. From a point out to sea, a ship sights the top of a lighthouse at an angle of elevation of 12°. It is known that the top of the lighthouse is 40 m above sea level. Calculate the horizontal distance of the ship from the lighthouse, correct to the nearest 10 m.

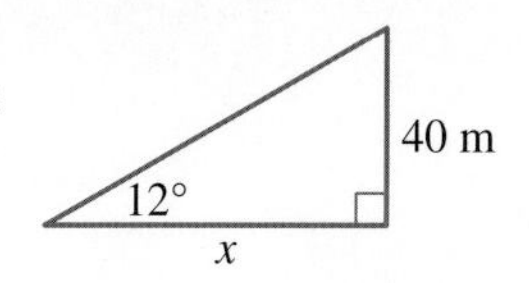

4. From a point 100 m from the foot of a building, the angle of elevation to the top of the building is 15°. Calculate the height of the building, correct to 1 decimal place.

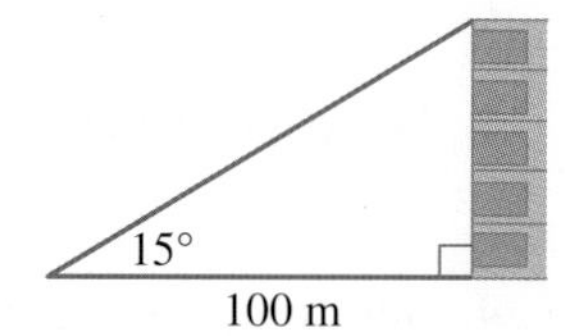

5. **WE20** From an aeroplane flying at an altitude of 4000 m, the runway is sighted at an angle of depression of 15°. Calculate the horizontal distance of the aeroplane from the runway, correct to the nearest kilometre.

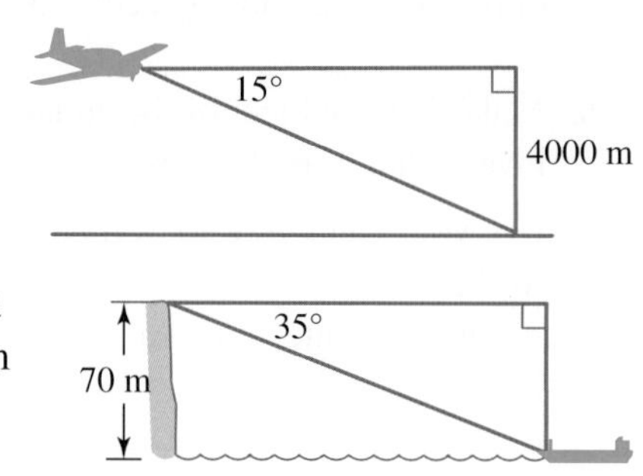

6. From the top of a cliff 70 m above sea level, a ship is spotted out to sea at an angle of depression of 35°. Calculate the distance of the ship from shore, to the nearest metre.

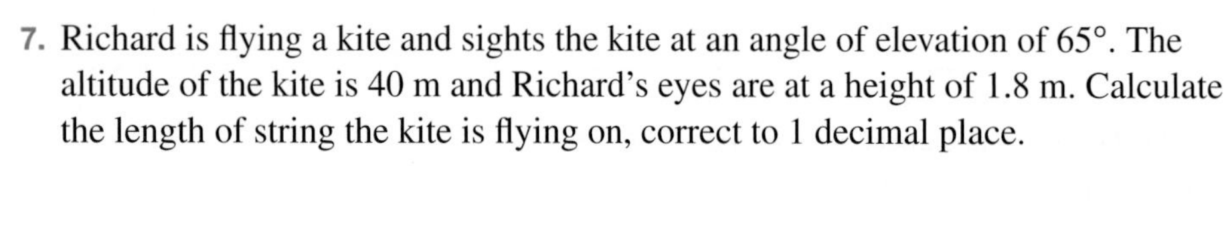

7. Richard is flying a kite and sights the kite at an angle of elevation of 65°. The altitude of the kite is 40 m and Richard's eyes are at a height of 1.8 m. Calculate the length of string the kite is flying on, correct to 1 decimal place.

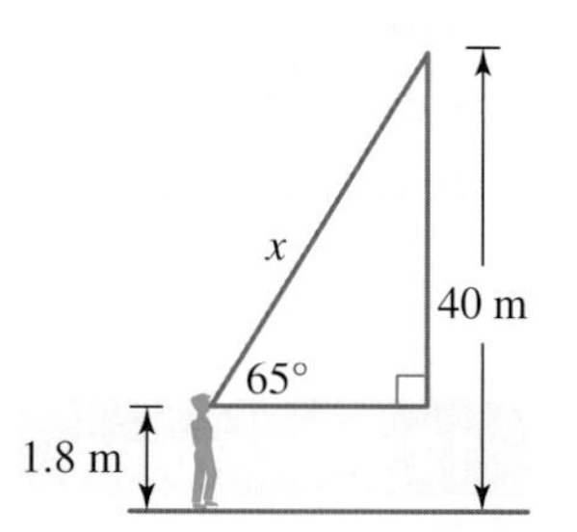

8. There is a fire on the fifth floor of a building. The closest a fire truck can get to the building is 10 m. The angle of elevation from this point to where people need to be rescued is 69°. If the fire truck has a 30-m ladder, determine whether the ladder can be used to make the rescue.

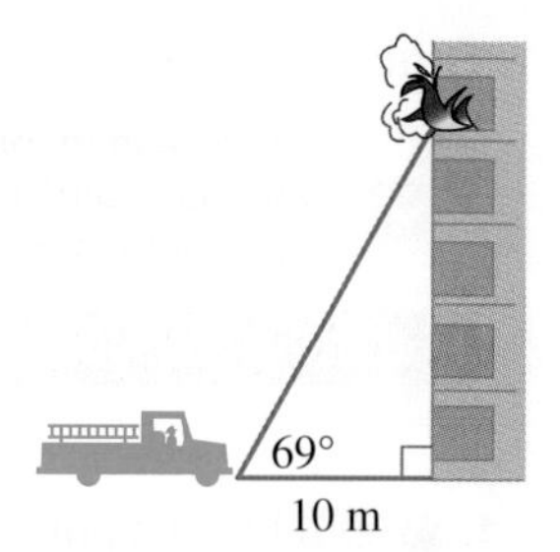

9. **WE21** A 12-m-high building casts a shadow 15 m long. Calculate the angle of elevation of the sun, to the nearest degree.

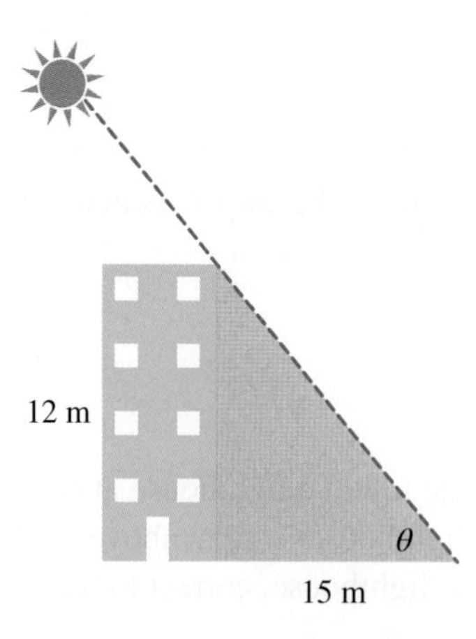

10. An aeroplane, which is at an altitude of 1500 m, is 4000 m from a ship in a horizontal direction, as shown. Calculate the angle of depression from the aeroplane to the ship, to the nearest degree.

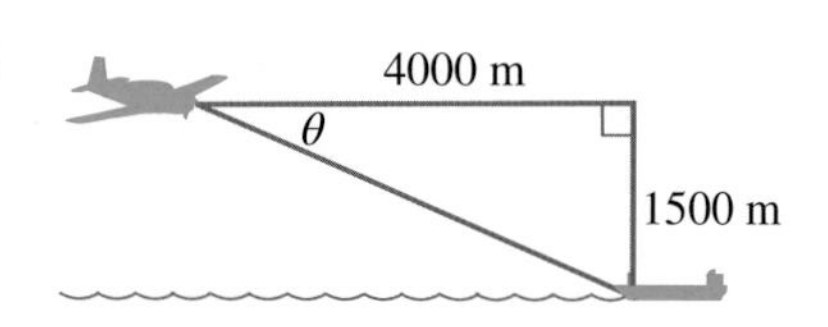

11. The angle of elevation to the top of a tower is 12° from a point 400 m from the foot of the tower.
   a. Draw a diagram of this situation.
   b. Calculate the height of the tower, correct to 1 decimal place.
   c. Calculate the angle of elevation to the top of the tower, from a point 100 m from the foot of the tower.

12. A beacon, which is 80 m above sea level, is sighted from a navy vessel at an angle of elevation of 5°. The vessel sailed towards the beacon and 30 minutes later the beacon is at an angle of elevation of 60°. Use the diagram to complete the following.

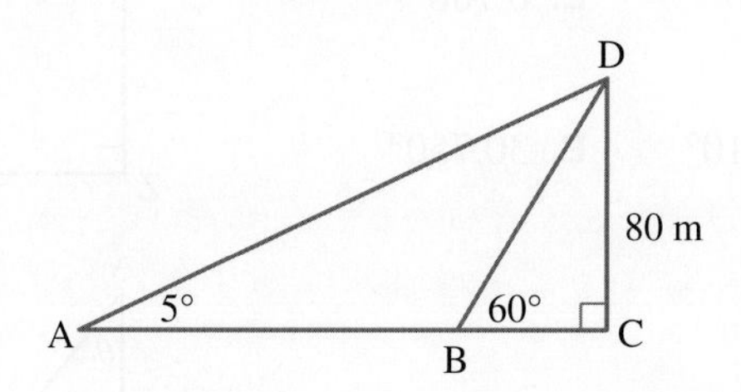

   a. Calculate the distance (AC) that the vessel was from the beacon, when the angle of elevation to the beacon was 5°.
   b. Calculate the distance that the vessel sailed in the 30 minutes between the two readings.

12.7 Review

12.7.1 Summary

doc-38020

Hey students! Now that it's time to revise this topic, go online to:

Review your results

Watch teacher-led videos

Practise questions with immediate feedback

Find all this and MORE in jacPLUS

12.7 Exercise

Multiple choice

1. **MC** The opposite side of the triangle shown is:

   A. θ **B.** XY **C.** XZ **D.** ZY **E.** YZ

2. **MC** The value of sin(58°) is closest to:

   A. 0.484 **B.** 0.530 **C.** 0.848 **D.** 1.600 **E.** 0.788

3. **MC** The value of θ in $\tan\theta = 0.75$ is:

   A. 36.870° **B.** 48.590° **C.** 34.930° **D.** 41.410° **E.** 30.750°

4. **MC** In the diagram shown, $\cos\theta$ is:

   A. 9
   B. $\frac{15}{12}$
   C. $\frac{12}{15}$
   D. $\frac{9}{12}$
   E. $\frac{9}{15}$

5. **MC** The value of t in the diagram shown is:

   A. 2.4 m
   B. 2.47 m
   C. 2.86 m
   D. 6.12 m
   E. 5.48 m

6. **MC** The sides and angle of the triangle shown are correctly described by:

   A. $\angle C = \theta$, AB = adjacent side, AC = hypotenuse, BC = opposite side
   B. $\angle C = \theta$, AB = opposite side, AC = hypotenuse, AC = adjacent side
   C. $\angle A = \theta$, AB = opposite side, AC = hypotenuse, BC = adjacent side
   D. $\angle A = \theta$, AB = adjacent side, AC = hypotenuse, BC = opposite side
   E. $\angle B = \theta$, AB = adjacent side, AC = hypotenuse, BC = opposite side

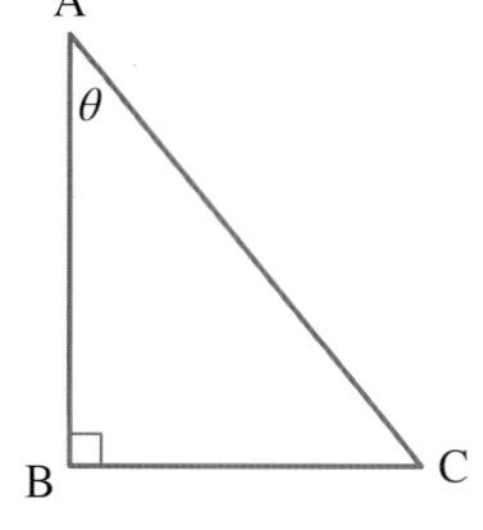

7. **MC** The length of the third side in this triangle is:

   A. 48.75 cm **B.** 0.698 m **C.** 0.926 m
   D. 92.6 cm **E.** 63.82 cm

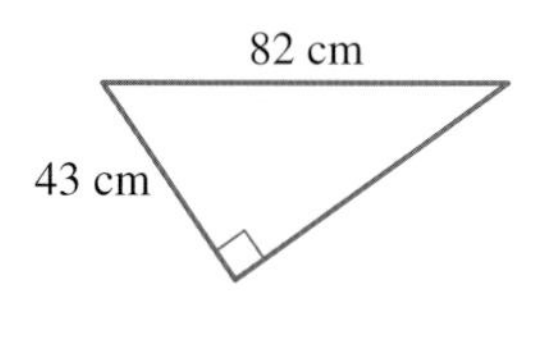

8. **MC** The value of t in the following diagram is:

A. 6.16 m
B. 12.26 m
C. 2.87 m
D. 14.58 m
E. 3.17 m

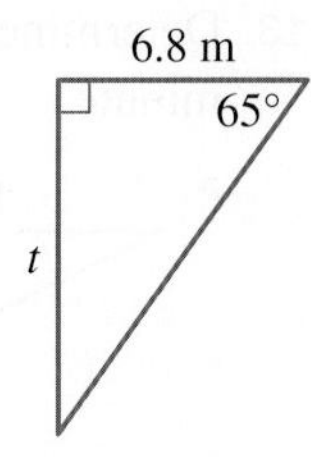

9. **MC** The value of m in the following diagram is:

A. 62.78 km
B. 5.27 km
C. 45.72 km
D. 38.54 km
E. 7.86 km

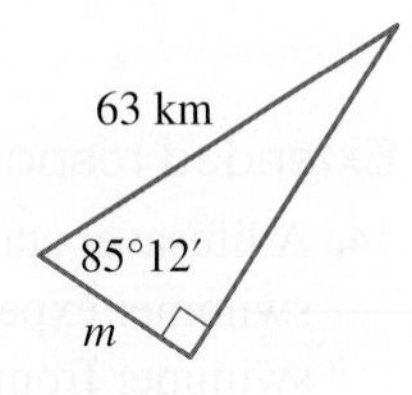

10. **MC** The top of a building is sighted at an angle of elevation of 40°, when an observer is 27 m away from the base.
The height of the building, correct to the nearest metre, is:

A. 22.66 m B. 17.36 m C. 23.55 m
D. 32.18 m E. 20.68 m

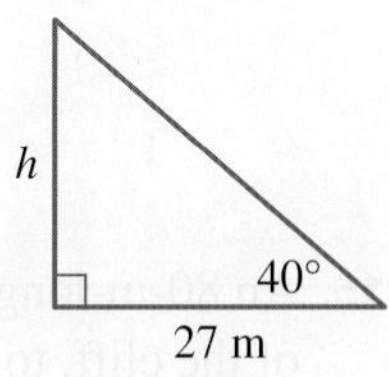

Short answer

11. Determine the length of the side marked with a variable in each case writing your answer correct to 2 decimal places.

a. 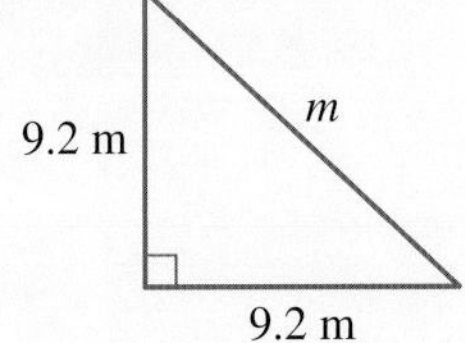

b. 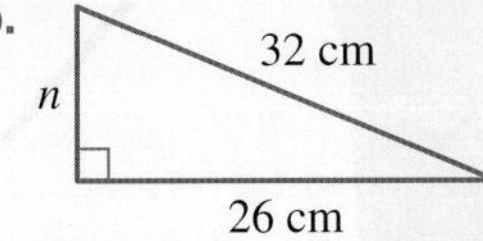

c. 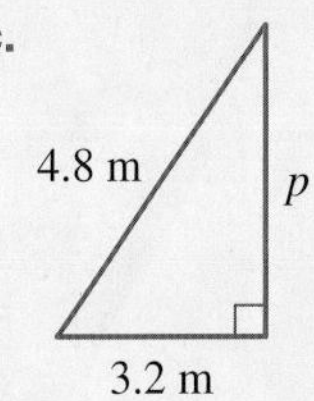

d. 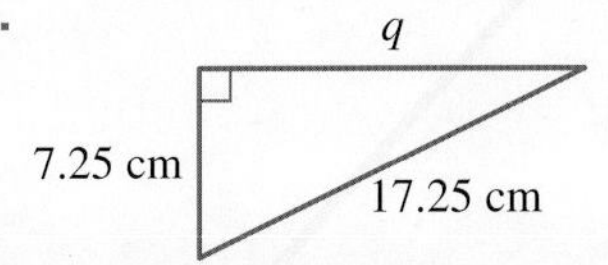

e. 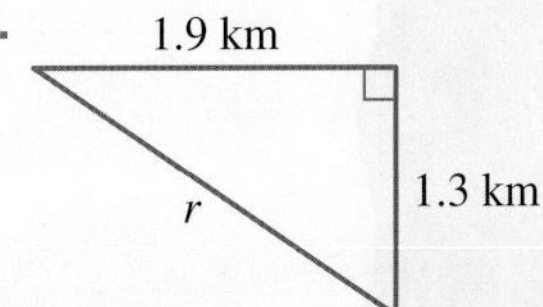

12. Determine the length of each side marked with a variable, correct to 1 decimal place.

a. 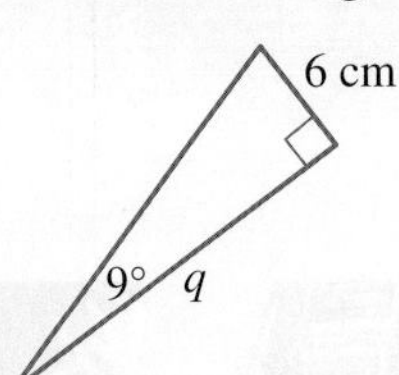

b. 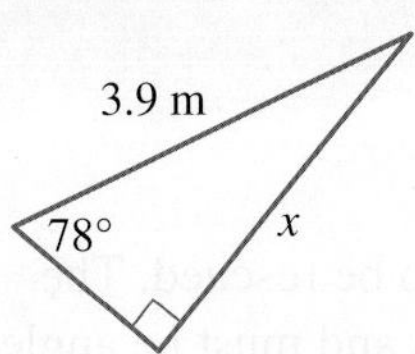

c. 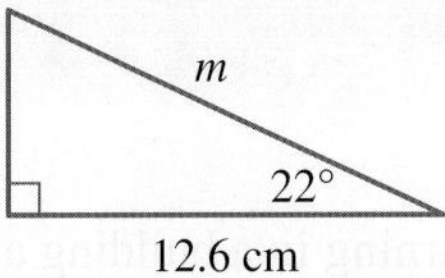

d. 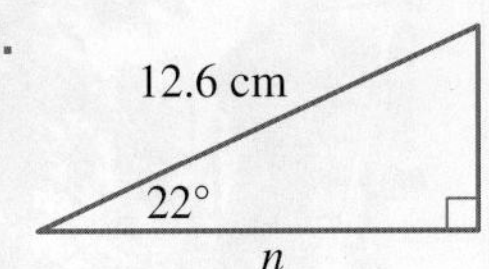

e. 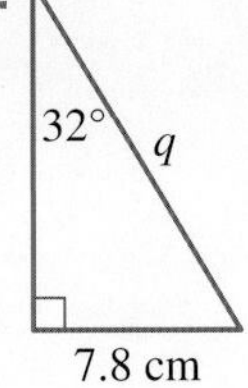

13. Determine the size of the angle marked θ in each of the following, giving your answer correct to the nearest minute.

a. 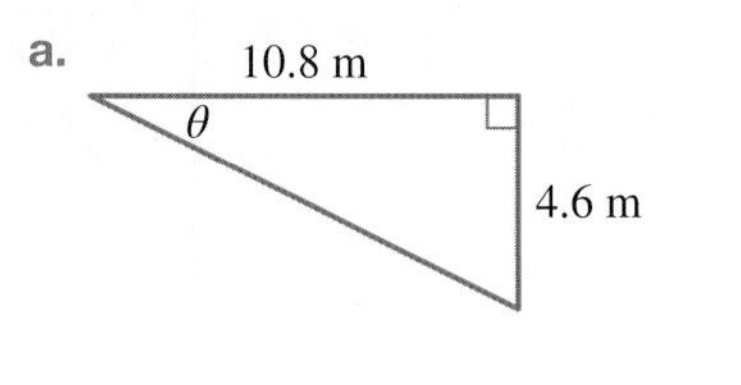

b. 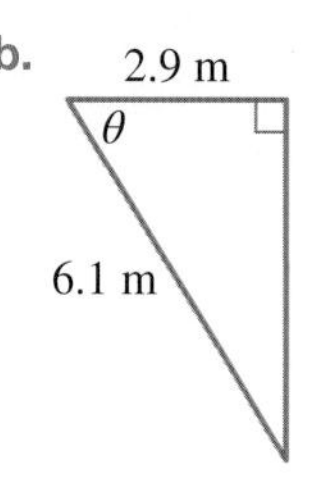

c. 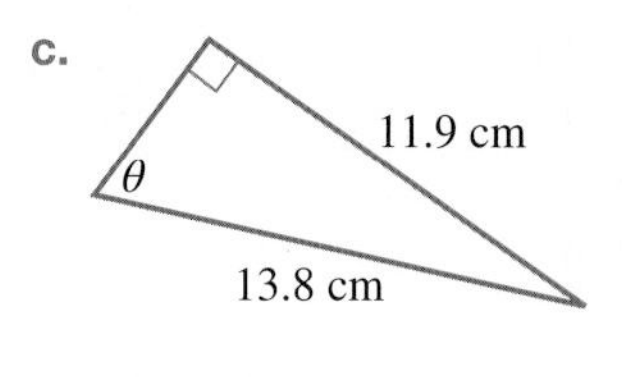

Extended response

14. A lifesaver standing on his tower 3 m above the ground spots a swimmer experiencing difficulty. The angle of depression of the swimmer from the lifesaver is 12°. Calculate the distance between the swimmer and the lifesaver's tower, correct to 2 decimal places.

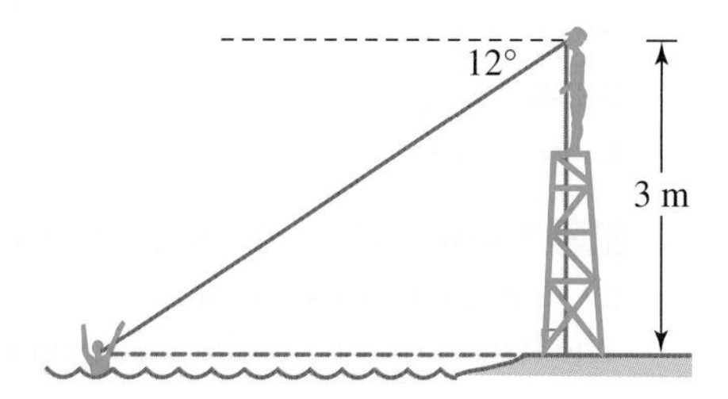

15. An 80-m-long rope runs from a cliff top to the ground, 45 m from the base of the cliff. Calculate the height of the cliff, to the nearest metre.

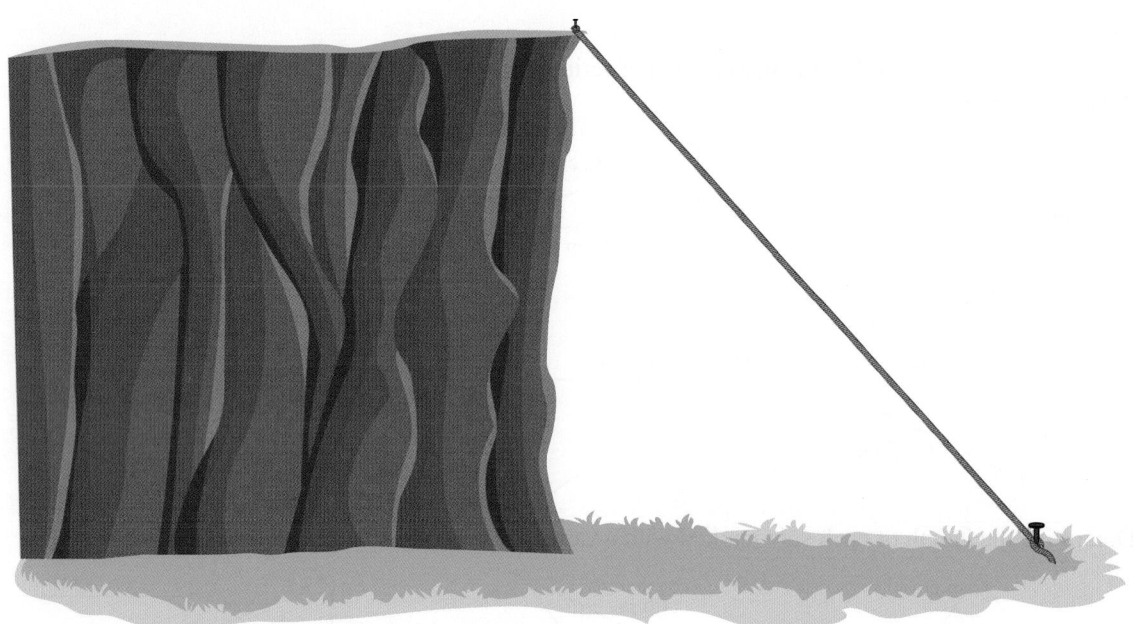

16. A fire is burning in a building and people need to be rescued. The fire brigade's ladder must reach a height of 60 m and must be angled at 70° to the horizontal. Calculate the height of the ladder needed to complete the rescue.

17. A rope that is used to support a flagpole makes an angle of 70° with the ground. If the rope is tied down 3.1 m from the foot of the flagpole, calculate the height of the flagpole, correct to 1 decimal place.

18. A child flies a kite on an 80-m string. He holds the end of the string 1.5 metres above the ground, and the kite reaches a height of 51.5 m above the ground in a strong wind. Calculate the angle the string makes with the horizontal, correct to the nearest degree.

19. There is 50 m of line on a fishing reel. When all the line is out, the bait sits on the bed of a lake and has drifted 20 m from the boat. Calculate the angle that the fishing line makes with the vertical, correct to the nearest degree.

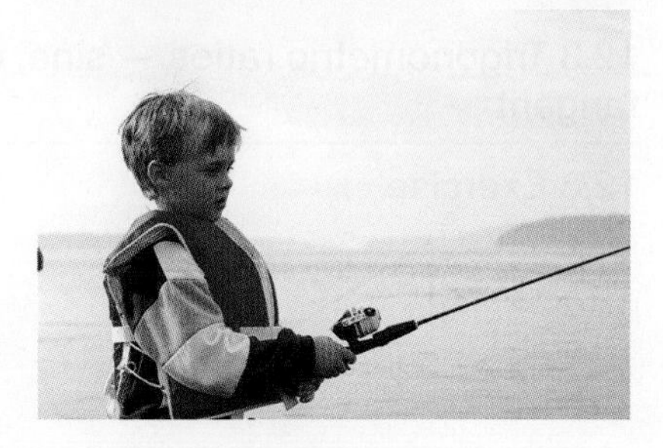

20. Hakam stands 50 m back from the foot of an 80-m telephone tower. Hakam's eyes are at a height of 1.57 m.

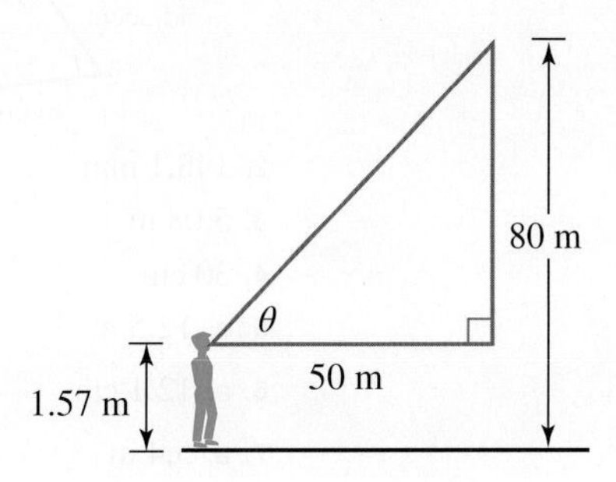

Calculate the angle of elevation at which Hakam must look to see the top of the tower, correct to the nearest degree.

Answers

Topic 12 Right-angled triangles

12.2 Pythagoras' theorem

12.2 Exercise

1. a. PR b. YZ c. AB
2. a. 13 cm b. 170 mm c. 61 m
3. a. 10.82 cm b. 6.93 m c. 14.20 km
4. a. 10.4 cm b. 1.9 m c. 3.9 m
5. a. 8.9 cm b. 22.1 cm c. 47.4 mm d. 37.3 m
6. A
7. 13 m
8. 3.23 m
9. 3.73 m
10. 7.5 m

12.3 Trigonometric ratios — sine, cosine and tangent

12.3 Exercise

1. a.

b.

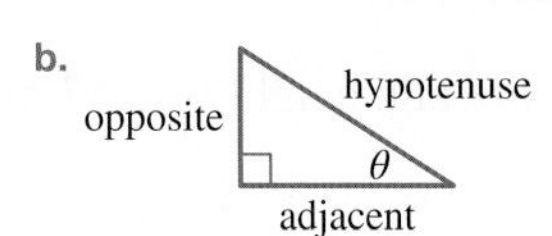

c.

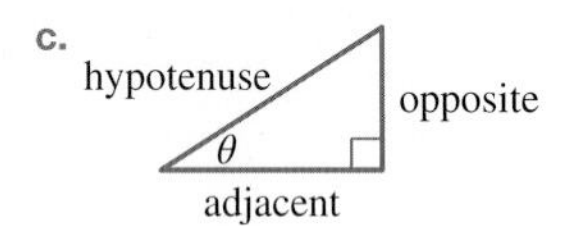

d.

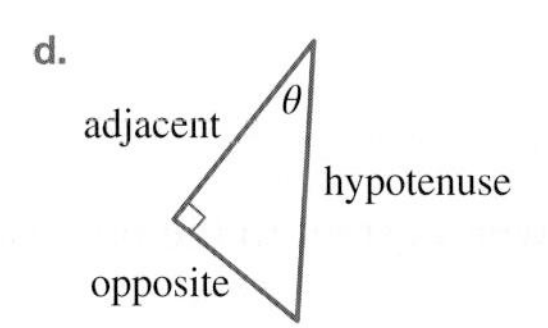

e.

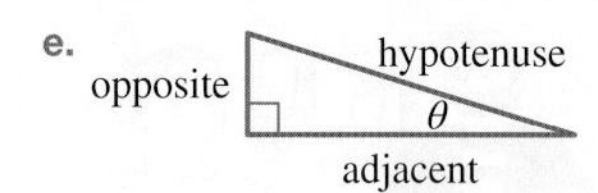

f.

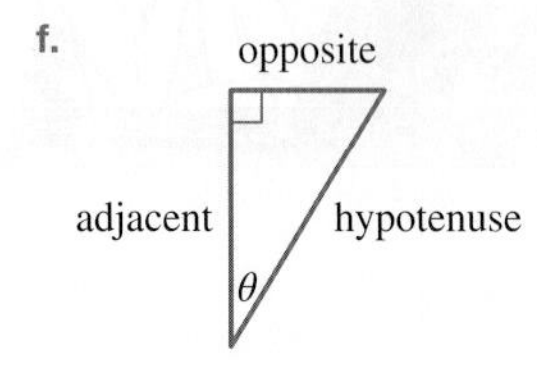

2. a. $\sin\theta = \dfrac{8}{10}$ b. $\cos\theta = \dfrac{6}{10}$ c. $\tan\theta = \dfrac{8}{6}$
3. a. 0.602 b. 2.092 c. 15.246 d. 51.893
4. a. 0.707 b. 0.247 c. 6.568 d. 5.896
5. a. 1.540 b. 17.663 c. 40.460 d. 0.657
6. a. 0.5 b. 0.966 c. 1 d. 548.643 e. 64 f. 1.301
7. a. 5.306 b. 1.374 c. 15.772
8. a. 0.42 b. 1.56 c. 0.09 d. 5.10 e. 2.87 f. 0.38
9. a. 7.77 b. 73.30 c. 0.87
10. 10°
11. 66°8′
12. a. 44° b. 80° c. 57°

12.4 Calculating unknown side lengths of right-angled triangles

12.4 Exercise

1. a.

b.

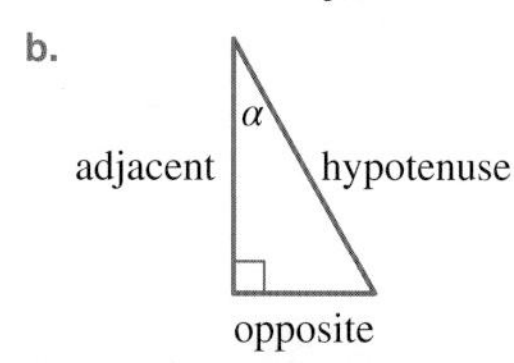

c.

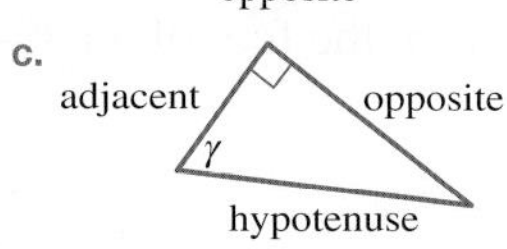

2. 148.1 mm
3. 5.08 m
4. 30 cm
5. a. 12.5 m b. 89.3 mm c. 10.1 m
6. a. 12.1 cm b. 55.2 m c. 9.4 km
7. a. 5.4 m b. 1.4 km c. 2.1 km d. 18.4 mm e. 3.2 cm f. 66.5 m
8. a. 5.4 m b. 5.4 km c. 0.2 m d. 41.6 km e. 84.4 m f. 13.2 cm
9. D
10. A
11. C
12. 6 m
13. 4.2 m
14. 20 km
15. a.

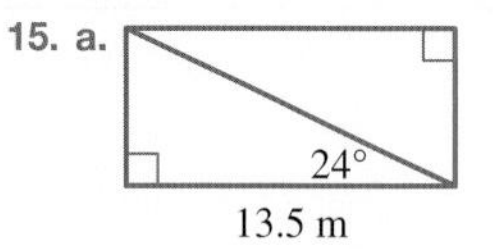

b. 6 m

16. a.

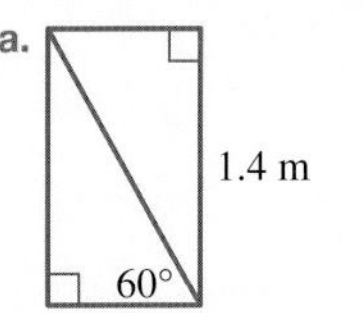

b. 1.6 m

17. 9.65 m

18. a.

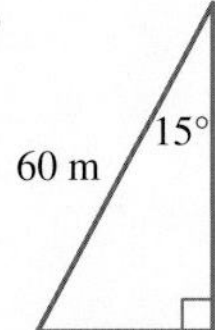

b. 58 m c. 15.5 m

12.5 Calculating unknown angles in right-angled triangles

12.5 Exercise

1. a. 30° b. 75° c. 81°
2. a. 32°48′ b. 45°3′ c. 35°16′
3. a. 53°8′ b. 55°35′ c. 45°27′
4. a. 50° b. 32° c. 33° d. 21° e. 81° f. 34°
5. a. 39°48′ b. 80°59′ c. 13°30′ d. 79°6′ e. 63°1′ f. 19°28′
6. A
7. C
8. 37°
9. 75°31′
10. 8°38′
11. 13°
12. 4°35′

12.6 Angles of elevation and depression

12.6 Exercise

1. 39.2 m
2. 3984 m
3. 190 m
4. 26.8 m
5. 15 km
6. 100 m
7. 42.1 m
8. 28 m, so a 30-m ladder can be used.
9. 39°
10. 21°
11. a.

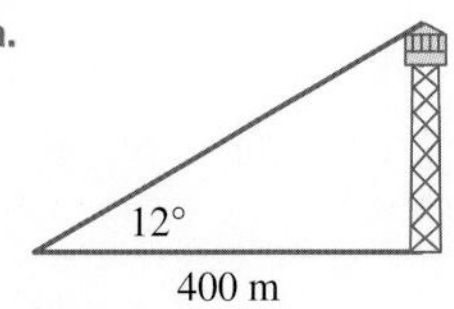

b. 85 m
c. 40°

12. a. 914 m b. 868 m

12.7 Review

12.7 Exercise

Multiple choice

1. D
2. C
3. A
4. C
5. B
6. D
7. B
8. D
9. B
10. A

Short answer

11. a. 13.01 m b. 18.65 cm c. 3.58 m d. 15.65 cm e. 2.30 km
12. a. 37.9 cm b. 3.8 m c. 13.6 cm d. 11.7 cm e. 14.7 cm
13. a. 23°4′ b. 61°37′ c. 59°35′

Extended response

14. 14.11 m
15. 66 m
16. 63.9 m
17. 8.5 m
18. 39°
19. 24°
20. 57°

13 Measurement

LEARNING SEQUENCE

Fully worked solutions for this topic are available online.

13.1 Overview

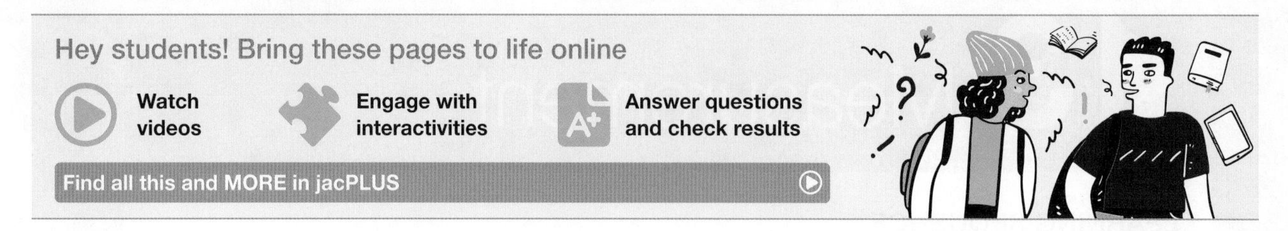

13.1.1 Introduction

We use measurement every day, often without even realising that we're doing so. For example, if you estimate the time that it'll take you to wash the dishes in the evening, that's a measure of time, and if you count the number of steps you take to walk to the corner shop, that's a measure of distance.

Understanding measurement allows us to understand the world around us, from the size of bacteria to the time it would take us to travel to the stars in the night sky. Think about how often you use measures every day!

KEY CONCEPTS

This topic covers the following key concepts from the VCE Mathematics Study Design:

- measurements and related quantities including derived quantities, metric and relevant non-metric measures
- conventions, properties and measurement of perimeter, area, surface area and volume of compound shapes and objects
- calibration and error in measurement, including tolerance, accuracy and precision.

Note: Concepts shown in grey are covered in other topics.

13.2 Units of measurement

LEARNING INTENTION

At the end of this subtopic you should be able to:
- record measurements using correct units
- convert one unit of measurement to another
- add measurements.

13.2.1 Units of measurement

Units of length are used to describe:
- the dimensions of an object, such as its length, width and height
- the distance between two points.

In the past, people used body part sizes as reference units to measure length, as shown.

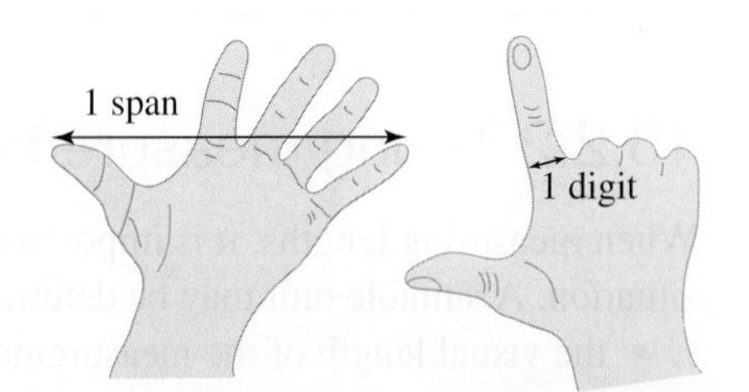

The **standard unit** of length in the metric system is the **metre**.

Metric units of length

The most common metric units of length are the millimetre (mm), centimetre (cm), metre (m) and kilometre (km).

10 millimetres = 1 centimetre
100 centimetres = 1 metre
1000 metres = 1 kilometre

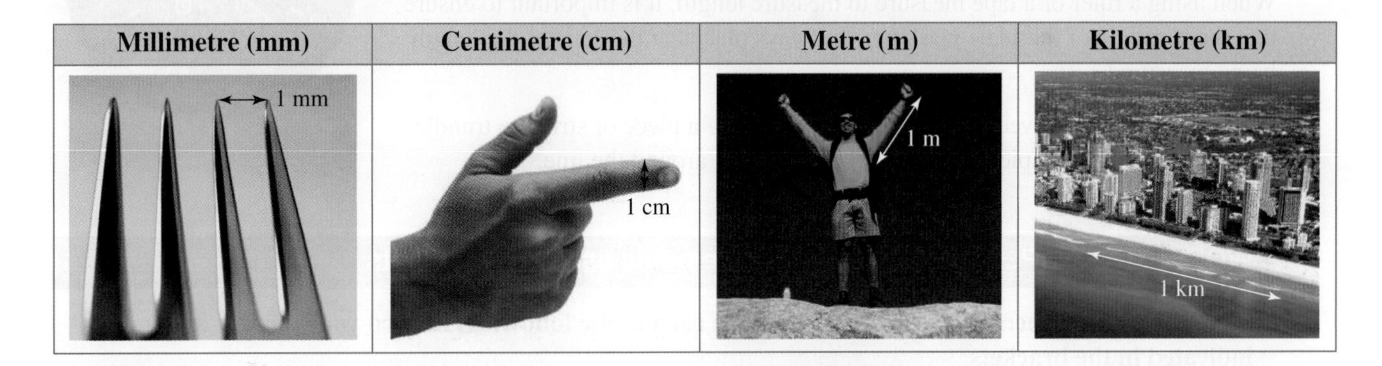

While a linear or curved distance can be measured using mm, cm, m or km, other units of measurement are required to measure liquid, weight and time.

In our modern society, we have standard measurements for volume (liquid, space, etc.), mass (weight) and time.

In the metric system, the standard unit of volume is the **litre** and the standard unit of weight is the **gram**.

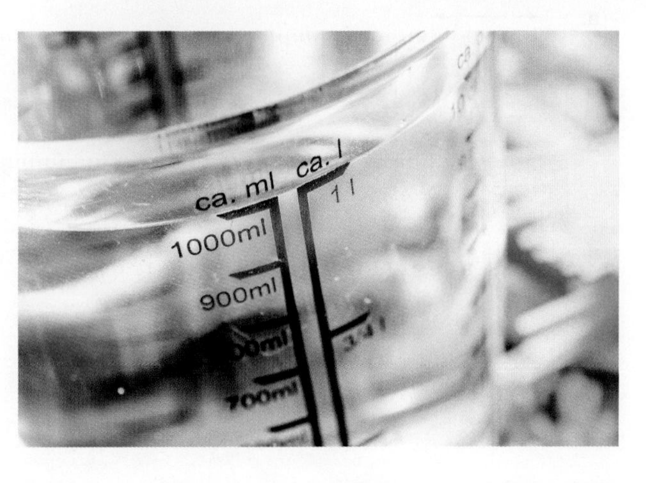

Metric volume measurements	Metric weight measurements	Universal time measurements
Millilitres (mL) Litres (L)	Milligrams (mg) Grams (g) Kilograms (kg) Tonnes (t)	Seconds (s) Minutes (min) Hours (h) Days Weeks Months Years Decades Centuries Millennia

13.2.2 Taking measurements

When measuring lengths, it is important to choose units that suit the situation. A suitable unit may be determined by:

- the visual length of the measurement
- the context in which the measurement is to be used; for example, builders, carpenters and plumbers work in millimetres and metres.

A variety of tools can be used to help measure length.

- A ruler can be used to measure short objects.
- A tape measure can be used to measure longer objects or distances.
- A car's odometer can record long measurements, such as the distance between two towns.
- A picture with a scale, such as a map or a microscope drawing, can be used to measure very large or very small lengths.

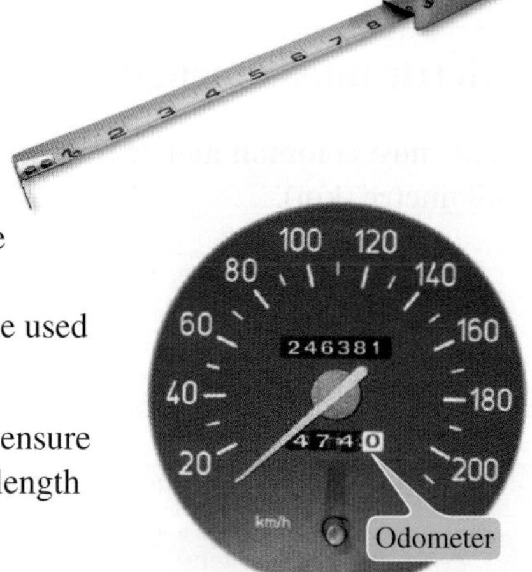

When using a ruler or a tape measure to measure length, it is important to ensure that the zero line of the scale markings is always placed at the start of the length being measured.

When measuring a curved line, some useful tools are a piece of string, a trundle wheel or the edge of a piece of paper, slowly rotated around the line.

WORKED EXAMPLE 1 Recording measurements using correct units

State the measurement marked by the arrow in each of the following. Record your answer in the unit indicated in the brackets.

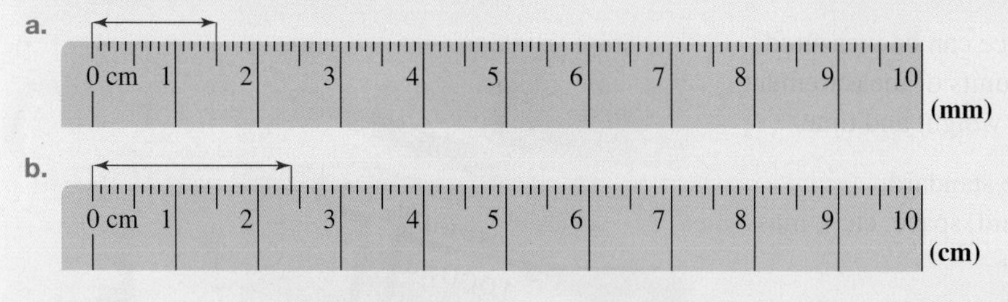

THINK	WRITE
a. 1. From left to right, count how many centimetres the arrow has passed.	**a.** 1 cm
2. Multiply that number by 10 to account for the number of mm in each cm (1 cm = 10 mm).	$1 \times 10 = 10$ mm
3. Add the extra millimetres (mm).	$10 + 5 = 15$ mm
b. 1. Count the number of centimetres the arrow has passed. The whole numbers represent the number of cm.	**b.** 2 cm
2. The extra 4 mm represent $\frac{4}{10}$ cm or 0.4 cm.	0.4 cm
3. Add the whole and part cm.	$2 + 0.4 = 2.4$ cm

The measurement of volume, weight and time has become extremely accurate due to advances in technology. Volume is now measured using advanced laser systems, whereas in the past some of the volumes that occurred in nature could not be accurately measured. Digital scales are used to accurately measure very small or large weights, and atomic clocks can be accurate to the nearest second for millions of years.

13.2.3 Converting between units of measurements

The relationships (or ratios) between the metric units of length can be used to convert a measurement from one unit to another.

Unit conversion

Units of length can be converted as shown in the following diagram. The number next to each arrow is called the conversion factor.

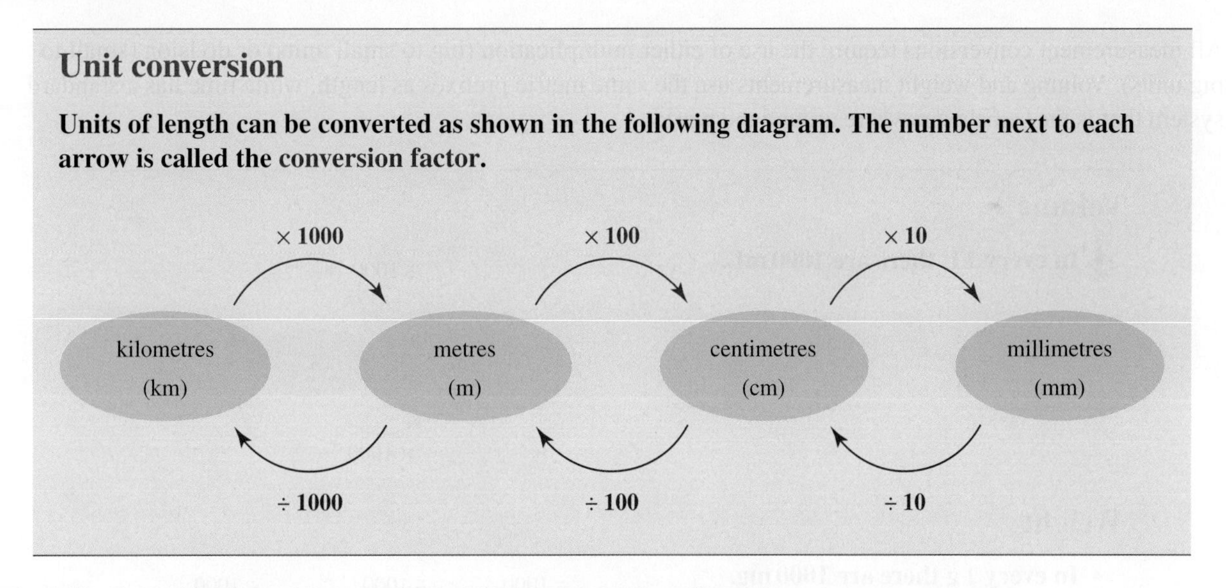

The following table shows how to convert a larger unit to a smaller unit.

In every 1 cm there are 10 mm.	In every 1 m there are 100 cm.	In every 1 km there are 1000 m.
5 cm = 50 mm 0.6 cm = 6 mm (Multiply by 10.)	5 m = 500 cm 0.5 m = 50 cm (Multiply by 100.)	25 km = 25 000 m 0.75 km = 750 m (Multiply by 1000.)

The following table shows how to convert a smaller unit to a larger unit.

In every 10 mm there is 1 cm.	In every 100 cm there is 1 m.	In every 1000 m there is 1 km.
400 mm = 40 cm 52 mm = 5.2 cm (Divide by 10.)	750 cm = 7.5 m 32 cm = 0.32 m (Divide by 100.)	25 000 m = 25 km 750 m = 0.750 km (Divide by 1000.)

tlvd-5058

WORKED EXAMPLE 2 Converting metric lengths

Convert the following lengths to the units shown.

a. 0.234 km = ___ m

b. 24 000 mm = ___ m

THINK	WRITE
a. To convert from kilometres to metres, multiply by 1000, because there are 1000 m for each 1 km.	a. 0.234 km = (0.234 × 1000) m = 234 m
b. To convert from millimetres to metres, first divide by 10 to convert from millimetres to centimetres, then divide by 100 to convert from centimetres to metres.	b. 24 000 mm = (24 000 ÷ 10 ÷ 100) m = (2400 ÷ 100) m = 24 m

All measurement conversions require the use of either multiplication (big to small units) or division (small to big units). Volume and weight measurements use the same metric prefixes as length, while time has a standard system that is the same everywhere around the world.

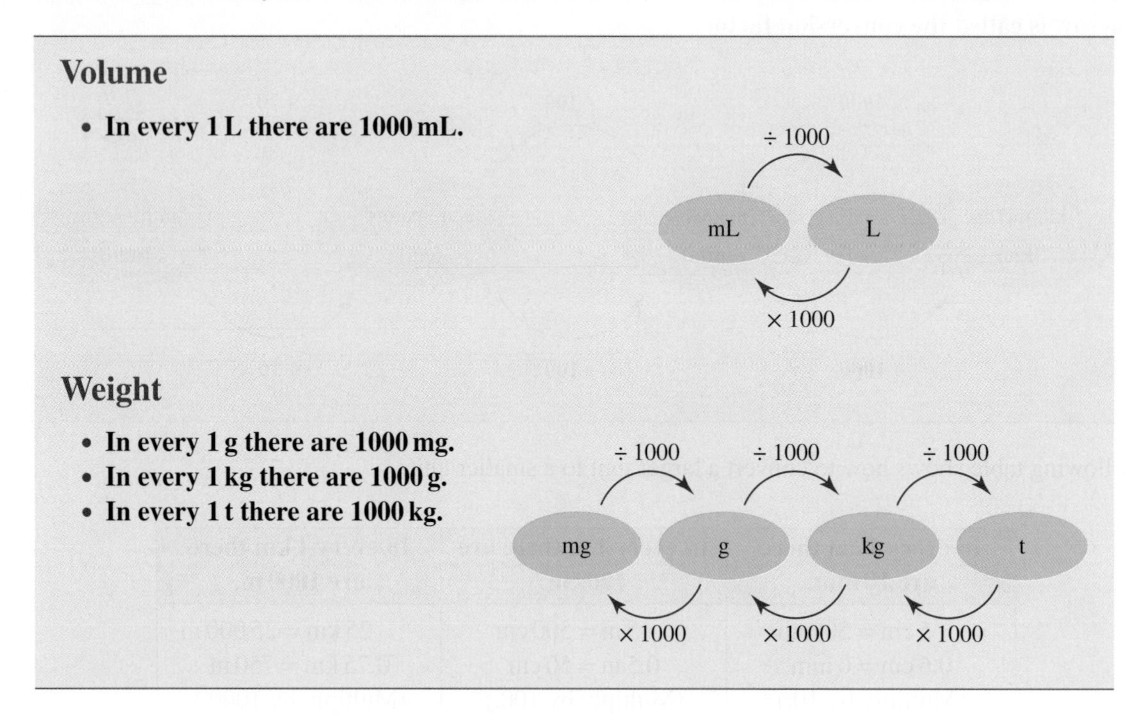

Time

- **60 seconds = 1 minute**
- **7 days = 1 week**
- **365 days = 1 year**
- **60 minutes = 1 hour**
- **12 months = 1 year**
- **10 years = 1 decade**
- **24 hours = 1 day**
- **52 weeks = 1 year**
- **100 years = 1 century**

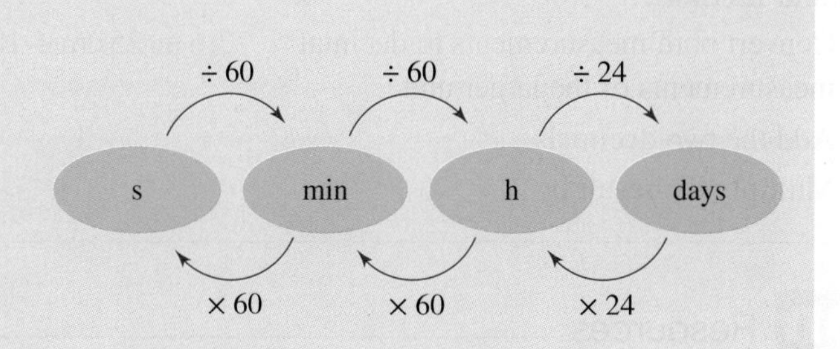

Note: There are 366 days in a leap year, which occurs every 4 years. If the last two digits form a multiple of 4, then the year is a leap year. For example, 2018 is a not a leap year because 18 is not a multiple of 4.

tlvd-5059

WORKED EXAMPLE 3 Converting volumes, weight and time

Convert the following volume, weight and time measurements.

a. **400 mL = ___ L** b. **1500 g = ___ kg** c. **5 h = ___ s**

THINK	WRITE
a. To convert from millilitres (mL) to litres (L) divide by 1000.	a. 400 mL = (400 ÷ 1000) L = 0.4 L
b. To convert from grams (g) to kilograms (kg) divide by 1000.	b. 1500 g = (1500 ÷ 1000) kg = 1.5 kg
c. To convert from hours to seconds, first multiply by 60 to convert from hours to minutes, then multiply by 60 again to convert from minutes to seconds.	c. 5 h = (5 × 60) min = 300 min 300 min = (300 × 60) s = 18 000 s

All measurements need to be in the same unit before they can be added or subtracted.

WORKED EXAMPLE 4 Adding like-sized measurements

Calculate the perimeter of a rectangle 6 m 25 cm wide and 18 m 92 cm long using two different methods.

THINK	WRITE
First method:	
1. Collect the like-sized measurements.	2(6 m 25 cm + 18 m 92 cm) = 2(6 m + 25 cm + 18 m + 92 cm)
2. Add the like-sized measurements.	= 2(6 m + 18 m + 25 cm + 92 cm) = 2(24 m + 117 cm)
3. Convert 117 cm into metres and centimetres. Add 24 m and 1 m.	= 2(24 m + 1 m + 17 cm) = 2(25 m + 17 cm)
4. Multiply the result by 2.	= 50 m + 34 cm

Second method:

1. Convert both measurements to decimal measurements of the larger unit. $2(6\text{ m } 25\text{ cm} + 18\text{ m } 92\text{ cm}) = 2(6.25\text{ m} + 18.92\text{ m})$
2. Add the two decimals. $= 2(25.17\text{ m})$
3. Multiply the result by 2. $= 50.34\text{ m}$

Resources

Interactivities Metric units of length (int-4010)
Converting units of length (int-4011)

13.2 Exercise

Students, these questions are even better in jacPLUS

Receive immediate feedback and access sample responses

Access additional questions

Track your results and progress

Find all this and MORE in jacPLUS

1. **WE1** State the measurement marked by the arrow in each of the following. Record your answer in the unit indicated in brackets.

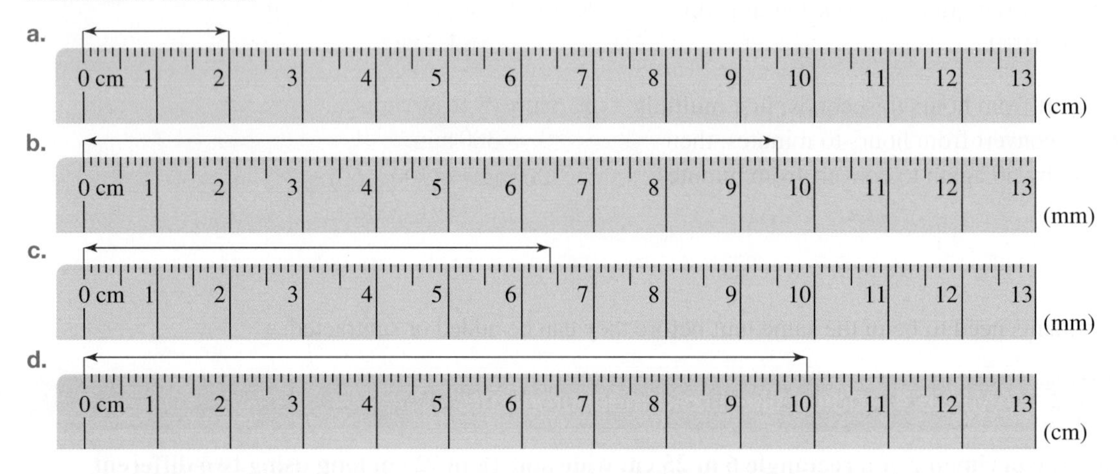

2. a. Convert 5600 m to km.
   b. Convert 21 cm to mm.
   c. Convert 55 km to m.
   d. Convert 2.2 m to cm.
   e. Convert 350 cm to m.
   f. Convert 25 mm to cm.
3. **WE2** Convert the following lengths to the units shown.
   a. 0.165 km = ______ m
   b. 780 cm = ______ m
   c. 0.158 m = ______ cm
   d. 13 m = ______ km
   e. $1\frac{1}{4}$ km = ______ cm
   f. 36 000 mm = ______ m
4. Convert the following measurements as indicated.
   a. Convert 5632 cm to km.
   b. Convert 204 310 cm to km.
   c. Convert 5.5 km to cm.
   d. Convert 2.2 m to mm.
   e. Convert 35 000 cm to km.
   f. Convert 250 mm to m.

5. Arrange the following lengths in descending order.

a. 89 000 cm, 7.825 m, 98 760 mm, 0.3217 km
b. 0.786 mm, 0.786 km, 0.000 786 m, 0.0786 cm

6. Arrange each of the following in ascending order.

a. 0.15 km, 135 m, 2400 cm
b. 25 cm, 120 mm, 0.5 m
c. 9 m, 10 000 mm, 0.45 km
d. 32 000 cm, 1200 m, 1 km
e. 1.5 m, 150 cm, 0.0015 km
f. 8.25 km, 825 m, 90 000 cm

7. **WE4** Calculate the perimeter of each of the following rectangles, in centimetres.

a. 15 mm wide and 5 cm long
b. 1.5 m long and 20 mm wide

8. Calculate the value of each of the following.

a. 855 mm + 1.8 m
b. 5.67 cm + 1156 mm − 0.25 m
c. 15 cm 15 mm + 27 cm 86 mm
d. 6 km 58 m + 84 km 47 m
e. 75 cm 10 mm − 15 cm 5 mm
f. 125 m 58 cm − 18 m 85 cm

9. Determine which volume measurement is the most appropriate for each of the following objects.

a.
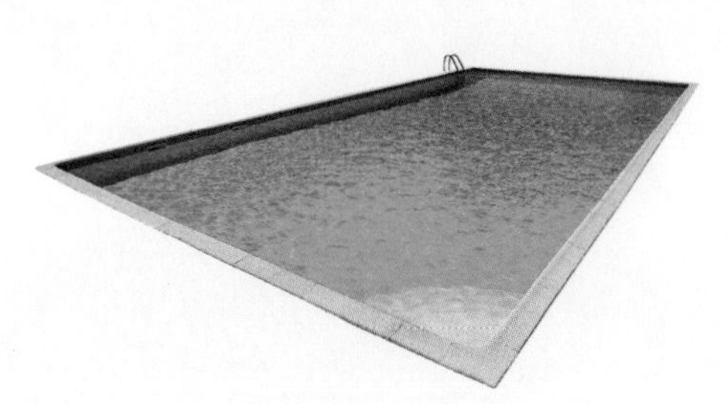

b.

10. Determine which weight measurement is the most appropriate for each of the following objects.

a.

b.
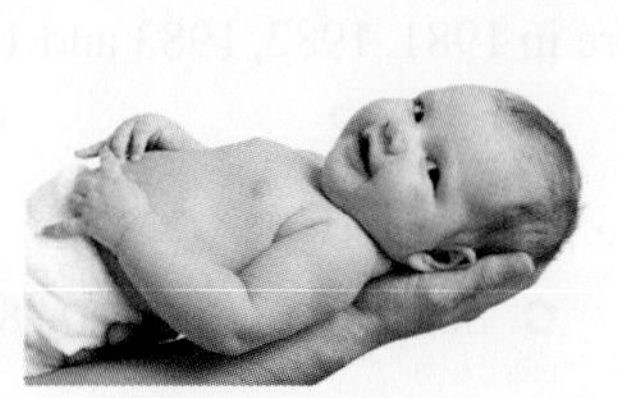

c.
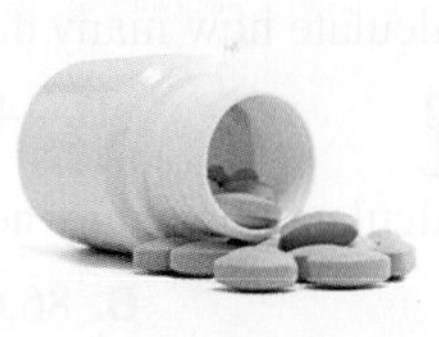

11. **MC** Of the following, the largest volume is:

A. 1500 mL
B. 3 L
C. 0.465 L
D. 858 mL
E. 1.521 L

12. **WE3** Convert the following volume measurements.

a. 2500 mL to L
b. 3 L to mL
c. 3.1 mL to L
d. 5.25 L to mL
e. 7.5 L to mL
f. $3\frac{3}{4}$ mL to L

13. Arrange the following weights in ascending order.

a. 2500 g, 2.4 kg, 400 000 mg, 0.000 024 t
b. 0.05 kg, 54 g, 52 000 mg, 0.0005 t

14. State whether each of the following is True or False.

a. 1555 mg > 2 g
b. 3.2 kg = 320 g
c. 45 000 mg < 0.145 kg
d. 500 000 mg > 500 g

15. Convert the following weight measurements.

a. 505 g = ______ mg
b. 4.5 mg = ______ g
c. 6.4 kg = ______ g
d. 5840 g = ______ kg
e. 75 821 mg = ______ kg
f. 0.000 07 kg = ______ mg

16. Fill in the blanks with the appropriate time measurement from the list below. (Use each term only once.)

second, hours, weeks, minutes, years, day

a. The speed of light is 299 792 458 m/_____.
b. Distance to most planets is measured in light _________.
c. A full-time job is considered to be 35–40 ______ per week.
d. There is always at least one low and one high tide every _______.
e. It takes about 2 ________ to cook instant popcorn.
f. A normal pregnancy takes between 37 and 42 ________.

17. Arrange each of the following in descending order.

a. 2 decades, 1.5 centuries, 250 years, 500 months
b. 750 minutes, 12.6 hours, 4250 seconds, 0.56 days, 0.1 weeks

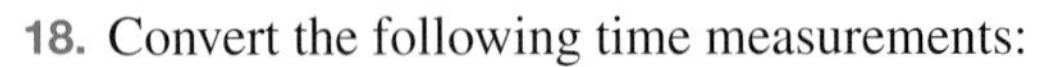

18. Convert the following time measurements:

a. 4.5 decades = _______ years
b. 825 years = ______ decades
c. 3.4 centuries = _______ years
d. 485 years = ______ centuries
e. 5 centuries = ________ decades
f. 120 decades = ______ centuries

19. Convert the following time measurements. Give answers to 2 decimal places where necessary.

a. 240 s = ______ min
b. 5.2 min = ______ s
c. 4.6 h = ______ min
d. 320 min = ______ h
e. 30 000 s = ______ h
f. 20.3 h = ______ s

20. **MC** Calculate how many days there were in 1981, 1982, 1983 and 1984 combined.

A. 1460
B. 1461
C. 1462
D. 1459
E. 1458

21. **MC** Calculate how many hours there are in 3.6 days.

A. 86
B. 86.6
C. 864
D. 86.4
E. 866

22. **MC** Calculate how many days there are in 396 hours.

A. 16.5
B. 16
C. 15.5
D. 15
E. 14.5

23. Explain why it would not be wise to use millimetres to measure the distance between Melbourne and Perth.

24. Wayne's Concreting Crew needs to pour concrete for a driveway. They have calculated that the driveway needs 20 cubic metres (20 m^3) of concrete.

a. If 1 cubic metre equals 1000 litres, calculate how many litres of concrete are needed.
b. Calculate how many cubic metres you can cover if you have 2584 litres of concrete.

13.3 Areas of composite figures

LEARNING INTENTION

At the end of this subtopic you should be able to:

- calculate the areas of composite figures by adding and subtracting the areas of regular shapes.

13.3.1 Sum of areas

A **composite figure** can be divided into two or more sections, each of which is a smaller regular shape with a known area formula. The shape shown here is a combination of a semicircle and a triangle.

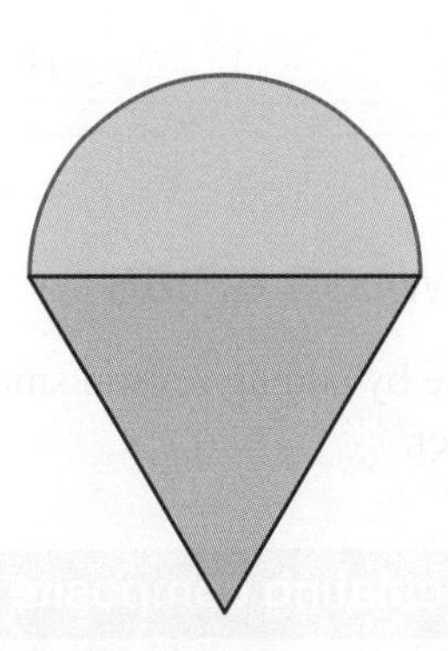

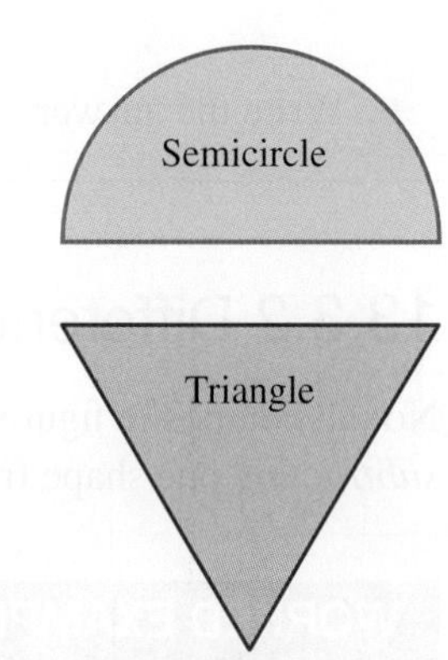

Determining the area of a composite figure

- **Divide the shape into smaller figures that have a known area formula.**
- **Calculate the area of each smaller figure.**
- **Calculate the area of the composite figure by *adding* the areas of the smaller figures.**

tlvd-5060

WORKED EXAMPLE 5 Calculating composite areas by adding

Calculate the area of the following composite figure.

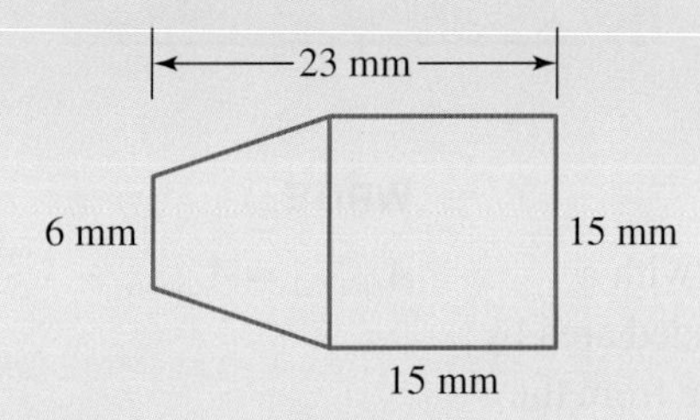

THINK	WRITE
1. This composite figure is made up of a square and a trapezium. Add the areas of the two to calculate the total area.	$A_{\text{total}} = A_{\text{trapezium}} + A_{\text{square}}$
2. Calculate the area of the square.	$A_{\text{square}} = l \times l$ $= 15 \times 15$ $= 225 \text{ mm}^2$

3. Calculate the area of the trapezium, where the height is 23 – 15 = 8 mm.

$A_{\text{trapezium}} = \frac{1}{2}(a+b)h$
$= \frac{1}{2}(6+15)\times 8$
$= 84 \text{ mm}^2$

4. Calculate the total area of the figure by adding the two areas.

$A_{\text{total}} = A_{\text{trapezium}} + A_{\text{square}}$
$= 84 + 225$
$= 309 \text{ mm}^2$

5. Write the answer.

The area of the figure is 309 mm^2.

13.3.2 Difference between areas

Not all composite figures are made by adding several smaller shapes together; some shapes are made by *subtracting* one shape from another.

tlvd-5061

WORKED EXAMPLE 6 Calculating composite areas by subtracting

Calculate the area of the shaded region.

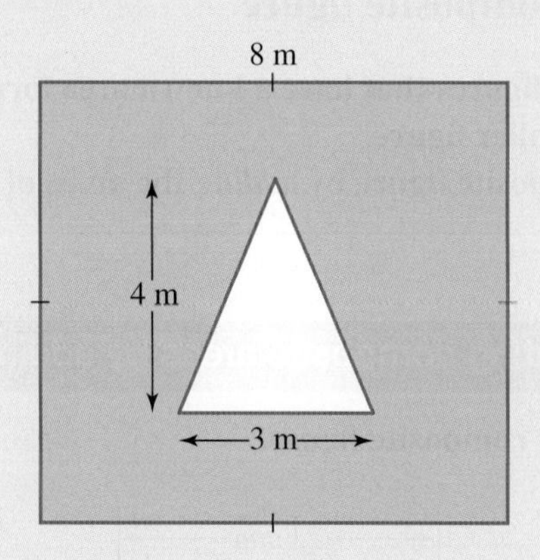

THINK

1. This composite shape is a square with a triangle cut out. Calculate the shaded area by subtracting the area of the triangle from the area of the square.
2. Calculate the area of the square.
3. Calculate the area of the triangle.

WRITE

1. $A_{\text{shaded}} = A_{\text{square}} - A_{\text{triangle}}$

2. $A_{\text{square}} = l \times l$
$= 8 \times 8$
$= 64 \text{ m}^2$

3. $A_{\text{triangle}} = \frac{1}{2} \times b \times h$
$= \frac{1}{2} \times 3 \times 4$
$= 6 \text{ m}^2$

4. Calculate the shaded area by subtracting the area of the triangle from the area of the square.

$A_{\text{shaded}} = A_{\text{square}} - A_{\text{triangle}}$
$= 64 - 6$
$= 58 \text{ m}^2$

5. Write the answer.

The shaded area is 58 m^2.

WORKED EXAMPLE 7 Calculating composite areas with circles

Calculate the area of the shaded region to 2 decimal places.

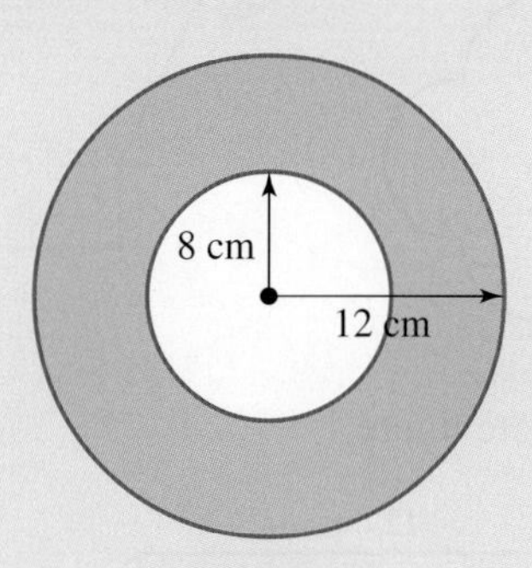

THINK	WRITE
This composite shape is a circle with a smaller circle cut out. Calculate the shaded area by subtracting the area of the smaller circle from the area of the larger circle.	$A_{\text{shaded}} = A_{\text{larger circle}} - A_{\text{smaller circle}}$
Calculate the area of the larger circle.	$A_{\text{larger circle}} = \pi r^2$ $= \pi \times 12^2$ $= 452.3893\ldots \text{ cm}^2$
Calculate the area of the smaller circle.	$A_{\text{smaller circle}} = \pi r^2$ $= \pi \times 8^2$ $= 201.0619\ldots \text{ cm}^2$
Calculate the shaded area by subtracting the area of the smaller circle from the area of the larger circle.	$A_{\text{shaded}} = A_{\text{larger circle}} - A_{\text{smaller circle}}$ $= 452.3893 - 201.0619$ $= 251.3274 \text{ cm}^2$
Write the answer correct to 2 decimal places.	The shaded area is 251.33 cm^2.

on Resources

Video eLesson Composite area (eles-1886)

13.3 Exercise

Students, these questions are even better in jacPLUS

Receive immediate feedback and access sample responses

Access additional questions

Track your results and progress

Find all this and MORE in jacPLUS

1. Name the two smaller shapes that make the following composite shapes.

a. 

b. 

c. 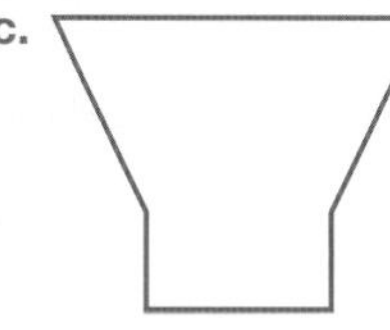

d. 

2.  WE5 Calculate the area of the composite figure.

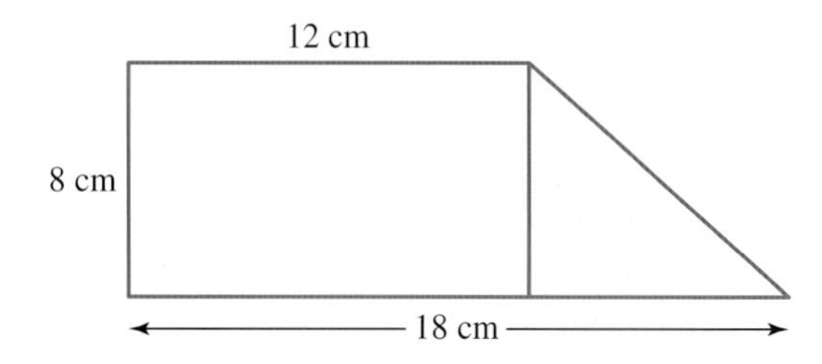

3. Calculate the area of the following shape to 2 decimal places.

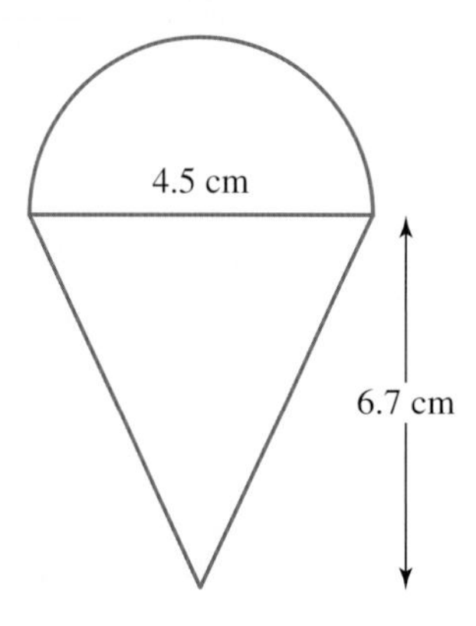

4. Calculate the areas of the following composite shapes.

a. 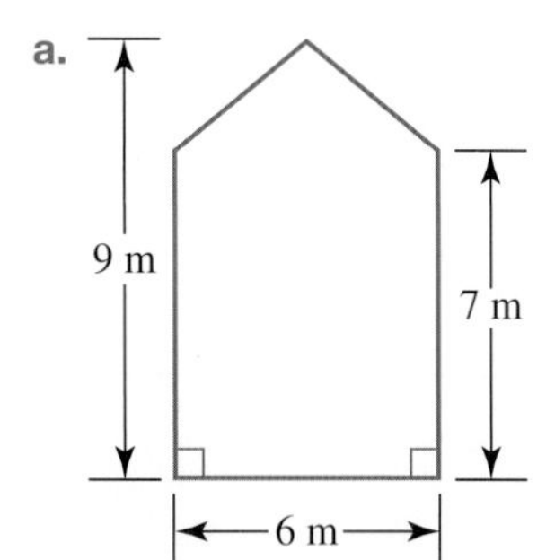

b. 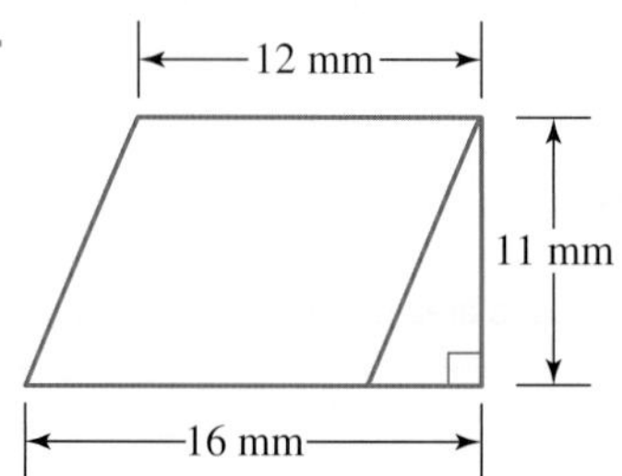

c.

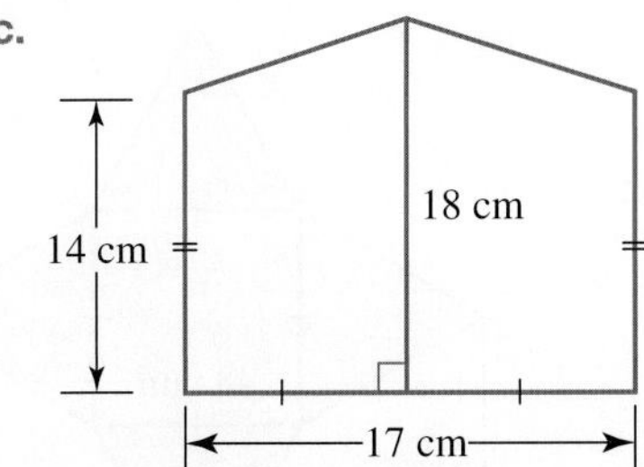

d.

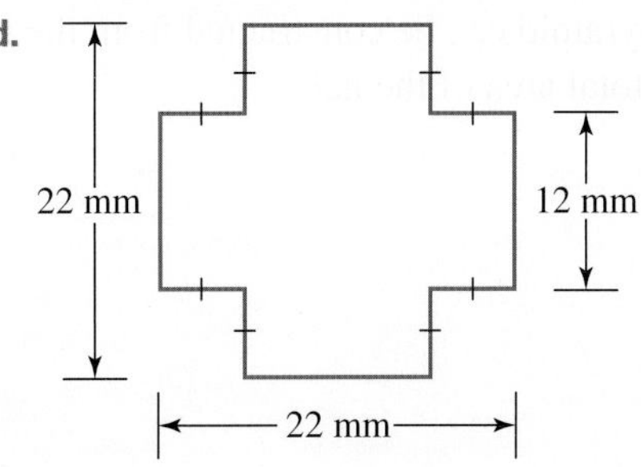

5. Calculate the areas of the following composite shapes.

a.

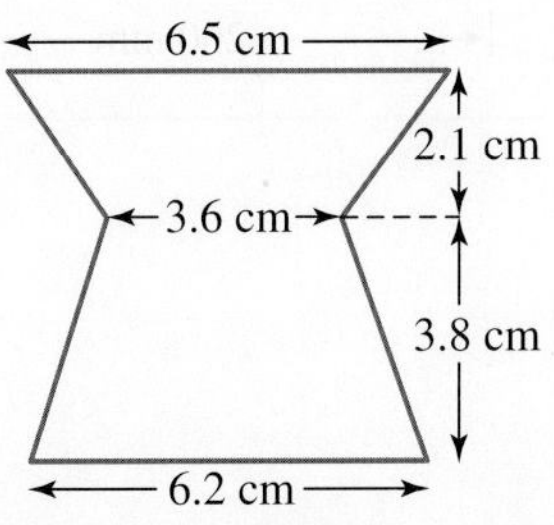

b.

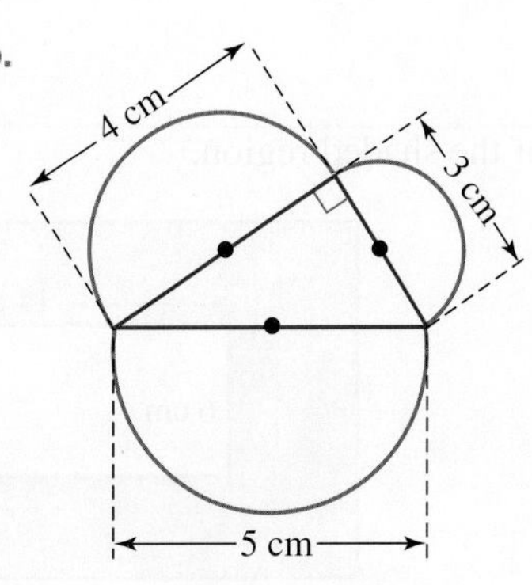

c.

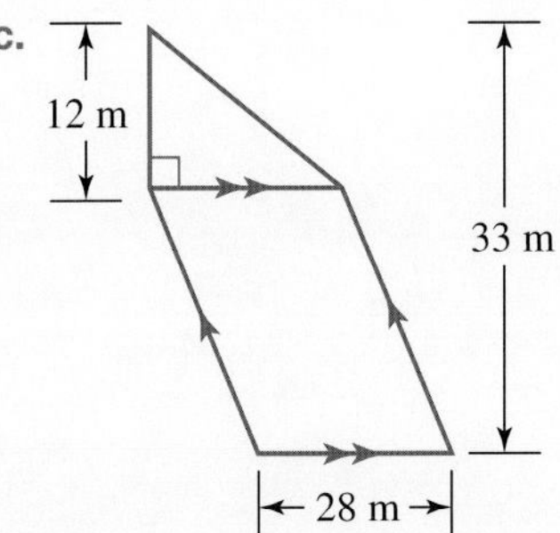

d.

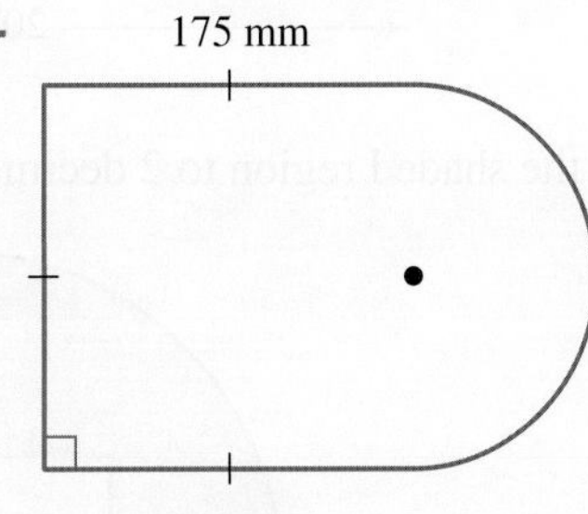

6. **WE6** Calculate the shaded areas of the following shapes.

a.

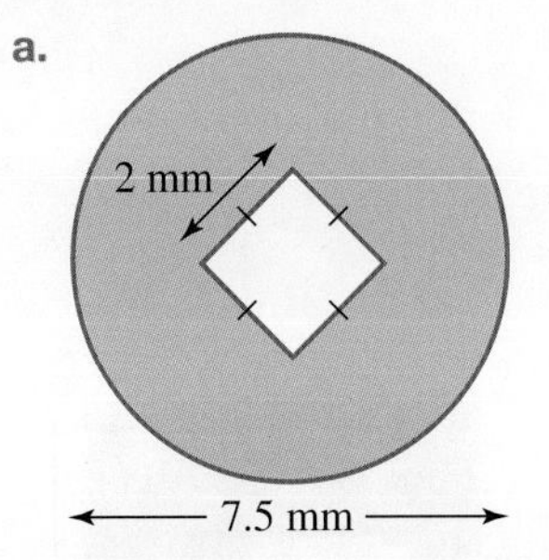

b.

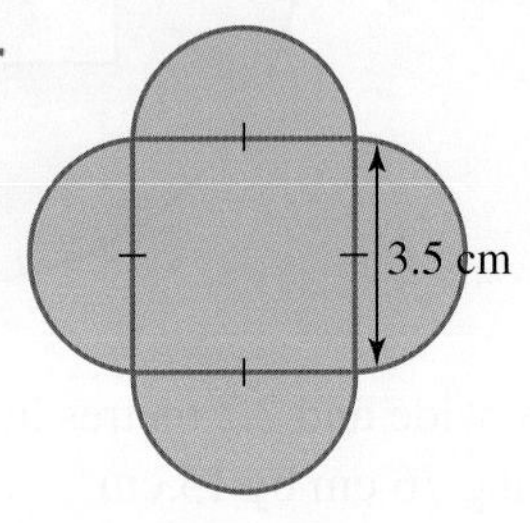

7. Consider the hexagon shown.
   a. Calculate the area of the hexagon by dividing it into two trapeziums.
   b. Calculate the area of the hexagon by dividing it into a rectangle and two triangles.
   c. Compare your answers to parts **a** and **b**.

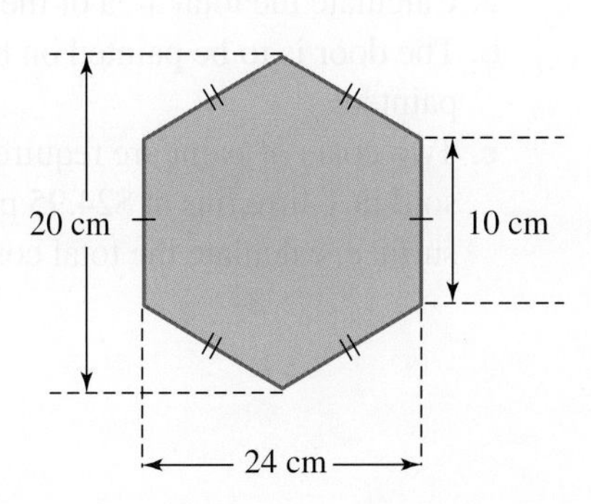

8. A triangular pyramid can be constructed from the net shown. Calculate the total area of the net.

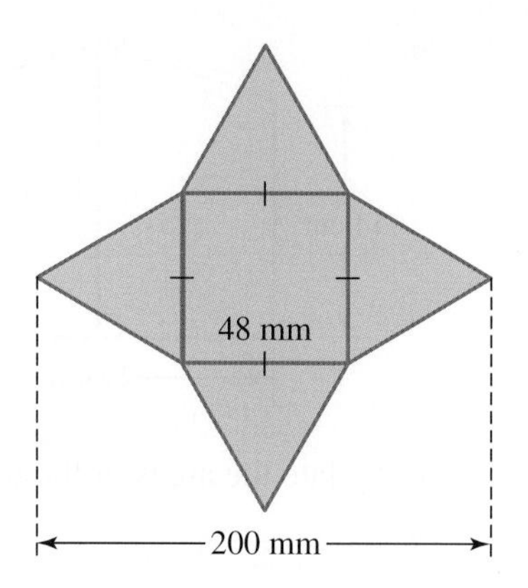

9. Calculate the area of the shaded region.

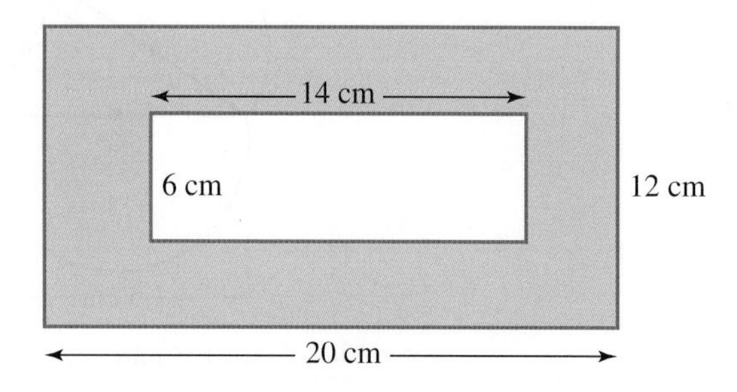

10. Calculate the area of the shaded region to 2 decimal places.

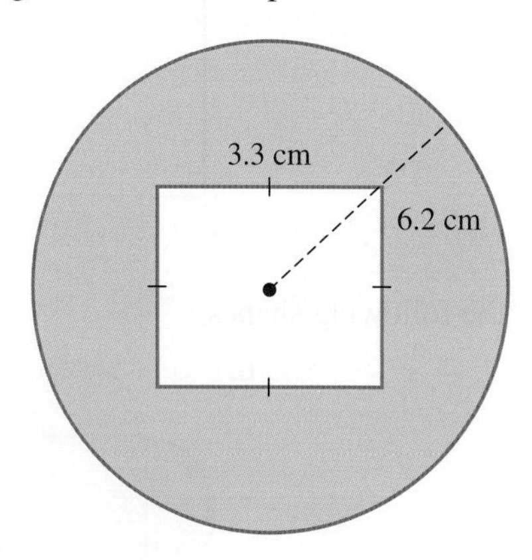

11. The door shown is 1 metre wide and 2.2 metres high. It has four identical glass panels, each measuring 76 cm by 15 cm.
    a. Calculate the total area of the glass panels.
    b. The door is to be painted on both sides. Calculate the total area to be painted.
    c. Two coats of paint are required on each side of the door. If the paint is sold in 1-litre tins at $24.95 per litre and each litre covers 8 m^2 of the surface, calculate the total cost of painting the door.

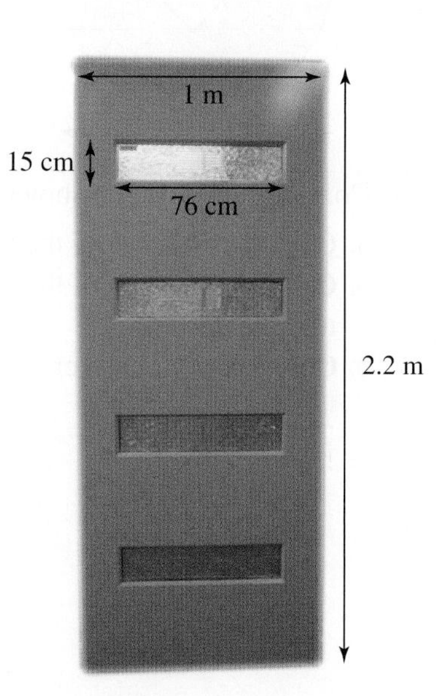

12. Calculate the area of the shaded region shown.

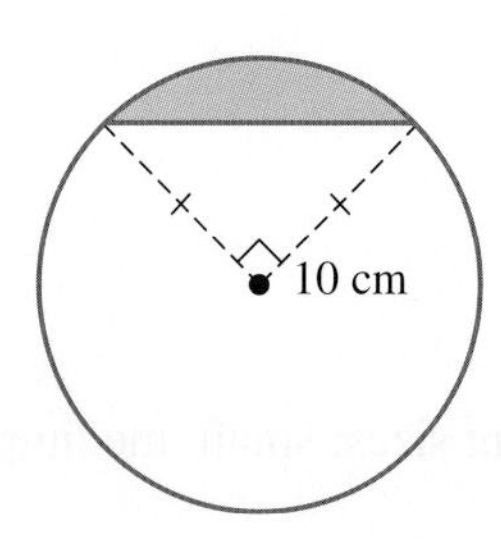

13. The leadlight panel shown depicts a sunrise over the mountains. The mountain is represented by a green triangle 45 cm high. The yellow sun is represented by a section of a circle with an 18 cm radius. There are 10 yellow sunrays in the shape of isosceles triangles with a base of 3 cm and a height of 12 cm, and the sky is blue. Calculate the area of the leadlight panel made of:

a. green glass
b. yellow glass
c. blue glass.

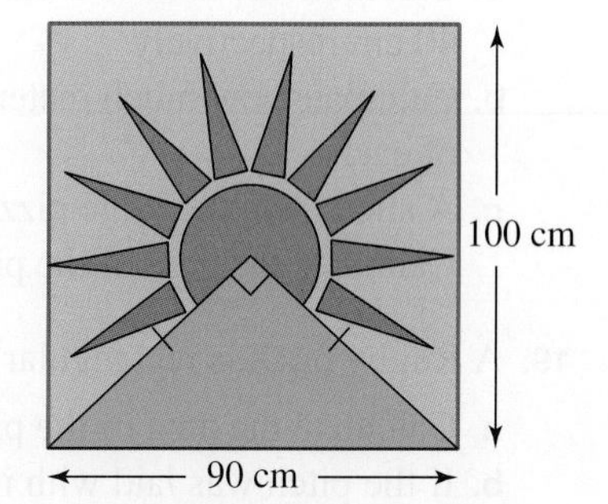

14. WE7 Calculate the area of the shaded region to 2 decimal places.

a.

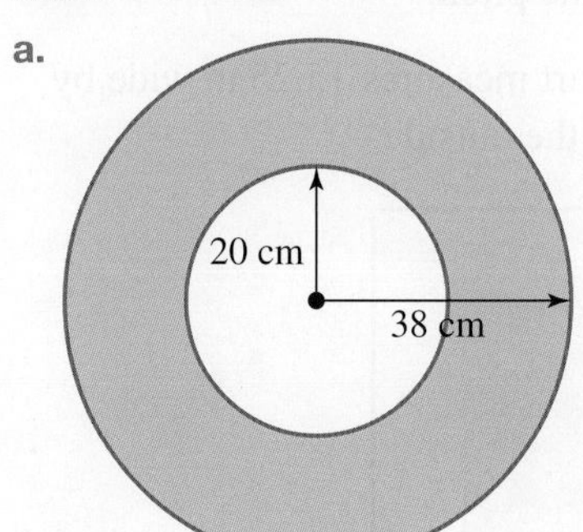

b.

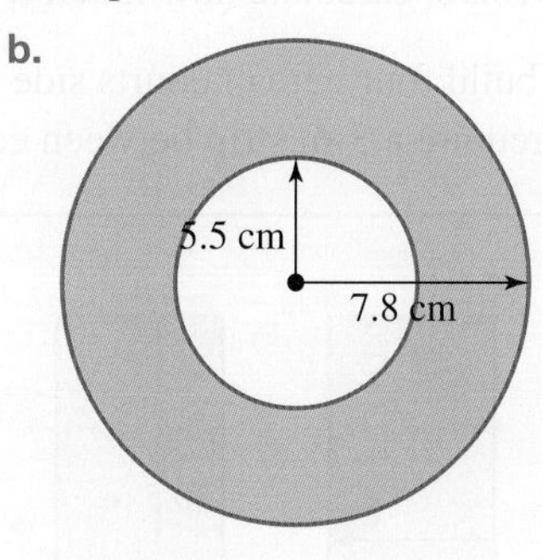

15. Calculate the area of the shaded region to 2 decimal places.

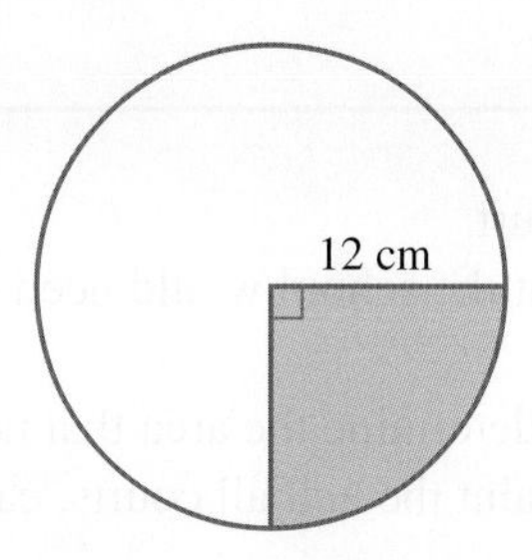

16. Calculate the area of the shaded region to 2 decimal places.

a.

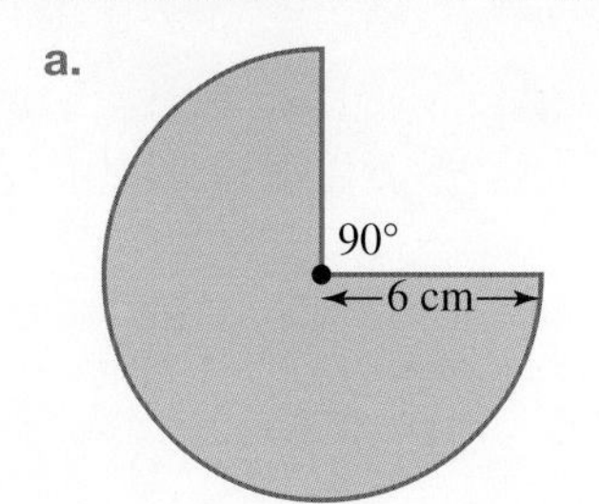

b.

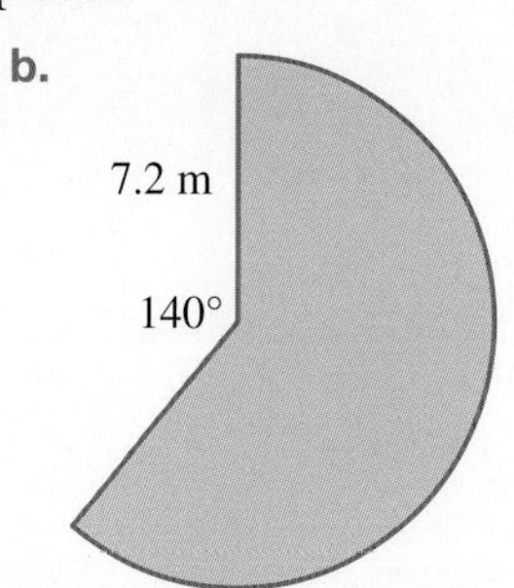

17. MC The area of a circle is equal to 113.14 mm^2. The radius of the circle is closest to:

A. 36 mm
B. 6 mm
C. 2133 mm
D. 7 mm
E. 8 mm

18. Circular pizza trays come in three different sizes: small, medium and large.
    a. Calculate the area of each tray if the diameters are 20 cm, 30 cm and 40 cm respectively.
    b. Calculate how much material must be ordered to make 50 trays of each.
    c. A slice from the large pizza makes an angle from the centre of 45°. Calculate the area of the pizza slice.

19. A Rugby pitch is rectangular and measures 100 m in length and 68 m in width.
    a. Calculate the area of the pitch.
    b. If the pitch was laid with instant turf, with each sheet measuring 4 m by 50 cm, calculate how many sheets of turf are required to cover the pitch.
    c. If each sheet costs $10.50, calculate how much it would cost to cover the pitch.

20. A school is looking to build four netball courts side by side. A netball court measures 15.25 m wide by 30.5 m long, and they require a 3-m strip between each court and around the outside.

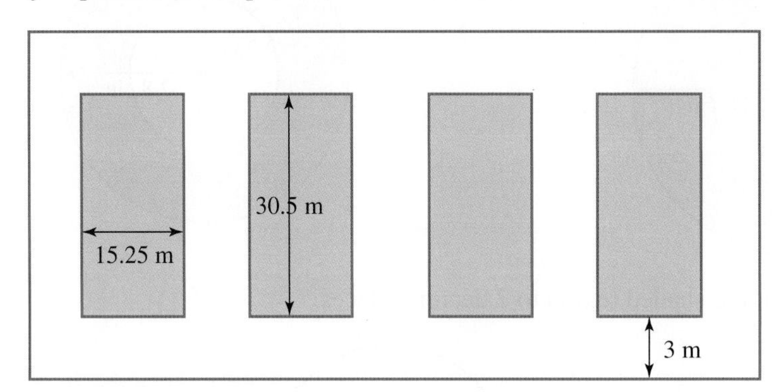

    a. Determine the area of each netball court.
    b. Calculate the total area, in metres, that the school would need for the four netball courts and its surrounding space.
    c. If the four netball courts are painted, determine the area that needs to be painted.
    d. If it costs $9.50 per square metre to paint the netball courts, calculate how much it would cost to paint the four courts.

13.4 Surface areas of pyramids and irregular solids

LEARNING INTENTION

At the end of this subtopic you should be able to:

- calculate the total surface area of cones and pyramids
- calculate the total surface area of composite shapes.

13.4.1 Review of total surface area

The total surface area (TSA) is the sum of all the areas that you can see in a solid. For example, the total surface area of a cube is the sum of the area of the six faces of the cube.

In previous years, you studied the formulas for the total surface area of a sphere and the total surface area of a cylinder.

Total surface area of a cylinder

$$\text{TSA}_{\text{cylinder}} = 2\pi r^2 + 2\pi rh$$

Total surface area of a sphere

$$\text{TSA}_{\text{sphere}} = 4\pi r^2$$

13.4.2 Total surface area of a cone

The total surface area of a cone, when the radius (r) and the slant height (s) are known, can be found using the formula shown below.

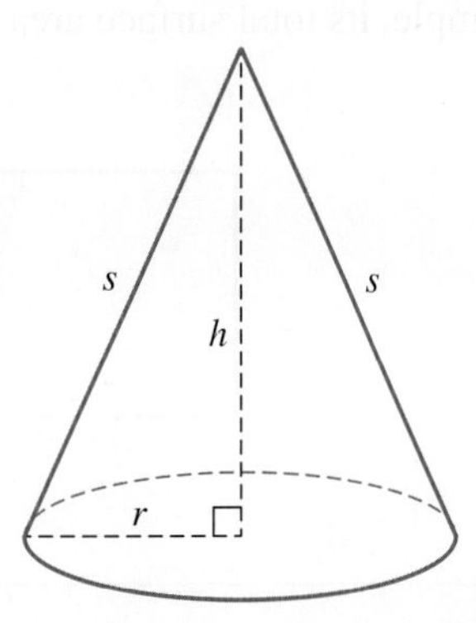

The slant height is the measurement along the side of the cone, not the vertical height (h) of the cone.

Total surface area of a cone

$$\text{TSA}_{\text{cone}} = \pi rs + \pi r^2$$

tlvd-5062

WORKED EXAMPLE 8 Calculating the surface area of a cone

Calculate the total surface area of the cone shown correct to 1 decimal place.

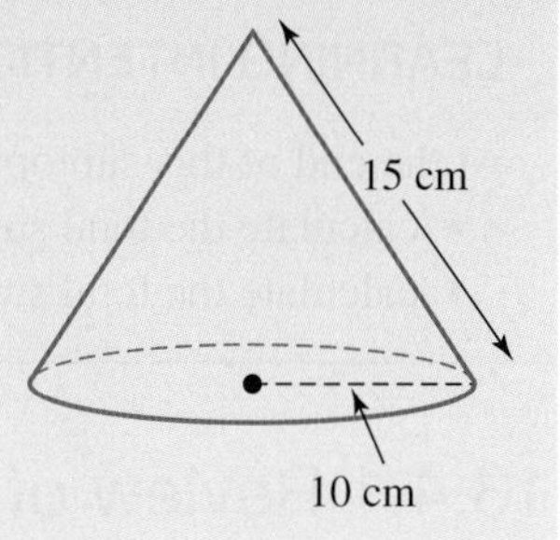

THINK	WRITE
1. Identify the shape as a cone and write the appropriate total surface area formula.	$\text{TSA}_{\text{cone}} = \pi rs + \pi r^2$
2. Identify the dimensions of the object.	$r = 10$ and $s = 15$
3. Substitute the values for the radius and the slant height into the TSA formula and calculate the TSA of the cone.	$\text{TSA}_{\text{cone}} = \pi \times 10 \times 15 + \pi \times 10^2$ $= 150\pi + 100\pi$ $= 250\pi$ $= 785.398\,163\,4$ ≈ 785.4
4. Write the answer.	The total surface area of the cone is 785.4 cm^2.

13.4.3 Total surface area of a pyramid

The total surface area of a pyramid can be calculated by adding the area of the base to the area of each of its sides.

Taking the square-based pyramid as an example, its total surface area can be found as shown.

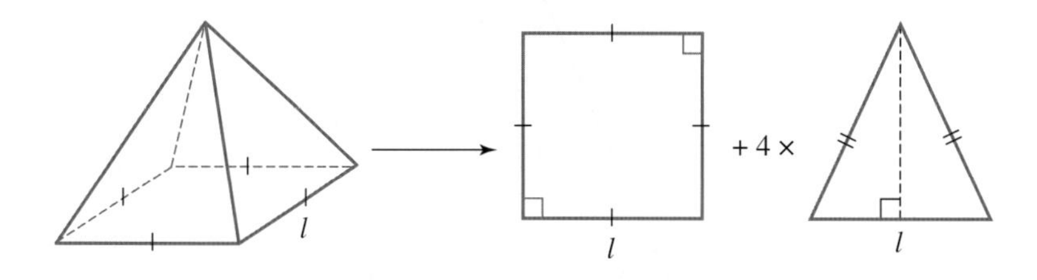

Total surface area of a square-based pyramid

$$\text{TSA}_{\text{square-based pyramid}} = A_{\text{base}} + 4 \times A_{\text{sides}}$$

This method can be adapted to rectangular-based pyramids and triangular-based pyramids.

WORKED EXAMPLE 9 Calculating the surface area of a pyramid

Calculate the total surface area of the pyramid shown correct to 2 decimal places.

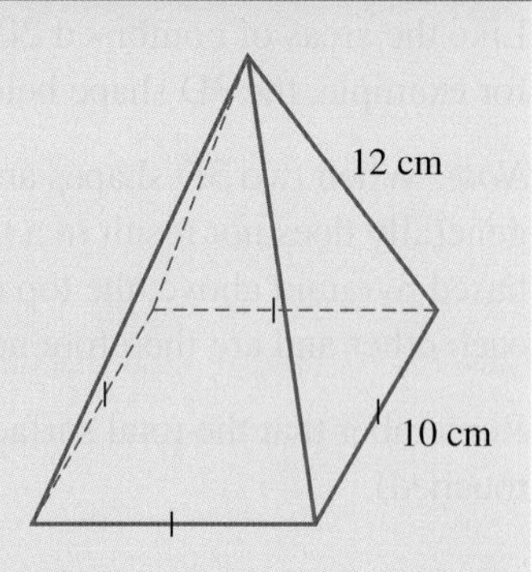

THINK

WRITE

1. The pyramid is made up of a square base and four triangular sides.

$$TSA_{\text{square-based pyramid}} = A_{\text{base}} + 4 \times A_{\text{sides}}$$

2. Calculate the area of the square base with a side length of 10 cm.

$$\begin{aligned} A_{\text{base}} &= l^2 \\ &= 10^2 \\ &= 100 \text{ cm}^2 \end{aligned}$$

3. To calculate the area of the triangular side, we need to determine the vertical height using Pythagoras' theorem.

Hypotenuse is 12 cm and half the width of the base is 5 cm.

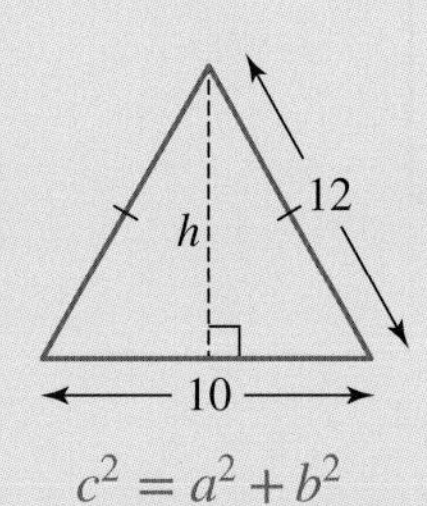

$$\begin{aligned} c^2 &= a^2 + b^2 \\ 12^2 &= 5^2 + h^2 \\ 144 &= 25 + h^2 \end{aligned}$$

Subtract 25 from both sides. To undo the square, take the square root of both sides.

$$\begin{aligned} h^2 &= 144 - 25 \\ h^2 &= 119 \\ h &= \sqrt{119} \end{aligned}$$

4. Calculate the area of the triangular side.

$$\begin{aligned} A_{\text{triangle}} &= \frac{1}{2}bh \\ &= \frac{1}{2} \times 10 \times \sqrt{119} \\ &= 54.5436 \text{ cm}^2 \end{aligned}$$

5. Calculate the TSA.

$$\begin{aligned} TSA_{\text{square-based pyramid}} &= A_{\text{base}} + 4 \times A_{\text{sides}} \\ &= 100 + 4 \times 54.5436 \\ &= 318.1744 \text{ cm}^2 \end{aligned}$$

6. Write the answer.

The total surface area of the pyramid is 318.17 cm^2.

13.4.4 Surface area of a composite shape

Like the areas of combined 2D shapes, some 3D shapes are made by combining a number of regular 3D shapes; for example, the 3D shape below can be made using a cube and a square-based pyramid as shown.

Note: When two 3D shapes are combined together, adding the total surface area of the two individual shapes generally does not result in a total surface area of the composite shape. Referring to the cube and the square-based pyramid above, the top of the cube (green) and the base of the square-based pyramid (pink) sit on top of each other and are therefore no longer an outside surface.

Remember that the total surface area is the area of all the **outside surfaces** (i.e. the surfaces that can be touched).

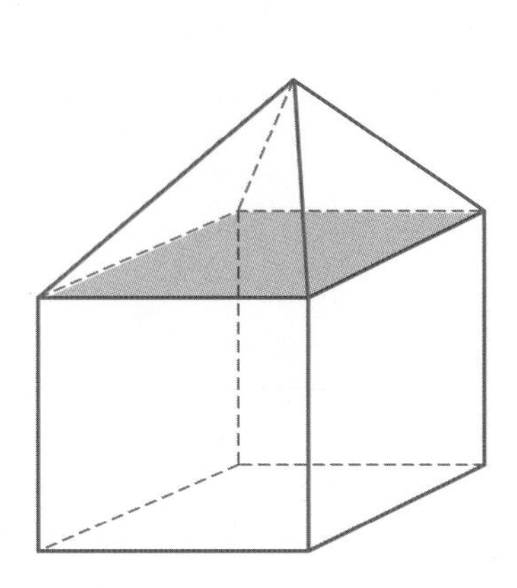

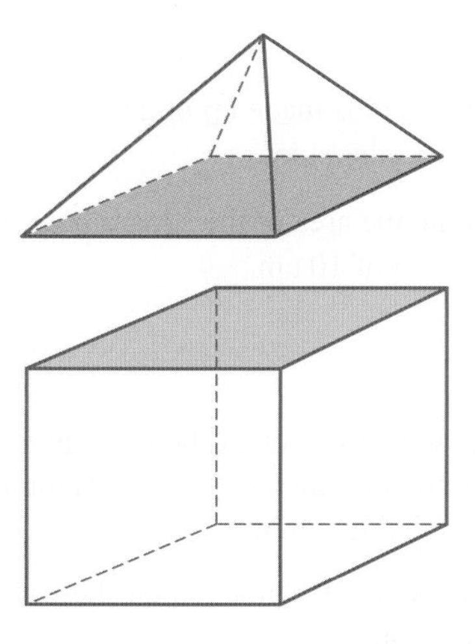

tlvd-5063

WORKED EXAMPLE 10 Calculating the surface area of a composite shape

Calculate the total surface area of the shape shown.

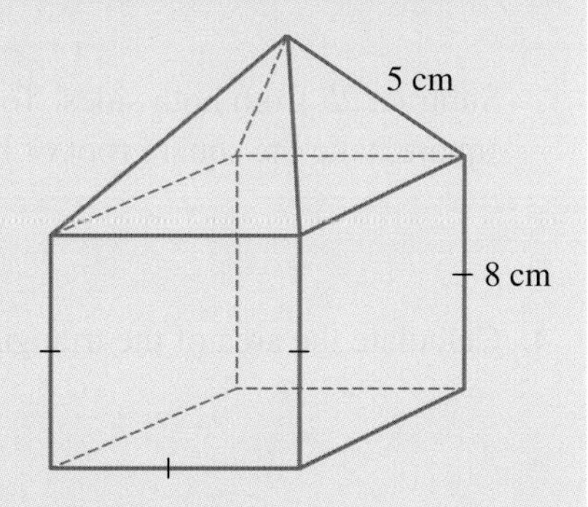

THINK	WRITE
1. The total surface area of the shape is made up of 5 faces of the cube and the 4 sides of the pyramid.	$\text{TSA} = 5 \times A_{\text{cube sides}} + 4 \times A_{\text{pyramid sides}}$
2. Calculate the area of the cube sides.	$A_{\text{cube sides}} = 8 \times 8$ $= 64$

3. Calculate the vertical height of the triangular side of the pyramid using Pythagoras' theorem.

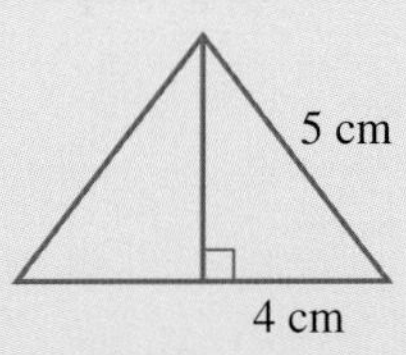

$$c^2 = a^2 + b^2$$
$$5^2 = a^2 + 4^2$$
$$25 = a^2 + 16$$

Subtract 16 from both sides.

$$a^2 = 25 - 16$$
$$a^2 = 9$$

Take the square root of both sides.

$$a = 3$$

4. Calculate the area of the pyramid sides.

$$A_{\text{pyramid side}} = \frac{1}{2} \times 8 \times 3 = 12$$

5. Calculate the total surface area of the shape.

$$\begin{aligned} \text{TSA} &= 5 \times A_{\text{cube sides}} + 4 \times A_{\text{pyramid sides}} \\ &= (5 \times 64) + (4 \times 12) \\ &= 320 + 48 \\ &= 368 \end{aligned}$$

6. Write the answer.

The total surface area of the shape is 368 cm^2.

Resources

Interactivity Area of composite shapes (int-4020)

13.4 Exercise

Students, these questions are even better in jacPLUS

Receive immediate feedback and access sample responses

Access additional questions

Track your results and progress

Find all this and MORE in jacPLUS

1. WE8 Calculate the total surface area of the cone shown, correct to 2 decimal places.

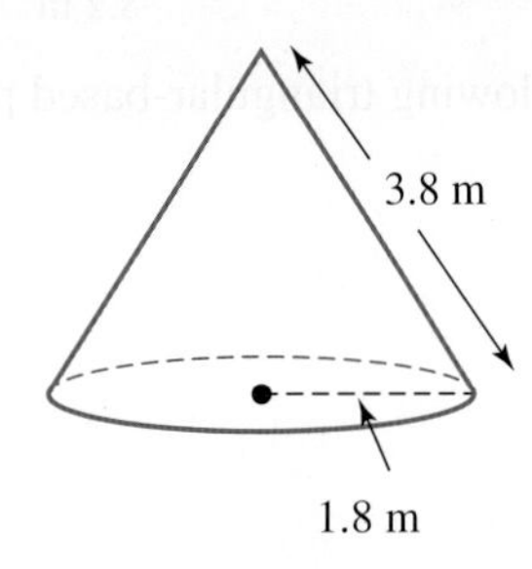

2. Calculate the total surface area of the following shape, correct to 2 decimal places.

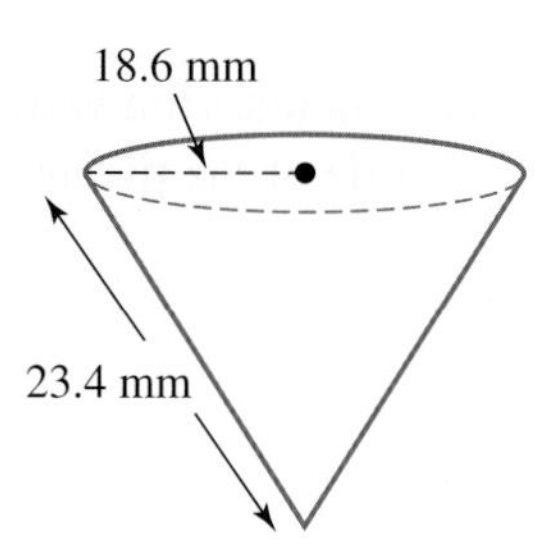

3. Calculate the total surface area of the following shapes, correct to 2 decimal places.

a.

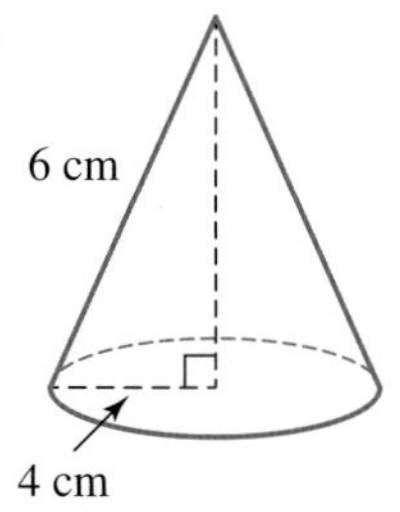

b.

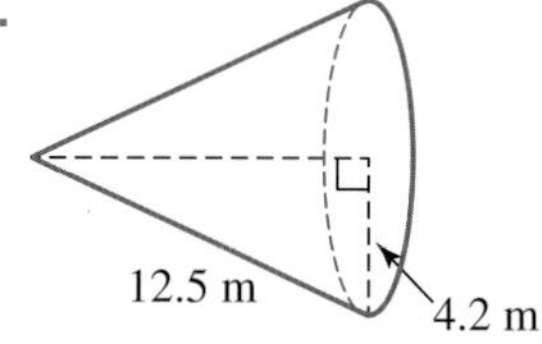

4. WE9 Calculate the total surface area of the following pyramid, correct to 2 decimal places.

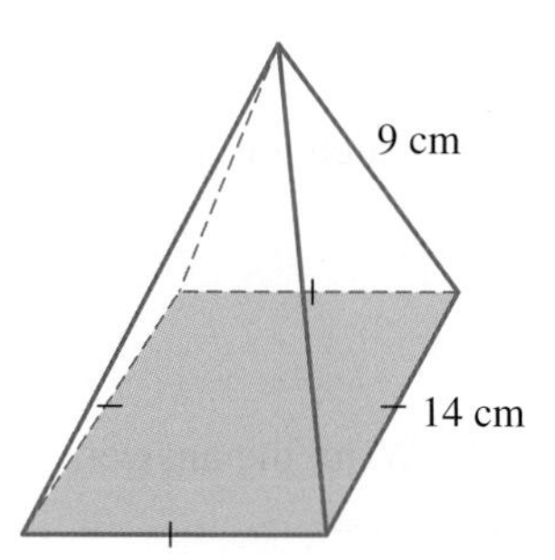

5. Calculate the surface area of the following shape, correct to 2 decimal places.

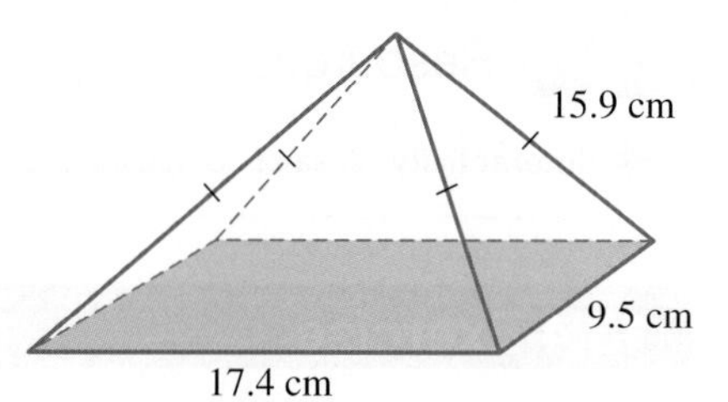

6. Calculate the total surface area of each of the following shapes, correct to 2 decimal places.

a.

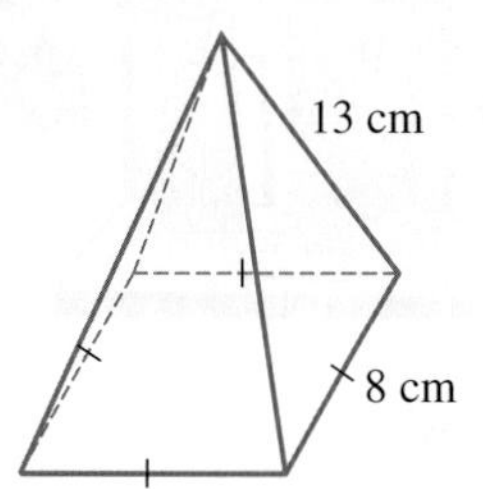

b.

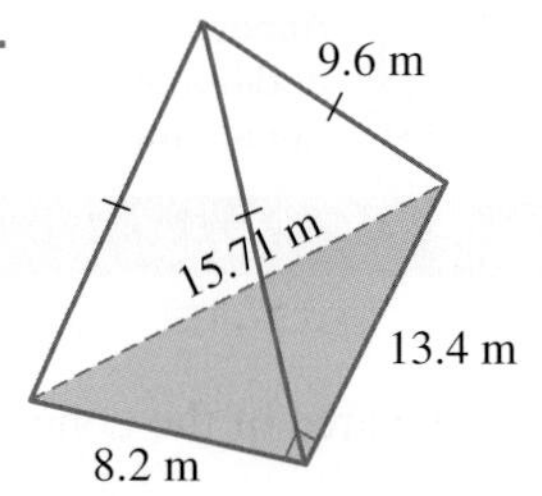

7. Calculate the total surface area of the following triangular-based pyramid, correct to 2 decimal places.

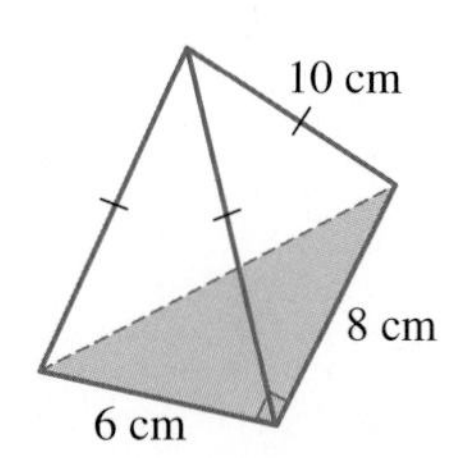

8. **WE10** Calculate the total surface area of the following shape, correct to 2 decimal places.

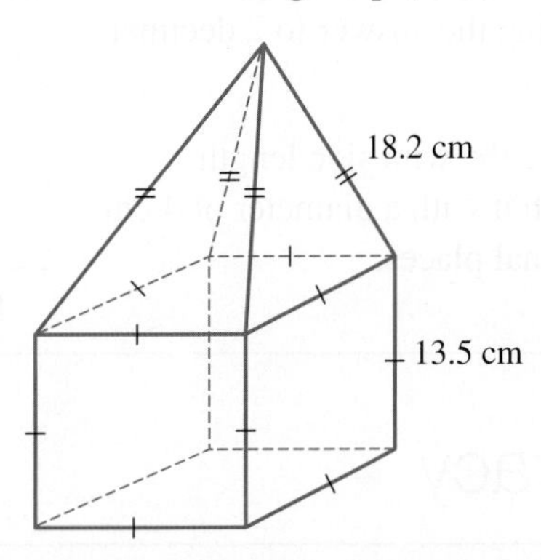

9. Calculate the total surface area of the following shape, correct to 1 decimal place.

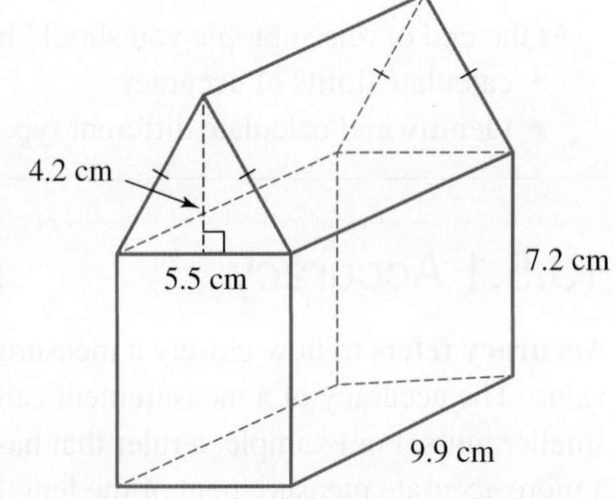

10. A 22-mm cube has a square-based pyramid sit on top of it with a slant length of 28 mm. Calculate its total surface area correct to 2 decimal places.
11. Calculate the total surface area of a hemisphere of radius 6 cm that sits on a cylinder that has a radius 6 cm and height of 12 cm.
Note: $TSA_{cylinder} = 2\pi r^2 + 2\pi rh$ and $TSA_{sphere} = 4\pi r^2$
12. Calculate the total surface area of two cones joined together at their circular bases if they both have a diameter of 24 cm and a slant length of 18 cm.
13. Calculate the total surface area of each of the shapes shown to 2 decimal places.

a.

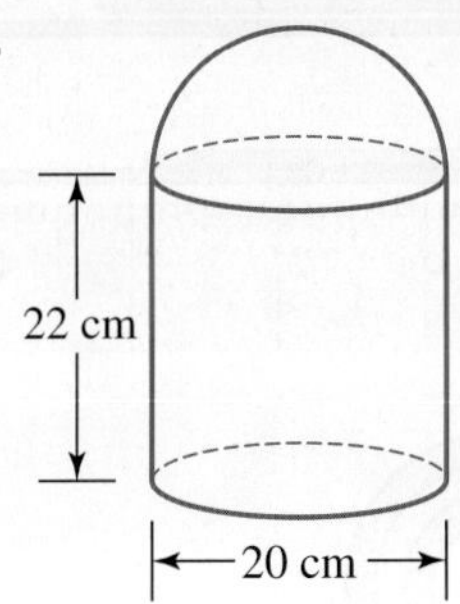

b.

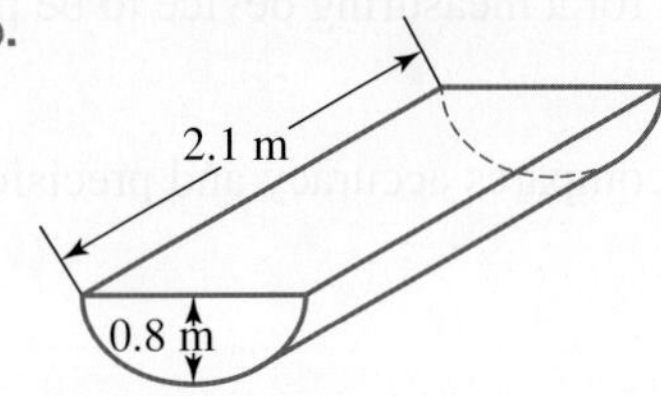

14. Calculate the total surface area of the shape shown, correct to 2 decimal places.

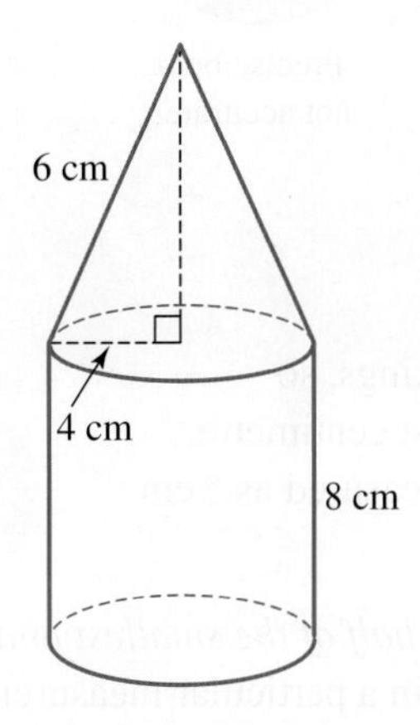

15. The drawing shown is part of a child's playground toy. Calculate the total surface area, giving the answer to 2 decimal places.

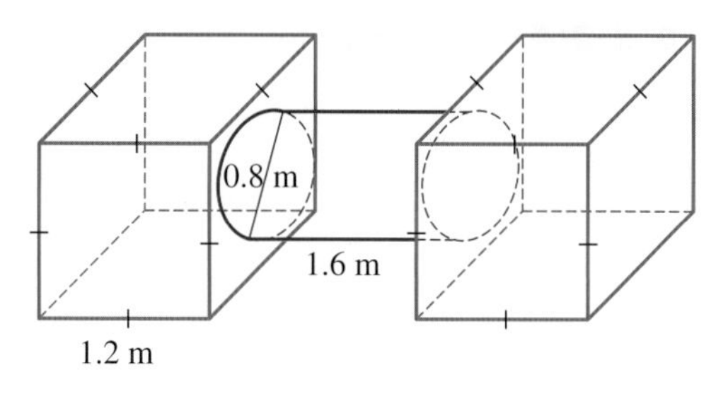

16. Calculate the total surface area of a cube with side length 15 cm that has a hole drilled through it with a diameter of 4 cm. Round the answer correct to 2 decimal places.

13.5 Limits of accuracy

LEARNING INTENTION

At the end of this subtopic you should be able to:
- calculate limits of accuracy
- identify and calculate different types of error.

13.5.1 Accuracy

Accuracy refers to how closely a measured value agrees with the actual value. The accuracy of a measurement can be improved by measuring with smaller units. For example, a ruler that has millimetre markings will give a more accurate measurement of the length of a paperclip than a ruler that has only centimetre markings, as shown.

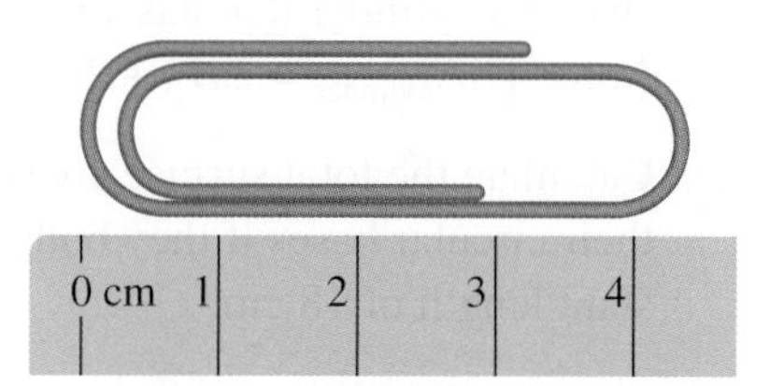

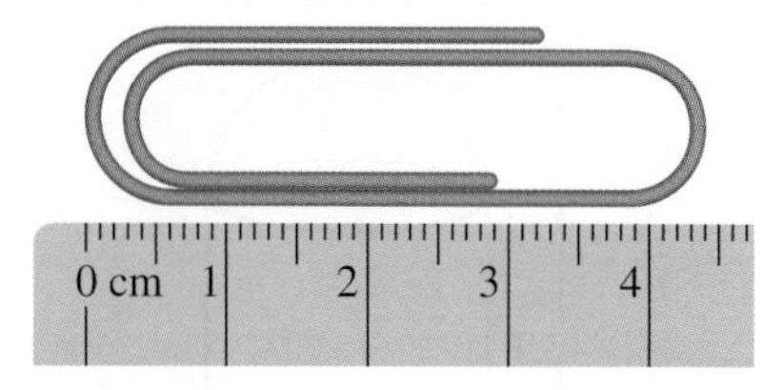

Precision refers to how closely the individual measurements agree with each other. It is possible for a measuring device to be precise but not accurate.

The following diagram compares accuracy and precision.

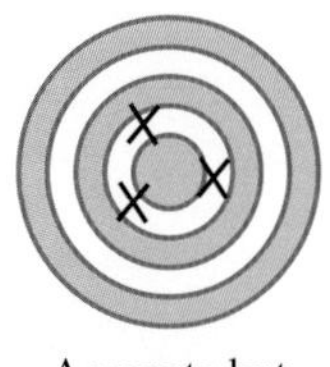
Accurate but not precise

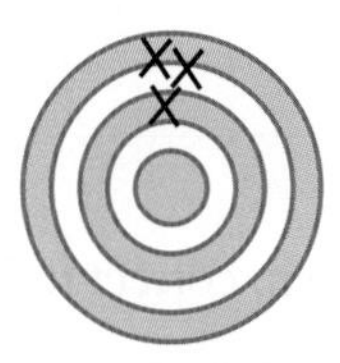
Precise but not accurate

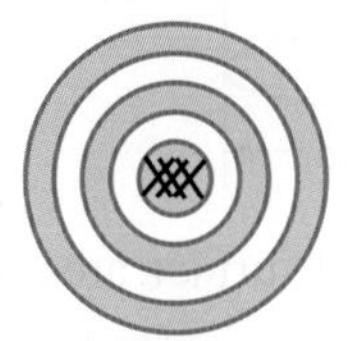
Accurate and precise

13.5.2 Limits of accuracy

The following ruler has only centimetre markings, so measurements are approximated to the nearest centimetre. Lengths between 4.5 cm and 5.5 cm are all measured as 5 cm with this ruler.

The **limit of accuracy** of a measuring tool is *half of the smallest unit* marked on the tool. This limit determines the highest and lowest values that will result in a particular measurement being recorded. The highest and lowest values indicate the **range** within which the actual value will lie.

The limit of accuracy of the ruler shown is 0.5 cm, which is half of the unit shown on the scale. If an object is measured with this ruler as being 5 cm long, the actual measurement will be between 4.5 cm and 5.5 cm; that is, half a unit above and below the nearest centimetre.

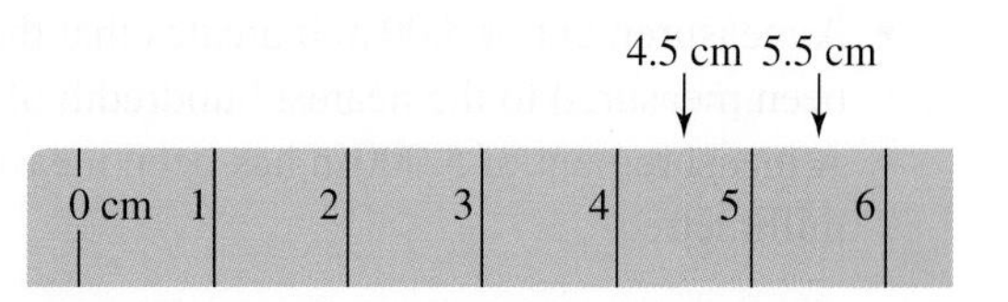

This can be expressed as 5.0 ± 0.5 cm; the ± 0.5 cm is known as the **tolerance**.

It is always possible to divide a measurement unit into smaller units. For example, metres can be divided into centimetres, centimetres can be divided into millimetres, millimetres can be divided into micrometres and so on. So the measurements that we take are never **exact**. They are **approximations** only.

All measurements are approximations, not just those relating to length. For example, the measurement of the mass of a teaspoon of butter is only as accurate as the kitchen scale with which it is measured, and the measurement of a baby's temperature is only as accurate as the thermometer's scale.

tlvd-5064

WORKED EXAMPLE 11 Calculating limits of accuracy

The speedometer shown measures the speed at which a car is travelling.

a. State the limit of accuracy for this speedometer.

b. Provide a range within which the car's actual speed (as indicated by the arrow) lies.

THINK	WRITE
a. Each marking on the scale represents 5 km/h, so the limit of accuracy is half of this measurement.	a. Limit of accuracy $= \frac{1}{2} \times 5$ $= 2.5$
b. The arrow is pointing to 80 km/h on the speedometer. The actual speed lies within the range of 80 km/h − 2.5 km/h to 80 km/h + 2.5 km/h.	b. $80 - 2.5 = 77.5$ and $80 + 2.5 = 82.5$ The actual speed lies within the range of 77.5 km/h to 82.5 km/h.

13.5.3 Types of error

Errors in measurement can be made due to a number of factors:

- An inappropriate instrument being used. For example, it would be inappropriate to use a centimetre ruler to measure the distance from Melbourne to Sydney.
- Human error when reading the measurement. For example, when measuring the length of a school football oval, it is possible for three different students to report three different results.
- Inaccuracy of the instrument being used. For example, it would be more accurate to measure 20 mL with a measuring cylinder instead of a beaker.

The difference between the measurement made (the estimated value) and the actual size of the object is called the **error**.

The error can be positive or negative and can be calculated using the formula:

$$\text{error} = \text{estimated value} - \text{actual value}$$

13.5.4 Accuracy of measurement

The number of decimal places a measurement contains can give an indication of the accuracy of the measurement. For example:

- A measurement of 5 m indicates that the length is closer to 5 m than it is to either 4 m or 6 m. It has been measured to the nearest metre.

- A measurement of 5.00 m indicates that the length is closer to 5.00 m than it is to 4.99 m or 5.01 m. It has been measured to the nearest hundredth of a metre or the nearest centimetre.
- A measurement of 5.000 m has been measured to the nearest thousandth of a metre or the nearest millimetre.

The number of decimal places a measurement can have is determined by the scale on the measuring device and the units the device has.

The more decimal places a measurement has, the more accurately it was measured.

13.5.5 The absolute error

The **absolute error** is the value of the error without the positive or negative sign. The calculation for the absolute error is done in a pair of vertical bars. This is called the **modulus** and means that you ignore the sign of the number. As the sign is ignored, the absolute error is never negative.

Absolute error

absolute error = | estimated value − actual value |

The significance of the error can be seen when it is compared with the actual value. For example, an error of 1 m may not be significant if the actual length of the object is 5 m. However, the same error would be significant if the actual length is 5 cm.

tlvd-5065

WORKED EXAMPLE 12 Calculating absolute error

a. Richie estimated his tennis grip to be 105 mm. When he arrived at the store to purchase the grip, the owner suggested they measure the circumference of his racquet handle to confirm his estimate. The actual measurement was 107 mm. Calculate the absolute error of Richie's measurement.

b. Geoff ordered a surfboard that was 2.25 inches thick. The thickness of the board was perfect for his weight of 70 kilos. His surfboard shaper thought he had written 2.75 inches on the order form, and shaped the board to that specification. Determine if this was a significant error and explain why or why not.

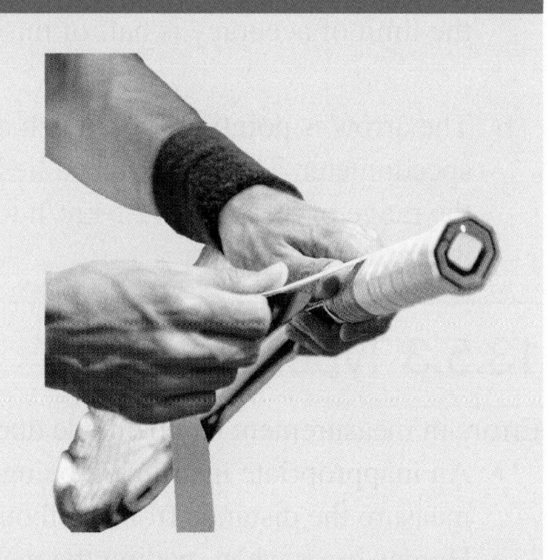

THINK	WRITE
a. Absolute error = |estimated value − actual value|	a. | 105 mm − 107 mm | = | −2mm | = 2mm
b. 1. Absolute error = |estimated value − actual value|	b. | 2.25 inches − 2.75 inches | = | − 0.5 inches | = 0.5 inches
2. Compare the absolute error to the original measurement.	Yes, 0.50 inches is almost a quarter of the original measurement and is therefore a significant error.

13.5.6 Percentage error

To compare the estimated value with the actual value, the **absolute relative error** is calculated. It expresses the absolute error as a fraction of the actual value.

Absolute relative error and absolute percentage error

$$\textbf{absolute relative error} = \left|\frac{\textbf{estimated value} - \textbf{actual value}}{\textbf{actual value}}\right|$$

The absolute relative error can be converted into the absolute percentage error by multiplying by 100%.

$$\begin{aligned}\textbf{absolute percentage error} &= \left|\frac{\textbf{estimated value} - \textbf{actual value}}{\textbf{actual value}}\right| \times \textbf{100\%} \\ &= \textbf{absolute relative error} \times \textbf{100\%}\end{aligned}$$

tlvd-5066

WORKED EXAMPLE 13 Calculating types of error

A student measured the width of a page of her workbook. Her measurement was 19.5 cm. If the actual width of the page was 20.5 cm, calculate:

a. **the error**
b. **the absolute error**
c. **the absolute relative error**
d. **the absolute percentage error.**

THINK	WRITE
a. The error is the difference between the estimated value and the actual value.	a. $\text{Error} = 19.5 - 20.5$ $= -1.0\,\text{cm}$
b. The absolute error is the error expressed without the positive or negative sign.	b. $\text{Absolute error} = \lvert -1.0 \rvert$ $= 1.0\,\text{cm}$
c. The absolute relative error is the absolute error expressed as a fraction of the actual value.	c. $\text{Absolute relative error} = \left\lvert\frac{\text{estimated value} - \text{actual value}}{\text{actual value}}\right\rvert$ $= \left\lvert\frac{19.5 - 20.5}{20.5}\right\rvert$ $= \frac{1.0}{20.5}$ $= 0.049$
d. The absolute percentage error is the absolute error expressed as a percentage of the actual value.	d. $\text{Absolute percentage error} = \text{absolute relative error} \times 100\%$ $= 0.049 \times 100\%$ $= 4.9\%$

13.5 Exercise

Students, these questions are even better in jacPLUS

Receive immediate feedback and access sample responses

Access additional questions

Track your results and progress

Find all this and MORE in jacPLUS

1. **WE11** For each of the scales shown:
   i. state the limit of accuracy
   ii. write a range to show the values between which the measurement shown on the scale lies.

a.

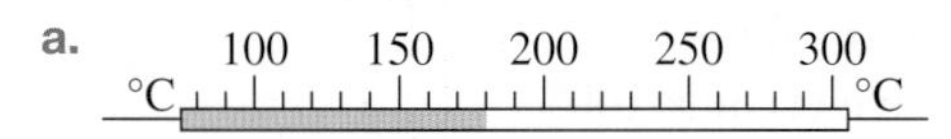

b.

c.

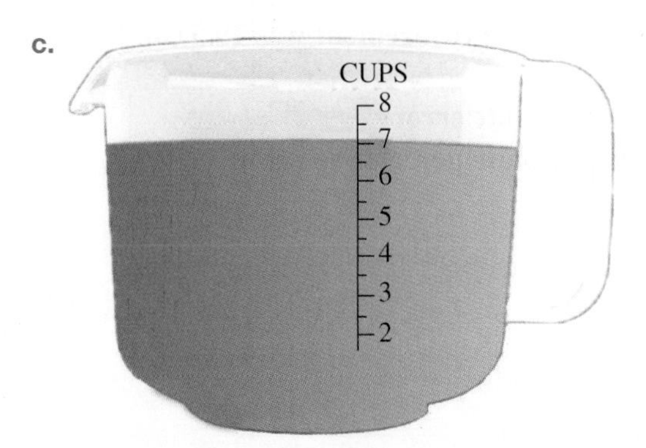

d.

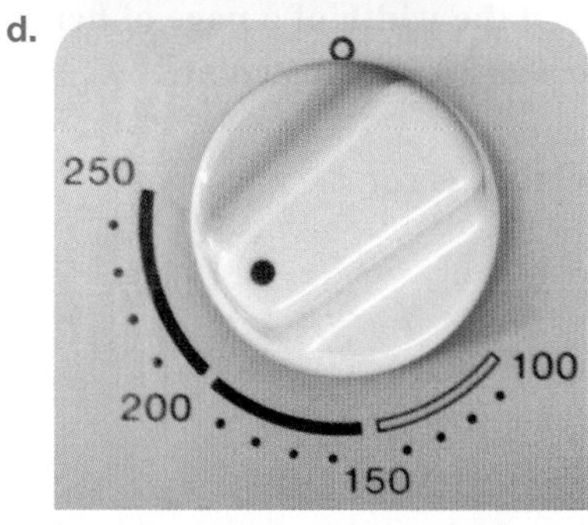

2. Write the measurements of the following lines as accurately as possible.

a.

b.

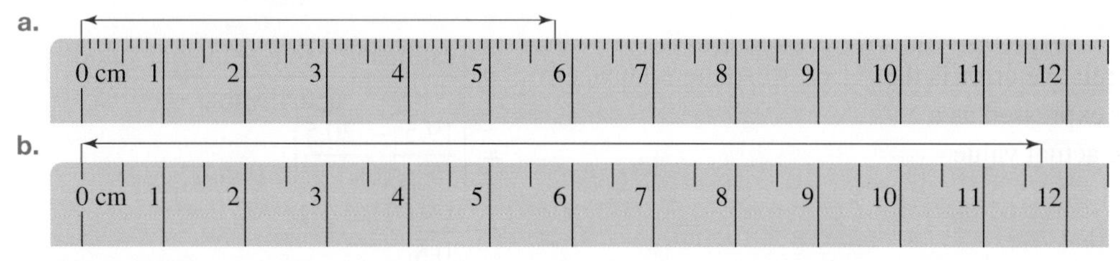

3. Each of the following lengths has been recorded to the nearest centimetre. Provide a range within which each measurement will lie.

   a. 5 cm b. 12 cm c. 100 cm d. 850 cm e. 5 m f. 2.5 m

4. For each of the following measurements, provide the two limits between which the actual measurement will lie.
   a. The length of a book is 29 cm, correct to the nearest centimetre.
   b. The mass of meat purchased is 250 grams, correct to the nearest 10 grams.
   c. The volume of cough medicine is 10 mL, correct to the nearest 5 mL.
   d. It takes me half an hour to walk to school, to the nearest minute.

5. **MC** The most accurate of the following measurements is:

   A. 7 **B.** 7.0 **C.** 0.7 **D.** 7.7 **E.** 7.07

6. **MC** The least accurate of the following measurements is:

A. 8.000 B. 8.0 C. 8 D. 8.008 E. 0.08

7. A measurement of 2.500 m indicates that the measurement is accurate *to the nearest millimetre* (2.500 m = 2500 mm). Determine the unit to which each of the following measurements is accurate.

a. 35.20 m b. 1.30 m c. 1.000 km
d. 8.0 cm e. 5.000 m f. 2.520 km

8. Four students recorded the following measured lengths during a Science class. The actual lengths were also recorded. Calculate the error made by each student.

Student	Estimated value	Actual value
a	10 cm	9.4 cm
b	15.5 cm	17.2 cm
c	250 km	228 km
d	37 mm	38.1 mm

9. **WE12** a. Jordan estimated the length of his bedroom as 410 cm..
However, when he measured it, he found out it was only 395 cm long.
Calculate the absolute error of Jordan's measurement

b. Jamie ordered a 30-cm pizza for a party with her friends. However, when it arrived, she found that the pizza was only 29.7 cm wide.
State if this is a significant error and explain why or why not.

10. Calculate the absolute error for each of the following.
a. The distance from home to school is estimated by a student to be 2 km. The actual distance is 1.85 km.
b. The weight stated on a packet of lollies is 250 g. The actual weight is 254 g.
c. The scheduled arrival time of the plane from Kansas City is 7:23 pm. The actual arrival time is 7:46 pm.
d. The estimated volume of water in a fish tank is 65 L. The actual volume is 72 L.

11. You measured your friend's height to be 165 cm. If the absolute error of your measurement is 4 cm, calculate how tall your friend is.
(*Hint:* There may be more than one answer.)

12. **WE13** Complete the following table by calculating the error, the absolute error, the absolute relative error and the absolute percentage error. Give your answers to the absolute relative errors correct to 2 decimal places where necessary, and give your answers to the absolute percentage errors to the nearest whole number.
(*Hint:* Make sure that both measurements are in the same unit.)

	Measurement					
	Estimated	Actual	Error	Absolute error	Absolute relative error	Absolute percentage error
a.	250 m	245 m				
b.	95 cm	1.02 m				
c.	2 cm	18.5 mm				
d.	48 seconds	1 minute				
e.	1.4 kg	1 kg 300 g				
f.	750 mL	0.9 L				

13. After ordering food in a local restaurant, a family estimates that their dinner will cost \$120. Compare their estimate with the actual total, as shown on the bill, and calculate their absolute percentage error to the nearest whole number.

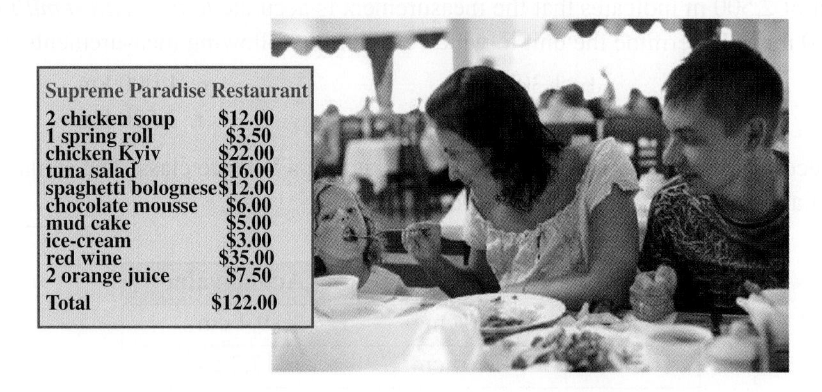

Supreme Paradise Restaurant	
2 chicken soup	\$12.00
1 spring roll	\$3.50
chicken Kyiv	\$22.00
tuna salad	\$16.00
spaghetti bolognese	\$12.00
chocolate mousse	\$6.00
mud cake	\$5.00
ice-cream	\$3.00
red wine	\$35.00
2 orange juice	\$7.50
Total	\$122.00

14. a. Explain how you would calculate the limit of accuracy of the ruler shown.

0 cm 1 2 3 4 5 6 7 8 9 10 11 12 13

b. A line measured with this ruler is recorded as 23 mm. Determine between which two measurements the actual length will lie.

15. Explain why any measurement is an approximation and can never be exact.

16. Explain the difference between a measurement of 4 km and a measurement of 4.000 km.

17. Medical researchers have found that estimating the weight of unborn babies is critical in determining the quality of their health at pregnancy. There are two methods that are used to estimate the weight of unborn babies: an ultrasound of the expecting mother, or specific measurements of her body. Both of these methods are performed over a five-month period, to ensure accuracy. Read the abstract of the scholarly report of the two methods at hindawi.com.

a. Explain what the abstract says about the mean absolute percentage error for each method.
b. State the mean absolute percentage error for each method.

18. The accuracy of clocks can vary significantly. Check the clock on your wall, your wristwatch, your mobile phone and your computer at home. These clocks could be set simultaneously, yet after weeks or months might differ by seconds or even minutes. Explain how you know which time is the most accurate.

An atomic clock uses the resonation of a caesium atom to measure time, and it is the most accurate clock in the world. If you go to **time.is** on your phone or computer, you will have access to an atomic clock that can compare your computer or mobile-phone time to the atomic clock. There is a problem with this method because of the inaccuracy of synchronisation of your computer or phone with the atomic clock. The accuracy of synchronisation is ± 0.425 s.

All computer or phone clocks are estimates compared to the atomic clock (actual time).

a. State if the accuracy of synchronisation is more or less significant if the estimated time is behind or ahead. Explain your answer.
b. If your computer clock's time was one minute ahead of the atomic clock, determine whether the accuracy of synchronisation would be significant. Explain your answer.
c. If your computer's clock time was one second behind the atomic clock, state if the accuracy of synchronisation would be significant. Explain your answer.

13.6 Scientific notation and significant figures

LEARNING INTENTION

At the end of this subtopic you should be able to:
- convert between scientific notation and basic numerals
- use scientific notation to represent real-world situations
- round to a specified number of significant figures.

13.6.1 Scientific notation

Scientists, economists, statisticians and mathematicians often need to work with very large or very small numbers.

Scientific notation (sometimes called **standard form**) is a way of writing a very large or very small numbers using less space. In scientific notation, the number is written as a multiple of a power of 10 and looks like $a \times 10^n$, where a is a number between 1 and 10. For example:
- $3600 = 3.6 \times 10^3$
- $0.036 = 3.6 \times 10^{-2}$
- $3\,600\,000\,000 = 3.6 \times 10^9$
- $0.000\,003\,6 = 3.6 \times 10^{-6}$

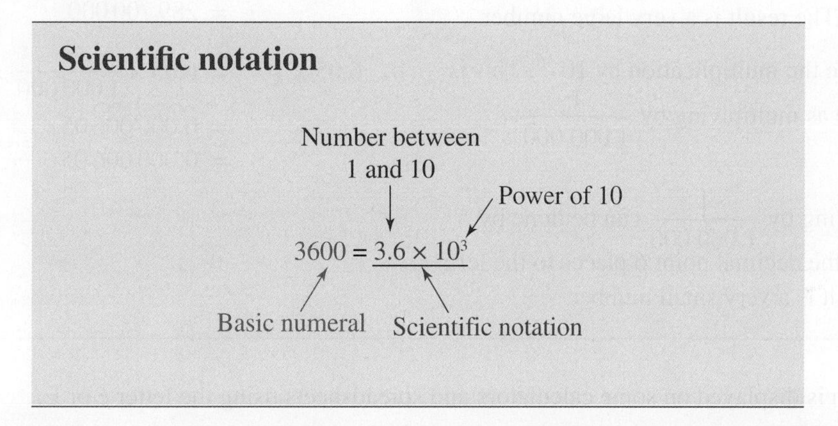

tlvd-5067

WORKED EXAMPLE 14 Converting to scientific notation

Write the following numbers in scientific notation.

a. **345**

b. **0.007**

THINK	WRITE
a. • The non-zero digit with the largest place value is 3 in the hundreds position, as shown in blue. • 3 will become the number between 1 and 10 in scientific notation, as shown in red. The rest of the digits in 345 become decimal places in scientific notation. • The digit 3 must still have a value of 300, so 3.45 needs to be multiplied by $100(10^2)$, as shown in purple.	a. $345 = 3.45 \times 10^2$

b. • The non-zero digit with the largest place value is 7 in the thousandths position, as shown in blue.
• 7 will become the number between 1 and 10 in scientific notation, as shown in red. The rest of the digits to the right (the trailing zeros) become decimal places in scientific notation.
• The digit 7 must still have a value of 7 thousandths, so 7.0 must be multiplied by $\frac{1}{1000}$ (10^{-3}), as shown in purple.

b. $0.007 = 7.0 \times 10^{-3}$

To write a number expressed using scientific notation as a basic numeral, simply complete the multiplication.

WORKED EXAMPLE 15 Converting from scientific notation to basic numerals

Write the following numbers as basic numerals.

a. $\mathbf{2.897 \times 10^8}$

b. $\mathbf{6.05 \times 10^{-6}}$

THINK	WRITE
a. Complete the multiplication by 10^8. This can be done by moving the decimal point 8 places to the right. The result is a very large number.	a. $2.897 \times 10^8 = 2.897 \times 100\,000\,000$ $= 289\,700\,000$ $= 289\,700\,000$
b. 1. Complete the multiplication by 10^{-6}. This is the same as multiplying by $\frac{1}{1\,000\,000}$.	b. $6.05 \times 10^{-6} = 6.05 \times \frac{1}{1\,000\,000}$ $= 0.000\,006\,05$ $= 0.000\,006\,05$
2. Multiplying by $\frac{1}{1\,000\,000}$ can be done by moving the decimal point 6 places to the left. The result is a very small number.	

Scientific notation is displayed on some calculators and spreadsheets using the letter e or E.

These calculators or spreadsheets show a number between 1 and 10, then an E, and then the power to which 10 is raised. For example:

- 3.45×10^{18} will appear as 3.45E18 on a calculator or spreadsheet.
- The display 5.9E – 03 represents the number 5.9×10^{-3}.

13.6.2 Measuring very large and very small units using scientific notation

Astronomers use very large numbers to express distances between planets, solar systems, stars and other celestial objects. If they are writing scientific papers for their research, scientific journals or the general public, the time it takes to repeatedly refer to these distances becomes prohibitive. Imagine writing the distance to the furthest galaxies in the solar system: 145 000 000 000 000 000 000 000 km. It would be much easier to express this distance in scientific notation as 1.45×10^{23} km.

Physicists have a similar problem writing numbers. They record weights that are very small and, like the astronomers, they need to describe these numbers to their readers.

For example, the mass of a stationary electron is 0.000 000 000 000 000 000 000 000 000 000 911 kg. (Try counting the 0s.) Instead, physicists use the scientific notation of 9.11×10^{-31} kg.

tlvd-5068

WORKED EXAMPLE 16 Using scientific notation in the real world

a. The approximate diameter of the Milky Way galaxy is 950 000 000 000 000 000 km. Use scientific notation to express this distance.

b. A stock trade on the New York Stock Exchange can be made in 740 nanoseconds or 0.000 007 24 seconds. Use scientific notation to express this time in seconds.

THINK

a. • 9 is in the largest place value, as shown in purple.
• 9 will become the number between 1 and 10, as shown in red.
• The rest of the digits become decimal places in scientific notation.
• The decimal point has moved 17 spaces to the left, so the power of ten is 17.

WRITE

a. $950\,000\,000\,000\,000\,000\text{ km} = 9.5 \times 10^{17}$ km

THINK

b. • 7 is in the largest place value, as shown in purple.
• 7 will become the number between 1 and 10, as shown in red.
• The rest of the digits become decimal places in scientific notation.
• The decimal point has moved 6 spaces to the right, so the power of ten is −6.

WRITE

b. $0.000\,007\,24$ seconds $= 7.24 \times 10^{-6}$ seconds

13.6.3 Significant figures

Significant figures are a method of indicating the precision of a measurement. The more significant figures a number has, the more precise the measurement. The words *significant figures* are sometimes abbreviated to sig. figs.

Rules for significant figures

1. **All non-zero digits are significant.**
   For example, the number 513 has 3 significant figures.
2. **Any zeros between two significant figures are significant.**
   For example, the zeros in the following numbers are all significant: 602, 1.5003, 600.1, 100.000 02
3. **Any zeros that are not between significant figures in a whole number are not significant.**
   For example, the zeros in the following numbers are not significant: 50 000, 30, 3510
4. **Zeros following the last significant figure in the decimal portion of a number are significant.**
   For example, the zeros in the following numbers are significant: 52.0030, 60.010 500, 0.340 00

5. **Zeros preceding the first significant figure in the decimal portion of a number are not significant. The zeros in the following numbers are not significant: 0.000 08, 0.007 80**
6. **Zeros in the decimal portion of a number that have no significant figures following them are significant, and therefore the zeros at the end of the whole number become significant, if they are present.**
   Some examples include: 80.000, 8.0, 7080.00

Scientific notation and rounding

If a number is expressed using scientific notation ($a \times 10^n$), then all the digits in a are significant, as shown in the table below.

Number	Scientific notation	Significant figures
34.56	3.456×10^1	4
3.456	3.456×10^0	4
346	3.46×10^2	3
3460	3.46×10^3	3
346 000	3.46×10^5	3
0.0034	3.4×10^{-3}	2
3.4×10^4	3.4×10^4	2

When working with significant figures in Maths or Science, do not round off to the required number of significant figures until the final answer.

When writing a number to a specified number of significant figures, the smallest significant digit may need to be rounded, depending on the value of the next digit. For example:

- 7.59 is written as 7.6, correct to 2 significant figures.
- 7.54 is written as 7.5, correct to 2 significant figures.

tlvd-5069

WORKED EXAMPLE 17 Writing significant figures

Convert the following measurements as specified, and then write the measurements correct to the number of significant figures indicated in brackets.

a. **0.000 078 1 cm to mm (2 significant figures)**
b. **550 900 g to kg (3 significant figures)**
c. **80.4586 s to min (4 significant figures)**

THINK	WRITE
a. • 10 mm = 1 cm • To convert from cm to mm, multiply by 10. • All of the non-zero digits in the number are significant; however, the three zeros to the right of the decimal point are not significant. • The second significant digit (8) will not need to be rounded, as 1 is less than 5.	a. $0.000\,078\,1$ cm $= (0.000\,078\,1 \times 10)$ mm $= 0.000\,781$ mm $0.000\,781$ mm $= 0.000\,78$ mm (2 sig. figs.)

b. • 1000 g = 1 kg
- To convert from g to kg, divide by 1000.
- All of the digits in the number are significant; however, we only need 3 significant figures, so these are the first three digits.
- 0 to the left of the decimal point will need to be rounded up to 1 because of the following 9.

b. $550\,900 \text{ g} = (550\,900 \div 10) \text{ kg}$
$= 550.900 \text{ kg}$
$550.990 \text{ kg} = 551 \text{ kg}$ (3 sig. figs.)

c. • 1 min = 60 s
- To convert from seconds to minutes, divide by 60.
- All the digits are significant; however, we only need the first 4 significant figures, so these are the first 4 digits.
- The fourth significant digit (4) will not need to be rounded, as 0 is less than 5.

c. $80.4586 \text{ s} = (80.4586 \div 60) \text{ min}$
$= 1.340\,97 \ldots \text{ min}$
$1.340\,97 \ldots \text{ min} = 1.341 \text{ min}$ (4 sig. figs.)

 Resources

 **Interactivity** Scientific notation (int-6456)

13.6 Exercise

Students, these questions are even better in jacPLUS

 Receive immediate feedback and access sample responses

 Access additional questions

 Track your results and progress

Find all this and MORE in jacPLUS

1. **WE14** Write the following numbers in scientific notation.
   a. 47.2 b. 3890 c. 0.56 d. −0.0067 e. 34 000 000
2. **WE15** Write the following numbers as basic numerals.
   a. 4.8×10^{-2} b. 7.6×10^{3} c. 2.9×10^{-4} d. 8.1×10^{0}
3. State which number(s) are not written in scientific notation.
   a. 10.75×10^{7} b. 3.54×10^{-3} c. 0.0091×10^{4} d. $7.23 \times 10^{\frac{1}{2}}$
4. For each of the following, state whether they're True or False.
   a. $0.000\,000\,51 \geq 5.1 \times 10^{-6}$
   b. $3.158\,400 \times 10^{7} < 3\,158\,400$
   c. $2.000\,000\,058 \times 10^{9} = 2\,000\,000\,058$
   d. $3.51 \times 10^{4} \geq 3.502 \times 10^{4}$
5. **MC** 0.000 67 written in scientific notation is:
   A. 67×10^{-5} B. 0.67×10^{-3} C. 6.7×10^{-4} D. 6.7×10^{-5} E. 0.67×10^{-4}

6. WE16 a. The nearest black hole to the Earth is thought to be only 1600 light-years away, which is equal to 15 137 000 000 000 000 km. Use scientific notation to express this distance.

b. The radius of an electron is 0.000 000 000 000 002 82 m. Use scientific notation to express this size.

7. Scientists used Earth's gravitational pull on nearby celestial bodies (such as the Moon) to calculate the mass of the Earth at approximately 5.972 sextillion metric tons.

5 972 000 000 000 000 000 000
metric tons

a. Write 5.972 sextillion in scientific notation.

b. State how many significant figures this number has.

8. Write the following numbers correct to the number of significant figures specified in brackets.

a. 0.057 89 (2 significant figures)
b. 0.050 58 (3 significant figures)
c. 55 830.000 01 (4 significant figures)
d. $2.978\,582 \times 10^{-17}$ (1 significant figures)
e. 138 000 (5 significant figures)
f. 1.054 (2 significant figures)

9. WE17 Convert the following measurements as specified, and then write the measurements correct to the number of significant figures indicated in brackets.

a. 1200 cm to m (3 significant figures)
b. 6.7481 kg to g (4 significant figures)
c. 0.000 08 h to min (3 significant figures)
d. 80.26 days to hours (4 significant figures)
e. 51 kg to g (1 significant figure)
f. 5.025 mm to m (2 significant figures)

10. MC Select the measurement that is the most accurate. (*Hint:* Count the number of significant figures.)

A. 2.5 m B. 2.51 m C. 2.059 m D. 2.5000 m E. 2.005 m

11. MC Select the measurement that is the most accurate. (*Hint:* First convert measurements to grams, then count the number of significant figures.)

A. 0.54 kg B. 500 g C. 501.0 g D. 0.541 kg E. 530 g

12. State the value of n in each of the following expressions.

a. $4793 = 4.793 \times 10^n$
b. $0.631 = 6.31 \times 10^n$
c. $134 = 1.34 \times 10^n$
d. $0.000\,56 = 5.6 \times 10^n$

13. State how many significant figures each of these numbers has.

a. 3.005×10^{-7} b. 3.005 c. 4000 d. 3.00×10^{-7}

14. An atom consists of smaller particles called protons, neutrons and electrons. Electrons have a mass of $9.109\,381\,88 \times 10^{-31}$ kg correct to 9 significant figures.

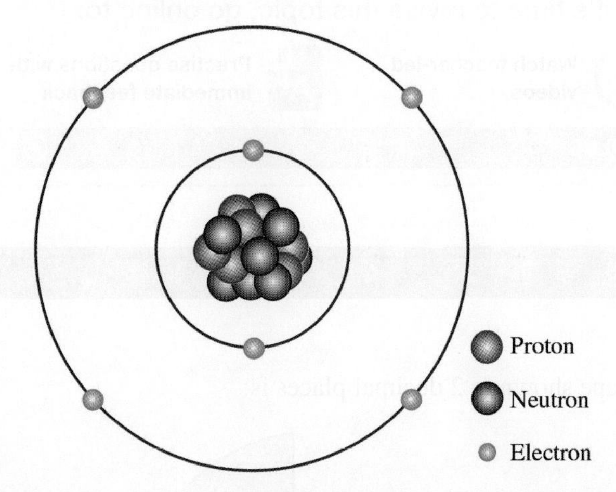

a. Write the mass of an electron correct to 5 significant figures.
b. Protons and neutrons are the same size. They are both 1836 times the size of an electron. Use the original mass of an electron (correct to 9 sig. figs.) and your calculator to calculate the mass of a proton correct to 5 significant figures.
c. Use the mass of an electron correct to 3 significant figures to calculate the mass of a proton correct to 5 significant figures.
d. Explain why it is important to work with the original amounts and then round to the specified number of significant figures at the end of a calculation.

15. Complete the following calculations, writing your answer to the number of significant figures given in the brackets.

a. $2.398 \text{ m} + 3.5 \text{ m}$ (2)
b. $57.86 \text{ cm} + 0.157 \text{ m}$ (3)
c. $52.89 \text{ m} - 1.3 \times 10^2 \text{ m}$ (2)
d. $10.253 \text{ cm} \times 0.251 \text{ m}$ (3)
e. $973.48 \text{ m} \times 3.5 \times 10^3 \text{ mm}$ (2)

16. Distance is equal to speed multiplied by time. If we travelled at 100 km/h it would take approximately 0.44 years to reach the Moon, 89.6 years to reach Mars, 1460 years to reach Saturn and 6590 years to reach Pluto.

a. Assuming that there are 365 days in a year, calculate the distance (as a basic numeral) between Earth and:

i. the Moon
ii. Mars
iii. Saturn
iv. Pluto.

b. Write your answers to part **a** correct to 3 significant figures.
c. Write your answers to part **a** using scientific notation correct to 3 significant figures.

13.7 Review

13.7.1 Summary

doc-38021

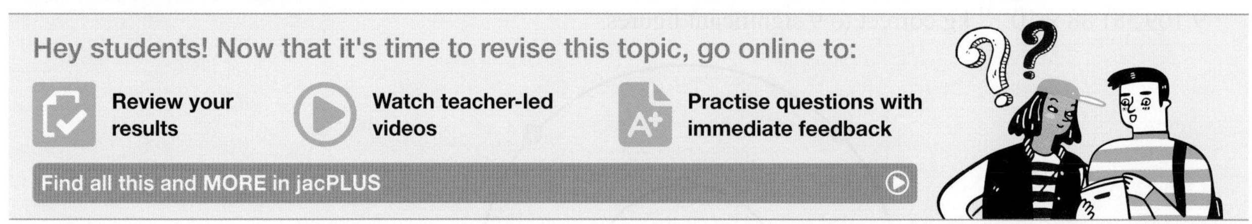

13.7 Exercise

Multiple choice

1. The shaded area of the shape shown to 2 decimal places is:

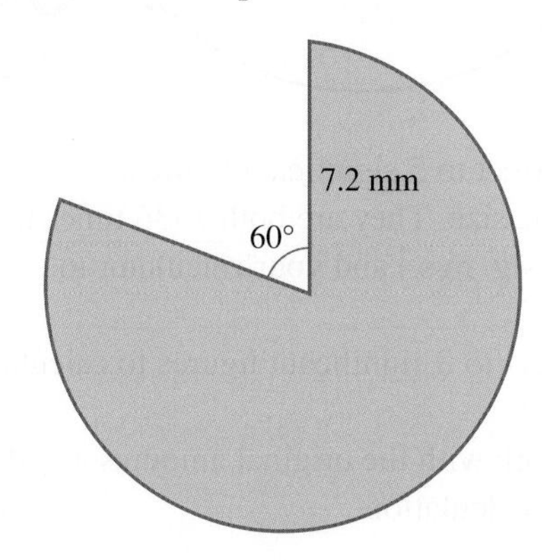

A. 162.86 mm^2 B. 155.52 mm^2 C. 27.14 mm^2 D. 135.72 mm^2 E. 153.25 mm^2

2. **MC** The area of the composite shape shown is closest to:
   A. 157.856 cm^2
   B. 225.543 cm^2
   C. 218.237 cm^2
   D. 209.56 cm^2
   E. 186.64 cm^2

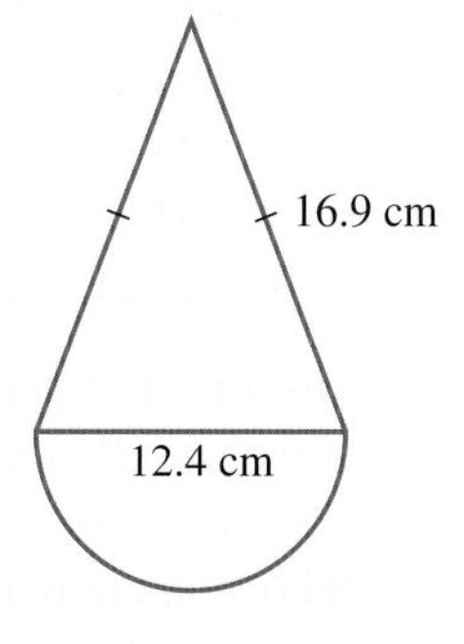

3. **MC** Select the most accurate measurement. (*Hint:* Count the number of significant figures.)
   A. 360.005 m B. 7000 m C. 36.500 m D. 3.71 m E. 0.700 000 0 m

4. **MC** The measurement marked by the arrow is:
   (Use the unit indicated in the brackets.)

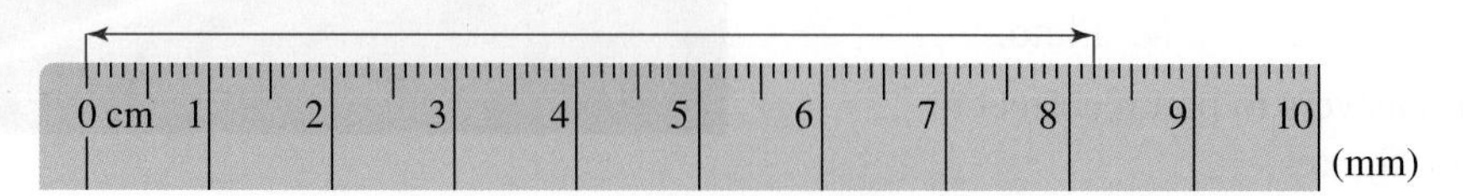

A. 8 cm B. 8.2 cm C. 8.2 m D. 8.2 mm E. 82 mm

5. **MC** The number with 4 significant figures is:
   A. 0.4 B. 3.023 C. 4.84 D. 0.0003 E. 0.0024

6. **MC** A resistor for an electrical circuit has a reading of $200 \pm 10\Omega$. The tolerance for this resistor is:

A. $\pm 200\Omega$ **B.** $\pm 2\Omega$ **C.** $\pm 20\Omega$ **D.** $\pm 10\Omega$ **E.** $\pm 100\Omega$

7. **MC** The value of each interval is:

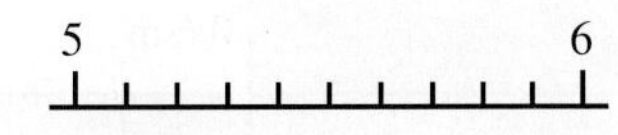

A. 0.1 **B.** 1 **C.** 5 **D.** 6 **E.** 0.5

8. **MC** The total surface area of the following shape is closest to:

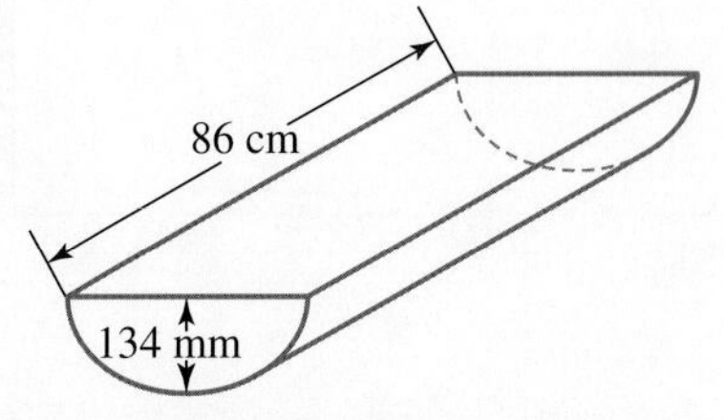

A. 0.78 m^2
B. 7804.84 cm^2
C. 649.28 cm^2
D. 0.65 m^2
E. 0.56 m^2

9. **MC** A cube of length 18 cm has a hole with a 6-cm diameter drilled through its centre. The total surface area of the new shape is closest to:

A. 1866.32 cm^2 **B.** 1548.16 cm^2 **C.** 2226.74 cm^2 **D.** 1944 cm^2 **E.** 2144.36 cm^2

10. **MC** The total surface area of a sphere with a diameter of 5.35 m, to 2 decimal places, is closest to:

A. 359.68 m^2
B. 22.48 m^2
C. 89.92 m^2
D. 33.62 m^2
E. 76.92 m^2

Short answer

11. Calculate the area of the following shape to 3 decimal places.

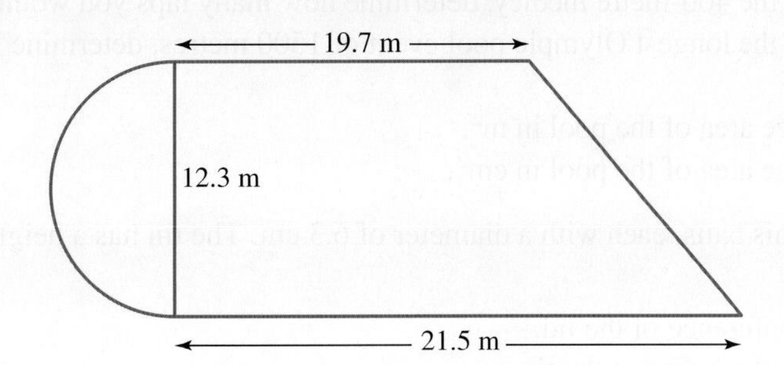

12. Calculate the total surface area of each of the following shapes to 1 decimal place.

a.

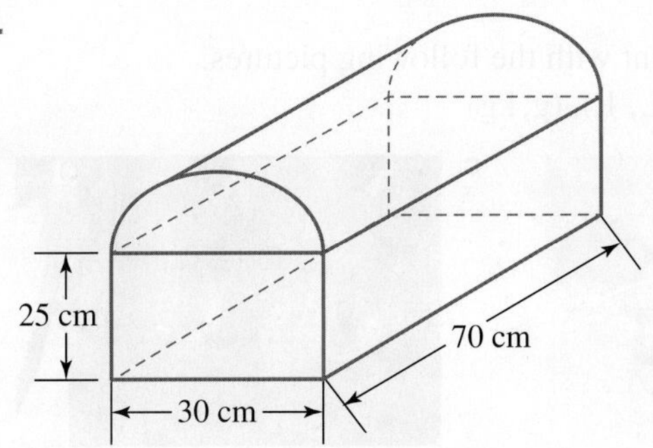

b.

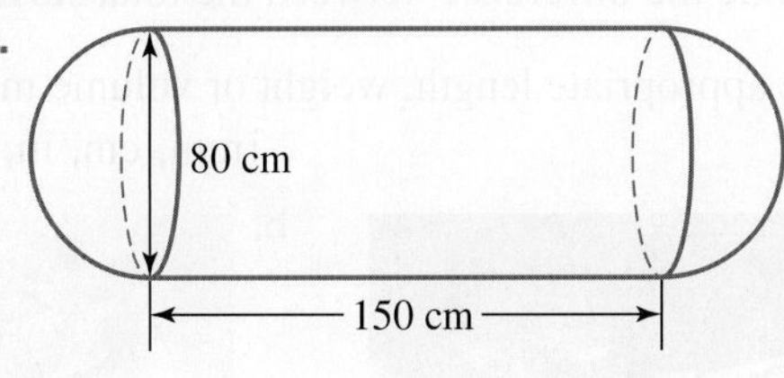

13. If a rectangular prism can be made from the following net, calculate:

a. the perimeter of the net
b. the total surface area of the prism.

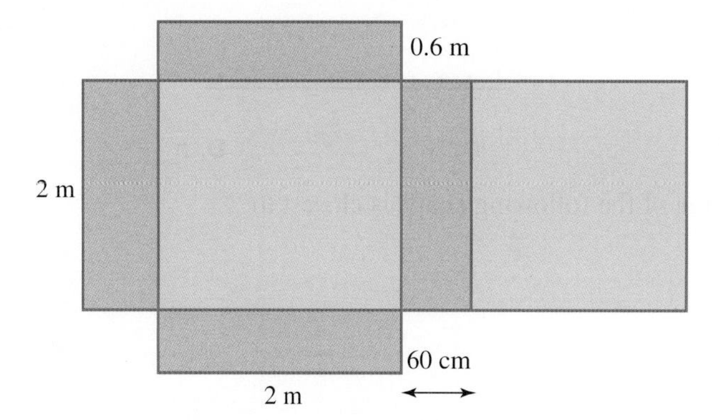

14. An Olympic swimming pool is 50 metres long, 25 metres wide and 1.8 metres deep.

a. If Haylee swam 25 laps of the pool, calculate how many metres she swam.
b. If you were to swim the 400-metre medley, determine how many laps you would swim.
c. If you were to swim the longest Olympic pool event of 1500 metres, determine how many laps you would complete.
d. Determine the surface area of the pool in m^2.
e. Determine the surface area of the pool in cm^2.

15. A tin contains four tennis balls, each with a diameter of 6.3 cm. The tin has a height of 27.2 cm and a diameter of 73 mm.

a. Determine the circumference of the lid.
b. Calculate the total surface area of the tin.
c. Calculate the total surface area of the four tennis balls.
d. Determine the difference between the total surface area of the four balls and that of the tin.

16. Match the appropriate length, weight or volume measurement with the following pictures.

(mm, cm, m, km), (mL, L), (g, kg)

a.

(mm, cm, m, km)

b.

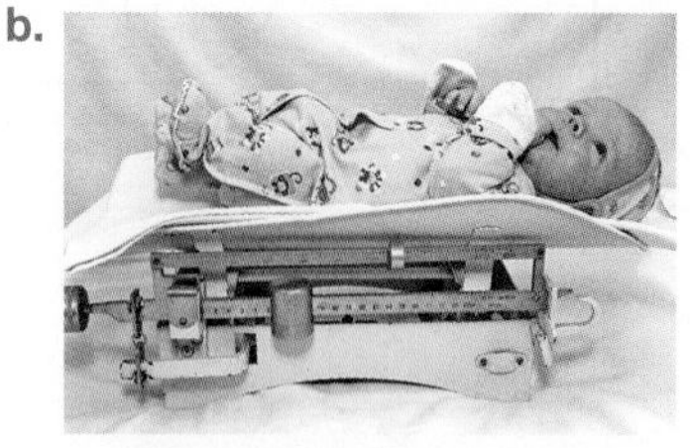

(g, kg)

c.

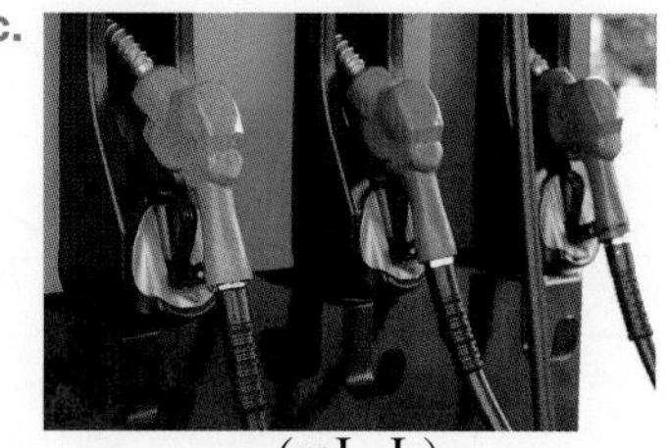

(mL, L)

Extended response

17. A speedometer in a car has markings every 5 km/h.
    a. State the limit of accuracy.
    b. If the speedometer indicated 55 km/h, state what range the speed could actually lie within.

18. Express the following numbers in scientific notation.
    a. 389 670 000
    b. 0.004 860 3
    c. 406 009 437
    d. 0.000 000 100
    e. 0.3750
    f. 100 300 000 000

19. Five students measured the length of a 42.7-m netball court. Complete the following table to determine which student had the most accurate answer. Round your answers to 4 decimal places where necessary.

Student	Measurement	Error	Absolute error	Absolute relative error
A	44.8 m	2.1 m	2.1 m	0.0492
B	41.9 m			
C	43.7 m			
D	42.0 m	−0.7 m	0.7 m	
E	43.3 m			

20. Nathan, a carpenter, tells his apprentice to build a rectangular prism with the dimensions as shown.

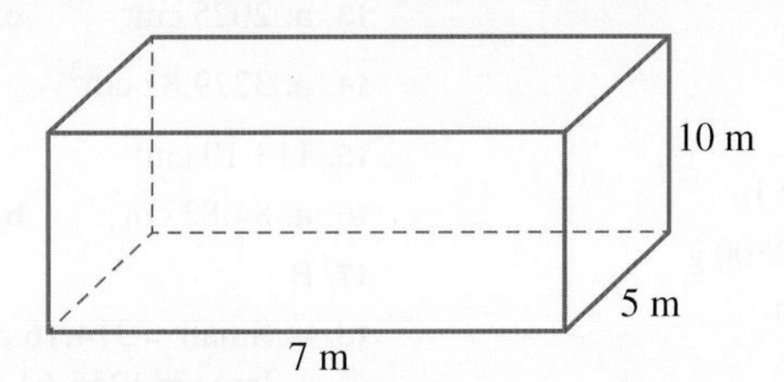

a. Calculate the correct volume of the prism.
b. If the length of the prism is measured incorrectly (an error of +1 m), determine what happens to the overall volume and calculate the new incorrect volume.
c. Calculate the error in volume, assuming he made +1 m errors on all three measurements.
d. State why the error is above 3 m^3, when the apprentice only made a +1 m error on each measurement.
e. If the apprentice made a +1 m error on the height dimension and a –1 m error on the width dimension, determine if the two errors would affect the original volume. Give the new volume.
f. Calculate the absolute error, absolute relative error (to 4 decimal places) and the absolute percentage error (to 2 decimal places) for the prism assuming the apprentice made a +1 m error on each dimension.

Answers

Topic 13 Measurement

13.2 Units of measurement

13.2 Exercise

1. a. 2 cm b. 95 mm c. 64 mm d. 9.9 cm
2. a. 5.6 km b. 210 mm c. 55 000 m
   d. 220 cm e. 3.5 m f. 2.5 cm
3. a. 165 m b. 7.8 m c. 15.8 cm
   d. 0.013 km e. 125 000 cm f. 36 m
4. a. 0.056 32 km b. 2.0431 km
   c. 550 000 cm d. 2200 mm
   e. 0.35 km f. 0.25 m
5. a. 89 000 cm, 0.3217 km, 98 760 mm, 7.825 m
   b. 0.786 km, 0.786 mm = 0.000 786 m = 0.0786 cm
6. a. 2400 cm, 135 m, 0.15 km
   b. 120 mm, 25 cm, 0.5 m
   c. 9 m, 10 000 mm, 0.45 km
   d. 32 000 cm, 1 km, 1200 m
   e. All values are equal.
   f. 825 m, 90 000 cm, 8.25 km
7. a. 13 cm b. 304 cm
8. a. 265.5 cm b. 96.27 cm
   c. 52.1 cm d. 90.105 km
   e. 60.5 cm f. 106.73 m
9. a. litres b. millimetres
10. a. grams b. kilograms c. milligrams
11. B
12. a. 2.5 L b. 3000 mL
    c. 0.0031 L d. 5250 mL
    e. 7500 mL f. 0.003 75 L
13. a. 0.000 024 t, 400 000 mg, 2.4 kg, 2500 g
    b. 0.05 kg, 52 000 mg, 54 g, 0.0005 t
14. a. False b. False
    c. True d. False
15. a. 505 000 mg b. 0.0045 g
    c. 6400 g d. 5.84 kg
    e. 0.075 821 kg f. 70 mg
16. a. second b. years c. hours
    d. day e. minutes f. weeks
17. a. 250 years, 1.5 centuries, 500 months, 2 decades
    b. 0.1 weeks, 0.56 days, 12.6 hours, 750 minutes, 4250 seconds
18. a. 45 years b. 82.5 decades
    c. 340 years d. 4.85 centuries
    e. 50 decades f. 12 centuries
19. a. 4 min b. 312 s c. 276 min
    d. 5.33 h e. 8.33 h f. 73 080 s
20. B
21. D
22. A
23. The distance from Melbourne to Perth is approximately 2721 km. Large distances are not easy to comprehend in mm, so we use the largest length measurement (km) to express long distances.
24. a. 20 000 L b. 2.584 m^3

13.3 Areas of composite figures

13.3 Exercise

1. a. Parallelogram and semicircle
   b. Square and four semicircles
   c. Trapezium and rectangle
   d. Triangle and semicircle
2. 120 cm^2
3. 23.03 cm^2
4. a. 48 m^2 b. 154 mm^2
   c. 272 cm^2 d. 384 mm^2
5. a. 29.225 cm^2 b. 25.63 cm^2
   c. 756 m^2 d. 42 651.41 mm^2
6. a. 40.18 mm^2 b. 31.49 cm^2
7. a. 360 cm^2
   b. 360 cm^2
   c. They are the same as you would expect.
8. 9600 mm^2
9. 156 cm^2
10. 109.87 cm^2
11. a. 4560 cm^2
    b. 3.488 m^2
    c. Need 1 litre of paint so it will cost $24.95.
12. 28.54 cm^2
13. a. 2025 cm^2 b. 943.41 cm^2 c. 6031.59 cm^2
14. a. 3279.82 cm^2 b. 96.10 cm^2
15. 113.10 cm^2
16. a. 84.82 cm^2 b. 99.53 m^2
17. B
18. a. Small = 314.16 cm^2; medium = 706.86 cm^2; large = 1256.64 cm^2
    b. 113 883 cm^2
    c. 157.08 cm^2
19. a. 6800 m^2
    b. 3400 sheets of turf
    c. $35 700
20. a. 465.125 m^2 b. 2774 m^2
    c. 1860.5 m^2 d. $17 674.75

13.4 Surface areas of pyramids and irregular solids

13.4 Exercise

1. 31.67 m^2
2. 2454.21 mm^2
3. a. 125.66 cm^2 b. 220.35 m^2
4. 354.39 cm^2
5. 541.01 cm^2

6. a. 261.91 cm^2 b. 179.97 m^2
7. 132.58 cm^2
8. 1367.60 cm^2
9. 344.3 cm^2
10. 3552.95 mm^2
11. 791.68 cm^2
12. 1357.17 cm^2
13. a. 2324.78 cm^2 b. 10.65 m^2
14. 326.73 cm^2
15. 20.30 m^2
16. 1513.36 cm^2

13.5 Limits of accuracy

13.5 Exercise

1. a. i. 5 °C
   ii. Between 175 °C and 185 °C
   b. i. 2.5 km/h
   ii. Between 47.5 km/h and 52.5 km/h
   c. i. $\frac{1}{4}$ cup
   ii. Between $6\frac{3}{4}$ cups and $7\frac{1}{4}$ cups
   d. i. 5°
   ii. Between 205° and 215°
2. a. 58 mm b. 11.5 cm
3. a. Between 4.5 cm and 5.5 cm
   b. Between 11.5 cm and 12.5 cm
   c. Between 99.5 cm and 100.5 cm
   d. Between 849.5 cm and 850.5 cm
   e. Between 4.995 m and 5.005 m
   f. Between 2.495 m and 2.505 m
4. a. Between 28.5 cm and 29.5 cm
   b. Between 245 g and 255 g
   c. Between 7.5 mL and 12.5 mL
   d. Between 29.5 min and 30.5 min
5. E
6. C
7. a. 3520 cm (accurate to cm)
   b. 130 cm (accurate to cm)
   c. 1000 m (accurate to m)
   d. 80 mm (accurate to mm)
   e. 5000 mm (accurate to mm)
   f. 2520 m (accurate to m)
8. a. 0.6 cm b. −1.7 cm
   c. 22 km d. −1.1 mm
9. a. 15 cm
   b. No, the error is only 1% of the advertised size.
10. a. 0.15 km b. 4 g
    c. 23 min d. 7 L
11. 161 or 169 cm
12. a. Error = 5 m
    Absolute error = 5 m
    Absolute relative error = 0.02
    Absolute percentage error = 2%
    b. Error = −7 cm
    Absolute error = 7 cm
    Absolute relative error = 0.07
    Absolute percentage error = 7%
    c. Error = 1.5 mm
    Absolute error = 1.5 mm
    Absolute relative error = 0.08
    Absolute percentage error = 8%
    d. Error = −12 s
    Absolute error = 12 s
    Absolute relative error = 0.2
    Absolute percentage error = 20%
    e. Error = 0.1 kg
    Absolute error = 0.1 kg
    Absolute relative error = 0.08
    Absolute percentage error = 8%
    f. Error = −150 mL
    Absolute error = 150 mL
    Absolute relative error = 0.17
    Absolute percentage error = 17%
13. Error = −$2
    Absolute percentage error = 2%
14. a. 0.5 mm (half of the smallest unit)
    b. 22.5 mm and 23.5 mm
15. The approximations can always be more exact by measuring to smaller and smaller increments toward infinity. There is no smallest increment.
16. The 4-km measurement is accurate to the km, while the 4.000-km measurement is accurate to the thousandth of a km.
17. a. Statistically there is very little difference between the two methods.
    b. Mother's measurements = 11.96% ± 9.48
    c. Ultrasound = 9.036% ± 7.61
18. a. The time, whether it is behind or ahead, does not affect the significance of the accuracy of synchronisation, because the absolute error calculation forces them to be equal.
    b. The absolute percentage error would be very small and therefore insignificant.
    c. The absolute percentage error would be large and therefore significant.

13.6 Scientific notation and significant figures

13.6 Exercise

1. a. 4.72×10^1 b. 3.890×10^3 c. 5.6×10^{-1}
   d. -6.7×10^{-3} e. 3.4×10^7
2. a. 0.048 b. 7600
   c. 0.000 29 d. 8.1
3. a, c and d
4. a. False b. False
   c. True d. True

5. C

6. a. 1.5137×10^{16} km b. 2.82×10^{-15} m

7. Sextillions have 21 zeros.
 a. 5.972×10^{21}
 b. 4 significant figures

8. a. 0.058 b. 0.0506
 c. 55 830 d. 3×10^{-17}
 e. 138 000.00 f. 1.1

9. a. 12.0 m b. 6748 g
 c. 0.004 80 min d. 1926 hrs
 e. 50 000 g f. 0.0050 m

10. D

11. C

12. a. 3 b. −1 c. 2 d. −4

13. a. 4 b. 4 c. 1 d. 3

14. a. 9.1094×10^{-31} kg
 b. 1.6725×10^{-27} kg
 c. 1.6726×10^{-27} kg
 d. Working with the original amounts and not rounding before the final calculation ensures that the answer will be more precise.

15. a. 5.9 m
 b. 73.6 cm
 c. −77 m
 d. 257 cm^2
 e. 3400 m^2

16. a. i. 385 440 km
  ii. 78 489 600 km
  iii. 1 278 960 000 km
  iv. 138 548 160 000 km
 b. i. 385 000 km
  ii. 78 500 000 km
  iii. 1 280 000 000 km
  iv. 139 000 000 000 km
 c. i. 3.85×10^5
  ii. 7.85×10^7
  iii. 1.28×10^9
  iv. 1.39×10^{11}

13.7 Review

13.7 Exercise

Multiple choice

1. D
2. A
3. E
4. E
5. B
6. D
7. A
8. D
9. C
10. C

Short answer

11. 312.791 m^2

12. a. 11105.53 cm^2 b. 57 805.30 cm^2

13. a. 16.8 m b. 12.8 m^2

14. a. 1250 m
 b. 8 laps
 c. 30 laps
 d. 1520 m^2
 e. 15 200 000 cm^2

15. a. 22.93 cm
 b. 707.50 cm^2
 c. 498.76 cm^2
 d. 208.74 cm^2

16. a. metres or kilometres
 b. kilograms
 c. litres

Extended response

17. a. 2.5 km/h
 b. Between 52.5 km/h and 57.5 km/h

18. a. 3.8967×10^8 b. 4.8603×10^{-3}
 c. $4.060\,094\,37 \times 10^8$ d. 1.00×10^{-7}
 e. 3.750×10^{-1} f. 1.003×10^{11}

19.

Student	Measurement	Error	Absolute error	Absolute relative error
A	44.8 m	2.1 m	2.1 m	0.0492
B	41.9 m	−0.8 m	0.8 m	0.0187
C	43.7 m	1 m	1 m	0.0234
D	42.0 m	−0.7 m	0.7 m	0.0164
E	43.3 m	0.6 m	0.6 m	0.0141

Student E has the most accurate answer. This is determined by finding the student who has the smallest absolute relative error.

20. a. 350 m^3
 b. 400 m^3
 c. 178 m^3
 d. The volume equation requires multiplication, not addition.
 e. Yes
 f. 50.86%

14 Volume

LEARNING SEQUENCE

Fully worked solutions for this topic are available online.

14.1 Overview

Watch videos

Engage with interactivities

Answer questions and check results

Find all this and MORE in jacPLUS

14.1.1 Introduction

It is important to have an understanding of metric measurements — length, area and volume — as they are used daily in many jobs and professions. Pool builders need to understand the measurements required to build a pool of a certain volume, carpet layers need to understand area for carpet laying, and nurses need to understand volume in order to administer medicine.

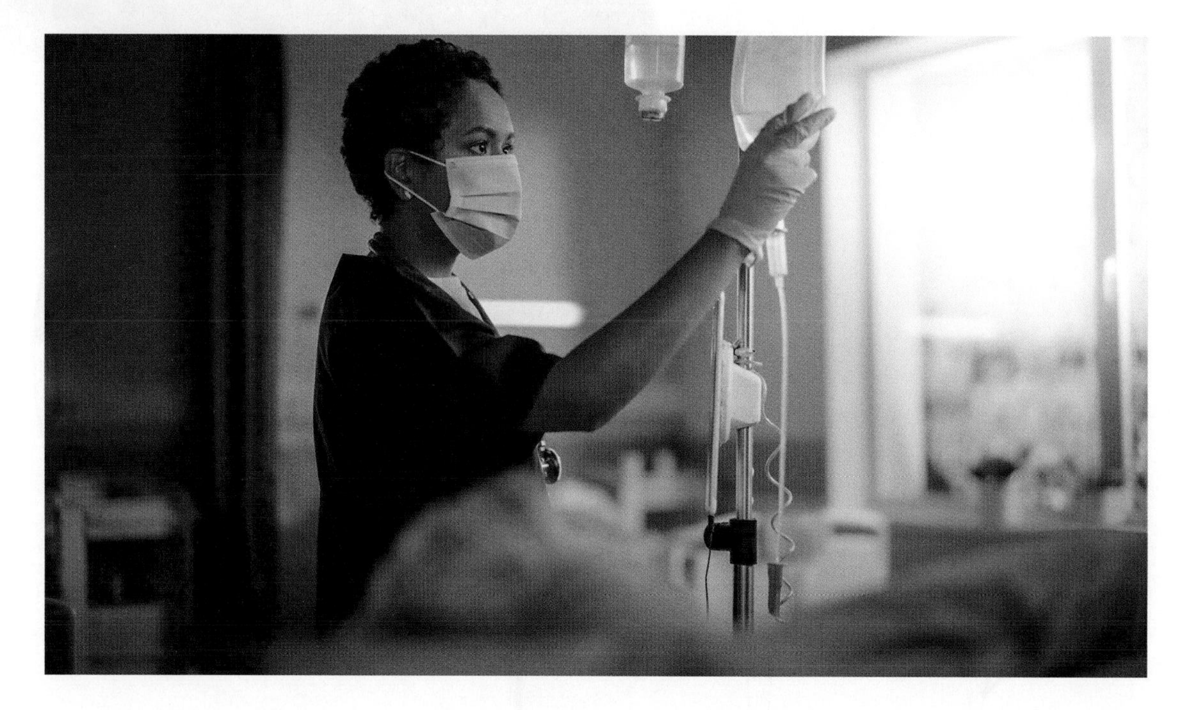

We need to be able to interpret our water bills, compare fuel consumption of cars, estimate volumes and areas when home renovating or landscaping, and use our knowledge of area and volume confidently in daily life.

KEY CONCEPTS

This topic covers the following key concepts from the VCE Mathematics Study Design:

- measurements and related quantities including derived quantities, metric and relevant non-metric measures
- conventions, properties and measurement of perimeter, area, surface area and volume of compound shapes and objects.

Note: *Concepts shown in grey are covered in other topics.*

14.2 Review of volume of prisms, units of volume and capacity

LEARNING INTENTION

At the end of this subtopic you should be able to:
- convert between metric units of volume
- calculate the volume of solids using metric cubes.

14.2.1 Units of volume

The volume of a three-dimensional object is the amount of space that is occupied inside it. As covered in Foundation Units 1 and 2, the units of volume can be converted using the diagram shown.

Converting units of volume

The following chart is useful when converting between units of volume.

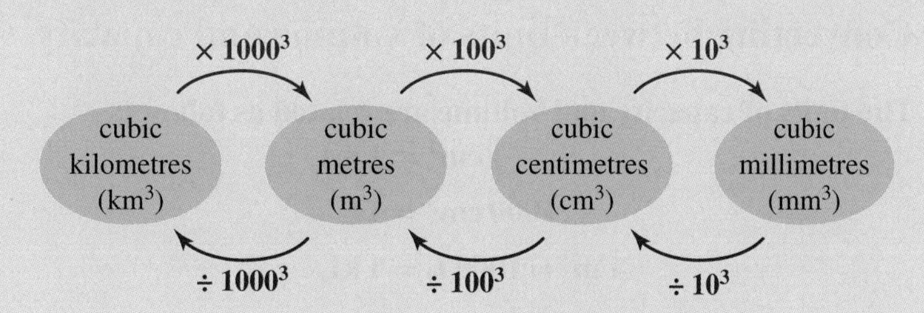

For example, $3\,\text{m}^3 = 3 \times 100^3 \times 10^3\,\text{mm}^3$

$= 3\,000\,000\,000\,\text{mm}^3$

14.2.2 Capacity

Capacity is another term for volume, which is usually applied to the measurement of liquids and containers.

The capacity of a container is the volume of liquid that it can hold.

The standard measurement for capacity is the litre (L).

Other common units are the millilitre (mL), kilolitre (kL), and megalitre (ML).

Converting units of capacity

The following chart is useful when converting between units of capacity.

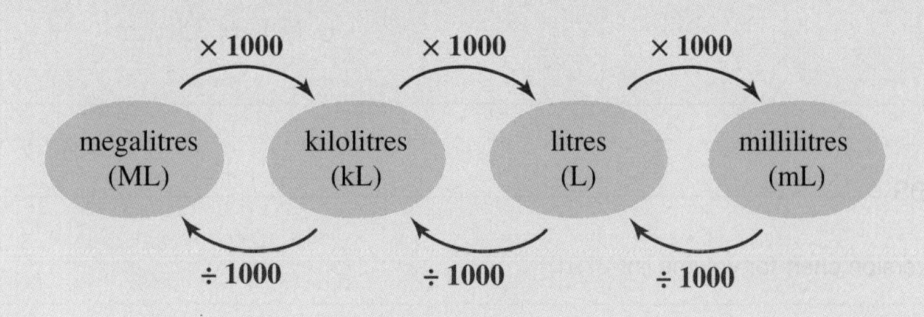

A container, such as a jar, has both volume and capacity, as it occupies space and can hold, contain or absorb something other than what it is made of. A solid block of wood has volume but not capacity, because it cannot hold or absorb anything.

The metric volume unit, 1 cm^3, was defined as having a capacity of 1 mL. Therefore, units of volume and capacity are related as shown below.

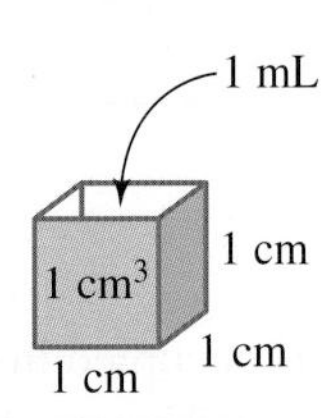

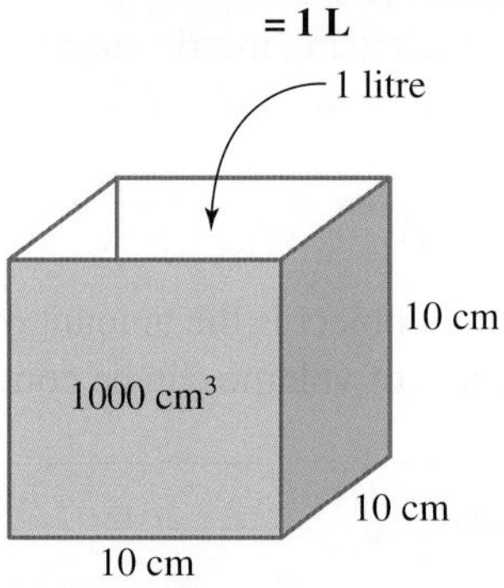

Converting between units of volume and capacity

The units of capacity and volume are related as follows:

$$1\ \text{cm}^3 = 1\ \text{mL}$$

$$1000\ \text{cm}^3 = 1\ \text{L}$$

$$1\ \text{m}^3 = 1000\ \text{L} = 1\ \text{kL}$$

tlvd-5070

WORKED EXAMPLE 1 Converting units of volume and capacity

Complete the following unit conversions.

a. 50 mL = ____ cm^3 **b. 150 mL = ____ L** **c. 0.35 L = ____ cm^3**

THINK	WRITE
a. 1 cm^3 holds 1 mL.	a. $1\ \text{mL} = 1\ \text{cm}^3$, so $50\ \text{mL} = 50\ \text{cm}^3$
b. There are 1000 mL in 1 L. So, to convert from millilitres to litres, divide by 1000.	b. $150\ \text{mL} = (150 \div 1000)\ \text{L}$ $= 0.15\ \text{L}$
c. 1. First, convert 0.35 L to mL. There are 1000 mL in 1 L. So, to convert from litres to millilitres, multiply by 1000.	c. $0.35\ \text{L} = (0.35 \times 1000)\ \text{mL}$ $= 350\ \text{mL}$
2. Use the relationship 1 cm^3 = 1 mL to complete the conversion.	$350\ \text{mL} = 350\ \text{cm}^3$ $0.35\ \text{L} = 350\ \text{cm}^3$

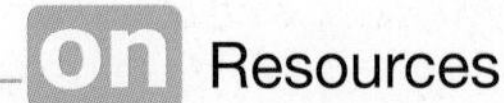

Interactivity Conversion chart for volume (int-3791)

14.2.3 Volume of prisms

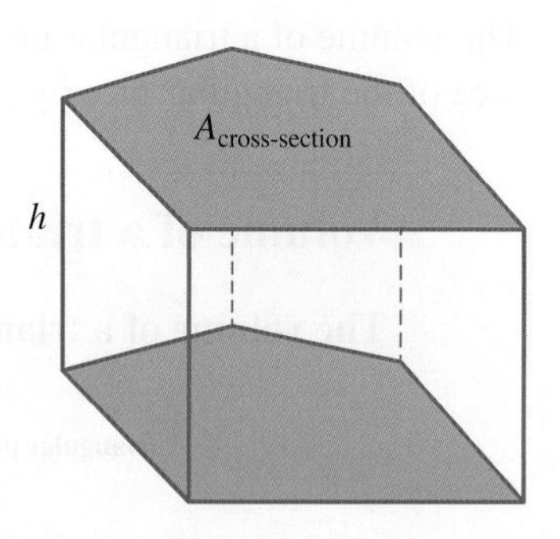

A prism is a three-dimensional shape with a constant two-dimensional polygon-shaped cross-section. The area of the cross-section is shaded in blue in the diagram shown.

The volume of a prism is calculated by multiplying the area of the cross-section by its height.

> **Volume of a prism**
>
> **The volume of a prism can be calculated by:**
>
> $$V_{\text{prism}} = A_{\text{cross-section}} \times h$$
>
> **where h is the vertical height.**

Volume of a rectangular prism

The volume (V) of a rectangular prism follows the general rule of all prisms, so you need to multiply the area of the cross-section by its height. Taking the bottom face as the cross-section:

$$V = A_{\text{cross-section}} \times h$$
$$V = (l \times w) \times h$$

This is the same as multiplying the three dimensions of the prism (length, width and height).

> **Volume of a rectangular prism**
>
> **The volume of a rectangular prism can be calculated by:**
>
> $$V_{\text{rectangular prism}} = \text{length} \times \text{width} \times \text{height}$$
> $$V_{\text{rectangular prism}} = l \times w \times h$$

WORKED EXAMPLE 2 Calculating the volume of a rectangular prism

Calculate the volume of the rectangular prism shown.

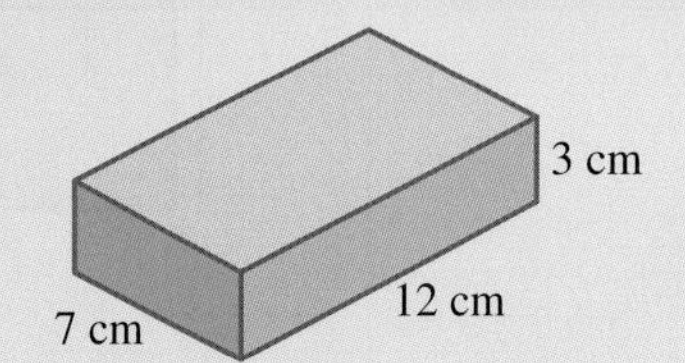

THINK	WRITE
1. Write the formula for the volume of a rectangular prism.	$V = l \times w \times h$
2. Substitute the values length = 12 cm, width = 7 cm and height = 3 cm into the formula.	$V = 12 \times 7 \times 3$
3. Calculate the volume and write the unit.	$V = 252\text{ cm}^3$

Volume of a triangular prism

The volume of a triangular prism can also be calculated by multiplying the area of the triangular face by its height.

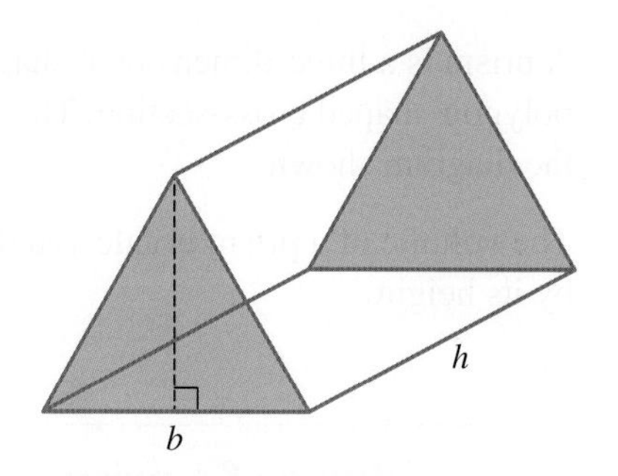

> **Volume of a triangular prism**
>
> **The volume of a triangular prism can be calculated by:**
>
> $$\begin{aligned} V_{\text{triangular prism}} &= A_{\text{cross-section}} \times h \\ &= \text{area of triangle} \times h \end{aligned}$$

WORKED EXAMPLE 3 Calculating the volume of a triangular prism

Calculate the volume of the triangular prism shown.

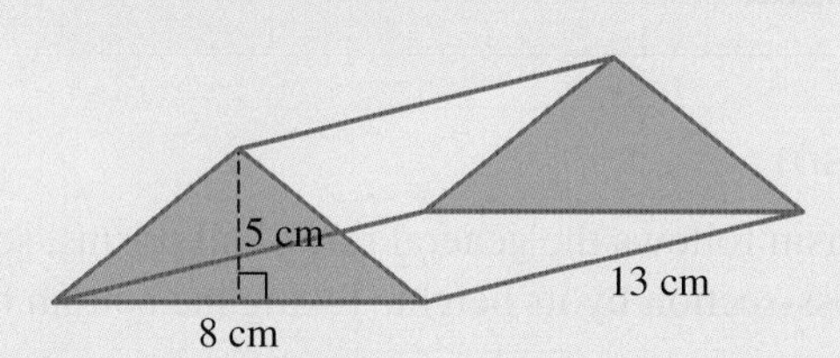

THINK	WRITE
1. Write the formula for the volume of a triangular prism.	$V = \text{area of triangle} \times h$
2. Recall that the area of a triangle is $A = \frac{1}{2} \times \text{base} \times \text{height}$. Be careful not to confuse the height of the triangle with the height of the entire object.	$V = \left(\frac{1}{2} \times 8 \times 5\right) \times 13$
3. Calculate the volume and write the unit.	$V = 260\ \text{cm}^3$

As covered in Foundation Units 1 and 2, the formulas to calculate the volume of prisms are shown below.

Shape	Diagram	Cross-sectional area	Volume formula
Cube		$A = l^2$	$V_{\text{square}} = A_{\text{square}} \times H$ $= \text{area of a square} \times \text{height}$ $= l^2 \times l$ $= l^3$
Rectangular prism		$A = lw$	$V_{\text{rectangle}} = A_{\text{rectangle}} \times H$ $= \text{area of a rectangle} \times \text{height}$ $= lw \times H$

(continued)

Shape	Diagram	Cross-sectional area	Volume formula
Triangular prism		$A=\frac{1}{2}bh$	$V_{\text{triangle}}=A_{\text{triangle}}\times H$ $=$ area of a triangle $\times$ height $=\frac{1}{2}bh\times H$
Hexagonal prism		$A=\frac{3\sqrt{3}S^2}{2}$	$V_{\text{hexagon}}=A_{\text{hexagon}}\times H$ $=$ area of a hexagon $\times$ height $=\frac{3\sqrt{3}}{2}S^2\times H$

14.2 Exercise

1. State whether the following measures represent volume or capacity.
   a. $3.75\ \text{cm}^3$
   b. 135 mL
   c. 1.25 L
   d. $4.75\ \text{m}^3$
2. WE1 Complete the following conversions.
   a. $224\ \text{L} = \_\_\_\_\_\ \text{cm}^3$
   b. 789 mL = ____L
   c. 0.348 L = ____mL
   d. $452\ \text{mL} = \_\_\_\_\ \text{cm}^3$
3. Complete the following conversions.
   a. 308 mL = _____L
   b. 1278 mL = ____L
   c. 0.045 L = ____mL
   d. 599 mL = ____L
4. Complete the following conversions.
   a. $555\ \text{cm}^3 = \_\_\_\_\_\ \text{mL}$
   b. $1890\ \text{L} = \_\_\_\_\ \text{cm}^3$
   c. $4.2\ \text{L} = \_\_\_\_\ \text{cm}^3$
   d. $12.58\ \text{L} = \_\_\_\_\ \text{m}^3$

5. Determine the volume of each of the following prisms.

a.

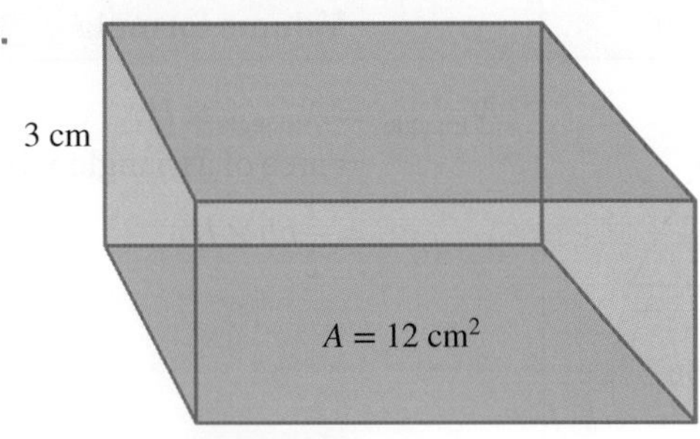

b.

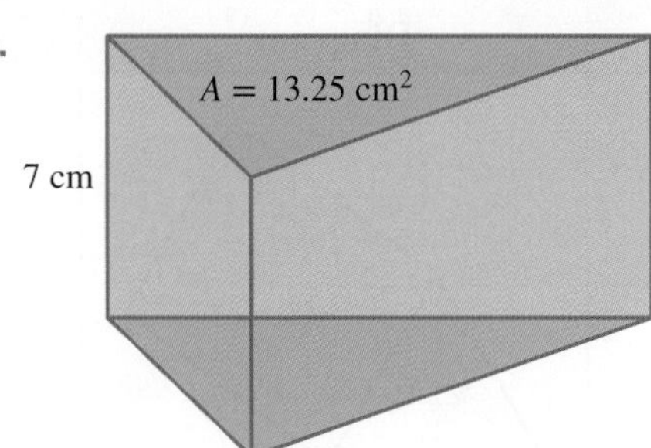

c.

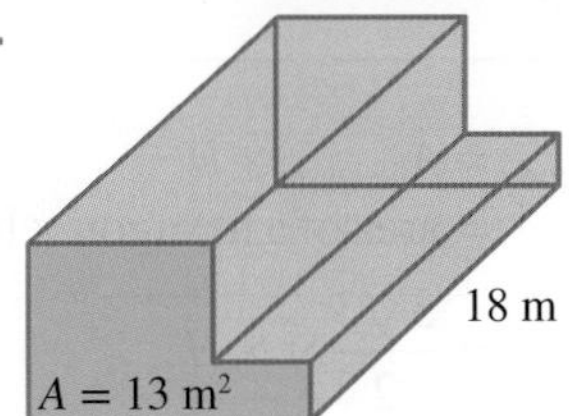

d.

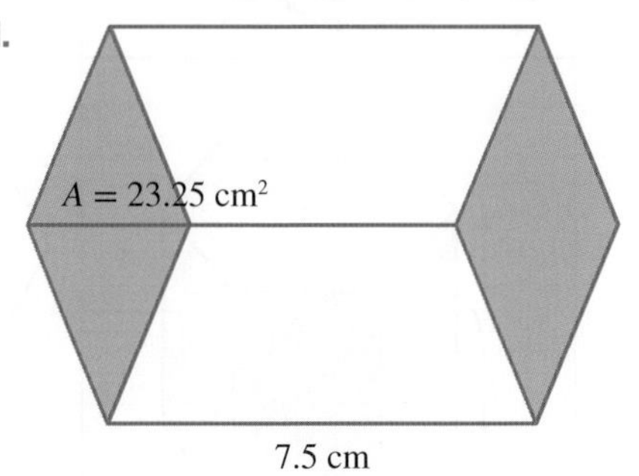

6. WE2 Calculate the volume of each of the following rectangular prisms.

a.

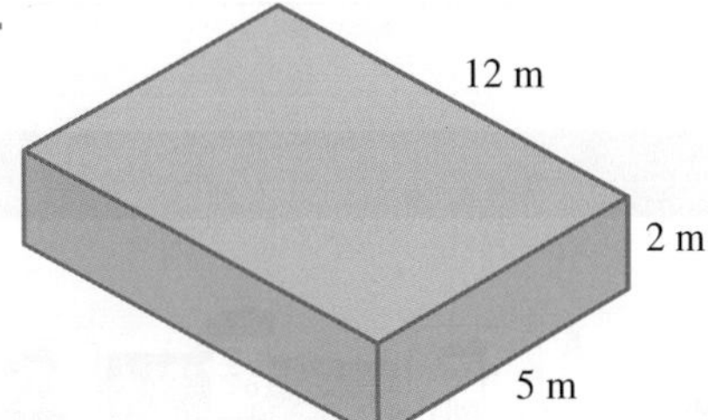

b.

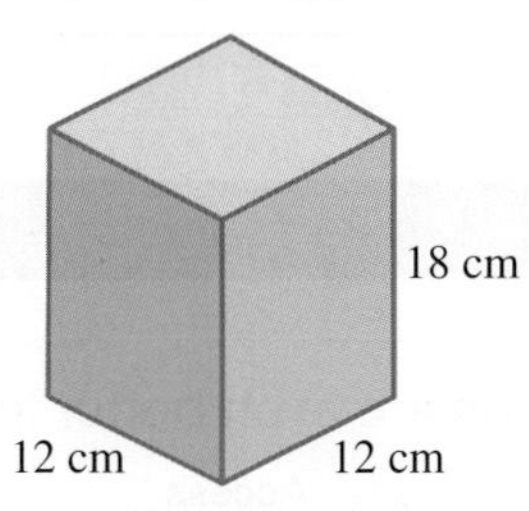

c.

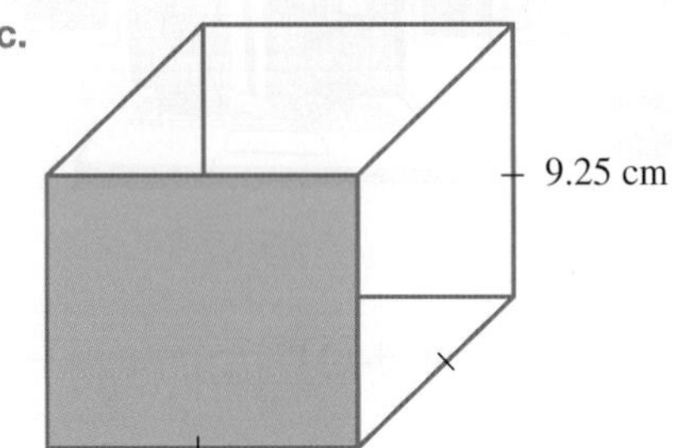

d.

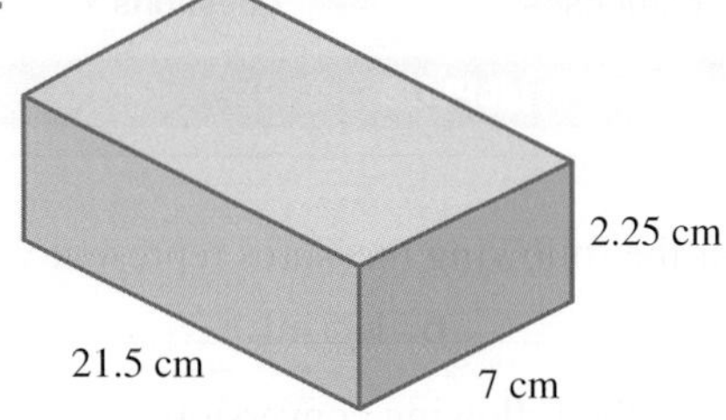

7. Calculate the volume of each of the following prisms.

a.

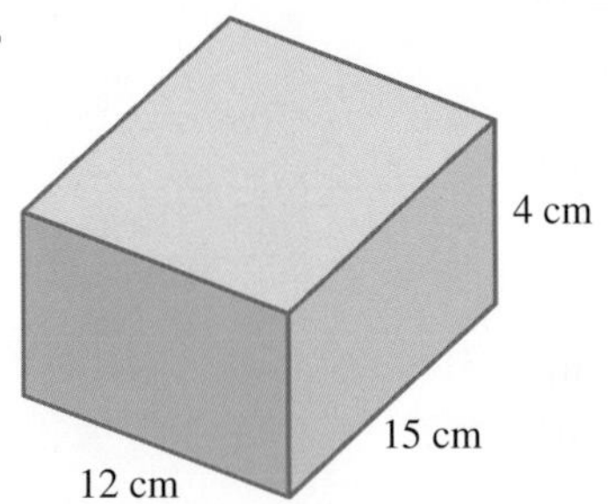

b.

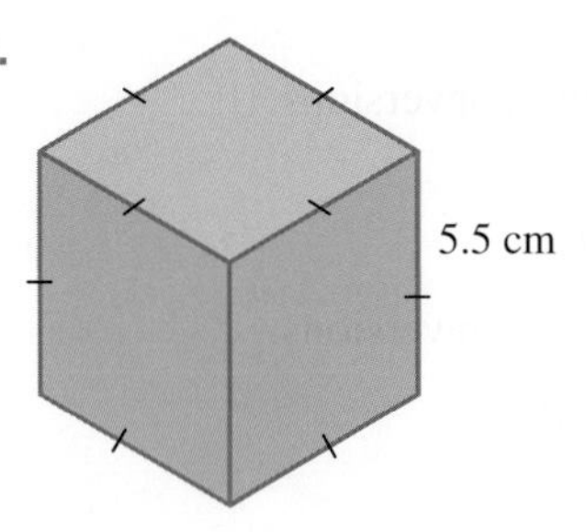

8. WE3 Calculate the volume of each of the following triangular prisms.

a.

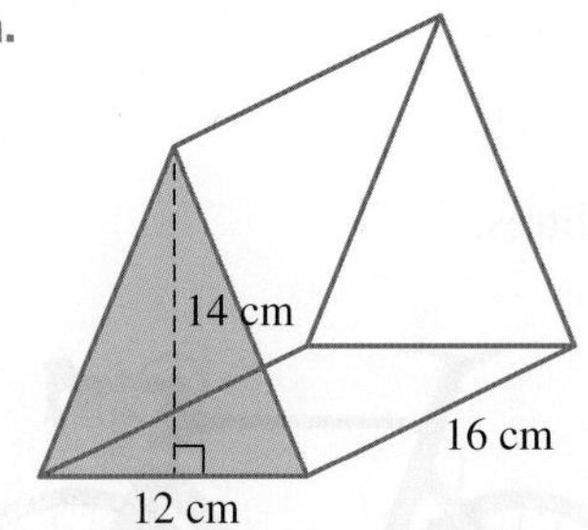

b.

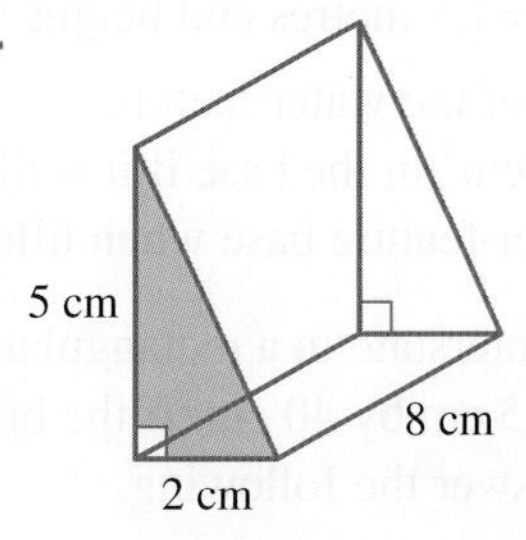

c.

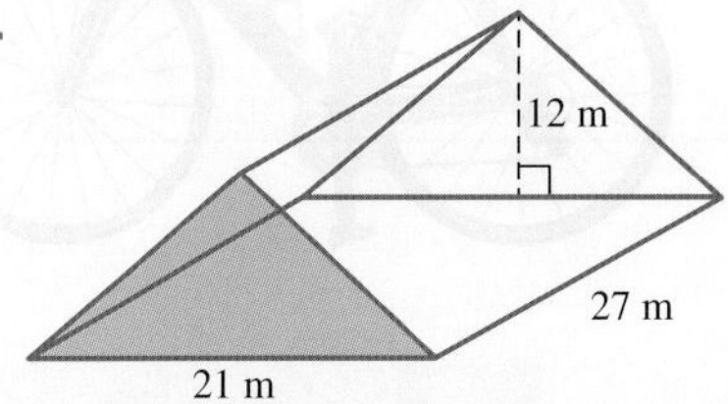

d.

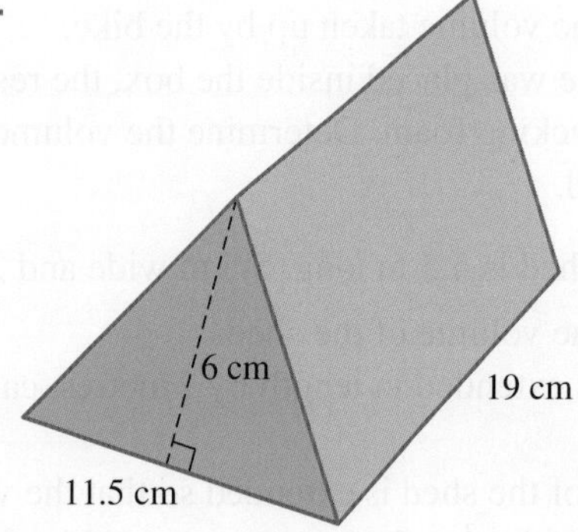

9. Calculate the volume of each of the following triangular prisms.

a.

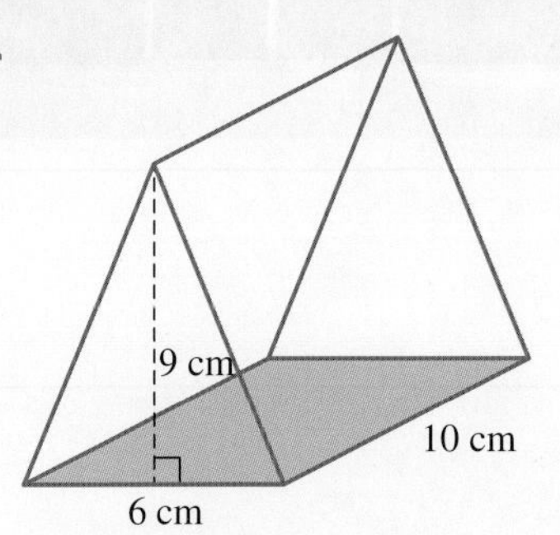

b.

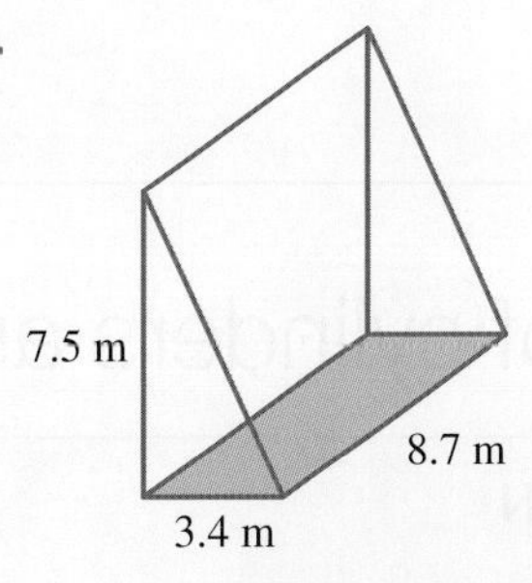

10. A pool is made in the shape of an open-top rectangular prism of length 12 m, width 5 m and depth 1.85 m. Calculate the total volume of the pool in cubed metres, and determine how many litres the pool holds when full.

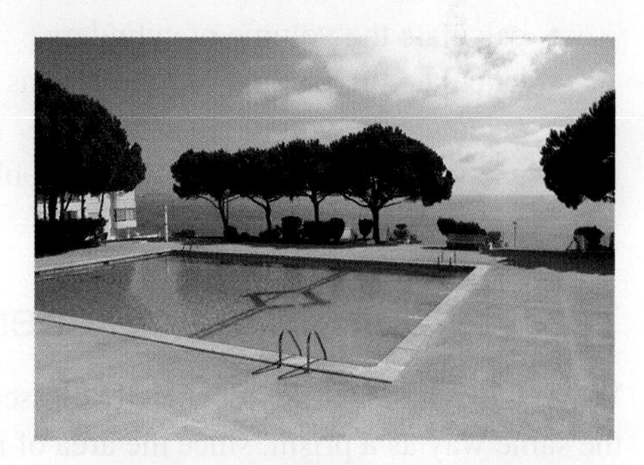

11. A tent is made in the shape of a triangular prism. If the tent has a vertical height of 2.2 metres, a width of 3.6 metres and a length of 4.2 metres, calculate the volume of the inside of the tent.

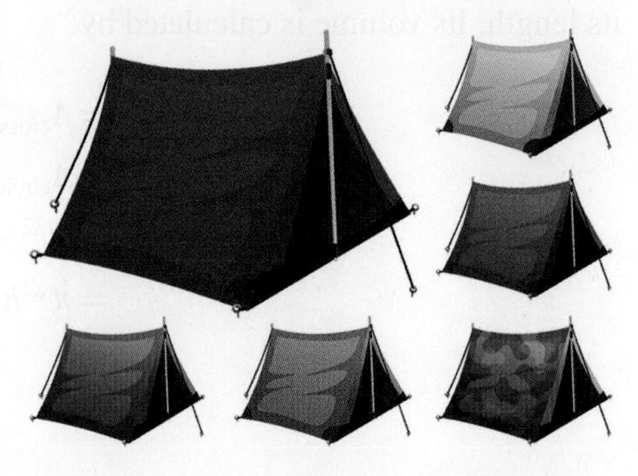

12. A garden water feature has a base in the shape of a rectangular prism with the following dimensions: length = 2 metres, width = 1.5 metres and height = 40 cm. Calculate:
   a. the volume of the base of the water feature
   b. the volume of water in cm^3 in the base if it is filled to a height of 30 cm
   c. the capacity of the water-feature base when filled to a height of 25 cm in litres.

13. A road bike is to be sent interstate in a rectangular prism–shaped box with dimensions 2 m by 85 cm by 40 cm. If the bike takes up 30% of the volume of the box, answer the following.
   a. Determine the volume of the box in m^3.
   b. Determine the volume taken up by the bike.
   c. After the bike was placed inside the box, the rest of the box was filled with packing foam. Determine the volume of packing foam that was used.

14. A rectangular shed is 5.5 m long, 5.5 m wide and 2.5 m high.
   a. Determine the volume of the shed.
   b. If the shed is extended in length by 2 metres, calculate its new volume.
   c. If the width of the shed is extended so that the volume of the shed is now 100 m^3, calculate the new width of the shed.

14.3 Volume of cylinders and spheres

LEARNING INTENTION

At the end of this subtopic you should be able to:
- calculate the volume of cylinders
- calculate the volume of right prisms and cylinders
- calculate the volume of spheres
- use volume to solve real-world problems.

14.3.1 Volume of a cylinder

While the cylinder is not technically classed as a prism, its volume can be calculated in the same way as a prism, since the area of its cross-section (the circle) is the same along its length. Its volume is calculated by:

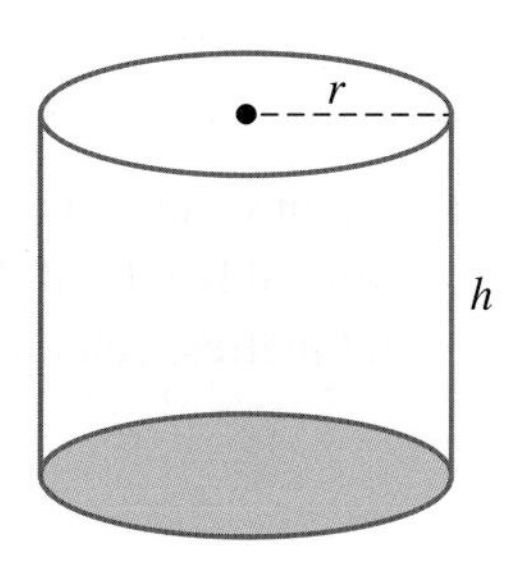

$$\begin{aligned} V &= A_{\text{cross-section}} \times \text{height} \\ V &= A_{\text{circle}} \times h \\ &= \pi r^2 \times h \\ &= \pi r^2 h \end{aligned}$$

Volume of a cylinder

The volume of a cylinder can be calculated by:

$$V_{\text{cylinder}} = \pi r^2 h$$

where r is the radius

h is the height.

WORKED EXAMPLE 4 Calculating the volume of a cylinder

Calculate the volume of the cylinder shown to 2 decimal places.

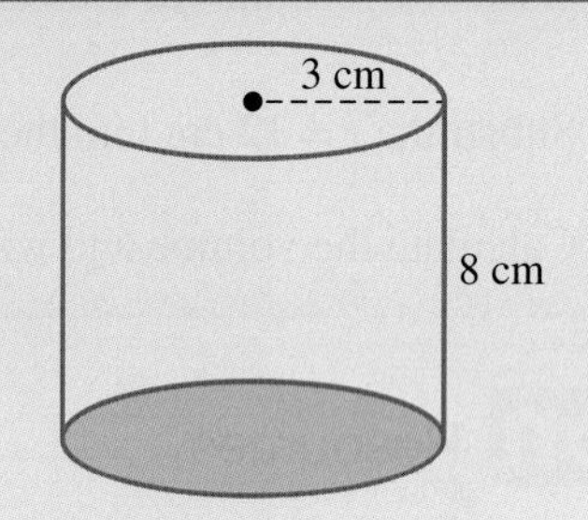

THINK	WRITE
1. Write the formula for the volume of a cylinder.	$V = \pi r^2 h$
2. Substitute the values radius = 3 cm and height = 8 cm into the formula.	$V = \pi \times 3^2 \times 8$ $= \pi \times 9 \times 8$
3. Calculate the volume and write the unit.	$V = 226.19\ \text{cm}^3$

Resources

Interactivity Volume of solids (int-3794)

14.3.2 Volume of spheres

A sphere is a three-dimensional circular shape, such as a ball. The volume of a sphere can be calculated using the following formula.

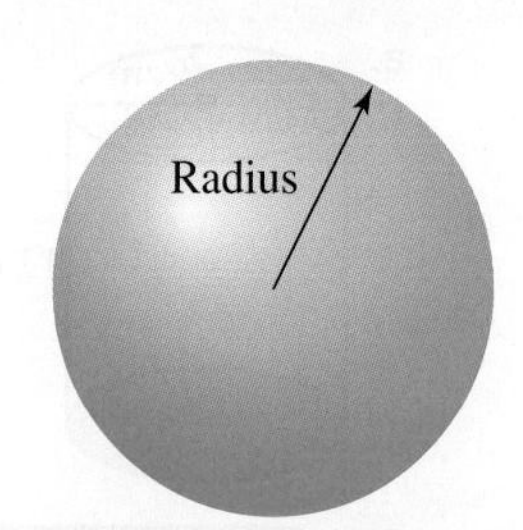

Volume of a sphere

The volume of a sphere can be calculated by:

$$V_{\text{sphere}} = \frac{4}{3}\pi r^3$$

where r is the radius.

tlvd-5073

WORKED EXAMPLE 5 Calculating the volume of a sphere

Calculate the volume of the following sphere.

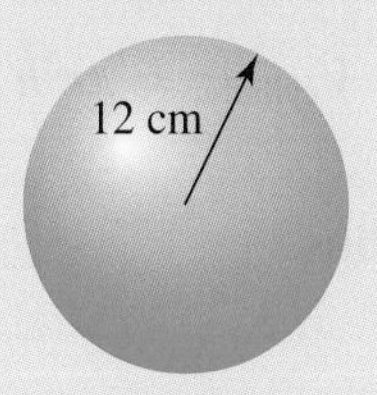

THINK	WRITE
1. Write the formula for the volume of a sphere.	$V_{\text{sphere}} = \dfrac{4}{3}\pi r^3$
2. Substitute $r = 12$ cm into the formula.	$V = \dfrac{4}{3} \times \pi \times 12^3$
3. Calculate the volume and write the unit.	$V = 7238.23 \text{ cm}^3$

 Resources

 **Interactivity** Volume (int-6476)

14.3 Exercise

1. **WE4** Calculate the volume of each of the following cylinders correct to 2 decimal places.

a. 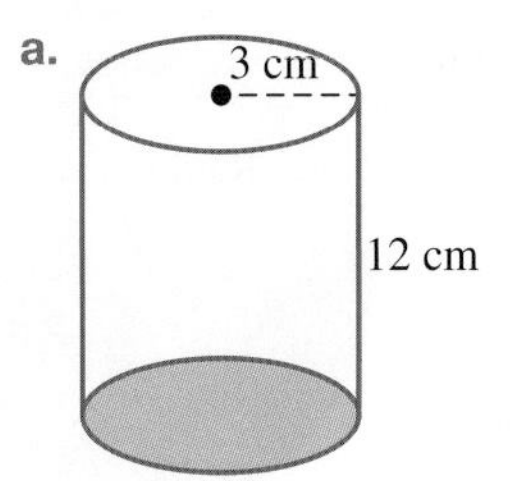

b. 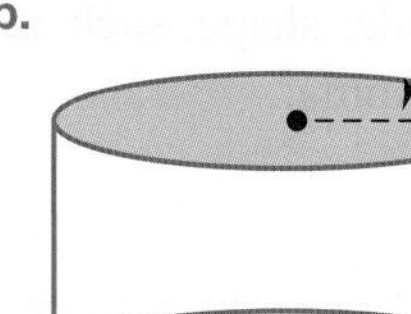

c. 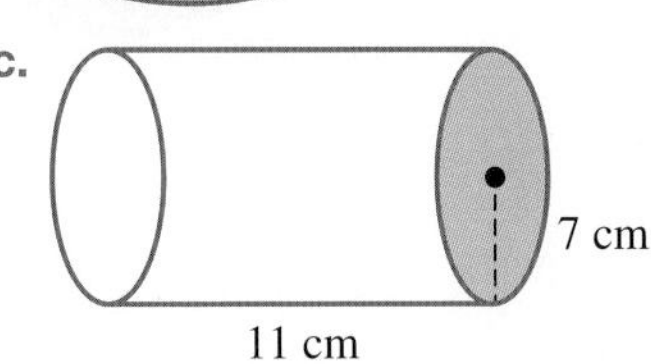

d. 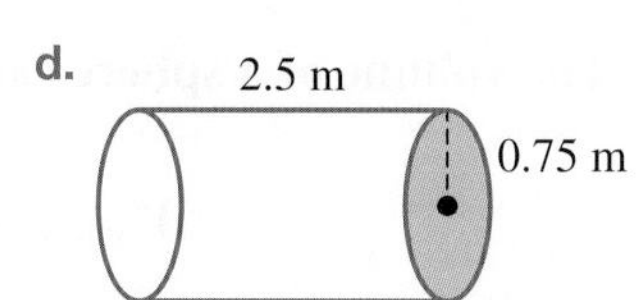

2. Calculate the volume of the cylinder to 2 decimal places.

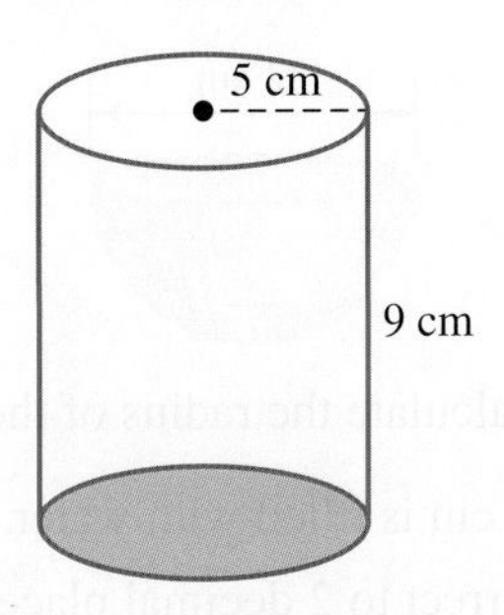

3. Calculate the volume of the following shape to 2 decimal places.

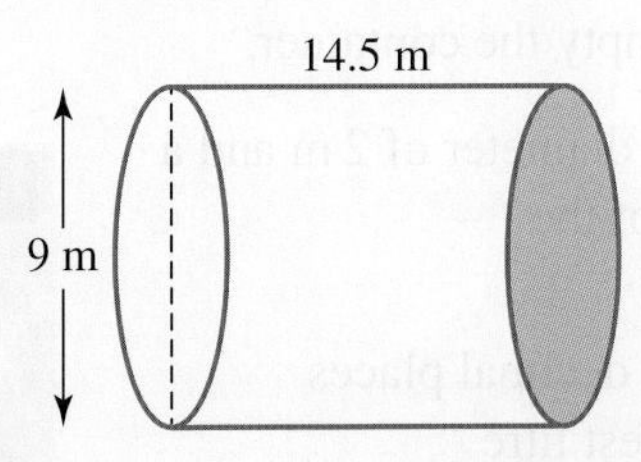

4. Calculate the volume of each of the following shapes correct to 2 decimal places.

a. 220 cm, 5.8 m

b.

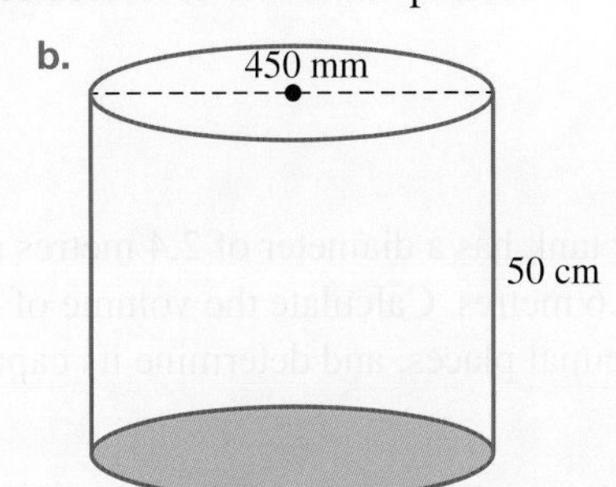

5. Calculate the volume of the spheres with the following properties correct to 2 decimal places.
   a. Radius of 5 cm
   b. Radius of 12 cm
   c. Diameter of 1.8 m
   d. Diameter of 27 cm
   e. Radius of 24.8 cm
   f. Diameter of 32.5 m

6. **WE5** Calculate the volumes of the following spheres.

a.

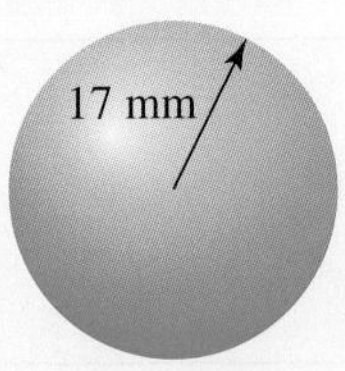

b.

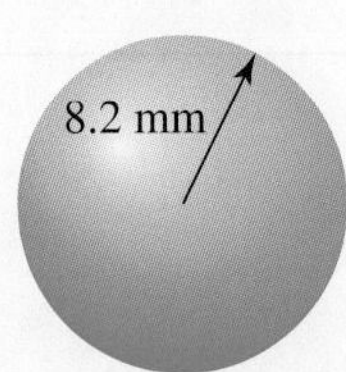

c.

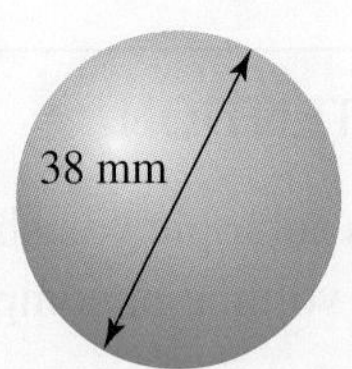

d.

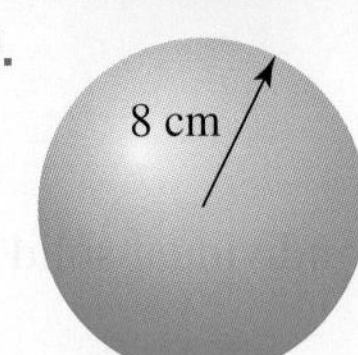

e.

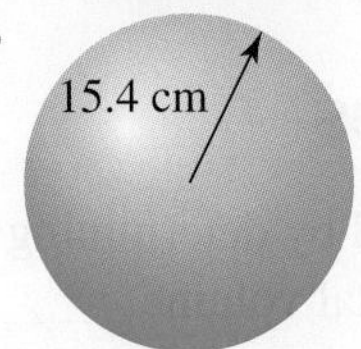

7. Calculate the volume of the hemisphere shown.

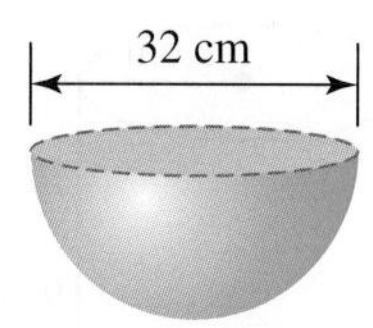

8. A sphere has a volume of 523.81 cm^3. Calculate the radius of the sphere.

9. A spherical container with a radius of 15 cm is filled with water.
   a. Calculate the volume of the sphere, correct to 2 decimal places.
   b. Determine the total capacity of the sphere, in litres, correct to 2 decimal places.
   c. If a tap at the bottom of the container allows the water to be emptied at a rate of 80 cm^3/s, calculate how long it would take to completely empty the container.

10. This cylindrical steel water tank has a diameter of 2 m and a length of 3 m. Ignoring the thickness of the material, calculate:
    a. the volume of the tank, correct to 3 decimal places
    b. the capacity of the tank, to the nearest litre ($1\ m^3 = 1000\ L$).

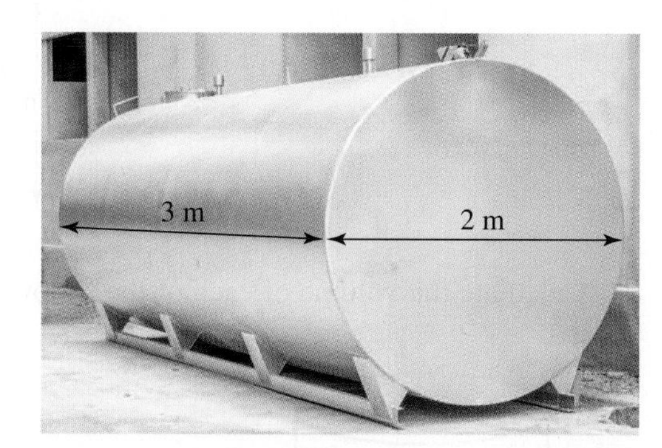

11. A cylindrical water tank has a diameter of 2.4 metres and vertical height of 2.6 metres. Calculate the volume of the water tank, to 3 decimal places, and determine its capacity when full.

12. A cylinder has a volume of 245 cm^3 and a height of 6 cm. Determine the length of the radius of the cylinder to 2 decimal places.

14.4 Volume of composite shapes

LEARNING INTENTION

At the end of this subtopic you should be able to:
- calculate the volume of composite shapes.

14.4.1 Volume of composite shapes

The total volume of a composite solid can be calculated by determining the volume of each individual solid that it contains and adding them together to calculate the total volume.

Some examples include:

tlvd-5074

WORKED EXAMPLE 6 Calculating the volume of a composite shape

Calculate the volume of this object, correct to 2 decimal places.

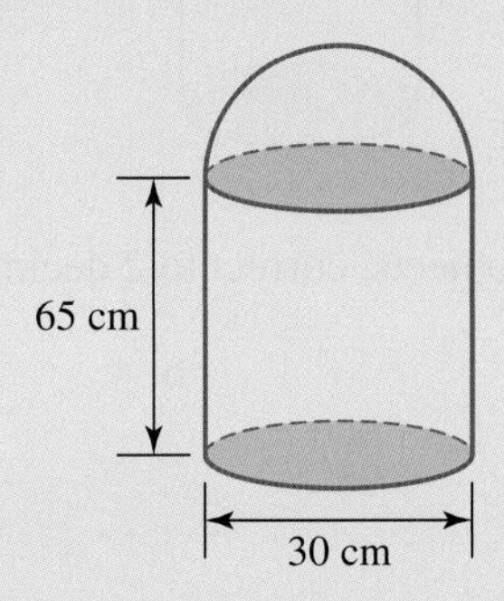

THINK	WRITE
1. The composite object is made up of a cylinder and a hemisphere.	
2. Write the formula for calculating the volume of a hemisphere and substitute the value of the radius into the formula. Calculate.	$V_{\text{hemisphere}} = \frac{1}{2} \times \frac{4}{3} \times \pi \times r^3$ $= \frac{1}{2} \times \frac{4}{3} \times \pi \times 15^3$ $= 2250\pi$
3. Write the formula for calculating the volume of a cylinder and substitute the values of the height and radius into the equation. Calculate.	$V_{\text{cylinder}} = \pi \times r^2 \times h$ $= \pi \times 15^2 \times 65$ $= 14\,625\pi$
4. Add the volumes of the hemisphere and the cylinder.	Total volume $= 2250\pi + 14\,625\pi$ $= 16\,875\pi$ $\approx 53\,014.38$
5. Write the answer with the appropriate unit.	$V_{\text{total}} \approx 53\,014.38\text{ cm}^3$

14.4 Exercise

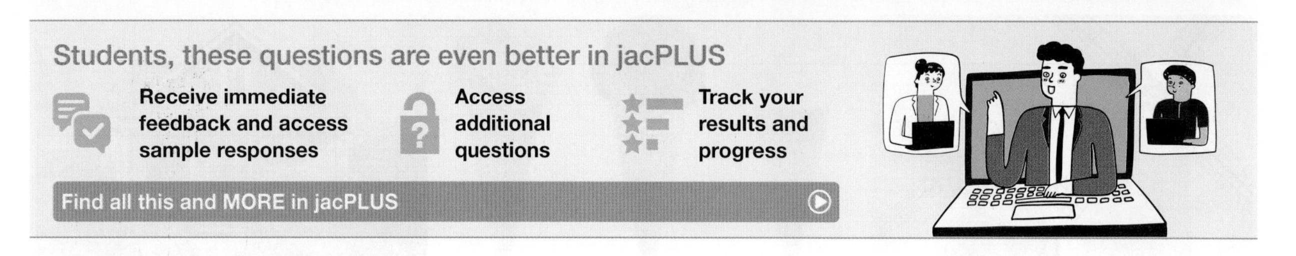

1. WE5 Calculate the volume of the solid shown, correct to the nearest cm^3.

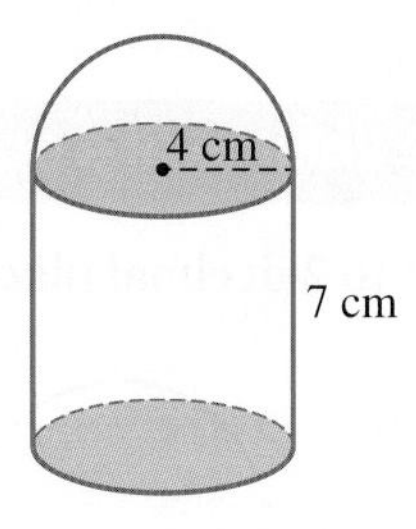

2. Calculate the volumes of the following objects, correct to 2 decimal places.

a.

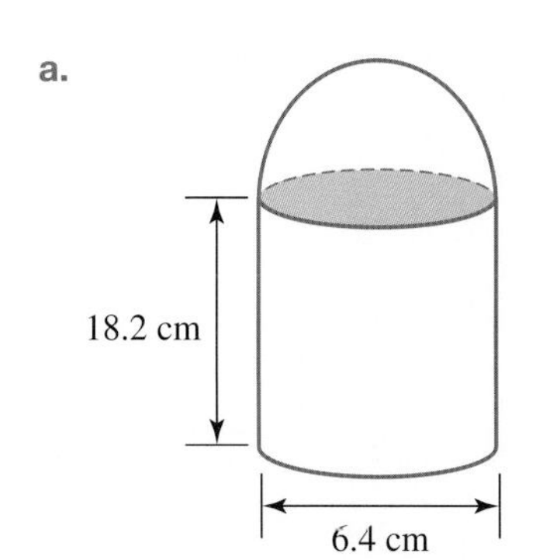

b.

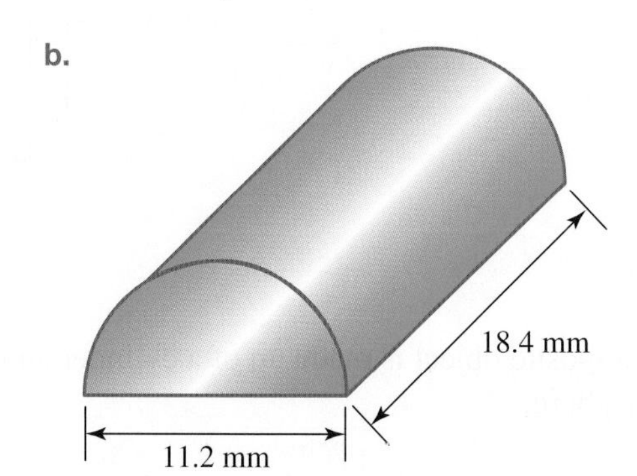

3. Calculate the capacity of the object shown in cm^3. Write the result to 4 significant digits.

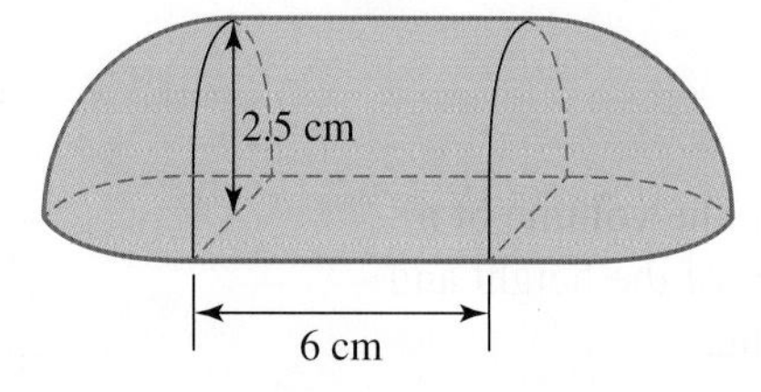

4. Calculate the volume of water that can completely fill the inside of this washing machine. Give your answer in litres, correct to 2 decimal places.

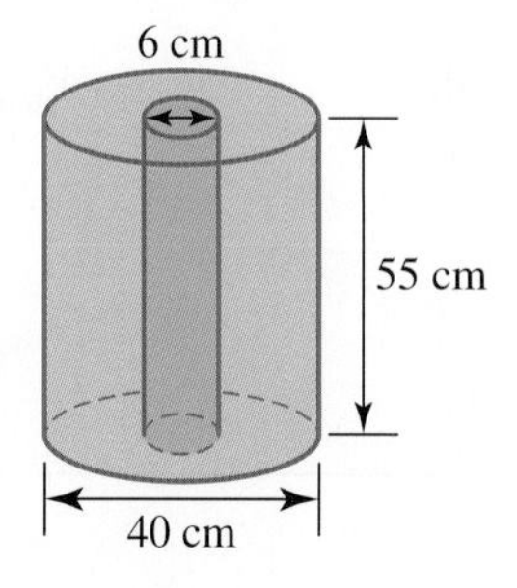

5. Calculate the volume of the following shape.

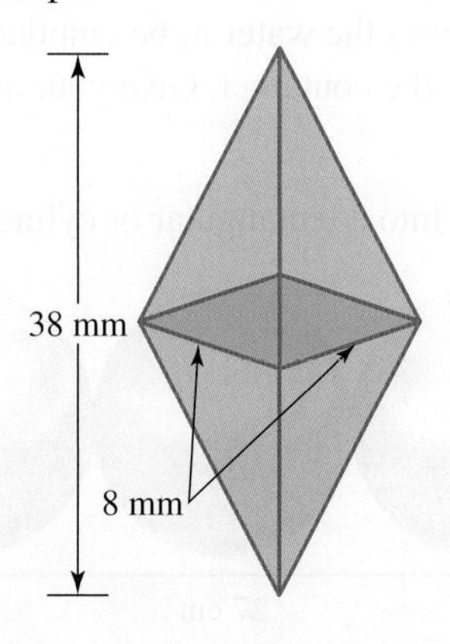

6. A cylinder fits exactly into a cube of side 8 cm.

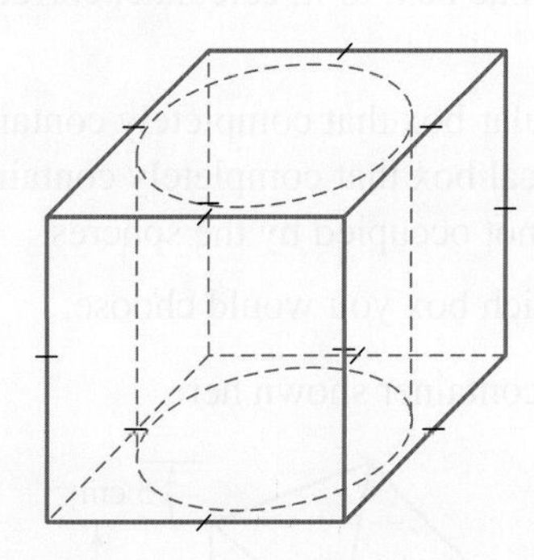

a. Determine the volume of the cylinder.
b. Determine the percentage of space inside the cube that is not occupied by the cylinder.

7. A house plan is drawn as shown. The house is going to be built on a concrete slab.

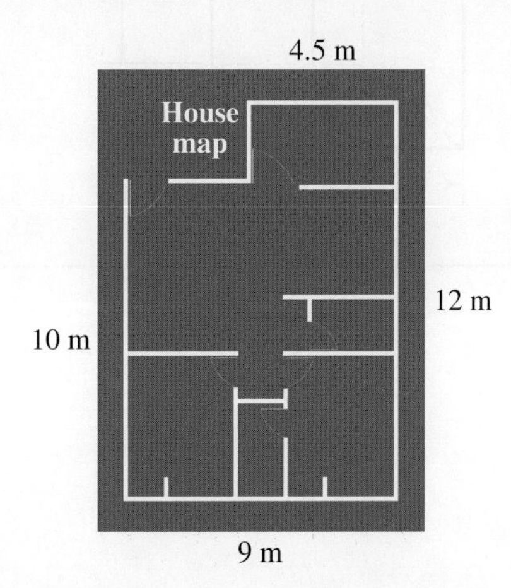

a. Calculate the area of the slab.
b. The slab is to be 15 cm thick. Calculate the volume of concrete needed for the slab. (*Hint:* Write 15 cm as 0.15 m.)
c. The cost of concrete is $\$180/m^3$. Calculate the cost of this slab.

8. A spherical container with a radius of 15 cm^3 is filled with water.
A tap at the bottom of the container allows the water to be emptied at a rate of 80 cm^3/s. Determine how long it would take to completely empty the container. Give your answer in seconds correct to 1 decimal place.

9. The sculpture shown is to be packaged into a rectangular or cylindrical cardboard box.

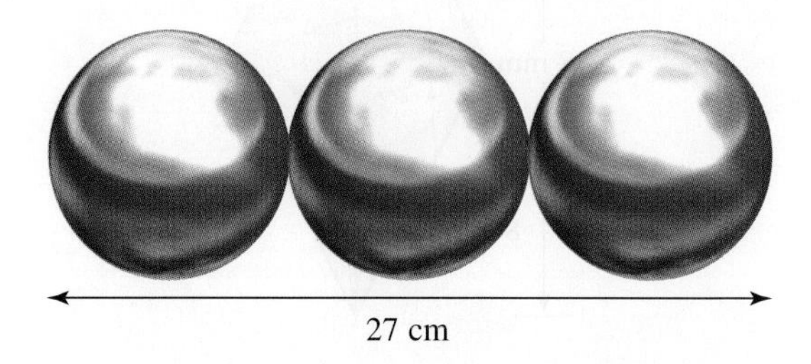

a. Assuming each sphere touches the one next to it, calculate, correct to 2 decimal places where necessary, the:
    i. volume of the smallest rectangular box that completely contains the spheres
    ii. volume of the smallest cylindrical box that completely contains the spheres
    iii. volume of space in both boxes not occupied by the spheres.

b. Use your calculations to justify which box you would choose.

10. Calculate the capacity in litres of the container shown here.

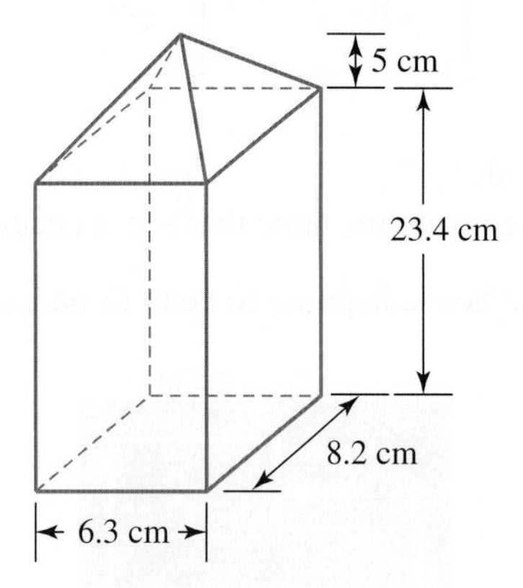

14.5 Review

14.5.1 Summary

doc-38022

Hey students! Now that it's time to revise this topic, go online to:

Review your results

Watch teacher-led videos

Practise questions with immediate feedback

Find all this and MORE in jacPLUS

14.5 Exercise

Multiple choice

1. **MC** The volume of a jug is 465 cm^3. Its capacity is:

   A. 564 mL **B.** 46.5 mL **C.** 4.65 L **D.** 0.465 L **E.** 5.54 L

2. **MC** The number of cubes with side length 1 cm that can fit into a larger cube with side length 4 cm is:

   A. 64 **B.** 16 **C.** 56 **D.** 32 **E.** 28

3. **MC** The volume of the shape shown is:

   A. 13 m^3
   B. 14 m^3
   C. 28 m^3
   D. 56 m^3
   E. 102 m^3

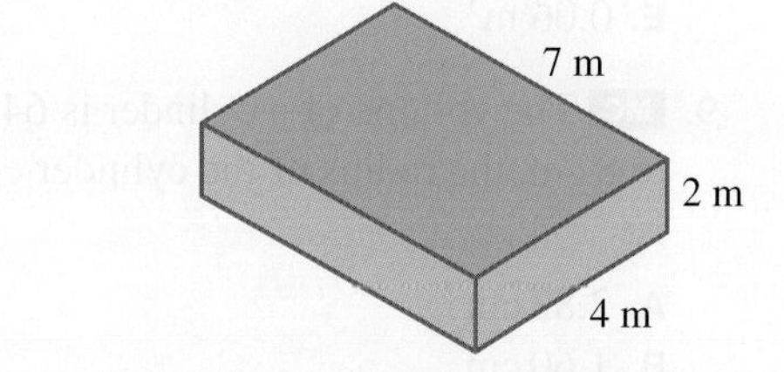

4. **MC** The volume of the triangular prism shown is closest to:

   A. 2406 cm^3
   B. 1203 cm^3
   C. 1895 cm^3
   D. 60 cm^3
   E. 1460 cm^3

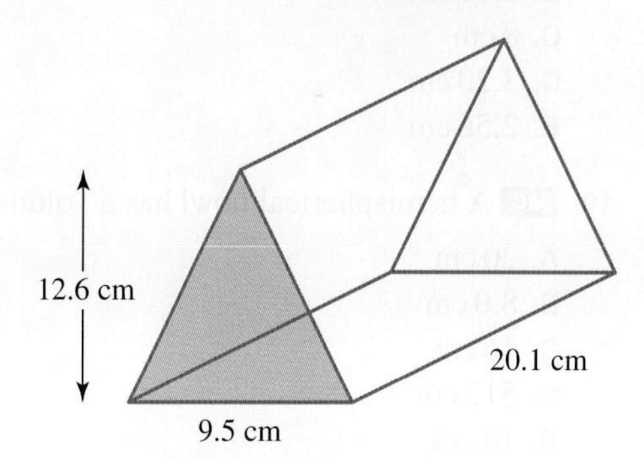

5. **MC** The volume of the cylinder shown is closest to:

   A. 285.5 cm^3
   B. 6589 cm^3
   C. 2097.5 cm^3
   D. 1647 cm^3
   E. 2863.5 cm^3

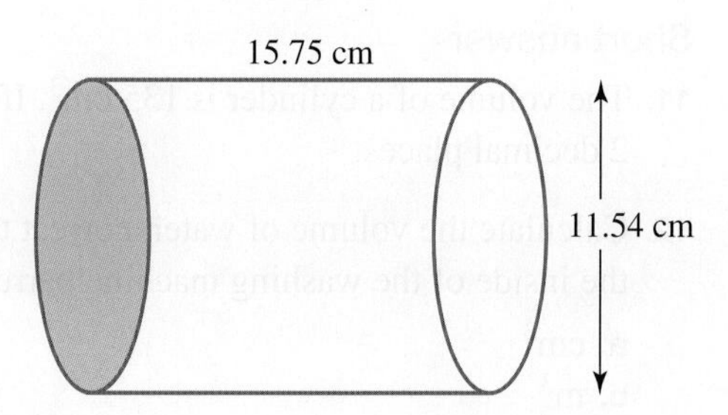

6. MC The volume of the following shape is closest to:

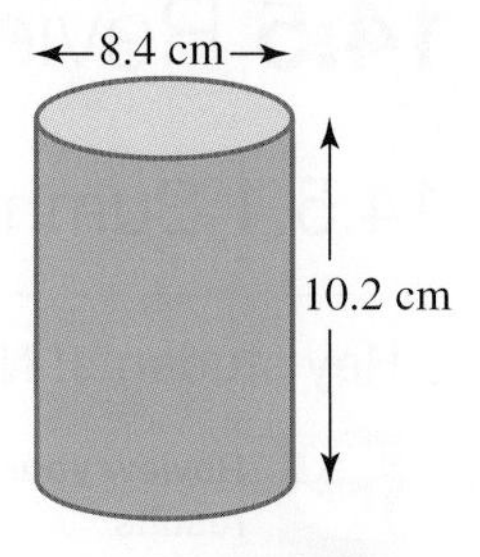

A. 85.68 cm^3
B. 134.59 cm^3
C. 1372.78 cm^3
D. 565.26 cm^3
E. 2010.62 cm^3

7. MC The volume of the following shape is closest to:

A. 1767.15 cm^3
B. 883.57 cm^3
C. 14 137.17 cm^3
D. 1325.36 cm^3
E. 7068.59 cm^3

15 cm

8. MC The volume of a disco ball that has a diameter of 55 cm is closest to:

A. 0.32 m^3
B. 0.07 m^3
C. 0.03 m^3
D. 0.09 m^3
E. 0.06 m^3

9. MC The volume of a cylinder is 64 cm^3. If the height is equal to 8 cm, the radius of the cylinder correct to 2 decimal places is:

A. 2.82 cm
B. 1.60 cm
C. 8 cm
D. 3.20 cm
E. 2.56 cm

10. MC A hemispherical bowl has a volume of 2145.52 cm^3. The diameter of the bowl is closest to:

A. 20 cm
B. 8.0 cm
C. 16 cm
D. 512 cm
E. 10 cm

Short answer

11. The volume of a cylinder is 135 cm^3. If it has a radius of 3 cm, calculate the height of the cylinder to 2 decimal places.

12. Calculate the volume of water, correct to 4 decimal places, that can completely fill the inside of the washing machine barrel shown in the diagram, in:

a. cm^3
b. m^3
c. litres.

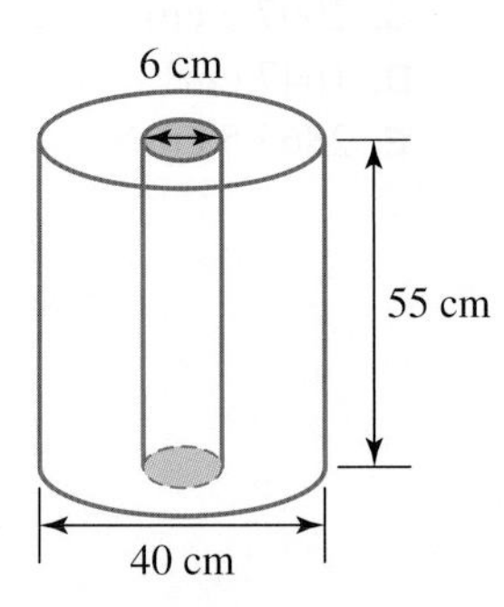

13. Calculate the volume of each of the following objects.

a. 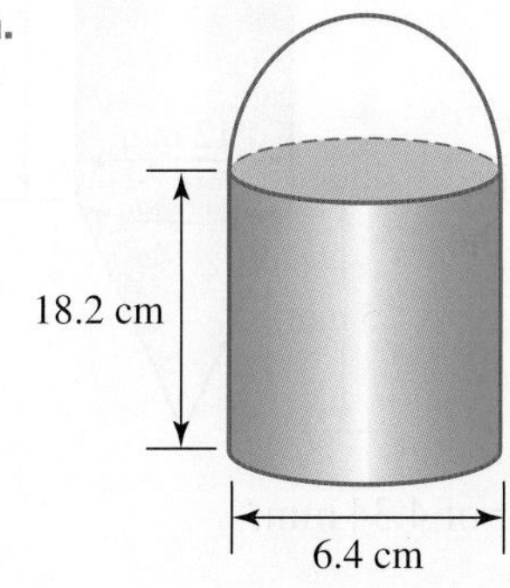

b. 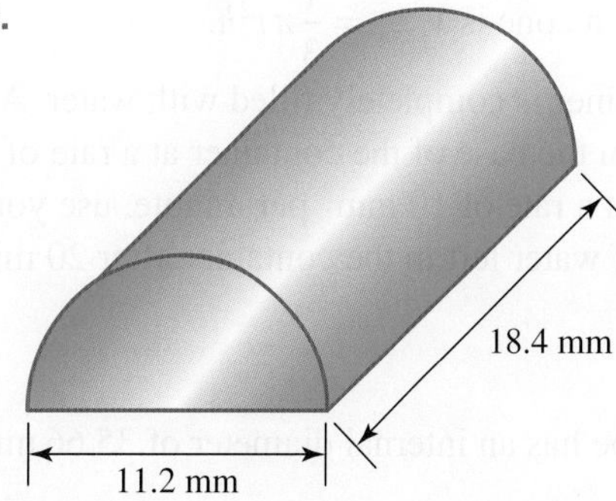

c. 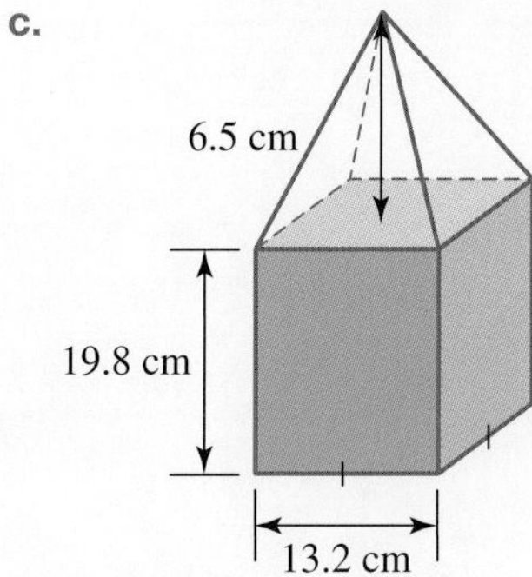

d. 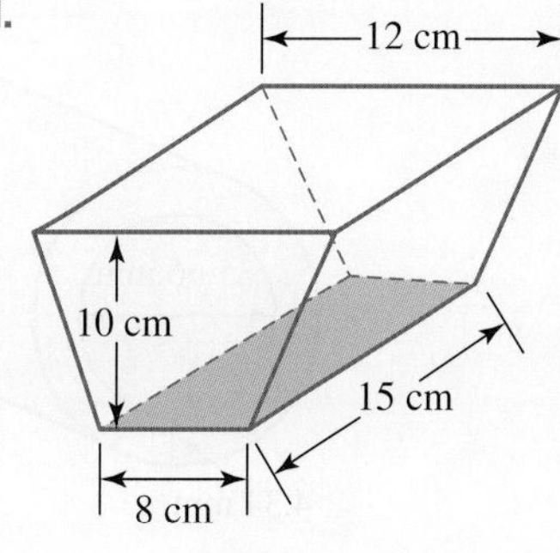

14. Find the capacity of the object shown (to the nearest litre).

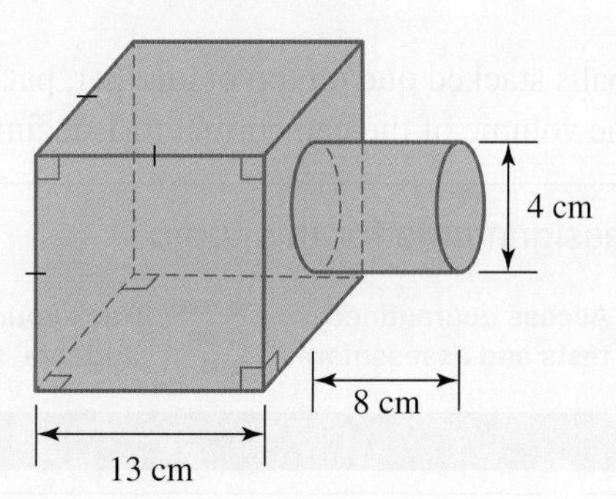

Extended response

15. A fish tank has dimensions 90 cm × 40 cm × 60 cm. Calculate how much water would be needed to fill:
    a. three-quarters of the tank in cm^3
    b. three-quarters of the tank in litres
    c. 90% of the tank in cm^3
    d. 90% of the tank in litres.

16. A car's engine has four cylinders, each of which has a diameter of 8.5 cm and a height of 11.2 cm. Calculate the capacity of the engine in cubic centimetres. Give your answer correct to 1 decimal place.

17. A fish tank has the dimensions 90 cm × 40 cm × 60 cm. Determine how much water would be needed to fill three-quarters of the tank.

18. a. Calculate the volume of the container shown, given that the formula for the volume of a cone is $V_{cone} = \frac{1}{3}\pi r^2 h$.

b. The container is completely filled with water. A tap is opened to release the water from the base of the container at a rate of 15 mm^3 per minute. If water is added at a rate of 10 mm^3 per minute, use your calculations to work out the amount of water left in the container after 20 minutes.

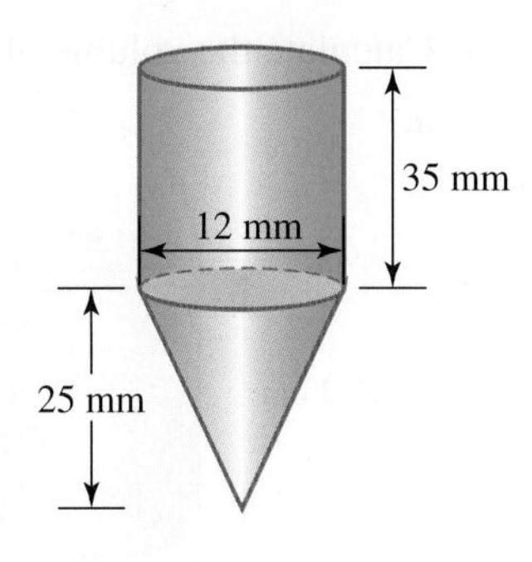

19. A copper pipe has an internal diameter of 35.66 mm and a thickness of copper of 4.34 mm.

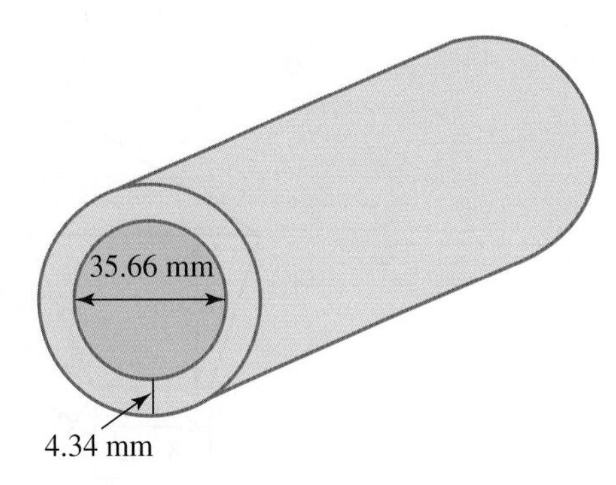

Calculate the volume of copper used per metre in making this pipe. Give your answer in cm^3, correct to the nearest whole number.

20. A cylindrical can holds 4 tennis balls stacked one on top of another, packed tightly. If the diameter of a tennis ball is 8.8 cm, determine the volume of the can correct to 1 decimal place.

Answers

Topic 14 Volume

14.2 Review of volume of prisms, units of volume and capacity

14.2 Exercise

1. a. Volume b. Capacity
   c. Capacity d. Volume
2. a. 224 000 cm^3 b. 0.789 L
   c. 348 mL d. 452 cm^3
3. a. 0.308 L b. 1.278 L
   c. 44.5 mL d. 0.599 L
4. a. 555 mL b. 1 890 000 cm^3
   c. 4200 cm^3 d. 0.012 58 m^3
5. a. 36 cm^3 b. 92.75 cm^3
   c. 234 m^3 d. 174.375 cm^3
6. a. 120 m^3 b. 2592 cm^3
   c. 791.4531 cm^3 d. 338.625 cm^3
7. a. 720 cm^3 b. 166.375 cm^3
8. a. 1344 cm^3 b. 40 cm^3
   c. 3402 m^3 d. 655.5 cm^3
9. a. 270 cm^3 b. 110.925 m^3
10. 111 000 L
11. 16.632 m^3
12. a. 1.2 m^3 b. 900 000 cm^3 c. 750 L
13. a. 0.68 m^3 b. 0.204 m^3 c. 0.476 m^3
14. a. 75.625 m^3 b. 103.125 m^3 c. 7.27 m

14.3 Volume of cylinders and spheres

14.3 Exercise

1. a. 339.29 cm^3 b. 678.58 mm^3
   c. 1693.32 cm^3 d. 4.42 m^3
2. 706.86 cm^3
3. 922.45 m^3
4. a. 88.19 m^3 b. 79 521.56 cm^3
5. a. 523.60 cm^3 b. 7238.23 cm^3
   c. 3.05 m^3 d. 10 305.99 cm^3
   e. 63 891.58 cm^3 f. 17 974.16 m^3
6. a. 20 579.53 mm^3 b. 2309.56 mm^3
   c. 28 730.91 mm^3 d. 2144.66 cm^3
   e. 15 298.57 cm^3
7. 8578.64 cm^3
8. 5 cm
9. a. 14 137.17 cm^3 b. 14.14 L
   c. 176.71 seconds
10. a. 9.425 m^3 b. 9425 L
11. 11.762 kL
12. 3.61 cm

14.4 Volume of composite shapes

14.4 Exercise

1. 486 cm^3
2. a. 654.12 cm^3 b. 906.39 mm^3
3. 91.63 cm^3
4. 67.56 L
5. Combined shape: 810.67 mm^3
   Each pyramid: 405.33 mm^3
6. a. 402.12 cm^3 b. 21.46%
7. a. 99 m^2 b. 14.85 m^3 c. $2673
8. $V = 14\,137.17$ cm^3
   176.7 seconds to empty
9. a. i. 2187 cm^3
      ii. 1717.67 cm^3
      iii. Empty space in the rectangular prism: 1041.89 cm^3
      Empty space in the cylinder: 572.56 cm^3
   b. Cylinder
10. 1.29 L

14.5 Review

14.5 Exercise

Multiple choice

1. D
2. A
3. D
4. B
5. D
6. D
7. B
8. D
9. B
10. A

Short answer

11. $h = 4.77$ cm
12. a. 67 559.9500 cm^3 b. 0.0676 m^3
    c. 67.5600 L
13. a. 654.1 cm^3 b. 906.39 mm^2
    c. 3827.47 cm^3 d. 1500 cm^3
14. 2 L

Extended response

15. a. 162 000 cm^3 b. 162 L c. 194 400 cm^3
    d. 194.4 L
16. 2542 cm^3
17. 162 L
18. a. 4900.88 mm^3
    b. $V_{left} = 4800.88$ mm^3
19. $V \approx 0.5$ cm^3
20. $V \approx 2140.9$ cm^3

15 Shapes

LEARNING SEQUENCE

Fully worked solutions for this topic are available online.

15.1 Overview

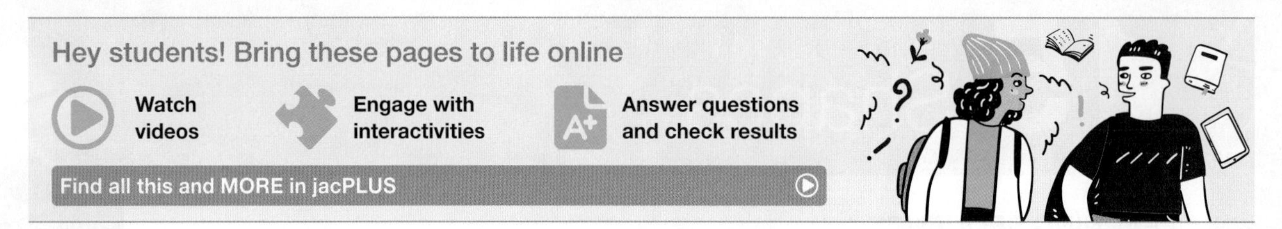

15.1.1 Introduction

Geometry is not only the study of figures and shapes; it is also the study of the movement of figures. Transformation is a general term used to describe a change to an object or its position. Patterns present in modern design, ancient design and architecture show how transformations can be applied.

Transformation is applied when tiling a floor or wall, where the same shape or shapes are moved to cover the surface. Some shapes, such as isosceles and equilateral triangles, have symmetrical properties. Symmetry is found in nature — some examples are honeycombs, starfish and orchids.

If we look around us, we will find symmetry and geometric patterns everywhere, all of which are examples of how transformations can be applied in everyday life.

KEY CONCEPTS

This topic covers the following key concepts from the VCE Mathematics Study Design:

- spatial and geometric constructions including transformations, similarity, symmetry and projections.

Note: Concepts shown in grey are covered in other topics.

15.2 Polygons

LEARNING INTENTION

At the end of this subtopic you should be able to:
- identify polygons and their features.

15.2.1 Introduction

A **polygon** is a closed shape where each edge is a straight line and all edges meet at vertices.

The number of edges in any polygon is equal to the number of interior angles.

Number of sides	3	4	5	6	7	8	9	10	n
Name of polygon	Triangle	Quadrilateral	Pentagon	Hexagon	Heptagon	Octagon	Nonagon	Decagon	n–gon
Example									

Regular polygons have edges that are all of equal length and interior angles that are all of equal size. Equilateral triangles and squares are examples of regular polygons.

If a polygon has at least one **reflex angle** (an interior angle greater than 180°), it is called a **concave polygon**.

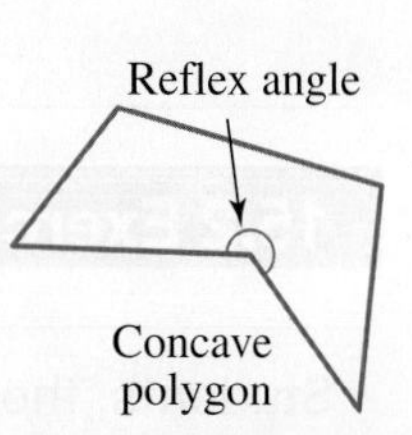

WORKED EXAMPLE 1 Determining regular polygons

State whether each of the following shapes is a polygon and, if so, whether it is regular.

a. 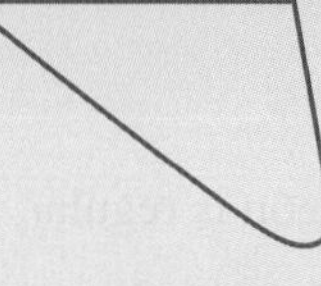

b. 

c. 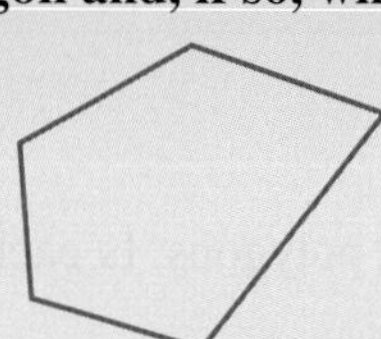

d. 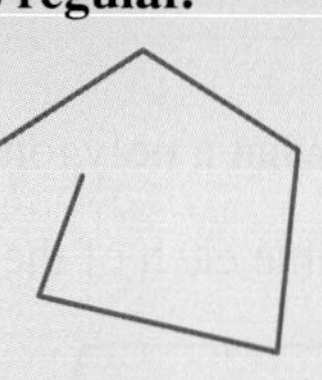

THINK	WRITE
a. This shape is closed, but it has only two edges and one of them is curved.	a. This shape is not a polygon.
b. This shape is closed. It has more than two straight edges and all of the edges are of equal length.	b. This shape is a regular polygon.
c. This shape is closed and has more than two straight edges, but the edges are not of equal length.	c. This shape is a polygon, but it is irregular.
d. This shape is not closed.	d. This shape is not a polygon.

15.2.2 Constructing polygons

Regular polygons can be constructed using a circle as a starting point.

tlvd-5075

WORKED EXAMPLE 2 Constructing polygons using circles

Construct a regular hexagon using a circle as a starting point.

THINK	WRITE
1. Use a pencil and a pair of compasses to draw a circle and mark the centre of the circle.	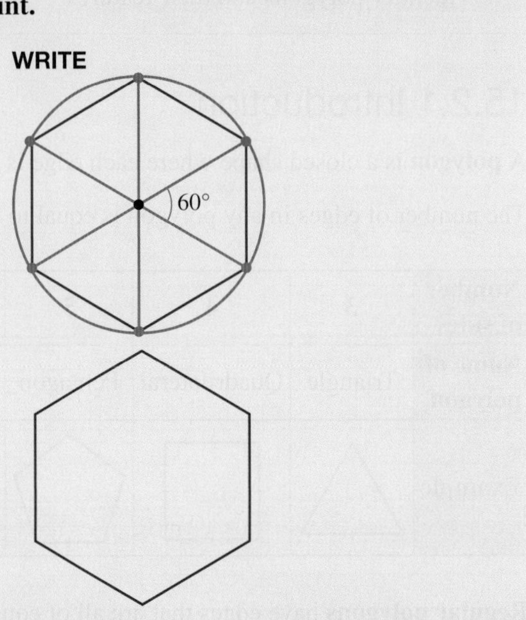
2. Because a regular hexagon has six edges of equal length, the six vertices must be evenly spaced around the circle; $360° \div 6 = 60°$. Start at one point on the edge of the circle, as shown in green, and mark every 60° around the circle, as shown in blue.	
3. Join the markings around the circle with straight-line segments, as shown in pink.	
4. Erase the circle and the interior blue lines, leaving the hexagon.	

15.2 Exercise

Students, these questions are even better in jacPLUS

Receive immediate feedback and access sample responses

Access additional questions

Track your results and progress

Find all this and MORE in jacPLUS

1. Explain what a polygon is.
2. WE1 Name each of the following polygons. In each case, state whether or not the polygon is regular.

a.

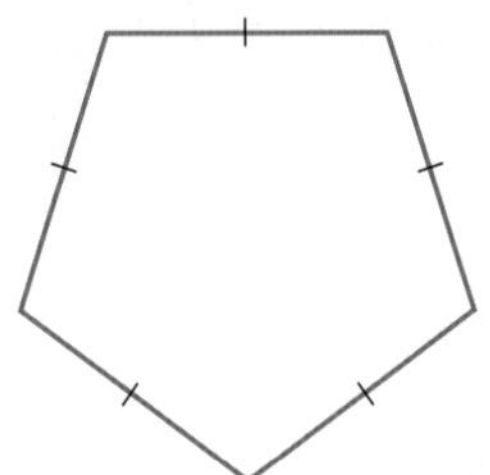

b.

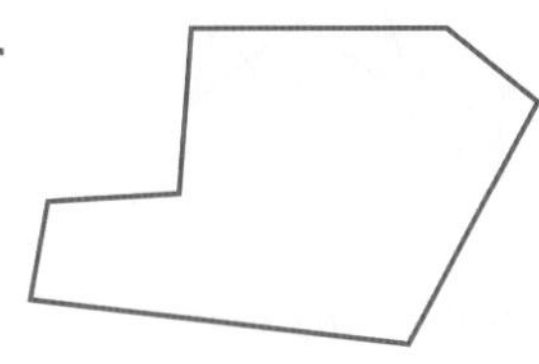

c.

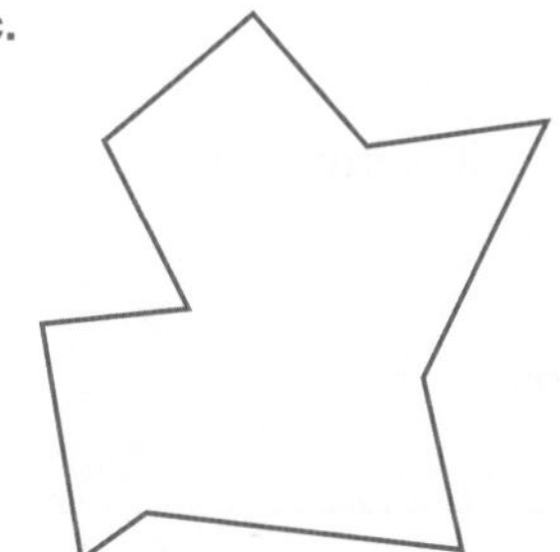

d.

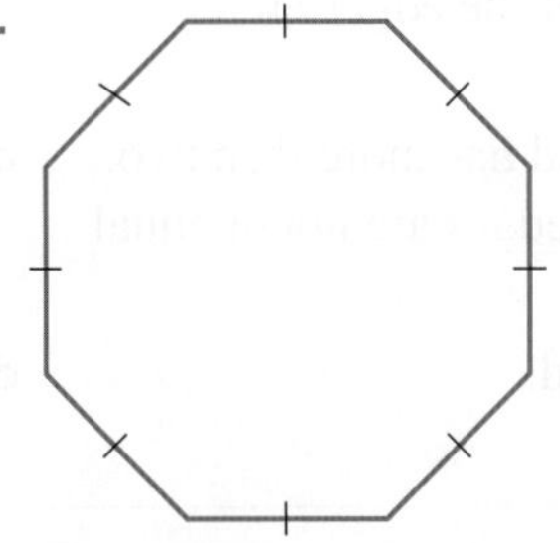

3. a. Explain what is the main feature of a concave polygon.
   b. Explain which of the polygons shown are concave polygons.

i. 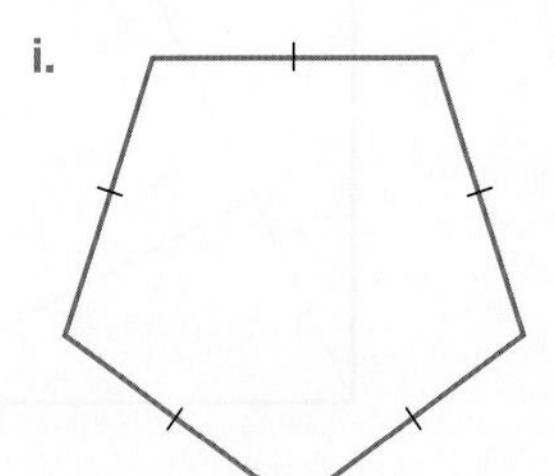

ii. 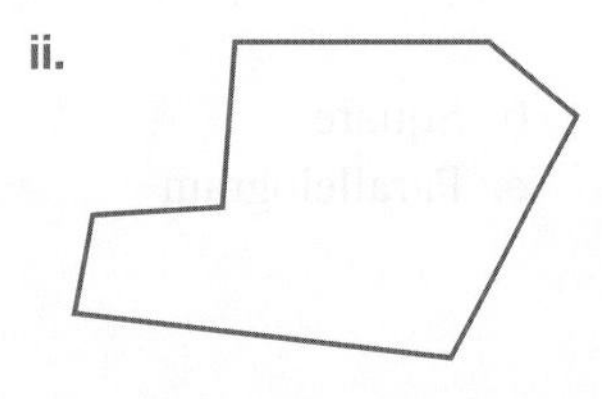

iii. 

iv. 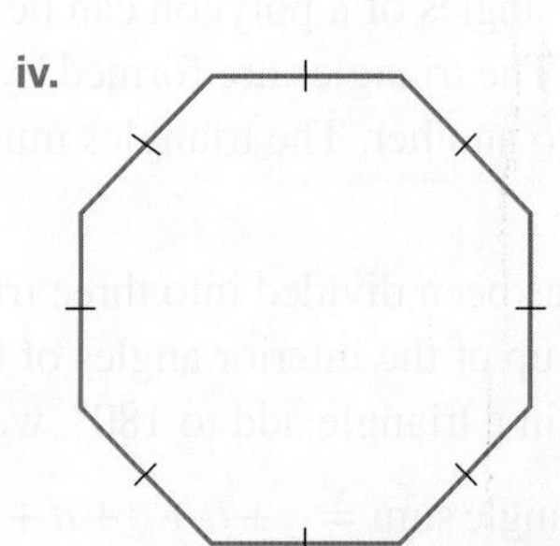

4. The stop sign shown is an example of an everyday object that is a polygon.
   a. State how many vertices the sign has.
   b. Name the polygon.
   c. Determine if the shape of the sign is a regular polygon. Explain your answer.

5. WE2 Construct the following polygons using a circle as a starting point.
   a. A square
   b. A regular pentagon
   c. A regular polygon with 8 vertices
6. Construct the following polygons using a circle as a starting point.
   a. A regular polygon with 9 vertices
   b. A regular polygon with 10 vertices
7. Consider the quilt pattern shown.
   a. Name the individual polygons from which the quilt is constructed.
   b. State if any of the polygons are concave.

8. Name one place at school or at home where you could find an example of the following polygons (they do not have to be regular).
   a. Triangle
   b. Quadrilateral
   c. Hexagon
   d. Octagon

9. Trace the shape shown and cut along the dotted lines to make three pieces. Use at least two of the pieces to make each of the following polygons. Draw the solutions in your workbook.

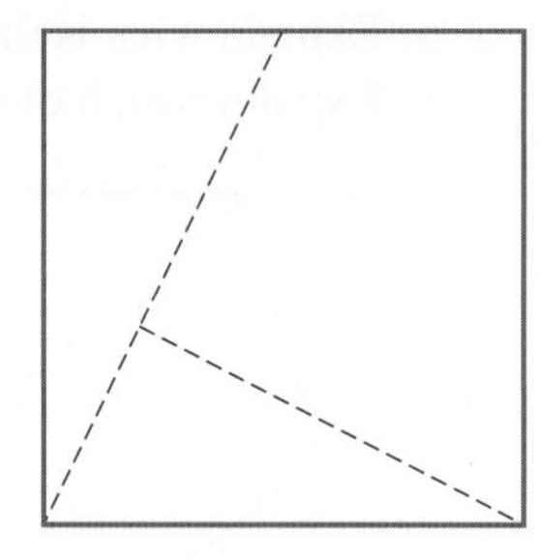

a. Triangle
b. Square
c. Rectangle
d. Trapezium
e. Parallelogram
f. Pentagon

10. The sum of the interior angles of a polygon can be found by dividing the polygon into triangles. The triangles are formed by drawing a line from one vertex on the polygon to another. The triangles must all have one vertex in common.

The pentagon shown has been divided into three triangles. The interior angles of the pentagon are made up of the interior angles of the triangles. Using the fact that the interior angles in a triangle add to 180°, we can calculate the following:

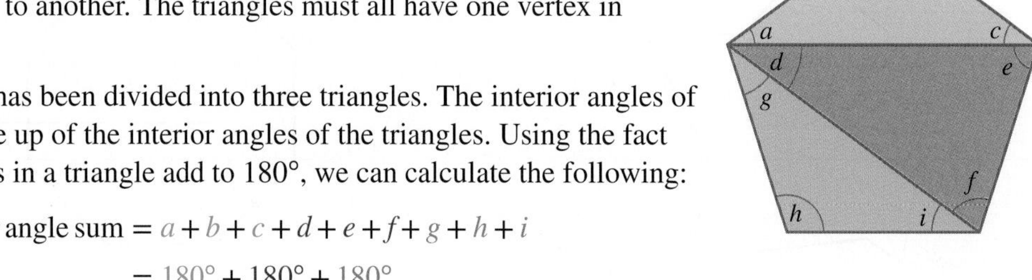

$$\begin{aligned}\text{interior angle sum} &= a+b+c+d+e+f+g+h+i \\ &= 180° + 180° + 180° \\ &= 540°\end{aligned}$$

a. Use your constructions from questions **5** and **6** to calculate the sum of the interior angles of each polygon.
b. Write a rule to predict the number of triangles a polygon can be divided into given the number of edges in the polygon.
c. Use your rule from part **b** to write a rule to calculate the sum of the interior angles of a polygon given the number of edges.

15.3 Transformations

LEARNING INTENTION

At the end of this subtopic you should be able to:
- translate points and objects on the Cartesian plane
- reflect points and shapes in the Cartesian plane
- rotate points and shapes in the Cartesian plane.

15.3.1 Transformations

Transformations are changes that move a point P from one position to another.

The position of the point, P, after a transformation is called the **image** of the point and is denoted P′.

Shapes can also undergo transformations. In this case, the transformations are applied to all points of the shape. For example, the image of the rectangle ABCD after a transformation is A′B′C′D′.

There are different types of transformations. In this topic we will cover **translations**, **reflections** and **rotations**.

15.3.2 Translations

A translation is the movement of a point or an object by shifting it up, down, left or right. After a translation, the object has the same shape and size but is in a different position. For this reason, translations are also known as 'slides'.

The Cartesian plane in the diagram shows some examples of translations.

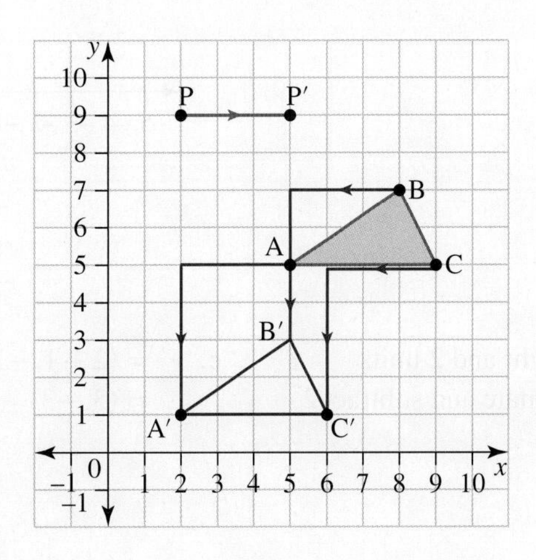

The point P has been translated 3 units to the right.

The triangle ABC has been translated 3 units to the left and 4 units down.

WORKED EXAMPLE 3 Translating points

State the coordinates and sketch the image of the point P(2, −1) after:

a. **it is translated 3 units to the left**

b. **it is translated 5 units up**

c. **it is translated 1 unit to the right and 2 units down.**

THINK	WRITE
a. 1. To translate 3 units to the left, subtract 3 from the x-coordinate.	a. $P' = (2 - 3, -1)$ $= (-1, -1)$
2. Sketch the image P′.	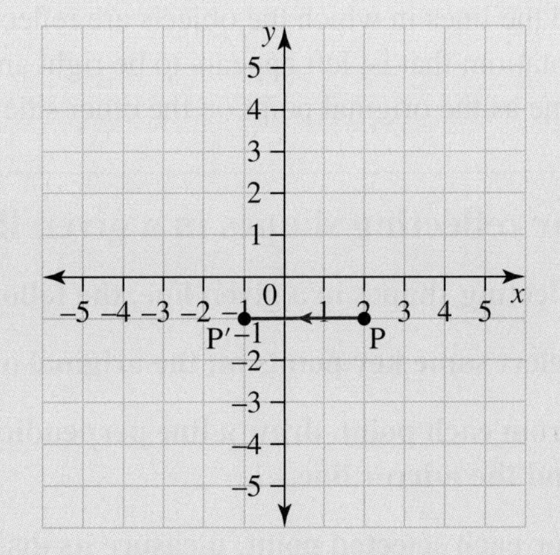

b. 1. To translate 5 units up, add 5 to the y-coordinate.

2. Sketch the image P′.

b. $P' = (2, -1 + 5)$
$= (2, 4)$

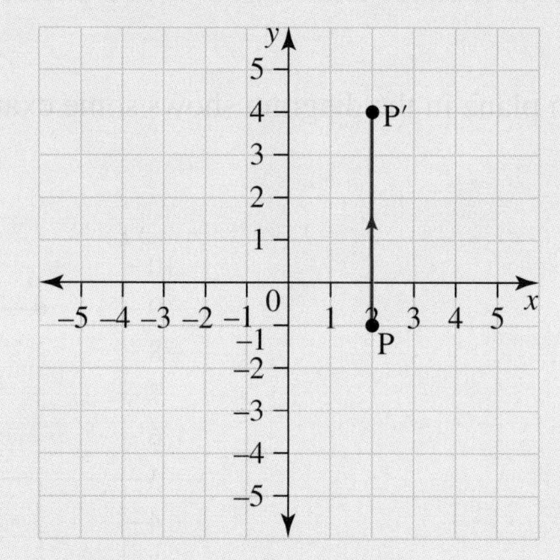

c. 1. To translate 1 unit to the right and 2 units down, add 1 to the x-coordinate and subtract 2 from the y-coordinate.

2. Sketch the image P′.

c. $P' = (2 + 1, -1 - 2)$
$= (3, -3)$

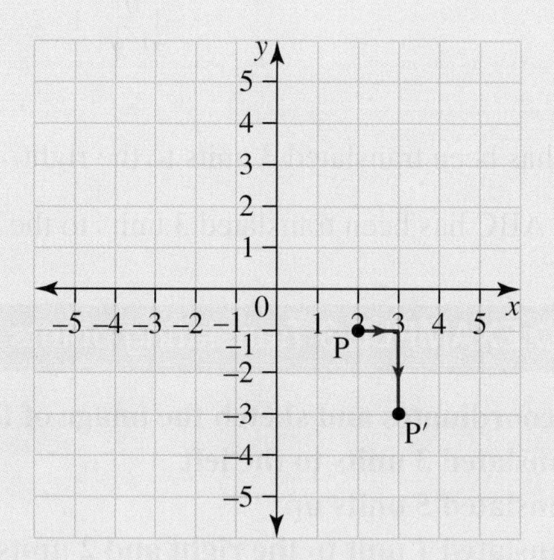

15.3.3 Reflections

A reflection is the image of a point or an object, as seen in the mirror. Reflections are often called 'mirror images', and the lines in which the objects are reflected are called 'mirror' lines. Mirror images always have reverse orientation; that is, left appears to be right and vice versa. Reflected points are the same distance from the mirror line as the original point on the other side of the mirror.

Steps for reflecting shapes in a given line

When reflecting shapes in a given line, the following steps can be of assistance:

Step 1: Select some key points on the original object (the vertices are usually a good choice).

Step 2: From each point, draw a line perpendicular (at a right angle) to the mirror line. Extend each line beyond the mirror line.

Step 3: For each selected point, measure its distance along the line from the mirror. The image of the point will be the same distance from the mirror line, on the other side of it.

Step 4: Complete the image using the reflections of the key points.

tlvd-5076

WORKED EXAMPLE 4 Reflecting in a mirror line

Sketch the image of the triangle after it is reflected in the mirror line.

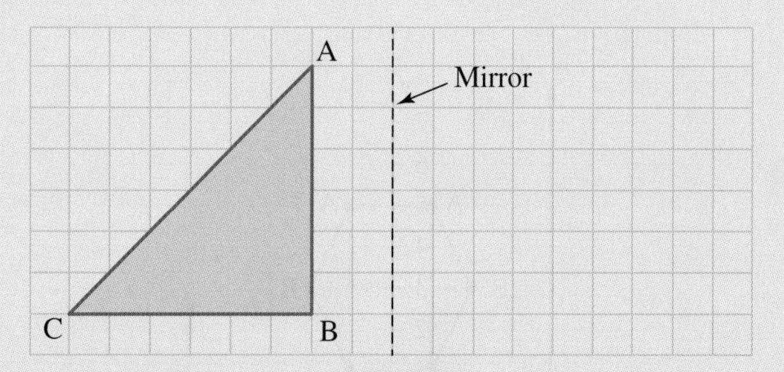

THINK

1. From each vertex of the given triangle, draw the lines perpendicular to and extending beyond the mirror line.

WRITE

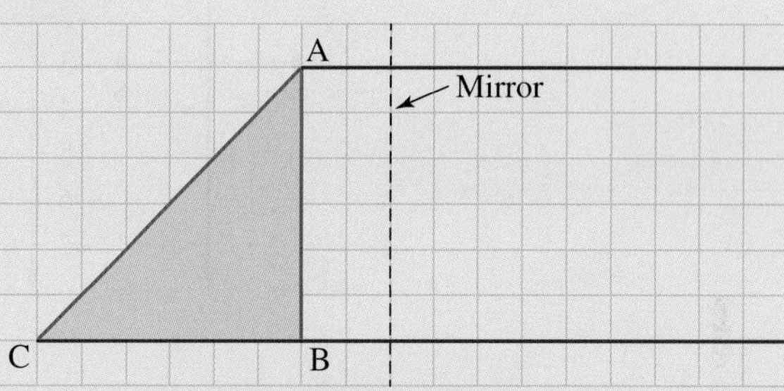

2. Points A and B are both 2 units to the left of the mirror line. Therefore, their images A′ and B′ will be 2 units to the right of the mirror line. Point C is 8 units to the left of the mirror line. Therefore C′ will be 8 units to the right of the mirror line.

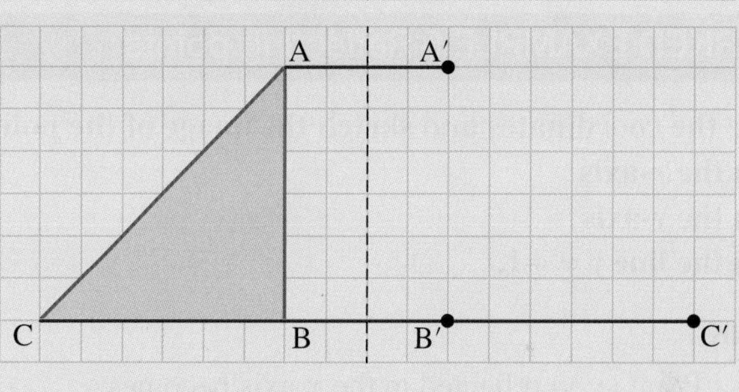

3. Join the vertices A′, B′ and C′ to complete the image.

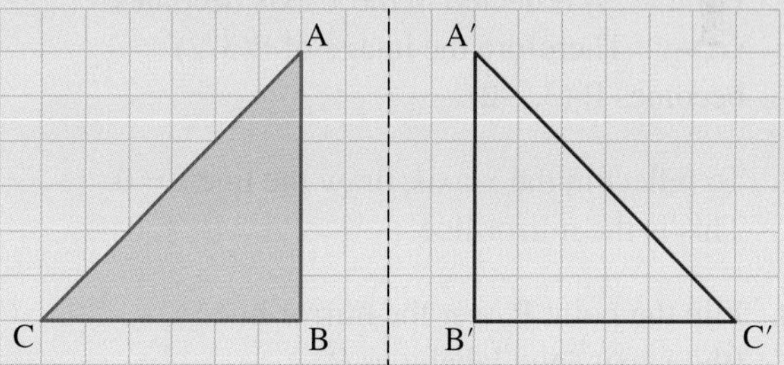

Reflections in the Cartesian plane

Points and objects in the Cartesian plane can be reflected in lines and axes.

Reflecting points in an axis

If a point P(x, y) is reflected in the x-axis, the x-coordinate stays the same and the y-coordinate changes sign. Therefore P′ becomes $(x, -y)$.

If a point P(x, y) is reflected in the y-axis, the y-coordinate stays the same and the x-coordinate changes sign. Therefore P′ becomes $(-x, y)$.

The Cartesian plane in the following diagram shows some examples of reflections. The point P has been reflected in the x-axis, the triangle ABC has been reflected in the y-axis and the point S has been reflected in the line $x = -4$.

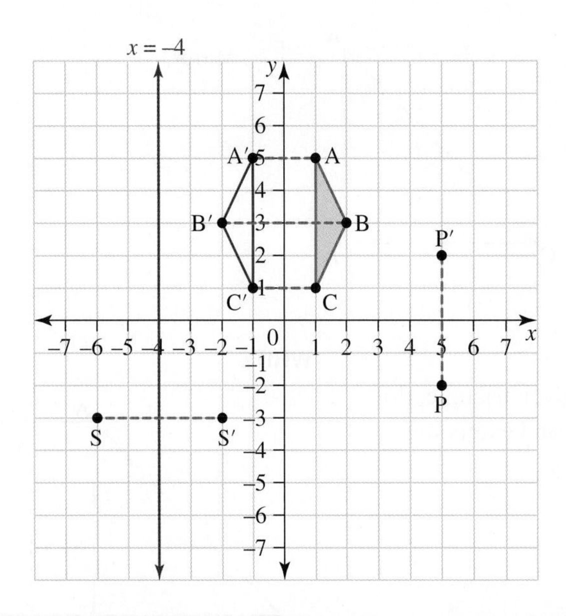

WORKED EXAMPLE 5 Reflecting points

State the coordinates and sketch the image of the point P(3, 2) after it is reflected:

a. in the x-axis

b. in the y-axis

c. in the line $y = -1$.

THINK	WRITE
a. 1. Point (x, y) reflected in the x-axis becomes $(x, -y)$. Therefore the image of P(3, 2) becomes P′(3, −2).	a. $P' = (3, -2)$
2. To reflect in the x-axis, draw the line $y = 0$. This is the mirror line.	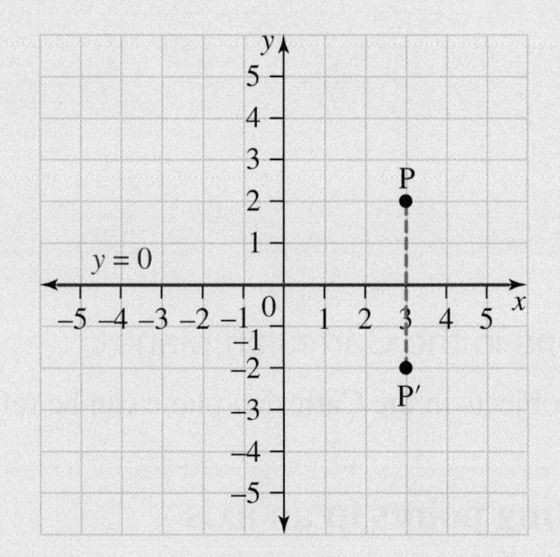
3. Flip the point P over the mirror line (the x-axis) and label it as P′.	

b. **1.** Point (x, y) reflected in the y-axis becomes $(-x, y)$. Therefore the image of P(3, 2) becomes P′(3, −2).

b. $P' = (-3, 2)$

2. To reflect in the y-axis, draw the line $x = 0$. This is the mirror line.

3. Flip the point P over the mirror line (the y-axis) and label as P′.

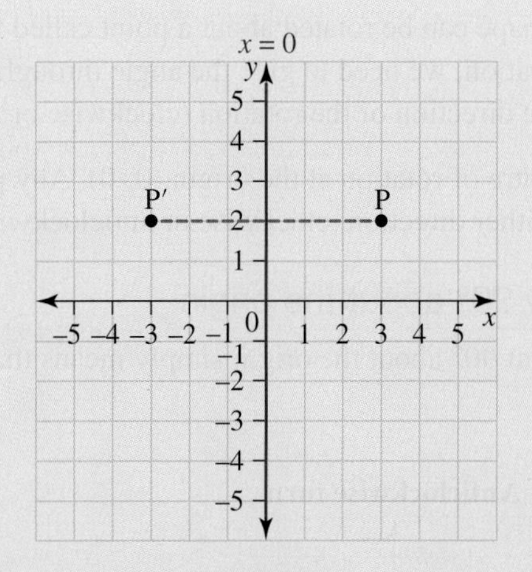

c. **1.** Draw the line $y = -1$. This is the mirror line.

c.

2. Draw a second line that is perpendicular to $y = -1$ and passes through P.

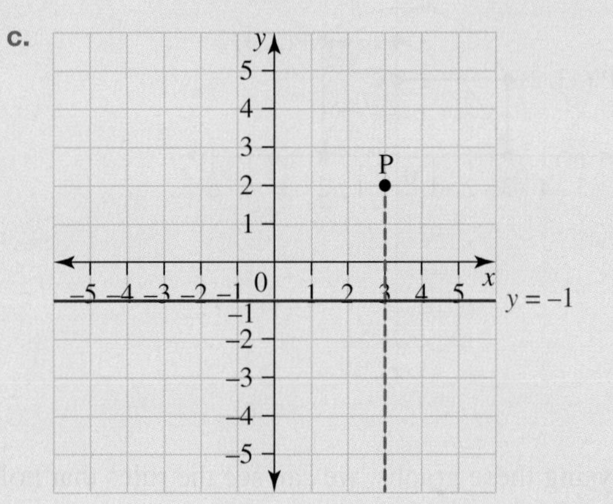

3. Place the image P′ on the dashed line so that the distance from the line $y = -1$ is equal for P and P′.

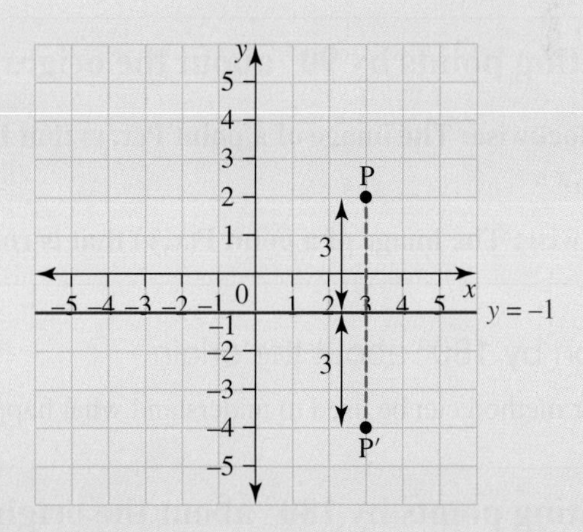

4. State the coordinates of P′.

$P' = (3, -4)$

15.3.4 Rotations

A rotation is simply a turn. When you turn your phone from portrait to landscape, you have performed a rotation.

A point or a shape can be rotated about a point called the **centre of rotation**. To specify the rotation, we need to give the angle through which the object is to be turned, and the direction of the rotation (clockwise or anticlockwise).

Consider a centre of rotation at the origin (0, 0). Any point may be rotated about the origin in either direction, clockwise or anticlockwise.

Rotation by 90° about the origin

Rotating a point 90° about the origin simply means that the angle at the origin between a point and its image must be 90°.

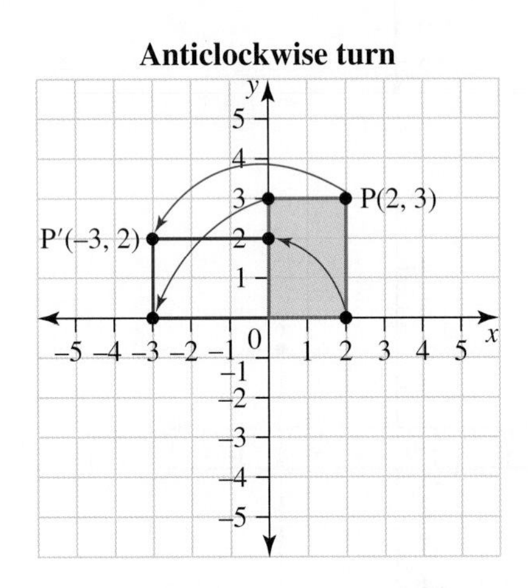

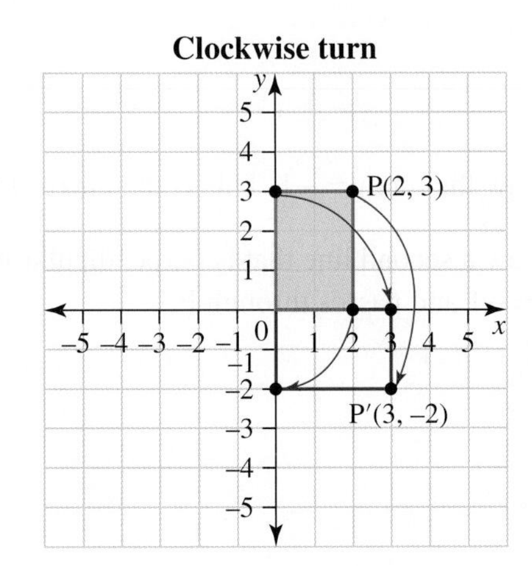

By analysing these graphs, we can see the rules that hold true for rotations by 90° about the origin.

Rotating points by 90° about the origin

Anticlockwise: The image of a point P(x, y) that is rotated by 90° anticlockwise about the origin is P′$(-y, x)$.

Clockwise: The image of a point P(x, y) that is rotated by 90° clockwise about the origin is P′$(y, -x)$.

Rotation by 180° about the origin

A similar method can be used to understand what happens when a point is rotated by 180° about the origin.

Rotating points by 180° about the origin

Regardless of the direction of rotation, the image of a point P(x, y) that is rotated by 180° about the origin is P′$(-x, -y)$.

Anticlockwise turn

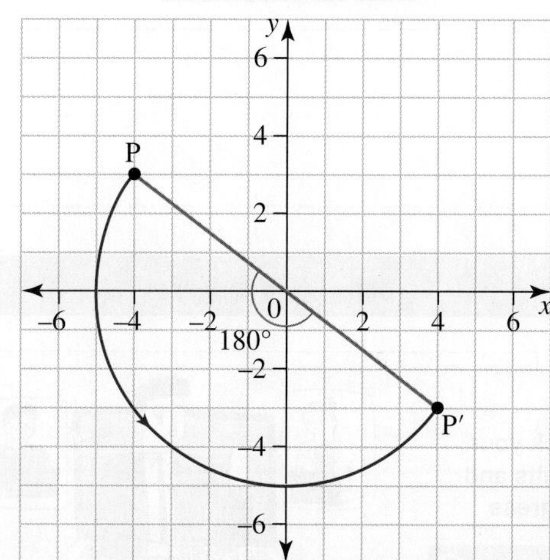

Clockwise turn

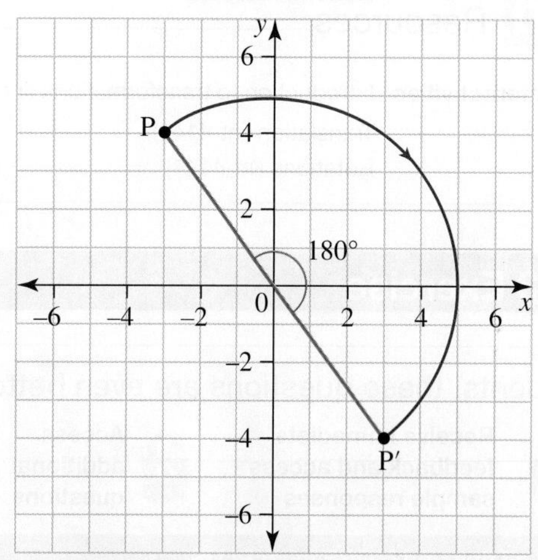

tlvd-5077

WORKED EXAMPLE 6 Rotating shapes

State the coordinates and sketch the image of the following shape after it is rotated by 180° clockwise about the origin.

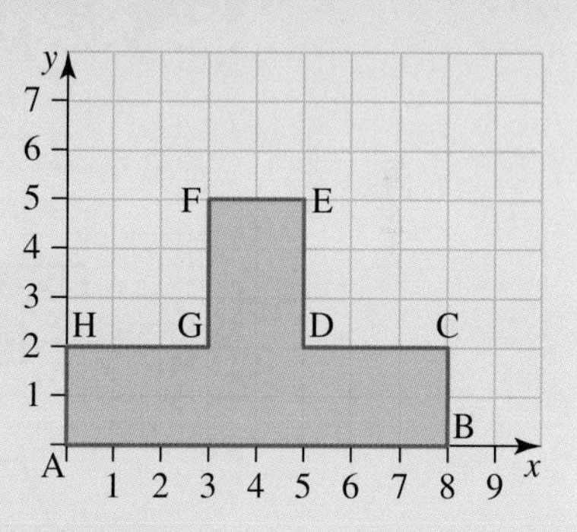

THINK	WRITE
1. Write the coordinates of the original shape.	The vertices of the original image are: A(0, 0), B(8, 0), C(8, 2), D(5, 2), E(5, 5), F(3, 5), G(3, 2) and H(0, 2)
2. A rotation of 180° clockwise about the origin changes the signs of the x- and y-coordinates. That is $(x, y) \rightarrow (-x, -y)$.	The vertices of the image are: $A'(0, 0)$, $B'(-8, 0)$, $C'(-8, -2)$, $D'(-5, -2)$, $E'(-5, -5)$, $F'(-3, -5)$, $G'(-3, -2)$ and $H'(0, -2)$
3. Sketch the image using the coordinates of the image that were determined in step 2.	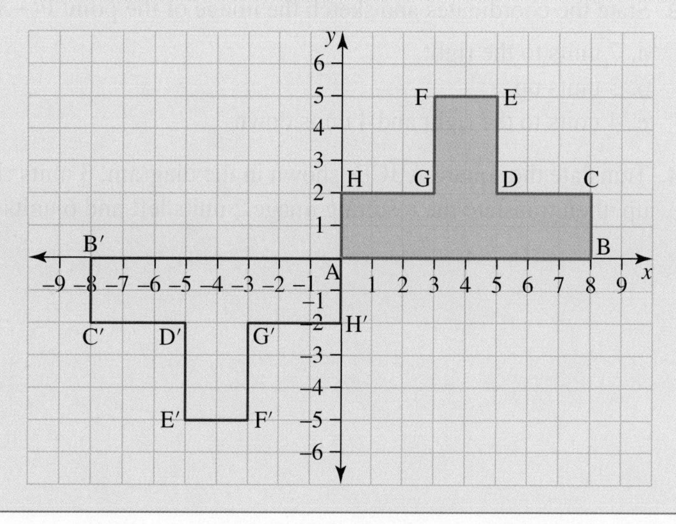

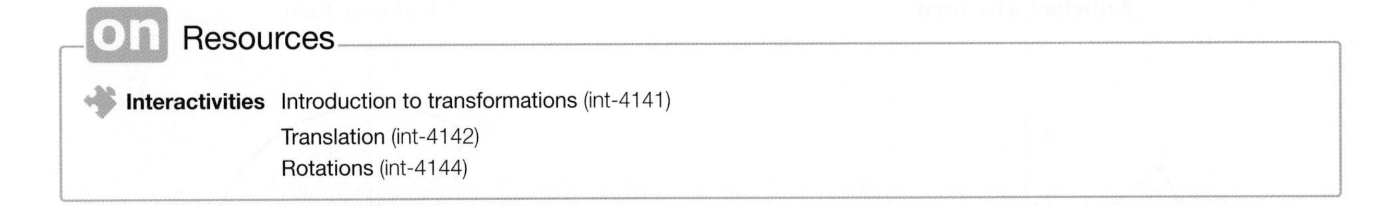

15.3 Exercise

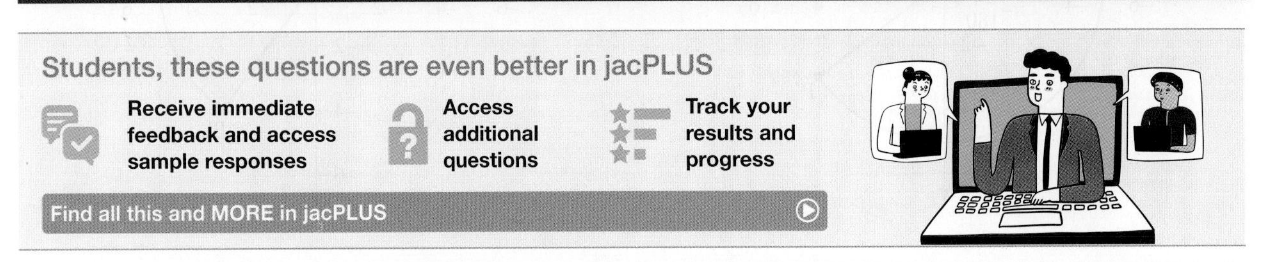

1. State the translation that has occurred to each of the shapes labelled a to d in the following figure.

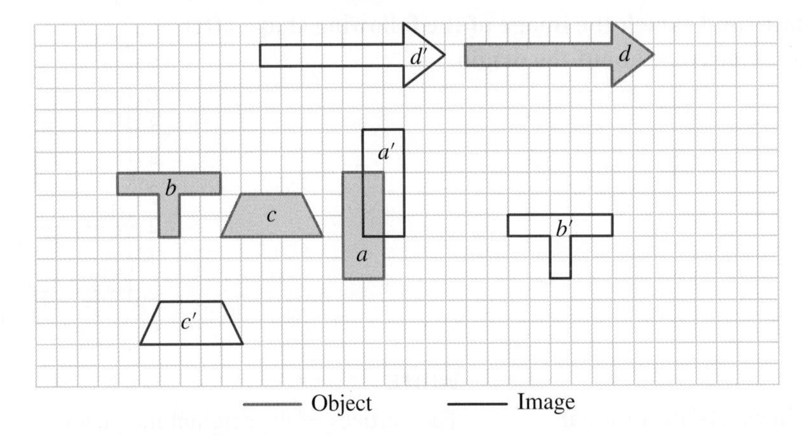

2. WE3 State the coordinates and sketch the image of the point P(1, 3) after it is translated:
   a. 2 units to the left
   b. 4 units up
   c. 1 unit to the right and 2 units down.

3. State the coordinates and sketch the image of the point P(−3, −5) after it is translated:
   a. 7 units to the right
   b. 3 units up
   c. 4 units to the right and 4 units down.

4. Translate the square ABCD, shown in the diagram, 3 units right and 1 unit up, then translate the resulting image 2 units left and 6 units up.

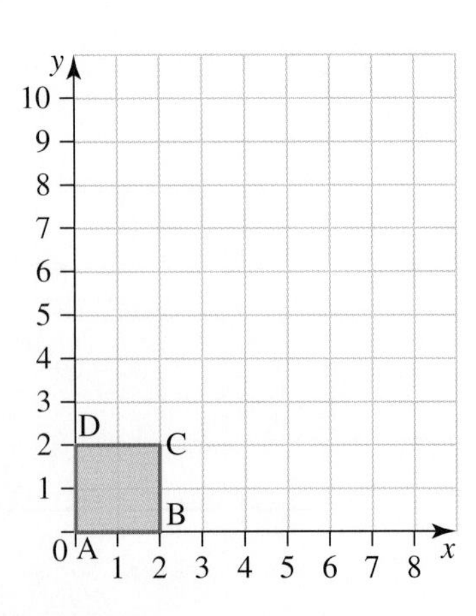

5. a. Translate the point P(3, 1) 3 units up and 2 units to the left. Label the new point P′ and give the coordinates of this point.
   b. Translate P′ 2 units down and 3 units to the left. Label the new point P″ and give the coordinates of this point.
   c. Determine what single translation would move P to P″.

6. **WE4** Sketch the image of the following shape after it is reflected in the mirror line.

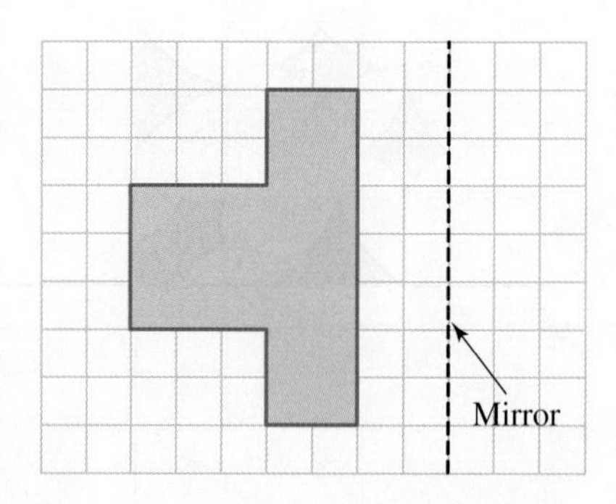

7. Sketch the image of the following shape after it is reflected in the mirror line.

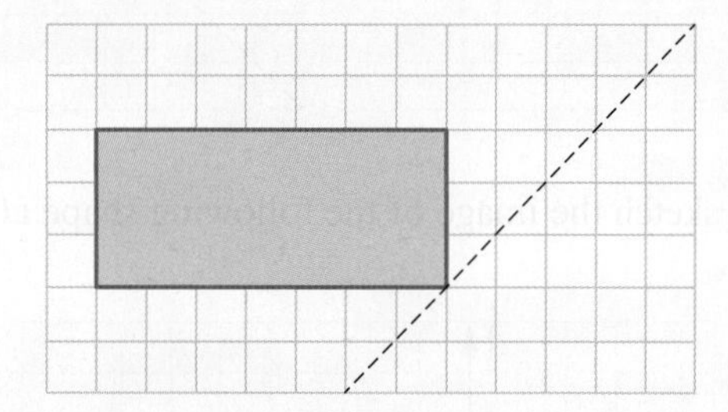

8. **WE5** State the coordinates and sketch the image of the point $P(1, 6)$ after it is reflected in the:
   a. x-axis
   b. y-axis
   c. line $y = 3$.

9. Consider the following shape.

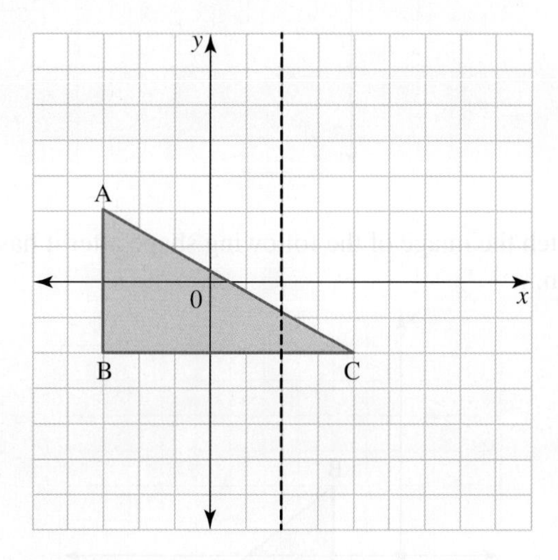

   a. Draw the image reflected in the mirror.
   b. Determine the coordinates of the vertices of the image.

10. A kaleidoscope uses more than one mirror to create wonderful patterns. To help design a kaleidoscope, reflect each object in each of the mirrors to complete the pattern in each of these figures.

a.

b.

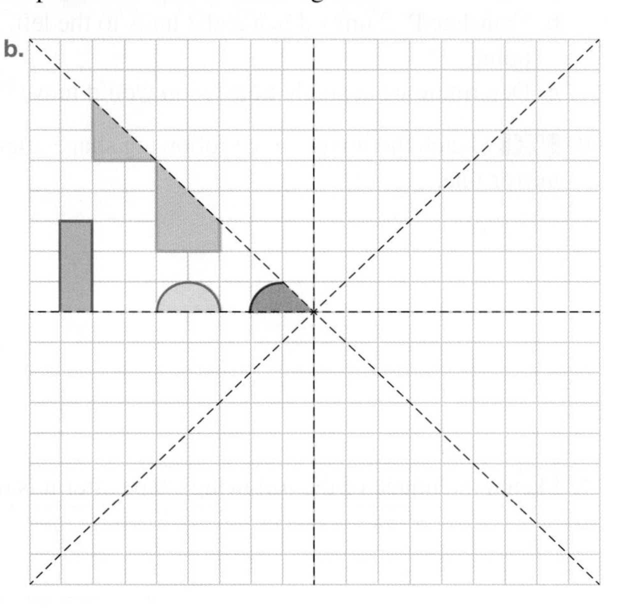

11. WE6 State the coordinates and sketch the image of the following shape after it has been rotated by 180° anticlockwise about the origin.

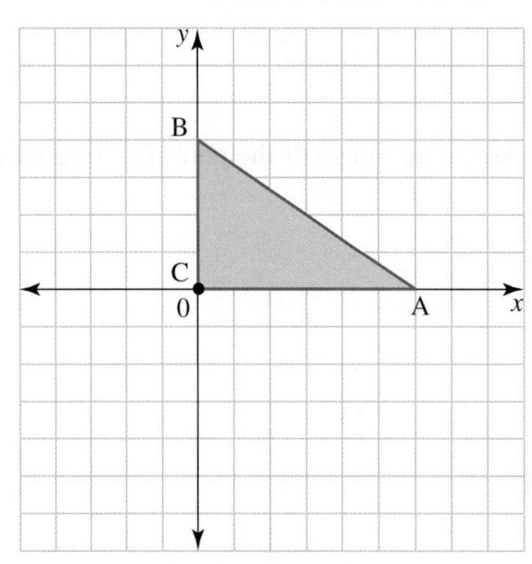

12. State the coordinates and sketch the image of the following shape after it has been rotated by 90° anticlockwise about the origin.

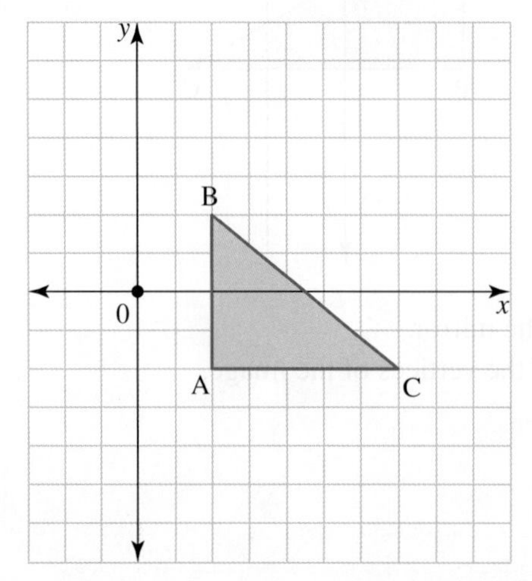

13. State the coordinates and sketch the image of the following shape after it has been rotated by 90° clockwise about the origin.

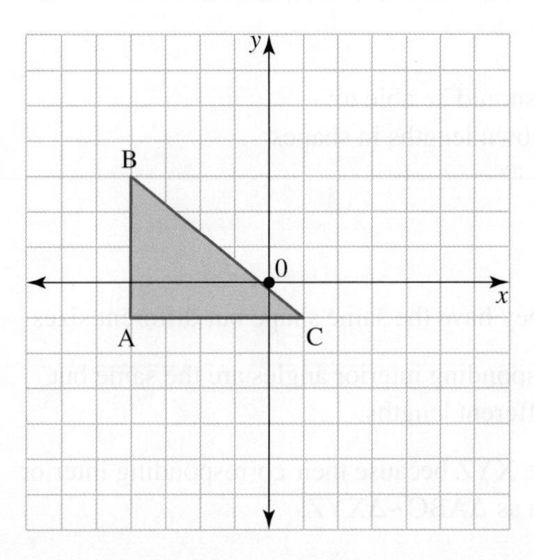

14. Reflect the triangle ABC in the x-axis, then rotate the image 90° clockwise about the origin; then, translate the second image 4 units right and 6 units down.

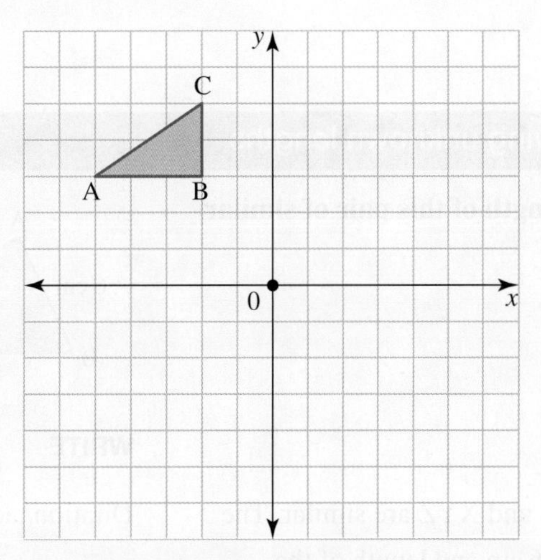

15. Rotate the tick ABC 90° clockwise about the origin, then translate the image 7 units right and 1 unit down; then, reflect the second image in the x-axis.

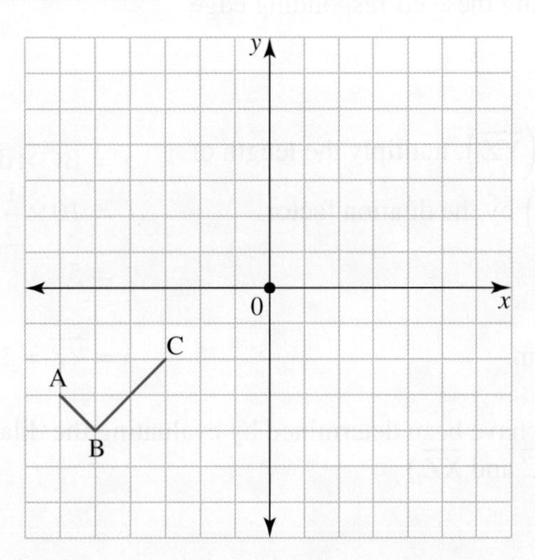

15.4 Similarity

LEARNING INTENTION

At the end of this subtopic you should be able to:
- use similarity to solve unknown lengths in shapes.

15.4.1 Similar figures

Figures are called **similar** when they have the same shape but different sizes.

Polygons are similar if their corresponding interior angles are the same but their corresponding **edges** have different lengths.

Triangle ABC is similar to triangle XYZ because their corresponding interior angles are the same; this is written as $\Delta ABC \sim \Delta XYZ$.

ΔXYZ is a reduction of ΔABC.

The **dilation** factor can be used to evaluate the unknown edge length of a similar figure.

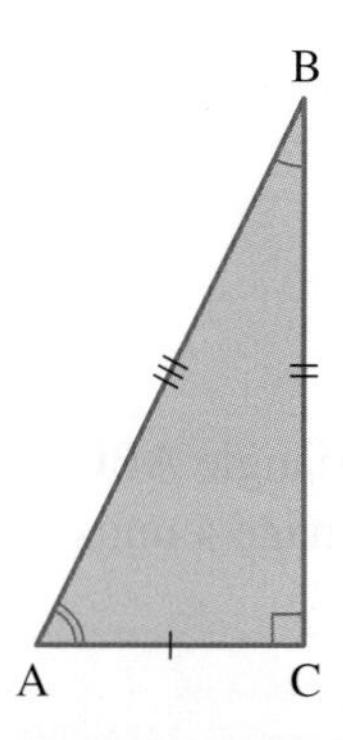

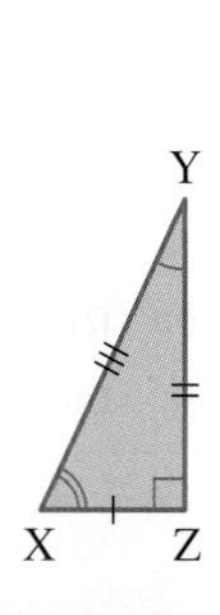

tlvd-5078

WORKED EXAMPLE 7 Solving unknown lengths

Evaluate the unknown edge length of this pair of similar triangles.

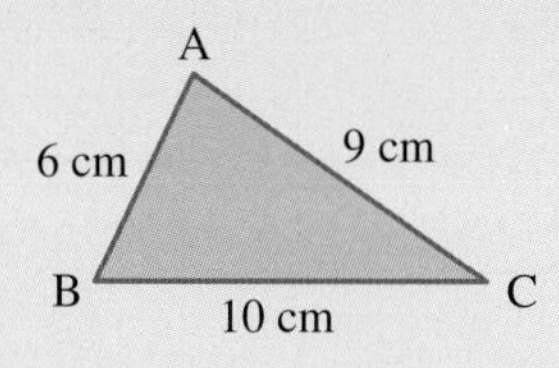

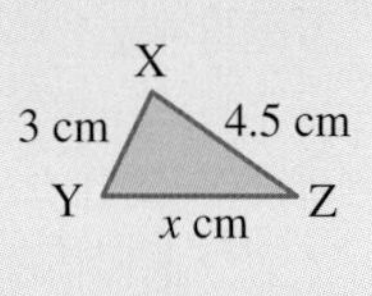

THINK	WRITE
1. We know that triangles ABC and XYZ are similar. The length of edge AB $\left(\overline{AB}\right)$ is 6 cm and length of the corresponding edge XY $\left(\overline{XY}\right)$ is 3 cm. The dilation factor can be found by dividing these corresponding edge lengths.	$\text{Dilation factor} = \dfrac{\overline{XY}}{\overline{AB}} = \dfrac{3}{6} = \dfrac{1}{2}$
2. To determine the value of x $\left(\overline{YZ}\right)$, multiply the length of its corresponding edge $\left(\overline{BC}\right)$ by the dilation factor.	$x = \overline{BC} \times \text{dilation factor} = 10 \times \dfrac{1}{2} = 5$
3. The length of edge YZ is 5 cm.	$x = \overline{YZ} = 5\text{ cm}$

(*Note:* The value of x could also have been determined by evaluating the dilation factor using other corresponding edges, such as $\overline{AC}$ and $\overline{XZ}$.)

15.4 Exercise

Students, these questions are even better in jacPLUS

Receive immediate feedback and access sample responses

Access additional questions

Track your results and progress

Find all this and MORE in jacPLUS

1. Explain what is meant by the term *similar figures*. Give some everyday examples.
2. Identify which shapes below are similar.

3. Determine if shapes *m* and *u* in question **2** are similar. Explain your answer.
4. Determine the dilation factor for the following similar figures.

a. 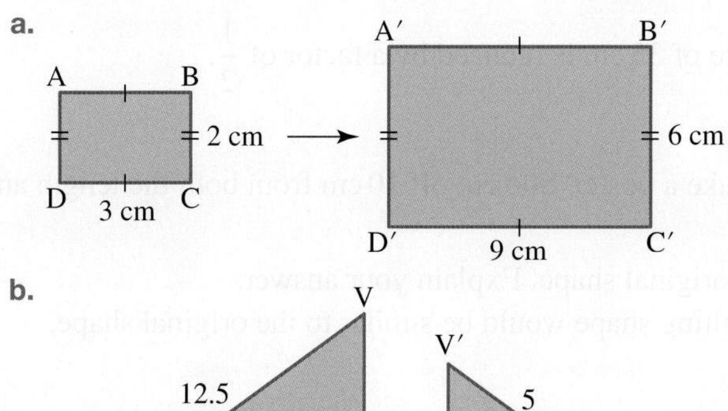

b.

5. Determine the dilation factor for the following similar figures.

a.

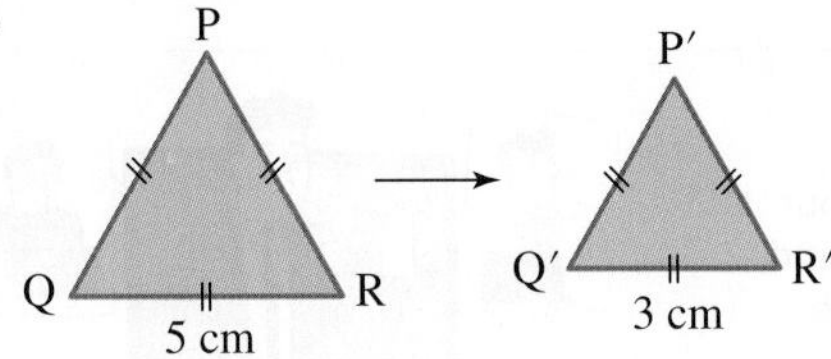

b.

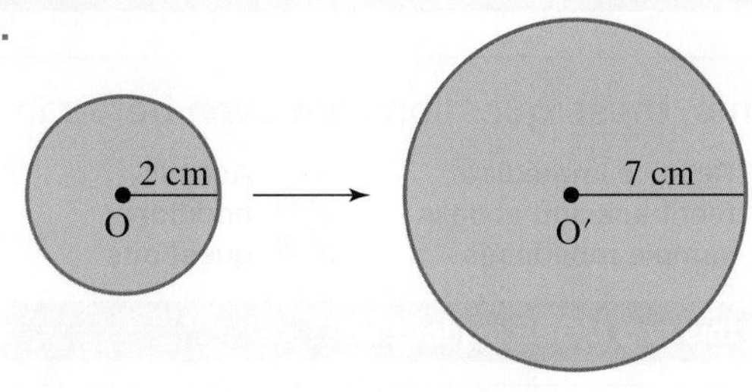

6. Use the dilation factors from question 4 to determine the length of the following sides.

a. $\overline{A'B'}$, if $\overline{AB} = 3$

b. $\overline{VW}$, if $\overline{V'W'} = 3$

7. WE7 Determine the length of the unknown side in each of the following pairs of similar figures.

a.

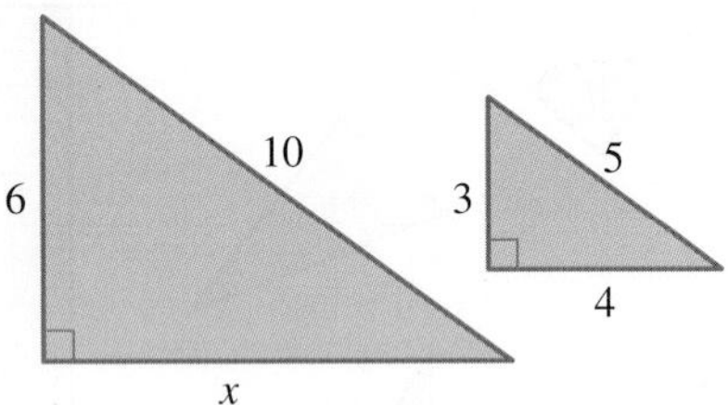

b.

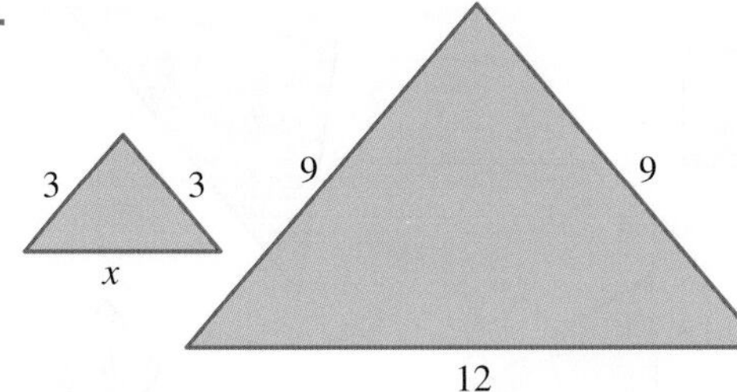

c.

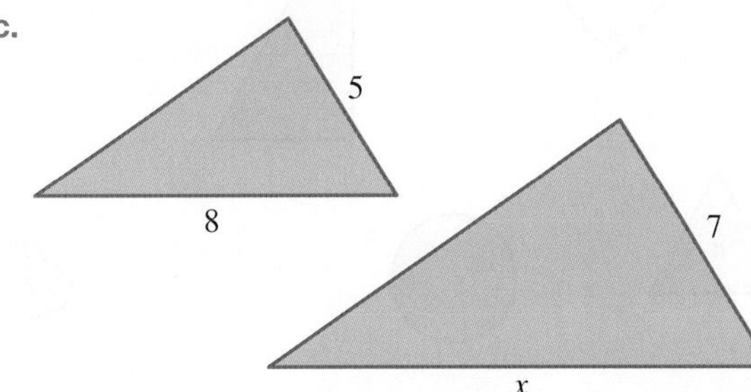

d.

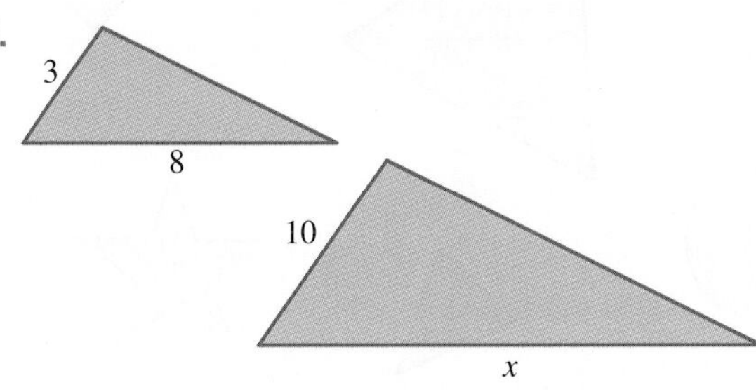

8. Identify the dilation factor in each of the following situations.

a. A classmate bought a toy aeroplane that is 15 cm long. In real life, the aeroplane would be 150 m long.

b. A friend gets a 16 cm × 10 cm photograph blown up to 20 cm × 12.5 cm.

c. A new laptop has a 10-inch screen, which is 2 inches smaller than the previous model.

d. Your parents get a new dining table, which is the same shape as the previous one. However, it is 10 cm wider than the previous table, which was 1 m wide.

9. For each of the following situations, apply the dilation factor and record the new dimensions.

a. A 50 cm × 25 cm photo is tripled in size to form a poster.

b. A triangular bag with a height of 40 cm and a base of 35 cm is reduced by a factor of $\frac{1}{2}$.

c. A 10 m length of cord has half its size added to it.

10. A classmate was cutting a large sheet of paper to make a poster. She cut off 10 cm from both the length and the width.

a. Determine if the resulting shape is similar to the original shape. Explain your answer.

b. Determine if there is a special case when the resulting shape would be similar to the original shape. Explain your answer.

15.5 Symmetry

LEARNING INTENTION

At the end of this subtopic you should be able to:
- identify axes of symmetry and rotational symmetry
- identify if a shape has line symmetry, and if so, state the number of lines of symmetry and location
- identify if a shape has rotational symmetry, and if so, state the order.

15.5.1 Axes of symmetry

If a line can be drawn that divides a shape into two parts, so that when folded along that line the parts would coincide exactly, the shape is symmetrical along that line.

The line itself is called the **line** (or the **axis**) **of symmetry**.

If a mirror is placed along the axis of symmetry, the half of the shape that faces the mirror, together with its reflection, will form the whole shape (that is, the mirror reflection will look exactly like the other half of the shape).

We can think of the axis of symmetry as the line that acts like a mirror.

Many things in nature are symmetrical. Some objects have more than one axis of symmetry, while others have none.

WORKED EXAMPLE 8 Determining axes of symmetry in the real world

Decide whether the front of the house shown is symmetrical and, if so, draw the line of symmetry on the image.

THINK	WRITE
A line of symmetry divides a shape into two parts so that if the shape is folded along it, the two parts will coincide. Imagine folding the picture along a line drawn through the top point of the roof and the midpoint of the base of the house. The two halves will coincide exactly. Draw this line of symmetry using a dotted line.	The shape of the house is symmetrical. The dotted line shows the axis of symmetry. 

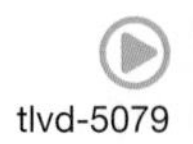 tlvd-5079

WORKED EXAMPLE 9 Determining axes of symmetry

Determine how many axes of symmetry this shape has.

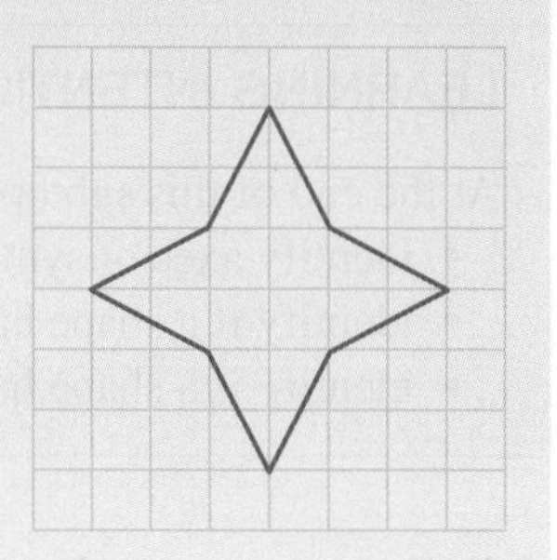

THINK

1. An axis of symmetry is a line that acts as a mirror. A vertical line through the top and bottom vertices of the given shape would divide it into halves that are mirror images of each other, so it is the axis of symmetry.

WRITE

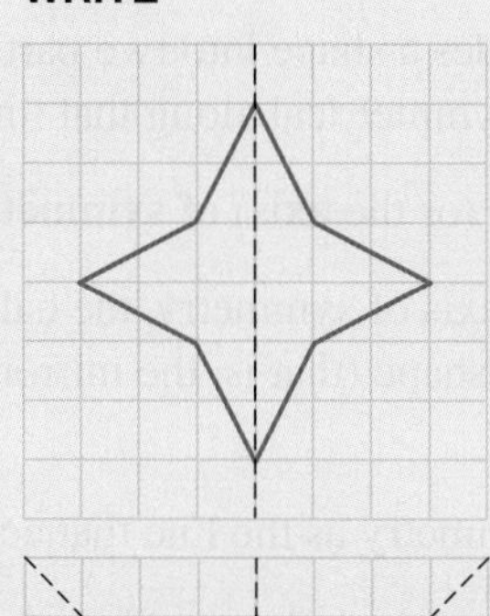

2. Look for other axes of symmetry. Indicate all the axes of symmetry with dotted lines. State their number.

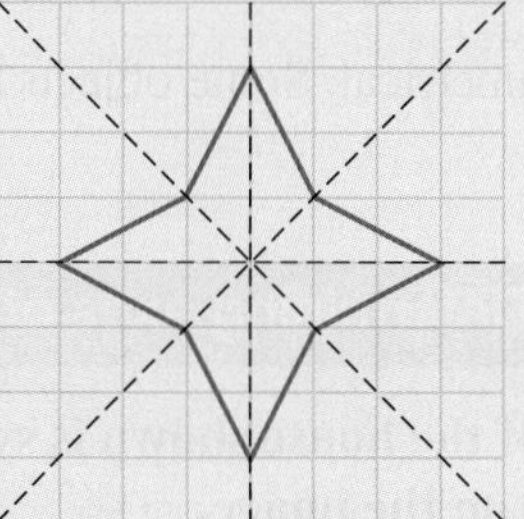

There are 4 axes of symmetry.

15.5.2 Rotational symmetry

A shape has rotational symmetry if rotating it less than 360° matches the original figure. The number of times a match occurs within a 360° turn is known as the **order** (or degree) **of rotational symmetry**.

A shape that has no rotational symmetry still has rotational symmetry of order 1, because the shape will match the original figure after a rotation of 360°.

Rotational symmetry can be seen in nature and in design, as shown in the following images.

Flower: Rotational symmetry of order 5

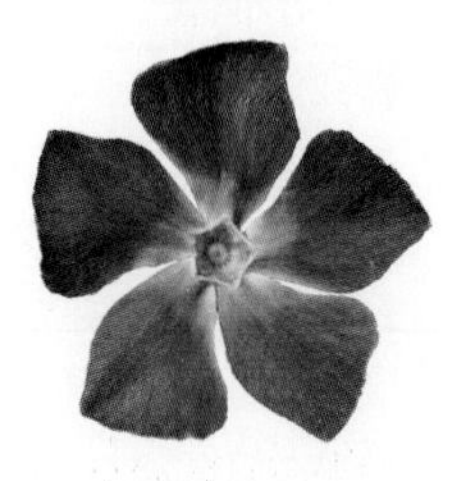

Hexagonal frame: Rotational symmetry of order 6

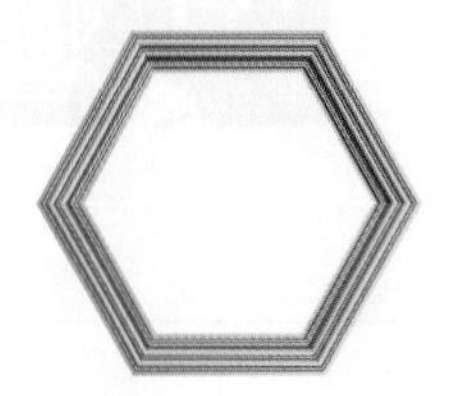

Mercedes Benz logo: Rotational symmetry of order 3

WORKED EXAMPLE 10 Identifying whether a figure has rotational symmetry

Determine if the figure shown has rotational symmetry and, if it does, state the order.

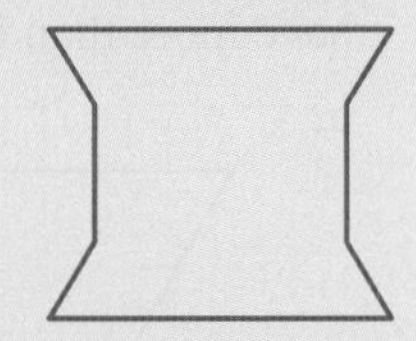

THINK	WRITE
1. The shape is rotated less than 180° to match the original and so it has rotational symmetry.	
2. Work out how many turns are needed to return to the initial position. Two turns are needed, at 180° and 360°.	The figure has rotational symmetry of order 2.

 Resources

Video eLesson Transformations (eles-0012)

15.5 Exercise

Students, these questions are even better in jacPLUS

 Receive immediate feedback and access sample responses

 Access additional questions

 Track your results and progress

Find all this and MORE in jacPLUS

1. WE8 Decide whether each of the following images is symmetrical and, if so, show the line of symmetry.

a. 

b. 

c. 

d. 

e. 

f. 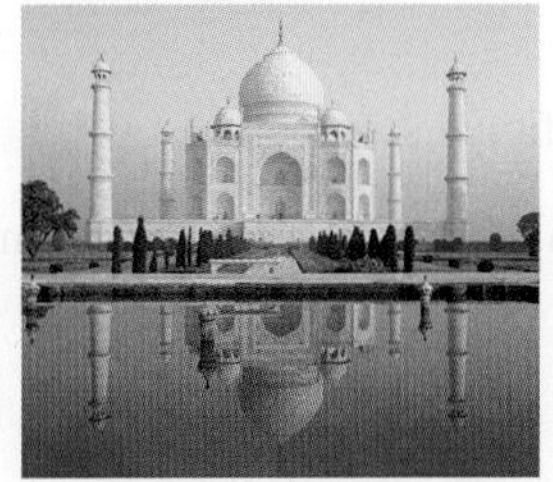

2. Copy each of the following shapes onto grid or squared paper. Carefully cut out each shape. Fold it to determine the axes of symmetry. State the number of axes of symmetry for each shape. (Some shapes will have more than one axis of symmetry.)

a.
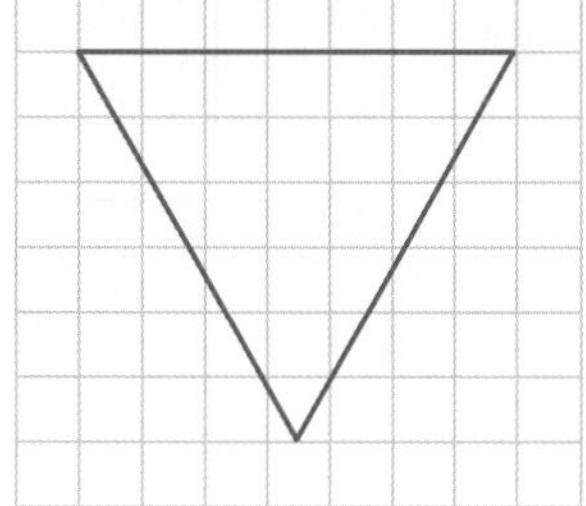

b.

c.
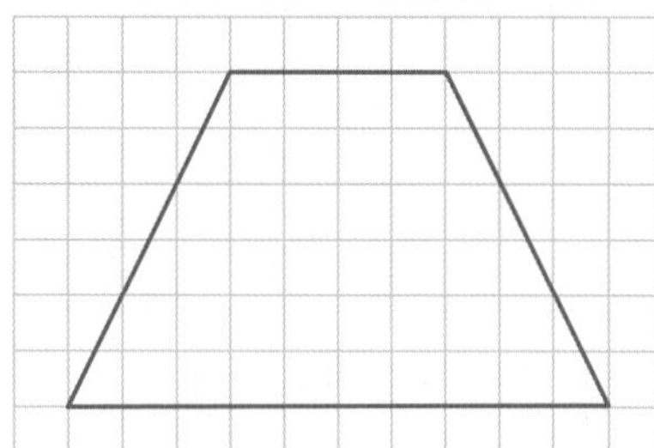

d.
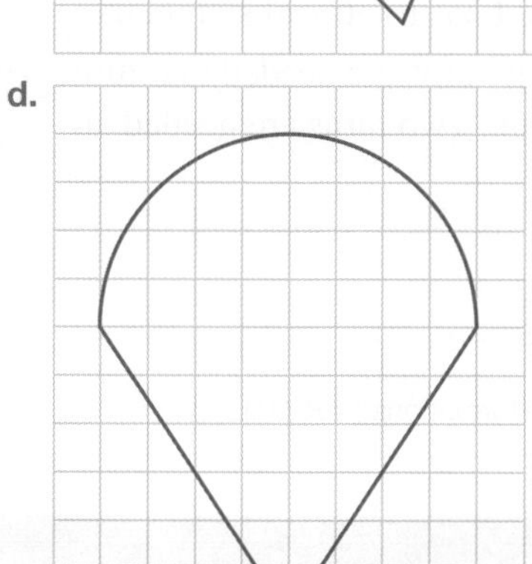

e.
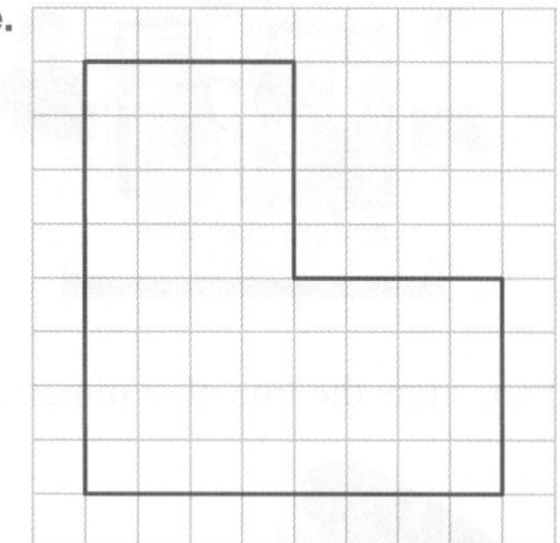

f.

3. WE9 i. Determine how many axes of symmetry each of the following shapes have.

a.
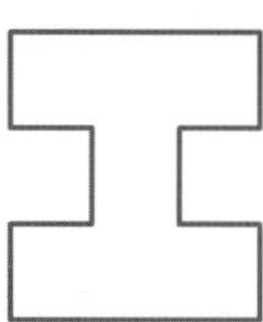

b.
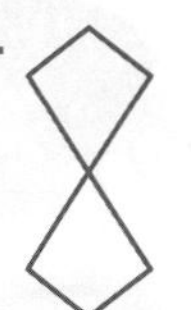

c.
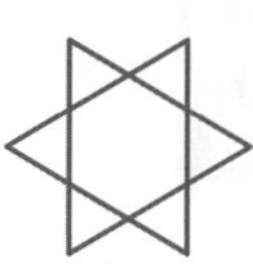

d.
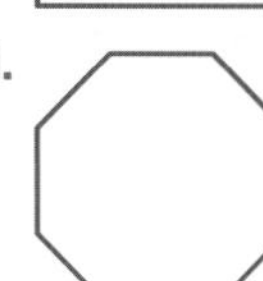

e.
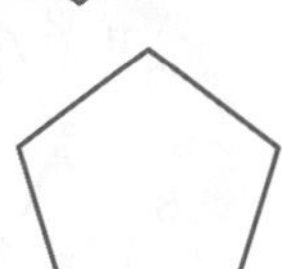

f.
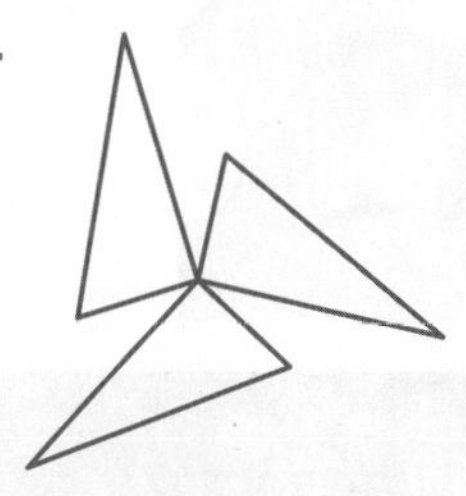

ii. State the order of rotational symmetry.

4. WE10 Determine which of the figures in question 2 have rotational symmetry and, for those that do, state the order.

5. Consider the following upper-case and lower-case letters of the alphabet.

A B C D E F G H I J K L M N O P Q R S T U V W X Y Z

a b c d e f g h i j k l m n o p q r s t u v w x y z

a. Determine which upper-case letters of the alphabet have a line of symmetry.
b. Determine which lower-case letters of the alphabet have a line of symmetry.
c. Determine whether any of the letters have more than one line of symmetry, and if so, which ones.

6. **MC** The number of axes of symmetry in this shape is:

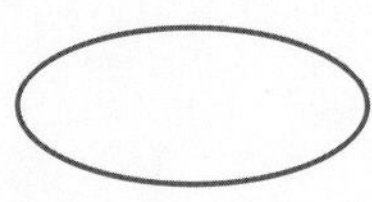

A. 0
B. 1
C. 2
D. 4
E. 3

7. For each of the following diagrams, use the dotted line as an axis of symmetry to complete the picture.

a.

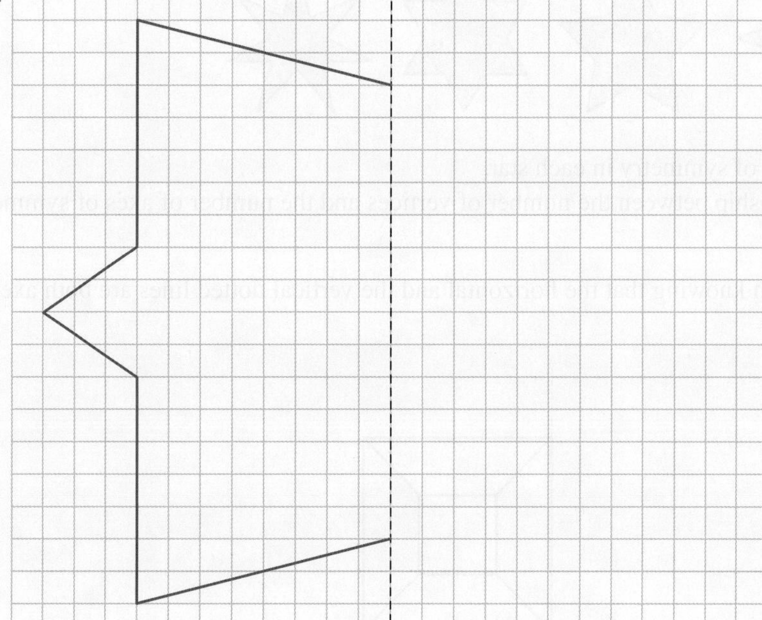

b.

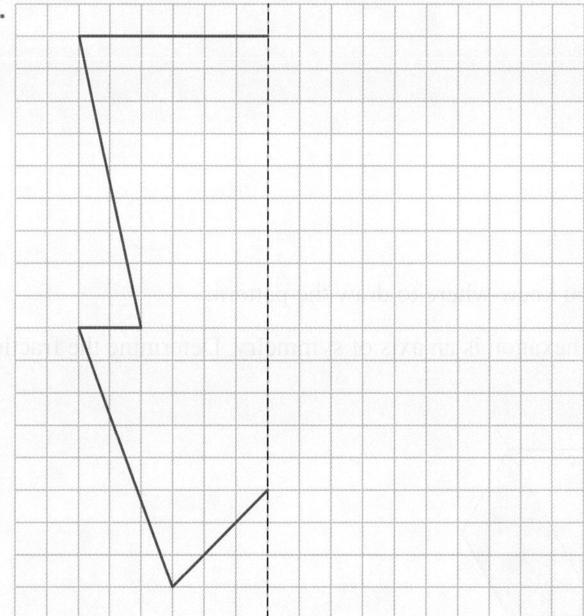

8. For each of the following, use the dotted line as an axis of symmetry to complete the picture.

a. 

b. 

9. The following shapes are regular polygons.

a. Determine the number of axes of symmetry in each star.

b. Determine if there is a relationship between the number of vertices and the number of axes of symmetry. Explain your answer.

10. a. Complete the following pattern knowing that the horizontal and the vertical dotted lines are both axes of symmetry for the pattern.

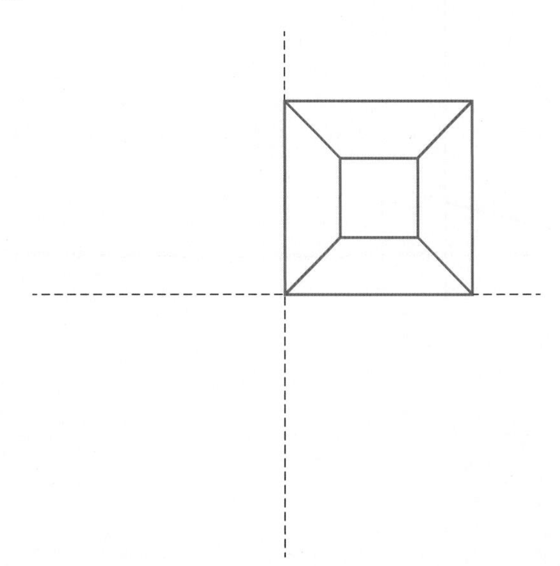

b. Using the concept of symmetry, explain how you knew where to draw the pattern.

11. Each of the dotted lines drawn on this rectangular hexagon is an axis of symmetry. Determine the fraction of the hexagon that is shaded.

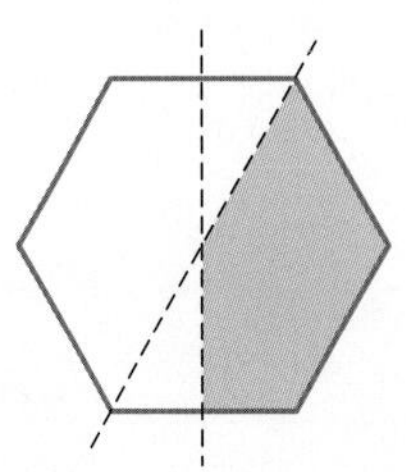

12. a. Determine how many axes of symmetry can be drawn on the umbrella shown.

b. Determine the number of axes of symmetry in the Melbourne Star.

15.6 Review

15.6.1 Summary

doc-38023

Hey students! Now that it's time to revise this topic, go online to:

Review your results

Watch teacher-led videos

Practise questions with immediate feedback

Find all this and MORE in jacPLUS

15.6 Exercise

Multiple choice

1. MC The polygon shown is:

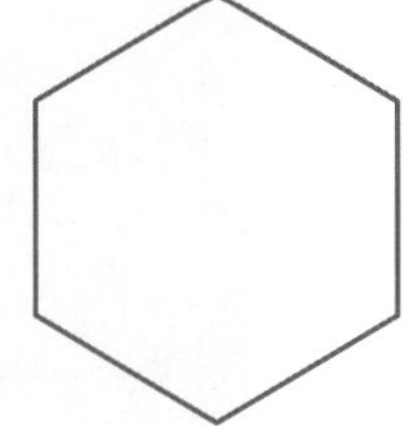

A. an octagon.
B. a pentagon.
C. a quadrilateral.
D. a heptagon.
E. a hexagon.

2. MC The number of vertices in the following polygon is:

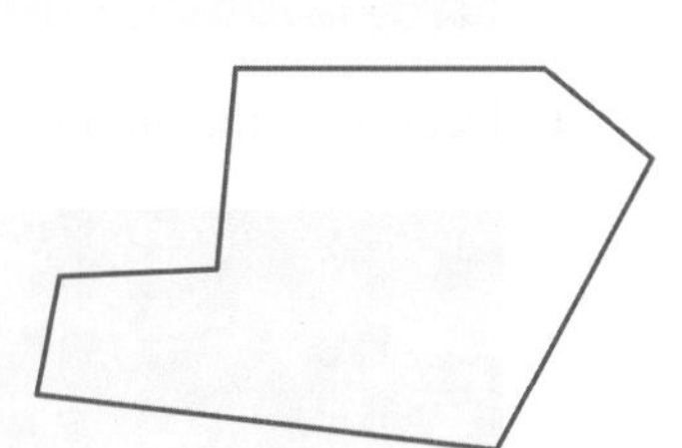

A. 4
B. 5
C. 6
D. 7
E. 8

3. MC The transformation from the blue polygon to the red polygon is:

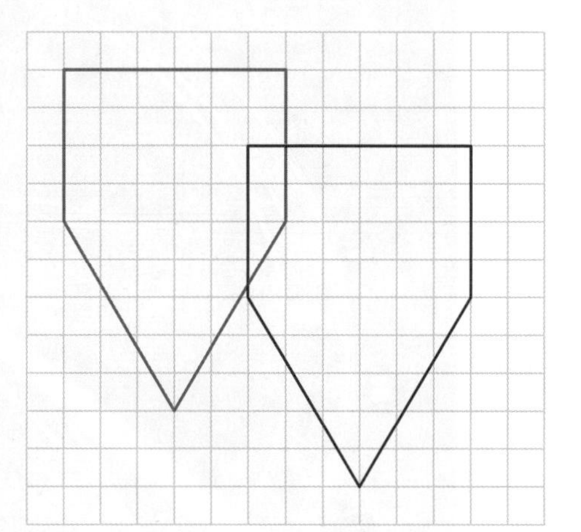

A. 2 right and 5 down.
B. 4 right and 3 down.
C. 5 right and 2 down.
D. 5 up and 2 left.
E. 2 up and 5 left.

4. MC A concave polygon has:

A. six sides.
B. at least one reflex angle.
C. the same number of vertices as edges.
D. more vertices than edges.
E. vertices of equal angles.

5. **MC** The following transformation of a blue polygon to a red polygon shows:
   A. a translation of 4 units right.
   B. a clockwise rotation of 180°.
   C. a translation of 16 units right.
   D. a reflection about the dashed line.
   E. a reflection about the dashed line and a translation of 2 units right.

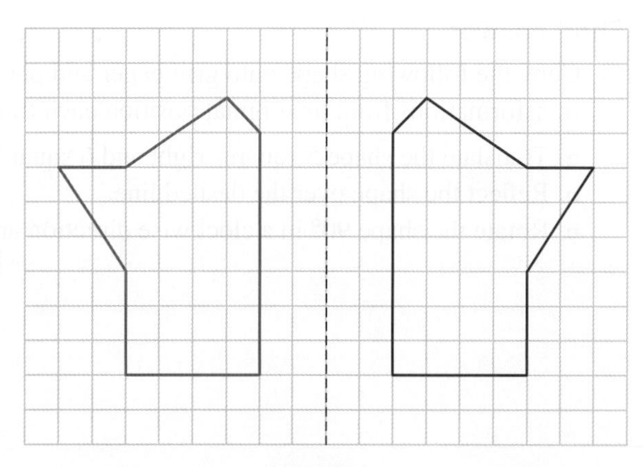

6. **MC** The value of x in the diagram is:
   A. 18 **B.** 16 **C.** 12
   D. 10 **E.** 8

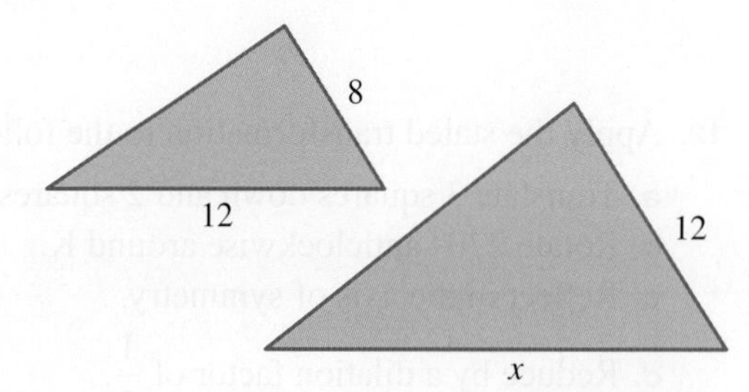

7. **MC** The dilation factor in the diagram is:
   A. 20 **B.** 8 **C.** 3
   D. 2.5 **E.** 3.5

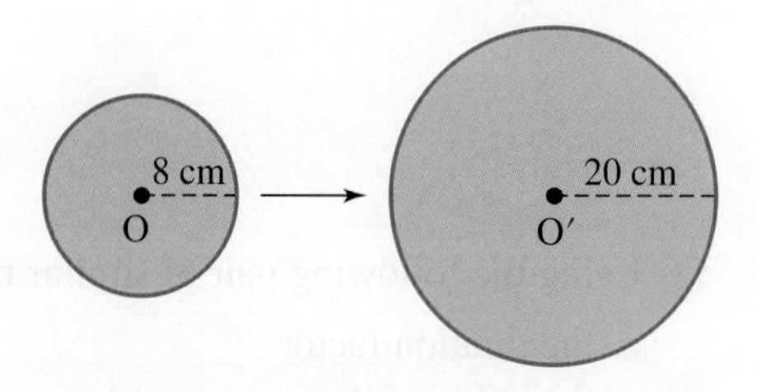

8. **MC** The dilation factor of a house plan that shows a wall 18 cm long, but whose actual length is 9 m, is:
   A. 0.5 **B.** 50 **C.** 2
   D. 5 **E.** 20

The following diagram refers to questions 9 and 10.

9. **MC** The number of axes of symmetry in the diagram is:
   A. 1 **B.** 2
   C. 3 **D.** 4
   E. 5

10. **MC** The order of rotational symmetry of the diagram is:
   A. 0 **B.** 1
   C. 2 **D.** 3
   E. 4

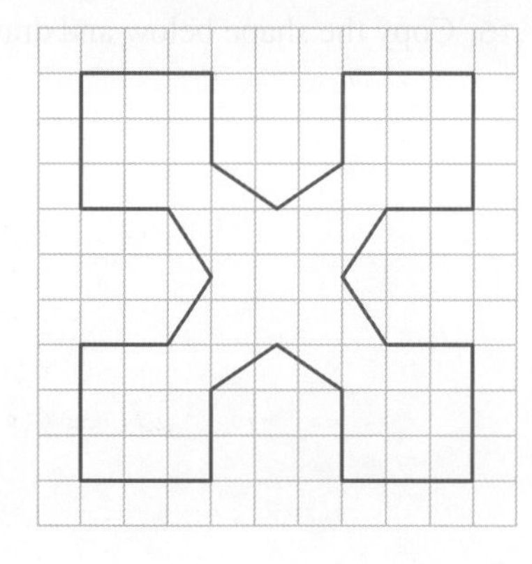

Short answer

11. Copy the following shape onto grid paper and perform the specified transformations from its original position each time.
    a. Translate the shape 5 squares right and 6 squares down.
    b. Reflect the shape over the dashed line.
    c. Rotate the shape 90° in a clockwise direction around the red dot.

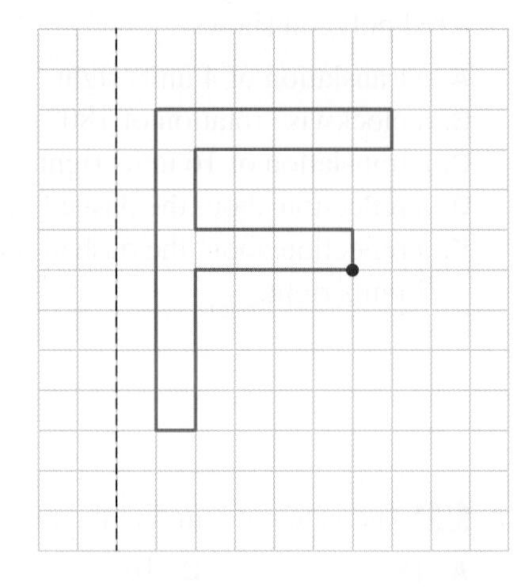

12. Apply the stated transformation to the following figure.
    a. Translate 3 squares down and 2 squares left.
    b. Rotate 270° anticlockwise around K.
    c. Reflect in the axis of symmetry.
    d. Reduce by a dilation factor of $\frac{1}{3}$.

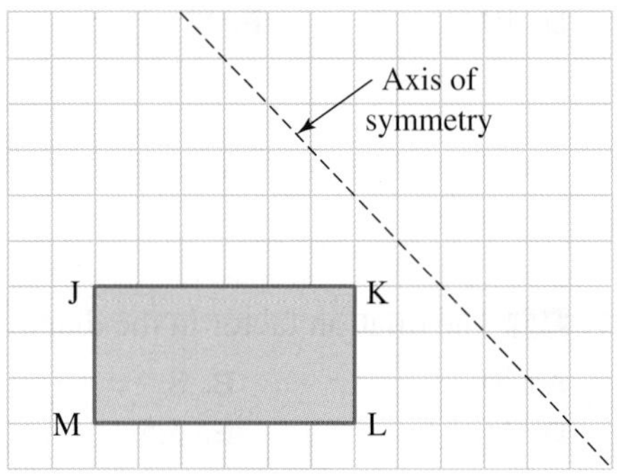

13. Using the following pair of similar figures, calculate:
    a. the dilation factor
    b. the value of the pronumerals.

3 cm
4 cm
x cm
12 cm

14. A new toy manufacturer makes a model car that is $\frac{1}{20}$ of the size of a real car.

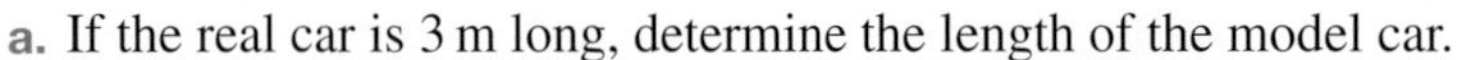

    a. If the real car is 3 m long, determine the length of the model car.
    b. The height of the model is 6 cm. Determine the height of the car in real life.

15. Copy the shape below and draw in its axes of symmetry.

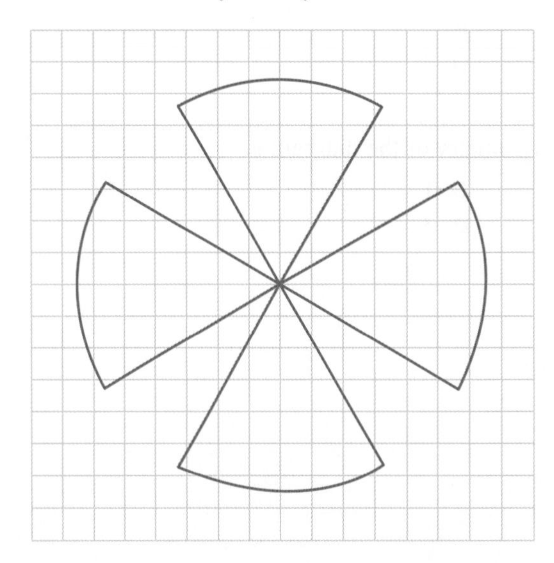

16. Show the following translations by copying the shapes and producing an image on a grid.

a. 3 units down and 2 units left

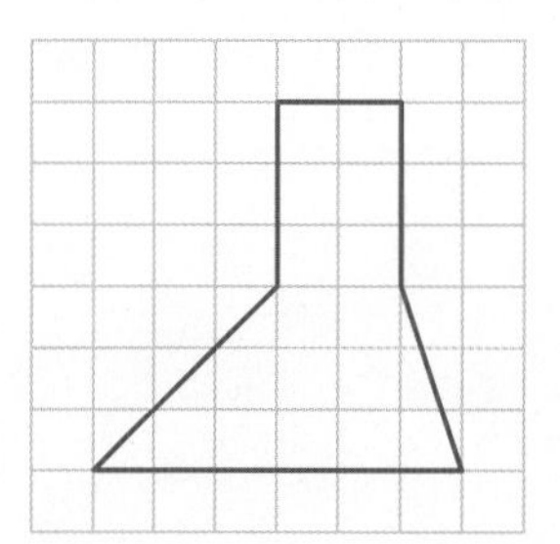

b. 2 units down and 5 units right

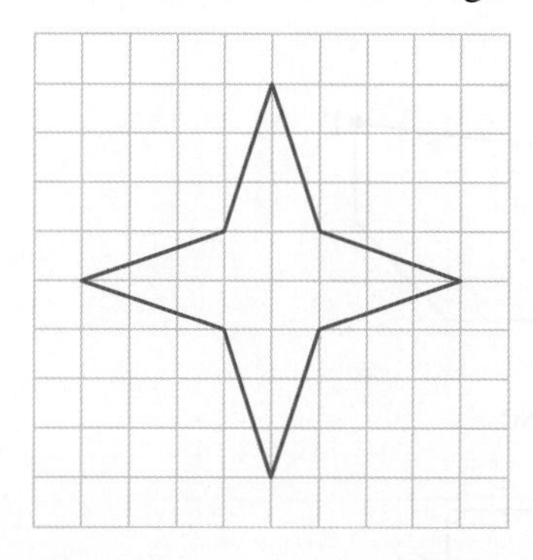

17. For each of the following shapes, draw the image reflected in the mirror line shown.

a.

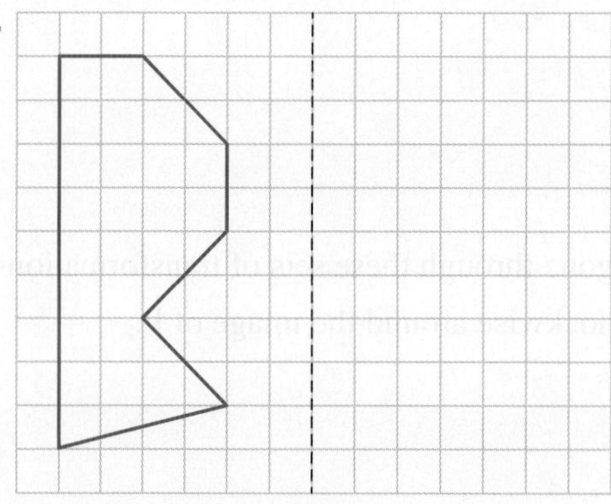

b.

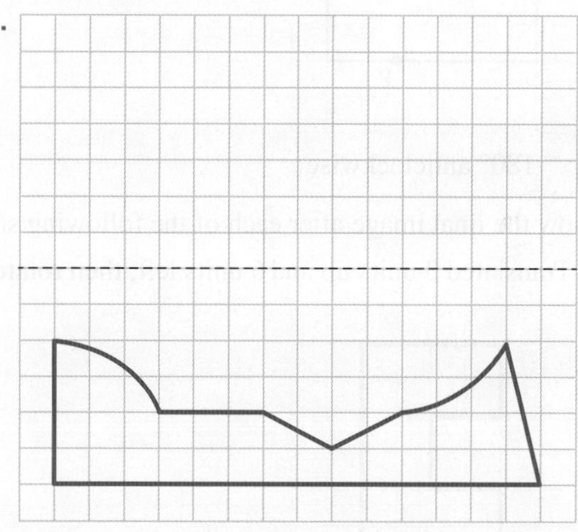

c.

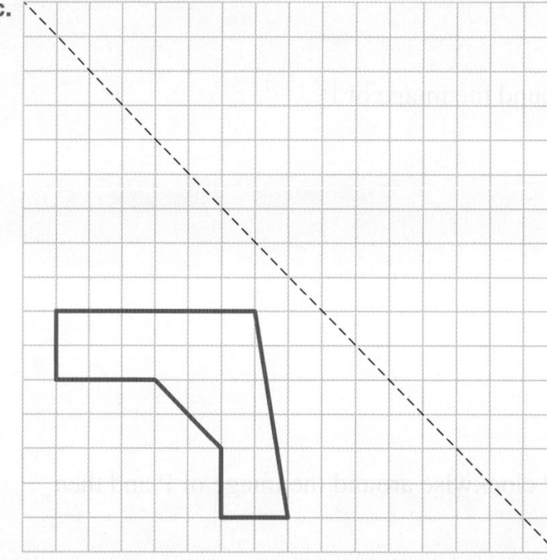

18. For each of the following shapes, show the image after the rotation about point P as specified below the figures.

a. 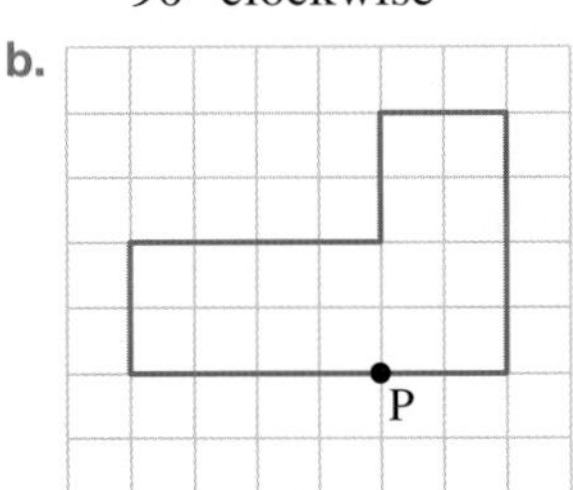

90° clockwise

b. 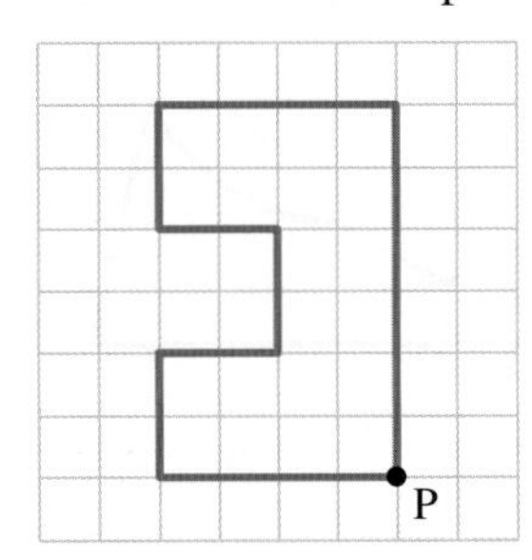

180° anticlockwise

19. Show the final image after each of the following shapes have gone through these sets of transformations.

a. Translated 3 units up and 6 units left; then rotated 90° anticlockwise around the image of P.

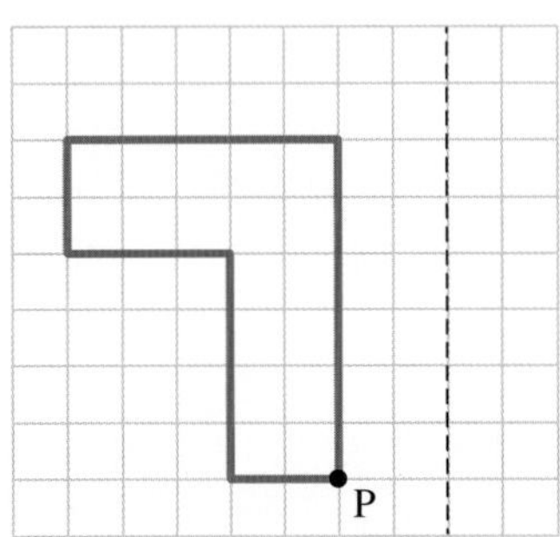

b. Reflected in the mirror; then rotated 90° clockwise around the image of P.

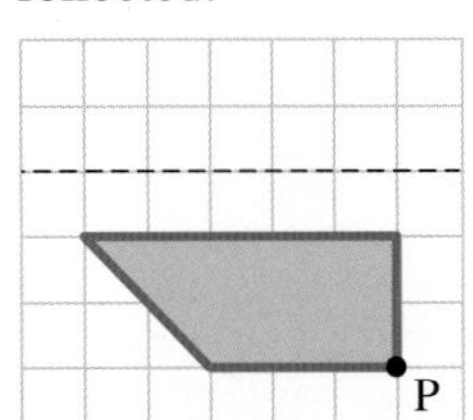

c. Translated 2 units down and 4 units right, rotated 180° clockwise around the image of P and then reflected.

P

Extended response

20. Braille is a code of raised dots that can be read by touch. It was developed by a 15-year-old blind French student named Louis Braille. The Braille alphabet is based on a cell three dots high and two dots wide.

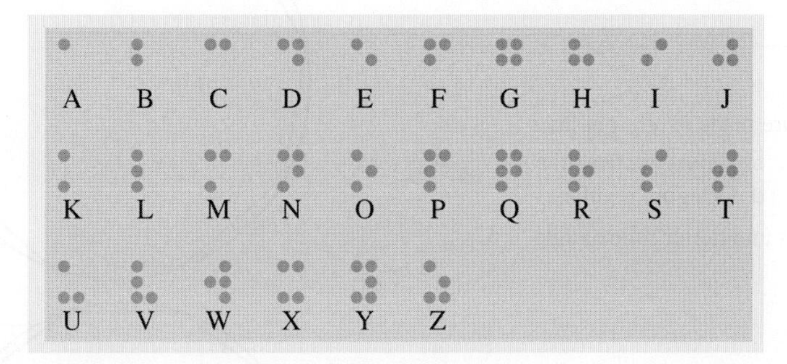

a. Compare the first 10 letters of the Braille alphabet with the second 10 letters. State what patterns you see.
b. State which Braille letters are reflections of each other.
c. State which Braille letters are rotations of each other.
d. There are no Braille letters that are translations of each other. Explain why.
e. A word like MUM has reflection symmetry. Determine two words that have reflection symmetry when written in Braille.

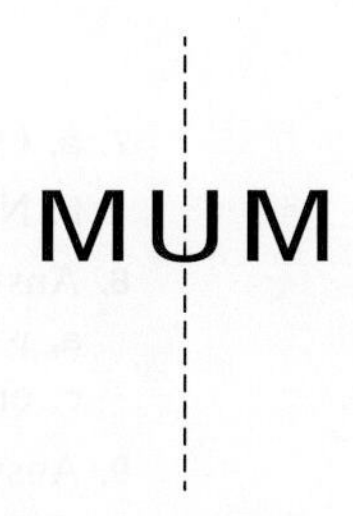

f. A word like SWIMS has rotational symmetry. Determine two words that have rotational symmetry when written in Braille.

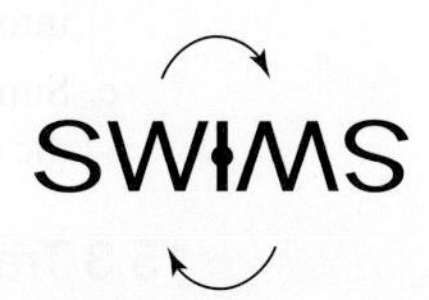

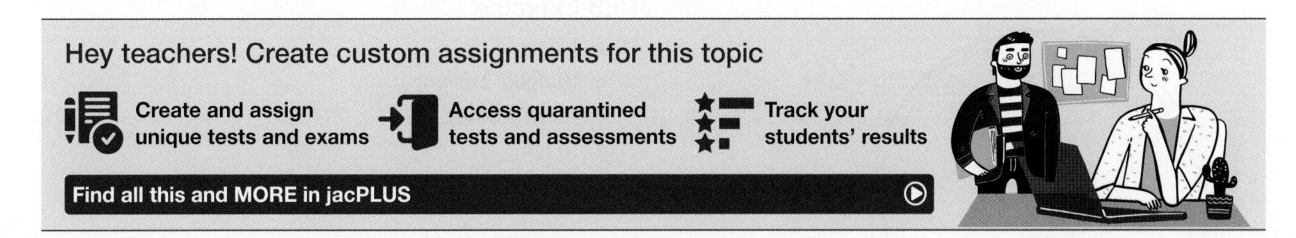

Answers

Topic 15 Shapes

15.2 Polygons

15.2 Exercise

1. A many-sided closed figure made of straight lines
2. a. Pentagon, regular b. Heptagon, irregular
   c. Decagon, irregular d. Octagon, regular
3. a. A concave polygon has an interior reflex angle.
   b. ii and iii
4. a. 8
   b. Octagon
   c. It looks regular because the edges look equal.
5. a.

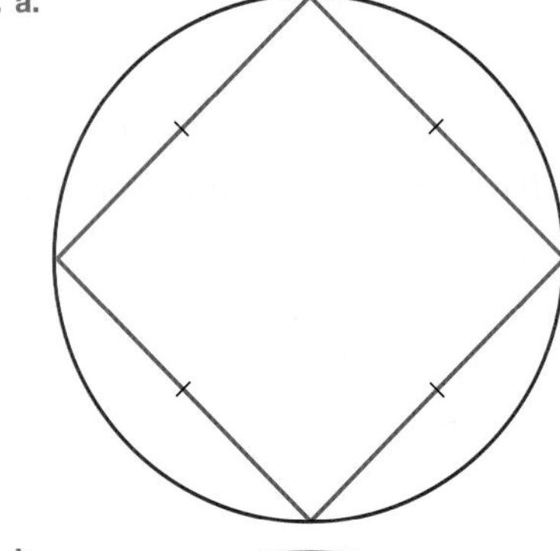

b.

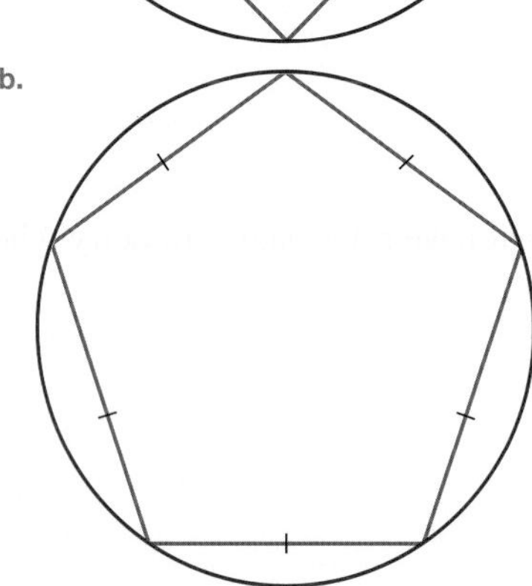

c.

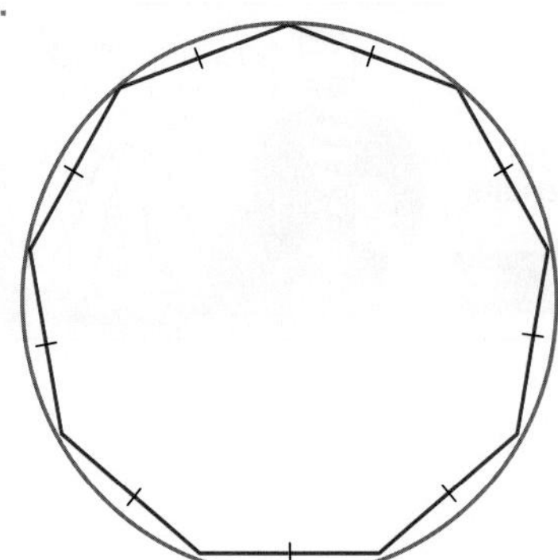

6. a.

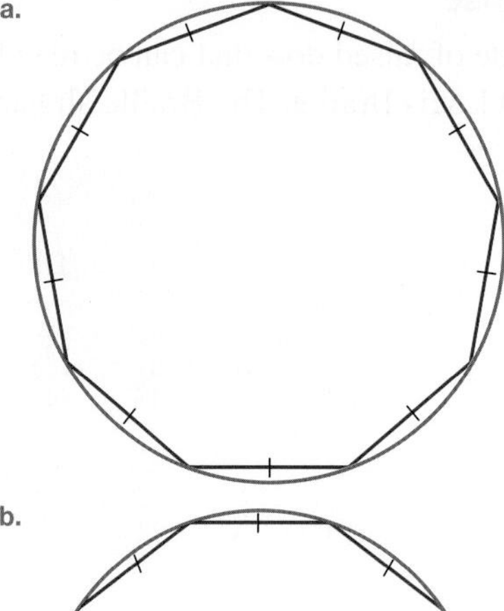

b.

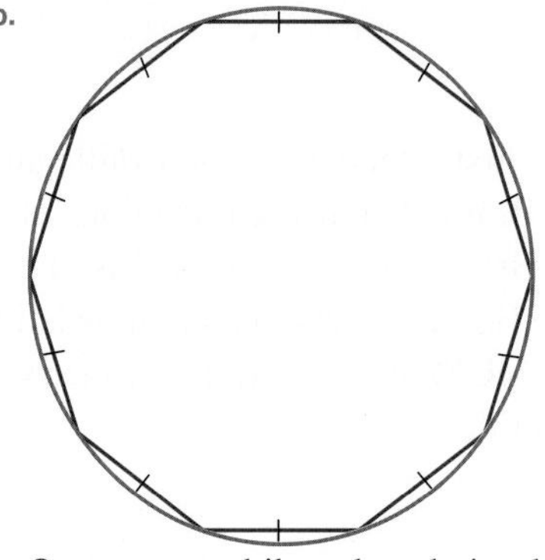

7. a. Octagons, quadrilaterals and triangles
   b. No
8. Answers will vary. Examples are:
   a. window b. tabletop
   c. on a book d. school sign.
9. Answers will vary.
10. a. Square: 360°; pentagon: 540°; octagon: 1080°; nonagon: 1260°; decagon: 1440°
    b. The number of triangles that a polygon can be divided into is the number of edges (or vertices) – 2.
    c. Sum of interior angles $= 180n° - 360°$, where n is the number of edges.

15.3 Transformations

15.3 Exercise

1. a. 1 right, 2 up
   b. 19 right, 2 down
   c. 4 left, 5 down
   d. 10 left
2. a. $P' = (-1, 3)$

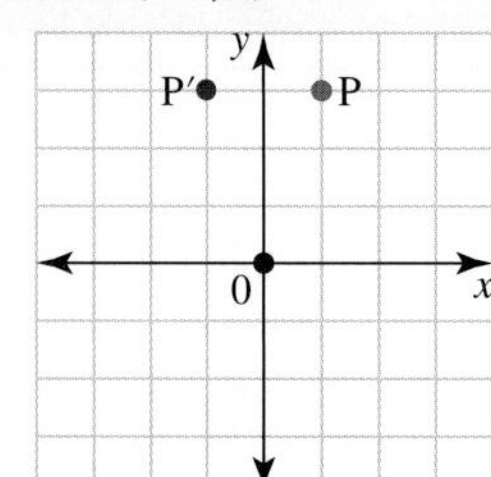

b. $P' = (1, 7)$

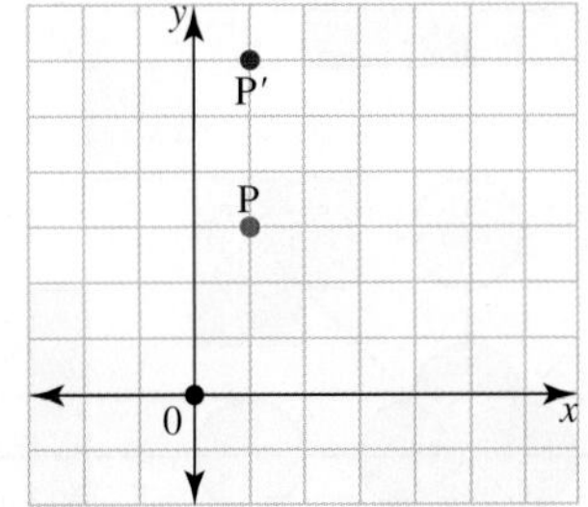

c. $P' = (2, 1)$

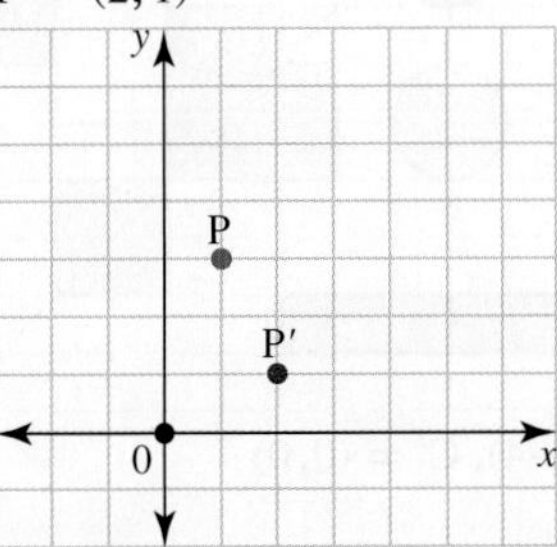

3. a. $P' = (4, -5)$

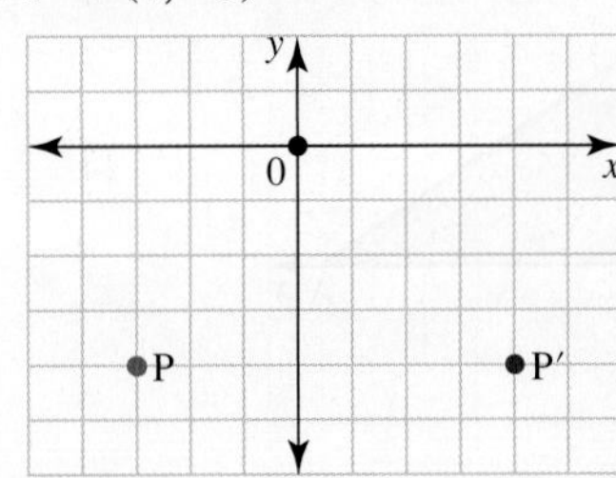

b. $P' = (-3, -2)$

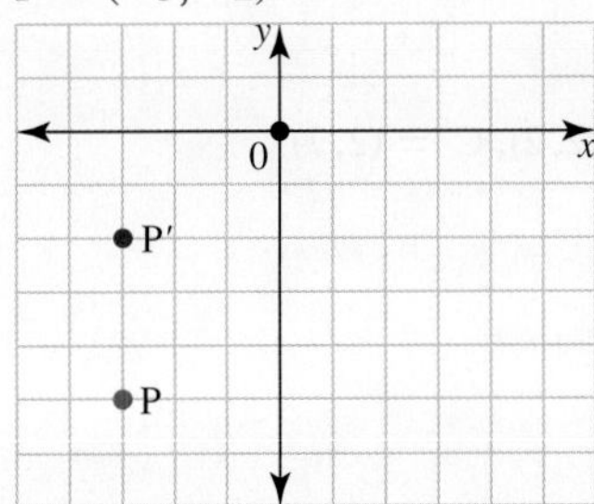

c. $P' = (1, -9)$

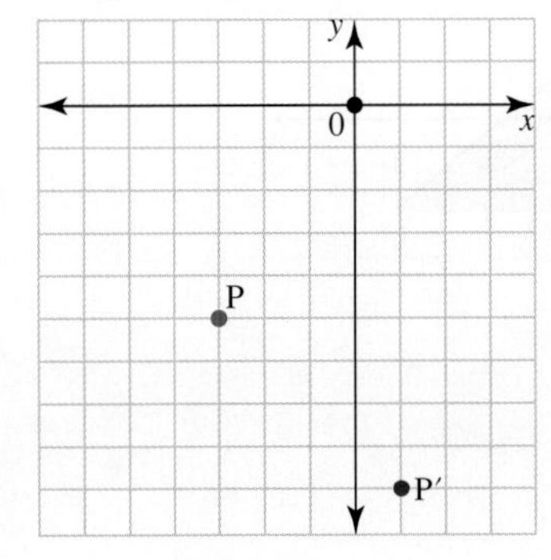

4.

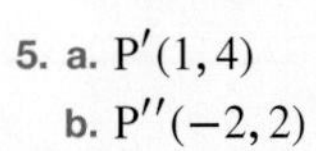

5. a. $P'(1, 4)$

b. $P''(-2, 2)$

c. A translation of 1 unit up and 5 units to the left

6.

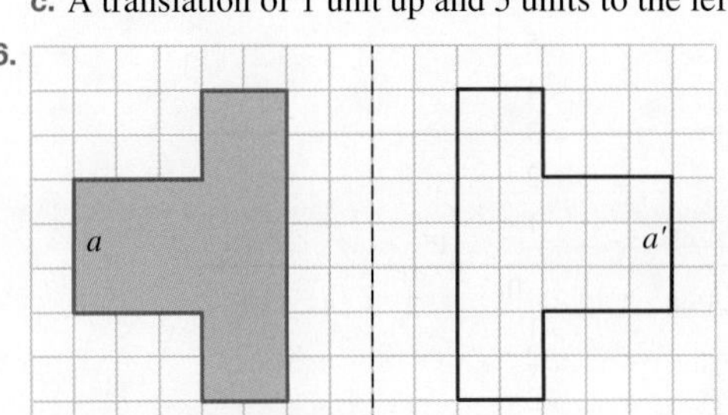

7.

8. a.

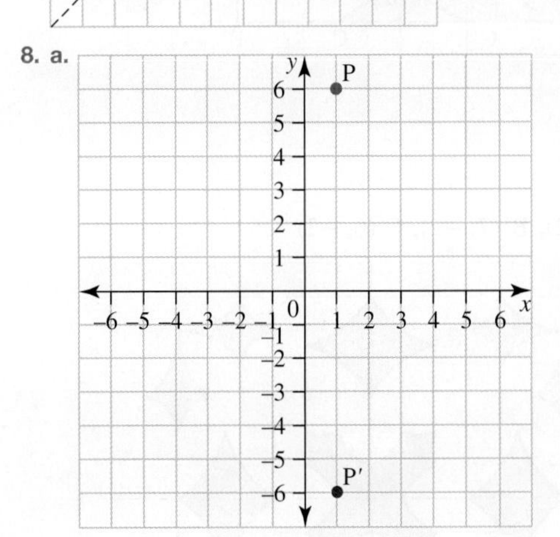

b.

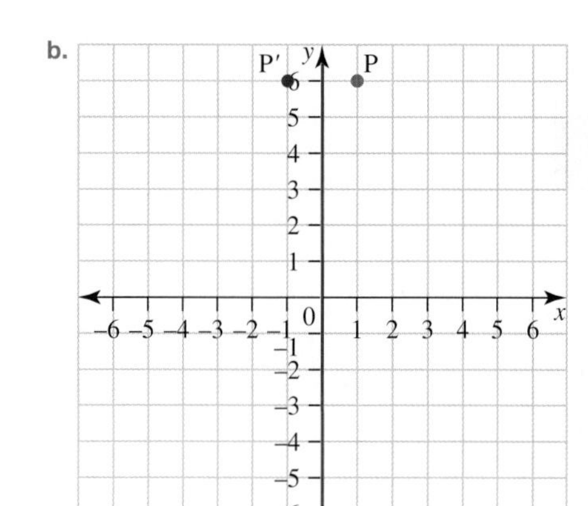

c.

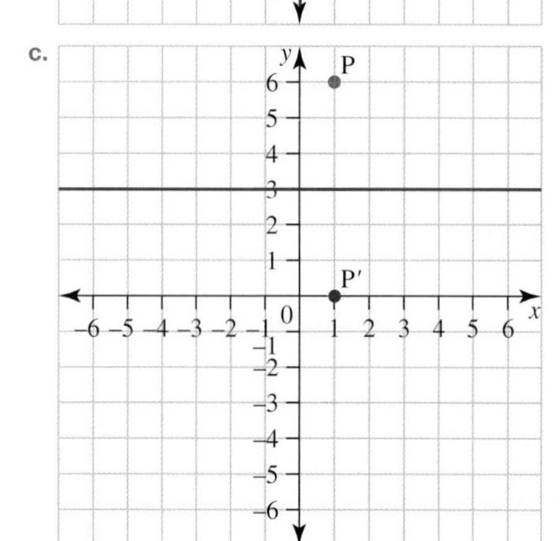

9. a.

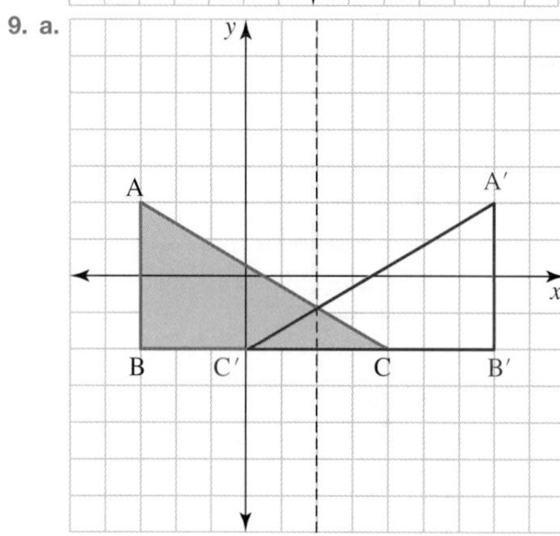

b. $A'(7, 2)$, $B'(7, -2)$, $C'(0, -2)$

10. a.

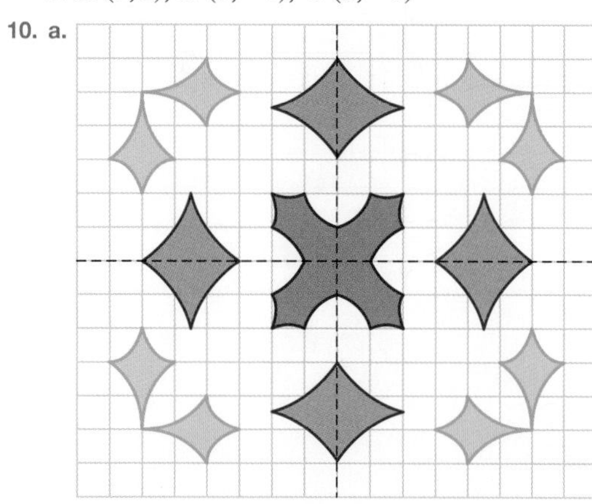

b.

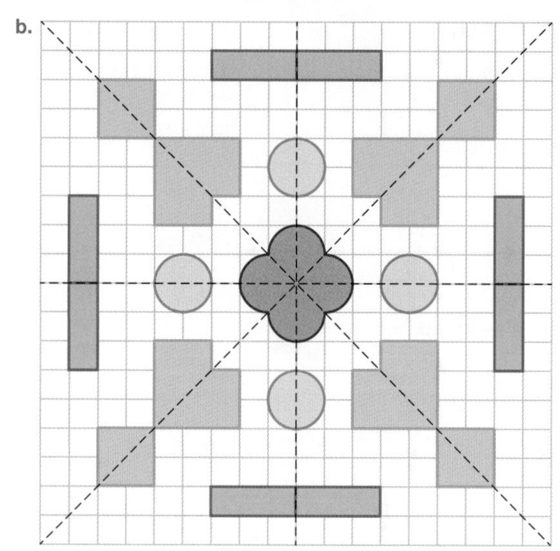

11. $A' = (-6, 0)$, $B' = (0, -4)$, $C' = (0, 0)$

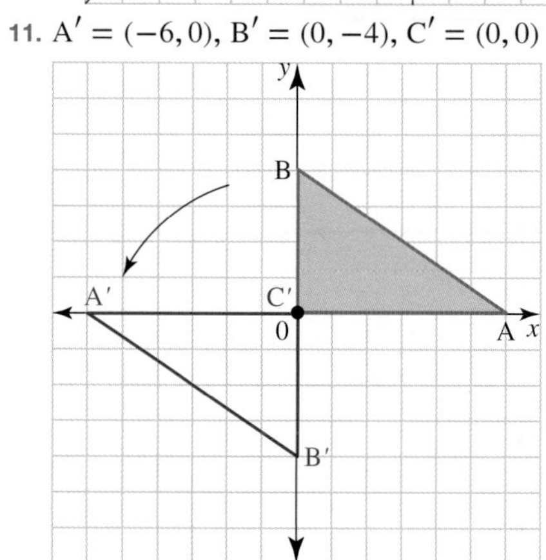

12. $A' = (2, 2)$, $B' = (-2, 2)$, $C' = (2, 7)$

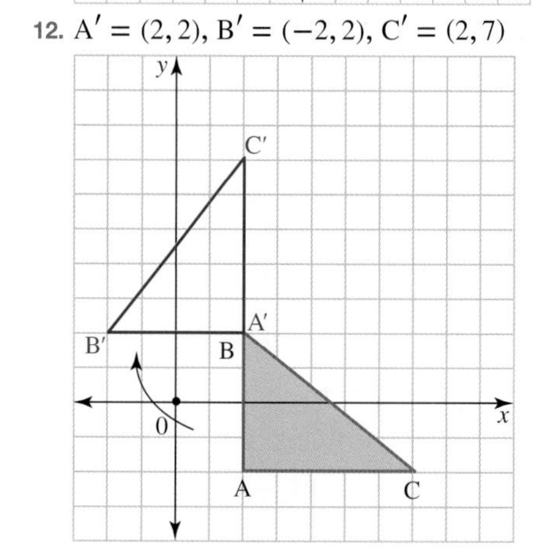

13.

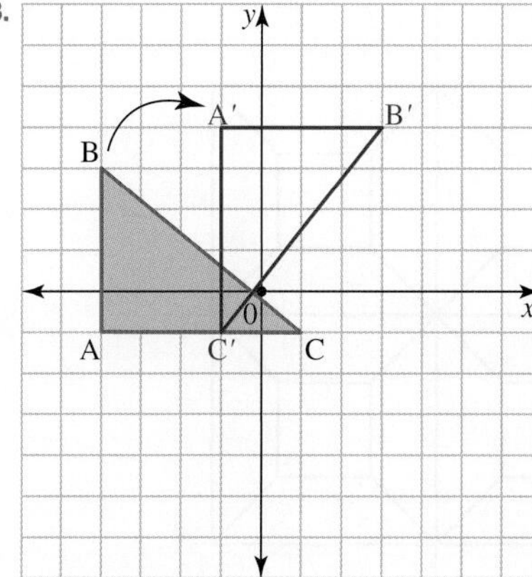

14.

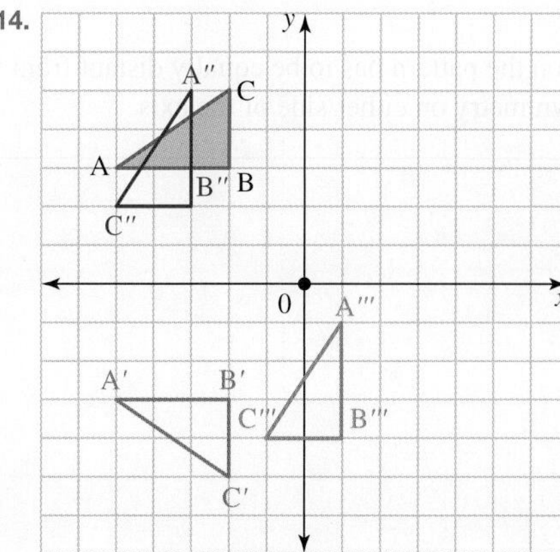

15.

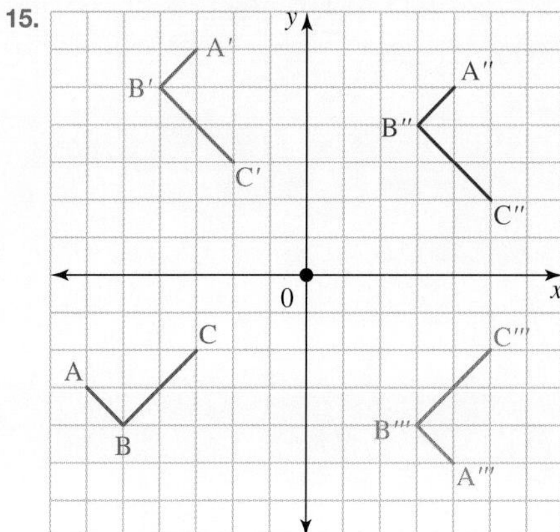

15.4 Similarity

15.4 Exercise

1. Similar figures have the same shape but different sizes.
2. a, f and r; m and u; g and n; s and o; d and p; q and h; e and j; i, d and p
3. Shapes m and u are similar because their angles are the same.
4. a. 3 b. $\frac{2}{5}$
5. a. $\frac{3}{5}$ b. $\frac{7}{2}$
6. a. 9 b. 7.5
7. a. $x = 8$ b. $x = 4$
   c. $x = 11.2$ d. $x = 26.7$
8. a. 1000 b. 1.25 c. $\frac{5}{6}$ d. 1.1
9. a. 150 cm × 75 cm
   b. Height of 20 cm and base of 17.5 cm
   c. 15 cm
10. a. Not similar
    b. If the original shape was a square

15.5 Symmetry

15.5 Exercise

1. a. No
   b. Yes

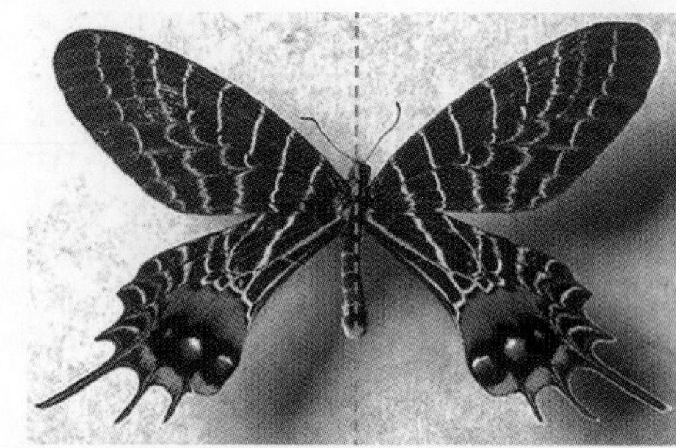

   c. No
   d. No
   e. No
   f. Yes

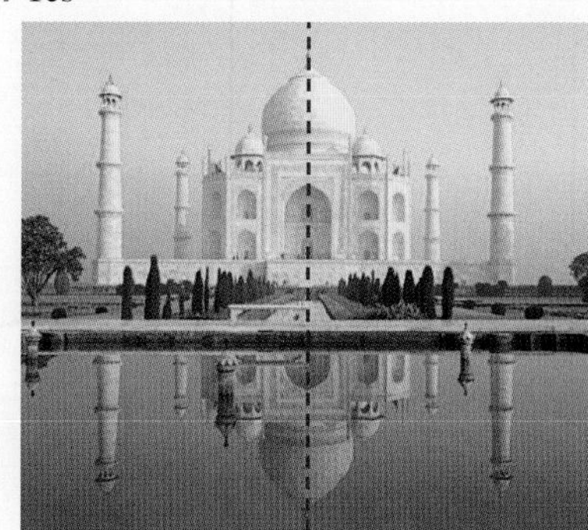

2. a. 3 b. 0 c. 1
   d. 1 e. 1 f. 0
3. i. a. 2 b. 2 c. 6
   d. 8 e. 5 f. 0
   ii. a. Order 2 b. Order 2 c. Order 6
   d. Order 8 e. Order 5 f. Order 3
4. a. Order 3 b. No c. No
   d. No e. No f. Order 2
5. a. A B C D E H I K M O T U V W X Y
   b. c i l o v w x
   c. H I O X l o x
6. C

7. a. See the figure at the bottom of the page.*

b. See the figure at the bottom of the page.**

8. Check with your teacher.

9. a. $4, 5, 6, 7$

b. Yes, the number of vertices is double the number of axes of symmetry, as each axis of symmetry goes through two vertices.

10. a.

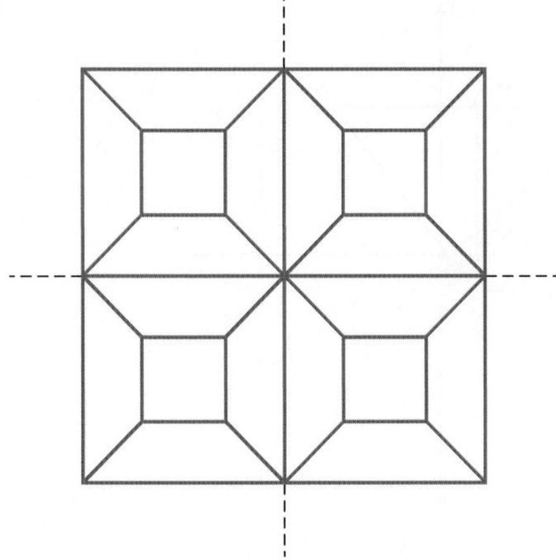

b. Every point on the pattern has to be equally distant from the axis of symmetry on either side of the axis.

---

*7. a.

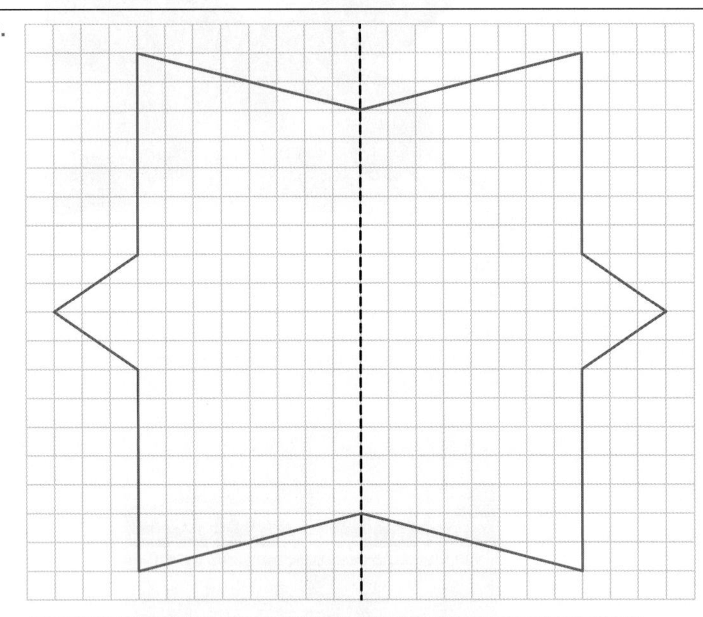

**7. b.

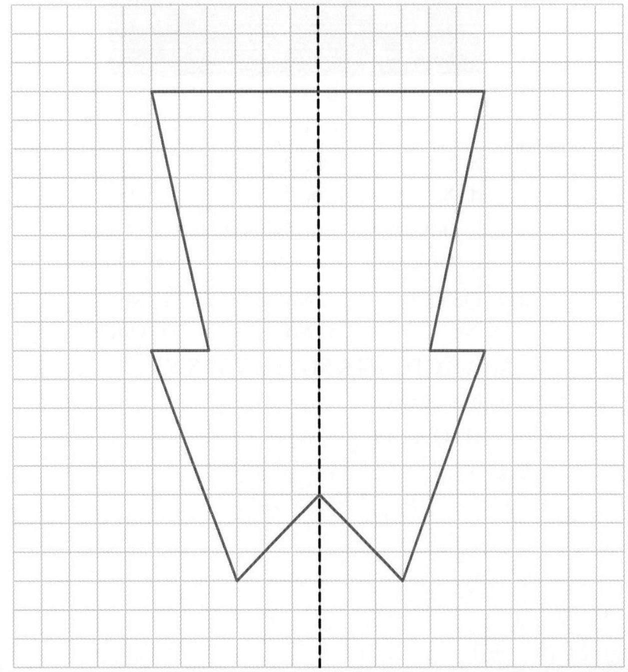

11. $\frac{5}{12}$

12. a. 8 axes of symmetry

b. 7 axes of symmetry

15.6 Review

15.6 Exercise

Multiple choice

1. E
2. D
3. C
4. B
5. D
6. A
7. D
8. B
9. D
10. E

Short answer

11. a.

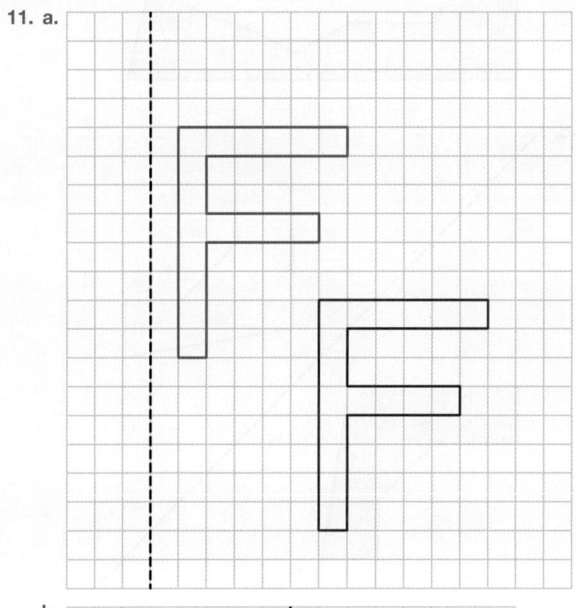

b.

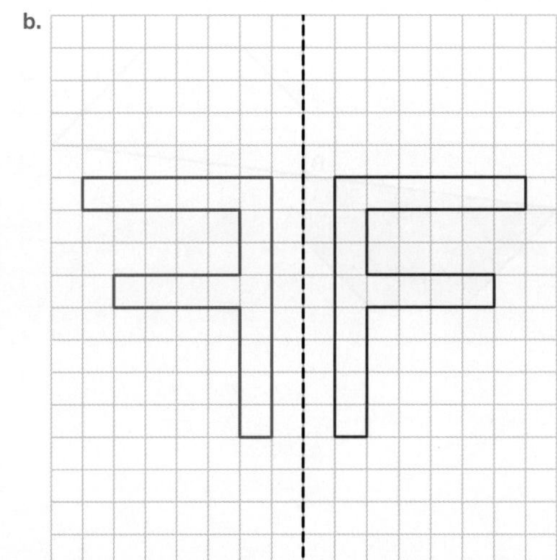

c.

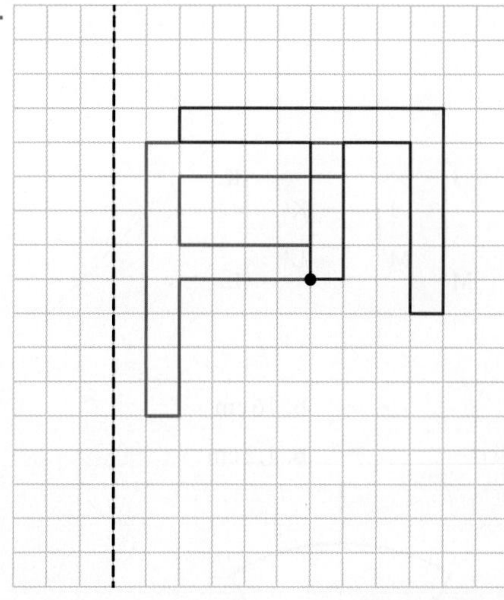

12. a.

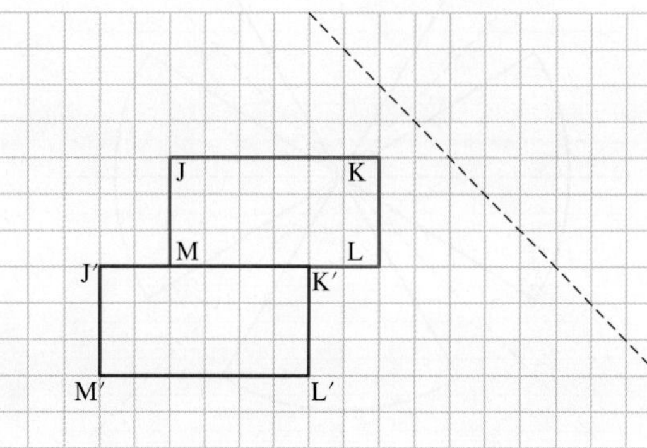

b.

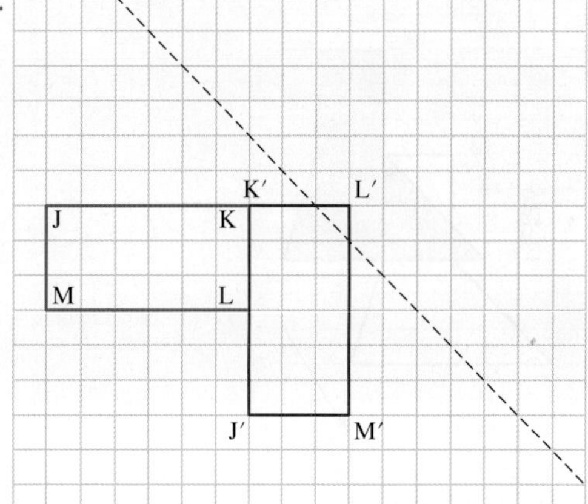

c.

d.

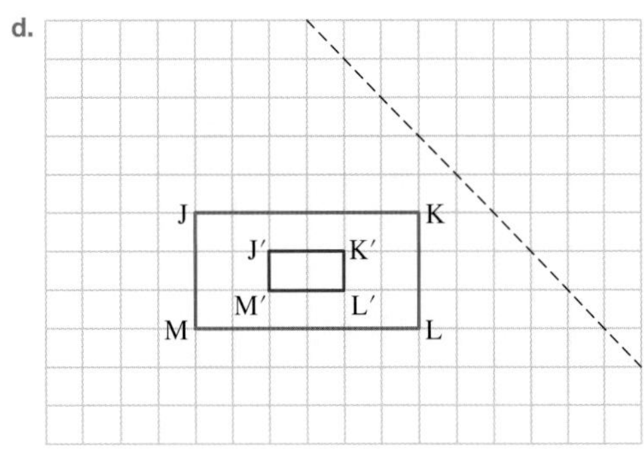

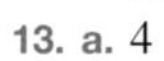

13. a. 4 b. 16 cm

14. a. 15 cm b. 1.2 cm

15.

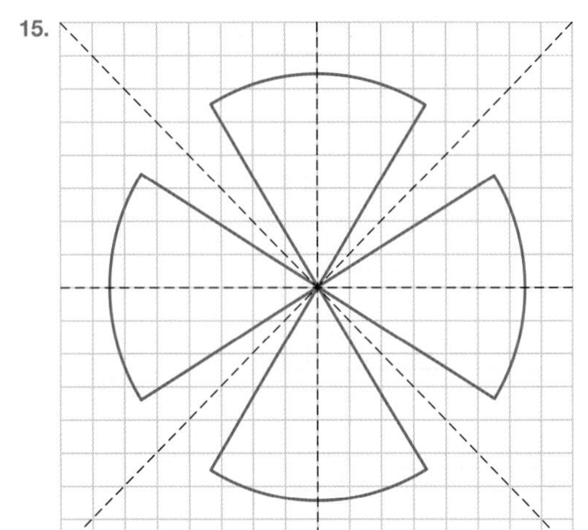

16. a.

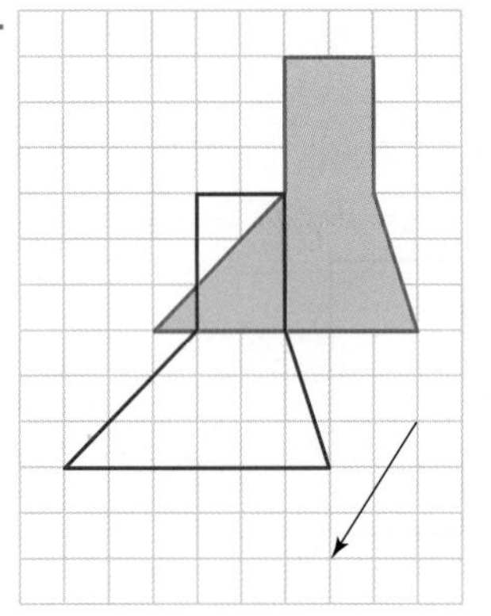

b.

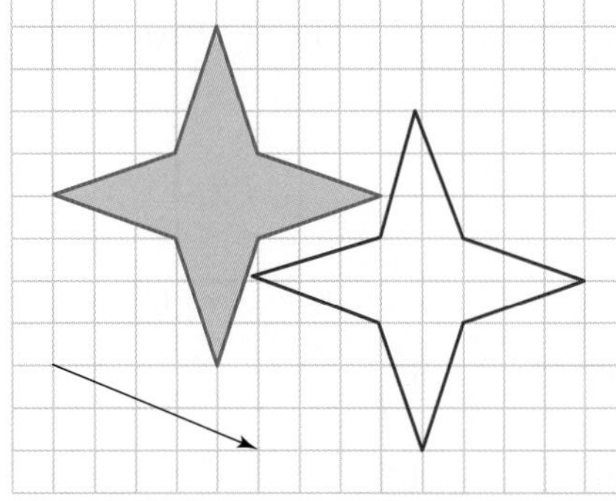

17. a.

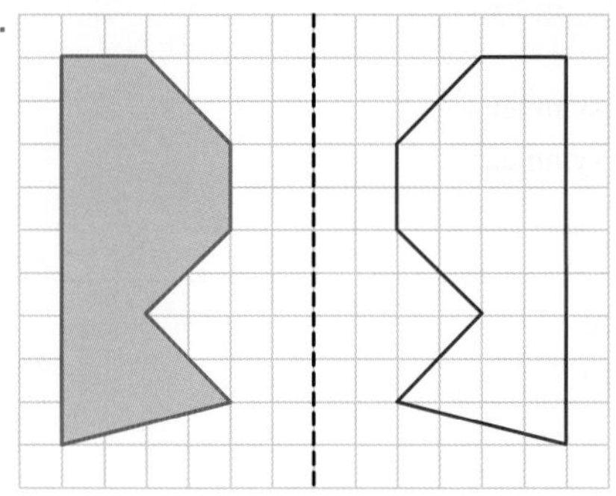

b.

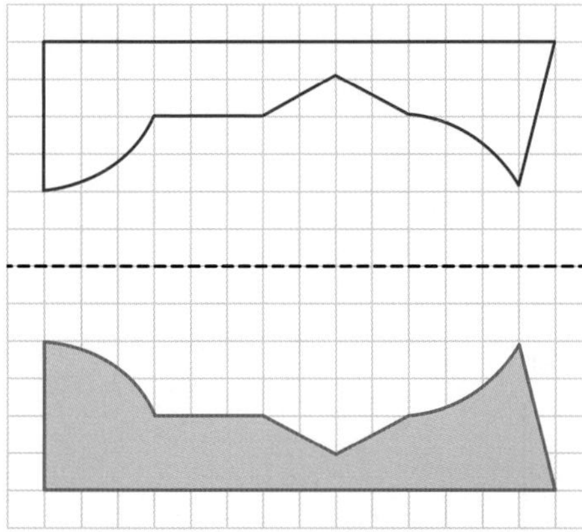

c.

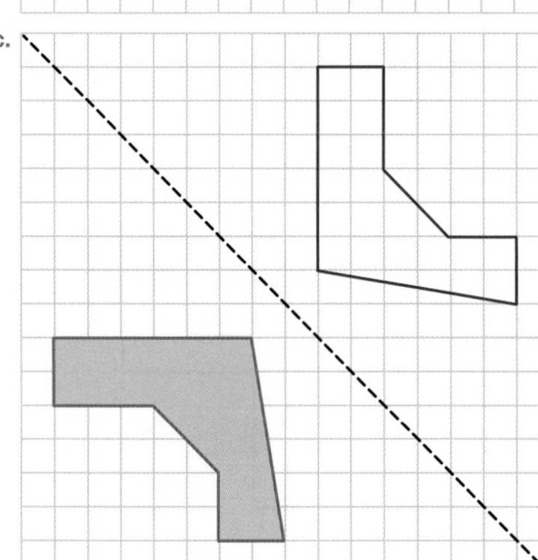

18. a.

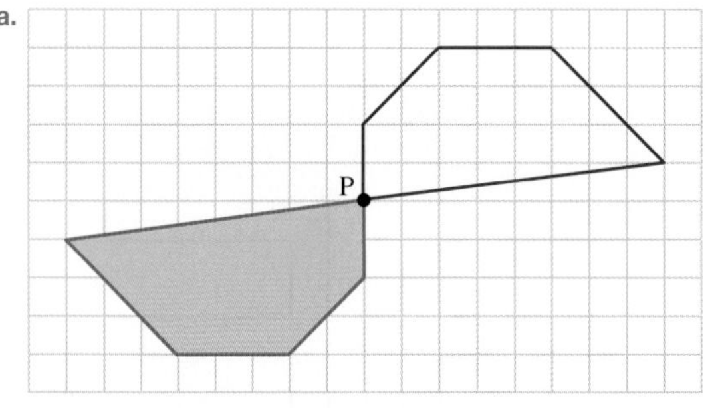

b.

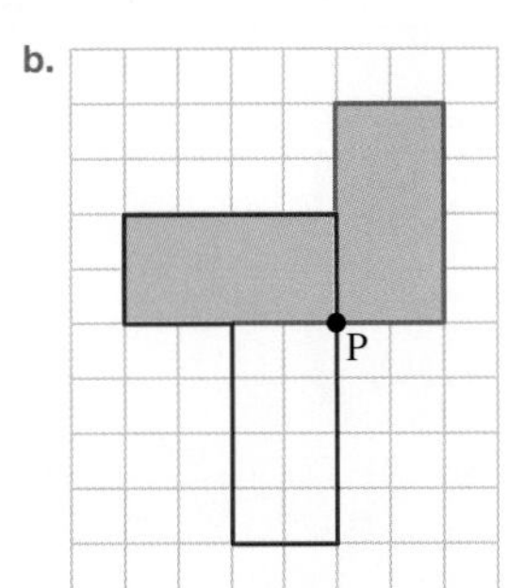

19. a.

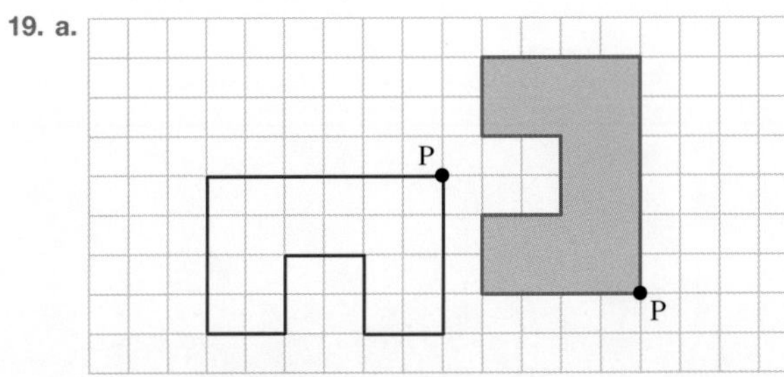

b.

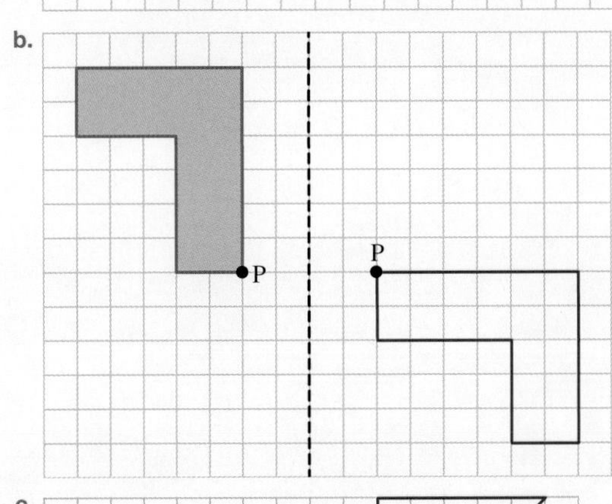

c.

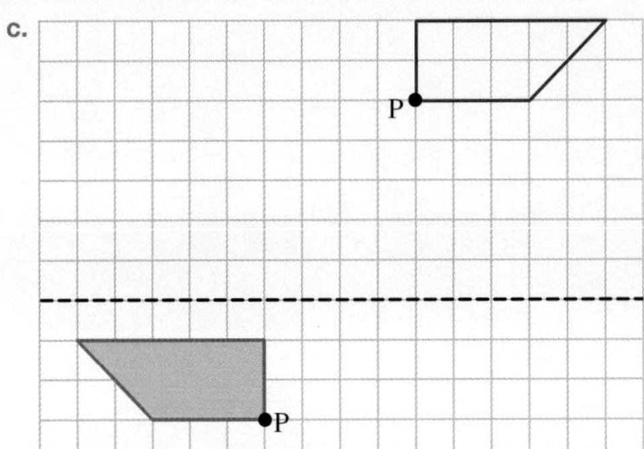

Extended response

20. a. Each of the second group of 10 letters has the same pattern as the corresponding letter in the first group, but with an extra dot underneath.

b. The following letters are reflections of each other: D and F, E and I, H and J, M and U, N and Z, P and V, R and W.

c. The following letters are rotations of each other: B and C; E and I; D, F, H and J; R and W.

d. Braille is read by touch and translations would not be determined by touch. The translated shape is the same as the original shape, just shifted on the plane; therefore, it will feel the same as the original.

e. Answers will vary, but a sample response is provided. The word *fixed* has rotational symmetry when written in Braille:

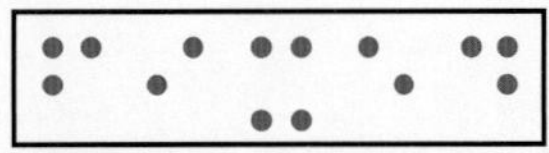

f. The letters W and R form a rotational symmetrical pair in Braille:

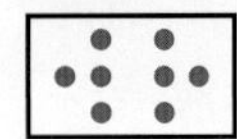

GLOSSARY

absolute error the value of an error without a positive or negative sign
absolute percentage error the absolute relative error converted to a percentage
absolute relative error to compare an estimated value with an actual value
accuracy how closely a measured value agrees with an actual value
adjacent in trigonometry, the side of a triangle adjacent to an angle is the line that together with the hypotenuse, forms the angle
angle of depression the angle measured down from the horizontal line (through the observation point) to the line of vision
angle of elevation the angle measured up from the horizontal line (through the observation point) to the line of vision
annual leave loading an additional payment received on top of the 44-week annual-leave pay
approximation an estimate based on the information available, when a precise answer is not necessary
array a collection of objects or values that have been ordered
axis of symmetry or **line of symmetry** the straight line that sits midway between an object and its image. An object is reflected along an axis of symmetry (mirror).
bar graph a display of the categories of data on one axis (usually the horizontal axis) and the frequency of the data on the other axis (usually the vertical axis)
back-to-back stem plot a display used to compare two different sets of data. Back-to-back stem plots share the same stem, with one data set appearing on the left of the stem and the other data set appearing on the right of the stem.
boxplot or **box-and-whisker plot** a graphical representation of the five-number summary
capacity the maximum amount of fluid contained in an object, usually applied to the measurement of liquids and measured in units such as millilitres (mL), litres (L) and kilolitres (kL)
categorical data data that are not numerical and are put into categories such as types of car
census collection of data from a population (e.g. all Year 10 students) rather than a sample
centre of rotation a specific point around which an object is rotated
certain the probability of an event happening is 1
class intervals a subdivision of a set of data; for example, students' heights may be grouped into class intervals of 150 cm – 154 cm, 155 cm – 159 cm
closed question a question that must be answered within given categories
coincident two lines are coincident if they lie one on top of the other
column graph a display of the categories of data on one axis (usually the horizontal axis) and the frequency of the data on the other axis (usually the vertical axis)
complement in probability, the likelihood that the event discussed will not happen
composite figure a closed figure that comprises two or more different common figures
composite solid a closed solid that comprises two or more different common solids
compound interest interest calculated on the changing value throughout the time period of a loan or investment. Interest is added to the balance before the next interest calculation is made. The amount of the loan or investment can be calculated using $A = P(1+i)^n$.
concave polygon a polygon with at least one interior angle greater than 180°
constant of proportionality the number multiplying x in the equation $y = kx$; equal to the gradient of the corresponding graph; also called the constant of variation
continuous data data that can take any value within a given range
conversion factors used to convert one unit of measurement to another
cosine ratio the ratio of the side length adjacent to an internal angle of a right-angled triangle with the hypotenuse: $\cos\theta = \dfrac{\text{adjacent}}{\text{hypotenuse}}$

council rates a fee home owners pay to support councils in providing parklands, libraries, rubbish collection and road maintenance
cumulative frequency the number of observations in a data set that are above or below the particular value
cumulative frequency curve the curve created when plotting data from a cumulative frequency table; also called an ogive
cumulative frequency table a method of recording cumulative frequency
deciles the nine values that split a ranked set of data into ten equal sections
decimal places the number of digits that are to the right of the decimal point
decimal point a point in a number that separates the whole number part from the fractional part
digital footprints data about ourselves that are generated by our digital activities
dimension lines thin continuous lines with arrowheads at both ends that show the dimension of a line
direct variation (proportion) the relationship between two quantities that are proportional to each other; produces a linear graph that passes through the origin
discount the amount that a price is reduced by
discrete data data that can take only certain values, usually whole numbers
edges connections between vertices
effective rate of interest the rate of interest that takes into account the reducing balance owing after each payment has been made: $R_{\text{ef}} = \dfrac{2400I}{P(m+1)}$
elimination method an algebraic method to solve simultaneous equations without graphing
equally likely outcomes outcomes that have the same chance of occurring
equation mathematical statement containing a left-hand side, a right-hand side and an equals sign between them
equivalent having the same value
error the difference between an estimated value and actual value
estimating approximating an answer when a precise solution is not required
evaluate to determine a value for an expression
exact a precise measurement
exchange rate the rate at which currencies can be interchanged. Buying rate refers to the rate at which banks will buy the currency from you. Selling rate refers to the rate at which banks will sell you a currency.
expanded form the form of a number where it is separated into each of its individual parts
expression mathematical statement made up of pronumerals (letters) and numbers. Expressions do not contain an equals sign.
extrapolation making a prediction from a line of best fit that appears outside the parameters of the original data set
favourable outcome the desired result in a probability experiment
finite decimal a decimal that reaches a point where it has no remainder
five-number summary a way of summarising data so that it can be put into a boxplot using the minimum, lower quartile, median, upper quartile and maximum values
formula a rule that is written using pronumerals
frequency the number of times a score occurs in a set of data
frequency histogram a display of numerical data similar to a bar chart where the width of each column represents a data value or a range of data values, and the height represents their frequencies
frequency polygon a line graph that can be drawn by joining the centres of the tops of each column of a histogram
goods and services tax (GST) a tax that is charged on most purchases
gram the standard unit of weight in the metric system
histogram a display of numerical data similar to a bar chart where the width of each column represents a range of data values and the height represents their frequencies
horizontal in trigonometry, the base line, parallel to the plane of the horizon, from which angles of elevation and depression are measured
hyperbola the shape of the graph formed by an inverse variation relationship

hypotenuse the longest side of a right-angled triangle. The hypotenuse will be opposite the right angle.
image a point or shape after it has been transformed
impossible the probability of an event happening is 0
infinite recurring decimal a decimal value with a repeating pattern
inflation the rise in prices within an economy over time
interest when investing money, interest is a payment earned for having money invested in a bank or financial institution. When borrowing money from a bank or financial institution, interest is charged as a fee for using their money.
interest rate the percentage by which an amount of money that is borrowed or invested is modified each time period
interpolation making a prediction from a line of best fit that appears within the parameters of the original data set
interquartile range (IQR) the difference between the upper and lower quartiles of a data set
inverse operations operations that undo a previous operation
inverse variation variation in which increasing one variable decreases the other
inversely proportional as one quantity increases, the other quantity decreases proportionally
irrational number a number that cannot be expressed as a ratio of two integers
limit of accuracy half of the smallest unit marked on a measuring tool
line of best fit a line drawn on a scatterplot that passes through or is close to as many points as possible; also called a regression line
linear equation an equation in which the variable has an index (power) of 1
linear function a function that is a straight line when drawn
litre the standard unit of volume in the metric system
lower fence the lower boundary beyond which a data value is considered to be an outlier: $Q_1 - 1.5 \times \text{IQR}$
mean commonly referred to as the average; a measure of the centre of a set of data. The mean is calculated by dividing the sum of the data values by the number of data values.
median the middle value of a data set when the values are placed in numerical order
Medicare levy a portion of taxpayer's funds used to pay for Medicare (healthcare for Australian residents)
metre the standard unit of length in the metric system
midpoint the average of the maximum and minimum values for the class interval
modal class the class interval with the highest frequency
mode the category or data value(s) with the highest frequency. It is the most frequently occurring value in a data set.
modulus a calculation done inside vertical bars, which means you can ignore the sign of the number
multiples of 10 numbers that can be divided by 10, such as 10, 20, 30 and 100
multiplicative factor the amount by which a number is multiplied or divided
nominal data categorical data that have no order associated with them
numerical data data that involve numbers or measurements
ogive the curve created when data from a cumulative frequency table is plotted; also referred to as a cumulative frequency curve
open questions a question that has no guidelines within which to answer
opposite in trigonometry, the side of a triangle opposite an angle is the only line that does not form a part of the angle
order of operations the correct sequence for performing the mathematical operations in an expression; often characterised by a mnemonic such as BIDMAS (Brackets; Index (or power); Division or Multiplication from left to right; Addition or Subtraction from left to right)
order of rotational symmetry the number of times a shape matches the original figure within a 360° rotation
ordinal data categorical data that are associated with some qualitative scale
outcome the result obtained when a probability experiment is conducted
outlier an extreme value that appears very different from most other values in a data set
outside surfaces the surfaces of a solid object that make up the total surface area
parallel lines lines that have the same gradient

parameters the boundary or limiting values associated with populations
partial variation variation involving quantities that produce a linear graph that does not pass through the origin
PAYG payment summary statement a statement of gross income and the PAYG tax deducted from that income throughout the financial year. It is given to the employee by the employer at the end of each financial year.
per cent the amount out of 100, or per hundred; for example, 50 per cent (or 50%) means 50 out of 100 or $\frac{50}{100}$
perpendicular lines lines that meet at right angles (90°)
polygon a closed shape where each edge is a straight line and all edges meet at vertices
population the entire set of subjects or objects being studied or investigated
powers of 10 numbers that include 10 (10^1), 100 (10^2), 1000 (10^3), 10 000 (10^4) and so on
precision how closely individual measurements agree with each other
principal the amount that is borrowed or invested
profit the difference between the amount a seller pays to produce the goods or service (the cost) and the amount for which the seller sells the same goods or service (the price)
projection lines thin continuous lines drawn perpendicular to the measurement shown, that do not touch the object
pronumerals a letter used in place of a number; another name for a variable
proportion sign the sign used to represent direct variation; equivalent to '$= k$'
proportion the relation or comparison of one thing to another
Pythagoras' theorem theorem stating that in any right-angled triangle, the square on the hypotenuse is equal to the sum of the squares on the other two sides. This is often expressed as $c^2 = a^2 + b^2$.
qualitative describes data that do not involve numbers or measurement
quantiles any number of equal parts that a set of data can be split up into
quantitative describes data that can be measured. Numerical values can be assigned to quantitative data.
quartiles these divide a set of data into quarters. The lower quartile (Q_1) is the median of the lower half of an ordered data set. The upper quartile (Q_3) is the median of the upper half of an ordered data set. The middle quartile (Q_2) is the median of the whole data set.
questionnaire a set of questions completed for a statistical investigation
random number generators devices or programs that generate random numbers between two given values
range a measure of spread determined by calculating the difference between the lowest and highest values
rate at which *y* varies with *x* the constant of variation or constant of proportionality
rate in the dollar the rate at which council taxes are charged
rate of interest the percentage of the principal that is paid out in a given time period as interest
ratio of *y* to *x* if y and x are directly proportional ($y = kx$), the constant of variation or constant of proportionality, k, is equal to the ratio of y to x
ratio a comparison of two or more quantities that are measured in the same units, written with a colon between the values
reflection exact image of an object, as seen in a mirror, as far behind the mirror as the object is in front of it. Reflections always have reversed orientations; right appears left and left appears right.
reflex angle an angle greater than 180°
regression line or **line of best fit** a line drawn on a scatterplot that passes through or is close to as many points as possible
regular polygon a polygon with edges of equal length and interior angles of equal size. Equilateral triangles and squares are examples of regular polygons.
relative frequency the frequency of a particular score divided by the total sum of the frequencies
repeating pattern the same number or numbers being repeated in the same order; for example, 333333 ... or 454545 ...
rotation the turning of an object about a certain point — the centre of rotation
rational number a number that can be expressed as a ratio of two integers
sale price the price of an item after the discount has been subtracted
sample a smaller selection of subjects or objects taken from a population
sample space in probability, the complete set of outcomes or results obtained from an experiment. It is shown as a list enclosed in a pair of braces, { }, and is denoted by the symbols ξ or S.

scale a series of marks indicating measurement increasing in equal quantities

scale drawings drawings of real-life objects where all dimensions of the drawing are kept in the same ratio as the actual object

scale factor a measure of the relative size of two similar figures. It is the amount of enlargement or reduction and is expressed as integers, fractions or scale ratios.

scatterplot a visual display of bivariate data

scientific notation used to express very large or very small numbers. To express a number in scientific notation, write it as a number between 1 and 10 multiplied by a power of 10.

sector graph A graph where a circle is cut into sectors. Each sector then represents a section of the data set. Each sector is the same proportion of the circle as the part of the data set it represents.

significant figures the number of digits in a whole number or decimal that are reliably known

similar describes figures that are the same shape but different sizes

simple interest interest calculation based on the original amount borrowed or invested; also known as 'flat rate' as it is a constant amount: $I = Pin$

simulations outcomes of events that are modelled to represent what would happen in real life

simultaneous occurring at the same time

simultaneous equations equations whose lines intersect or coincide

sine ratio the ratio of the side length opposite an internal angle of a right-angled triangle with the hypotenuse: $\sin\theta = \dfrac{\text{opposite}}{\text{hypotenuse}}$

spread how far a data set is spread from the centre or from each other

standard deviation a measure of the spread of continuous numerical data around the mean

standard form *see* **scientific notation**

standard unit a unit used most commonly for measurement

stem-and-leaf plot or **stem plot** an arrangement used for numerical data where each value is grouped according to its numerical place value (the 'stem') and then displayed horizontally as a single digit (the 'leaf')

substitute to replace a variable in a formula with a number

successful trial a trial where the result is the outcome that was desired

summary statistics the boundary or limiting values associated with samples

super guarantee percentage a percentage of employees' annual salary paid into a recognised superannuation fund

superannuation a percentage of annual salary that is set aside for retirement

surd an irrational number containing a radical sign such as a square root or cube root, for example $\sqrt{10}$

survey collection of data from a sample of a population

symmetrical identical on opposite sides of a line or plane; a parabola is symmetrical about a line called the axis of symmetry

system of equations a set of equations that can all be satisfied by the same values for variables

tangent ratio the ratio of the side length opposite an internal angle of a right-angled triangle with the side length adjacent to it: $\tan\theta = \dfrac{\text{opposite}}{\text{adjacent}}$

term deposit lending money to a bank for a set amount of time

term the time period of a loan or investment

translation horizontal (left/right) or vertical (up/down) movement

transposing rearranging a formula to make a particular pronumeral the subject

tree diagrams branching diagrams that list all the possible outcomes of a probability experiment

trial an experiment performed in the same way every time

trigonometry a branch of mathematics that analyses the relationships between angles and sides of triangles

upper fence the upper boundary beyond which a data value is considered to be an outlier: $Q_3 + 1.5 \times \text{IQR}$

variable another word for a pronumeral (algebra); a symbol in an equation or expression that may take many different values (data)

vary inversely if two quantities vary inversely, then increasing one variable decreases the other

INDEX

S

T

U

V

W